IET RADAR, SONAR AND NAVIGATION SERIES 16

Series Editors: Dr N. Stewart
Professor H. Griffiths

Target Detection by Marine Radar

Other volumes in this series:

Volume 1 **Optimised radar processors** A. Farina (Editor)
Volume 3 **Weibull radar clutter** M. Sekine and Y. Mao
Volume 4 **Advanced radar techniques and systems** G. Galati (Editor)
Volume 7 **Ultra-wideband radar measurements: analysis and processing** L. Yu. Astanin and A.A. Kostylev
Volume 8 **Aviation weather surveillance systems: advanced radar and surface sensors for flight safety and air traffic management** P.R. Mahapatra
Volume 10 **Radar techniques using array antennas** W. Wirth
Volume 11 **Air and spaceborne radar systems: an introduction** P. Lacomme (Editor)
Volume 13 **Introduction to RF stealth** D. Lynch
Volume 14 **Applications of space-time adaptive processing** R. Klemm (Editor)
Volume 15 **Ground penetrating radar, 2nd edition** D. Daniels
Volume 16 **Target detection by marine radar** J. Briggs
Volume 17 **Strapdown inertial navigation technology, 2nd edition** D. Titterton and J. Weston
Volume 18 **Introduction to radar target recognition** P. Tait
Volume 19 **Radar imaging and holography** A. Pasmurov and S. Zinovjev
Volume 20 **Sea clutter: scattering, the K distribution and radar performance** K. Ward, R. Tough and S. Watts
Volume 21 **Principles of space-time adaptive processing, 3rd edition** R. Klemm
Volume 101 **Introduction to airborne radar, 2nd edition** G.W. Stimson
Volume 102 **Low-angle radar land clutter** B. Billingsley

Target Detection by Marine Radar

John N. Briggs

The Institution of Engineering and Technology

Published by The Institution of Engineering and Technology, London, United Kingdom

First published 2004

The Institution of Engineering and Technology
Michael Faraday House
Six Hills Way, Stevenage
Herts, SG1 2AY, United Kingdom

www.theiet.org

British Library Cataloguing in Publication Data
Briggs, John
Target detection by marine radar. – (IET radar series no. 16)
1. Tracking radar 2. Radar in navigation
I. Title II. Institution of Electrical Engineers
623.8'933

ISBN (10 digit) 0 86341 359 5
ISBN (13 digit) 978-0-86341-359-9

Typeset in India by Newgen Imaging Systems (P) Ltd, Chennai

Contents

Target Detection by Marine Radar

John N. Briggs

The Institution of Engineering and Technology

Published by The Institution of Engineering and Technology, London, United Kingdom

First published 2004

The Institution of Engineering and Technology
Michael Faraday House
Six Hills Way, Stevenage
Herts, SG1 2AY, United Kingdom

www.theiet.org

British Library Cataloguing in Publication Data
Briggs, John
Target detection by marine radar. – (IET radar series no. 16)
1. Tracking radar 2. Radar in navigation
I. Title II. Institution of Electrical Engineers
623.8'933

ISBN (10 digit) 0 86341 359 5
ISBN (13 digit) 978-0-86341-359-9

Typeset in India by Newgen Imaging Systems (P) Ltd, Chennai

Foreword

The opportunity arose, with the end of hostilities in 1945, to make available to Merchant Shipping the new technology of radar, which had developed so rapidly in the secrecy of war. In the United Kingdom, the Government made a design available to manufacturers and this was followed by Performance Specifications for Radar for Merchant Ships in 1946. This specification and its later revisions were prepared by the Ministry of Transport in consultation with representatives of shipowners, lighthouse and harbour authorities, marine manufacturers, the General Post Office and the Admiralty. The GPO at the time were responsible for the use of radio frequencies and the Admiralty contributed the technical expertise with a new Transport Experimental group at ASE (Admiralty Signals Establishment) in Eastney which subsequently developed into the Civil Marine Division of ASWE (Admiralty Surface Weapons Establishment). Together with these Performance Specifications a system of type testing of designs was put in place to certify their compliance.

Radar on Merchant Ships was initially installed for commercial purposes. The early customers were ferries, which could then maintain better schedules in fog, and large fishing vessels. Radar was treated with great suspicion by the mariners of the day and was usually the preserve of the master, who locked it so that it could only be used when he was on the bridge. Ports also started using radar for the commercial purposes of berthing ships in fog; one example is the Port of Liverpool, in 1948.

With improving technology and ease of use, the Merchant Ship radar became more accepted but it was some time before the use of radar for safety purposes was recognised. In fact the melancholy situation in the 1950s was that the introduction of radar had not resulted in any reduction of the number of serious collisions at sea. The collision between the passenger ships *Andrea Doria* and *Stockholm* off Nantucket in 1956 is a well recorded example of the misinterpretation of radar information in a passing situation in fog. Accordingly the International Conference on The Safety of Life at Sea in 1960 revised the International Regulations for Preventing Collisions at Sea by adding rules to take account of the use of radar and recommendations on the use of radar information as an aid to avoiding collisions at sea.

Today radar with plotting is accepted by mariners as the primary tool to assist in collision avoidance. The International Conference on Safety of Life at Sea in 1974 adopted provisions making radar a mandatory carriage requirement for Merchant

Ships in a phased programme starting in 1980 and finally completed in 2002. All Merchant Ships over 300 gross tonnage now carry a radar and many carry two. Many small craft also carry radar voluntarily as manufacturers have produced cost effective designs for their needs.

The Maritime and Coastguard Agency is the direct successor of the Marine Safety Division of the 1946 Ministry of Transport and the consultation processes used then in agreeing radar specifications are still used today through its Safety of Navigation Committee. The task is somewhat more complicated today as specifications are seldom written for national needs but are designed to be internationally agreed so that standards can be maintained all over the world. The ensuing type approval process has also led to international agreement with one approval often being acceptable to Administrations world-wide.

No standard setting of the regulatory bodies of the world would be successful, however, if the basic physics of the radar and the resulting technology constraints were not fully understood by those responsible for the drafting. The author of this book is to be congratulated in his descriptions of the physical processes at work and the methods by which the technology can be used in the specialised world of civil marine radar.

Kim Fisher FRIN FIEE
MCA
Chairman UKSON

Preface

Radar is a legal necessity for the safe navigation of merchant ships, is voluntarily carried by many leisure craft, within vessel traffic services (VTS) is indispensable to the operation of major ports and harbours, and has several important other maritime applications. This book tells how these civil surveillance radars detect their targets. Brief historical outlines help explain the way modern practice developed from the first faltering steps of the 1930s. We describe and illustrate today's radars, as well as passive and active beacon targets. But natural features such as coastlines and – vital for collision avoidance – vessels of all shapes and sizes are not specifically designed to reflect radar transmissions. Radar operation however relies on their ability to return echoes. This ability is therefore examined in detail from both practical and theoretical standpoints.

Our reason for adding to the many existing radar treatises is, for the first time, to concentrate solely on the civil marine scene, unadulterated by consideration of aircraft flying at 40 000 ft, stand-off jammers, Doppler effects and the host of other factors, civil and military, which do not concern civil marine operations.

Radar design is constrained by the laws of man and of physics, so we explain the framework lying behind the numerous international regulations governing marine radar and highlight the fundamental technical constraints. Detection is an engineering problem founded on scientific and statistical principles, so we have to include more mathematics than we might wish, but which can be skipped on first reading. Without excluding anything of practical significance, we have simplified our account as much as possible, helping the many users, managers and regulators whose backgrounds lie elsewhere, even perhaps outside the marine industry.

We include full technical analysis of the many factors in play within the radars and within targets, the great and sometimes under-estimated parts taken by the weather and the environment, and not least, the ability of operators to set the controls to influence performance for good or ill. Our analysis shows why, when and whether radar will pick up targets as diverse as yachts, low-lying coasts or super-tankers. Numerous graphical and other diagrams and worked examples help the reader to grasp the principles underlying radar operation and to quantify the practical importance of the many factors in play. The analysis culminates in full instructions for use of a set of spreadsheets, available on the IEE website (www.iee.org), which give detection ranges, probability of detection and many other performance parameters for the user's own equipment, illustrated by a set of case studies.

The accuracy with which targets are positioned on the radar screen and with which their progress is tracked and predicted depends upon how definitely they have been detected, so we devote a chapter to the general question of accuracy, which underlies the ability of plotting aids such as ARPA to predict closest point of approach and give other warnings vital to the navigator. Not being master mariners, we leave interpretation of the displayed traces to the specialist navigation textbooks. In the final chapter, Chris Baker looks at some of the ways in which marine radar may develop to meet the challenges of the next two decades.

We have striven to be comprehensive, minimising the need to hunt up references elsewhere. Copious cross-references, a logical layout, the glossary and the index should quickly guide the busy reader to specific items.

Although we authors take full responsibility for the errors and omissions that surely remain, this book would have been impossible without the help, advice and encouragement of many experts.

J.N.B. would like to thank several old GEC-Marconi colleagues whose friendship often goes back too many years to admit. They include John Ashley, Niall Davies, Steve Holland, George Hurd who made extensive comments on an early draft, Bob James, David Ogleby who identified the need for such a book when scheming VTS systems, Richard Parsons, Janet and Peter Sykes, and Dudley Taylor whose computer expertise was vital at times. Several friends with senior engineering management experience elsewhere in the marine radar industry freely gave valuable help, including among others David Hannah, Richard Trim and Professor Phil Williams. J.N.B. is particularly in debt to Phil for valuable comments on the whole draft text and for permission to quote freely from his privately published CD work 'Civil Marine Radar' with its wealth of technical detail and historical aspects describing the evolution of Decca radars. In the early days Professor John Kemp, then Editor of the Journal of Navigation and Julian Parker, then Secretary of the Nautical Institute gave valuable encouragement to persevere with what seemed a daunting endeavour. The Royal Institute of Navigation generously supplied a scarce copy of *The Uses of Radar at Sea*. Jonathan Ansell of Easat Antennas freely gave much background information on modern VTS and scanner developments, while Peter Munro of Munro Engineering kindly read and commented on Chapter 8.

Officials have been equally helpful. Martin Hart and his colleagues at the Maritime and Coastguard Agency checked and corrected the regulatory sections of Chapter 1 and looked up a number of technical points. J.N.B. also benefited from extensive discussions with Roy Lee on regulatory aspects. Dr J. M. Williams took much trouble to explain the fundamentals of ships' radar reflectivity, which Chapter 10 attempts to summarise. Bill Paterson, then Director of Engineering, Northern Lighthouse Board, facilitated the atmospheric refraction experiments described in Chapter 5. Dr Nick Ward, Principal Development Engineer, Trinity House Lighthouse Service, kindly dug out a number of IALA papers and supplied illustrations.

A number of leading manufacturers and other organisations also kindly contributed illustrations, as acknowledged in the text. We do not however endorse particular suppliers' products – all the leading manufacturers offer highly developed equipment capable of excellent performance. Other illustrations are from the authors' collections.

J.N.B. would like to thank Sarah Kramer, Commissioning Editor at the IEE, and the IEE team for their forbearance, also Professor Hugh Griffiths, Radar Series Editor. His predecessor, Professor Ramsay Shearman, gave much guidance and prevailed on Professor Baker to contribute to the book.

Finally, and most important of all, J.N.B. must record his thanks to his long-suffering wife Betty for putting up with the disruption of the supposed leisure of retirement with 'yet more work'.

Chris Baker is especially indebted to Professor Hugh Griffiths (University College London), Dr Andy Stove (THALES Sensors), Dr Steve Harman and Mr Graham Binns (QinetiQ) and Professor Ramsay Shearman for their invaluable advice and numerous corrections to Chapter 16. In particular C.J.B. wishes to thank J.N.B. for the invitation to write the chapter and for having the patience to integrate it with the earlier chapters.

Chapter 1
Introduction

'It is on men that safety at sea depends and they cannot make a greater mistake than to suppose that machines can do all their work for them.'

Mr Justice Cairns, giving judgement in the Trentbank–Fogo case, 1967, quoted in A. N. Cockcroft and J. N. F. Lameijer, A Guide to the Collision Avoidance Rules, Stanford Maritime, London, 1976, p. 39

1.1 Purpose and scope

This introductory chapter explains the book's aims and methods, introduces the persons who operate radar, gives a brief historical background and outlines the regulatory framework under which marine radar is used.

1.1.1 Purpose

The art, science and skill of the navigator is to get a ship from A to B without hitting anything. Ships come in many shapes and sizes, Figure 1.1, from pleasure yachts to half-million tonne tankers. A and B may be berths in harbours, but B may be a pilot cutter or a shoal of fish. Among the things to be avoided are coastal features, other vessels, heavy flotsam and ice. Radar is an essential tool used to locate these hazards, assisting the navigator to make timely manoeuvre decisions.

Radar, an acronym for radio detection and ranging, detects objects of interest by transmitting radio signals in known directions from a narrow-beam antenna or scanner which scans the horizon, then timing the instants of reception of returned echoes from these 'targets'. (This Second World War term dates from development of radar for gunnery control and other military tasks; like much of the early jargon, it has stuck.) Each detected target is displayed mapwise at its correct range and bearing. Additional to the anti-collision task, prudent seamen also still use radar to cross-check their position against known shore targets, despite the advent of satellite navigation systems as the prime onboard position finder. Ashore, vessel traffic service (VTS) radars depict the traffic situation in and around busy harbours and traffic lanes, and

Figure 1.1 Ships come in all sizes

range surveillance radars check that offshore gunnery ranges are clear to commence firing.

The radar's first task is to pick out targets against a background of unwanted electrical fluctuations within the radar receiver and unwanted reflections from objects of no interest to the observer, such as waves and rain. In the jargon, targets must be detected in presence of noise and clutter. Only when this has been done can the radar go on to display target range and bearing or generate tracks of the target's movement to provide the operator with useful information.

Much of this book treats the technical circumstances enabling a target to be displayed but one must never forget that even the finest radar is merely a tool which can be ignored, misused or misunderstood. It is always up to the operator to decide whether and how to use the radar and its information. The human–machine interfaces are the radar controls and the display screen. We shall step outside the realm of electronic engineering when considering how operators should utilise radar when performing their safety-related tasks, but steer clear of purely navigational matters, for which readers should turn to specialist books, such as those by Burger [1] for deep-sea ships, Bole and Dinelley [2], which concentrates on plotting aids and Wylie [3], which though old and out of print, remains surprisingly relevant in many respects, with many illustrations of radar displays of targets and clutter and much observational experience soundly based on theory. See Bartlett [4] for small vessels.

The present book stems from a paper published in the *Journal of Navigation* [5]. It is primarily offered to those, perhaps neither radar specialists nor mariners, who need an understanding of how well civil marine shipborne and shore-based radars are likely to detect their targets under practical conditions. Targets may be 'natural' objects such as ships and coastlines which happen to reflect radar echoes, or may be specially prepared devices intended to return good radar signals. These include passive radar reflectors, racons and radar target enhancers (RTE), the latter two being secondary radars, speaking only when spoken to, and hence 'active' targets. The book discusses all aspects of the detectability problem and helps the reader answer such questions as:

- Is higher transmitter power worth its cost to me?
- What is the best height for my scanner?

- Which radar band should I choose?
- Would carriage of an RTE significantly improve detection of my craft?
- Which supplier's VTS proposal is best for my harbour?
- Do trials results match theoretical expectations?
- How does weather affect performance?

Radar is useless unless it can detect the targets of interest to the user, yet manufacturers in their data sheets rarely claim any specific detection ranges for their products. This is because the environment and many targets fluctuate in ways which are difficult to quantify or measure. Reports from seafarers (e.g. as quoted by Bell and Starling Lark [6]) stress the variability of observed detection range and the difficulty of detecting small targets in bad clutter; we shall explain why this is so. Although precision is impossible, we provide methods of calculating the likely detection performance of all the radar/target/clutter combinations likely to be met in practice, with spreadsheets enabling readers to calculate the performance of their own particular systems on a personal computer.

1.1.2 Scope

We consider only radars used for civil shipborne navigation and collision avoidance and for VTS and related tasks such as sea surveillance of gunnery ranges, drug interdiction and sea traffic research studies. These all employ centimetric, non-coherent, pulsed, low pulse repetition frequency surface to surface surveillance radars with directional rotating common transmit–receive scanners. 'Non-coherent' means that our radars use only the amplitude of echoes, ignoring signal phase. In general, they also operate unambiguously, with only one pulse in play between radar and target at any instant.

Other forms of radar exist, differing to lesser or greater extent, but are currently seldom used in marine contexts. Over the horizon radars and some others, although of course sharing the same underlying physical principles, differ so radically that only parts of this book apply directly. Much of the book is applicable to warship surface surveillance radar used for detection of surface targets, although we have completely excluded all the specifically military problems, such as jamming, which so dominate military radar design. Instead, we concentrate entirely on the civil field, where a modern treatment of detectability is lacking. Accuracy of positioning and tracking, which demands good detectability, is discussed towards the end of the book. The many other important radar qualities, such as size, cost and ease of use, are considered only where they influence detectability.

The linked components of the radar – transceiver, scanner, display – are frequently referred to as a system. We however prefer to treat the whole radar as part of a wider system which also contains the other main interacting elements: the marine environment, the target and the operator. We point out some of the problems facing designers and indicate solutions. Likewise, we hope better understanding of the factors in play will help operators get the best from their equipment, and so contribute something to the safety of life at sea.

1.2 Radar users and uses

There are many specialist users such as warships, oil industry support vessels, buoy tenders, search and rescue craft and coastal surveillance systems. But the main operating personnel, whom we shall call operators, are probably associated with the following.

1.2.1 Merchant ships

About 96 per cent of the world's international trade, some 6 billion tonnes per year, is transported in 35 000 or more merchant ships. Their radars are operated by the master, officer of the watch (OOW) and, if embarked, the pilot. These officers primarily use the radar or radars as an important navigational tool to do the following.

1. Assess the traffic situation – the position and tracks of other vessels, usually employing an Automatic Radar Plotting Aid (ARPA) or the simpler Automatic Tracking Aid (ATA); anticipate likely traffic movements, and make timely manoeuvre decisions under the Collision Regulations to give a safe clearing range of typically 1.5–4 km. The assessment may be that own ship is the 'stand on' ship under the regulations, the decision then being usually to maintain course and speed, keeping a close eye on the target's manoeuvres. Accurate tracking demands particularly good detection performance.
2. Monitor movements of other shipping for collision avoidance.
3. Monitor own ship's progress relative to sea-marks or coastal features, particularly in port approaches. Coastal echoes can be difficult to interpret, so racons or reflectors are used to pinpoint otherwise radar-inconspicuous lighthouses and buoys.
4. Detect ice, uncharted wrecks or other obstructions.
5. Rendezvous with pilot cutters, etc.
6. Maintain anchor watch, both for own ship dragging and for movements of other vessels.
7. Independently confirm position data provided by specialist instruments such as a global navigational satellite system (GNSS), currently comprising the [differential] global positioning system ([d]GPS, operated by the United States) and [differential] global navigation satellite system ([d]GLONASS, operated by the Russian Federation), which interpret data obtained by radio down-link from constellations of special-purpose satellites. Both GPS and GLONASS form part of the global maritime distress and safety system (GMDSS) and will be supplemented or superseded by later satellite chains such as Galileo. Lang [7] has published an interesting discussion of satellite reliability. Satellite signals are weak enough to be vulnerable to certain radio and television transmissions. Four-watt jammers, available to terrorists on the Internet, apparently can jam to over 200 km. Worse, they may spoof receivers into indicating false position. It therefore remains a basic tenet of seamanship, as well as being a legal requirement, to use all available navigational aids for position finding. Well-found passenger

ships have grounded through sole reliance on very precise (but faulty) satellite navigation, ignoring the more woolly but independent radar coastline echo.
8. Make landfall. Since the advent of satellite navigation, this long-range task has lost some of its former importance.

The relationship between radar suppliers and deck-officer operators is often tenuous. The radars may be procured by the shipowner, owner's purchasing manager, ship management company or shipbuilder, perhaps as a just-compliant product within a third party's overall bridge electronics deal rather than on technical merit. Repair and maintenance is perhaps in the hands of independent contractors, and the operators may be provided by a manning or ship management agency. One unfortunate result is that designers receive insufficient feedback on how well their products actually perform in the rough and tumble of actual service, rather than under controlled laboratory test conditions.

1.2.2 Leisure craft

Beside general navigation, the skipper or helmsman operates radar for obstruction avoidance, monitoring position of competitors when racing and sometimes to detect sea or precipitation clutter – to find smoother water, fairer weather or a good racing wind. Special radars have evolved for leisure craft.

1.2.3 Fishing vessels and small commercial vessels

The watchkeeper is usually the skipper or mate, who operates the radar for most of the above tasks. In addition fishing vessels have to detect dan buoys which mark nets, monitor activities of other fishing vessels, perhaps hope to detect flocks of birds feeding on fish and use clutter returns to help keep an eye on the weather.

The bigger yacht radars are often also fitted to small commercial vessels such as tugs and harbour workboats which are too small to have to carry radar within the international Safety of Life at Sea Convention (SOLAS, Section 1.4.5) but which choose to do so for operational convenience and safety. The International Maritime Organisation (IMO) has produced guidelines on appropriate performance for two radar sizes.

1.2.4 High speed craft

Collision avoidance is usually the first priority for fast ferries and other high speed craft (HSC). They use slightly modified conventional marine radar, augmented by night-vision infrared optical sensors for waterlogged obstructions which are poor radar targets.

1.2.5 Vessel traffic services

VTS systems provide a service to shipping, while technically similar vessel traffic management and information services (VTMIS) provide vessel movement data to

Figure 1.2 *Dover Maritime Response Co-ordination and Control Centre (MRCC), UK Maritime and Coastguard Agency. On 80 m cliff commanding extensive views of the English Channel. Takes data from radar outstation at Hastings. Operates in conjunction with Port of London Authority and French CROSS radar networks. Reflector scanner operating in dual frequency diversity mode at 9 GHz. Eight workstations, three being dedicated to emergency responses. Reproduced by permission of UK Maritime and Coastguard Agency, Norcontrol-IT and Easat*

the port management. For our purposes we shall lump them together. VTS operators, sometimes called watchstanders, use radar to gain awareness of the traffic situation, confirm manoeuvres and positions reported-in by radio or by the radio-based automatic identification system fitted to all ships over 300 gross tons (AIS; from 31 December 2004) and maintain tracks on all significant targets within the surveillance area. At the shorter ranges, ships' aspect is often of considerable interest in confirming reported manoeuvres. Moves are afoot in IMO to introduce long range identification and tracking of ships by AIS for security purposes, which may necessitate some form of integration with VTS radar displays.

An increasing number of systems akin to VTS are being installed to monitor shipping movements within traffic separation schemes (TSS) for safety purposes. Figure 1.2 shows one such station and Figure 1.3 shows a typical tower-mounted scanner.

At the other end of the scale, coastal surveillance radars are used by coastguard and voluntary organisations to keep a safety watch on inshore craft near popular leisure resorts, particularly where currents can so often cause problems to the inexperienced, see Figure 1.4(*a*)–(*c*).

1.2.6 Military applications

For general navigation, in naval parlance Marine and Pilotage (M&P), many warships carry navigational radar generally similar to those on merchant ships. The radars we shall describe are devoid of any form of anti-jamming facilities (electronic protection, formerly called electronic counter-countermeasures, ECCM) and are unsuited to warfare. However, outfits similar to VTS and marine radars are used for safety surveillance of coastal gunnery and missile-firing ranges, and although little publicised,

Figure 1.3 *High gain reflector scanner for sea surveillance service at 9 GHz. Inverse cosec^2 type, aperture 5.5 m, gain 40 dBi, switchable circular and horizontal polarisation. Steel tower also carries transmitter–receivers and communications equipment, plus infrared and CCTV cameras. Reproduced by permission of Easat Antennas Ltd, Stoke on Trent, UK*

increasingly for drug, intruder and piracy interdiction by coastguard or gendarmerie services. The majority use frequencies in or near the marine 9 GHz band. Equipments vary from powerful (~250 kW) sets with massive reflector scanners derived from military practice and displaying in a VTS-like range control centre, down to the smaller deep-sea ship sets or even yacht radars, deployed in small rough-terrain vehicles. Performance can be predicted from the following chapters, but we shall not cover the specialist precision tracking pulse or frequency-modulated (FM) radars which follow the flight of the projectile and of its sub-munitions.

1.3 The past and future

1.3.1 The history of marine radar

Some historical knowledge may help us understand why things are done as they are. After the First World War, liners were getting bigger and faster; would there be another *Titanic* disaster? The ship's whistle was acknowledged to be inadequate as the primary anti-collision aid in fog – despite the old *Mauritania's* whistle, reputed of 12 miles range, deafening the young author when sounding the Friday 'knocking-off' time at the BT-H Works at Rugby. Strategic coastal lighthouses started to sport lights and fog signal emitters of heroic size. By the 1930s, the technology of short-wave radio had matured. Several countries continued Guglielmo Marconi's pioneering experiments in target detection by radio, particularly aimed at marine collision avoidance. It was realised that to cover surface targets, the best wavelengths were a few centimetres. These 'microwaves' were very short indeed by the standards of the day and barely

(a)

(b)

Figure 1.4 *Small 9 GHz marine radars on surveillance duty. Sea Safety Group volunteers operate an expanding chain of coastal stations around the UK coasts. All Illustrations reproduced by permission of Captain A. P. Starling Lark and Sea Safety Group. (a) Watchroom of SSG Redcar Station, North-east England. A Furuno radar (left) augments the visual lookout. Combination with a fluxgate compass speeds reporting of exact latitude and longitude of casualties to the Search and Rescue services, to which the fully trained volunteers are officially accredited. Roof-mounted scanner at 40 m above sea level has 15 nmi horizon range and can detect target as close as the adjacent beach. (b) View from the watch-room towards the busy Tees Bay with its commercial shipping. (c) Twin displays. SSG Great Yarmouth Station. Quite small radars may suffice for inshore safety surveillance. Furuno 1832 radar, 4 kW transmitter power, 60 cm scanner. Feeds two 10 inch rectangular monochrome raster-scan displays. One observer can quickly guide Royal National Lifeboat Institution lifeboat to a casualty while a colleague maintains general surveillance, essential when sudden bad weather catches many leisure craft unawares*

practical with the available technology. An experimental centimetric apparatus on the French liner *Normandie* was able to detect presence of ships at several miles range shortly before that great ship was destroyed by fire, and Radar Type XAF was at sea from 1938 in USS *New York*, but lack of transmitter power stymied progress.

(c)

Figure 1.4 Continued

Meanwhile the storm gathered over Europe. Britain, fearing airborne bombing and invasion, hastily and secretly set up a shore-based air surveillance system, constrained to conventional radio wavelength ($\sim$12 m) by the power problem. These Chain Home radars, described with a good engineering outline by Latham and Stobbs [8], resembled broadcast radio transmitter stations, with their huge static wire antenna arrays. They drew on nascent television technology. Immediate success in the Second World War triggered concerted research into the crucial need for powerful microwave generators. Necessity is the mother of invention and a superb solution was found with remarkable speed. Building on low-power cylindrical split-anode magnetrons developed in Japan, the cavity magnetron of 1940 opened the way to practical centimetric radar, first on fighter aircraft and then warships. Submarine periscopes became detectable as early as 1943. Perhaps exceeded in technological effort only by atomic energy and code-breaking, wartime Anglo–American radar research yielded an excellent understanding of basic principles. Declassified before the Cold War, the work was published around 1950, notably in the 28-volume MIT book series,[1] much of whose content is still valid today. Late in the War thoughts turned to peacetime uses and the British were able to write a confident design aim for 'Post War Radar', including the first commercial marine radars. As early as May 1945, a 9 GHz prototype navigational radar Type 268 was demonstrated [9] to civil marine interests aboard HMS *Pollux*, using a submerging submarine, HMS *Umbra*, as a variable-sized target – the wheel had turned full circle.

The first commercial (1946) marine radars typically came complete with own cabin and roof-mounted scanner. Initially they were not very reliable and it took time for deck officers to accustom themselves to the strengths and limitations of this radically new aid to navigation. Nevertheless, the basic concept was sound,

[1] Massachusetts Institute of Technology Radiation Laboratory, McGraw-Hill. Some volumes have been re-published by IEE in UK and all are available in CD-ROM form from McGraw-Hill.

early problems were gradually beaten and radar became mandatory for all ships. Wavelengths and frequencies are discussed in Chapter 2, but we mention here that frequencies around 3 GHz, wavelength 10 cm (S band, 3000 MHz), was found best for detection in severe rain, in other conditions 9 GHz (3 cm, X band, 9400 MHz) being preferable. Some ships got the best of both worlds and carried both. Early suppliers were mainly large electrical firms who had been active in wartime radar. We shall not often mention individual firms but must make an exception for Decca (now Northrop Grumman Sperry Marine), who later deservedly won a pre-eminent position. Their war work included the Decca Navigator, but they did not enter the radar market until 1950.

Until the late 1960s, purely analogue circuits were used, the raw echoes being presented on a cursive plan position indicator (PPI) using a cathode ray tube having a dim monochrome long-persistence phosphor. Operators had to draw tracks representing the movement of each target by grease-pencil on the Perspex face of a reflection plotter. Radars of those days contained several electro-mechanical devices – motors, relays, etc. – and the only semiconductors were a few rectifiers and the microwave detector 'crystals'. The hard work was done by about 50 valves (thermionic tubes in the United States, where the anode is known as a plate). Each valve was supported by a couple of dozen passive components, mainly resistors, capacitors and inductors, hand-soldered to insulating tags, and all carried on metal trays, connected by cable bundles. Each valve anode consumed about 2.5 W at 250 V and its heater dissipated another 2 W. The total heat was considerable and the dust attracted by the high voltage could cause arcing failures. Digital technology was in its infancy. Hardware meant screws. Software? What have woolly jumpers to do with radar? The term did not exist. Detail design by pencil, paper and sliderule was laborious and of variable quality, necessitating much 'cut and try' prototype testing.

In 1977, after a spate of accidents causing pollution, political pressure caused the United States Coast Guard to issue, after discussion, regulations requiring all ships entering US waters to carry and use a collision avoidance system, to include continuous evaluation of the echoes of all ships posing a collision risk. This démarche quickly resulted in the international adoption of ARPA, the computer-based automatic radar plotting aid so widely used today. The remainder of the radar soon went solid state, although retaining a few analog circuits. Displays now often carry chart material from an electronic chart system (ECS).

1.3.2 Secondary radars

Ship superstructures tended to be angular. To a radar engineer they were corner retro-reflectors riveted together, so were inherently good targets, helping to give the early sets useful range. From the first it was realised that lighthouses ought to be made radar-conspicuous, using transponders called racons (radar beacons), which early British experiments showed to be superior to ramarks radar markers. After initial setbacks, the first chain went into regular service around the British Isles in the late 1960s. Early sets weighed some 300 kg and consumed between 45 and 450 W, see Figure 1.5, but advent of solid-state electronics soon brought weight below 20 kg and consumption

(a) (b) (c)

Figure 1.5 *Early 9 GHz racons. (a) Bell Rock lighthouse. This historic structure, which first exhibited its light on 1 February 1811, carried one of the first racons to enter regular service, 1968. (b) Lighthouse engineer inspecting antenna. Transmit and receive full-height WG16 waveguides, four slots giving ±10° elevation beamwidth, flares giving omnidirectional azimuth response ±2 dB, gain 7 dB. Perspex λ/2 radome. Reproduced by permission of the Northern Lighthouse Board. (c) Circuits rack. Duplicated transistorised receive–transmit units. Tuneable magnetron transmitters. Automatic monitoring and changeover unit above. Power consumption 45 W from diesel generator. Sub-units transportable by ship's boat. Reproduced by permission of the Northern Lighthouse Board*

Figure 1.6 *Early low-power racon. GEC-AEI Sea-Watch 300; here in transponder duty on support ship's mast, 1973*

to 1 W, extending use to buoys; Figure 1.6. The market is well under 1000 a year, so racons remain rather costly. According to ITU,[2] the world population is about 6000, of which rising 60 per cent are dual band, the remainder being 9 GHz types.

In the 1980s racon technology was used as a basis, initially in Japan, for search and rescue transponders (SART) for the specialist task of marking liferafts. Being a mandatory carriage requirement, some 50 000 are carried by shipping, although it is hard to find evidence that they have saved many lives.

The echoes of small craft have for long been augmented by metallic reflectors, which are necessarily rather bulky. Active reflectors – containing electronic circuits – have now appeared, called radar target enhancers (RTE). Problems remain, particularly provision of adequate radar cross section for heeling yachts. The potential user base is very large, exceeding that for small-craft radar, and prices are falling as products becomes better established and the regulatory authorities encourage carriage.

1.3.3 VTS

Progressive port authorities quickly adapted ships' radar to monitor shipping movements from shore, initially to regulate traffic during and after fog. Although marine radars are still sometimes used for this task, 'harbour radar' has evolved into a specialist VTS discipline. Powerful radars with big scanners feed extensive data-handling adjuncts. Stations may have to combine echoes received by land-line or radio link from a dozen or more remote radar heads, and add data from other sources such as closed circuit television and shipping databases to provide the operator with full situational awareness of the traffic. The market is small – perhaps 100 stations a year – and

[2] International Telecommunications Union document ITU-R SE34(99)1 Annex 3 (SE(99)TEMP 171 rev1). Compatibility studies between existing and proposed new radio services in the band 2700–3400 MHz.

equipments are often custom-built to port requirements, so unit cost tends to be high. Developments are concentrated on data fusion from multi-head systems, with some convergence with air traffic control practice. Range surveillance systems are broadly similar to VTS. Argument raged for a long time whether VTS should control or advise. The author soon found that use of VTMS (M for management) was likely to stir such a hornets' nest that the term is banished from this book. IMO is harmonising[3] the training and certification of VTS operators and supervisors.

As already noted, there is a trend to colocate maritime rescue coordination centres (MRCC) with VTS stations to enhance the resources available to handle marine incidents.

1.3.4 The current generation of radars

All large fishing vessels, and all ships within SOLAS – broadly, all merchant ships – have to carry a 9 GHz radar. The bigger merchant ships have to have a second set, preferably at 3 GHz. It is remarkable that after half a century's development in a competitive international market the WW2 design concept is still followed, demonstrating fitness for purpose rather than complacency. Although there have of course been numerous improvements in detail, centimetric-wavelength short pulses are still generated by a magnetron and radiated by a rotating scanner as a narrow beam. The tiny echoes bounced back from targets are amplified in a form of radio receiver and displayed as a map representation of the area surrounding the radar. Range continues to be calculated from elapsed time and bearing from scanner pointing angle.

Instead of functioning as a stand-alone device, shipborne radars now tend to be treated as units within an integrated navigation-aid system, linked together to form an integrated bridge system or IBS. Radar designers seized on computer technology as it developed, particularly for processing the signal delivered from the receiver. This has enabled use of colour raster-scan television-style daylight-viewing screens, capable of displaying auxiliary data such as track lines and alpha-numerics, culminating in a complete superimposed electronic chart. Ever more computing capacity is now being harnessed, primarily to improve detection and tracking of multiple targets in clutter. Radars are now seen as sensors within an integrated data system, whether on the bridge in an integrated navigation system (INS), at the VTS centre or at the range control building. Most big-ship radars are no longer supplied as stand-alone items, but as part of a complete electronics package. In all cases the overall aims are improvement of the operator's situational awareness and enhanced detection of weak targets.

Marine and VTS equipments are simpler than some military radars, and very much cheaper. But they are not a poor relation. On the contrary; long continuous development in a fiercely competitive market gives the user choice of the optimal engineering solution for almost every requirement of performance, reliability, ease of use, size and price. The world market for deep-sea radar is only moderate – less than 25 000 a year. Technological development work is costly and radar improvements

[3] IMO MSC Circular 1065, IALA standards for training and certification of vessel traffic services (VTS) personnel.

perforce ride on the back of mass produced items for communications, computers and television.

At the other end of the scale, miniaturisation and price reductions have enabled smaller craft to benefit from radar, and a large specialist yachting market has developed, particularly in the United States and the Far East. There may be[4] about 30 000 of 3 GHz radars at sea, predominantly on large ships and certain large fishing vessels, and 800 000 at 9 GHz, on vessels of all sizes.

Marine radar engineers, of course, keep fully abreast of developments in related fields. Current work programmes concentrate on integration of displays with electronic charting and with integrated bridge systems generally, which all make wide use of digital technology, and this task is by no means yet complete. Emphasis is being placed on improving the echo strength of small vessels, some travelling fast, using passive or active reflectors. Radar is beginning to be employed to determine wave information for routeing assistance systems, to find the optimal route to minimise voyage time while avoiding seas bad enough to cause cargo damage, particularly when containers are carried on deck, or even to hazard the vessel itself.

Increasing pressure on the electromagnetic spectrum from telecommunications interests has recently forced the radar industry to pay more attention to unwanted out-band transmitted interference and further tightening of permissible emissions can be expected.

1.3.5 Future possibilities

Technical advances, evolving operational spectral and other constraints, plus desire for new facilities may induce substantial design changes in the fairly near future. Professor Baker, who has been studying the possibilities, has very kindly, and bravely, contributed a concluding chapter reviewing a wide range of options for change. Some are evolutionary, continuing existing trends. More revolutionary re-design would discard much of the present 'pulse magnetron' configuration in favour of low power, long pulse modulated transmissions, active scanners, microwave data processing and other unfamiliar concepts. What will happen only time can tell. Such developments would demand extensive revision of IMO and other regulations, not only those relating to radar itself, but extending to SARTs, racons, RTEs and reflectors. If only for that reason, new concepts will not supersede conventional technology for at least a decade and equipment to the current concept will surely be around in a quarter of a century's time. The underlying laws of physics are eternal and well understood, so most of the content of this book will always remain valid, come what may.

Irrespective of the form radar may eventually take, it will be more necessary to combine marine or VTS radar data with that gathered from other sources, particularly the emerging AIS system. Sollosi [10] gives a useful summary of the VTS function and the problem of fusion of radar and AIS data.

[4] From ITU-R, IALA standards for training and certification of vessel traffic services (VTS) personnel.

1.4 The regulators

1.4.1 Overview

Mr Kim Fisher gives a good outline of the development of marine radar and its regulation in the Foreword. At first, marine radar was seen as experimental and the whole merchant marine was far less closely regulated than today. Innovation flourished. When designs had settled down and it became obvious that radar was useful and here to stay, some of the leading maritime nations began to require new designs to be type-approved by their admiralty laboratories to national specifications. A manufacturer faced the wearisome and expensive grind of seeking approval to slightly differing specifications in the United Kingdon, United States, Germany and other states. Racon specifications tended to be written by the suppliers in detailed consultation with leading lighthouse authorities. Meanwhile, IMO (see Section 1.4.3) had been founded, initially as the Inter-Governmental Maritime Consultative Organization (IMCO). Nowadays a formal international regulatory system is well established, as we shall see.

Ports and harbours differ widely in size, traffic, weather, legislative basis, affluence and navigational difficulty. Harmonisation of working practices is slowly being tackled internationally through the International Association of Marine Aids to Navigation and Lighthouse Authorities (IALA; see Section 1.4.6) and others, leading to closer standardisation of VTS equipment and procedures.

Today's shipping industry certainly does not lack international regulation. The following summary giving the spheres of operation of the main regulatory authorities and instruments involved in radar issues only skims the surface; for more on marine regulation see Maclachlan [11]. The regulations are revised from time to time and readers should always confirm the current position. For example, at the time of writing (March 2004), IMO and IEC were radically reviewing the radar performance standards.

1.4.2 UNCLOS

The United Nations Convention on the Law of the Sea codifies international law and customs of use of the sea which have evolved over the centuries and sets down the rules binding States on international marine affairs. It covers such concepts as the right of innocent passage and defines the limits of coast States' jurisdiction. Changes are made within the IMO framework at Diplomatic Conferences of Contracting Governments.

1.4.3 IMO

The following sections benefit from advice kindly supplied by Mr Rob Andrews[5] and Mr I. Eckert.[6]

The International Maritime Organisation, with headquarters in London, is the United Nations agency charged with regulating maritime technical matters, with

[5] (UK) Maritime and Coastguard Agency.

[6] I. Eckert of the Bundesamt für Seeschiffahrt und Hydrographie, Hamburg. BSH is the German Federal Maritime and Hydrographic Agency.

strong emphasis on safety of life. IMO is a sister of the International Civil Aviation Organisation (ICAO), and the two cooperate on air–sea rescue and other topics of mutual interest. The newer industry of aviation was closely regulated from inception, but people had been going to sea from time immemorial, working to rules and customs thrashed out piecemeal as problems arose. It is therefore natural for IMO to take a more consensual approach than ICAO, regulations usually only being made binding after the more advanced users have developed them as best practice.

Speaking broadly, IMO is a club of over 160 member administrations (in general, Nation States). The International Chamber of Shipping and 50 or more other official and non-governmental accredited organisations having the privilege of observer status also contribute to its decisions, which always follow wide consultation and debate. IMO only does what members agree and cannot act on its own initiative. To get a change made, one must persuade an accredited body, such as the International Chamber of Shipping or one's Government, to table the matter at, say, IMO's Marine Safety Committee. If MSC agrees the question merits attention, it will instruct the appropriate sub-committee (on which the administrations are represented) to investigate; observers may be invited to speak on topics where they have specialist expertise and concerns. Most radar subjects fall to the Sub-committee on Safety of Navigation (NAV) but a few are more appropriately handled by other Sub-committees; maybe Communications and Search and Rescue (COMSAR), Design and Equipment (DE), responsible for hulls and fittings or even by the Legal Committee. The[7] Navigational Safety and Maritime Security section, formerly the Maritime Safety Division, has traditional duties pertaining to the work of the sub-committees on NAV and on COMSAR, and has assumed responsibility for regulatory matters relating to the prevention and suppression of terrorism against shipping.

Any administration or observer can circulate discussion papers, in practice usually after taking soundings of interested parties within their country or membership. Papers are then debated at a session of, say, NAV. Having reached consensus after perhaps a couple of sessions' work, NAV tables its reply to MSC. If it approves, MSC either amends current requirements or proposes a formal Assembly resolution, for adoption at the next IMO Assembly. These diplomat-level meetings are held every 2 years. This rather ponderous but democratic procedure gives all interested parties internationally the opportunity to consider fully the proposed measure and its side-effects. Adopted resolutions are recognisable from their numbering system, thus: A.222(12), meaning Assembly Resolution number 222, adopted at the twelfth Assembly – formerly roman numerals were used, e.g. A.222(XII). Resolutions take the form of minimum operational performance standards (MOPS) or carriage requirements. Having written the performance standard, ISO or IEC (see Sections 1.4.8 and 1.4.9) prepare test or technical standards.

1.4.4 National consultations

Each state has its own way of forming its position on any topic before IMO. In the United Kingdom, the Maritime and Coastguard Agency (MCA) is responsible for

[7] This paragraph is reproduced from *Seaways, The International Journal of The Nautical Institute*, April 2002, 'Nautelex', Brian Bailey.

matters affecting marine safety. MCA is an executive agency of the Department for the Environment, Transport and the Regions – titles seem to change after every general election – whose ministers are of course responsible through parliament to the electorate. Two groups gather opinion. The Safety of Navigation Committee (UKSON) is responsible for policy and political issues. Its membership includes government departments, shipowners' and seafarers' representative associations and less formal bodies. The Marine Navigational Equipment (MNE) sub-committee handles technical matters, with membership drawn from shipowners, navigators, lighthouse authorities, seafaring trade union interests, equipment manufacturers and the training colleges. There is overlap and the committees liaise closely. This arrangement has stood the test of time and the author has always found it possible to get a fair hearing for a cogently expressed view.

Operators frequently grumble that 'the powers that be' have foisted yet another inappropriate measure on them. The remedy is partially at least in their own hands; to make their views known through the channels outlined above. Communication has never been easier.

1.4.5 SOLAS and the Colregs

Most IMO assembly resolutions are incorporated within major IMO legal instruments, in particular SOLAS (sometimes written Solas). This International Convention on Safety of Life at Sea is binding on all Contracting States after ratification by a pre-set number of them. It has a number of chapters and is revised from time to time. Chapters 5 and 10 include the Carriage Requirements for radar and other on-board navigation equipment for ships and for HSCs, respectively. These define what must be carried on each size and class of ship, including the navigational radar. SOLAS also states if and when exceptions may be permitted by national authorities and defines specifications and type-testing procedures applicable to the equipments.

Similarly the Convention on the International Regulations for Preventing Collisions at Sea, variously known as IRPCS, the collision regulations, Colregs, or the rules, is a mandatory set of 'rules of the road' for navigation of all craft, including requirements for full and appropriate use to be made of radar. Cockcroft and Lameijer [12] print and fully explain the Colregs.

1.4.6 IALA

The International Association of Marine Aids to Navigation and Lighthouse Authorities[8] brings together services and organisations concerned with the provision of marine aids to navigation systems and allied activities including vessel traffic services, at sea and on inland waterways. It is responsible to IMO for matters concerning operation of aids to navigation such as lights and buoys, electronics and VTS matters being handled by its radionavigation and VTS Committees, respectively.

[8] 20ter rue Schnapper, 78100 Saint Germain-en-Laye, France. iala-aism@wanadoo.fr.

The General Lighthouse Authorities (almost invariably government departments) of most leading maritime states are members. Specialist organisations such as the International Association of Ports and Harbours are represented and most of the leading suppliers of lighthouse and VTS equipment are Industrial Members. Because harbours differ so widely and so many are state-controlled, international regulation of their activities has been fairly light.

1.4.7 Enforcement

IMO has no policemen. Enforcement of its instruments is entrusted to domestic legislation of the national Administrations adopting IMO conventions, through Flag State (in which the vessel is registered) and Port State (states visited by the vessel) controls. For example, the Colregs are given force of law in the United Kingdom by Statutory Instrument.[9] It is the duty of Flag State administrations to ensure their shipping carries radar conforming to SOLAS requirements and that ships' officers are properly qualified. Classification Societies provide design, build and maintenance standards for shipping, primarily for insurance purposes, to answer that centuries-old question: Is this ship a good insurance risk – is she seaworthy? Among many other tasks, their surveyors need to assure themselves that the radar installation complies with IMO carriage requirements.

1.4.8 ISO

The purpose of the International Organisation for Standardisation is to issue internationally acceptable technical standards to facilitate world trade. It has over 100 member nations and over 180 technical committees and has published around 10 000 standards on all manner of things, although electrical equipment is generally handled by the IEC (see below). Membership structure is similar to that of IMO. ISO/TC-8 is the ISO Technical Committee on Ships and Marine Technology, responsible for design, construction, structural elements, outfitting parts, equipment, methods and technology, and marine environmental matters, subject to IMO requirements. Users are encouraged to feed back their experience into the standardisation process. TC-8 links IMO to the marine industry. Of its ten sub-committees, SC6 deals with navigation.

1.4.9 IEC

The International Electrotechnical Commission issues detailed test and technical specifications for electrical and electronic equipment. ISO or IEC, as appropriate, clarify IMO's requirements, tie them in with related requirements of bodies such as the International Hydrographic Organisation (IHO, responsible for coordination of

[9] Statutory Instrument SI 1996/75, Merchant Shipping Safety. The Merchant Shipping (Distress Signals and Prevention of Collisions) Regulations (HMSO, London).

nautical charts) and ITU (see below), and define methods of testing and required test results to harmonise with IMO performance standards, without extending the requirements. Older IEC specifications having three digit serial numbers have been renumbered by addition of 60 000; for example, IEC936 becomes IEC60936:1999, the suffix indicating date of latest revision.

1.4.10 ITU

As for any other radio device, radar transmissions are governed by the Radiocommunication Sector (ITU-R) of the International Telecommunication Union. This United Nations body regulates all classes of radio service to assure interoperability and prevent interference and has taken on former Comité Consultatif International de Radiocommunication's (CCIR's) responsibilities. Radar lies within the Radiodetermination Service. ITU-R allocates blocks of the radio spectrum to the various services, with limits on permissible out-of-band transmission and forms of modulation. It holds World Radio Conferences (WRC) or World Administrative Radio Conferences (WARC) every couple of years. Its regulations are administered by national regulatory authorities. Further details are given in Hall *et al.* [13]. There is considerable and useful overlap of delegates to IMO, ISO, IEC and ITU working groups.

1.4.11 National regulations

Many IMO Instruments have to be backed by parliamentary approval and national legislation to gain legal force within the Flag or Port State. National regulations and approvals outside the IMO framework may also apply to some aspects of radar equipment and its use; for example siting and height of VTS scanner towers are often constrained by visual amenity considerations. Transmission licences are usually issued by national post telephone and telecommunications (PTT) authorities; in the United Kingdom by the Radiocommunications Agency, an executive agency of the Department of Trade and Industry; in the United States by the Federal Communications Commission (FCC).

1.4.12 National and supra-national groups; the European Community

Many states have national standards organisations, for example, the British Standards Institute. Frequently their standards antedate ISO and IEC but have now usually been harmonised with the international equivalents. The national organisations, by a consultation and voting process, endorse draft IEC and ISO standards before enactment.

A tendency is developing for regional organisations such as the North American Free Trade Agreement (NAFTA) and particularly the European Community (EC) to interest themselves in marine regulation, particularly of fishing vessels and their equipment. The EC has spawned further standards authorities, which publish the EN (European Norm) series of standards as follows (EN IEC standards are technically identical to IEC standards):

- European Committee for Standardisation (CEN)
- European Committee for Electrotechnical Standardisation (Cenelec)
- European Telecommunication Standards Institute (ETSI).

After a parliamentary process, the EC imposes its will on member states by directives. One such[10] lists many international testing standards for radar, SARTs and much else. Subject to derogations in special cases, directives have the force of law in the states. Domestic laws have to be amended to comply, directives overriding any conflicting national legislation. The European Marine Equipment Directive (MarED) harmonises procedures within member states. It permits independent laboratories to type-test marine equipment against relevant specifications and recommendations issued by bodies such as IMO, ISO, ITU, CENELEC and ETSI. Compliant equipment is identified by a wheelmark. Wheelmarked items are recognised by all member states, irrespective of the approving state, doing away with the former multiplicity of national tests. Subject to 'grandfather clauses' allowing certain existing equipment, ships flagged by member states are obliged to have wheelmarked obligatory navigational equipment, including the radars.

Optional additional equipment may be carried, such as radar on vessels below IMO's minimum tonnage limit for obligatory radar carriage, currently 300 gt. The EC requires these equipments to comply with the R&TTE Directive, which includes ITU regulations. Within Germany, for example, SOLAS, the MarED and the R&TTE Directive have been made mandatory by inclusion within national law, their yachting radars only having to comply with the R&TTE Directive.

The European Union (EU) has a Maritime Transport Directorate, (DG, TREN) and wishes also to upgrade its position as observer-only at the IMO. Current EU maritime policy concentrates on the enforcement by member states of international regulations and to fill the gaps in existing legislation. It also aims to make its waters the safest in the world. It remains to be seen how these aspirations will be received by the other IMO States.

Following major shipping disasters in European waters, the EU has adopted very substantial packages of legislation to improve maritime safety and reduce pollution, and in 2003 created the European Maritime Safety Agency to contribute to the enhancement of the overall maritime safety system in the EC. Among its tasks will be to establish a Community vessel traffic and information system. The EMSA will also facilitate cooperation between the member States. For details see Seaways,[11] March 2004, from which this paragraph is extracted.

1.4.13 The courts

The final interpreter of the law is the lawcourt. Although criminal prosecutions are few, civil litigation is often used to determine who was to blame and should pay for marine

[10] EC Council Directive 96/98/EC of 20 December 1996 (OJ No. L 46, 17.2.1997, p. 25) as amended.

[11] "Making European waters safer? The European Maritime Safety Agency programme for 2004." Seaways, the International Journal of the Nautical Institute, March 2004, p10.

accident damage. The courts' interpretation of all rules and regulations is governed by an accumulated body of national and international case law. In recent years, the thrust in many jurisdictions has been to stress the responsibility of senior management to provide appropriate tools for the job and to ensure by formal procedures that they are safely and correctly used by trained personnel.

Imagine a fishing vessel (FV) has collided with a coaster. The evidence of its officer of the watch, confirmed by the voyage recorder, is accepted that although diligently observing the radar, which was correctly set for the severe sea clutter, the FV was undetectable. Apportionment of blame might depend on the court's findings on these questions.

- Was the radar capable of detecting the target?
- Ought the coaster owners to have known by how much the long feeder to the high scanner would degrade the radar's performance?
- Ought they to have known that this radar outfit with its small high scanner was unsuited to the small targets and rough weather prevalent in the coaster's trading area?
- Should the skipper have known that his inferior radar reflector, incorrectly mounted, would jeopardise his vessel's detectability?

1.5 The regulations

1.5.1 Radar for ships within SOLAS

The following particularly clear overview of the main performance standards for marine radar, written by Mike Pope, of the radar manufacturer Sperry Marine cannot be bettered. It is reproduced with permission [14], with formal specification titles and IMO references added.

Performance standards for marine radars are promulgated by the IEC. Historically, these performance standards have been published as separate stand-alone documents and I believe that a future task will be to combine them into a single performance specification. Herewith a brief description of the current standards.

1. IEC 60936-1, *Maritime navigation and radiocommunication equipment and systems – Radar – Part 1: Shipboard radar – Performance requirements – Method of testing and required test results.* This is the fundamental radar specification all manufacturers must meet (for equipment to be fitted on Solas vessels). It defines all of the important parameters that the equipment must comply with to achieve type approval. It defines the range scales to be offered (others may be offered, but it must include those specified), minimum range requirements, range and bearing discrimination, minimum antenna scan rate, azimuth stabilisation details, bearing scale details, range and bearing measuring tools and accuracies, range performance, display modes of operation and other specifications. (Pope later comments that IEC 60936-1 Annex D contains tighter emission regulations for radars in the 3 and 9 GHz bands to meet new ITU regulations. WRC 2003 is to consider, and probably tighten, the boundary between out of band emissions and in band peak equivalent power, PEP).

2. IEC 60936-2, *Maritime navigation and radiocommunication equipment and systems – Radar – Part 2: Shipboard radar for high speed craft (HSC) – Performance requirements – Method of testing and required test results.* This standard is, in effect, an extension of the above, and describes the additional requirements for radar which are to be fitted to HSC. The minimum range and range discrimination requirements are more demanding and the antenna rotation rate is higher (40 rpm minimum). This standard also details the scenarios that the associated ATA or ARPA must comply with.
3. IEC 60936-3, *Maritime navigation and radiocommunication equipment and systems – Radar – Part 3: Radar with chart facilities – Performance requirements – Method of testing and required test results.* This standard is new, published in 2002. It details the testing standards and test results required for radars with charting facilities. It also defines what information can be displayed: in effect, it is only selected parts of the system electronic navigation chart (SENC) that may be shown. The most important point to make here is that it is a radar, not a chart display system, and it is vital that the radar information should not be masked or degraded in any way when the chart information is added to the display.
4. IEC 60872-1, *Maritime navigation and radiocommunication equipment and systems – Radar plotting aids – Automatic radar plotting aids (ARPA) – Methods of testing and required test results.* This standard details the minimum number of targets we have to track, the tracking accuracy to be achieved, alpha-numeric data to be displayed for the tracked targets, details on guard zones and acquisition zones, operational warnings, trial manoeuvre details, interfacing, symbols and other specifications. [Based on IMO Resolution A.823:1995, performance standards for automatic radar plotting aids (ARPAs).]
5. IEC 60872-2, *Maritime navigation and radiocommunication equipment and systems – Radar plotting aids – Automatic tracking aids (ATA) – Methods of testing and required test results.* The ATA specification is very similar to that of the ARPA, except that the ATA has to track a minimum of 10 targets compared to the ARPA's 20. Also, for the ATA, trial manoeuvre and history dots are not required.
6. IEC 60872-3, *Radar plotting aids – Electronic plotting aids (EPA) – Methods of testing and required test results.* This is the simplest of the three plotting standards. EPA is a manual plotting system. Again, the standard defines how many targets are to be manually plotted, how the information is to be displayed, symbols to be used, accuracy to be required and other specifications.
7. IEC 60945, *Maritime navigation and radiocommunication equipment and systems – General requirements – Methods of testing and required test results.* This standard deals with issues such as environmental testing for heat, cold, humidity, vibration and corrosion. It also deals with electromagnetic emissions and susceptibility to electromagnetic interference, illumination of controls, compass safe distance, equipment manuals and acoustic noise.

IMO's 2004 radar review will doubtless lead to the amendment of IEC 60936 and IEC 60872.

1.5.2 Radar for craft outside SOLAS

These radars, carried voluntarily, may comply with a recent performance specification, IEC 62252 ED1 Maritime navigation and radiocommunication equipment and systems – Radar for craft not in compliance with IMO SOLAS Chapter V – Performance requirements and methods of test and required test results. Summarising, three classes of radar, A, B and C, are recognised and may use the 3 or 9 GHz bands. Scanner first sidelobes should not exceed −20dB (class A) otherwise −18dB. When mounted at 7.5m, with normal propagation and no clutter, the radars should pick out the following targets on 8 out of 10 scans:

Ground rising to 60 m: Class A 9 nmi; classes B and C 5 nmi.
Ground rising to 6 m: Class A 5 nmi; classes B and C 3 nmi.
Radar reflector, RCS 400 m^2 at height 7.5 m: Class A 5 nmi; classes B and C 3 nmi.
Radar reflector, RCS 10 m^2 at 3.5 m: Class A 2 nmi; classes B and C 1 nmi. Minimum ranges: Class A 50 m, class B 60 m, class C 75 m.
Radar reflector, RCS 5 m^2 at 3.5 m: Class A 1 nmi, classes B and C not applicable. In sea clutter a target of 200 m^2 (class A) otherwise 400 m^2 should be detected at 100 m to 1 nmi on 5 of 10 scans.
Display effective diameter: Class A 150 mm, class B 85 mm, class C 75 mm.

1.6 Theory and calculations

1.6.1 Sources

No book of this sort could be written without reference to several good textbooks such as those by Barton [15], Kingsley and Quegan [16] and Skolnik [17]. These and numerous papers treat detection as part of radar or communications theory. Our text references all sources of important quantitative data, but it is impossible to reference every qualitative statement, many of which have been absorbed by the author over the years. Readers needing more depth should go to the relevant textbook chapters. The textbooks are generally well indexed and contain extensive bibliographies, routing interested readers to the specialist literature, which is voluminous but often difficult for the layman. Unfortunately, most textbooks either generalise by including aircraft, military jammers and other problems of no interest to us, or demand deep prior knowledge of electronics and mathematics. We shall consider only factors likely to have some practical significance, mentioning matters of purely academic interest only where necessary as a step to solution of some practical problem. We have aimed always to give sufficient theory to highlight what actually happens, the conditions under which results are valid and the likely residual error in calculations, and have tried to be consistent in terminology. Readers wishing to brush up on their basic understanding of electronics may find Bishop [18] or Hagon [19] useful.

Detectability depends on aspects of radar engineering, navigation, meteorology, oceanography and statistics. We have tried to give a straightforward account assuming no prior knowledge of these subjects. All of them abbreviate their common technical statements using jargon, which we shall explain as we go along. For example, the navigator might ask ‘At what range should our X-band radar raise a panamax bulker loaded to her WNA marks?' and the engineer might answer '12 miles assuming four-thirds Earth and RCS 40 dB square metres.' The value of this jargon is shown by the long-windedness of these technical sentences in plain speech: 'At what maximum range should our radar lying in the marine 9 billion cycles per second frequency band display the radar reflection of a ship whose size is the maximum allowable in the Panama Canal and designed for carriage of bulk cargo, when laden to the winter North Atlantic load-line mark on her hull?' and 'At 22 km, assuming the rate of change of refractive index with height of the atmosphere is such that if the Earth were assumed to have $\frac{4}{3}$ its actual radius radar rays would travel in straight lines, and she reflects equivalently to a metal sphere of silhouette area 10 thousand square metres.' The more important jargon terms are defined in the Glossary, Appendix 1.

1.6.2 Mathematics and units

We have tried our best to keep the inevitable mathematics of the calculations as simple as we can. This approach leads to a few differences from the standard treatment of some topics, indicated in the text. Some of the other possible treatments give slightly higher accuracy at the cost of more difficult mathematics, but the uncertainties surrounding the environment between the radar and its targets, as well as those within some of the targets, usually swamp any approximations of our approach. Refer to Chapter 13 for comments on accuracy of calculations. Wide use will be made of decibel (dB) notation, explained in Chapter 2, Section 2.1.7.

Scientific writing uses SI (Système International des Unités) units,[12] formerly called the rationalised metre–kilogram–second (MKS) system, whose base units are: length (metre, symbol m), mass (kilogramme, kg), time (second, s), electric current (ampere, A), temperature (kelvin, K = degree Celsius + 273.3); also amount of substance (mole, symbol mol) and luminous intensity (candela, cd), which do not concern us. The radian (rad) of plane angle and the steradian (sr) of solid angle are supplementary units. Derived units for all other physical quantities use the above in a manner which minimises constants of proportionality. Examples are the newton, the measure of force, dimensions $\mathrm{m\,kg\,s^{-2}}$, and the volt, the measure of electric potential, $\mathrm{m^2\,kg\,s^{-3}\,A^{-1}}$.

We retain some everyday non-SI units: $1^\circ = 2\pi/360\,\mathrm{rad} \sim 0.001\,745\,\mathrm{rad}$, $1\,\mathrm{foot} = 0.3048\,\mathrm{m}$. The nautical mile remains in widespread marine use, including radar display scaling (although kilometres are preferred for river radar), and is often abbreviated n.m. but to avoid confusion with nanometres (10^{-9} m) we prefer nmi. 1 cable is 0.1 nmi, its use generally inferring an approximation. Relationships are: 1 nmi = 1 min of latitude. Napoleon's savants defined the metre as 10^{-7} the distance from the North Pole to the Equator, through Paris; by which $1\,\mathrm{nmi} = 10^7/(60 \times 90)\,\mathrm{m} = 1851.85\,\mathrm{m}$. (The metre has since been redefined as the distance travelled by light in vacuum in 1/299 792 458 s and 1 nmi as 1852 m exactly).

12 Units and Symbols for Electrical and Electronic Engineering, an IEE Guide (Institution of Electrical Engineers, London, 1997).

If once upon a time a foot represented 0.01 s of latitude, the modern foot is 1.3 per cent short. The statute mile is 5280 ft, 0.869 nmi or 1.609 km. Our calculations will usually prefer km to nmi to minimise tedious conversion factors.

Bearings are either True (relative to the North Pole) or relative to some stated reference such as ship's head (i.e. centreline). With wide use of gyrocompasses, Magnetic North is less used. A point (of the compass) is $360/32 = 11.25°$, used by mariners for rough estimates. Grads, 400 to a circle, milliradians and circular mils (6400 to a circle) are not used in civil marine.

We do not follow a forthcoming ISO standard, ISO 19018, which defines many navigational terms in a consistent manner. It includes the nautical mile (NM, but abbreviated as M on charts) as the fundamental length; 1 NM = 1852 m. 1/10 NM is 1 cbl (cablelength, also named cable). The unit of speed is the knot (kn): 1 kn = 1 NM/hr. The standard uses abbreviations RM, H up, C up, N up for relative motion, ship's head up, ... course up, ... north up of displays. For details see Junge,[13] on which this paragraph is based.

For convenience, we often use practical units; ranges in kilometres, wavelengths in centimetres, antenna beamwidths in degrees, transmitter powers in kilowatts, pulse-lengths in microseconds. But except where definitely stated, our formulae always express angles in radians, lengths (wavelength, range, etc.) in metres and times in seconds. That is, we use SI except where other measures are widespread – an example is rainfall rate, quoted in mm h^{-1} rather than kg m^{-2} s^{-1}. *Always remember to make appropriate conversions*; for example, in calculations 10 kW (10 000 W) transmitter power *must* be written 10^4 W or preferably in scientific notation as 10×10^3 W, and 1.0° scanner beamwidth *must* be put as 0.01745 ... rad.

The basic units often give inconveniently large or small numbers, for example, frequency 60 000 000 s^{-1}, or current 0.0007 A. These zeros and 'damned dots' are a fruitful source of error – the result of a calculation may be dead accurate apart from being a thousand times too big! Prefixes are applied every 10^3: giga (G) = 10^9, mega (M) = 10^6. The smaller multipliers use lower-case letters: kilo (k) = 10^3, milli (m) = 10^{-3}, micro (μ = mu) = 10^{-6}, nano (n) = 10^{-9}, pico (p) = 10^{-12}. Other multipliers in common use are the centimetre (cm), 10^{-2} m; decibel (dB), 10^{-1} bel, and for atmospheric pressure the hectopascal (hPa), 10^2 Pa.

Inadvertent use of mixed units is productive of error in calculations and unfortunately not all information sources clarify their units. The author has been caught out often enough to make no apology for insertion of units after equations – especially when non-SI – for avoidance of doubt, when purists would deem them unnecessary. Where no units are stated, SI units are to be understood.

The electrical units – volt, amp, ohm, watt, farad, henry, etc. – form a coherent set within the SI system, so when we write $V = IR$ we do not need to add that the answer is in volts when current I is in amperes and resistance R in ohms. The fly in the ointment is frequency, where 2π often sneaks in. Radian frequency ($\omega = 2\pi f$) merely

[13] H. Junge, "Harmonisation of navigational terms. Synopsis of ISO/19018/Final/draft." Seaways, the International Journal of the Nautical Institute, April 2004 p26.

transfers 2π to formulae involving wavelength, so we generally retain conventional (cyclic) frequency, generally referred to simply as 'frequency', f (Hz) and put up with a sprinkling of 2πs in expressions.

We write $\log(x)$ for the common logarithm $\log_{10}(x)$, having base 10, of a quantity x; and $\ln(x)$ for the natural logarithm of x, base ε. So $\ln(x)$ is identical to $\log_\varepsilon(x)$, where the Euler number $\varepsilon = 2.71828\ldots$; $\log_{10}(x) = 0.4343\ldots \ln(x)$; $\ln(x) = 2.3025\ldots \log(x)$. Only when there is specific need to emphasise the base would we add the subscripts. Scaling of graphs within figures is always linear except when axes are specifically indicated as, for example, 'log scale'. Many of our graph axes run between non-zero quantities, with suppressed zero.

1.6.3 Basis of performance calculations

Equipment manufacturers rarely divulge the exact strategies they use for the various steps of the detection process, particularly now that maintenance is by exchange of printed circuit boards or other lowest replaceable units containing high levels of functionality and many functions are carried out within custom-made, digital application-specific integrated circuits (ASICs) whose workings are not divulged. We therefore proceed by noting the performance available within the published 'data sheet' limits of transmitter power, scanner gain, prf, etc. and making reasonable allowance for shortfalls likely to arise in practical designs. The validity of the approach depends in part on the competitive nature of the industry; what *can* be done soon *is* done, and woe betide those firms who do not keep up!

1.6.4 Spreadsheet calculation

The equations needed for calculation of system performance are presented in forms convenient for generation of personal computer (PC) spreadsheets. Chapter 14 gives full listings and operating instructions. Having entered radar, target and environmental parameters, the spreadsheets, available on the IEE website (www.iee.org), deliver results such as maximum detectable ranges and indications of how well detectability is maintained as range closes, with graphs linking such parameters as detectability and range. It is then easy to explore the effect of radar, environmental and target parameters on performance. This approach informs judgement of how worthwhile the various parameters may be in a given situation or the robustness of a configuration against clutter and other environmental uncertainties. The spreadsheets should also help design trials of radars and aids to detection for minimum error from unwanted environmental effects, which was not always possible in the past.

1.6.5 Approximate methods

Full calculation is often complex and may not be justified for the task in hand, so we often include alternative approximate methods, sometimes using graphs. Rough approximations should not be despised, their uses include the following.

- Highlighting the major factors in play, separating the wood from the trees.
- Explaining principles to others.
- Getting 'orders of magnitude' for preliminary work.
- Not least, checking for blunders.

1.7 The layout of this book

The text assumes a working knowledge of elementary electro-technology. It tells the story of the detection of targets, setting out the factors in a logical sequence. To avoid harking forward, some early chapters summarise topics whose details come later. The calculation examples frequently included should not be taken out of context; extreme values or drastic simplifications are sometimes chosen to illustrate particular points under discussion, so calculated results may not apply to the complexities of real life. Where costs are quoted, they are broadly indicative 2004 values, at exchange rates £1 ~ $US1.80 ~ €1.50.

Chapter 2 qualitatively describes how marine, VTS and similar surveillance radars illuminate targets, stressing that the radar and scanner, the operator, the target and the environment work together as a system. Here and throughout, we give examples of typical values of frequency, losses and other parameters, which do not describe or endorse any particular make or model of equipment.

Chapter 3 discusses reception of echoes and unwanted clutter, signal processing, detection and display. Readers with a good understanding of radar technology but coming fresh to marine work should note how the emphasis differs from defence and other radars, for example; there are no jammers and few other man-made impediments, but MTI technology is not appropriate.

Chapter 4 describes what would happen in hypothetical 'free space' where the environment plays no part. The inverse-square law of schooldays physics is at the core of the radar range equation, used to calculate echo strengths.

Chapter 5 looks at the influence of the environment on propagation of signals from the radar scanner to the target and back. As propagation depends on atmospheric factors and sea surface roughness, some elementary meteorological and oceanographic background is included. Although we speak of the 'sea', the surface might equally be that of a ship canal, river or fresh-water lake.

Chapter 6 develops a multipath factor to describe how direct and indirect rays combine to form the resultant signal actually received from simple point targets at short and long ranges. The range equation is extended to include atmospheric and precipitation attenuation.

Chapter 7 discusses the simple theory of reflection from insulators and conductors and then outlines the performance of small targets such as aid-to-navigation reflectors which approximate geometrical points. The effects of target tilt are included. Some examples are given of target pairs and their effect on the uniformity of reflection with viewing aspect.

Chapter 8 describes active point targets, including racons, SARTs and radar target enhancers. The influence of target characteristics on the response received by the radar is presented for each class of active target. Allowances are made for tilt and the passive echo of the host platform.

So far the emphasis has been on point targets. Chapter 9 develops a multipath factor to cover the very important classes of extended targets such as ships and coastlines which are too big to behave like points, developing a method of calculation which accounts for the changing effective cross-section when the incoming target rises over the horizon, and if it exceeds scanner beamwidth and pulselength.

Chapter 10 examines the radar cross section of ships etc., which is often very uncertain. Some of the values reported in the literature are reviewed, followed by an attempt to clothe the experimental results with an elementary theoretical justification.

By the end of this chapter all the necessary information has been assembled for calculation of the mean echo strength of all classes of target, including variation with range, weather and the other environmental factors.

Chapter 11 examines noise, precipitation and sea clutter, which compete with echoes to make detection more difficult.

In Chapter 12, the detection process is shown to depend on fluctuations of echoes and on the noise and clutter background. Statistical analysis finds the probability of detection and of false alarms. Approximations are included, whose accuracy is appropriate to the inevitable uncertainties of actual target cross-sections, environmental parameters and inherent radar performance.

Chapter 13 considers the effect of signal strength on positional accuracy and target tracking, with special reference to track formation using electronic plotting aids, and the particular problems faced by VTS and coastal surveillance systems.

Chapter 14 introduces a method of calculation for all the factors in play, suited to spreadsheet calculation of probability of detection and many associated parameters on a personal computer.

Chapter 15 contains worked case studies which highlight the factors of significance in representative practical situations.

In Chapter 16, Professor Baker reviews future possibilities for improvement of marine radar.

Appendix A1 is a glossary of specialist terms, while Appendix A2 augments some statistics detail of Chapters 11 and 12.

1.8 References

1 BURGER, W.: 'The radar observer's handbook' (9th (revised) edn.)
2 BOLE, A. G. and DINELLEY, W. O.: 'Radar and ARPA manual' (Heinemann and Newnes, 1990)
3 Capt. WYLIE, F. J. (Ed.): 'The use of radar at sea' (Hollis & Carter for the Royal Institute of Navigation, London, 1952, 1st edn.; 1978, 5th (revised) edn.)
4 BARTLETT, T.: 'Radar afloat (official background reader to the RYA radar course)' (Fernhurst Books)

5 BRIGGS, J. N.: 'Detection of marine radar targets', *Journal of Navigation*, 1996, **49** (3)
6 BELL, S. W. and STARLING LARK, A. P.: 'Radar detectability and collision risk, nautical briefing' (Nautical Institute, 1995), Tables 2 and 6
7 LANG, G.: 'If GPS fails', *Seaways, The International Journal of the Nautical Institute*, 2002, reporting a Royal Institute of Navigation Technical Committee Meeting of 26 June 2002
8 LATHAM, C. and STOBBS, A.: 'Radar, a wartime miracle' (Sutton Publishing, 1996)
9 COX, P: 'Memories of surface warning radar' *Transmission Lines, The Newsletter of the Defence Electronics History Society*, Bournemouth UK, 2003, **8** (4)
10 SOLLOSI, M.: 'The automatic identification system and vessel traffic services', *IALA Bulletin*, 2003, **1**, p. 20
11 MACLACHLAN, M.: 'The shipmaster's business companion' (The Nautical Institute, 2003, 4th edn.)
12 COCKCROFT, A. N. and LAMEIJER, J. N. F.: 'A guide to the collision avoidance rules' (Stanford Maritime, London, 1976)
13 HALL, M. P. M.: in HALL, M. P. M., BARCLAY, L. W. and HEWITT, M. T. (Eds): 'Propagation of radiowaves' (Peter Peregrinus for The IEE, 1996, 1st edn.), Preface and Chaper 1, Section 1.5
14 POPE, M. (of Sperry Marine): 'Marine radar technology. Current status and future directions', *Seaways, the International Journal of the Nautical Institute*, 2002, p. 11
15 BARTON, D. K.: 'Radar evaluation handbook' (Artech House, London)
16 KINGSLEY, S. P. and QUEGAN, S.: 'Understanding radar systems', (McGraw-Hill, New York, 1997), ISBN 0-07-707426-2
17 SKOLNIK, M. I.: 'Introduction to radar systems' (McGraw-Hill, New York, 1983), ISBN 0-07-066572-9
18 BISHOP, O.: 'Understand electronics' (Newnes, Oxford, 2nd edn. 2001)
19 HAGEN, J. B.: 'Radio frequency electronics, circuits and applications' (Cambridge University Press, 1996)

Chapter 2

The system and the transmitter

'Power corrupts, but lack of power corrupts absolutely.'

A Parody of Lord Acton

2.1 The operator and the system

2.1.1 Scope of chapter

This chapter outlines radar operation in general terms, and then describes the transmission systems of the relatively large radars used in deep-sea ships, vessel traffic service (VTS) systems and firing-range surveillance. Receiving systems are described in Chapter 3. Chapter 4 onwards detail the various facets of the detection problem, including quantitative analysis.

Figure 2.1 shows the whole radar/target/environment system. A person is studying the traffic situation at the display console. No mere passive observer, this officer adjusts the radar controls to optimise the display of targets of most current importance. Stressing this interaction, we refer to the person as the operator. The display itself, sometimes still called the indicator or scope, with associated controls forms the human–machine interface (HMI) between the radar and operator and one task of this book is to consider how the machine can best help the human perceive the targets – apprehend them within the mind to gain situational awareness. Figure 2.2(a) shows a traditional deck-mounted console, while Figures 2.2(b) and (c) depict alternative formats suited to building into operator workstations.

2.1.2 Operators afloat

Radars on merchant ships, including vessels subject to IMO's high speed craft (HSC) code, are primarily operated by the officer of the watch (OOW), who is the ship's master or a qualified deck officer. The radar(s) are one of the principal tools which aid navigation of the intended route, avoid collision with other craft and confirm positions determined by satellite and other means. The OOW may be the sole person

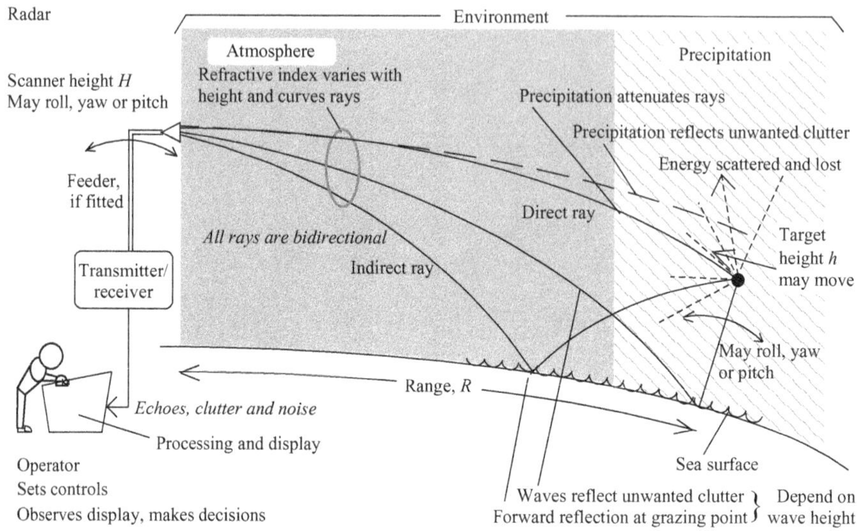

Figure 2.1 The radar system. The operator controls the radar to best observe the target of interest within its environment. The system elements interact; all affect detectability. Radars and targets may be afloat or groundfast

on the bridge during daylight, but at night or in thick weather must be assisted by a seaman lookout, perhaps posted at the bow in telephone contact, but usually on the bridge, keeping visual and aural watch for ships and other hazards but never using the radar. HSC always have two navigators on duty. Figure 2.3 shows a typical bridge layout with displays and controls available to either officer's chair.

A pilot with special local knowledge may be hired to advise the master and often conns the ship, using the radar as would the OOW. The Master or OOW then monitors the proceedings, partly by observation of the radar secondary viewing display, but remains in charge. A seaman helmsman or quartermaster may actually steer the ship under orders, but never uses the radar.

Pilots in some VTS systems are provided with portable laptop computers incorporating modems and radio links giving copies of the current VTS display for the local area, independent of the ship's radar, perhaps revealing targets masked from the ship by bends in the waterway, and annotated with VTS alpha-numeric data. It is less usual for shore radars to transmit data direct to the portables, dispensing with a VTS centre but providing all ships with a common overall high-quality view of the traffic situation. AIS radio-based systems such as the Tideland Signal AIMS Base are now available, offering radarless VTS, claiming all the precision, accuracy and reliability without the costs and maintenance. It will be interesting to see how well they catch on.

Pilots and OOWs hold Certificates of Competency or 'tickets', awarded by a national authority. Radar operation and display interpretation are taught and examined prior to award, standards according with IMO's Convention on Standards of Training,

(a) (b) (c) (d)

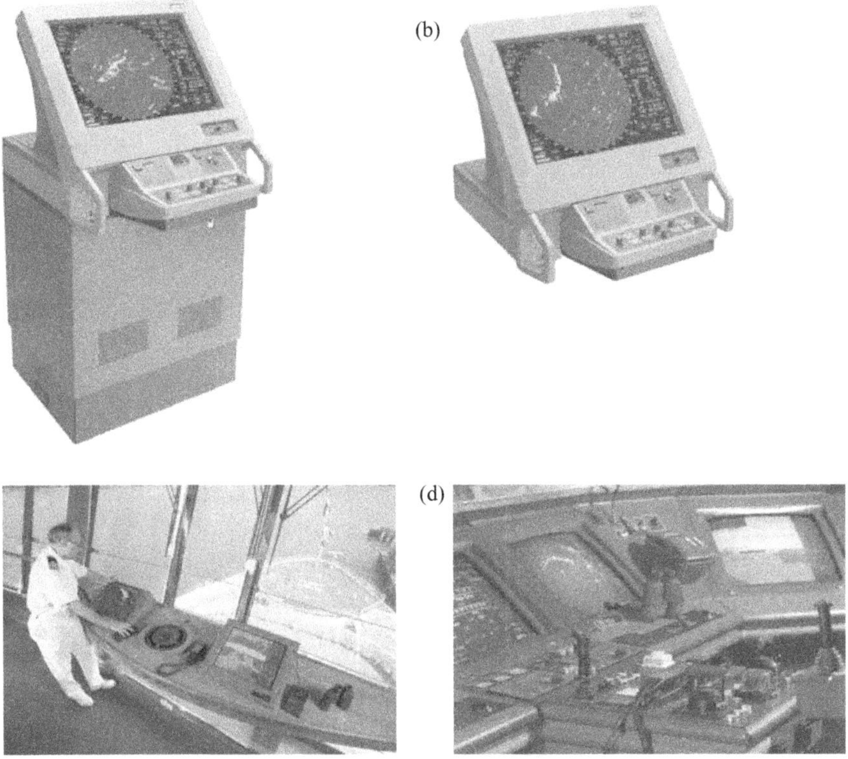

Figure 2.2 *Deep-sea marine radar. BridgeMaster E Series. All reproduced by permission of Northrop Grumman Sperry Marine Ltd, New Malden UK. (a) Traditional deck-mounted console. Controls immediately below display screen, transmitter and receiver in base cabinet. For standing operator, substantial bracing handles for heavy weather. Menu-drive controls below display. (b) Desk-top display for use seated or standing. (c) Flat panel display on RCCL cruise ship Brilliance of the Seas. Bow 3 and 9 GHz scanners for berthing, main scanners above bridge. (d) Main navigation workstation. Radar and chart displays, with engine and steering controls to hand by navigator's chair, Brilliance of the Seas. (e) (overleaf) Bridge wing workstation, again with radar and chart displays, Brilliance of the Seas*

Certification and Watchkeeping (STCW). Deck officers frequently transfer from ship to ship and may be presented with unfamiliar models of radar, so the IMO Marine Radar Performance Specification includes detailed requirements for uniformity of display depiction and of controls and their labelling.

As well as gaining hands-on experience at sea, navigators are taught on full mission or radar simulators ashore at nautical colleges. Simulators can replicate numerous scenarios, exercising the most effective operation and interpretation of

(e)

Figure 2.2 Continued

Figure 2.3 High speed craft command workstations. Typical layout with displays and controls available to either officer's chair. As always, a clear view forward is essential. Reproduced by permission of Kelvin Hughes Ltd, Ilford, UK

the radar. Students sometimes emerge ashen-faced from close-quarter situations they hope never to encounter at sea. Complete ship's bridge simulators take the process further by inclusion of life-like and interactive views of the surroundings, with a full suite of navigational controls.

Naval officers are trained and examined in navigation much as their merchant navy cousins. Although naval bridge teams are larger, warships often take civilian pilots in unfamiliar harbours. Skippers and mates of fishing vessels (FV) are often part-owners, or at least share voyage profits. Time is money to them, and they make full use of radar on passage. FVs are unmanoeuvrable while fishing and careful watch is kept for collision risks from approaching shipping. Owners of private leisure craft are not usually required to carry radar or be trained in its use, but will want to get the best out of an expensive gadget they have chosen to buy out of their own pockets.

2.1.3 Integrated bridge systems

Beside radar, operators gain situational awareness from the view from the window, radio traffic now including AIS, sound signals and maybe night vision equipment; VTS may include radio direction finders and closed circuit television. The importance of the radar display varies sharply between, say, night in thick weather and heavy traffic, and daytime in fair weather with little traffic, when the display may legitimately go almost unregarded.

Formerly, the navigation aids on a ship's bridge were almost autonomous, with minimal interconnection. Links to the radar were confined to heading and speed feeds from the compass and log for the True Motion and North-Up display modes. The radar(s), compass, log and other instruments each had their own displays, positioned in a rather uncoordinated manner. Nowadays the trend is to provide each member of the bridge team with definite seated work-stations, each having ergonomically designed controls and displays appropriate to the member's function, see Figures 2.2(d), (e) and 2.4(a)–(c). The screens may be capable of displaying some electronic chart and other data as well as the radar picture. The radar, less display, then forms a sub-system of an integrated bridge system (IBS), being sometimes termed a black box radar.

The ship's voyage data recorder (VDR; bright orange, but sometimes called a black box nevertheless) is used for incident investigation and training purposes. Among much else, it is required by IMO to record all the information currently presented to the operator on the master display of one radar, including range rings, radar status data (e.g. range scale), navigation alarms, etc., but not gain and other control settings.

2.1.4 Operators ashore

VTS may cover conflicting traffic flows in a navigationally difficult sea area. The traffic area of port VTS usually extends well to seaward of the harbour area. A small team of operators, sometimes called watchstanders, is led by a supervisor who may be the Harbourmaster. Methods of operation vary with port size, traffic patterns, local practice and the legal regime. There may be half a dozen sectors, each with its radar or radars, target data being handed from operator to operator as the ship transits the area. Beside radar, the operators use other sensors and information to build up situational awareness of the current and intended movements, anticipating conflictions and advising traffic to take appropriate actions; for example requesting or requiring a small vessel to keep clear of the deep channel while a supertanker passes. Except perhaps *in extremis*, for legal and other reasons VTS operators are not generally responsible for fine detail of movements or collision-avoidance manoeuvres – they

(a)

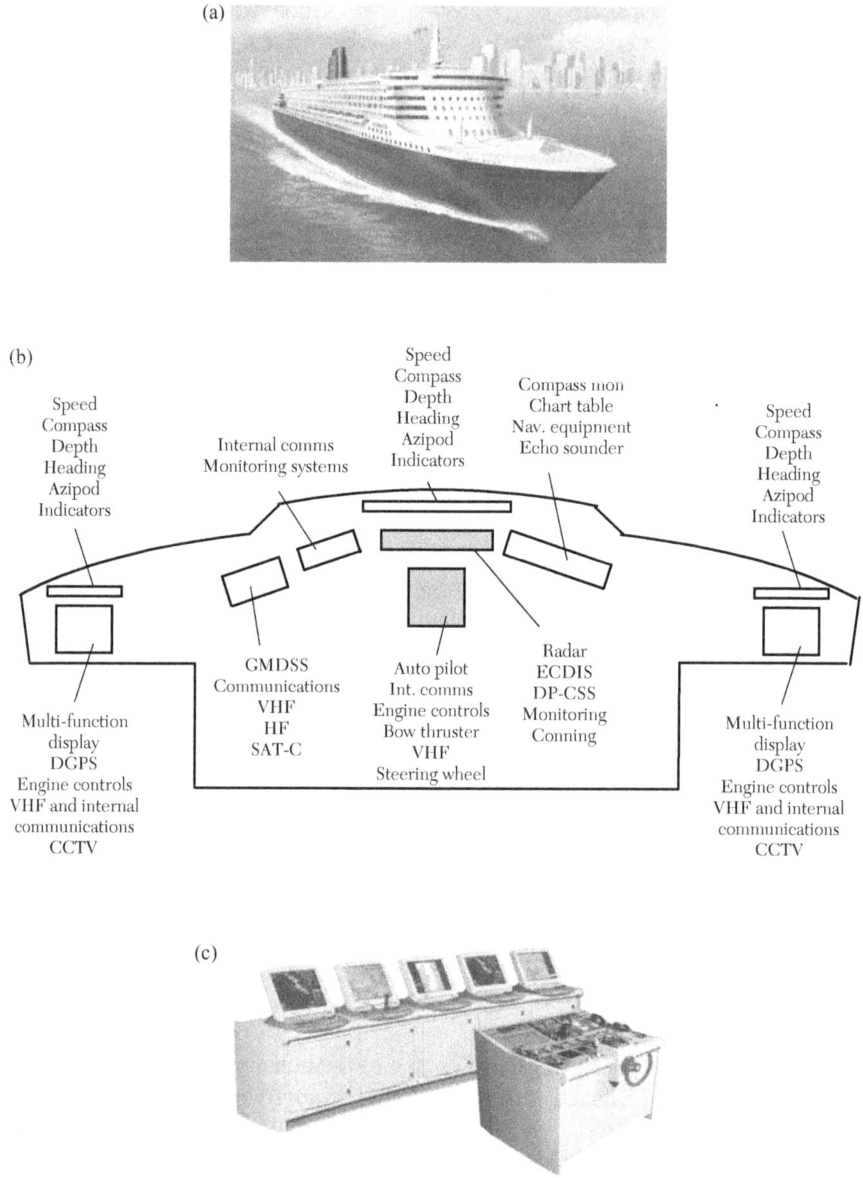

Figure 2.4 Liner RMS Queen Mary2, Cunard Line. All courtesy Kelvin Hughes Ltd, Ilford UK. (a) The largest passenger ship afloat, 150 000 gt. Entered service between Southampton and New York 2004. The bridge occupies prime space, the top floor forward. (Artist's impression.) (b) Bridge console contents. This comprehensive outfit omits to mention the all-important Mark 1 Eyeball. (c) Main radar and pilotage consoles, shaded in (b)

do not seek to drive the ship. Training standards meet the need of the particular port. Some states have national standards, others do not, since individual VTS systems vary so widely in complexity. Internationally recognised unified training standards are however being introduced through IALA and IMO. A surveillance system with many similarities to VTS was mentioned and illustrated in Chapter 1, Section 1.2.5.

There is a tendency to provide shore pilotage assistance from VTS centres, along the lines of Air Traffic Control. Only one aspect of this vexed question concerns us, registration of the targets displayed on the ship and shore radars, viewed by the OOW and by the shore pilot, respectively.

- Instead of using the ship's radar to detect local targets, the VTS radar must display all of a group of distant targets in correct register to the piloted ship, demanding particularly high performance. A degree or so bearing error, trivial on the ship's display, might translate into a quite unacceptable relative positional error.
- Displays used by the shore pilot and the OOW should both contain exactly the same set of targets. Given a pair of weak targets A and B ahead of the piloted ship, there is rich possibility of confusion should the ship detect A but not B, while the VTS, with its different aspect, detects B but not A.

Surveillance radars on coastal gunnery and missile firing ranges primarily ensure the hazard zone is clear of non-participating vessels, secondarily control movements of military craft participating in exercises. The civilian or military operators are trained and drilled in radar operation, interpretation and safety procedures. As members of the Range Safety Officer's team they operate to standing instructions which stress safety to all. Modern ranges take safety seriously. A UK Ordnance Board officer once told the author that they classed as a 'frequent occurrence' a life-threatening hazard predicted to arise once per 10 000 years. On the other hand, one has heard of a tanker master finding a deep indentation in the deck plating after passing a certain Mediterranean rocket range.

Fixed or mobile surveillance radars, often adapted ships' radars, are increasingly employed by Coastguard or Police forces on anti-terrorist or drug interdiction missions, again after suitable training. Feeds may also be taken from VTS installations, where the security dimension is becoming an important factor in system design.

2.1.5 Basic radar operation

Conventional marine and VTS radars generate a steady train of pulses – bursts of oscillation – of microwave power. An antenna transmits the energy in a continuously rotating beam as shown in Figures 2.5(a) and (b). Any object in its path scatters the radiation reaching it. A very little returns to the radar. Object bearing is that of the antenna, range being measured by the delay before reception.

Let us look at the process in a little more detail, giving some typical shipborne radar performance parameters – like many of those quoted later on, these are approximate and vary from radar to radar. The pulses have quite high power of 10 kW but very

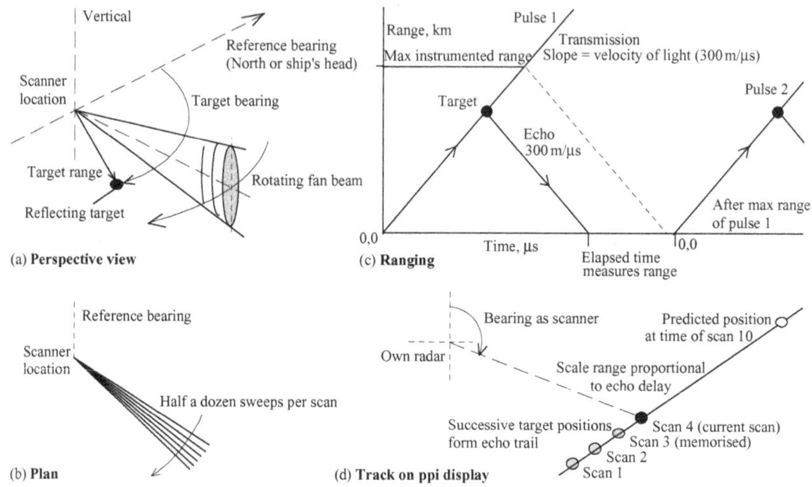

Figure 2.5 Radiolocation and ranging

short duration, 1 μs or less. A pulse is transmitted at the speed of light, 300 m/μs, sweeps out and strikes any scatterer on or above the sea surface lying in its path, indicated by the direct path of Figure 2.1. Some of the incident energy is absorbed within the scatterer. The remainder is scattered through a broad solid angle. The tiny part returning to the antenna forms an echo. Knowing that transmission and echo each propagate at the speed of light, the elapsed time to reception measures echo range, Figure 2.5(c), with uncertainty inversely proportional to the pulselength. The two-way scaling is 150 m/μs or 6.67 μs/km. In radar work, time and range are often interchangeable. Each transmitter pulse is in effect 'time stamped' for measurement of echo delay.

After waiting long enough to receive the echo from a possible scatterer at the longest range of interest, another pulse is transmitted, the time between successive transmissions being the sweep time or pulse repetition interval, typically 0.001 s or 1 ms. A steady train of such pulses is emitted, the pulse repetition frequency (prf) being $1/0.001 = 1000$ pulses per second (pps); pps is preferred to Hz to stress the extremely non-sinusoidal waveform. Sometimes prf varies with control settings. A few ancillary displays may operate ambiguously, with two transmissions simultaneously in flight.

The directional antenna radiating the pulses is called a scanner. Its beam rotates continuously at 25 rpm and typically covers 25° in elevation to cater for roll of the platform (ship carrying the radar), but is only 1° wide. Any particular scatterer is therefore scanned every $60/25 = 2.4$ s for a period of $2.4/360 = 0.0067$ s, being illuminated by a packet of $0.0067/0.001 = 6.67$ successive sweeps, say half a dozen, Figure 2.5(*b*). Any echoes received during this period are assumed to come from objects lying on the known azimuth bearing currently being illuminated, azimuth accuracy approximating the beamwidth, Figure 2.5(d).

The positions of all detected objects in range and bearing (polar or R, θ coordinates) are therefore determined on each scan. Their echoes are laid down to scale as plots on a display screen called a plan position indicator (ppi) which informs the operator of their positions relative to the radar. Plots are refreshed by the new measurements taken on each scan. By following the progress of a plot over several scans, the operator can determine the object's track or course made good relative to the radar. Historic plots may be shown as trails, roughly indicating target course and speed during the last few scans, Figure 2.5(d).

Targets are all objects, such as ships, of current interest to the operator. Although the Collision Regulations are written round aspect (relative bearing of target centreline) as indicated visually by navigation lights, often the radar discrimination is too coarse separately to display the individual scatterers comprising the target object and thus its aspect. Heights cannot be determined by radar. Radar is valued for its ability to position targets in range as well as bearing, and its general independence of cooperative equipment at the target. Although good signal processing facilities do the donkey work in presenting the clearest possible display, only the operator can decide that vital question – what to do?

2.1.6 Target detectability

Targets can only be displayed and tracked when the echo power or signal can be distinguished or detected with reasonable certainty from competing clutter, electrical noise, and such man-made interference as the transmissions of other radars. Figure 2.6 shows a ship's radar display with clearly visible coastal features. Areas of speckling over the sea surface are clutter caused by rain squalls and would mask any small target echoes within. We will now briefly examine these and other factors which affect detectability. They are discussed in detail later in this book.

When examining its passage through the atmosphere, the transmitted beam is often regarded as a bundle of linear elements called rays. The atmosphere subjects the rays to loss or attenuation and to variable curvature in the vertical plane on both transmit and receive legs. Figure 2.1 indicates that there are both direct and indirect ray paths between scanner and target, the indirect path being formed by intermediate reflection at the sea surface. Interaction between the two rays causes constructive or destructive multipath interference. At long range, the horizon intrudes on the scanner–target path.

Having reached the target, the proportion of unit incident transmitter power reflected back towards the radar governs the apparent reflecting strength of a target and is called its radar cross section (RCS, defined in Chapter 7, Section 7.1.1, and sometimes called cross section area, CSA). Most targets, such as ships and coastlines, are inanimate or passive. Racons, RTEs and SARTs (radar beacons, radar target enhancers or active reflectors; search and rescue transponders, Chapter 8) are active devices which include a reception and retransmission process. Although much modified by environmental effects, transmissions reaching any target basically follow an inverse square law and returning echoes or responses again follow this law, so echo

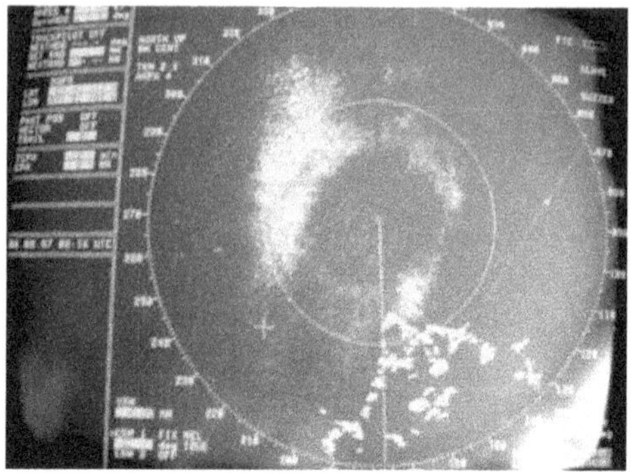

Figure 2.6 Ship's basic radar display. Atlas 9 GHz monochrome (green) raster cathode ray tude display, 12 nmi range scale, North-Up. VRM set to 5.35 nmi, measuring range of a ship target bearing 228°. Own ship heading 177°. Rugged cliffs of Cape Wrath, NW Scotland, to South; the lighthouse is not conspicuous. Rain squalls to NW and SSE would mask small ship echoes. Two blind arcs astern (North) from masts. Alpha-numeric data around edge of screen. Lighthouse tender Pharos. Author, reproduced by permission of Northern Lighthouse Board, 1997

power, S, at the radar tends to follow an inverse fourth power law of range, R; ($S \propto 1/R^4$, discussed in Chapter 4).

Clutter arises from scatterers such as a volume of precipitation or an area of sea-waves, not interesting to the operator. Their returns clutter the display and so hinder perception of targets. Although we will often use signal loosely, properly speaking signals convey information, wanted or unwanted. Echoes are signals but transmitter pulses are not, for they contain no information. Each echo is tiny (10^{-6} to 10^{-12} W) and may fluctuate in strength, say as own ship or the target ship rolls. The signal to noise-plus-clutter ratio, often shortened to SNR, is of great importance. Unavoidable imperfections within the radar receiving system also generate clutter-like background power, called noise. Because noise and clutter are random in nature, detection is never clear cut. There is always some probability of detection (P_D) less than unity, associated with a finite probability of false alarm (P_{FA}). As to be expected from information theory, high SNR raises P_D for any given P_{FA}.

For detection on a single sweep with acceptably high P_D (>0.5) and low P_{FA} ($<10^{-6}$), echo amplitude must exceed the adjacent noise and clutter by a large margin (SNR at least 10 : 1 power ratio or 10 dB) – there must be adequate contrast. Candidate events are winnowed by thresholding, only returns above a predetermined strength passing to the signal processor following the receiver, where they are assigned to the appropriate one of an array of detection cells or bins in range and bearing. Detection is improved by having several sweeps per beamwidth and averaging or integrating them. Echoes, being associated with a definite position, are statistically correlated and their counts build up more rapidly than those of clutter, noise and interference, which are more random in nature and decorrelated. Targets are declared valid when there is

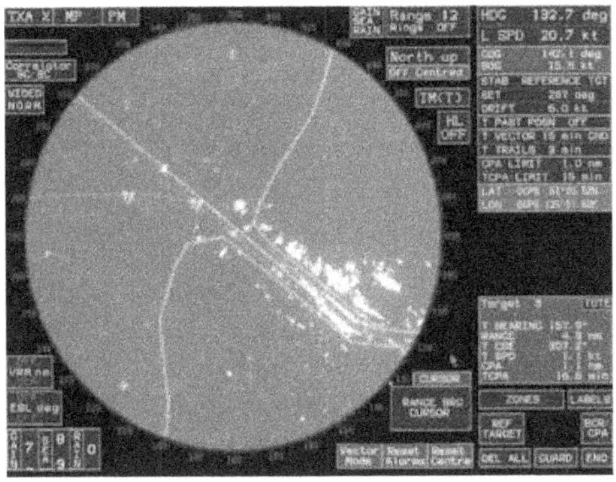

Figure 2.7 *Display with radar, ARPA and ECDIS data. Flat panel eleven-colour display. Reproduced by permission of Kelvin Hughes Ltd, Ilford UK*

more than a certain number of counts, blips or hits per scan on which SNR exceeds a threshold value. Additionally, sometimes returns are required to meet detection criteria on two, occasionally more, successive scans, called scan to scan correlation.

The operator, helped by built-in circuits and software, selects appropriate radar settings to optimise, as far as possible, echoes rather than clutter on the display; that is, seeks to optimise displayed SNR. Maximising SNR is the key to many aspects of radar design and performance, preserving the information within echoes while keeping noise and clutter down to acceptable limits.

Current target positions are displayed as plots, points of light, generally on a raster display, formed from the raw polar data by a digital scan converter (DSC). The process of joining sequential plots by a line to display a target track or vector is called track-forming. Ancillary devices called automatic radar plotting aids (ARPA) or automatic tracking aids (ATA) automate this process and can generate numerous tracks unless overloaded when an excessive number of clutter returns arising from low SNR give high P_{FA}. Plotting aids can extrapolate tracks to estimate future target positions, closest point of approach to own radar (CPA), time of CPA (TCPA) and can activate alarms should a target enter a guard zone around own ship. Accurate prediction demands high SNR (Chapter 13). VTS radars operate similarly and often include more elaborate forms of ARPA. They may combine data from several radar heads or from other sensors, particularly electronic charts; Figure 2.7. This data fusion, although technically difficult, can refine display quality and improve SNR. Marine radars will soon be required to associate radar plots with AIS reports.

2.1.7 Radar construction

The plethora of 50 valves (vacuum tubes), resistors, capacitors and inductors of early radars has given way to in excess of 50 integrated circuits but few other components. Each IC has an area of a couple of square centimetres and contains many – sometimes many thousand – transistor elements, each functionally equivalent to a valve. The

ICs draw some 10 mA apiece, at the low voltage of 5 V. They are assembled on a few multilayer printed circuit boards and there is muck less bulk. Cable bundles are few. Digital technology predominates, and computers are also widely used during design. Extensive climatic and durability tests are performed at extremes of temperature, humidity, supply voltage, vibration and mechanical shock, especially on the scanner and upmast transceivers which inhabit an extreme (Class X) environment. These developments have transformed reliability, greatly enhance performance and have reduced price.

Instead of renewal of a failed component, repair is generally by change of a complete sub-unit such as a board. Service engineers no longer need to know the minutiae of circuit arrangements, so manufacturers no longer divulge physical or software design details in service manuals. By jealously guarding their intellectual property rights much more than formerly, they make it well-nigh impossible for an outsider to infer the detection strategies used in specific models, or to describe in detail how they detect their targets. Use of proprietary application specific integrated circuits (ASICs) further obscures detail of operation. But even the subtlest designs must obey the laws of physics, enabling us to clearly state the boundaries of available performance, which we can be pretty sure all modern radars closely approach. Descriptions here and in later chapters are therefore to be regarded as basic concepts, intended to show in broad terms how and why transmissions are generated, reach targets, reflect as echoes, are detected and are displayed to the operator.

2.1.8 Decibels

Radar calculations often involve outlandishly large or small quantities – we have already encountered transmitter power 10 000 W and received echoes of 1 pW (10^{-12} W). It is often convenient to express such quantities logarithmically, using decibels to give handier values which are added and subtracted rather than multiplied or divided. The dictionary[1] definition cannot be bettered:

> Decibel: A logarithmic unit (one-tenth of a bel, abbreviation dB) used to express the ratio between two levels of sound intensity, electrical power, etc, one of which is usually a (stated or unstated) reference level . . .

Power ratio in decibels, P_{dB}, of power P_1 (watts) relative to power P_2 (watts) is defined as:

$$P_{\mathrm{dB}} = 10 \log_{10}\left(\frac{P_1}{P_2}\right). \tag{2.1a}$$

From this,

$$\frac{P_1}{P_2} = 10^{P_{\mathrm{dB}}/10}. \tag{2.1b}$$

Note that decibels are power *ratios*. Power itself may be expressed as dBW or dBm, meaning dB relative to 1 W or to 1 mW (we use dBW exclusively); RCS as dBm^2,

[1] Shorter Oxford English Dictionary.

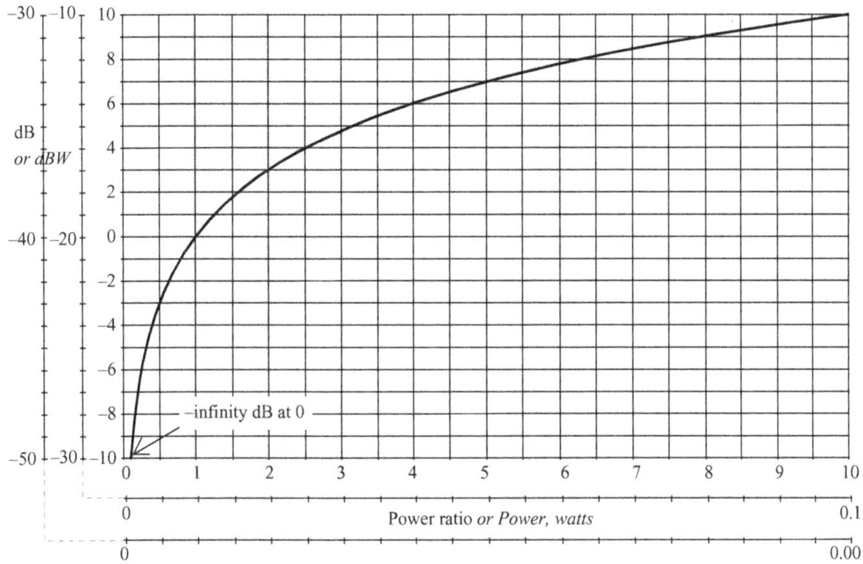

Figure 2.8 *Decibels. Relates power ratio to dB and power to dBW. Scaling can be extended indefinitely using fresh pairs of axes as shown, the curve always remaining the same shape. Any 2 : 1 change in power is always a 3 dB change, a 10 : 1 change being 10 dB*

dB relative to 1 m^2, sometimes written dBsm or dbsm. Unlike some authors, we do not express distance in dB form (dB relative to 1 m). Time is never expressed in dB and millibels, kilobels, etc., are never used. To add powers which are expressed in dBW, they must first be converted to watts by Eq. (2.1b). Figure 2.8 plots the earlier expressions.

Where voltages are denoted V and resistances R, and remembering that $P \propto V^2$, substitution into Eq. (2.1a) gives

$$P_{\mathrm{dB}} = 10 \log \left[\frac{V_1^2/R_1}{V_2^2/R_2} \right] \mathrm{dB}. \tag{2.1c}$$

If $R_1 = R_2$ *but not otherwise*

$$P_{\mathrm{dB}} = 10 \log \left(\frac{V_1^2}{V_2^2} \right) \mathrm{dB} \tag{2.1d}$$

or

$$P_{\mathrm{dB}} = 20 \log \left(\frac{V_1}{V_2} \right) \mathrm{dB}. \tag{2.1e}$$

Logarithms cannot be taken of negative numbers so Eq. (2.1d) is more general than Eq. (2.1e). For d.c. or a.c. phasor quantities V_1 and V_2 are the *magnitudes* (always positive) of the voltages across the (equal) resistances R_1 and R_2. Corresponding forms are used for currents.

Decibel conversion is straightforward using Figure 2.8, a PC or pocket scientific calculator's 'Log' and '10^' functions. Tens of dB multiply by 10, 100... etc. Negative dBs can be split thus for calculation: -127.3 dBW $= -120 - 7.3$ dBW $= 1/10^{12} \times 1/5.370$ W $= 0.1862 \times 10^{-12}$ W.

Examples: 0 dB = 1 : 1, 1 dB ~ 1.26 : 1, 0.1 dB ~ 1.023 : 1, −3 dB ~ 0.5 = 0.5 : 1, 5 dB ~ 3.18 : 1, 10 dBm^2 = 10 m^2, 30 dB = 1000 : 1, −123 dBW ~ 0.5×10^{-12} W; 0 dBm = 1 mW = −30 dBW; 0 W (zero power) = $-\infty$ dB, emphatically *not* 0 dB.

2.2 Components of the radar

The following description concentrates on big-ship radars, which lie between the large VTS sets and small-craft radars.

2.2.1 Transmission

Each transmitted pulse is a pulselength burst of sinusoid having the very high frequency necessary for efficient propagation close to the sea surface, typically 3 or 9.4 GHz (3 or 9.4×10^9 Hz). Corresponding wavelength is 10 or 3.2 cm, much shorter than conventional radio practice. The radar is therefore said to operate at microwave frequency or centimetric wavelength. The microwave sinusoid is the carrier (or bearer) of frequency f_c, modulated by a train of rectangular unidirectional baseband or video pulses, shaped as Figure 2.9(a), at the prf frequency f_m, typically 1 kHz. The microwave magnetron power oscillator is switched on for the duration of each pulse by a modulator device. Modulation superimposes the train on the carrier; Figure 2.9(b).

Speaking generally, although the energy of a sine wave signal is concentrated at a single (fundamental) frequency, all pulse trains have energy components spread between the fundamental (the prf) and its harmonics. A pulse train, prf $= f = 1/T$, having pulses of any desired shape, can be synthesised by summation of a judiciously chosen d.c. component, plus a Fourier series of sine waves of frequencies $f, 2f, 3f, \ldots, nf$, of appropriate amplitudes and phasing. (It is permissible to speak of phasing of these differing frequency components because they are harmonically related.) Where the ratio of the pulse on time, τ, to the pulse repetition interval is $k = \tau/T$, the frequency of the nth harmonic, f_n, is

$$f_n = nf = \frac{n}{T} = \frac{nk}{\tau}. \tag{2.2a}$$

As the radar modulation is a pulse train rather than a sinusoid, f_m is a spectrum of prf harmonics centred on f_c, Figure 2.10, with two equal and opposite sidebands at frequencies

$$(f_c + f_m) \quad \text{and} \quad (f_c - f_m). \tag{2.2b}$$

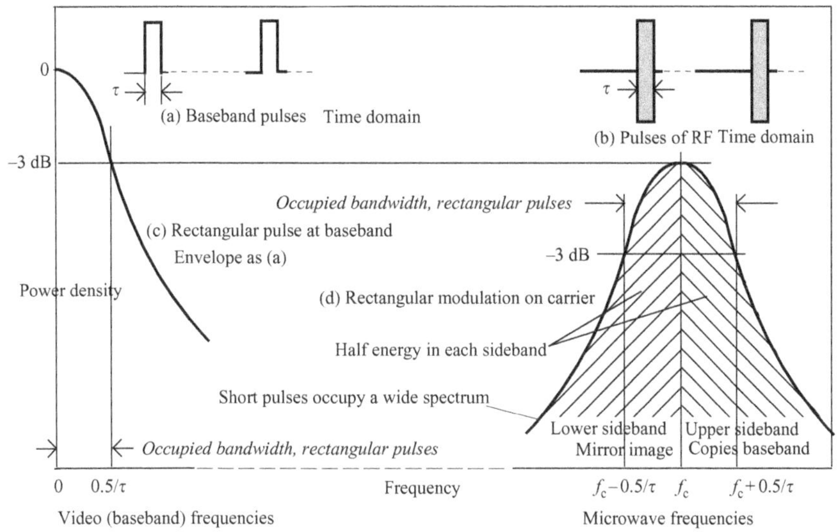

Figure 2.9 Echo spectrum. In frequency domain, lower and upper sideband voltages are mirror images centred on carrier frequency f_c

Each sideband contains half the pulse energy so occupied bandwidth is doubled to $\pm 0.5/\tau$ as shown in the transmitted spectrum envelope of Figure 2.9(d). Short pulses, especially those having sharp edges, of necessity occupy a wide spectrum. Energy density (watts per hertz) in the far skirt regions, although low, may be enough to interfere with users at other frequencies. Imperfections in the magnetron may introduce unwanted further spectral components.

$$\text{Occupied modulated bandwidth} = \frac{1}{\tau}. \tag{2.2c}$$

The baseband spectrum of Figure 2.9(c) has bandwidth extending from zero to $0.5/\tau$.

$$\text{Occupied baseband bandwidth} = \frac{0.5}{\tau}\ \text{Hz}. \tag{2.2d}$$

Conversion between time (Figures 2.9(a) and (b)) and frequency (Figures 2.9(c) and (d)) domains is possible using mathematical Fourier transforms. The harmonic voltages of a rectangular pulse of height E_R are given by an infinite series comprising a d.c. term kE_R, fundamental and harmonics.

$$V = kE_R + \frac{2}{\pi}E_R \sum_{n=1} \frac{1}{n} \sin \pi nk \cos nx, \tag{2.2e}$$

where $x = 2\pi T$.

Radar pulse trains have k so low ($\sim$0.001 max) that the d.c. term can be neglected. The amplitude of the nth harmonic, V_n (of frequency $f_n = nk/\tau$ Hz) is found by

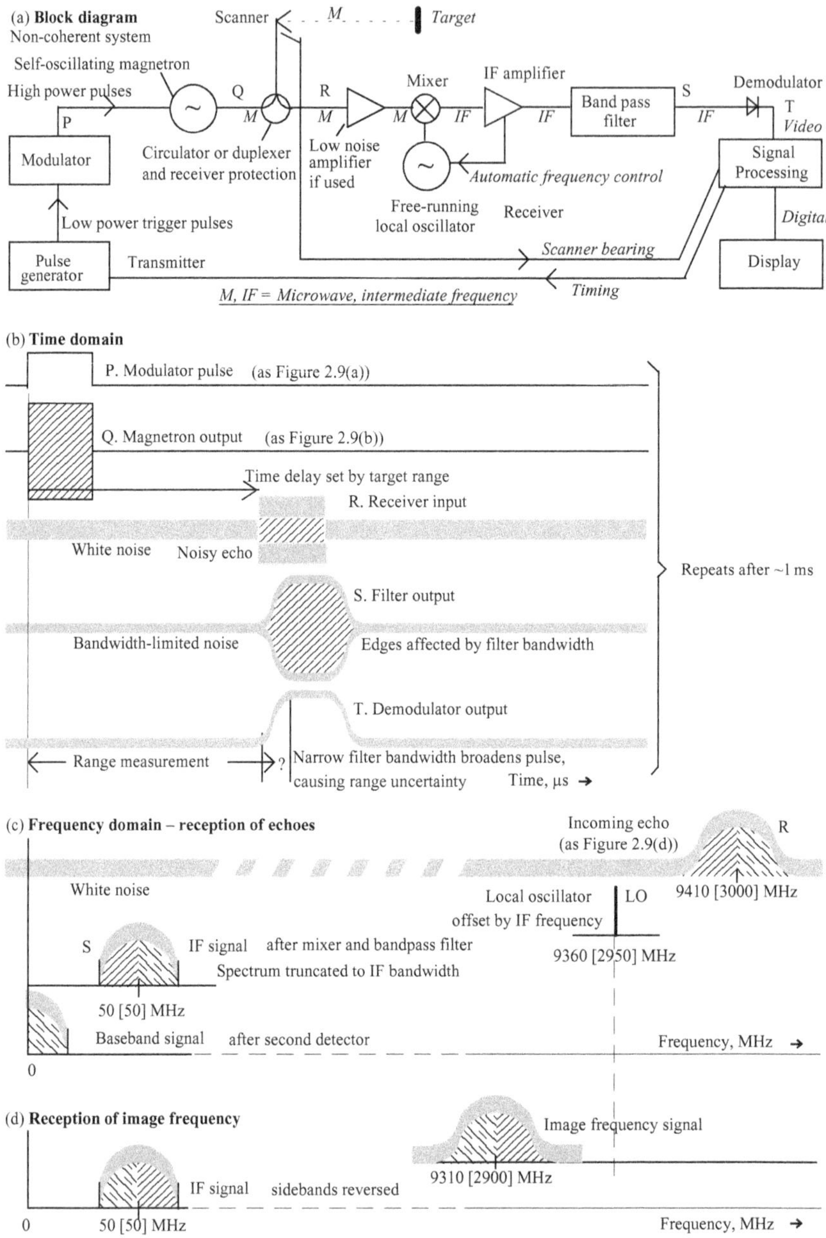

Figure 2.10 Radar block diagram and signal flow. Time and frequency domains. Typical 9 (and 3) GHz band frequencies indicated. Block diagram (a) represents the usual non-coherent system

setting the cosine term to its maximum value of 1.0:

$$V_{n\,\max} = \frac{2}{\pi}\frac{E_{\mathrm{R}}k}{f_n\tau}\sin(\pi f_n\tau). \tag{2.2f}$$

Figure 2.10(a) is a block diagram of a typical radar, and depicts signals in time and frequency domains. The components are grouped within two or three physical modules. Low-power trigger pulses fire the magnetron via the modulator. The magnetron block has too much power abstracted from it (low Q-factor) to define magnetron frequency exactly. The magnetron output feeds the scanner. Echoes are routed to the receiver and thence to the demodulator, which removes the carrier, leaving a baseband or video pulse train similar to that generated by the pulse generator, but with delay proportional to target range and, at a given range, height (voltage) dependent on echo strength. The video train is processed to decide which pulses are likely to represent echoes rather than noise or clutter, then fed to the display for viewing by the operator.

Figure 2.10(b) shows events during a single sweep. The pulse generator delivers a train of trigger pulses at prf near 1000 pps to suit the operator's choice of range scale. (A slight timing jitter to help suppress interference from other radars is not shown). The pulses fire the modulator, whose output, P, is a train of powerful (25 kW) pulses at the selected length. The magnetron is 40 per cent efficient and generates 10 kW (40 dBW) bursts of oscillation, Q, centred on the microwave frequency of its resonator in the 3 or 9 GHz bands. The pulsed nature of the transmission causes a fairly broad frequency spectrum to be radiated by the scanner.

2.2.2 *Reception*

A circulator or duplexer (device routing bidirectional signals) directs the return signal R to the receiver, whose input circuit is preceded by devices to protect its sensitive components from burn-out by the powerful transmitter pulses. The first stage is usually a low noise amplifier working at microwave frequency, which lifts the echo voltage well above unwanted noise injected by later parts of the receiver. Microwave amplifiers are expensive and inconvenient, so the main amplification is done at a lower frequency called the intermediate frequency (IF, 50 MHz).

Figure 2.10(c) shows the frequency relationships within conventional marine radars. The signal R is shifted bodily down to IF, here 50 MHz, at point S by a mixer, sometimes called a first demodulator or first detector, the receiver being a superhet (supersonic heterodyne). In more detail, when the weak microwave signal is superimposed with a strong sine wave from a continuously running local oscillator, LO, whose frequency is offset from transmitter frequency by the intended IF frequency; this oscillation beats with the echo. Components are generated at the sum and difference of the two frequencies, the latter being accepted as the IF. A symmetrical arrangement of diodes is used as a balanced mixer, which introduces no LO noise. Balanced mixers have a noise factor around 8 dB so receivers without LNAs have system noise factors around 9 dB.

2.2.3 Non-coherent system

The LO in the system described is a free-running semiconductor microwave oscillator with output power of a few milliwatts. Its frequency must remain approximately tuned to any drifts in magnetron frequency otherwise the mixer output would drift out of the passband of filters further down the receiver. Tuning is primarily by an automatic frequency control (AFC) circuit which applies a correction voltage proportional to IF frequency error to the LO. The correction is typically derived from the changing phase of the double balanced mixer output. A manual fine tune control and associated tuning indicator are sometimes provided.

Within AFC limits, LO frequency 'does its own thing' – it is not exactly harmonically related (is non-coherent) to transmitter frequency. In Figure 2.11(a) the block diagram is redrawn to emphasise the frequency-determining elements, here shown for the 3 GHz band. While the modulator is firing, the magnetron output (Q) centre frequency is determined by the anode resonator block dimensions, surrounded by a spectrum based on that of the modulator pulse, P. The echo (R) spectrum is more or less identical (target RCS is only slightly frequency-sensitive), although of course echo power is drastically lower.

Extra IF bandwidth has to be retained to cover the residual tuning error, degrading SNR, since noise power is proportional to bandwidth. On long-range scales, where receiver bandwidth is least, residual tuning errors may cause some loss of receiver sensitivity, further spoiling SNR. If the limited range of the AFC is exceeded, it may throw off to a large error, grossly degrading receiver performance, so the AFC loop must be reset when the magnetron or other component is renewed.

After the LNA, the echo is multiplicatively mixed with the microwave continuous-wave LO oscillation, which preserves the spectrum, shifting it bodily to IF. The echo is amplified in a multistage IF amplifier, containing bandpass filters. Filter centre frequency is the nominal difference between LO and magnetron frequencies and bandwidth has to be wide enough to accept the main components of the pulse spectrum, its occupied bandwidth. The IF output, S, is applied to a diode demodulator where it is rectified to give the baseband video pulse, T, whose spectrum approximates modulator pulse P, with amplitude proportional to echo strength at R. This envelope detection process preserves the envelope of the IF and microwave signals (compare Figure 2.10 S and T).

The non-coherent system just described is wasteful of precious signal because the information resident within the echo phasing is discarded. The following systems improve SNR by preserving echo phase information but are more complicated. Use is currently confined to a few VTS systems.

2.2.4 Coherent-on-receive system

Figure 2.11(*b*) depicts one form of coherent-on-receive system, which seeks to retain the cost and efficiency advantages of the magnetron. The transmitter and receiver are basically as the non-coherent system except for the local oscillator,

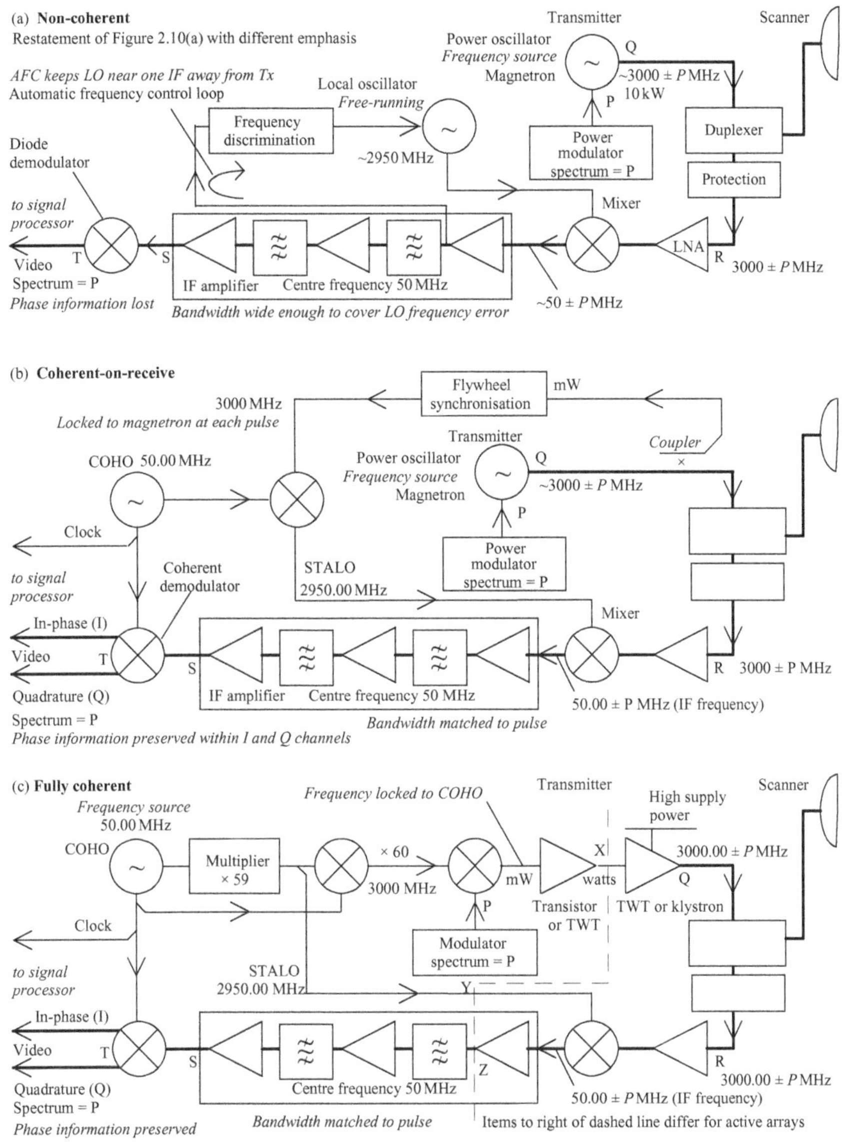

Figure 2.11 Frequency management strategies. Typical 3 GHz band frequencies. Most radars are non-coherent

whose frequency is synthesised as follows. A coupler (a pair of parallel waveguides linked by small slots, or parallel stripline conductors) extracts a sample of the transmitter frequency actually generated, from which the pulse sidebands are then removed. The frequency of this sample is remembered between pulses by either a flywheel oscillator or digitally and is maintained exactly at magnetron frequency.

It is mixed with the output of a coherent oscillator (COHO) which runs at IF centre frequency to give a stable local oscillator (STALO) signal. The figure shows typical frequencies.

After the LNA, the echo is mixed with the STALO to give a pulse spectrum centred on IF centre frequency; this feeds the IF amplifier. Filter bandwidth can be matched to the modulator pulselength without need of additional allowance for LO tuning error. The demodulator is a pair of mixers taking direct and 90° phase-shifted drives from the COHO to give in-phase and quadrature (I and Q) video outputs, preserving both the amplitude and phase information contained in the echo. Successful operation is critically dependent on accurate capture of magnetron frequency during the transmitter pulse, followed by drift-free memory during the relatively long inter-pulse reception phase. Relative to non-coherent operation, integration loss is roughly halved.

2.2.5 Fully coherent system

Figure 2.11(*c*) depicts a fully coherent system. The COHO is the primary frequency source, multiplication (here by 59 and 60) synthesising the STALO and transmitter frequencies, respectively. The continuous-wave transmitter feed is modulated by the pulse at low power (facilitating precise pulse-shaping and control of transmitter spectrum), followed by a multistage power amplifier, which replaces the magnetron. The STALO frequency is positively locked one IF frequency away from echoes at all times. The remainder of the receiver follows the coherent-on-receive system. Amplifier-type transmitters are bulky, inefficient and expensive, but give operational flexibility. Although not shown, it is straightforward to stagger transmitter frequency from pulse to pulse to decorrelate sea clutter and hence improve detection for a given SNR; or to reduce transmitter power; for short-range operation, for example. The amplifier may contain a travelling wave tube (TWT) feeding a klystron amplifier tube. Both are high voltage thermionic valves. Coherent systems are particularly suited to active array scanners, not in current marine service, see Chapter 16.

2.2.6 Ambiguity; image frequency, prf constraints

It is desirable to maximise the number of sweeps – minimising sweep time – taken into the detection process to maximise the effective SNR and get maximum probability of detection for the chosen false alarm rate. But if sweep time is less than the range delay of the furthest target returning a detectable echo, it becomes uncertain or ambiguous whether a plot is from a close target reflecting the last pulse transmitted, or from a more distant echo of an earlier transmission. Ambiguity can be resolved by severely jittering (staggering) the pri but signal to clutter ratio then tends to suffer. Operation at relatively low prf is universally preferred, the radar being non-ambiguous out to the maximum instrumented range of the display (on big ships often 96 nmi, 178 km, constraining maximum prf to 840 pps on long-range scales). On short-range scales higher prf can be used. Beside minimising risk of receiving second time around echoes, relatively low prf is often retained to enable plotting aids to continue to track targets which are too distant for current display, ready for the operator to return to long-range operation; to permit a second display to show the distant scene while the primary display examines the short-range scenario, or to service guard zones. Low-power radars can utilise high prf without risking significant second-time around echoes from distant targets, except under conditions of anomalous propagation or anaprop, Chapter 5, Section 5.2.5. It is questionable whether the 96 nmi instrumented range frequently provided serves much purpose, only inland mountains being likely to rise above the radar horizon. The operator is sometimes provided with a long/short pulselength switch. Long pulse/narrow bandwidth reduces noise and improves detection of weak targets if clutter is slight; short pulse/wide bandwidth improves range resolution and illuminates less clutter around the target, improving clutter rejection and helping to depict target aspect.

Figure 2.10(d) shows that, depending on detail design, the receiver may be responsive or open to a band of unwanted frequencies lying on the other side of LO frequency from the transmitter. Usually this image frequency band contains nothing synchronised with the transmitter and merely contributes some additional noise, which is included within the overall noise figure. Swept-frequency racons are sometimes received at image frequency, see Chapter 8. Modern receivers include double balanced mixers which reject image frequency.

2.2.7 *Typical station configuration*

Figure 2.12 shows a typical radar station and its links with other bridge equipment, Figure 2.13 being a particularly futuristic realisation of the bridge components. A pair of transmitter/receivers, if in the same band, may be connected to a single scanner by a combining device called a diplexer (*not* duplexer), diplex operation being the simultaneous transmission/reception of two signal channels using a common component such as a scanner. In principle it is possible to combine or fuse the receiver output data streams before feeding a single display. The difficulties exceed the advantages on shipboard, although data fusion is occasionally used in VTS. The bigger ships are mandated by IMO to carry two radars, primarily for reliability. One set has to use the 9 GHz band. IMO encourage the other to be at 3 GHz, giving the operational advantages of each band. Good seamanship usually requires one of a pair of radars or displays to be kept on a long range scale for landfall verification and to give early warning of traffic movement, the other running on a shorter scale for collision avoidance.

Here the installations have their own scanners, but either's receiver output may be switched to the other's display, and it is not always very obvious to the operator which band is active. A very few merchant ships voluntarily duplicate their equipment with typically two each of 3 and 9 GHz radars. Many roll-on roll-off (ro-ro) ferries and some other ships carry additional small, usually 9 GHz, radars low down forward and sometimes astern to assist berthing. Lack of omnidirectional azimuth coverage unsuits them to general navigation.

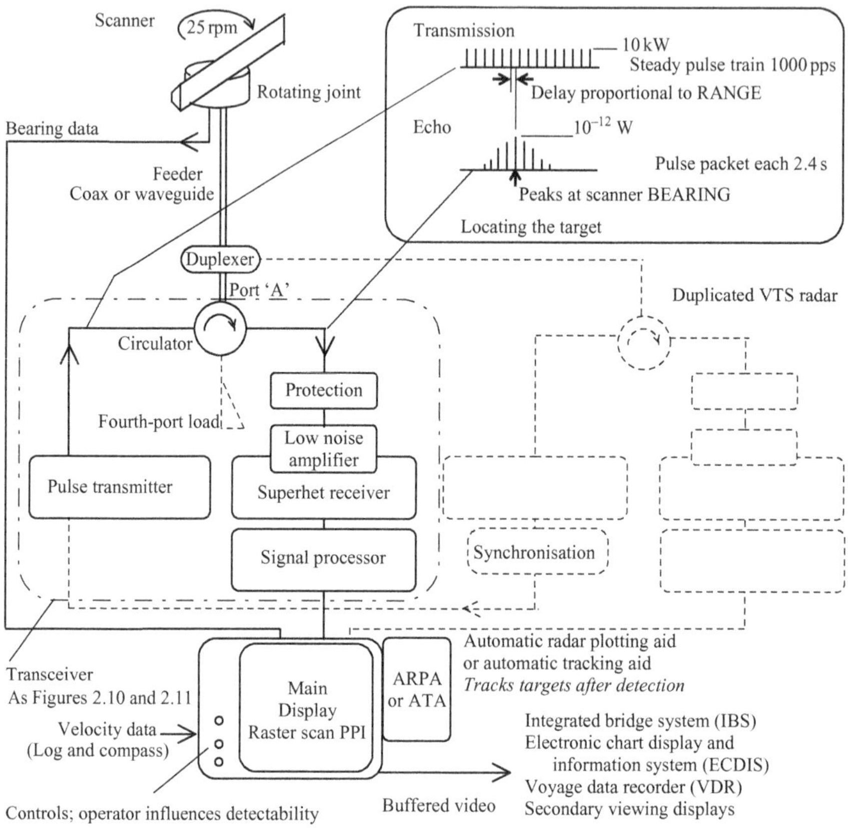

Figure 2.12 Typical station configuration. Twin-radar installations usually use two scanners which usually remain functionally independent even when sharing a display. The display may contain other facilities, particularly when forming part of an integrated bridge system (IBS)

The remainder of this chapter details how radars illuminate targets. Reception is detailed in Chapter 3.

2.3 Transmitter

2.3.1 Overview

The transmitter, sometimes abbreviated Tx, is usually built as a unit with the receiver, Rx, to form a transmit–receive unit, transceiver or Tx/Rx. If not aloft with the scanner, the transceiver is located below deck, either within a stand-alone cabinet or integrated within the display unit. The necessary transmission line or feeder to the scanner introduces loss, noise and reflection clutter.

Figure 2.13 *A modern trend in bridge workstation design. Reproduced by permission of Kelvin Hughes Ltd, Ilford UK*

A steady train of powerful, short, pulses of electromagnetic energy is required. The train is sometimes said to be uncoded, for it carries no data. In radio terminology, the emission type is P0N. Equipment limitations preclude generation of truly rectangular pulses – and they would have undesirably wide spectral width – but it is convenient for most purposes to think of them as rectangular or 'square'.

2.3.2 Magnetron power source

Except in coherent systems, the generator is usually a cavity magnetron valve (tube). The magnetron has always been the cheapest and most efficient power generator and is a transit-time high power oscillator. Within a sealed envelope it contains a central cylindrical heated cathode, surrounded by pairs of anode poles connected to parallel inductor–capacitor (LC) tuned cavities, the inductor centre points being earthed. The high-vacuum working space between cathode and the anode system lies in the strong axial field of a permanent magnet, the tube/magnet assembly thus forming a packaged magnetron. Application of a negative pulse of about 10 kV to the cathode makes it emit electrons. They are subject to crossed fields, radial electric and axial magnetic, and take spiral paths. Some electrons return to bombard the cathode, increasing its emission to several amperes, but most fly to the anode poles. When a random grouping of electrons causes Poles 1 to fall slightly below earth potential, the centre-tapped tuned circuit forces Poles 2 positive. Electrons hitting −ve Poles 1 deliver more energy than absorbed by those hitting +ve Poles 2, so the oscillation builds up. After a microwave half-cycle the tuned circuits swing the pole polarities to Poles 1 +ve, Poles 2 −ve. This instant is arranged to coincide with the transit time of the spinning beam to Poles 2, feeding more energy into the tuned circuits.

There are usually four pole pairs and the electrons cluster into four spokes. Intervening C-shaped cavities form the tuned circuits of the cavity resonators within the copper anode block. Frequency is therefore determined during manufacture and is ordinarily not externally controllable. Frequencies lie within sub-bands, usually

30 MHz wide, often located near the centre of the IMO operating band. The outermost 15 MHz 'guard bands' are always avoided to minimise radiation of outband spectral components.

The power is coupled out by a short coaxial line launching into an integral waveguide. Magnetrons are simple and have long life, and have stood the test of time. Their remarkable efficiency makes for cool running and high reliability. Permissible duty cycle or on/off ratio is low and despite the high peak power, mean output power is only around 10 W (and waste heat needing to be dissipated is not much more), similar to that of shipborne very high frequency (VHF) radio with its quite different modulated continuous wave transmission. Actual transmitted pulse shapes are not generally disclosed by radar suppliers. Because target detectability is improved when there are many pulses within the packet, occasionally prf is raised so far on short-range scales that, to keep within the magnetron's maximum permissible duty cycle, pulselength has to be reduced below that essential for good range resolution.

The scanner is never perfectly matched. When the feeder is long, quite minor frequency change sharply changes the phase of the mismatch power returned to the magnetron. To minimise risk of provoking unwanted 'long line effect' modes of oscillation (Section 2.6.1), with their poor spectra and low efficiency, usually either a ferrite circulator is used as duplexer (Chapter 3, Section 3.2.3) or a ferrite isolator (a non-reciprocal device; ordinary reciprocal devices and components behave equally to either direction of energy flow) is inserted at the magnetron output to improve load match. Similarly, the rate of rise of drive voltage has to be controlled. More expensive coaxial magnetrons have tighter spectrum control and are better suited to coherent systems.

2.3.3 *Modulator*

The magnetron is driven from a modulator, designed to introduce insignificant noise. Typically the modulator contains inductors and capacitors in a pulse-forming network (PFN) – or formerly a coaxial cable containing distributed inductance and capacitance – which accumulates low-voltage energy between pulses. When a solid-state thyristor, silicon controlled rectifier (SCR) or high voltage insulated gate field-effect transistor (FET) switch is fired by a small trigger pulse, the PFN rapidly discharges in a controlled manner through a step-up transformer, whose high voltage secondary is bifilar wound to carry the magnetron heater current. In some designs, the inductors and capacitors in the line overswing to double voltage. Pulse length is changed by switching inductor/capacitor combinations. Widely varying lengths pose difficult design problems and a delightfully named tail biter diode is often used to suppress secondary short pulses; alternatively FET switches terminate the pulse in a more definite manner. Formerly modulators used hard-vacuum triodes, or thyratron valves containing gas at low pressure – the author's first radar task was production test of hydrogen thyratrons. Occasionally saturable reactors (pulsactors, Melville lines) were employed, where a control current pulse switched magnetically stored energy. These were heavy, complex and best avoided!

2.3.4 *Influence of transmitter on system*

The modulator/magnetron arrangement of non-coherent transmitters constrains many characteristics of the whole radar.

- Only a few discrete pulselengths are available, modulator design constraints precluding smooth variation.
- Magnetrons are bang-bang devices. Transmitter power is either full or zero. Unlike radio, there can be no low-power mode, although peak power may be a couple of decibels less than nominal on the shortest pulselength.
- Power builds up very rapidly (~10 ns) at the start of the pulse and pulse-end decay is nearly as fast, broadening the frequency spectrum and necessitating output filters to minimise interference to other spectrum users.
- Transmitter frequency is built into the magnetron cavity and cannot be adjusted by the operator or service engineer.
- Frequency changes by 10 MHz or so due to heating, pulling (load match change, VSWR preferably being held below 1.3), pushing (drive voltage change) and ageing. The receiver has to match drift using automatic frequency control (AFC) or manual retuning and bandwidth has to include a margin for error, letting through more noise and effectively reducing receiver sensitivity and SNR.
- Maximum duty or pulselength/prf combination is dictated by prf to prevent overload.
- The cathode may take several minutes to heat from cold, necessitating a heater-on hot-standby mode.
- Small size and high efficiency permit compact installation, enabling mounting at the scanner, obviating wasteful feeders.

2.3.5 *Spectrum problems*

It is difficult to tame magnetron output pulse edges and the output spectrum is undesirably rich in harmonics, often being especially dirty on short pulses. The anode tuned circuit is tightly coupled to the output to extract maximum power. Selectivity is perforce low, like a muffled bell, so the tuned circuit cannot fully suppress out-of-band frequency components. Other oscillation mechanisms become significant when the valve ages, especially when rate of rise of drive voltage is outside specification, and tend to cause spectral lines some tens of megahertz from centre frequency, called moding. Oscillation amplitude is limited by onset of saturation and cut-off effects which cause harmonics of the microwave carrier frequency to be generated. A rounded Gaussian pulse (shape similar to Chapter 3, Figure 3.5) would deliver a cleaner spectrum, lying almost wholly within the marine band, but magnetrons are unsuited to this pulse shape. Successful modulator design demands particularly close liaison with the magnetron supplier.

At first the microwave spectrum was not intensively used. There were a few industrial and, later, domestic microwave heaters at 2.45 GHz and some low power industrial activities at 10.688 GHz. Astronomical research receivers near 10 GHz demanded quiet conditions. Otherwise the civil and military radar fraternities had the field to themselves. Although radar receivers are sensitive, the highly directional antennas invariably employed mitigate mutual interference. Civil marine sets employ prf stagger to break up 'running rabbit' interference – patterns of dots slowly traversing the display – from other sets employing similar frequency and mean prf. The military have to be prepared to counter hostile jamming, so can generally put up with considerable inadvertent interference. The situation had similarities to the spark transmitter days of early marine radio telegraphy.

For many years there was therefore little objection to transmission of rectangular pulses, with their profligate spectrum. Rectangular pulses are particularly convenient to generate by discharge of a delay line into the transmitter valve, and facilitate good range resolution.

As pointed out by Williams [1], the 1990s telecommunications explosion placed intense pressure on the lower microwave frequencies. Governments recognised the spectrum as a finite and valuable resource, to be auctioned to the highest bidder for billions of pounds. Covetous eyes focused on the centimetric bands. It is likely that radar will soon have to share the frequency bands with telecomms. While a good case can be made on safety and commercial grounds for pulse marine radar, together with sufficient spectrum for accurate range determination, it is hard to justify the pollution of adjacent frequencies by unnecessary transmission of rectangular pulses, not to mention moding lines and harmonics. Occupied bandwidth must be minimised.

To a telecommunications receiver, it is peak power that matters, in other words the equivalent isotropic radiated power (EIRP), the power in the beam, the product of transmitter power and antenna gain (their sum when using dB). Radio circles prefer the term peak emitted power, PEP, which assumes radiation from a dipole of gain 2.15 dBi. Typical big-ship radars of say 20 kW peak power and scanner gain 1000–2000 (30–33 dBi) have very high EIRP, ~30 MW (75 dBW). As receiver noise in telecomm receivers is only a few decibels above the thermal noise floor of −204 dBW/Hz, they may encounter severe interference. Since 2003, Appendix S3 of the ITU-R Radio Regulations[2] in essence requires out of band shipborne radar emissions to be 60 dB below EIRP, for example, 30 W in the example, the 'relative' phrasing of the regulation (inserted at military behest) giving little incentive to reduce EIRP itself. VTS radars are required to be considerably better. This situation is likely to be tightened within a few years; current relative friendliness to high EIRP may be rebased to require out of band emissions less than some specified 'absolute' low number of watts per megahertz.

Some future spectrum control possibilities are outlined in Chapter 16. Currently, control is often by a bandpass filter at the magnetron output. Problems include space availability, designing to handle the peak power, provision of sufficient attenuation through to harmonic frequencies, energy loss in the passband and maintenance of good impedance match. If the filter appears highly reactive away from centre frequency, it may pull the magnetron frequency or provoke moding. Filtering has

[2] ITU-R SM 329-7; Category A, Shipborne radar; Category B, VTS radar.

little practical effect on the shape, bandwidth or detectability of the received echo apart from introduction of ~1 dB insertion loss.

Radiated spectrum depends on several linked features.

- Modulator detailed design, particularly the slope (~100 V/ns) and dynamic impedance of its output pulse.
- Magnetron detailed design. New 'third generation' designs being developed claim sharply reduced out-of-band spectral components.
- Magnetron age. Spectrum deteriorates near end of life.
- In-band load mismatch presented by the duplexer or output filter. Four-port circulators with matched loads (Figure 2.11 and Chapter 3, Section 3.2) are superior to three-port types.
- Out-of-band load mismatch. It is difficult to design filters to retain good match in the stop bands.
- The out-of-band scanner efficiency. If poor, out-of-band radiation is reduced.

2.4 Transmitted frequency

2.4.1 Frequency and wavelength

Wavelength, λ, is tied to frequency, f, by the velocity of propagation, c, which is the speed of light 299.7×10^6 m/s, conveniently remembered as approximately 300 m/μs or 1 ft/ns. When the refractive index of the medium is n (1.0 for vacuum, ~1 for air)

$$\lambda \sim \frac{299.7 \times 10^6}{fn} \text{ m.} \tag{2.3}$$

When it is necessary to confirm wavelength is in free space rather than some other dielectric, λ is replaced by λ_0. A frequency of 3.0 GHz (3×10^9 Hz) has wavelength $\lambda_0 = 0.0999$ m, ~10 cm. Operational frequency or wavelength can be indicated in various ways and fashions have changed over the years.

- All marine and VTS radars are microwave (defined by wavelength <0.3 m) and centimetric (wavelength variously defined as 30 to 2 cm or 10 to 1 cm).
- IMO's current preference, used in this book, is nominal frequency, expressed in gigahertz. The two marine bands are called 3 and 9 GHz. The latter's actual allocation is centred on 9.4 GHz, confusingly 9.00 GHz lies well outside the band.
- The actual ranges of permissible carrier frequencies, in megahertz or gigahertz: 2900–3100 MHz or 2.9–3.1 GHz and 9200–9500 MHz or 9.2–9.5 GHz (formerly 2920–3100 and 9320–9500 MHz).
- The corresponding wavelength, usually nominal rather than actual, and – in robust defiance of SI principles – always expressed in centimetres: 10 and 3 cm (actually centred on 100 and 32 mm). Convenient when discussing scanners and other wavelength-related phenomena.
- Code letter designations, introduced for secrecy in the Second World War, have stuck: marine S and X bands. Unfortunately the higher frequency band used by a few VTS sets is designated J in the United Kingdom but Ku in the United States, alas not the only radar example of two peoples divided by a common language.

Table 2.1 Radar bands, frequencies and wavelengths

Former name	L	S	C	X	J (Ku in USA)
Band frequency limits (MHz)	1000–2000	2000–4000	4000–8000	8000–12 500	12 500–18 000
Marine radar allocation (MHz)	None	2900–3100	None, see notes	9200–9500	See notes
Marine radar allocation (GHz)		2.9–3.1		9.2–9.5	
IMO name	–	**3 GHz**	–	**9 GHz**	–
Common preset frequencies (MHz)		3050		9375, 9410, 9440	
Equivalent wavelength, nominal (cm)	20	10	6	3	2
NATO equivalent band name	D	E/F	G	I	J
Radio band	UHF	UHF/SHF	SHF	SHF	SHF

Notes: Some past use has been made of 5470–5650 MHz within C band. There is no internationally recognised marine radar allocation in J/Ku band. VTS radars may lie within the 13.4–14.0 or 15.7–17.7 GHz ranges.

- NATO letter designations, rarely used in civil marine circles, perhaps because the transition from E to F band bisects the 3 GHz marine band: F and I bands.
- Radio band designations of hoary antiquity but seldom encountered in radar work: part of the civil marine 3 GHz band lies within the ultra high frequency (UHF) band (300–3000 MHz), the remainder of the 3 GHz band and the 9 GHz band are super high frequency (SHF), 3.0–30 GHz.

Of course there is no abrupt change of physical properties at band boundaries, which are all artificial concepts. Table 2.1 lists the commonest designations of the two radar bands and the J/Ku band occasionally used for VTS. The current 3 cm band extends between 9200 and 9500 MHz but racons are limited to 9300–9500 MHz; the 9200–9300 portion is not widely used. ITU-R allocates 13.4–14.0 and 15.7–17.7 GHz for radar service of any sort, and 14.1 GHz has been used for continuous wave short-range berthing aids which primarily measure velocity. The table also includes a couple of bands (L and C) which have been used for transponders. VTS has often used frequencies centred on 9170 MHz. As we have seen, transmission frequency of non-coherent radars is not under the control of the user – the tune control works not on the transmitter but on the receiver.

In equations, unless otherwise specifically directed, frequency must ALWAYS be expressed in hertz, for example, 9.4×10^9 *Hz and wavelength in metres, for example, 0.032 m.*

2.4.2 Choice of band

Marine radar has to use centimetric frequency, lower frequencies giving poor illumination of the sea surface and higher frequencies suffering excessive precipitation clutter. Choice of the best band for a given application is determined by several factors, and should be made after detailed cost/performance/benefit analysis, using the methods of Chapter 15. The intervening chapters present all the factors in play. In general, the following considerations are significant.

- *Regulatory:* ships have to carry 9 GHz and are encouraged to carry 3 GHz as the second set, for its superior performance in precipitation.
- *Azimuth resolution:* for acceptable resolution, scanner aperture should exceed 3 m at 3 GHz or 1 m at 9 GHz in marine service or 4 m at 9 GHz and 3 m at 14 GHz for VTS.
- *Long-range precipitation attenuation:* small at 3 GHz, moderate to severe at 9 GHz, prohibitive at 14 GHz.
- *Precipitation clutter:* small at 3 GHz, moderate to severe at 9 GHz, very severe at 14 GHz.
- *Sea clutter:* detection is improved by scanner of narrow azimuth beamwidth, favouring high frequency for a given physical size.
- *Compactness:* favours high frequency.
- *Cost:* for roughly comparable performance, 9 GHz is the cheapest and 14 GHz the most expensive.

To counter clutter, also second-time-round echoes when there is anomalous propagation to extra-long range, some VTS installations use two radars operating on frequencies spaced a few per cent apart, called frequency diversity, or frequency may be quasi-randomly varied (staggered) from pulse to pulse, using either a special tuneable magnetron or a coherent system with klystron or travelling wave tube. See Chapter 12, Section 12.8, and Chapter 16, Section 16.3.1.

2.5 Choice of other parameters

Transmitter pulselengths vary between about 50 ns and 1 μs. Short pulses give best range resolution (nominally 7.5 m for 50 ns) and illuminate less clutter, but may not work well with some scanner designs. They occupy a wider spectrum so the receiver bandwidth has to be wider and collects more noise. Long pulses require less receiver bandwidth, minimising noise, and are generally preferable when distant targets have to be detected in benign clutter. Choice is not always clear-cut and is examined in later chapters.

Transmitter power is typically 10–30 kW (40–45 dBW) for deep-sea ships. Some trawlers favour up to 60 kW (47.8 dBW), to improve detection of distant flocks of gulls hovering over tasty shoals of fish. Yachts needing only short detection range may use power as low as 1 kW (30 dBW). 'Transmitter power' means peak power, the power delivered during the pulse. After leaving the magnetron, the signal traverses several microwave components including the duplexer which links the scanner to the transmitter and receiver. Each component introduces some loss. The radar's datasheet power may mean: (a) power at the magnetron output, or (b) the available power at the flange of the transceiver cabinet or at the scanner flange when the scanner is integral with the transmitter. Unless specified otherwise it is prudent to assume (a) – the highest value in the datasheet! – so allowance must be made for internal loss, typically 0.5 dB. Feeder loss is specific to the installation and always excluded from datasheet ratings.

Transmitter power is only one of the parameters affecting performance. Although 25 kW sounds grander than 20 kW, it represents barely 1 dB more (44, 43 dBW). For equal performance it might be cheaper to choose a set with receiver noise figure 1 dB lower, or to shorten the path of the feeder to reduce its one-way loss by 0.5 dB (see below). Do not be fooled into believing transmitter dBW are somehow more important than dBs elsewhere.

2.6 Feeder

Combination of the transmitter and receiver input stage with the scanner aloft eliminates feeder loss, but may impede servicing access and reduce reliability in the hostile masthead environment. Feeders do not always get proper attention in system design. At 3 GHz, a coaxial cable is usual, but at 9 and 14 GHz cable is generally too lossy and tubular metal waveguide is preferred.

2.6.1 Waveguide

Figure 2.14 shows an electromagnetic wave propagating energy through a vacuum, air or other dielectric medium. There are mutually perpendicular electric (E) and magnetic (H) fields, the direction of energy flow being mutually perpendicular to both.

When a plane wave such as that of Figure 2.14(a) strikes a perfectly conducting metal sheet placed normal to the direction of propagation,

- the magnetic field component of the wave induces circulating surface currents of finite amplitude in the sheet;
- with zero resistivity, there can be no voltage in the plane of the sheet. So immediately clear of the surface there must be a cancelling electric field in phase opposition to the incident wave.

These conditions make the sheet re-radiate as an antenna and the incident beam is specularly reflected just like light at a mirror. The angles of incidence and reflection are equal. Ray paths can be drawn geometrically, the rays travelling in straight lines when the medium is uniform. (Sometimes, e.g. when a ray grazes an obstacle and diffracts around its surface, behaviour ceases to be properly represented by geometrical optics and can only be described by the much more complex rules of wave theory.)

Figure 2.15 shows a plane wave reflecting from a flat conducting metal sheet. The angles of incidence and reflection are each θ. Phase fronts are shown, denoting

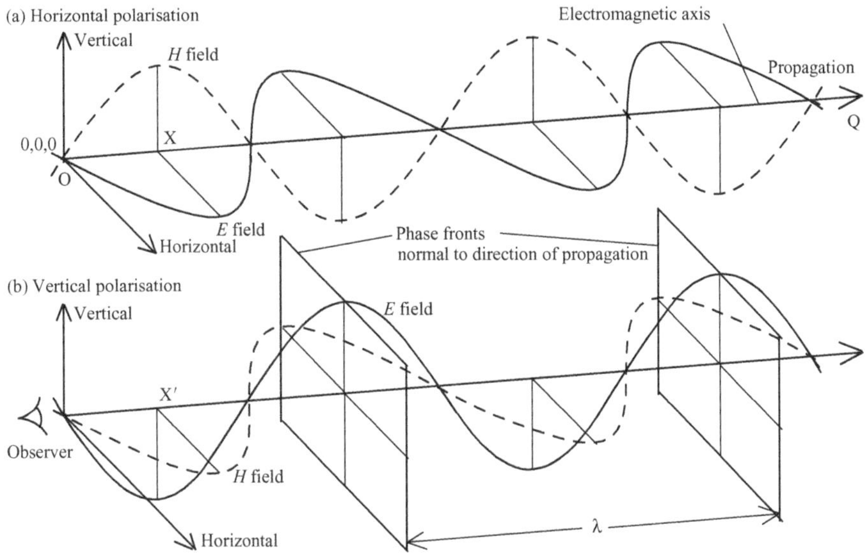

Figure 2.14 *Propagation. (a) Shows the electric (E) and magnetic (H) fields of a horizontally polarised ray. Reversing polarity of either field reverses the direction of propagation, (b) is for vertical polarisation. Vector addition of (a) and (b) gives slant polarisation. Vector addition of (a) and (b) with $\pi/2$ rad phase shift gives circular polarisation, the hand depending on which planar component leads*

positions having common phase and one wavelength apart. They are always normal to the direction of propagation. Clearly, once a pattern of incident and reflecting fronts has been launched, a second conducting sheet could be inserted parallel to the first sheet at position YY (spaced distance a) without affecting the pattern in the bounded space. The ray perpendicular to the fronts represents the oblique direction of propagation of the plane wave, assumed to have been launched from some point to the left of the diagram. This wave reflects back and forth as it proceeds towards the right within the bounded space. The velocity in the oblique direction is that of light, c, so the horizontal component of velocity is lower and depends on the obliquity. So once a ray has been launched between two parallel conducting plates, it propagates freely. Electric field exists between the plates, while magnetic fields link with circulating currents on their inner surfaces.

This geometry requires a whole number of projections of a half-wavelength between the plates, therefore $\sin\theta = n\lambda_0/(2a)$, where n is an integer. The following waveguide properties arising from this geometry can be proved rigorously from Maxwell's fundamental propagation equations, see, for example, Mehler [2].

The plates can be bounded by another pair, giving a rectangular cross section, making a pipe called a waveguide. The distance between this second pair, b, affects the characteristic impedance (discussed later), and is usually made about $a/2$, partly

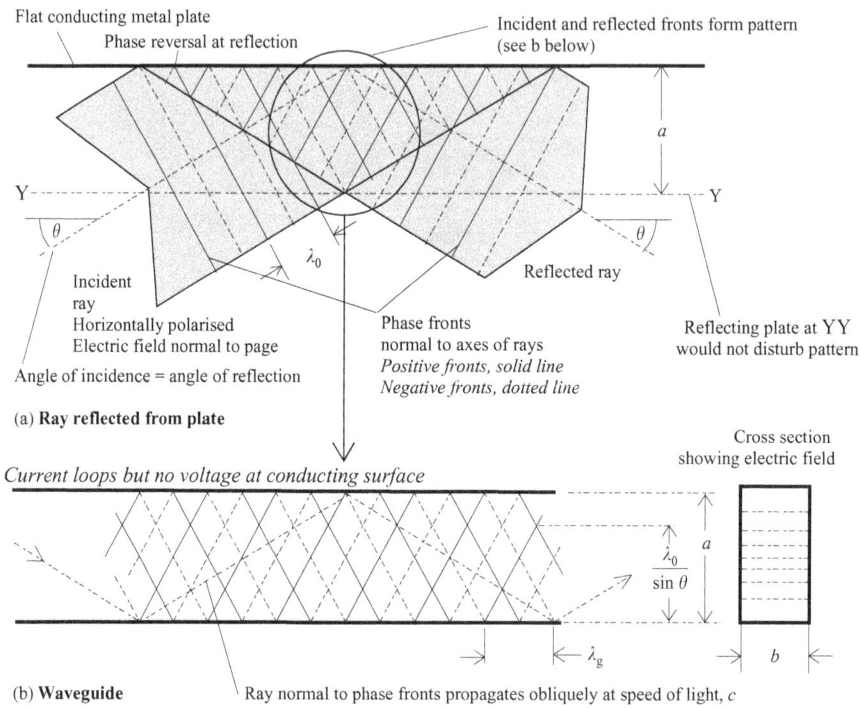

Figure 2.15 Reflection from flat sheet, and propagation in waveguide. Properties can be described using geometrical optics, with total reflection at the narrow walls

to inhibit unwanted transmission modes in the wrong plane. The desired mode of transmission is then called the H_{01} mode. If frequency is too high, over-moding may occur, unwanted high-order modes existing as if the guide were compartmented into a honeycomb of parallel smaller tubes. The electric field is everywhere perpendicular to the broad face, varying sinusoidally from zero at the sides to a maximum at the centreline. The magnetic field forms closed loops. Intensities at any given point vary sinusoidally with time.

The pattern of electric and magnetic fields travels along the guide. Its phase velocity is faster than the speed of light:

$$\text{phase velocity} = \frac{c}{\sqrt{1-(f/f_c)^2}}. \tag{2.4a}$$

From the geometry, as signal wavelength rises, the ray (velocity c) becomes more oblique and forward progress is slower. Eventually, when $\lambda_0 = 2a$, θ reaches $\pi/2$ rad, perpendicular to the guide axis, making propagation impossible. Below this cut-off frequency, f_c, a given height waveguide will not propagate but reflects the signal back to the source. The velocity of information propagation in the guide is called the group velocity and is the forward component of the velocity vector. It is lower than the velocity of light. From the geometry;

$$\text{group velocity} = c \times \sqrt{1 - (\lambda_0/2a)^2}. \tag{2.4b}$$

Within the guide, wavelength, λ_g, is increased relative to free-space wavelength λ_0. It follows from the geometry that:

$$\lambda_g = \frac{C\lambda_0}{\text{group velocity}} = \frac{\lambda_0}{\sqrt{1 - (\lambda_D/2a)^2}}. \tag{2.4c}$$

For example, at 9.4 GHz ($\lambda_0 = 3.136$ cm), waveguide WG16 (WR90, R100) having $a = 22.86$ mm has group velocity $= 0.728c = 218.1$ m/μs and $\lambda_g = 4.448$ cm.

Waveguide, but not coaxial cable, is dispersive; the velocity of propagation varies with frequency. If the pulse has a wide spectrum (short pulselength), dispersion in a lengthy waveguide operating not much above cut-off may spread the echo between a couple of range cells in the data extraction system, reducing detectability. In more normal circumstances dispersion can be ignored. For example, 9300 MHz ($\lambda_0 = 0.03216$ m) in WG16 has group velocity 213.226 m/μs, taking 468.94 ns to transit a 100 m feeder. Changing to 9500 MHz, the velocity rises to 216.922 m/μs, reducing transit time by 7.94 ns to 460.99 ns, spreading the echo over 1.2 m range equivalent. Any mismatch at the scanner changes phase by nearly 514 rad at the magnetron, causing the 'long line effect' referred to in Section 2.3.2.

The cut-off frequency has free-space wavelength $\lambda_0 = 2a$. Cut-off sets a definite lower limit to the workable frequency. Immediately above cut-off, operation is very frequency dependent. On the other hand, if frequency is too far above cut-off, over-moding occurs. These factors define the waveguide cross section, chosen from one of a range of standard sizes, see Table 2.2. In WR nomenclature, the number is the broad internal H-plane dimension a, expressed in units of 0.01 inch $= 0.254$ mm. Waveguide has only about 15 per cent useful bandwidth, which is however sufficient to cover a marine band. Preferred operating frequency $\sim 1.5 \times$ cut-off. Velocity of propagation in coaxial or waveguide feeders is usually about three-quarters the speed of light.

Current flow is confined to the inner surface, the small skin depth (limited by electromagnetic interactions within the metal) precluding any exterior voltage. Waveguide tubes may be rigid rectangular or semi-rigid elliptical in section. The elliptical semi-rigid type has few joints and is less susceptible to water entry. Although too expensive and lossy for long runs, fully flexible and twistable/flexible short lengths are useful to accommodate misalignments and vibration.

Peak power rating is set by arcing voltage. Higher frequency waveguides have smaller cross section, with more loss and lower rated power. In theory, at atmospheric pressure size WG 10 for 3 GHz breaks down at 2.6 MW and WG 16 (9 GHz) at 256 kW, but it is imprudent to operate at more than half breakdown power. Bends generally withstand lower power than straight lengths and need particular consideration on 9 and 14 GHz high-power VTS installations. Pressurised dry air or nitrogen is sometimes used to obviate water ingress and increase flashover voltage. Flashover may severely

Table 2.2 *Feeder sizes and ohmic loss, copper plain rectangular waveguide, one-way*

Band	3 GHz	9 GHz	9 GHz, low loss	14 GHz
Waveguide size (see note)	WG10, R32, WR284	WG16, R100, WR90	WG15, R84, WR112	WG18, R140, WR62
Outside dimensions (mm)	76.2 × 38.2	25.4 × 12.7	31.75 × 15.875	17.83 × 9.93
Internal, $H \times E$ planes (mm)	72.14 × 34.04	22.86 × 10.16	28.70 × 12.83	15.8 × 7.899
Loss: plain straight guide (dB/m)	0.02	0.11	0.09	0.27
Loss: likely in service (dB/m)	0.04	0.18	0.15	0.35

Note: Waveguide sizes quoted are RCSC, IEC-R 153 and RETMA respectively and are derived from inch units.

damage both transmitter and receiver. In marine radar service, power handling is rarely a problem, unless bad mismatch is introduced by a dent, broken joint or water ingress. When using coaxial cable, high power necessitates large diameter with substantial bending radius.

2.6.2 Mismatch

Mismatch occurs whenever energy is fed from one component to another of differing impedance. All transmission lines have characteristic impedance, Z_0, a vector quantity. For coaxial cable $Z_0 \sim \sqrt{L/C}$ where L is the central conductor's inductance per unit length and C is the capacitance per unit length between the central conductor and the metal sheath. Figure 2.16 shows these distributed constants conventionally, as a series of lumped elements. Although less simply defined, waveguides also have characteristic impedance, rising with the b dimension. The device at the far end of the feeder is called the load. The load on a transmitter is the scanner; to echoes it is the receiver. If the load is resistive and equals Z_0, it accepts all the energy flowing along the feeder. None is wasted or rejected so the return loss is infinite. This is the desired matched load condition.

The load impedance may differ from Z_0. One limiting case is a short-circuit. At the shorted load terminals current flows but no voltage can exist. Power $= VI$ and no power is accepted as $V = 0$. The power rejected must go somewhere; it is reflected back toward the source. The load is mismatched. A quarter-wave back along the feeder the incident and reflected voltage components are in phase and add to give a high-voltage node. Slightly further back they are in opposite phase giving a low-voltage antinode. If we measured the voltage along the feeder we should find alternate

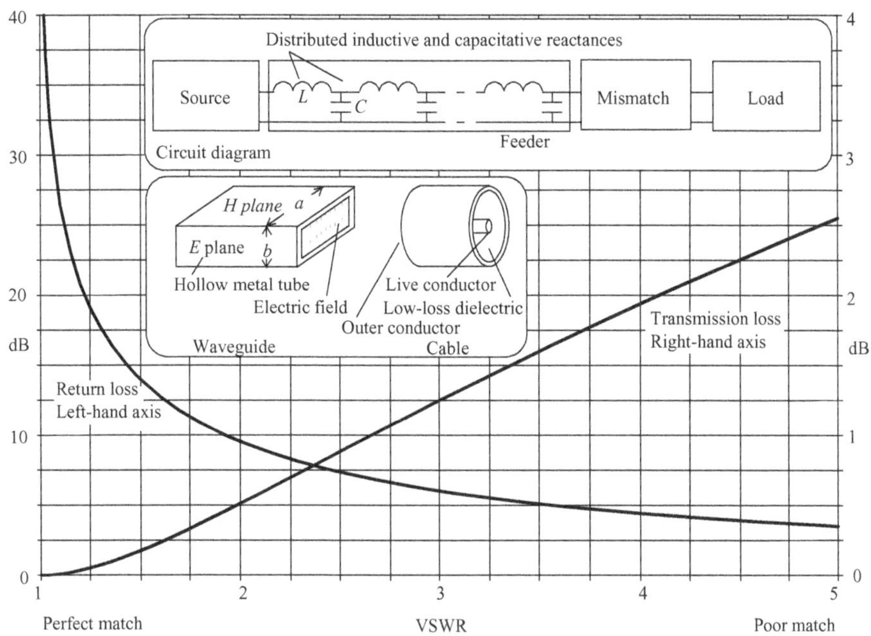

Figure 2.16 Feeder mismatch. Transmission and return losses plotted against VSWR. Within waveguides, electric fields are in the E plane and magnetic fields in the H plane (mnemonic: easy and hard to bend!)

high- and low-voltage points. A standing wave exists, as when a train of sea waves hits a harbour wall. For our short circuit, the ratio between the node and antinode voltages, called the voltage standing wave ratio (VSWR, sometimes denoted S), is infinite for a short or open circuit (the former British practice sensibly used the reciprocal, giving VSWR between 0 and 1). For a matched load VSWR falls to 1.0. VSWR is a measure of match quality. Risk of arcing from the high voltages at mismatch nodes may limit the power handling capacity of feeders.

At spot frequency, a mismatch can be cancelled by another mismatch nearby, usually within the component itself, and this is routinely done during manufacture within scanner/rotating joint and other assemblies. It becomes increasingly hard to maintain match over wide frequency bands or wide physical separation of the original and correcting mismatches. In practice, there are always several residual mismatches, spaced many wavelengths apart. Their phasing becomes extremely frequency dependant and there is considerable mismatch uncertainty, preventing accurate prediction of mismatch loss.

As bad VSWR prevents some of the incident power getting to the load, it represents a transmission loss, usually tolerably small unless the mismatch is severe. Often more important is the return loss – the reflection into the receiver when a transmitter feeds a mismatched scanner. Losses are related to VSWR through a voltage reflection coefficient, ρ:

$$\rho = \frac{\text{VSWR} - 1}{\text{VSWR} + 1} \text{ numerical,} \tag{2.5a}$$

$$\text{transmission power loss} = -10\log(1 - \rho^2)\text{ dB,} \tag{2.5b}$$

$$\text{reflection or return power loss} = -20\log\rho\text{ dB.} \tag{2.5c}$$

Figure 2.16 connects loss and reflection with VSWR and depicts these equations.

2.6.3 *Feeder losses*

Feeder resistive (ohmic) and mismatch losses affect both transmission and reception paths and can significantly degrade overall performance. Unless there is a pre-amplifier or head amplifier aloft, feeder ohmic loss, L, beside attenuating echoes introduce additional noise, often expressed as a noise temperature $= (1 - 1/L)T_a$ Kelvin (T_a is absolute ambient temperature, typically 290 K). With no head amplifier, feeder loss precedes the source of noise in the receiver and the degraded SNR cannot be recouped by subsequent amplification. Ohmic loss can be particularly severe where VTS transceivers are at the foot of tall masts. It is all too easy to throw away expensive transmitter power and low-loss receiver performance in long and lossy feeders. If a lengthy feed is inevitable, consideration should be given to placement aloft of the main parts of the transceiver, including the head amplifier. Feeder losses are one of the factors affecting choice of scanner height, discussed in Chapter 15, Section 15.2.2. Losses rise if the waveguide becomes damaged or water seeps in. Kinks sharply increase loss and must be avoided.

Table 2.2 includes a rough guide to practical ohmic insertion losses in undamaged, internally dry, copper waveguide and includes an allowance for internal tarnish and a typical proportion of bends. Losses in semi-rigid and aluminium guide are similar. The UK Maritime and Coastguard Agency[3] offers useful hints on good installation practise for feeders.

The following reduce feeder losses.

- Microwave circuit devices such as scanners having good match.
- Feeder cables of the correct characteristic impedance or using matching transformer sections.
- Protection from accidental damage.
- Short length. Avoid ‘round the houses plumbing’.
- Cable or elliptical waveguide not curved to less than its recommended radius.
- Minimal number of waveguide bends.
- Do not attempt to bend or twist straight waveguide, use a special section.
- Never cross waveguide planes.

[3] Survey of Merchant Shipping Navigational Equipment Installations: Instructions for the Guidance of Surveyors, UK Maritime and Coastguard Agency.

Mismatch causes a standing wave in the feeder, VSWR rising as mismatch worsens. For example, when VSWR $= 2$, the associated voltage reflection coefficient, ρ, is 0.33 (Eq. (2.5a)) and transmission loss is 0.51 dB (Eq. (2.5b)).

Two-way, this throws away $\sim$1.0 dB (26 per cent) of radar performance. Reflection loss is 9.65 dB, so as much as 11 per cent of the transmitter power reflects back into the receiver – a kilowatt or more, hugely bigger than the echoes it is designed to handle. Protection devices may prevent damage, but the receiver will have a bout of severe indigestion, taking several microseconds to recover full sensitivity after transmission, spoiling performance out to a kilometre or so range.

2.6.4 Ringing

Transmitter power reflected by mismatch introduces unwanted ringing clutter. As the receiver input is never perfectly matched, it re-reflects. The reflection bounces back and forth along the feeder before finally petering out, false echoes or ringing appearing each time. At the usual velocity of propagation near 200 m/μs, a 50 m feed would deliver reflections to the receiver each 0.5 μs or 75 m range equivalent. These successive strong reflected pulses can mask true echoes, spoiling the radar's short-range performance. Chapter 11, Section 11.8, details how they interact, depending on their spacing.

2.7 Scanner, qualitative description

2.7.1 Plane and circularly polarised rays

An antenna or aerial is a device which, when fed by an a.c. signal, radiates electromagnetic energy into the surrounding space, or converts intercepted electromagnetic energy to an a.c. signal. All antennas used in marine radar applications are reciprocal devices working equally well either way, although it is usually easier to visualise the transmit mode. A scanner is an antenna (plus turning gear, etc.) whose energy is radiated as a narrow beam which continuously rotates to successively illuminate the whole horizon. It receives target echoes while collecting as little clutter as possible within the same beamwidth (defined in Section 2.7.2), thus filtering out signals from other bearings and defining the bearing of echoes. We shall use some of the principles developed here when discussing targets in Chapters 7 and 10, for they behave in some ways as receiving and re-transmitting antennas. The scanner is a key component of both transmission and reception paths. It dominates many aspects of system performance and merits thorough examination. Simplified quantitative analysis follows in Section 2.8 and reception in Chapter 3, Section 3.1. Readers requiring deeper analysis should consult the extensive specialist literature. Silver [3], although old, is recommended.

We consider the radiation to or from antennas as a bundle of linear rays filling the solid angle through which the antenna radiates, rather as hypothetical lines of force help to explain operation of a magnet. Figure 2.14(a) shows a ray travelling from left to right at the speed of light, $c \sim 300$ m/μs, along the electromagnetic axis OQ. At any point X, the electric (E, V/m) and magnetic (H, A/m) alternating field components are mutually orthogonal to the direction of propagation, the ray energy being EH W/m^2. The ratio E/H is set by the characteristic impedance of free space $= \sqrt{\mu_0/\varepsilon_0} = \sqrt{4\pi c/10^7} = 377\,\Omega$. Reversing the polarity of either field reverses the direction of propagation. In Figure 2.14(a) the electric field lies in the horizontal plane so the ray is horizontally polarised (HP).

Figure 2.14(b) shows a vertically polarised (VP) plane-polarised ray, having its nulls at the same instants as A. Addition of the E or H vectors of A and B at any instant, say X and X', gives a resultant plane-polarised ray with slant polarisation, the slant angle depending on the relative powers of the components. Therefore, any slant-polarised ray can be resolved into HP and VP components.

Changing the relative phasing of B relative to A, so that a B null lies at an A peak, causes the resultant pair of E and H vectors to rotate spirally one revolution per wavelength. The ray is said to be circularly polarised (CP) if the component powers are equal. More generally, unequal components give elliptical polarisation. The polarisation is said to be right-hand if an observer at the source looking along the direction of propagation sees the vectors rotating clockwise, like a nut on a conventional right-hand threaded screw. This is the usual (American) IEEE convention. The (British) IEE convention was the opposite, a fruitful source of confusion. We avoid the term ‘clockwise’, partly because the scenario would imagine an observer within the clock, looking out along the direction of data flow, from which position the hands rotate anticlockwise! Changing the relative phasing of one component by π rad (inverting its polarity) reverses the hand of polarisation to left hand. Thus any circular or elliptically polarised ray may be resolved into a pair of orthogonal plane components.

An isotropic radiator, Figure 2.17(a), is an antenna which radiates uniformly through the whole $360° \times 360°$ (4π sr) of the surrounding space. The concept is useful, but because polarisation cannot be retained throughout the sphere it is impossible actually to make such a device. Placed at the centre of a large hollow sphere of radius R, the whole internal surface area $4\pi R^2$ would be weakly illuminated, an analogy being a torch’s bare bulb suspended from wires with the reflector removed.

At some instant the radiator transmits with, say, the E field at a positive maximum, half a cycle later there is a negative maximum and so on. The phase fronts spread out at the speed of light in expanding spherical shells as they leave the radiator. Within the small solid angle occupied by a ray, fronts may be regarded as planar. The isotropic antenna can pick up signals from a source anywhere on the sphere, but it is necessarily insensitive, for it must ‘watch’ over the whole sphere all the time. Neither can it give any indication of direction.

From consideration of conservation of energy, it follows that when a source radiates power P watts from an isotropic radiator in unobstructed free space, at range R metres the signal power gives a flux density d watts per square metre of transverse cross section, which is the power divided by the surface area of the sphere:

$$d = \frac{P}{4\pi R^2}\ \text{W/m}^2. \tag{2.6}$$

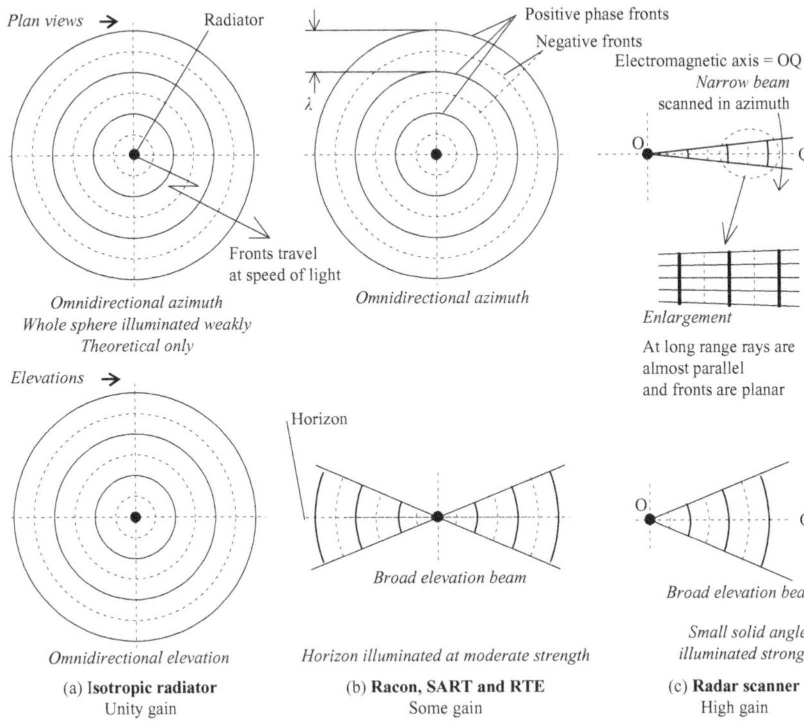

Figure 2.17 Antenna radiation patterns. Diagrammatic plans (upper row) and side elevations of (a) theoretical isotropic radiator, covering whole sphere; (b) active device antenna covering 360° azimuth ×~25° elevation; (c) radar scanner concentrating energy into ~1° × 25° fan beam. Transmit mode. Receives within same solid angles

The isotropic radiator is used as a reference against which practical antennas are compared, and is defined as having unity gain, $G = 1$. In decibel form, $G = 0\,\text{dB}$ (usually written 0 dBi, gain relative to isotropic). Its radiation in azimuth and elevation patterns are shown as circles in Figure 2.17(a).

2.7.2 *Directional radiation*

Fitting a paraboloidal reflector to the torch bulb concentrates its light on a small part of the surface of the sphere, which it brightly illuminates at the expense of the remainder, now dark. A small solid angle has gained light, although of course the total output remains unaltered, or diminishes if the reflector is lossy. Placing a reflector behind the isotropic radiator has a similar effect; a beam is formed within which energy density is stronger than before, Figure 2.17(c). The radiator/reflector combination forms a directional antenna, with gain of power density within the beam. The gain is high when the beam is narrow in azimuth and elevation, covering a small solid angle. The annulus of sphere wall at the horizon can be scanned by slowly rotating the bulb or antenna, gradually building up a picture of the whole horizon. The beamwidth is the angular width of the main lobe of the beam, measured between its half-power (-3 dB) points, where electric field strength is $1/\sqrt{2}$ (71 per cent) that at beam axis. Figure 2.17(*b*) shows an intermediate arrangement with only elevation directivity and low gain, used for racons, etc.

Unless otherwise stated, the gain of a scanner always means the one-way gain on beam axis, denoted by $G_{\max}$ or plain G. Gain is a pure number, the ratio of actual power density on beam axis to power density at the same range from an isotropic radiator (gain 1.0 by definition). There are 4π steradians in a sphere so gain of a lossless scanner is given by:

$$G_{\max} = \frac{4\pi}{\theta\phi}$$

where θ, ϕ are the azimuth and elevation beamwidths in radians. Putting $\theta\phi = \Omega$ sr, the solid angle of the beam,

$$G_{\max} = \frac{4\pi}{\Omega}. \tag{2.7a}$$

To cover the surface of a sphere needs 41 253 tiles, each 1° square, so when beamwidths are expressed in degrees and the solid angle illuminated is $\Omega_{\text{deg}} = \theta_{\text{deg}}\phi_{\text{deg}}$,

$$G_{\max} = \frac{41\,253}{\Omega_{\text{deg}}}. \tag{2.7b}$$

A handy rule of thumb for typical actual scanners is:

$$G_{\max} \sim \frac{25\,000}{\Omega_{\text{deg}}}. \tag{2.7c}$$

This is often expressed in decibel form:

$$G_{\max} = 44 - 10 \log \Omega_{\text{deg}} \text{ dBi}. \tag{2.7d}$$

The ratio 25 000/41 253 = 0.606 represents a typical loss of ~2.2 dB (one-way) within the scanner, accounting for unavoidable resistance, mismatch, etc. Loss tends to be high when antennas have particularly good sidelobe levels.

We shall see in Section 2.8.1, Eq. (2.11b) for a slotted array having constant power density across the aperture width, a, that beamwidth, θ, in the plane of the aperture is

$$\theta = \frac{\sqrt{\pi}}{2}\frac{\lambda}{a} = 0.886\frac{\lambda}{a}.$$

The factor $\sqrt{\pi}/2$ arises from the power lost to sidelobes. A hypothetical uniformly illuminated antenna without sidelobes would have

$$\theta = \frac{\lambda}{a}.$$

For an aperture of $a \times b$, mouth area A, solid beamwidth angle $\Omega = \lambda^2/A$.

Substituting in Eq. (2.7a)

$$G_{\max} = \frac{4\pi}{\Omega} = \frac{4\pi A}{\lambda^2}. \tag{2.7e}$$

The gain, aperture area A, wavelength and directivity D (the ratio of the energy per unit area transmitted relative to that from an isotropic radiator at the same range, neglecting losses), are therefore related.

$$D = \frac{4\pi A}{\lambda^2} = \frac{4\pi}{\Omega}. \tag{2.7f}$$

Losses reduce efficiency below unity:

$$\text{efficiency} = \frac{G}{D}. \tag{2.7g}$$

When calculating detectability, the scanner is often assumed to have a rectangular radiation pattern; full gain at all angles between the half-power points and zero elsewhere, allowing for the reduced gain at the edges of this rectangle by introduction of a hypothetical beamshape loss, see Section 2.7.15.

2.7.3 Beam characteristics

Ideally, an antenna would direct all its energy within the desired beamwidths, but this is impossible on theoretical grounds. If the energy distribution across the aperture is uniform, called uniform illumination, a proportion is transmitted in unwanted subsidiary beams at other bearings, called sidelobes. The loss of energy within the main beam is not usually very important. But the sidelobes may illuminate some scatterer, its reflection being in turn picked up by the sidelobe in receive mode and falsely displayed as a target at main beam azimuth. Sidelobes are minimised in practical scanners by shaping or weighting the energy density or illumination so that the edges of the aperture receive less than the centre, the aperture not being fully filled. There are penalties – for a given effective aperture size, the physical size rises, beamwidths go up and gain goes down. Cross-polarisation effects may cause butterfly lobes, sidelobes at 45° to the principal planes.

When both beamwidths are fairly small and approximately equal, radiating a near-circular cone, as in a few VTS scanners, the beam is called a pencil beam. A fan beam antenna has elevation beamwidth sharply larger than azimuth to cater for roll and pitch and radiates an elliptical solid angle (Figure 2.5(a); radiation as Figure 2.17(c)). As stabilised mounts are not used, to cater for roll all marine radars must have fan beams, the vertical beamwidth (sometimes called divergence) being an order of magnitude larger than the horizontal. A few VTS scanners use inverse cosec^2 beams, discussed later, others being intermediate in shape between fan and inverse cosec squared. Unfortunately, narrow beams demand great physical size. The beam electrical centreline is called the beam axis or boresight, dating from WW2 fire-control service, when radar axis was referenced to the gun-barrel using a sight inserted in the bore.

All scanners need narrow azimuth beamwidth for the following reasons.

- To obtain enough gain for adequate signal to noise ratio from weak distant targets.
- To minimise illumination of sea and precipitation clutter.
- For precision determination of target position, particularly for track prediction on ARPA or ATA.
- To reveal target aspect, particularly in VTS service.
- To discriminate between close-spaced targets, revealing presence of tugs, etc. near another ship and to prevent track-seduction (swapping) when two targets manoeuvre close together. This also demands low sidelobes.

2.7.4 Rotation

The scanner is continuously rotated clockwise, typically driven from a three phase induction motor through spur gears or toothed belts. Speed may fluctuate a few per cent with wind load. Typical speeds are:

- *Marine radar:* 20 rpm IMO minimum, with a tendency toward rather smaller scanners rotated rather faster, typically 25 rpm.
- *High speed craft (HSC):* 40 rpm (IMO minimum) to 60 rpm for fast display update, the penalty being few sweeps per scan.
- *VTS:* 10–20 rpm, low speed helping to decorrelate sea clutter from one scan to the next, also reducing the drive motor load of these large structures. Windage may be further reduced by aerofoils.

Instantaneous bearing of the beam and hence of currently received targets is reported to the signal processor by a shaft encoder having much better resolution than the beamwidth. VTS radars and auxiliary docking radars on some ferries often also use bearing data to inhibit transmission within blind arcs or blanking angles containing obstructions or persons. The feeder or transceiver is connected to the radiator by a rotating joint forming part of the scanner and drive motor assembly.

2.7.5 Size and beamwidth

Table 2.3 lists some parameters of typical scanners. Apertures are still sometimes expressed in feet nominal width, 1 ft $= 0.3048$ m. Nine gigahertz band marine scanners have azimuth apertures between 0.45 m (1.5 ft, yachts) and 3.7 m (12 ft, biggest ships) giving azimuth beamwidth, θ, between about 5° and 0.7°, respectively. Three gigahertz band has $\frac{1}{3}$ the number of wavelengths, hence $3 \times \theta$ for the same aperture, which is generally here 2.7–4.0 m (9–13 ft), $\theta \sim 2°$. In each case elevation beamwidth $\phi \sim 20°$–$25°$ to cope with rolling. Typically, the effective azimuth aperture, a, is about 75 per cent of the nominal overall width of the scanner, the scanners of Table 2.3 varying between 67 and 85 per cent. Typical radiation patterns are given in Figures 2.29–2.31. Losses are discussed in Section 2.7.15.

2.7.6 Marine radar scanners

Azimuth beamwidth is required by IMO for ships within the SOLAS convention to be less than 2° at −3 dB points and less than 5° at −20 dB points, with sidelobes less than

Table 2.3 Typical scanner parameters, all horizontally polarised

Ref.	Aperture, a (Metres)	(Feet)	Band (GHz)	Type	Shape	Beamwidth, ° θ Azimuth	ϕ Elevation	Gain G (dBi)	First sidelobe (dB down)
	Small craft								
A	0.45	1.5	9	Slot array	Fan beam	4.7	25	24	20
B	1.2	4	9	Slot array	Fan beam	2	25	27	26
	Deep-sea ships								
C	2.7	9	9	Slot array	Fan beam	0.7	20	33	31
D	3.6	12	3	Slot array	Fan beam	2	27	26	23
	VTS service								
E	3.6	12	9	Slot array	Fan beam	0.6	19	32	21
F	5.5	18	9	Slot array	Fan beam	0.41	20	37	35
G	5.4×0.64	17.7×2.1	9	Reflector	Inv. cosec^2	0.4	4.5	39	28

−23 dB within ±10° of beam axis and less than −30 dB further out (all relative to main beam). Elevation beamwidth has to be sufficient for ±10° ship's roll. Horizontal polarisation has to be available.

Originally separate small 'hoghorn' (waveguide horn) radiators illuminated transmit and receive parabolic reflector strips which were supported by horizontal flat plates, aptly called cheeses. Today's marine radars combine transmit and receive functions in a single slotted waveguide array (SWG) scanner, Figures 2.18(a) and 2.19(a)–(c), necessitating only a single rotating joint and feed, and giving better performance. The radiator is a horizontal waveguide fed from one end. Aperture, a, is somewhat less than the physical width. Up to 80 or so slots form the radiating elements, alternately cut at opposite slants and spaced S metres apart.

Some smaller vessels cannot accommodate scanners whose turning circles exceed about 1.3 m, giving undesirably wide azimuth beamwidth even at 9 GHz.

Using the concept of wavelets introduced by Huygens for optics in the seventeenth century, the simplified geometrical description of operation of a slotted array scanner is as follows. The slot elements disturb the electric and magnetic field patterns on the inner waveguide wall, causing radiation in a similar manner to radio dipole or loop antennas. Wavelets of energy radiate from each element, increased slot width and obliquity increasing the coupling, the proportion of power extracted. What little energy remains at the far end of the guide is absorbed in a built-in resistive plug or load, preventing end-reflections from spoiling the wavelet phasing. If each element radiated the same power, so the aperture was uniformly illuminated, an excellent main beam would form at the expense of poor sidelobes. Elements near the ends are therefore slanted less to make them radiate less, tapering the illumination across the aperture to improve sidelobe performance. If S were $\lambda_g/2$, beam axis would be

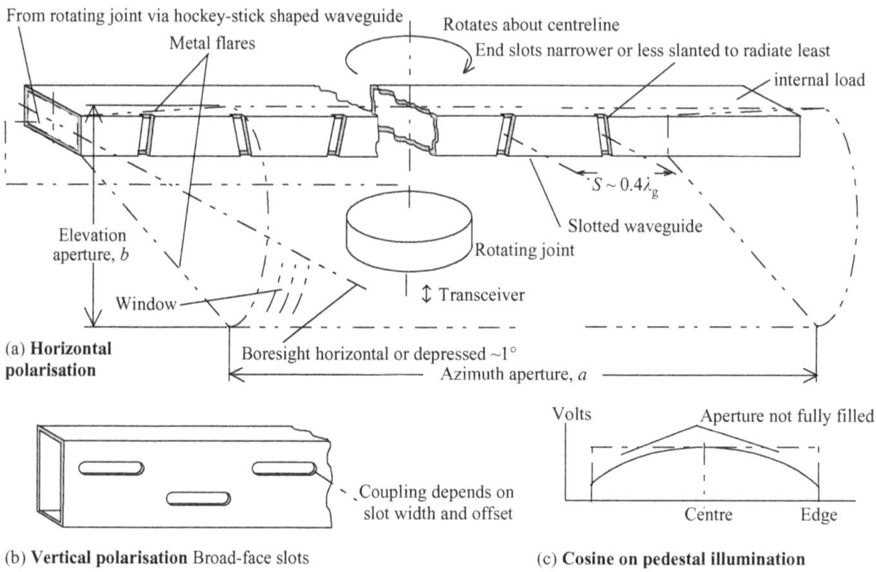

Figure 2.18 *Slotted waveguide scanner. Perspective view. Horizontally polarised fan beam, narrow in azimuth, broad in elevation. Slot radiation set by inclination and width. Flare mouth height sets elevation beamwidth. Slotting for vertical polarisation shown at (b)*

normal to the mechanical axis. Sidelobe performance is improved by making S rather less, typically $0.4\lambda_g$, squinting the beam $\sim 3^\circ$ from the mechanical axis. Care is taken to control the inevitable coupling which always exists between each element and its neighbours, to prevent unwanted phase and amplitude variations, often by stub filter waveguides, visible in Figure 2.19(c). If S exceeded $\lambda_0/2$, the smooth pattern would be disrupted by grating lobes, as with optical gratings.

2.7.7 Radiation patterns

Variation of gain with angle within a single plane (azimuth or elevation) is described graphically by the radiation pattern in that plane. Combination of patterns of both planes in a map-like representation gives a target pattern map (TPM). In azimuth, each slot forms a small antenna, transmitting in a broad beam, shown in Figure 2.20(a). On the elementary geometrical theory of operation, the element wavelets merge to give the fully formed beam at the Rayleigh distance, defined as that giving 45° variation of phase across the beam. This range, $\sim 2a^2/\lambda$, a few tens of metres for the smallest marine scanners and a kilometre or so for the largest VTS scanner, separates the near-field, Fresnel or focussing region, from the far-field or Fraunhofer region of most interest to us. Here the phase fronts are planar overall and previously-unfocussed transmitted energy combines to become a single coherent energy wave. Rays hitting a target in this region are considered as parallel. Fronts, TU in the figure,

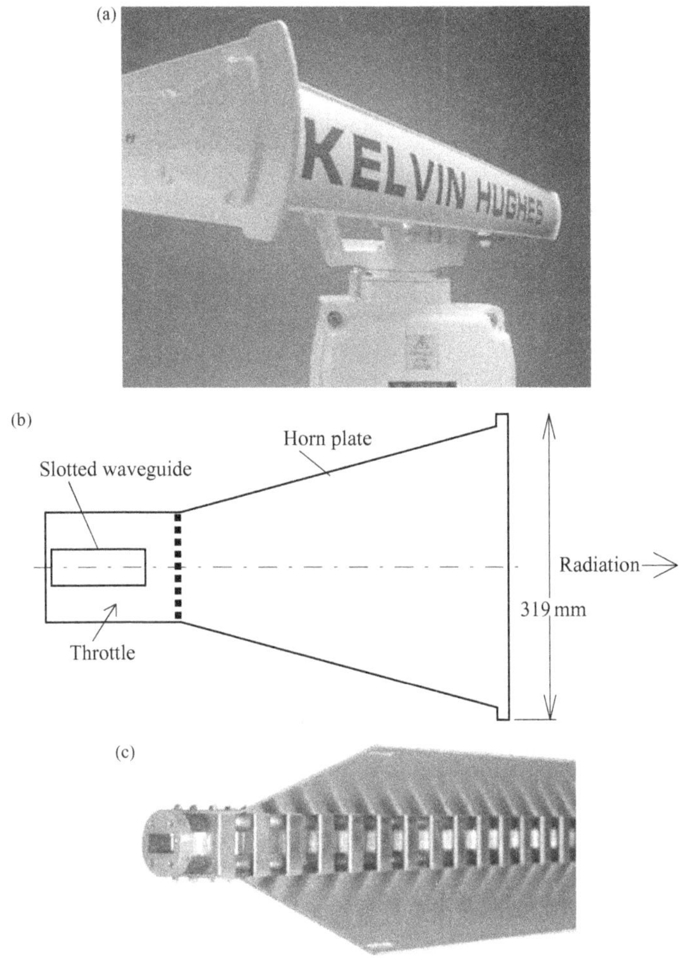

Figure 2.19 Conventional slotted waveguide scanner. Elevation beamwidth controlled by flares. All reproduced by permission of Kelvin Hughes Ltd, Ilford UK. (a) Outer view. Wing plates improve aerodynamics. Weatherproof window; turning gear, rotating joint and transceiver below. (b) Cross section, 3 GHz band. (c) Internal view, showing 9 GHz end-fed slotted waveguide, horn plate flares and a few of the slots and their associated filters. Aluminium construction

are approximately parallel to the waveguide face. Direction of propagation along the boresight OQ is of course normal to the front.

If the frequency and therefore the waveguide wavelength change, the wavelets no longer emerge quite at the same phase, so the beam axis in the horizontal plane is somewhat frequency dependent or dispersive (centre-fed designs are not dispersive).

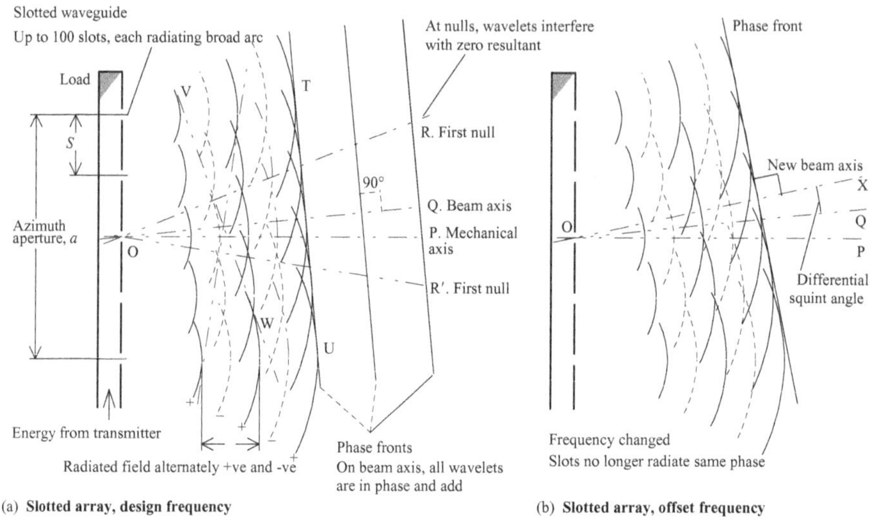

Figure 2.20 Radiation from slotted array. Plan views. (a) At design frequency, wavelets from many elements add coherently on boresight to form a beam but suffer destructive interference when sin $\theta \sim a/\lambda$, defining the position of first nulls and so the beamwidth. (b) When frequency is offset, λ_g changes and changes squint angle; a problem for offset-frequency racons and some VTS radars. Reception uses reciprocal processes

This differential squint, shown in Figure 2.20(b), is typically about 1° per 100 MHz in the 9 GHz band, more or less independent of aperture. It may necessitate realignment of the display after renewing the magnetron by one of different frequency. Differential squint also places limitations on frequency stagger and on operation of frequency offset racons, considered in Chapter 8. Few ships' scanners exceed 4.0 m (13 ft) aperture. To improve azimuth resolution, VTS scanners routinely reach 5.5 m (18 ft), some more.

Apertures as big as 11 m have been produced, but require very stiff support to prevent sag and attendant performance loss. Because the slots are driven in series, problems arise with frequency dispersion and propagation time when transmitting short (~50 ns) pulses. Velocity of propagation is about 0.2 m/ns (two-thirds the speed of light), so for the 27 ns it takes the leading and trailing edges to traverse a 5.5 m aperture, only some of the slots are energised, reducing gain, introducing squint and spoiling sidelobe performance (see next section); 11 m apertures are unsuited to pulselength <70 ns. At pulse edges, rate of change of power is high, depending on the rise and fall times of the modulator/magnetron combination, so spectrum is broad. The upper and lower frequency components are dispersed by the frequency-dependent nature of λ_g, further degrading the radiation patterns. Only during the central part of the pulse can the scanner meet its designed performance.

In elevation, metal flares traditionally act as a tapered waveguide opening out to match the slots to the impedance of free space. Mouth height, aperture b, defines the elevation beamwidth. Uniform illumination is desirable, to give the narrowest possible beamwidth and hence highest gain and least wind resistance, elevation sidelobes being unimportant. In practice the relatively wide beam together with the shortness of the flares introduces some illumination taper.

Weather protection is either by a static radar-transparent radome (radar dome) overall cover in small sizes, or a low loss dielectric window over the mouth of the aperture on the bigger open scanners.

2.7.8 Recent developments

Figures 2.21(a) and (b) show a design recently introduced for marine radar service. Radiations from a conventional slotted waveguide pass through a polarisation grid to a low-loss dielectric polyrod (polyethylene rod), which forms an endfire radiator, swapping axial length for height, after the fashion of a Yagi radio antenna. This enables better control of the elevation radiation pattern and gives a low profile assembly having only about one-third the usual height, sharply reducing wind resistance and permitting smaller and lighter turning gear.

Some small-craft radars (e.g. Raymarine Pathfinder series) have for several years used centre-fed arrays of patch elements, formed as printed circuits on low-loss substrates. Each small rectangular patch radiates as a dipole. Elevation beamwidth is controlled by the number of rows of patches (typically 3), dispensing with metal

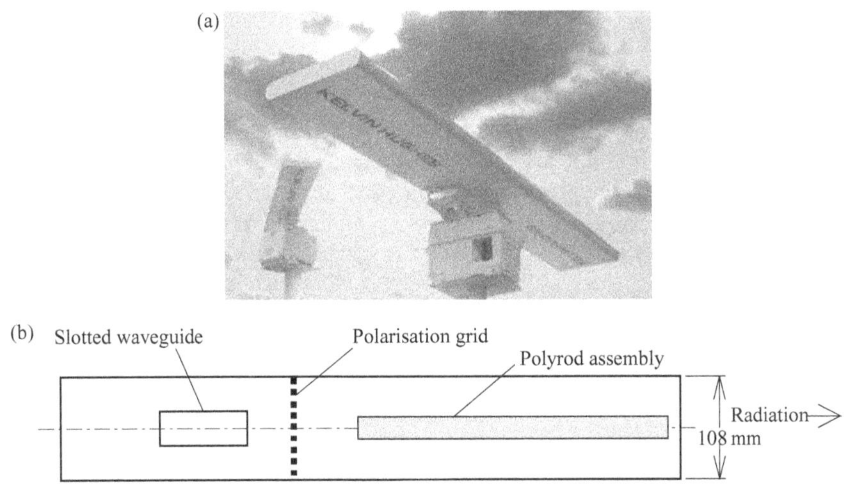

Figure 2.21 Low profile scanner using polyrods. For marine radar service, claimed to combine reduced face area with improved performance. Both reproduced by permission of Kelvin Hughes Ltd, Ilford UK. (a) Shows 9 and 3 GHz scanners. Reduced height permits lighter turning gear. (b) Cross section. The dielectric polyrod swaps height for depth

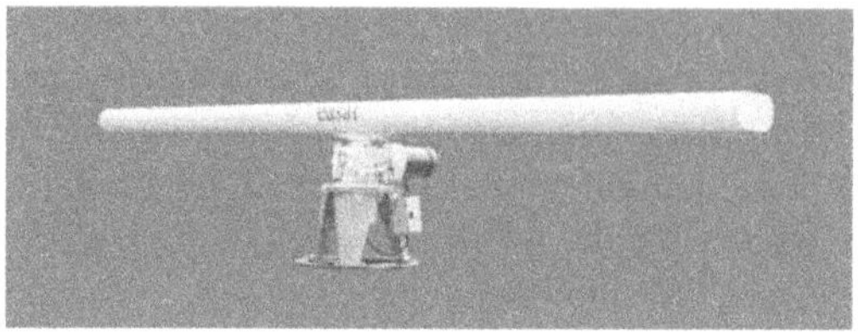

Figure 2.22 Linear phased array scanner. Aperture 7.4 m, 9 GHz band. Can handle very short (10 ns) pulses without distortion. Azimuth beamwidth 0.3°, inverse cosec^2 elevation beam for good detection of short-range as well as distant targets. Passive phased arrays are increasingly being used for wide-aperture VTS scanners. Reproduced by permission of Easat Antennas Ltd, Stoke on Trent UK

flares. The patches are fed by equi-phased splitter conductor patterns, facilitating dipping the input conductor into a static waveguide to form a simple and effective rotating joint. Such scanners are suited to mass production and short pulse operation but may be lossier than slotted waveguide types. Pathlengths to all patches being equal, differential squint is avoided. Phased arrays are increasingly also used for wide-aperture VTS scanners, see Figure 2.22.

Some VTS scanners are also now combining patch arrays with dielectric lenses, facilitating production of large sizes from thin modules. See also Chapter 16, Section 16.3.2, dielectric resonator antennas.

2.7.9 Obstructions

Sometimes the beam scans across local obstructions such as masts, samson posts or funnel. Because of the relatively long wavelength, their shadows are less sharp than with light. Some radiation refracts into the shadowed arc at the expense of some attenuation outside the arc, visible in Figure 2.6, Section 2.1.6. Calculation is difficult, but Wylie [4] gives some examples for vertical metal masts in the 9 GHz band, horizontal polarisation, which may be summarised as the following one-way approximate signal losses:

- 0.9 m diameter mast at 3/6/9 m from scanner axis: 7.5/6/4 dB through 5°, 6/3.5/1 dB through 10°, 5.5/1.5/0.5 dB through 15°;
- 0.3/0.6/0.9 m diameter masts at 9 m: dip to 2/6.3/7.5 dB on mast axis;
- 0.9 m diameter mast at > 10 m: about 1.5 dB at geometrical edge of obstruction and > 6 dB on axis.

Reflections from obstructions can also cause false echoes, see Chapter 10, Section 10.12.6.

2.7.10 Sidelobes

Figure 2.20(*a*) shows that away from boresight there are cuts such as *VW* where positive and negative wavelets are equal and cancel, with no resultant energy, giving nulls at bearings normal to the cut. The first nulls are OR and OR'. As the angle from boresight increases, the energy falls to zero at the first null, partially recovers to give the principal or first sidelobe, followed by further sidelobe nulls and smaller peaks. Although the mechanism differs in detail, reflectors and all other directional antennas also have sidelobes – indeed it is theoretically impossible to avoid them entirely. The outer parts of reflectors have to be illuminated weakly to reduce spillover and sidelobes induced by edge currents. Alas, beamwidths increase and gain falls. Scanners are precision devices and good sidelobe performance depends critically on element coupling and the straightness of the structure.

Within limits, the designer can trade sidelobe strength against gain on axis. Many alternative aperture weighting strategies exist, the subject being something of a mathematicians' delight, see, for example, Meikle [5]. Often the illumination function has a variant of the 'cosine on a pedestal' form; Figure 2.18(c):

$$\text{relative voltage density at angle } \nu \text{ from boresight} \propto 1 - (1 - k)\cos(\pi\nu)$$

where k is a constant.

The radar can only assume that echoes come from an object on boresight. Azimuth sidelobes therefore give false echoes. The art and science of scanner design is to minimise these sidelobes, by careful computation, illumination taper and precision slot-cutting. Principal sidelobes on big-ship scanners are at least ~23 dB below main beam, see Table 2.3 C, D. The higher sidelobes generally diminish by another 10 dB. Good scanner sidelobe performance can be ruined by close obstructions which deflect energy to unwanted bearings, causing sidelobe interference.

Figure 2.23(a) illustrates a scanner having uniform elevation illumination, 25° beamwidth and heeled 10° by roll or pitch. Even under this adverse condition the first null strikes the surface at only three times scanner height, H, with the second null at about $0.6H$. One hopes targets will have been detected and any collision avoidance manoeuvre completed at much longer range. Null ranges are very sensitive to roll and pitch angle, so all targets are likely to be illuminated from time to time and elevation nulls are not usually too troublesome to ships' radar. Elevation sidelobes illuminate extra sea or precipitation clutter and waste expensive transmitter power.

2.7.11 VTS reflector scanners

Many VTS stations use slotted arrays; either standard marine types or derivatives having wider aperture, somewhat narrower elevation beamwidth and sometimes with choice of polarisation. When more gain or narrower azimuth beamwidth is required, e.g. for distant detection of approaching shipping or to monitor movements of possibly illicit small craft, although patch arrays are being introduced (Figure 2.22, Section 2.7.8), reflector scanners are generally used, Figure 2.24. As also indicated in Figure 2.25(a), a small primary radiator (sometimes confusingly called the feed) is mounted in front of the reflector. It may use a plain or corrugated horn or short vertical slotted waveguide, broad-face slots giving horizontal polarisation. The radiator is protected by a window and placed at the focus O of an open reflecting metal dish reflector shaped as part of a shallow paraboloid (parabolic in both planes) having focal

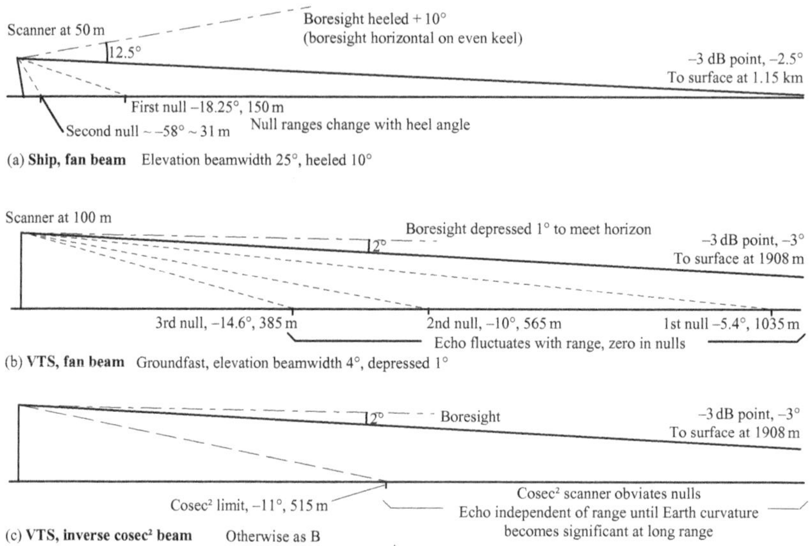

Figure 2.23 Short-range cover. Shows position of scanner elevation events for: (a) Ship with high scanner: despite adverse heel, broad (20°–25°) elevation beamwidth covers surface to short range. (b) VTS with high scanner, 4° beamwidth, fan beam, troublesome nulls. (c) Inverse cosec^2 beam obviates nulls

Figure 2.24 Dual-band coastal surveillance reflector scanner: 3 and 9 GHz bands, 36 and 44 dBi gains. Bulky feed assembly contains polariser for linear and circular operation. For exceptionally arduous operation at the Bay of Fundy, N Atlantic coast of Canada. Heater to prevent icing. Note the very stiff girderwork necessary to retain rigidity and microwave performance in high winds. Reproduced by permission of Easat Antennas Ltd, Stoke on Trent UK

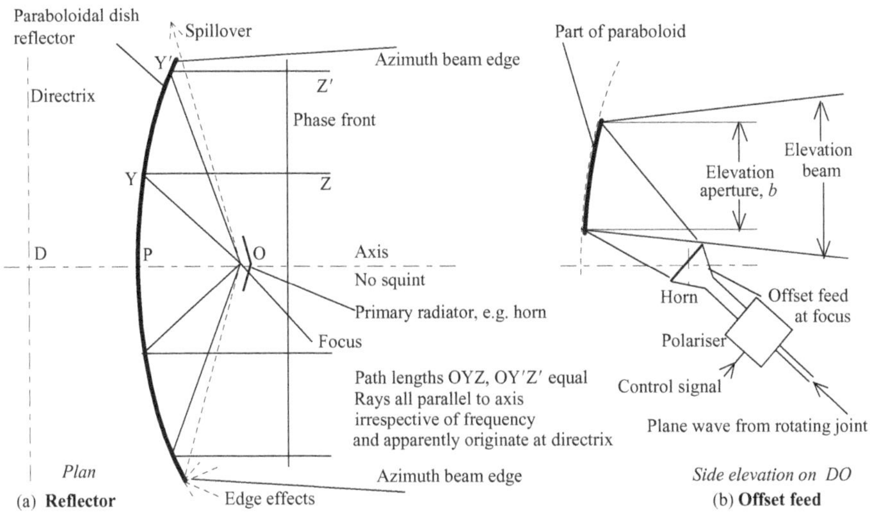

Figure 2.25 Reflector scanner. No azimuth squint. Offset feed avoids aperture blocking

length OP. Paraboloids have the property that all rays from the focus, after reflection at the surface, are (a) parallel and (b) have the same path length to the mouth, giving a plane-fronted pencil beam, very narrow in azimuth and narrow in elevation. The geometry is equivalent to radiation from the directrix one focal length behind the dish (PD = OP).

The primary radiator is usually offset below the dish (Figure 2.25(b)) to keep it out of the beam. Blocking would spoil the pattern, introducing blocking loss. The blocking 'hole' in the illumination would act as a small negative illuminator, generating sidelobes at its edges. Reflector scanners have no azimuth differential squint and if designed for wide bandwidth, differential elevation-plane squint is minimal. The field strength contours are elliptical and the dish is made elliptical also; corners of a rectangular dish would be too weakly illuminated to pull their weight. Beamwidth is increased and gain reduced for a given aperture – power taken from the sidelobes in effect broadens the beam, reducing the aperture efficiency. Reflector scanners, beside having sidelobes near the boresight, suffer from spillover radiation, some energy from the feed missing the dish and forming far-out 'sidelobes' at around $\pm 120°$ off axis. It is usual to minimise elevation radiation above boresight, which does not intercept the target and collects unwanted precipitation clutter in receive mode. Below boresight, the near-in sidelobes are often fused or integrated with the main beam to minimise loss of gain at sidelobe null angles, as with the inverse cosec^2 type described shortly.

Although some radio telescopes of great precision have reflector areas up to about $5000\,\text{m}^2$, the economic limit for VTS is much smaller; around 6 m × 1 m aperture in the 3 and 9 GHz bands, and maybe two-thirds these dimensions at 14 GHz. Beamwidths

Table 2.4 Fan beam scanner elevation features

Illumination feature	Uniform			Cosine		
	ν	w	g (dB)	ν	w	g (dB)
$\frac{1}{2}$ power beamwidth	ϕ	0.886	3	ϕ	1.184	3
3 dB points	0.5ϕ	± 0.443	3	0.5ϕ	± 0.592	3
1st null	1.288ϕ	1.0	inf	1.27ϕ	1.5	inf
1st sidelobe	1.58ϕ	1.42	13.3	1.59ϕ	1.9	23
2nd null	2.22ϕ	1.97	inf	2.11ϕ	2.5	inf
2nd sidelobe	2.77ϕ	2.45	17.8	2.45ϕ	2.9	31

go down to about 0.35° azimuth at 9 GHz, gain exceeding 43 dB. Illumination is always tapered to minimise azimuth sidelobes, at cost of a bigger dish for given gain and beamwidth. Computer design techniques such as finite element analysis, coupled with the additional data derived from near-field testing, enable modern reflector scanners to deliver considerably higher gain than previously possible, with lower sidelobes.

All parts of the beam are fed in parallel, so very short pulselengths may be used. Designs are also less frequency dependent, facilitating improvement of performance by frequency diversity, described in Chapter 12, Section 12.8.

2.7.12 Elevation performance; inverse cosecant squared reflectors

VTS elevation half-power beamwidths, ϕ, are about 10°–20° (slotted arrays) and 4° (reflectors). When mounted high, the boresight is generally depressed a degree or so to improve illumination of the horizon.

Table 2.4 compares theoretical elevation performance for uniform and cosine-weighted illumination models. The angles are almost identical for all sidelobe events and of course are somewhat dependent on wavelength variations across the frequency band. Performance of practical scanners is rarely published in detail, but is probably intermediate between these models. Beamwidth w links the featured angle with the wavelength/aperture ratio, thus, for example, $\frac{1}{2}$ power beamwidth $= w\lambda/a$, where factor $w = 0.886$ for uniform illumination or 1.184 for cosine illumination. As an aside, this indicates aperture of a cosine-illuminated scanner has to be $1.184/0.886 = 1.336$ that of a uniformly illuminated scanner of the same beamwidth, the price paid for improved sidelobe performance. At the featured angle, gain is g dB below gain on axis.

Alternatively, either the vertical curvature of the reflector may be made non-parabolic, or the feed radiation may be tailored, so that below the lower half-power point the pattern follows a nominally inverse cosecant squared pattern down to about $-10°$, 'inverse' because the cosec^2 law applies for angles of depression rather than elevation as in air traffic control service. Echo strengths of surface targets then remain constant as range changes, particularly when the scanner is high and range fairly short, reducing the dynamic range demanded of the receiver. Figure 2.23(b) shows the nulls of a conventional VTS fan-beam scanner of elevation aperture 4° and depressed 1°. The first null lies at about $11H$, about 1.1 km when $H = 100$ m. Changing to an inverse cosec^2 type, Figure 2.23(*c*), eliminates nulls down to very short range. More energy is transmitted away from beam axis so cosec^{2}scanners sacrifice a couple of decibels boresight gain on fan beam types of the same aperture. Choice of most cost-effective scanner for a specific application is not always clear-cut.

2.7.13 Polarisation

The voltages at the edges of near-vertical slots in the narrow face of a horizontal slotted array generate H fields in elevation and E fields in the azimuth plane, so the radiation, mutually perpendicular to both planes, is horizontally polarised (HP); Figure 2.12(*a*). IMO requires all marine 9 GHz radars to have HP to give compatibility with racons and SARTs, which use this polarisation. A few 3 GHz band sets (Sperry Radar, c1970) preferred vertical polarization (VP), Figure 2.14(b), where the field planes are interchanged, the guide broad face being slotted as Figure 2.18(b). Here slot coupling is a function of offset from centreline. These slots are usually resonant, approximately $\lambda_g/2$ long. Although marketed for marine radars at one time, circular polarisation (CP) facilities seem now to be confined to VTS. HP picks up least sea clutter from the regular waves typical of open waters but VP is better in coastal areas where seas are more confused. Circular polarisation, consisting as it does of equal HP and VP components, has intermediate sea clutter performance but is much better against rain clutter because the near-spherical shape of raindrops causes them to reflect at the opposite hand, which is rejected by the scanner in receive mode. At one time, CP was provided by dropping a grid of slanted wires over the primary radiator, but it is now usual to insert a waveguide network immediately behind the feed horn; Figure 2.24 (Section 2.7.11) shows the bulk of such networks. A waveguide switch selects HP or VP via plane-rotating elements. For CP, a second switch passes half the power through each rotator, with ±45° phase shift. Williams [6] has found that many targets depolarise echoes sufficiently for a cross-polarised receive mode to enhance echo to clutter ratio. An antenna transmitting say HP and receiving VP is designated HV mode, the ordinary mode being HH.

Polarisers add complexity and some additional loss. Plane polarisation is used in clear conditions, because CP reduces RCS of many targets such as ships by several decibels. On reception, only the transmitted hand of polarisation is accepted. Echoes from scatterers having an odd number of internal reflections, for example, from a sheet or trihedral, as well as raindrops, come back with the other hand and are rejected.

The measure of circularity is ellipticity

$$= 20 \log (\text{ratio of outputs in the planes of maximum and minimum gain}), \text{dB}. \quad (2.8a)$$

For wrong-hand echoes, for example spherical raindrops, the clutter cancellation ratio to CP is

$$20\log\left[\frac{e^2+1}{e^2-1}\right]\ \text{dB} \tag{2.8b}$$

where e is the average ellipticity weighted by the two-way antenna voltage characteristic. For example, for 25 dB clutter cancellation, $20\log[(e^2+1)/(e^2-1)] = 25$, so $e = 1.057$.

2.7.14 Surface tolerance loss

Reflector manufacturing tolerances, sag under the weight of the structure and wind loads which combine to distort the shape from true by δ m rms, (root mean square; the square root of the sum of the squares of instantaneous values; standard deviation δ) introduce a phase error:

$$\text{phase error} = \frac{2\pi\delta}{\lambda}\ \text{rad} \tag{2.9a}$$

$$\text{consequent loss per plane, one-way} = \exp\left(\frac{2\pi\delta}{\lambda}\right)^2\ \text{numerical.} \tag{2.9b}$$

With imperfection in both planes the surface tolerance loss is squared:

$$\text{two-plane loss} = 343\left(\frac{\delta}{\lambda}\right)^2\ \text{dB each way.} \tag{2.9c}$$

The energy removed from the beam axis appears as sidelobes and a widening of the main beam. Slotted arrays suffer similar loss when distorted by wind load or heavy vibration. Static surface tolerance loss is included within published scanner data sheet gain. It is difficult to measure or quantify additional loss caused by wind loading and dynamic forces, but maintenance of good performance demands a particularly rigid structure, pushing up weight and cost. Figure 2.24 (Section 2.7.11) well shows the substantial cross-bracing of reflector and feed necessary to retain performance of large reflector scanners for service in exposed locations.

Figure 2.26 depicts Eq. (2.9c) and illustrates how rapidly loss increases with operating frequency. For example, for 1 dB one-way loss in the 9 GHz band, $\delta = 1.7$ mm rms. The dish surface therefore has to be contoured with great accuracy on a stiff support structure to maintain gain, the desired pattern and to control sidelobes. Size, weight and cost of the scanner itself, its drive motor and gearbox, and the support tower all rise rapidly as the reflector is made bigger. Cost of big scanners should be carefully balanced with necessary performance before purchase. For a given gain, high frequency scanners are smaller, but the need for increased accuracy impacts on cost. Large reflector scanners require ~10 kW motor drive power. The reduction gears can be noisy enough to require noise-deadening mountings, especially when mounted on steel monopole masts in residential areas.

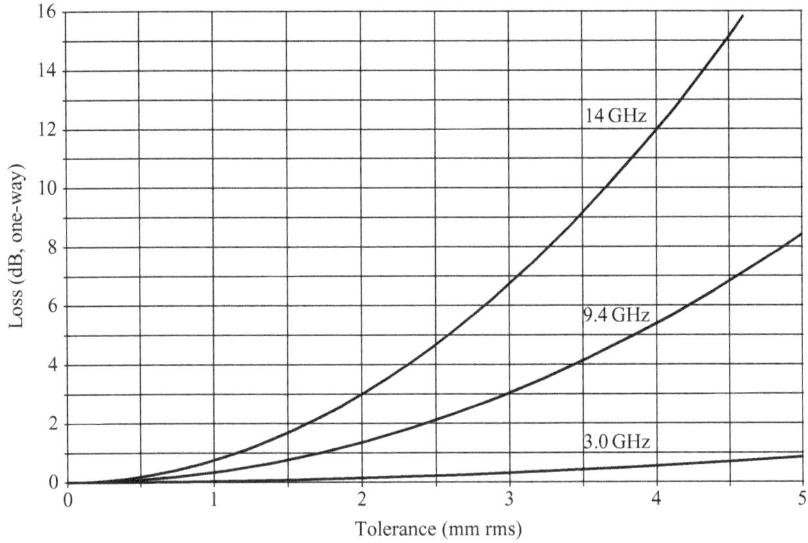

Figure 2.26 Surface tolerance loss. Reflector scanner. Much finer tolerances are necessary at high frequency, driving up cost and weight

2.7.15 Beamshape and scanning losses

As mentioned in Section 2.7.2, beamshape loss arises when the scanner is assumed to transmit rectangular beams for purposes of system performance calculation. This approximation assumes full gain to all targets within the beam, whereas sometimes the target will be near its edge and will see less gain. If the scanner axis remains vertical, the beamshape loss due to the assumption of a rectangular beam can be treated by taking the beam to be narrower, see Meikle [5, p. 128]. From the integration of the Gaussian beamshape approximation, the loss is theoretically $\sqrt{\pi/(8\ln 2)}$ numerically (1.23 dB), in practice about 2 dB and depending somewhat on the signal processing system if using coherent integration. For the more usual non-coherent integration it is about 0.69 actual, a loss of 1.62 dB two-way. Alternatively, beamshape loss can be regarded as being of the order of (area of circumscribing rectangle)^2/(area of ellipse)^2 $\sim$ $(4/\pi)^2$ $\sim$ 2.1 dB. If the platform rolls or pitches, sometimes the target comes above or below the beam axis. Beamshape loss therefore to some extent depends on circumstances. An appropriate general two-way value is 4 dB.

If the scanner rotates a significant part of a beamwidth between the instants of transmission and reception, a scanning loss arises. For a target at range R, beam flight time $= 2R/c$. When R is the maximum instrumented range of a standard marine radar, 96 nmi $=$ 178 km, flight time $=$ 1.187 ms. In this time a 25 rpm scanner turns 0.18°. For a narrow-beam 9 GHz scanner of 0.7° beamwidth this represents 0.25 beamwidths, representing about 0.5 dB scanning loss. At the shorter range of practical targets, loss falls rapidly and scanning loss can be safely neglected except perhaps for long-range, high-gain VTS scanners.

2.7.16 Summary of scanner losses

The losses mentioned here are among those discussed in Chapter 13, Section 13.3. Unless otherwise stated, 'gain' of a scanner means gain on boresight, where it is highest, denoted by G_{max}.

Scanners suffer several losses on theoretical performance, typically including all or some of the following.

- Ohmic loss with current flow on the metallic surfaces, causing energy loss as heat.
- Dielectric loss of 0.2 dB in the radome or window.
- Rotating joint loss 0.3 dB.
- Radiated as sidelobes 0.3 dB.
- Absorbed in internal load.
- Surface tolerance loss (Section 2.7.14).
- Beamshape loss (Section 2.7.15, but often treated as a processing loss, Chapter 13, Section 13.3.5).
- Path obstructions such as masts cause several decibels loss each way within their subtended angle and a few decibels nearby (Section 2.7.9).

Apart from the last two, these losses, totalling about 2.5 dB two-way, are generally included within suppliers' quoted overall gain figures, so rarely need further consideration. Radomes and windows are finished with a self-cleaning slippery surface to help shed dirt, water and ice. Minor grime has little effect but thick build-up of dirt (soot especially) or ice may introduce further attenuation and mismatch losses, the build-up absorbing and/or reflecting energy back to the source, often treated as a service loss. Many ordinary paints are lossy and have rather high dielectric constant so should never be applied to radomes and windows.

2.7.17 Testing antennas

Large-aperture antennas only deliver a beam of final shape in the far field, which for a 9 GHz scanner of 1° beamwidth commences at ~450 m, precluding far-field measurement of performance in an anechoic chamber. Formerly open-air test ranges were employed, the scanner being rotated while mounted well clear of the ground on a tower while receiving test transmissions from another tower a kilometre or more distant. Testing was laborious and it was extremely difficult to control reflections from the terrain, buildings and vegetation. Now, near-field measurements are taken within an anechoic chamber, unwanted reflections being minimised by pyramids of radar absorber; see Figure 2.27. Transmissions are from a range of a few metres, the antenna on test being rotated in azimuth and tilted in elevation. Amplitude and phase of received signals are digitally recorded, a computer program recovering the far field azimuth and elevation radiation patterns. Tests with the antenna removed calibrate-out any residual reflections from the chamber walls. The procedure is automated, reasonably quick – usually less than a day – and out of the weather. Tests are repeated at several frequencies and all polarisations of interest. Results are precise, repeatable

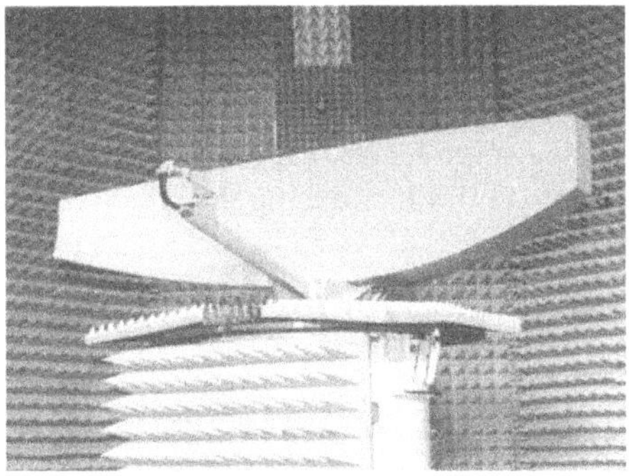

Figure 2.27 *Reflector antenna on test. Walls of anechoic chamber lined with lossy pyramids. Further absorbent pyramids screen turning gear from near-field testgear beyond left of picture: 2.8 m aperture 9 GHz naval sea and air surveillance antenna for frigates and offshore patrol vessels. Box girder construction. Reproduced by permission of Easat Antennas Ltd, Stoke on Trent UK*

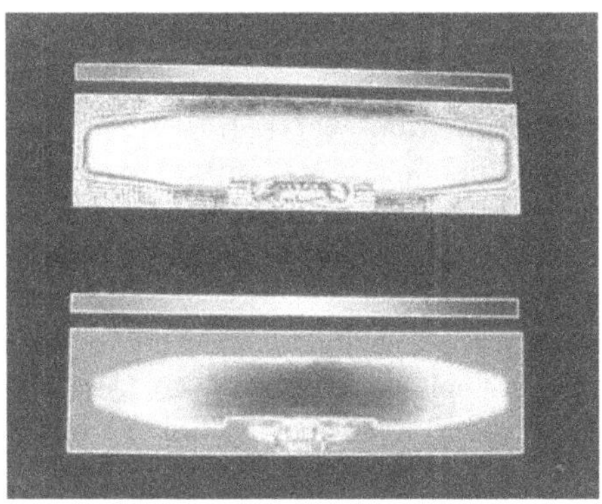

Figure 2.28 *Holographic plots of anechoic chamber measurements. Originals in colour. Upper plot shows phase, accurately quantifying the surface error with respect to the theoretical optimum curvature. Confused areas away from reflector are phase noise in zero-signal areas. Lower plot is amplitude, showing the RF power distribution across the reflector. Minimal power near edges to reduce spillover and far-out and back lobes. Graphically indicates tapered distribution, particularly in azimuth. Scaling bars above each plot. Reproduced by permission of Easat Antennas Ltd, Stoke on Trent UK*

and traceable to international standards. The small antennas of active devices are far-field tested within anechoic chambers at a few metres range.

Figure 2.28 shows phase and amplitude distributions from near-field microwave radiation measurements taken in an anechoic chamber and used for holographic analysis. The upper plot shows a contour map of the signal phase, accurately depicting the surface error with respect to the theoretical optimum curvature. The lower plot is derived from the signal amplitude. It shows there is little power near the reflector edges, minimising spillover and far-out and back sidelobes. It will be seen that illumination is far from uniform, reducing the effective size and gain.

2.8 Quantitative scanner analysis

2.8.1 Elevation performance, marine and VTS slotted arrays

Elevation sidelobe performance is not usually stated in datasheets, but can be inferred from the following analysis, which is based on the worst case of uniform illumination. From the geometry, off-axis voltage, V, relative to axial voltage V_{max} is

$$V = \frac{\sin x}{x} V_{max}.$$

Hence, off-axis gain is

$$G = \left(\frac{V}{V_{max}}\right)^2 G_{max} = \left[\frac{\sin x}{x}\right]^2 G_{max} \text{ numerically,} \tag{2.10a}$$

where ν is the elevation angle from axis (rad) $[-1 < \nu < 1]$, ϕ the beamwidth between half-power points (per datasheet) (rad), G the numerical gain at angle ν, G_{max} the numerical gain on axis (per datasheet), g_{el} the loss of gain at angle ν, $x = C\nu/\lambda_0$ where C is a constant $\sim$2.78312, a the aperture (m) and w the aperture fill factor.

Actual gain, G, should be used when calculating performance against targets lying well out of the scanner equatorial plane, replacing G_{max} by a term G_{max}/g_{el}; the loss recurs on the receive leg. Like all losses in this book, $g_{el} \geq 1$ numerically.

$$g_{el} = \frac{G_{max}}{G \text{ numerically}} \quad \text{or} \quad g_{el} = G_{max} - G \text{ dB.} \tag{2.10b}$$

By definition, when $\nu = \pm\phi/2$, $G = G_{max}/2$, $g_{el} = 2$ and loss is 3 dB. By substitution in Eq. (2.7a), here $\sin^2 x = 0.5x^2$, from which $x \sim \pm 1.392\nu/(\phi/2) = 2.78312\nu/\phi$. That is, $C = 2.78312 \sim \frac{1}{2}\pi^{3/2}$. Substituting in Eq. (2.7a) at angle ν

$$\frac{G_{max}}{G} = g_{el} \sim \left[\frac{2.783\nu/\phi}{\sin(2.783\nu/\phi)}\right]^2 \text{ numerically}$$

or

$$g_{el} \sim 20 \log \left[\frac{2.783\nu/\phi}{\sin(2.783\nu/\phi)}\right] \text{dB.} \tag{2.10c}$$

Nulls occur when the sine term is zero. Here, where n is an integer,

$$2.78312\frac{\nu}{\phi} = \pm\pi, \pm 2\pi, \ldots, \pm n\pi. \tag{2.11a}$$

At the first nulls, $\nu = \pm 1.1288\phi$. Here $V/V_{max} = 0$ and $x = \pi$. But here $\sin\theta = \lambda_0/a$. If $a \gg \lambda_0$, $\sin\theta \sim \theta$. Substituting,

$$\pi = \frac{2.78312\lambda_0}{aq}, \qquad \text{so } a = \frac{w\lambda_0}{\theta}. \tag{2.11b}$$

For the fully filled case considered here, $w \sim 0.886 = \frac{1}{2}\sqrt{\pi}$ and efficiency $D = (0.886)^2 = 0.785$, depressed below unity by the power lost to sidelobes.

Sidelobe peaks occur when the sine term is ± 1. The principal sidelobe is when $n = 1$. At peaks:

$$2.78312\frac{\nu}{\phi} = \pm\frac{3\pi}{2}, \pm\frac{5\pi}{2}, \cdots \quad \text{and} \quad \nu \sim \pm 1.69\phi, \pm 2.82\phi, \ldots. \tag{2.11c}$$

Substituting in Eq. (2.10c), gain at principal sidelobe peaks $= 0.045$. That is 4.5 per cent of the radiated power is lost to each of the four sidelobes (two each in azimuth and elevation); a few per cent is also lost to the secondary sidelobes, together accounting for the value of D obtained above.

At the -20 dB points, $\sin(2.78312\nu/\phi) = 0.1(2.78312\nu/\phi)$. Assuming uniform illumination, the width is just over twice the half-power beamwidth:

$$\nu_{-20\,\mathrm{dB}} = \pm 1.025\phi. \tag{2.11d}$$

Figure 2.29 shows calculated radiation patterns of uniformly illuminated apertures having elevation beamwidths typical of ships' scanners. Actual patterns may differ

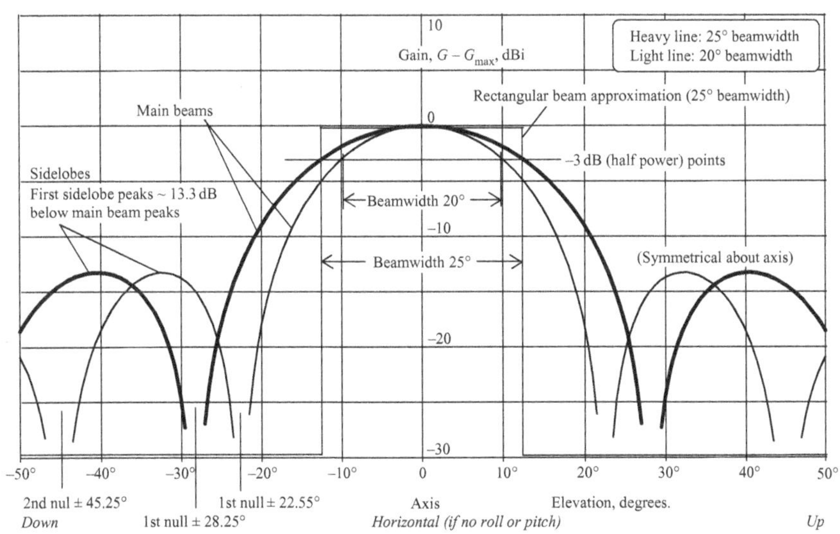

Figure 2.29 Elevation patterns, ships' scanners. For beamwidths 20° and 25°, assuming uniform illumination. One-way transmission. Roll or pitch reduces the effective gain on target elevation

slightly because illumination may not be quite uniform. Gain remains reasonably constant within the half-power points, then falls rapidly.

When assuming a rectangular beamshape (sometimes called square) in performance calculations:

$$G = G_{\max} \text{ (i.e. } g_{el} = 1) \qquad \text{if } -\phi < \nu < \phi,$$

otherwise

$$G = 0 \text{ (i.e. } g_{el} = 0) \text{ numerically.} \tag{2.11e}$$

Effective aperture height, b, can be inferred from published beamwidth assuming the aperture is fully filled:

$$b = \frac{1}{2}\sqrt{\pi}\frac{\lambda_0}{\phi} = 0.886\frac{\lambda_0}{\phi}. \tag{2.12}$$

Substituting for typical beamwidths:

- at 3 GHz, $\lambda_0 = 0.1$ m: $b = 0.254$ m (20°), 0.203 m (25°);
- at 9 GHz, $\lambda_0 = 0.032$ m: $b = 0.081$ m (20°), 0.065 m (25°).

Using flares, actual scanner heights exceed these apertures because illumination is not fully uniform and window sealing flanges have to be accommodated. As noted earlier, polyrod designs may trade front to back length for height, permitting sharp height reduction.

Severe roll or pitch swings targets lying near to the roll or pitch plane through the scanner elevation polar diagram. Assuming uniform aperture illumination in the elevation plane and sinusoidal roll motion, Eq. (2.10c) can be used to calculate gain reduction throughout the roll or pitch cycle. If the roll is sinusoidal, reduction is 1 dB one-way when the roll component in the target plane is ϕ peak–peak.

2.8.2 Inverse cosecant squared VTS scanners

In the cosec^2 elevation region, for half-power beamwidth ϕ rad:

$$\text{voltage at } \nu \text{ rad off axis} = \left(\frac{\operatorname{cosec} \nu}{\operatorname{cosec} \phi/2}\right) \text{ numerically.} \tag{2.13a}$$

At the half-power point, again $\nu = \phi/2$, and gain loss $g_{el} = 2$. Putting cosec (angle) $= 1/\sin$ (angle), in the cosec^2 region:

$$\text{gain relative to half-power point gain} = \left(\frac{\sin x}{\sin \phi/2}\right)^2 \text{ numerically;} \tag{2.13b}$$

$$\text{gain loss (positive number) } g_{el} = \frac{G_{\max}}{G} = \frac{2(\sin x)^2}{(\sin \phi/2)^2} \text{ numerically.} \tag{2.13c}$$

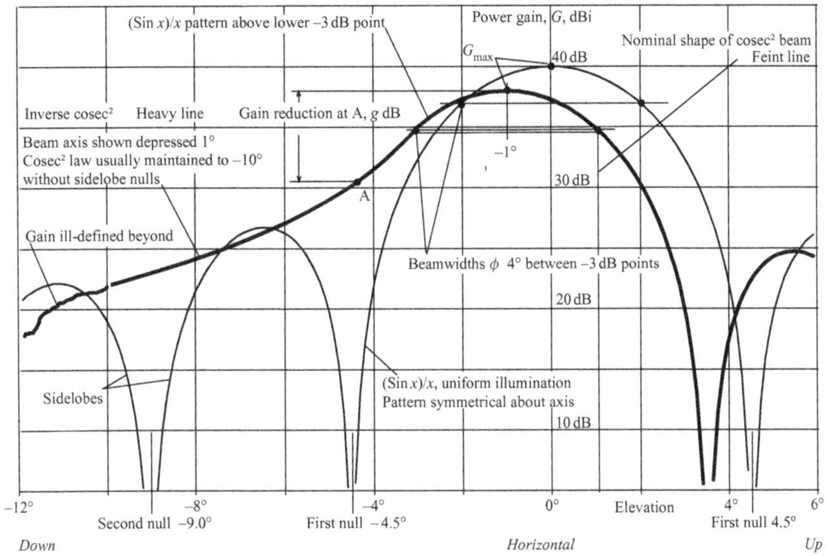

Figure 2.30 Elevation patterns, VTS scanners. For uniformly illuminated and inverse cosec squared beams, each having beamwidth 4°. Although in practice the inverse cosec² pattern may contain ripples, it avoids the deep nulls at −4.5° and −9°. It illuminates short-range targets more uniformly, at cost of slightly lower G_{max} for a given aperture

The expression for the one-way elevation off-axis loss relative to G_{max} of an inverse cosecant squared scanner in both its regions, taking upward angles as positive, is:

$$\text{If } \nu > -\frac{\phi}{2}, \quad g_{el} = 20\log\left[\frac{\sin(2.78312\nu/\phi)}{2.78312\nu/\phi}\right],$$

otherwise

$$g_{el} = 3 + 20\log\left[\frac{\sin(2.78312\nu/\phi)}{(\sin\phi/2)}\right] \text{dB}. \tag{2.13d}$$

The first term represents (in dB) the region of uniform illumination (Eq. (2.10c)), the second being the cosec² region (Eq. (2.13c)). Figure 2.30 compares elevation beamshapes of uniformly illuminated and inverse cosecant squared large VTS scanners, the latter shown depressed 1° for clarity. The pattern locus lies a few decibels below the sidelobe peaks of a same-size uniformly illuminated scanner, because of the energy transferred to the former null angles. In practice the cosecant squared law is not always closely followed; when calculating short-range performance actual gain/angle values should be substituted if known. Above the lower half-power point both types have the same nose shape.

An inverse cosec² scanner receives approximately constant echo power from a surface target at the shorter ranges for thee following reason. Assuming the scanner

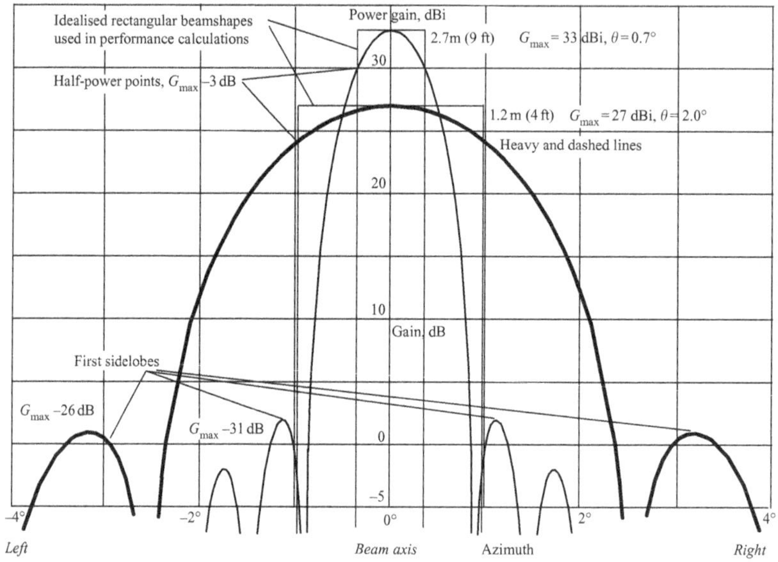

Figure 2.31 Typical azimuth patterns. Typical 9 GHz band marine radar scanners. At 3 GHz, gains for the same apertures would be 5 dB lower and beamwidths 3 times wider. Large aperture is needed for high gain. Linear gain scaling would make the patterns look much thinner

is mounted at height H above a flat Earth, from the geometry

$$R = H \operatorname{cosec} \nu. \tag{2.14}$$

Neglecting multipath effects, received echo power $\propto G^2/R^4$. Substituting $G \propto \operatorname{cosec}^2 \nu$, echo power $\propto (\operatorname{cosec}^2 \nu)^2/(H \operatorname{cosec} \nu)^4 \propto 1/H^4$, irrespective of R.

This incidentally shows that *at short range* doubling scanner height reduces signal by 12 dB because ν is higher and gain lower.

2.8.3 Azimuth radiation pattern

When considering operation of any type of scanner with target off-axis by angle ν, and having half-power beamwidth θ, loss factor g_{az} is defined similarly to g_{el}. If the target is off beam in both planes, loss factor is $g_{az}g_{el}$ numerically.

Figure 2.31 shows azimuth radiation patterns of two of the typical ships' scanners listed in Table 2.3, with main beams and close-in sidelobes. When a rectangular beam is assumed, following Eq. (2.11c):

$$G = G_{max} \text{ (i.e. } g_{az} = 0\,\text{dB)} \qquad \text{if } -\frac{\theta}{2} < \nu < \frac{\theta}{2}, \tag{2.15}$$

otherwise

$$G = 0 \text{ (i.e. } g_{az} = \infty\,\text{dB).} \tag{2.16}$$

2.9 References

1 WILLIAMS, P. D. L.: 'Civil marine radar – a fresh look at transmitter spectral control and diversity operation', *The Journal of Navigation*, 2002, **55**, pp. 405–18
2 MEHLER, M. J.: in HALL, M. P. M., BARCLAY, L. W. and HEWITT, M. T. (Eds): 'Propagation of radiowaves' (The IEE, 1996), Chapter 2
3 SILVER, S. (Ed.): 'Microwave antenna theory and design, MIT Radiation Laboratory Series, vol. 12' (McGraw-Hill 1949, re-published The IEE, 1984)
4 WYLIE, F. J.: The use of radar at sea' (Hollis & Carter for the Royal Institute of Navigation, 5th edn.), Figures 60–62
5 MEIKLE, H.: 'Modern radar systems' (Artech House, London, 2002), Chapter 5
6 WILLIAMS, P. D. L.: 'Civil marine radar' (Privately published, 2002), Chapter 9

Chapter 3

Radar receiver

'Little Sir Echo, how do you do?'

3.1 Scanner – receiving

As reciprocal devices, scanners receive incoming signals from the same volume of space illuminated on transmission. The receive and transmit gains, radiation patterns, sidelobes and polarisation are identical. Gain benefits both the transmit and receive legs, so increased gain is doubly beneficial to detection of weak targets.

Most marine radars use linear polarisation in which the electric field exists in one plane, sometimes vertical but usually horizontal. Antennas can always efficiently receive the polarisation which they transmitted. Passage through the atmosphere does not significantly affect ray polarisation; most targets and sea clutter also more or less preserve the incident polarisation when they reflect. Any depolarisation to plane polarised signals is allowed for by a reduction of the nominal radar cross section (RCS) of the target or clutter. Targets and sea clutter are therefore received by plane polarised scanners without ostensible polarisation loss.

A circularly polarised (CP) electric vector proceeds corkscrew fashion. The nearly spherical shape of raindrops causes CP rays to be reflected at the opposite hand so, in principle, rain clutter is rejected by the scanner. In practice the drops wobble between oblate and prolate spheroidal as they fall and are imperfect spheres, so scanners have some residual ellipticity and CP does not entirely eliminate rain clutter. To CP, target depolarisation reduces RCS of many ship targets by a few decibels, so stronger echoes are received with linear polarisation. Overall, a good CP scanner can give very useful improvement of signal to rain clutter ratio, 20–30 dB, completely outweighing the reduction in target RCS. Snowflakes are by no means spherical so CP is inferior to plane polarisation except in rain.

3.2 Receiver input

3.2.1 Rotating joint or sliprings

The scanner is connected to the remainder of the radar by a rotating joint or by sliprings. Although used on transmit as well as receive, any defects in these components have more effect on the receive function. Waveguide rotating joints contain short vertical lengths of coaxial transmission line on the axis of rotation, with transitions to waveguide above and below. Metal to metal contact is obviated by choke sections, an open-circuit a quarter-wavelength distant appearing as a 'conductor', on the principle of a mismatched feeder. Coaxial feeders also use contacting or non-contacting coaxial rotating joints. When the receiver mixer and pre-amplifier are aloft, the sliprings also carry LO, IF and power feeds. Alternatively, a transformer may be used with rotating primary and static secondary windings to obviate noise from dirty sliprings. Rotating joint losses are incorporated within published scanner overall performance figures when the scanner forms part of the scanner/turning gear assembly, the usual case except in some VTS systems. Rotating joints and sliprings, particularly the contacting variety if in poor mechanical condition, can introduce cyclic or random signal fluctuation or noise which interferes with the detection process. The effect is difficult to quantify.

Ohmic losses in the scanner, rotating joint and feeder (if used) introduce thermal noise as well as attenuation.

3.2.2 Receiver protection

The receiver must be protected from transmitter pulse breakthrough. Not only would it be swamped, paralysing it from reception for many microseconds and spoiling short-range performance, but the high power would also burn out the input stage, which will safely withstand only about a watt instantaneous maximum, or 10^{-7} J (joule; $=1$ erg) integrated through the pulselength. Alternative protection and duplexing arrangements are shown in Figure 3.1. Low-power radars may use a PIN diode switch in the receiver path. Positive, Intrinsic, Negative are the forms of silicon comprising the device. The transmitter pulse drives the diode to a low-impedance state, short-circuiting the receiver terminals to keep transmitter power out. More powerful modern radars may have twin PIN diodes, tuned respectively to magnetron frequency and magnetron principal spurious output frequency, for example, E2V Technologies' Dupletron arrangement.

Within older powerful radars, breakthrough is controlled primarily by a transmit–receive (TR) cell in the receiver input. The cell is a section of waveguide a centimetre or so long, sealed by quartz windows at its ends and containing a 'gas' of particularly short de-ionisation time ($\sim$2 μs), such as water vapour at absolute pressure $\sim$0.01 bar. Relatively weak incoming echoes pass unscathed but the strong transmitter pulse strips electrons from the gas molecules. The resulting negative electrons and positive ions conduct freely, short-circuiting the guide to shut the receiver off from the transmission circuit; Figure 3.1(*a*). A mild source of radio-activity and/or a 'keep-alive' d.c. low-current discharge speed initiation, but may introduce some additional noise. About

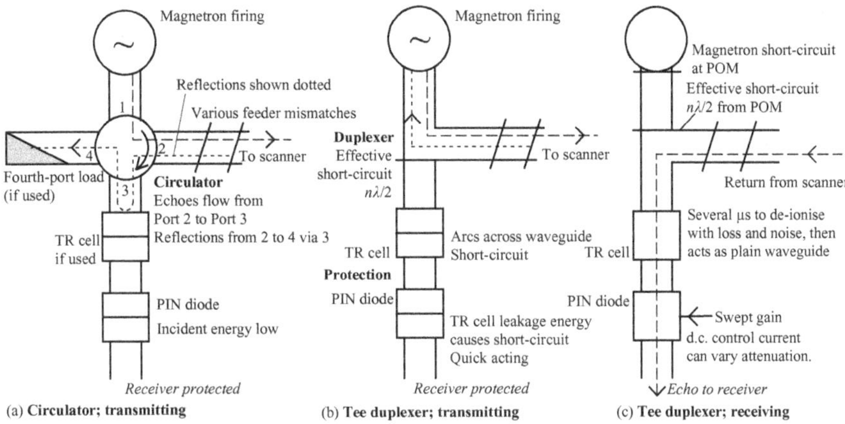

Figure 3.1 Duplexer and receiver protection. If the magnetron is buffered from load 'long line' mismatch by an isolator (not shown), POM is referenced to the isolator rather than the magnetron

50 mW leaks through, preceded by a stronger leakage spike during the 10 ns taken for the cell to fire, so the TR cell may be backed by a fast spike protection PIN switch.

After the transmission, the gas takes a few microseconds to de-ionise, meanwhile protecting the receiver against feeder ring signals (Chapter 2, Section 2.6.4). While de-ionising, TR cell attenuation (incorporated within the swept gain attenuation) and the noise generated by the ionised gas affect reception of short range echoes. Unless the TR cell is failing, recovery is complete within a few tens of microseconds, before full receiver performance is needed to detect weak targets at moderate range. The TR cell and in particular the PIN switch also protect the receiver from powerful transmissions picked up from other radars as interference.

About 0.5 dB of transmitter power is lost to the arc. Un-ionised, insertion loss in the receiver arm is similar. These design losses are included within the published radar performance parameters. TR cells gradually deteriorate. The gas 'cleans up', being partially adsorbed by the metal walls, particularly when severe scanner or feeder mismatch has been working the cell hard. De-ionisation time increases and protection is impaired. Spike energy breakthrough may then permanently damage the detection efficiency of the receiver, whose noise factor deteriorates, spoiling detectability of distant weak targets. This slow and insidious process may not be recognised by the operator, and is a main reason why it is essential to make routine performance checks; Section 3.13. Such losses are allowed for in calculations by a service loss term, typically 2 dB two-way.

3.2.3 Duplexer

The scanner is connected to the transmitter and receiver by a duplexing device, often a circulator. This non-reciprocal multiport device uses molecular spin properties of a ferromagnetic material within the field of a built-in permanent magnet.

Figure 3.1(*a*) shows input to port 1 (transmitter) emerges at port 2 (scanner); input at 2 emerging at port 3 (receiver). A fourth port, not always provided, may be connected to a resistive load and improves the match seen by the magnetron, benefiting its spectrum. When the transmitter fires, energy reflected from the inevitable scanner-circuit mismatches are routed to the receiver port, firing the receiver protection device, which then presents a short-circuit, reflecting the mismatch energy to the fourth-port load, whose good match absorbs it all. Circulator loss (0.25 dB) is usually included within transceiver published parameters. The circulator routes echoes to port 3, feeding the receiver.

Alternatively, a waveguide tee junction duplexer may be used. A TR cell or PIN diode placed an integral number of half-wavelengths away causes the duplexer receiver port to appear short-circuited, routing all the transmitter power to the scanner; Figure 3.1(b). The position of minimum impedance (POM) of the magnetron is similarly spaced, so when it presents a bad match after firing, it too appears as a short circuit, routing echoes to the receiver; Figure 3.1(c).

3.3 Receiver and filter

3.3.1 Overview

After traversing the rotating joint, feeder and protection circuits, the echo and clutter signals enter the receiver proper, all at transmitter frequency except for a few specialised racon targets discussed in Chapter 8, which can respond at an offset frequency. Each scatterer delivers a packet of several pulses each scan; Figures 3.2(a) and (b).

If number of pulses in packet $= n$, scanner azimuth beamwidth $= \theta$ rad and scanner rotation rate $= r$ rpm,

$$
\begin{aligned}
n &= \text{prf} \times \text{scan time} \times \frac{\theta}{2\pi} = \left(\text{prf} \times \frac{60}{r}\right) \times \frac{\theta}{2\pi} \\
&= 9.55 \times \text{prf} \times \frac{\theta}{r} \text{ pulses per packet.} \qquad (3.1)
\end{aligned}
$$

For example, Figure 3.2(b) is drawn for 1000 pps, 30 rpm and 1° beamwidth, $\theta =$ 0.0175 rad, $n = 5.57$. The prf is not synchronised with scanner rotation and non-integral n signifies individual packets fluctuate between 5 and 6 pulses. The figure excludes environmental effects such as target roll which may make individual pulses fade quasi-randomly, either sweep to sweep or scan to scan.

It is often sufficient to assume that all pulses within the beamwidth have full scanner gain on transmit and receive, with scanner gain zero elsewhere; Figure 3.2(c). This rectangular assumption introduces the beamshape loss discussed in Chapter 2, Section 2.7.15, targets slightly off-axis being credited too much gain.

The functions of the receiver are to amplify echoes to a level easily handled by the signal processor, convert them to baseband or video frequency, preserve information content represented by pulse shape and spectrum, introduce as little distortion and

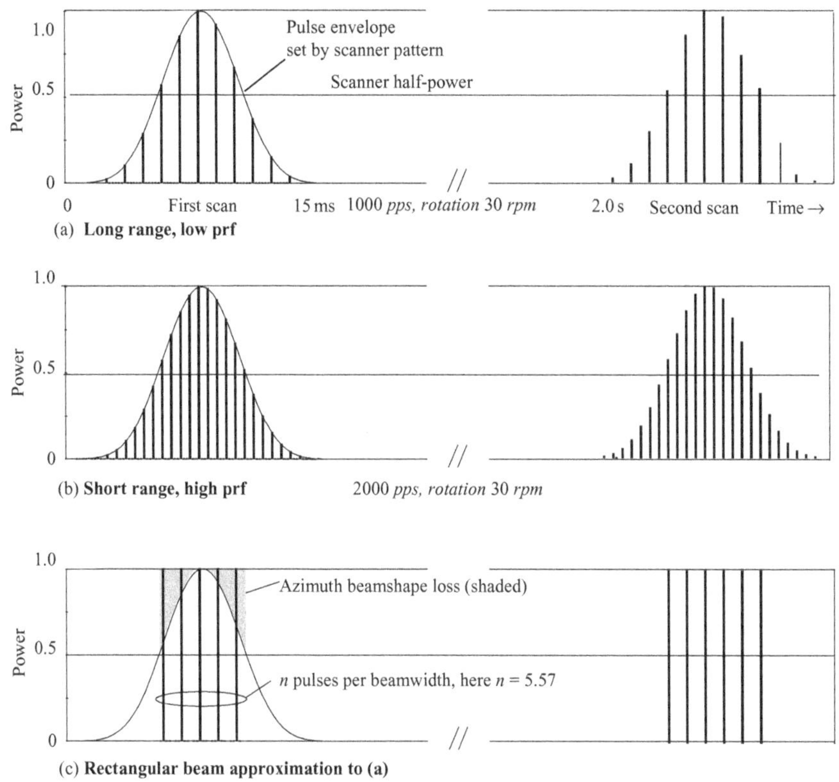

Figure 3.2 Echo pulse packets. Packet received as each scan sweeps across target. Number of pulses in packet proportional to prf. Rectangular approximation assumes all returns in beamwidth encounter full scanner gain, all other returns being ignored

additional noise as possible and control bandwidth by filtering, so delivering the optimum signal to noise ratio (SNR).

Target and clutter scatterer motions have similar velocities, giving similar Doppler frequency spectra and precluding use of the moving target indication (MTI) technique so common in aeronautical and military radar. The signal to clutter (but not signal to noise) ratio is almost independent of receiver bandwidth, but not of transmitter pulse-length, which alters the illuminated footprint. The echo spectrum replicates that of the transmitter, possibly trivially modified by dispersion in a waveguide feeder, Chapter 2, Section 2.6, which would cause the higher frequency components to lead slightly.

Echoes may be as low as -120 dBW (10^{-12} W, 1 pW), 16 orders of magnitude feebler than transmitter power, a huge ratio equivalent to a penny to a major State's gross domestic product for a century.

Chapter 4 will show that, speaking very generally, echo strength follows the inverse fourth power law, rising as range is halved by a factor of $2^4 = 16$ (12 dB).

For example, a 0.001 m^2 target at 1 km returns the same echo power as a 10 m^2 target at 10 km. To reduce the dynamic range of echoes, overall gain is made to rise with time from the instant of pulse transmission, full gain being reached at fairly long range, say 10–20 km, equivalent to 65–130 μs. Swept gain, discussed further in Chapter 12, is introduced by insertion early in the receiver of a swept gain attenuator, often formed by a PIN diode attenuator preceding or immediately following the initial microwave amplifier, to prevent later stages developing cross-modulation output components when a small echo is surrounded by high clutter spikes.

Conventional printed circuit boards are too lossy at microwave frequency, so the input circuits use microstrip, a form of transmission line printed as a metal-strip conductor pattern, with surface-mounted components, on a low-loss sapphire or alumina substrate backed by a metallic ground plane. The substrate is about the size of a couple of postage stamps and a millimetre thick. Microstrip propagates rather like an opened-out coaxial cable.

The TR cell recovery characteristic, if applicable, is also utilised and sometimes the protective PIN diode is biassed. Swept gain is adapted to the current clutter input, augmented by the operator's swept gain or sensitivity time control (STC, Chapter 12, Section 12.7.2). The control is advanced to reduce sea clutter, which primarily occurs at short range.

The microwave amplifier is specially designed for low noise, usually using a pair of GaAs FET (gallium arsenide field effect transistor) stages. It feeds the mixer. Complete receiver front ends, sometimes integrated with the transmit microwave components, are often procured by radar manufacturers from specialist suppliers. After the initial microwave low noise amplifier (LNA, alternatively known as a microwave integrated circuit, MIC) and the mixer, the IF signal is further amplified, then detected to give an output at baseband, as shown in Chapter 2, Figure 2.10(c).

Although several controls, Figure 3.3, are provided to optimise target detection, we cannot always assume that they have been set to suit the target in question. For example, the operator may be primarily concerned with a nearby echo, while keeping an eye on more distant traffic, accepting some reduction in its detectability.

3.3.2 Receiver noise

If targets reflected steady echoes and no noise or interference returns intruded, we could increase receiver amplification until targets of interest at indefinitely long range registered a strong echo, or we could reduce transmitted power and save money. As usual, life is less easy. Thermal noise can never be avoided. Together with unwanted clutter from precipitation and the sea, noise limits the smallest echo detectable by any given radar. The noise limit is dictated by the fundamental laws of physics. Design imperfections may prevent achievement of the theoretically possible performance, but under no circumstances can that performance be exceeded. Nonetheless, the Writer of the laws of physics has been kind to the marine radar community, enabling small and relatively cheap effective radars to be produced with sufficient performance for most purposes.

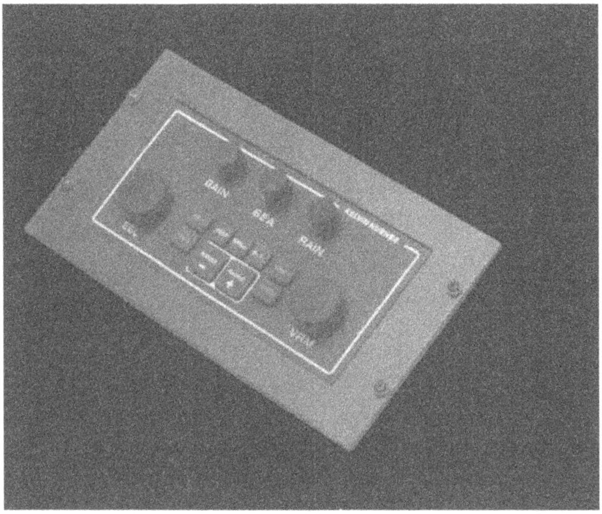

Figure 3.3 *Main controls of a modern radar. Simplicity of use, with all settings shown on alpha-numeric panels of display. Reproduced by permission of Kelvin Hughes Ltd, Ilford UK*

Beside noise fluctuations, the echo usually fluctuates also. The detection task boils down to maximising detection of wanted targets while minimising unwanted false alarms from the intruders; both usually fluctuating in strength.

The fundamentals of thermal noise summarised here are fully discussed in Chapter 11, Section 11.2. Thermal noise arises when current flows in a resistor of any kind, including semiconductors and feeders, so all the receiver components contribute some noise. The first amplification stage dominates the noise performance of the whole system because its noise is amplified most. Great care is therefore taken to minimise noise sources prior to amplification, and the noise generated within the first stage of the receiver.

Noise is a random variation of voltage, hence power, uniformly distributed through all frequencies as 'white noise' unless restricted, as it always is in practice, an observation bandwidth being implicit in all statements of noise. The sine wave, pulse or pulse train waveforms already considered follow definite patterns. Once we have 'cracked the code' we can state with confidence the instantaneous voltage at any time, past or future. Noise is different; it is random and there is no 'code'. Noise is often best described in terms of power density – power per unit bandwidth. Figure 3.4 shows (a) a sample of noise within a wide bandwidth, (b) the same sample after filtering to a narrow bandwidth (with the same noise power density), and then (c) after re-amplification to the former power level, event X being represented by Y after a bandwidth-dependent delay. The number of separate noise events per second is the reciprocal of the bandwidth; the rounded form of (c) shows the event rate to be less than (a). Another sample would look the same in general form but quite different in detail, just as no two sea waves are identical. The instantaneous voltage amplitude has Gaussian or normal statistical distribution; Figure 3.5 is the probability density function, discussed further in Chapter 11, Section 11.2. Occasionally spikes have several times the rms amplitude. Noise is the antithesis of a sine wave's $\sqrt{2}$ times rms

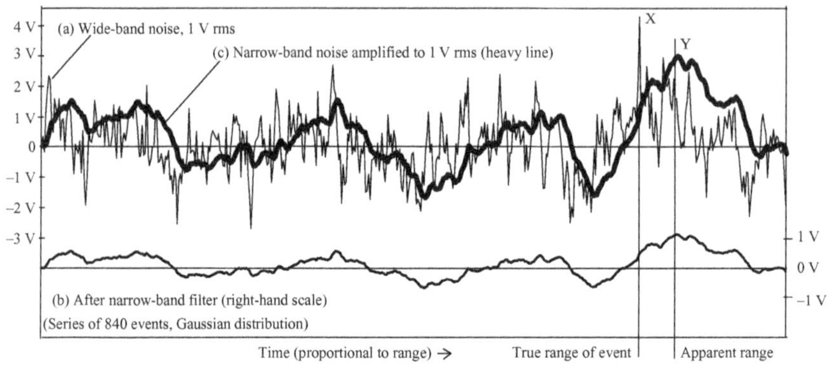

Figure 3.4 *Noise voltage waveform. Passage of wide-band noise (a) through a narrow-band low-pass filter (b) reduces amplitude. When amplitude is restored (c), the original amplitude distribution is restored, although the narrow bandwidth makes changes sluggish. Events such as X occasionally considerably exceed the rms, irrespective of bandwidth. Bandwidth reduction imposes a short delay, to Y*

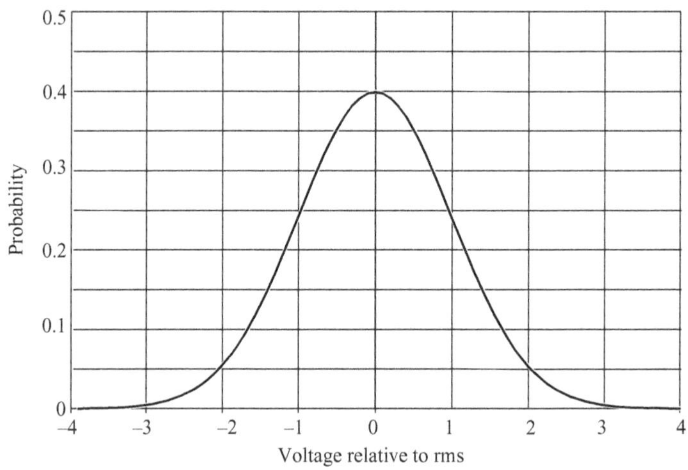

Figure 3.5 *Probability, Gaussian noise or clutter. There is a small but finite probability that noise and clutter events may substantially exceed the rms*

voltage. Until the statistics are detailed in Chapter 11, noise powers and voltages are to be understood to be average or rms values, subject to Gaussian distribution unless specifically stated otherwise.

Because noise is random, all that can be said is that, averaged over a time interval much longer than the event rate:

- power, measurable on a wattmeter, will average a certain amount;
- the statistical distribution of instantaneous voltage will be Gaussian.

It can never be quite certain whether any event such as X represents a noise spike or a target echo. Precipitation clutter amplitude is noise-like with Gaussian distribution, although the spectrum is set by the transmitter. Sea clutter is also rather similar. Under adverse conditions clutter power far exceeds noise.

At the receiver input average noise power, P_n, is

$$P_n = nkT_0B \text{ W} \tag{3.2a}$$

where k is Boltzmann's constant (1.381×10^{-23} J/K), T_0 is absolute temperature (conventionally taken as 290 K, 17°C. K stands for Kelvin; measured from the absolute zero of temperature. Degrees Celsius or centigrade = Kelvin − 273.3; degrees Fahrenheit = 1.8 K − 460) and B is the bandwidth under consideration, hertz. Excess over the lowest noise power theoretically attainable for the bandwidth is described by the receiver noise factor, n. The equation is frequently put in decibel terms, where noise figure $N = 10 \log n$:

$$P_n = N + 10 \log B - 204 \text{ dBW}. \tag{3.2b}$$

Because clutter rather than noise often dominates detection performance, the bother of cryogenic cooling to reduce T_0 has not been found worthwhile. Increasing amplifier gain to increase a weak signal also increases the noise and does not improve SNR. Great care is therefore taken to minimise the noise factor of the first amplifier stage. Although noise sets a definite limit to the useful sensitivity of all amplifiers, even before bandwidth is limited, noise power is never remotely high enough to damage the amplifier. Putting $B = 10^6$ MHz (1000 GHz) gives $kT_0B = 4 \times 10^{-9}$ W, whereas the most sensitive electronic devices can generally withstand 10×10^{-3} W without difficulty.

Beside internally generated noise, the scanner gathers galactic noise, approximately equal to the noise contribution of a resistor of the same value as the radiation resistance of the scanner (a few hundred ohms) and usually insignificant.

Mixers have rather poor noise figure so are often preceded by a specialised microwave low noise amplifier with $N \sim 3.5$ dB. The noise contributions of the mixer and succeeding stages, and of the scanner, feeder protection system and swept gain attenuator at its maximum gain setting, are usually accounted for by assuming the input stage has a rather poorer noise figure and generates all the noise. The noise figure so defined is properly the system noise figure, 'system' often being omitted. Although radar data sheets usually quote the system noise figure less the installation-specific feeder, sometimes called the overall noise figure, occasionally the first stage noise figure is given; if so, the overall figure will be slightly (~1.5 dB) higher. Attamable noise figures are mildly frequency dependent, the 14 GHz band being a couple of decibels worse than the 3 GHz band, with the 9 GHz band intermediate. Placing the low noise amplifier adjacent to the scanner in effect prevents the feeder receive-leg attenuation from degrading the system. Reducing receiver gain by insertion of swept gain attenuation of course increases the system noise figure. This is immaterial since only relatively large echoes need be detected at such times. If radar has to share its frequency allocations with telecommunication services, additional telecomms modulated continuous wave signals will be received. Their format will appear noise-like and may degrade the radar system noise figure by about 1 dB.

As an example of practical performance, a receiver having 5 dB overall noise figure and bandwidth 13 MHz ($10 \log 13 = 11.1$) has equivalent input noise power $5 - 144 + 11.1 = -127.9$ dBW.

3.4 Superhet receiver and mixing

3.4.1 Superheterodyne principle

Direct microwave amplifiers are inconvenient and expensive. Modern radars usually have a low noise microwave pre-amplifier, containing a simple microwave amplifier, gain ~10 dB, followed by frequency conversion or mixing to an intermediate frequency (IF) of around 50 MHz where the lower frequency makes amplification much easier, cheaper and controllable. Finally, the signal is demodulated to give a unidirectional pulse replica of the echo at video frequency or baseband, with spectrum resembling that of the transmitter modulator. Video voltage is of the order of a volt, suitable for digital processing; Chapter 2, Figure 2.10 refers.

3.4.2 Mixing

Mixing shifts the signal bodily down to a much lower frequency, where it is easier to process. The mixer is sometimes called a down-convertor, first demodulator or first detector. Figure 3.6 illustrates one of its simpler forms. At (a) a weak microwave echo signal, f_M, is shown connected by centre-tapped transformer T1 so that when its primary drives secondary terminal (x) +ve, the other terminal (y) goes −ve. T1 feeds a network of diodes and thence another centre-tapped transformer, T2, and an output bandpass filter. T1 centre-tap is driven with a strong continuous sine wave from a microwave free-running local oscillator (LO) whose frequency is f_{LO}. Its power is a few mW, much higher than the signal power. (b) shows that when the LO output is positive diodes D1 and D4 are driven into the linear conduction region, connecting T1 to T2 directly; D2 and D3 are reverse biased and cannot conduct. But on negative LO half-cycles (c), D1 and D4 are biased off and D2, D3 conduct, reversing the signal connection to T2. The diodes act as reversing switches and do not rectify the signal. Transformer T1 is not wire-wound but realised in microstrip or within a waveguide ‘magic tee’. The LO waveform always contains some phase and amplitude noise, manifested as jitter in the switching instants and variation in the ‘on’ resistance of the diodes, but fully symmetrical mixer circuits such as the one shown, called balanced mixers, keep LO noise out of the signal channel. Because originally mixer diodes used crystalline compounds like the old ‘cat’s whisker’ radio detectors, they are sometimes still called detector crystals, not to be confused with quartz oscillator crystals.

Balanced mixers have a noise figure around 8 dB, so receivers without LNAs have slightly worse system noise figures, say 9 dB.

Figure 3.6(d) is the microwave signal burst voltage waveform f_m, (e) is the LO voltage f_{LO} and (f) is the signal at T2 output. A much lower frequency component f_{IF} is clearly visible, emphasised at (g) by passage through a simple low-pass filter. Because the LO merely switches, IF amplitude must be proportional to the microwave

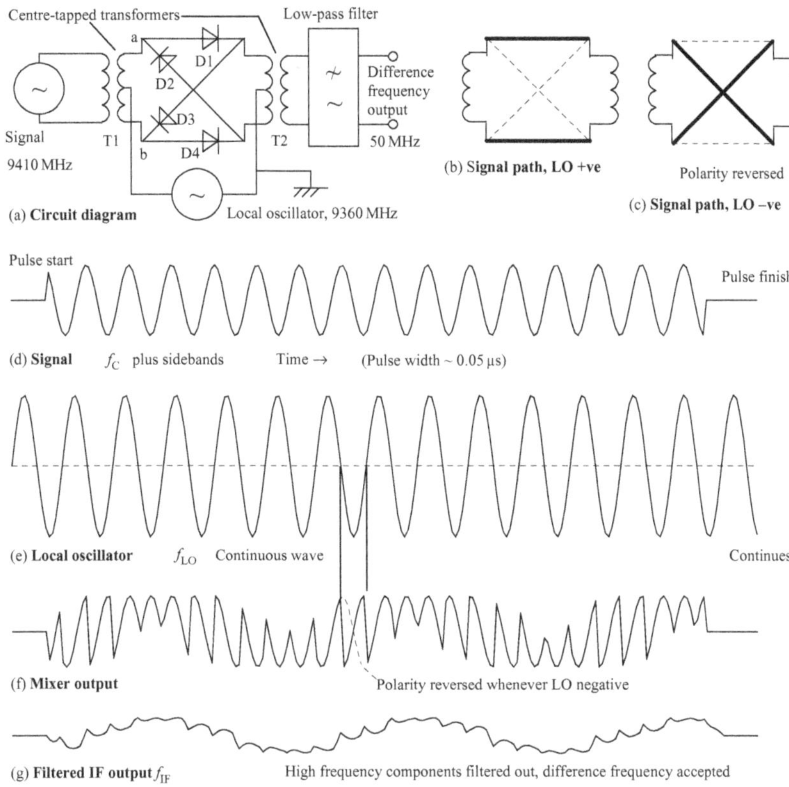

Figure 3.6 Mixer. The weak microwave signal beats (is mixed) with a microwave local oscillator to give a difference-frequency signal at intermediate frequency suitable for amplification and filtering. Diagrammatic

signal strength. The two waveforms are multiplied within the mixer (hence sometimes called a multiplicative mixer), beating together to produce sum and difference frequencies. The microwave sum frequency is discarded. The difference at IF, shown bold in the following equation, is accepted. Note that with the narrow pulselength shown there are less than three cycles of IF.

If the signal and LO reference voltages are, respectively, $A\sin(2\pi st + \phi)$ and $1.0\sin(2\pi rt)$,

$$\text{output voltage} = \text{signal} \times \text{reference} = A\cos(2\pi st + \phi) \times \sin(2\pi rt)$$

$$= \frac{A}{2}[\sin(2\pi(r+s)t + \phi] - \boldsymbol{\frac{A}{2}[\sin(2\pi(r-s)t + \phi)]}. \quad (3.3a)$$

Note that signal amplitude, A, and phase, ϕ, information are preserved. However, the latter is in fact lost when LO frequency is non-coherent and r is not exactly known.

f_{LO} may be higher or lower than f_m. Where operator $|\cdots|$ denotes the modulus of the expression:

$$f_{IF} = |f_m - f_{LO}|. \tag{3.3b}$$

3.4.3 *Local oscillator*

The LO may be a Gunn oscillator containing a special-purpose gallium arsenide (GaAs) diode fed at d.c. via a radio-frequency blocking inductor. The Gunn diode characteristic includes a current fall as voltage is raised. This negative resistance cancels the loss in a tuned circuit, causing continuous oscillation at a frequency determined by the tuned circuit and the diode transit time. Alternatively, a FET transistor may be used in an oscillating positive feedback amplifier. Formerly, reflex klystron valves were used. In non-coherent systems the LO frequency is subject to similar but smaller frequency shifts to some of those in the magnetron. The LO is electronically tuned, often by application of a quasi d.c. voltage to a varactor variable-capacitance diode in its resonant circuit, keeping it approximately in step with magnetron frequency. Error is detected within a double balanced mixer integrated circuit, which performs as a frequency-determining phase sensitive detector in the processing section. Beside this automatic frequency control (AFC) servo loop, a manual tuning control and indicator are sometimes provided. Error cancellation is not perfect. The AFC has a limited range, which if exceeded, may throw off to a large error, grossly degrading receiver performance. The AFC loop must be reset when the magnetron or LO is renewed. Performance calculations implicitly assume correct AFC operation.

Because in non-coherent systems the phasing of the LO relative to the magnetron was neither controlled nor measured at time of transmission, echo phase cannot be measured relative to that of the transmission and this information component is irretrievably lost. This is immaterial when there is only one pulse, but integration gain of non-coherent systems relative to their coherent cousins is lower, increasing more slowly with N, the number of pulses in the packet. The integration loss (relative to perfect integration) approaches $\sqrt{N}$ numerically or $5 \log N$ dB, so is greatest when there are many pulses in the packet, on short-range scales where prf is the highest; Chapter 12, Section 12.6.

3.5 IF amplifier, demodulator and video sections

3.5.1 *IF section*

The intermediate frequency subsystem or strip contains several amplification stages with gain stabilised by feedback, utilising transistor integrated circuits, interspersed with the bandpass filters. The demodulator or second detector converts the signal to a unidirectional video or baseband pulse, which after further amplification feeds the processing system.

For the typical frequencies shown in Figure 3.6(a) and Chapter 2, Figure 2.10, the IF is a line spectrum at 50 MHz. This bearer represents the microwave carrier.

A sideband component of the modulation f_m, say +10 MHz on the 9410 MHz centre frequency, would cause an IF sideband at 60 MHz, also offset 10 MHz from the IF bearer, confirming that the spectrum and pulse shape of the microwave signal are preserved, shifted bodily down in frequency by the mixing process. There is a tendency to raise intermediate frequency above 50 MHz as better amplifier chips become available, as signal processors become more elaborate and when very short pulses may be transmitted.

Echo strength varies very widely between, say, a small buoy at long range and a large ship at short range. Unless the receiver has enough dynamic range, strong signals overload its later stages, causing saturation and unwanted stretching of the pulse, destroying some of the information in the echo and introducing range error. Very strong echoes may give the receiver such a bout of indigestion that reception is paralysed for a few microseconds, spoiling reception of nearby targets.

Automatic gain control (AGC) sets gain at long range to maintain maximum tolerable noise and clutter, which optimises detection of weak targets, helping to reduce dynamic range to manageable proportions. At short range, even small targets return strong echoes and full sensitivity is not needed. A manual gain control enables the operator to optimise performance on the target of greatest current interest. Dynamic range is usually improved by inclusion of a non-linear logarithmic amplifier (Section 3.5.7).

3.5.2 Filter

Unless the target is long enough to cover a significant range bracket, the echo reaching the radar receiver reproduces the pulse shape and spectrum of the incident signal, at drastically reduced amplitude. All the timing information, essential for determination of target range, is carried in the sidebands. The receiver must therefore deliver substantially all the sideband energy to the demodulator.

When a noisy signal passes through a wideband amplifier the noise is also amplified – SNR can only be reduced by reduction of bandwidth ($P_n \propto B$, Eq. (3.2a)); magnifying a photograph does not improve its contrast. For efficient echo detection SNR must be optimised, using a frequency selective filter to pass frequencies containing some noise but also most of the signal, hence preserving most of the timing information within it, while blocking those frequencies with noise but little signal. Two or more successively narrower IF band-pass filter stages are usually included, so bandwidth reduction is gradual, reducing the risk of strong signals causing ringing and ghost echoes behind the displayed echo pulse. Stand-alone filter blocks in the signal path are often aided by frequency-dependent feedback within the amplifiers.

If filter bandwidth is too wide, all the frequency components of the signal are indeed accepted but with much noise, noise being proportional to bandwidth. If bandwidth is too narrow, signal and noise are both attenuated, letting the baby out with the bathwater; the pulse is also broadened, spoiling range accuracy. Matched filters in which the attenuation/frequency characteristic matches the pulse spectrum give best signal to noise ratio but may slope the echo pulse edges too much to determine the exact instant of arrival, introducing navigationally undesirable range error to displayed plots, degrading ARPA or ATA track prediction and demanding extra AFC accuracy. The amplitude/frequency transfer function is the complex conjugate of the Fourier transform (in frequency domain) of the signal pulse shape.

However, there is little loss of SNR when the filter response shape differs considerably from the matched case. Extra IF bandwidth also has to be retained to cover the residual tuning error, degrading SNR. On long-range scales, where receiver bandwidth is least, residual tuning errors may cause some loss of receiver sensitivity, further spoiling SNR. IF bandwidth is therefore often made rather wider than the matched value, some loss of SNR being accepted in return for preservation of as much signal data as possible to better preserve timing information. Such a filter is termed a matching filter. The minimum necessary bandwidth, B_{n}, which must be retained to preserve the shape of the echo pulse, length τ, to permit range determination is similar to Chapter 2, Eq. (2.2d):

$$B_{\mathrm{n}} \sim \frac{1}{\tau}. \tag{3.4}$$

In practice, somewhat wider receiver bandwidth is often used at the expense of a dB or so poorer SNR

- to preserve signal data and avoid loss of ranging accuracy;
- to facilitate operation of twin displays set to differing range scales;
- to ease local oscillator tuning accuracy.

When pulselength is restricted to permit highest available prf on short range scales, IF bandwidth is occasionally made somewhat less than necessary bandwidth.

In system performance calculations, where the filter shape is in general not known, the receiver passband is usually assumed rectangular, called an abrupt or hard filter. Frequency components away from band centre are then credited with too much gain, introducing a bandshape loss in the same fashion to the beamshape loss of Section 3.3.1 (Figure 3.2). Figure 3.7 shows the similar spectra of rectangular and Gaussian (minimum spectral width) pulses and a hard filter response. The beamshape loss is shaded for the rectangular pulse; the Gaussian pulse loss is similar.

For most transmitted pulse shapes the matched filter bandwidth is quite close to $1/\tau$. For purposes of system performance calculation, rather than entering detailed analysis of pulse shape and filter shape, the following simplifying assumptions are usually made, with little error:

- filter is hard, with zero loss in passband and infinite loss at all other frequencies. Although not manufacturable, the concept is useful;
- signal pulse is rectangular;
- filter output contains all the signal energy;
- filter output contains noise power equivalent to filter bandwidth;
- filter bandwidth is assumed to be $1/\tau$ (some prefer to use $1.2/\tau$) unless actual value is known;

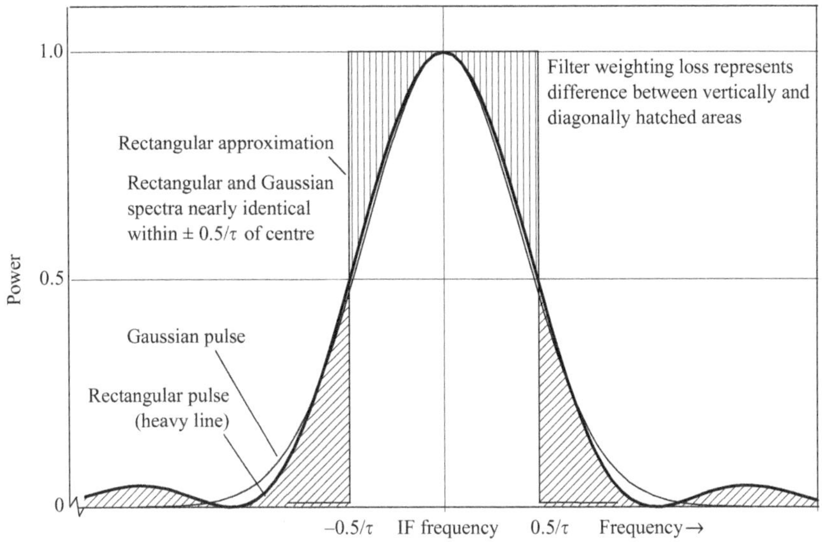

Figure 3.7 Filter responses

- reduced noise pick-up just within the passband does not fully offset additional noise just beyond, so filtration introduces a filter weighting loss to account for the earlier approximations; Figure 3.7. Loss is somewhat dependent on assumed bandwidth as well as shape but is between 1 and 3 dB when bandwidth is near $1/\tau$.

3.5.3 Linear and square-law demodulators

The second demodulator determines whether the IF amplifier output contains a candidate echo. Figure 3.8(a) shows a diode current/voltage characteristic curve. An alternating voltage applied to the anode gives positive half-cycles at the cathode, negative half-cycles being blocked, rectifying the input signal to d.c. with a residual a.c. component. Following radio practice, the process is often called detection. We prefer the alternative term demodulation, reserving detection for the overall determination of whether a signal contains a candidate echo. Figures 3.8(b) and (c) are circuit diagrams of simple half-wave and full-wave demodulators. The latter uses centre-tapped transformer T1 to collect information from negative as well as positive components of the IF waveform. When the a.c. is an IF pulse burst (d), a unidirectional pulse is produced at (e). The a.c. component is removed by a smoothing capacitor as shown at (f). The capacitor accepts charge on peaks and holds the voltage intermediately. It acts as a low pass filter, inherently introducing a small time delay. Radar second detectors are generally driven well into the diode linear region, and here video output voltage is proportional to microwave signal voltage, forming a linear demodulator.

Direct microwave rectification with no mixer or IF stage gives a video-frequency pulse direct from a microwave pulse. Although inefficient, this may suffice within a racon for direct demodulation of the transmissions of interrogating radars (Chapter 8, Section 8.2.5). Near the origin, diode current rises proportional to V^2, so diode

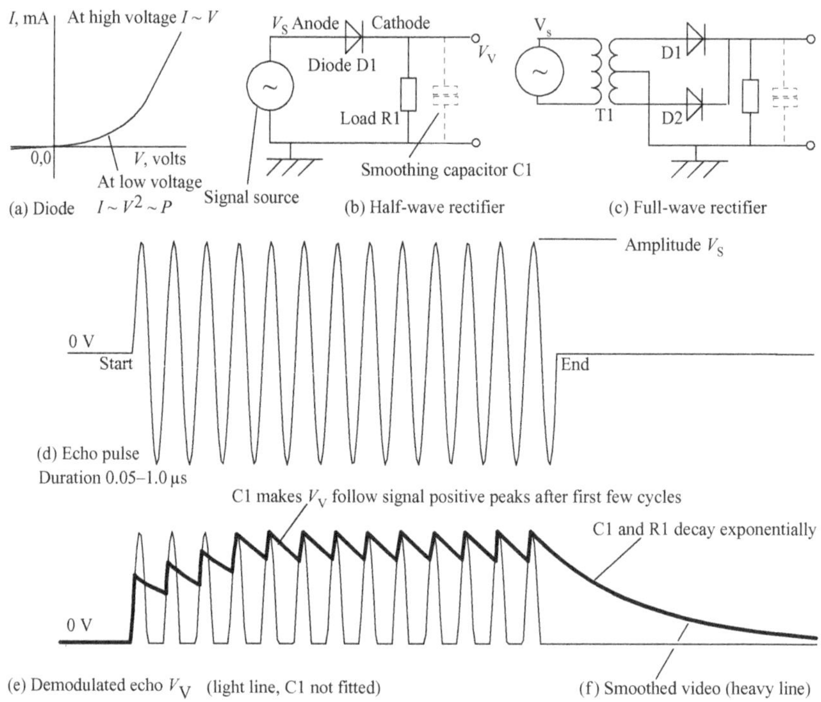

Figure 3.8 Diode demodulator. The smoothed output forms the video signal

rectifiers are called square-law demodulators. Video *voltage* is proportional to input signal *power*, typical sensitivity into a 1 kΩ load is 1 mV per 1 µW input, efficiency approximating 0.1 per cent. Efficiency is improved somewhat by a small forward current bias (~50 µA d.c.), putting the signal on the most curved part of the V/I characteristic and reducing the microwave characteristic impedance to a few hundred ohms, matching the feed impedance. Receivers having direct microwave detection are sometimes called crystal-video receivers. When biassed, typical 9 GHz minimum detectable signal is −72 dBW for 2 MHz video bandwidth.

Demodulators of coherent systems take the form of multiplicative mixers, the output frequency being the difference between IF and COHO frequencies, Chapter 2, Sections 2.2.4 and 2.2.5, Figures 2.11(*b*) and (*c*). A pair of demodulators is used, accepting in-phase and quadrature IF and COHO components (derived via 90° phase shifters) and delivering in-phase (I) and quadrature (Q) video components, their joint amplitude being proportional to echo amplitude, while their relative amplitudes convey the signal phase information.

3.5.4 Factors affecting detection

Despite swept gain, the receiver input voltage/output voltage relationship or transfer characteristic must support events when the noise and clutter are at their maximum;

that is the receiver as a whole must have sufficient dynamic range to support strong as well as weak signals without clipping off (limiting) which destroys the information in large events. As well as provoking reflections from feeder mismatch (Chapter 2, Section 2.6.2), excessively strong echoes or clutter can also drive the receiver into paralysis or induce damped oscillation, dumping or repeating nearby targets. Severity of these overload effects is a matter of detail design and difficult to quantify.

3.5.5 Detection cells

The signal processing system divides the surveillance area into a matrix of detection cells, each covering an area of roughly pulselength × (range × azimuth beamwidth). Big cells cover more surface area so pick up more sea clutter. They also cover more atmospheric volume, getting more precipitation clutter. Thermal noise is indirectly increased by small cell length, for receiver bandwidth has to be wider for short pulses.

Small cells, matched by narrow pulselength and large scanner aperture (narrow beamwidth) are desirable to:

- give crisp display of weak targets, somewhat improving the operator's perception;
- improve signal to clutter ratio unless the target is big enough to overflow the cell;
- resolve adjacent targets, which otherwise merge; often a main reason for giving VTS stations such large scanners;
- increase position accuracy, necessary for accurate ARPA/ATA track prediction;
- but narrow cells have fewer hits available for integration, especially in the case of fast crossing targets whose angular velocity changes azimuth bearing per scan by more than a cell width.

3.5.6 Effect of range scale selection

At long range in benign clutter, the receiver can be operated at high gain to receive weak echoes from distant targets. Bandwidth is minimised to suppress noise and raise SNR, necessitating maximum available transmitter pulse length to satisfy the necessary bandwidth criterion, some range accuracy being sacrificed. PRF is reduced (a) to give unambiguous operation with only one pulse in flight at a time, (b) to avoid transmitter on/off ratio or duty cycle overload.

Radars have 8–10 range scales, in 2 : 1 ratio. At short range, echoes are stronger and more noise is tolerable, so bandwidth is raised and pulselength reduced to improve range resolution. The reduced SNR reduces the likelihood of reception of very distant echoes, easing the range ambiguity problem and enabling transmitter prf to be raised, partly restoring the duty. The additional pulses per packet partially restore sensitivity by integration, somewhat offsetting the increased noise. The short pulses also illuminate less clutter. PRF is not in fact always increased to the fullest possible extent as it is often desirable to retain a long-range scale on the secondary viewing display for early warning of new targets during a coastal passage, restricting prf to avoid ambiguity. Also the ARPA or ATA must continue surveillance of distant undisplayed targets to maintain their tracks. In severe clutter considerably exceeding noise, it is desirable to minimise range cell size by retaining short pulse/wide bandwidth operation.

Table 3.1 Range scale comparison

Range scale setting	Long	Short
Pulselength	Long	Short
Cell size	Large	Small
Range discrimination	Poor	Good
PRF	Low	High
Pulses/scan	Low	High
Receiver bandwidth	Low	High
Receiver noise	Low	High
Signal to noise ratio	High	Low
Signal to clutter ratio	Low	High

Individual radars differ in their matching of pulselength/bandwidth to range scale, and may include intermediate pulse length/bandwidth steps. Table 3.1 summarises the broad effects on detection performance of the chosen range scale.

3.5.7 Video amplifier

The demodulator restores the signal to baseband. The output instantaneous video voltage is proportional to the IF input voltage. The spectrum of any target is converted back to that of the modulator, so necessary bandwidth reverts from the IFs $1/\tau$ to $0.5/\tau$. Response need not extend down to d.c. but can roll off at ~50 kHz, simplifying detail design and rejecting semiconductor low-frequency flicker noise. Phase information is only preserved in coherent systems, for which twin I and Q video channels are required. Video amplifiers are usually contained on a couple of semiconductor chips.

A conventional linear amplifier has output voltage proportional to the input voltage. Negative feedback is employed to improve linearity and stabilise gain. Supposing the dynamic range has to be 100 dB and the maximum output the amplifier can give before overload (i.e. within its dynamic range) is 1 V, then a weak echo, 100 dB down, has amplitude only 10 μV. This voltage range of $10^5 : 1$ demands a 17-bit analog to digital convertor ($2^{17} = 131072 : 1 = 102.3$ dB) followed by 17-bit (plus parity bits) processing in the signal processor, which is profligate in computing power.

Usually a logarithmic amplifier or log amp is included. Here output voltage is proportional to $a \times$ (logarithm of input voltage), so 100 dB (10^{10}) range causes only $10a : 1$ output voltage variation, which can be handled by a smaller processor; when $a = 1$, 100 dB input change (equivalent to $10^5 : 1$ voltage variation) gives $\log 10^5 = 50$ output change, which is less than 2^6 or 64, enabling 6-bit plus parity processing. Logarithmic amplifiers sum the outputs of several successive limiting stages; Figure 3.9(a). They can handle extremely wide dynamic range, but their chips have to be precisely made to avoid error, placing the radar designer at the mercy of the

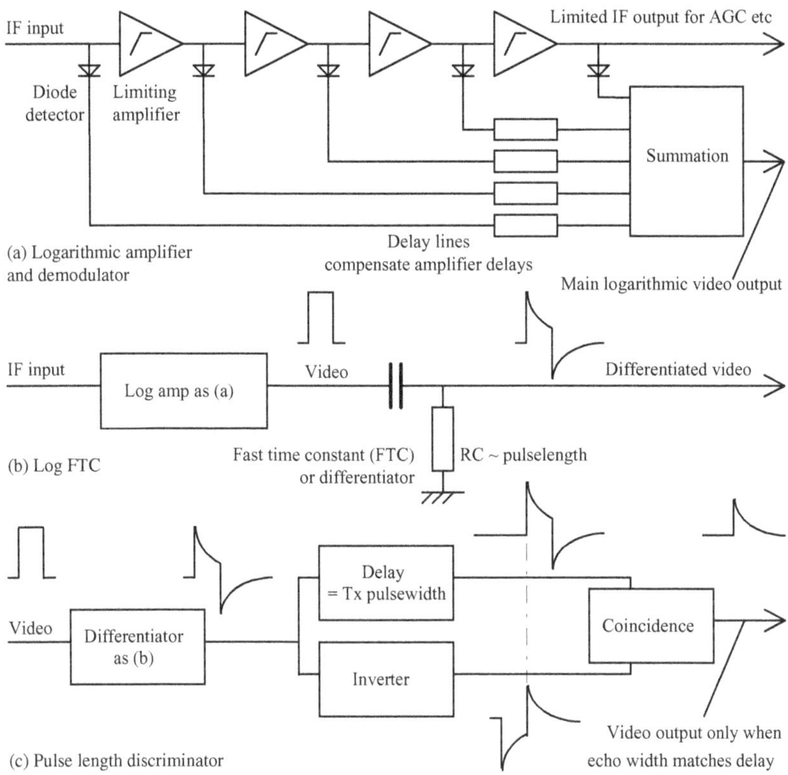

Figure 3.9 *Receiver ancillary circuits showing waveforms. (a) Logarithmic amplifier. As the signal voltage increases, the last stage saturates, followed by the others in turn. (b) Fast time constant. Breaks up solid blocks of clutter by extracting changes in signal + clutter level. (c) Pulse length discriminator. Breaks up solid clutter by extracting signals matching transmitter pulselength. Not applicable to large targets*

chip supplier. Especially when used with a differentiator, they help to reveal echoes within clutter. If used in a constant false alarm rate (CFAR) system which maximises probability of detection, the first limiter stage must always be saturated by noise. The gain control may actually control the clipping level rather than amplifier gain. The human ear has a logarithmic scale, enabling us to handle an extreme loudness range.

When several displays are driven from the output of the main display's logarithmic amplifier, each may be given its independent gain, swept gain and differentiator controls, which may be set to suit each operator's requirements.

3.5.8 Fast time constant, differentiator

In areas of rough water or precipitation producing weather clutter (sea and particularly precipitation clutter), solid clutter may return a substantial and near-equal signal

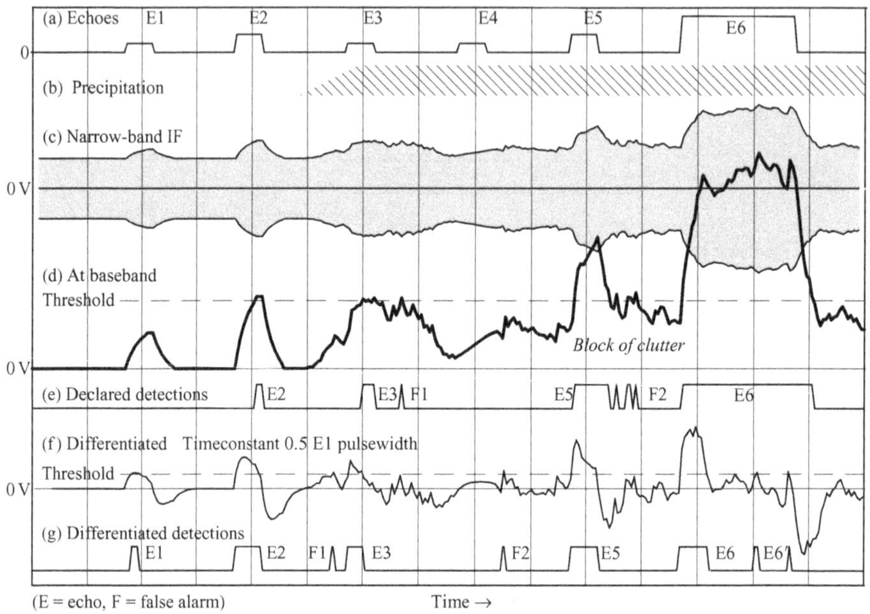

Figure 3.10 *Differentiator in precipitation clutter. Time domain for a single sweep. Differentiation partially suppresses blocks of clutter to reveal the edges of echo pulses*

to all detection cells in the area, masking any echoes they may contain, as shown in Chapter 2, Section 2.1.6, Figure 2.6 rain areas. The echoes can be enhanced by looking not at signal amplitude, but its rate of change with time during each sweep. Rate of change is mathematical differentiation and in analog systems is achieved by passing the signal through a fast time constant (FTC) circuit, such as a series capacitor–resistor high-pass network. Alternatively, an analog resistor–inductor combination or a digitised functional equivalent may be used. Figure 3.9(b) shows a simple FTC circuit following a log amp. Figure 3.9(c) shows a simple pulselength discriminator, which can enhance detection of small area targets, but which is rarely applicable in marine practice.

Figure 3.10 depicts detection of six targets having differing ranges and echo strengths (a) represents the echoes, E1–E6. Part of the range bracket lies in rather heavy precipitation (b) giving solid clutter. The narrow-band IF signal at (c) shows the echoes, the envelope being rounded by the narrow bandwidth. E3–E6 are buried in clutter; each echo being equivalent to echo S in Chapter 2, Section 2.2.1, Figure 2.10(b). After the second detector (d), similar to T in Figure 2.10(b), the baseband signal's detection threshold has to be set high to reduce probability of false alarm (P_{FA}) to an acceptable level. Waveform (e) shows echo E4 is lost and there are false alarms at F1 and F2. The high threshold completely prevents detection of weak echo E1 and moderate echo E2 is only just seen, although both lie outside the clutter area.

Waveform (f) shows baseband signal (d) after differentiation. Peak echo amplitude has fallen, but clutter has fallen further permitting a lower threshold (g). All echoes except E4 are now detectable, with a couple of false alarms. Strong echo E6 occupies considerable time, equivalent to radial length. It might represent an islet or a racon response. When the radar is operated conventionally, all parts of the echo are detected and displayed as an axial trace. Engaging the differentiator suppresses nearly everything but the leading edge, perhaps with a faint suspicion of the remainder from enhanced noise, E6′. Display quality is reduced. The extended nature of the echo is no longer apparent, neither can the shape of its paint reveal the aspect of a large target. At the end of the echo, voltage overshoot (not shown) in the baseband amplifier circuit may cause a faint blip which might be mistaken for another target.

The figure is drawn for differentiator timeconstant half a pulselength to emphasise the principle. In practice the timeconstant is usually about one pulselength. All radars provide a differentiator or FTC on/off control, usually with control of timeconstant, giving the operator more scope to balance detectability against display quality. As differentiation degrades echo plots, it should be disengaged in clear conditions. Its value lies in its ability to break up solid clutter. Sea clutter is less solid than that from precipitation – the structure of individual sea-waves is often visible as striations on the display – so differentiation is less effective against sea than precipitation clutter.

3.6 Signal processing basics

3.6.1 The task

The data stream at the receiver output contains a serial jumble of thermal noise and clutter returns, mixed up with any target echoes. All modern radars use extensive digital signal processing to remove as much noise and clutter as possible before reaching the display. Processing optimises detection of wanted echoes and minimises false alarms. Information theory sets definite limits to what can be achieved.

The processing strategies of individual radar suppliers are not normally disclosed and doubtless differ in detail, but the competitive nature of the market dictates that they all achieve performance close to the theoretical limit for the target and clutter scenario. This section gives a broad outline of the signal processing section of the radar; quantitative analysis following in Chapter 12. Automatic tracking aids, discussed in Chapter 13, use the detected echo positions to form tracks of individual targets. From these likely future positions are extrapolated and closest points of approach are predicted.

The processor's task is to extract the maximum information – matter informative to the user – from the undigested stream of data; that is, to maximise the probability of detection, P_D, and minimise the probability of false alarms, P_{FA}. Data once discarded is irretrievably lost, so must be retained until as much information as possible has been wrung from it. P_D and P_{FA} are interlinked with the relative strength of the wanted signal to unwanted noise, the all-important signal to noise and clutter ratio (SNR). It is impossible to have high P_D and low P_{FA} with low SNR. The processor plays a big part in the effort to maximise SNR. Detectability is affected by the random or partly random manner in which echoes, noise and clutter fluctuate with time, necessitating statistical treatment in Chapter 12.

Prior to the processor, we have already encountered several features which use the precepts of information theory to maximise the echo at the expense of its competitors:

- high transmitter power, attention to feeder loss and low receiver noise figure improve the ratio of echo to thermal noise;
- narrow scanner beamwidths – high gain – minimise the amount of clutter illuminated;
- receiver bandwidth is restricted to reduce thermal noise;
- swept gain reduces necessary dynamic range of components;
- available dynamic range may be increased by logarithmic amplifiers, automatic gain control, etc.;
- differentiation breaks up solid clutter.

3.6.2 P_D and P_{FA} for target perception

The detection process in its widest sense contains several steps, with operator intervention in each. The target in question may not be the one of primary interest to the operator, who therefore may not optimise the controls for it. The steps are:

- generation of a prospective target plot in the signal processor – the operator chooses the radar band if both 3 and 9 GHz are available, then sets gain and other controls;
- display of the plot – the operator chooses an appropriate range scale;
- display of track – the operator may choose whether a plot is to be tracked;
- operator perceives paint as being a valid plot or track, not clutter.

The last step is mental, with electrotechnical implications. The eye has good pattern recognition capability and under optimum conditions can perceive faint target plots, if they persist, in quite severe clutter. When viewing a display containing noise or clutter, the operator consciously or subconsciously applies a mental threshold, whose consistency depends on motivation, fatigue and circumstance; the threshold may be higher in mid-ocean than when anxiously searching for a liferaft. Conditions are rarely optimum: the operator has other tasks, may be fatigued or stressed, cannot devote full attention to the display, or may not have the luxury of time for long observation of a barely visible plot before having to make the navigation decision. Signal processors bypass most of these limitations.

To optimise perception, the radar has to display the target with high probability of detection (P_D) but also low false alarm rate (P_{FA}). 'High' and 'low' are not always easy to quantify, but the following are often accepted as reasonable for display of the integrated return of a single scan:

$$P_D > 0.5 \qquad P_{FA} < 10^{-6}.$$

P_D 0.5 is sometimes called 50 per cent blip/scan ratio, defined as the number of times an echo (blip) is detected divided by the number of sweeps within the packet.

False alarm probability of one in a million may sound very low, but it must be remembered that a noise paint can arise from any of the hundreds of thousands of detection cells on a half dozen sweeps per scan. Psychological or human factors (HF) such as fatigue, use of dim cursive displays, etc. can be partly allowed for in calculations by adjustment of necessary P_D.

3.6.3 Digital conversion, detection cells

So far, the radar signals and processes have been analog: 'Designating,[1] pertaining to or operating with signals or information represented by a continuously variable quantity, such as spatial position, voltage, etc.' Older radars were entirely analog, their detection strategy making use of the characteristics of long-persistence display tubes, described in Section 3.10. In the digital technology used in all modern radars, quantities take discrete values: 0, 1, 2, 3, ..., expressed as whole binary numbers 000, 001, 010, 011, ..., but nothing in between.

Although analog circuits can perform arithmetical operations such as addition, subtraction and multiplication, it is far more convenient, quick, flexible, cheap, stable and accurate to use digital processing, adopting the technology of the computer industry. The first operation is to digitise the baseband data stream by an analog to digital converter; Figure 3.11. The analog voltage is repetitively sampled to give digital number 'snapshots' best representing the voltages at those instants. If the sampling or clock rate is less than the video bandwidth (undersampling), significant data is lost. If the clock frequency exceeds video bandwidth (oversampling), there is little loss but more and faster processing capacity is necessary. For pulselength say $\tau = 0.1\ \mu s$, video bandwidth $\sim 0.5/\tau$ or 5 MHz so at least 5×10^6 samples/s are taken. Samples may have any amplitude from the noise background up to the echo of a large target. Together, finite clock frequency and digitising quantum or least significant bit (LSB) introduce a quantising loss.

Apart from a little dead time before each sweep, every instant of the nominal 2.5 s scan time represents a slightly different basic element of surveillance area which may contain a target. In all there are some 10^7 area elements or detection cells, alternatively called range cells or bins, each identified by a unique address related to range and bearing and together forming a digital frame store. Economies may be made by reusing cells within areas containing insignificant content, but large memory is always needed. The integration process demands a set of detection cells for each beamwidth's worth of bearing, if the cell footprint is not to degrade that of the scanner beamwidth/pulselength. The azimuth might be divided into 512 bearing $\times$ 256 range cells $=$ 131 000 cells, doubled for scan to scan correlation, but modern frame stores may contain over a million addresses. Only in the last years of the twentieth century did such large memory become economically viable. It is usual for cells to have equal angle resolution.

Straddling loss occurs when small targets do not sit central to a cell, but sit astride a pair of cells to record unduly low strengths in each.

[1] Shorter Oxford English Dictionary.

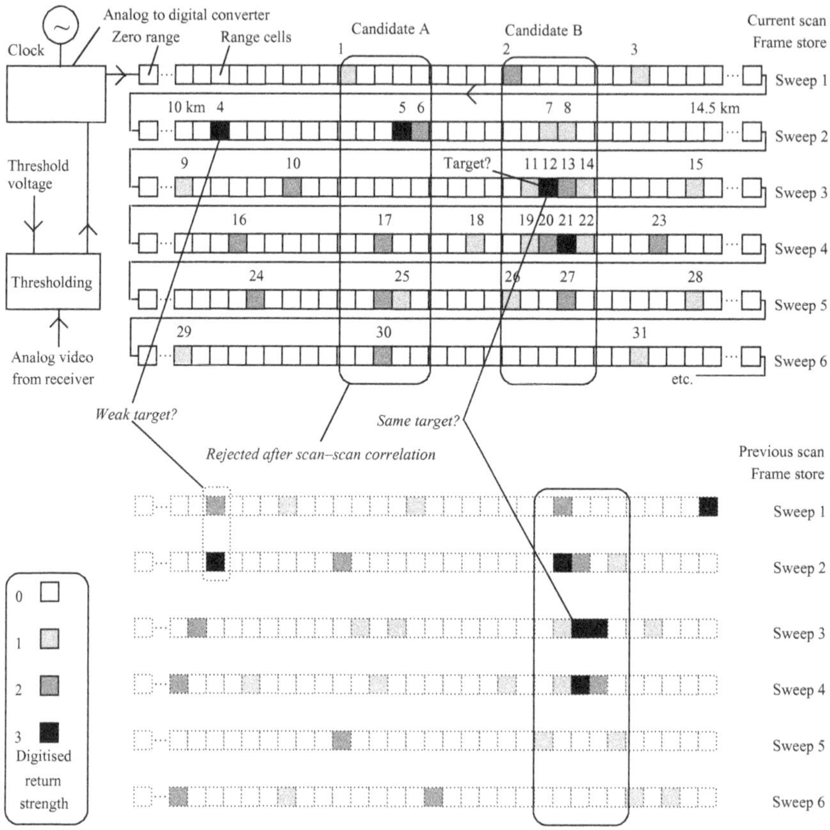

Figure 3.11 *Signal processing. Digitised data is routed to detection cells within the frame store*

3.6.4 *Logical process of target detection*

Figure 3.11 includes part of a simplified digital frame store, drawn as a matrix in range and bearing. For clarity a cut-down system is depicted. Thirty cells of range are shown, representing, say, from 10 to 14.5 km for pulselength 1 μs. There are six sweeps in bearing, each sweep advancing by $\frac{1}{6}$ scanner beamwidth, representing 0.65 km mean azimuth bracket at the range in question. Each cell contains thermal noise plus the echo and clutter content of about 150 m × 110 m average sea area and the atmosphere above.

For simplicity, cell capacity is shown as a mere 2 bits, content representing 0, 1, 2 or 3 analog baseband voltage units. The figure assumes modest noise or clutter, so most of the cells have count zero; a few have strengths of 1–3.

Do cells having strength 1–3 represent targets? First, we examine the sweeps of the current scan by eye, for the human mind is good at pattern recognition. Looking at any single sweep in isolation, we may feel little confidence in drawing the line above which we accept a return as valid, or put formally, set the threshold for detection declaration. Figure 3.10(*d*) shows such a threshold applied to the video signal.

Examining the six sweeps as a set, pulse to pulse integration, gives a clearer picture. Returns of strength 1 seem so far below the strongest returns that we feel sure they must represent noise or clutter and are too feeble to have significant probability of being targets. We discard entirely all cells whose signal strength lies below threshold. Bearing in mind the steepness of the skirts of Gaussian distribution (Figure 3.5), it seems much more profitable to concentrate on the stronger returns, which are statistically much less likely to represent clutter spikes. Cells labelled 1, 5, 6, 17, 25 and perhaps 30 seem to form a significant cluster (candidate A). Another cluster, candidate B, contains cells 2, 7, 8, 11–14, 19–22, and perhaps 26 and 27. Cells 3, 4, 9, 10, 15, 16, 23, 24, 28, 29 and 31 seem random but might possibly represent small targets. In other words, choice of a low threshold risks numerous false alarms, but a high threshold risks rejection of genuine targets.

Comparison with the previous scan's frame store appears to rule out candidate A because there seems no previous concentration of returns at or near the range and bearing in question – scan to scan correlation is poor. If A is a target, it must be subject to severe fading. Has a weak target at cell 4 been discarded? The claims of candidate B are reinforced: the previous scan contained a cluster at about the same bearing and range. Unless A is a very fast approaching target, only B is probably a valid target and we might well declare it, meanwhile watching future scans for confirmation.

All that can be concluded from the evidence is:

- there is high probability of a target at B;
- there is moderate probability of a fading target at A;
- several other weak targets possibly exist;
- increased SNR would reduce the uncertainties.

3.6.5 Machine detection

The data processing system more or less replicates the mental processes outlined above in a formalised manner. To replace the mind's more subjective assessments, processing uses algorithms, defined and objective mathematical procedures, to squeeze the most information from the evidence quickly, tirelessly and rationally.

We have seen that in isolation it is not possible to state with confidence whether a particular cell content represents signal or a noise/clutter spike. Optimum threshold level is usually computed to be several times average cell occupancy, which on the premise that most cells are devoid of targets, is a measure of the prevailing noise and clutter. The first step is to reject as clutter or noise all cells whose counts lie below threshold, called thresholding. The threshold stage can be analog, thus reducing the load placed on the digital convertor and signal processor. The threshold voltage is set in part by the processor, by reference to the clutter environment (generally assessed by the average cell contents) and partly by the operator's use of the gain control. In the example, the average signal, noise and clutter count is 53 in 180 cells, average 0.29. As the least significant bit in our simplified model is 1, the threshold might be set at 1 (1 V : 0.29 V represents 10.75 dB, a reasonably typical value). Using 0 would let through all the noise and 2 would probably exclude too many genuine echoes.

Signals or hits exceeding threshold form first-stage single-pulse target candidates. Clearly, detection is optimised by setting the threshold as low as possible consistent with accepting as high a false alarm rate (FAR) as the operator or system can abide. The noise/clutter density varies with range and other circumstances – Figure 2.6 shows how a squall may occupy only part of the playing area. It is therefore beneficial to juggle the threshold (or receiver gain, which amounts to almost the same thing) to give constant false alarm rate (CFAR). This is accomplished by a combination of methods:

- swept gain, reducing gain at short range in a partially predetermined and partially adaptive manner;
- automatic gain control, averaging relative gain over the whole surveillance area;
- manual gain control, the operator setting IF gain (or threshold/logarithmic clipping levels) to give optimum display of a particular target;
- dedicated CFAR algorithms, which measure the average noise/clutter occupancy of cells near the candidate target, setting local FAR to maximum permissible by swept gain (or swept clipping level with logarithmic amplifiers);
- clutter mapping, where local clutter over each part of the playing area is regularly assessed, threshold being adaptively set for optimum FAR at each part.

3.6.6 Clutter map

Particularly with VTS groundfast radar, it is feasible to memorise or map the locations of heavy clutter and adapt detection thresholds to match that locally prevailing, editing out stationary clutter from ground features, and much weather clutter. Much of the benefit of differentiation is obtained without loss of display quality. Clutter mapping is less effective when the radar platform moves rapidly through varying clutter density and is not yet common on marine radars. Even in the most crowded harbour, most cells contain only noise and clutter and are devoid of targets. The clutter map is constructed from the average count (discarding high values which are probably targets) of blocks of cells much bigger than a ship dimensions, taken over a succession of scans. The map is stored digitally and steadily updated at a rate sufficient to track clutter movement, e.g. of a rainstorm driven by the wind. The detection threshold at any part of the surveillance area is automatically adapted to suit the average local clutter by reference to the map.

The map has to be very detailed to store coastal features and the strong clutter from breaking waves on the shoreline. One stratagem confines the playing area within the low-tide line, dumping echoes from elsewhere. This needs to be done with discretion; a VTS once puzzled why its radar had lost a ship, known to be within the harbour. The ship turned out to have drifted into the intertidal zone before grounding at high water in fog. Embarrassingly, as the playing area had been defined by the low-water line, the ship could not be displayed.

3.6.7 Detection decision process

The detection threshold is set automatically to suit the prevailing clutter, but always with possibility of operator intervention. But it must be stressed that even the most advanced processing can never give certainty from corrupted and incomplete data having low SNR, as shown by Section 3.6.5.

Figure 3.12(a) reproduces the cell matrix of Figure 3.11 more compactly. Figure 3.12(b) is after thresholding at 1 voltage unit. Cells having value 0 or 1 are dropped, all others being accepted as hits. Actual strengths are not stored, so the processor is compact, at cost of discarding some data. The detection criterion is some minimum number of hits per scan, sometimes called the m out of n criterion, here three hits out of six sweeps per scan. Candidates A and B each score three hits so would be declared as targets.

The next increase of complexity is scan correlation, storing the previous scan's results, only declaring a target when both register some lower number of hits per scan, say 2. Targets with poor blip/scan ratio are likely to be rejected by scan to scan correlation. By adding further memory, the strength information can be stored, enabling the strength of each hit to be accounted for as shown in Figure 3.12(c). Again candidate B is declared, but with somewhat higher P_D. The full potential of correlation is not always realised, the designer sometimes being content to suppress interference from other radars using a combination of pulse–pulse timing jitter and a 2 pulse correlator ($n = 2$).

Coherent systems also store phase data. Echo phasing jitters, particularly when SNR is low. Comparison of amplitudes during a scan with constancy of phasing sweep to sweep raises P_D. Scan to scan correlation can be used as before.

In general we see that, for a given P_{FA}, P_D is raised if the following occur.

- Many digitising bits are used. There is less quantising loss. The penalty is additional and faster memory capacity.
- SNR is raised.
- Cell size is reduced, needing bigger scanner aperture and short pulses, high bandwidth with the attendant additional noise, as well as additional processing power.
- Scan to scan correlation is used, again needing more processing power.
- The system is coherent, requiring a more complex transceiver.

Physically big targets sprawl over several adjacent detection cells. Particularly with short pulselength, even small targets may straddle a pair of cells. Probability of the candidate being valid increases when its immediate neighbours exceed average clutter and when the previous scan contained a cluster at or near the same position. Neither own ship or a target moves significantly during the 10 ms or so of the packet of sweeps within a scan, although their intensities may fluctuate. Receiver noise spikes meanwhile occur at random position. The processor has however to allow for fast targets traversing a couple of cells in the inter-scan period.

Modern radars display targets with remarkably little clutter on the screen, even in adverse conditions. But it is important to realise that because the signal processor has

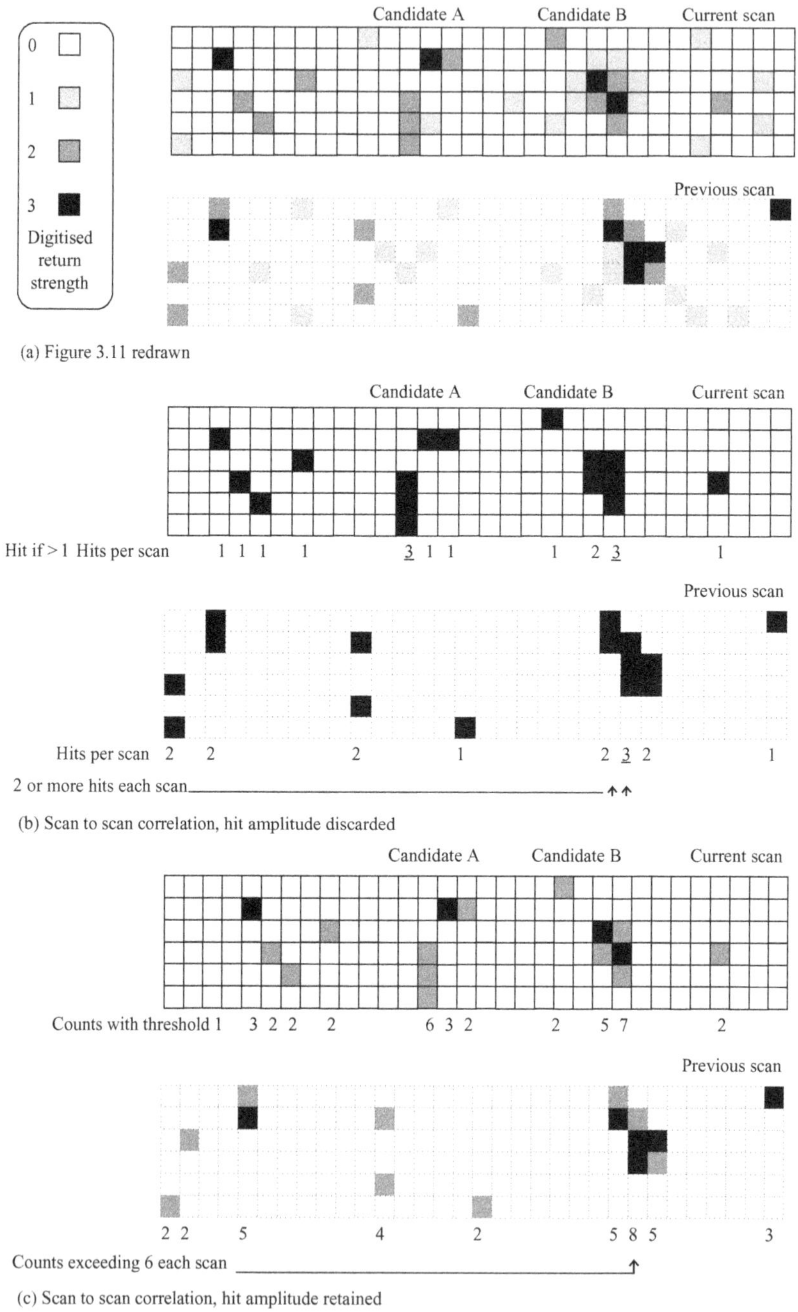

Figure 3.12 Digital detection. Adding sophistication improves probability of detecting valid targets while discarding false alarms

kept clutter off the screen does not mean that clutter ceases to degrade performance. Severe clutter necessitates raising the threshold to restore FAR to an acceptable level, so weak targets cease to be detected. Presence of noise or strong clutter always degrades performance.

Declared targets are fed to a target register which memorises target locations and supplies a bright-up pulse to the display at the appropriate range and bearing.

Radar data may be exported to other on-board instruments. For example, wave height is of particular interest to high speed craft which are not licensed to operate in extreme conditions. The Wavex [1] directional wave monitoring system calculates significant wave height from sea clutter signals extracted from the craft's standard navigational radar.

3.7 Additional features

3.7.1 Within single radar

Relatively large ships viewed at short range may give paints approximating the target waterline plan, yielding valuable aspect data. The paints tend to be too small for easy viewing so some radars include a target expansion facility to double the apparent target size.

The navigator needs to see at a glance the approximate course and speed of all targets. The target registers for the previous few scans are applied to the display at successively reduced brilliance, leaving a trail similar to a wake to indicate approximate course and speed. Control of the trail timeframe or persistence can be provided as an operator control.

When own ship yaws or manoeuvres, or the operator changes range scale or off-centres the display, the target register redirects itself to the correct part of the screen without smearing, the display remaining legible. The target may manoeuvre also, making it futile to integrate over many scans. Manoeuvrability assumptions have to be built into the scan–scan integration algorithm. How many adjacent range cells may be integrated depends on an assumption of maximum speeds of own ship and the target. A 30 knot ship ought to accept a HSC at say 70 knot (the upper limit of IMO's HSC Code, although some military patrol boats are faster). Maximum closing speed is then 100 knot ($\sim$49 m/s). A HSC radar ought to accept another HSC, maximum closing speed 140 knot (72 m/s). It would be counter-productive to accommodate aircraft at 200 knot. Setting the algorithm for unnecessarily high speeds lets in more noise cells and increases probability of false alarm.

3.7.2 Multiple sensors, track combiners

The diversity principle seeks to improve detectability by observing the target in two or more decorrelated ways, where their clutter returns on any given sweep usually differ. Twin identical radars can have their scanners mounted at different heights or positions, or a single scanner can alternately use differing polarisations or transmitter frequencies.

Track combination is not current practice on shipborne dual radar installations, perhaps because the additional complexity outweighs the benefits. However, combination of tracks from two radar heads to form and display a single track from the composite data is standard practice on the larger VTS installations, enabling a target to be tracked seamlessly as it traverses the whole surveillance area. Considerable care has to be taken with accuracy of the basic radars, to ensure that a single target is not regarded as two, or two targets are not incorrectly merged or 'associated' as one.

The first step is plot association, that is, make the decision whether a pair of plots on the two radars are from a single target or from a pair of close-spaced targets. This is not as easy as it sounds, particularly when SNRs are low, just when track combination would be most beneficial.

Track combination is performed on digitised data within a dedicated computer. Cost–benefit analysis informs the choice from amongst the following alternative arrangements.

1. Each radar transmits a digitised but otherwise unprocessed data stream to the VTS, where tracks are formed in a central computer. Quasi-raw (relatively unprocessed) data can be displayed if required, and is preferred by some operators in certain clutter conditions, although it is nowadays hard to justify the assertion that an operator, however skilful, can improve on the performance of a well-designed trackformer. A disadvantage is volume of data, necessitating wide link bandwidth. On the other hand, the central computer can be duplicated for reliability at moderate cost and it is easy to adjust the tracking and combining algorithms to suit prevailing circumstances. Much grief can be (and has been) caused if data is discarded prematurely, e.g. if the link restricts echo dynamic range or overloads when clutter is severe.
2. Targets can be detected locally with detection thresholds set for fairly high P_{FA}. This moderates the quantity of reported false alarms. Plots are reported scan by scan to the centre at moderate bandwidth, for subsequent further processing and track-forming. Weak targets may fail to be reported and be lost.
3. Tracks can be formed locally, only track combination from each outstation being central. Little transmission bandwidth is required to report formed tracks, but the operator is denied knowledge of local clutter conditions and cannot search for plots on which tracks have not been formed, such as new entrants to the area. Failure of a trackformer disables its outstation.

3.8 Display principles

The whole purpose of the radar is to communicate information to an operator by means of the display. The display, screen, or scope is the human-machine interface and is arranged to suit the operator – it must be user friendly. It should display as much as possible of what the operator needs – targets and associated alpha-numeric data – with as little extraneous clutter and noise as possible. Echoes displayed on the screen are often called paints or blips.

3.8.1 Display format

On ships, IMO demands relatively large screens, 250 mm, to 340 mm active diameter on the bigger tonnages, to reduce eyestrain when alternately looking out of the window, viewing other bridge instruments and refocussing on the screen at moderate viewing distance, and to facilitate two persons viewing together, for example, the master and OOW when evaluating a manoeuvre. The very large ‘conference displays’ of some military systems are not used – perhaps merchant ships cannot muster enough deck officers! Screen area must have some influence on mental perception of echoes, but data is lacking.

Current echoes or declared target plots are displayed as spots of light at scale range and bearing from the scanner position. This map-like form of display is called a plan position indicator (PPI); Figure 3.13. Targets are shown in polar or R, θ (range, bearing) coordinates. The simplest PPI presentation keeps own ship or station at the centre of the display while all target paints, including coastlines, move past in relative motion. Ships’ displays may be aligned to suit the traffic situation: North-Up, Ship’s Head Up or Course Up (differing from head up if the ship makes leeway). Own ship may be offset from the centre to extend the view ahead. In the important True Motion mode, the display is ground stabilised and all targets and own ship move as across a map, resetting when approaching the display rim. VTS displays are of course always referenced to a fixed datum. They are sometimes aligned to the view from the station window – for example, the West Coast port of Vancouver, whose station lies to the North of the harbour, has found South-Up displays convenient.

The operator measures target position relative to the scanner by intersecting a variable range marker (VRM) and electronic bearing line (EBL) over the plot of interest, digital readouts giving range and bearing relative to own scanner. Alternatively approximate position can be judged from a set of electronically generated range rings at say 1 nmi intervals and a bearing scale or pelorus (compass rose, nowadays

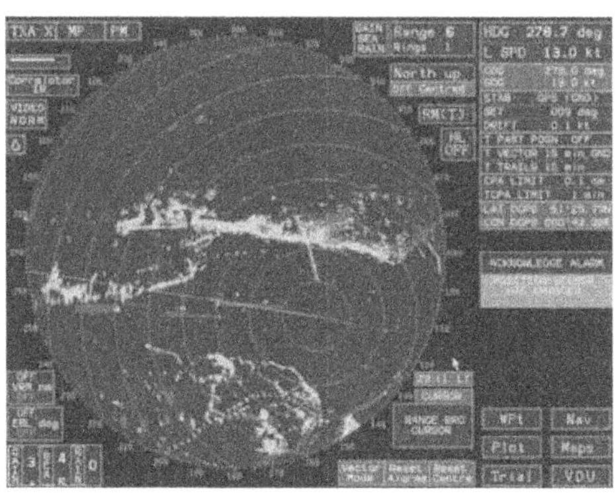

Figure 3.13 Display including radar data only: coastlines, targets (some with trails), offset with range rings, heading marker and copious alphanumeric information. Original in full colour. Reproduced by permission of Kelvin Hughes Ltd, Ilford UK

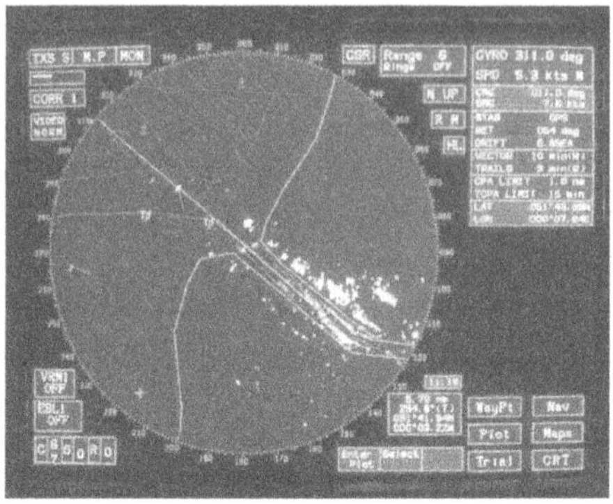

Figure 3.14 Display with chart data superimposed. Original in full colour. Reproduced by permission of Kelvin Hughes Ltd, Ilford UK

electronic rather than engraved, avoiding parallax) around the circumference. It is usually possible to lay a line between a pair of targets of interest and read off their relative range and bearing. A cursor can also be placed over any feature of interest, its coordinates being displayed alpha-numerically. Choice of presentation, although very important to the operator, rarely affects the detectability problem and we shall not consider it further.

Displays of relatively unprocessed radar broadband video signals are called raw radar, but it is now usual to show processed or 'synthetic' information on the screen. Formerly, each radar had its own stand-alone display, the only non-radar data being a few short alpha-numeric statements of course, speed etc. Nowadays, partly because the display occupies prime real estate on the bridge, selected parts of the system electronic navigation chart (SENC) are routinely added to radar displays, Figure 3.14, and plots are exported as a radar layer within the ship's electronic chart and display system (ECDIS). There is scope for addition of AIS data and plots transmitted by VTS stations, although care must be taken not to present the operator with an inassimilable mass of material – information overload is dangerous, the mind tending to dump data in a rather arbitrary manner. The trend is for black box radars to export plots and tracks to multifunctional displays at operator workstations forming part of an integrated bridge system (IBS). Data fusion has been discussed by Lee [2] and the same fusion and association problems arise as when radar tracks are combined, Section 3.7.2.

The radar market is much too small to support development and manufacture of special screens, so high quality computer and air traffic control devices are used, available in a restricted size range up to about 580 mm diagonal.

3.8.2 Cathode ray tube

Many displays still use cathode ray tubes (CRT), a specialised triode valve or electron tube. Within the rear neck of an evacuated glass/metal conical envelope are three electron guns for the three primary colours, a single gun sufficing for the older monochrome displays. Each gun has a red-hot metal cathode operated at a high negative voltage of about -10 kV. Cathodes are coated with a mixture of oxides facilitating electron emission, under control of nearby 'grid' electrodes. Positive-going grid bright-up voltage signals permit cathode rays of high-velocity electrons, $\sim 100\,\mu A$, to fly to the glass screen at the front of the tube, drawn by an earthed metal cylindrical anode (plate) within the neck. The rays are electrostatically focussed to sharp dots and can be steered by the time base system to any part of the screen by application of voltage to pairs of deflector plates (or by current to pairs of coils) surrounding the neck. The screen is made as flat as possible consistent with forming part of a pressure vessel and carries sets of electro-luminescent phosphor coating picture elements or pixels on its inner face which fluoresce in primary colours or the predetermined monochrome hue when struck by the beam, the display being intensity modulated by the grid signal. Colours, if used, are standardised by IMO (for navigational data) and IHO (International Hydrographic Organisation for charting features), differing for day and night illumination.

Modern screens have very short afterglow and are scanned raster-fashion like TV screens, the bright-up signals coming from the frame stores, which here perform an R, θ to Cartesian coordinate conversion, giving a flicker-free display, Section 3.9. Older CRTs laid down raw radar R, θ video pulses sweep to sweep and scan to scan. Long-afterglow phosphors integrated the real time scan to scan to give a reasonably steady image, Section 3.10.

3.8.3 Other display devices

CRTs are bulky and are rapidly being replaced in television and PC monitors by various kinds of flat screen low voltage semiconductor arrays, including liquid crystal displays (LCD), which are produced by the million. Application of a low voltage switches certain organic molecules between transparent and opaque states. In LCDs, a very thin film of the active liquid is sandwiched between glass sheets bearing a near-transparent pattern of conductors to form pixels, controlled by transparent thin film transistor (TFT) arrays. Backlit and colour variants are available. LCDs consume very little power. Long widespread in laptop computers, they were first taken up by the radar industry in small craft radars. Problems of available brilliance range, adequate area and wide angle of view are being overcome and price is falling. As a result, high resolution LCDs are replacing CRTs in new big-ship marine radars. Figure 3.15 shows part of a modern screen and shows the high resolution available. Advantages include low bulk, screen flatness, low power consumption and avoidance of high voltage. Sufficiently large screens or screen arrays for VTS stations are now becoming feasible, at a price. TFT screens drive each pixel continuously, not once per screen scan, improving brilliance and eliminating flicker.

3.9 Raster scan display

The modern flicker-free television-type raster scan display, Figure 3.16, has the phosphor applied as a fine pattern of (usually) coloured short persistence triplets

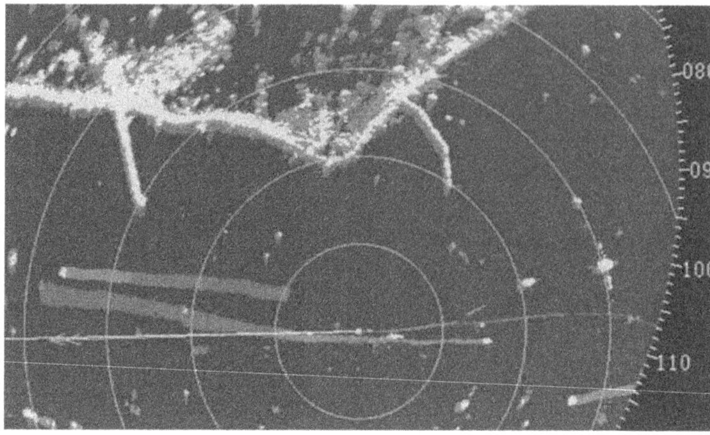

Figure 3.15 Portion of display. Target trails shown as varying tones in background. Original in full colour. Reproduced by permission of Kelvin Hughes Ltd, Ilford UK

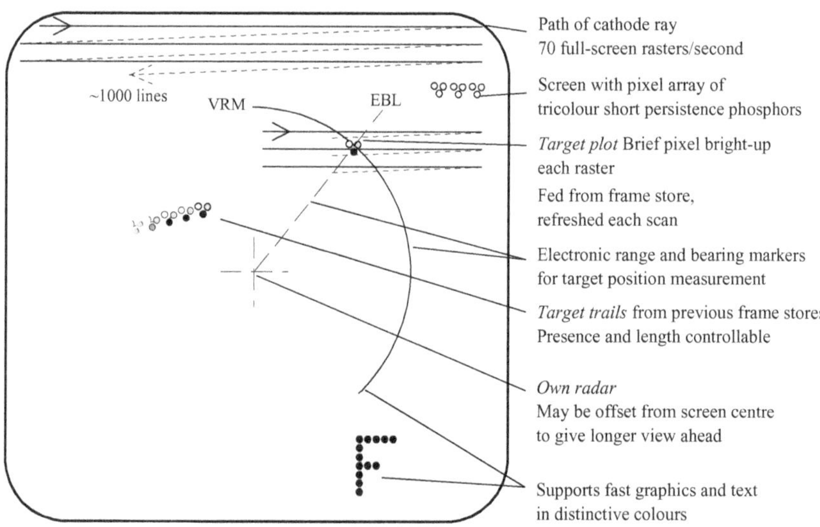

Figure 3.16 Raster display. Shown on 12 nmi range scale with 1000 pps. Target position is found by intersecting the variable range marker and electronic bearing line on its paint

of phosphor pixels in primary colours. Raster is Latin for rake. Operation is digital in character. The beam is steered in a quick zigzag path, raking the screen with about 70 complete rasters per second, like TV or PC displays, giving sufficient brilliance for viewing in full daylight. Use of a raster coarsens the basic R, θ footprint at close range. Kelvin Hughes have experimented with spiral scans to maintain footprint so detection cell area is proportional to range, but this has not so far fully caught on, perhaps because television display technology is unsuitable. Monochrome displays are sometimes used, see Chapter 2, Section 2.1.6, Figure 2.6. Oftener, full colour is used. The palette may be selected to suit daylight or night viewing conditions. The electron beams take up one of usually 16 or 32 digitally determined intensities, giving the pixel 16 or 32 grey scale (brilliance) values. The digital frame store data is converted from R, θ to Cartesian format by a digital scan converter. The scaling and raster pitch are sufficiently fine for the eye to perceive the display as a flicker-free continuous range of tones and hues. Monochrome displays usually use a green phosphor, least tiring to the eye. The phosphors fluoresce during the electron pulse, but do not phosphoresce afterwards – they have short persistence, to cope with dynamic navigational situations and evolving alpha-numeric messages.

To support text and chart data, and to avoid degradation of the echoes, radar displays need higher resolution than TV. The PPI itself occupies a circle or square field, up to 340 mm diameter, the whole field being 380 mm square as minimum on the bigger ships. The remaining space of the rectangular tube forms a convenient area for display of secondary alpha-numerics such as course and speed, control menus, etc. The pelorus is generated electronically on the display face, avoiding parallax when reading target bearings. Trackforming (generation of vectors) is performed electronically before paints are laid down, so any nonlinearities of scale caused by changes of deflection sensitivity near the rim of the tube do not affect calculations of CPA etc.

The tube grids permit beam current to flow and a bright-up of appropriate colour to be generated only when the raster is at an address whose target register cell contains a target, or where secondary information is to be displayed. Each frame store cell is addressed each raster, although of course its input is updated only once per scan in an R, θ manner. Scrutiny of a moving echo shows it to advance one step per scan. Target trails (see Section 3.10.1) are generated from the frame stores and are under the operator's control. Trails are shown in Figure 3.15.

Necessary resolution is around a couple of million pixels. For example on the 12 nmi (22 200 m radius) scale, pulselength may be 0.1 μs, equivalent to 15 m range cell. The diameter therefore contains $22\,200 \times \frac{2}{15} = 2960$ cells and ideally the tube would have this number of lines with the same number of pixels per line, pixel size being 0.13 mm × 0.13 mm for 400 mm (16 inch) tube width. In practice, the eye cannot resolve such small pixels at normal viewing distance and pixels are made about 0.2 mm × 0.2 mm. About 1000 lines × 1500 pixels per line are used for ships' radar. Some large VTS displays have higher resolution, partly to support many small alpha-numeric target legends or 'tags'.

Being reconstituted from stored and manipulated data, the display is synthetic, although there is a confusing tendency to reserve this term for tracks, terming plots as 'raw', even when they are the output of a processor. The screen is scanned frequently, giving enough brilliance for daylight viewing. The digital format of the input facilitates hooking-in secondary displays using local area network (LAN) techniques. At one time differing colours were used to indicate echo strengths. This practice has been discontinued, since strength is not navigationally significant. The phosphor persistence is too short for smearing when own ship manoeuvres or the range scale is changed.

Beside brightness, important features of raster scans include their ability to display in different colours, and quickly to amend:

- the radar picture (plots and tracks);
- text, enabling targets to be tagged with identification symbols (also operator's menus for controls);
- radar graphics such as predicted tracks (from ARPA, etc.), range and bearing markers for position measurement, target alpha-numeric identification tags;
- non-radar graphics such as charts (from ECDIS, etc.).

By lessening the operator's mental task, raster systems give much more consistent detection performance and free the operator for the tasks that no machine can replicate – intelligent decision-making.

3.10 Cursive display

3.10.1 Raw radar

Before digital computer technology evolved, echo plots were perforce displayed raw on a long-persistence monochrome CRT in PPI format, Figures 3.17 and 3.18. The display is polar, showing the received echoes in real time as the scanner turns, the beam (or 'strobe') being deflected by a motor-driven rotating coil or a fixed orthogonal pair of electromagnets, the latter facilitating off-centring of own ship's position. Long-persistence phosphors are available in few colours; many displays

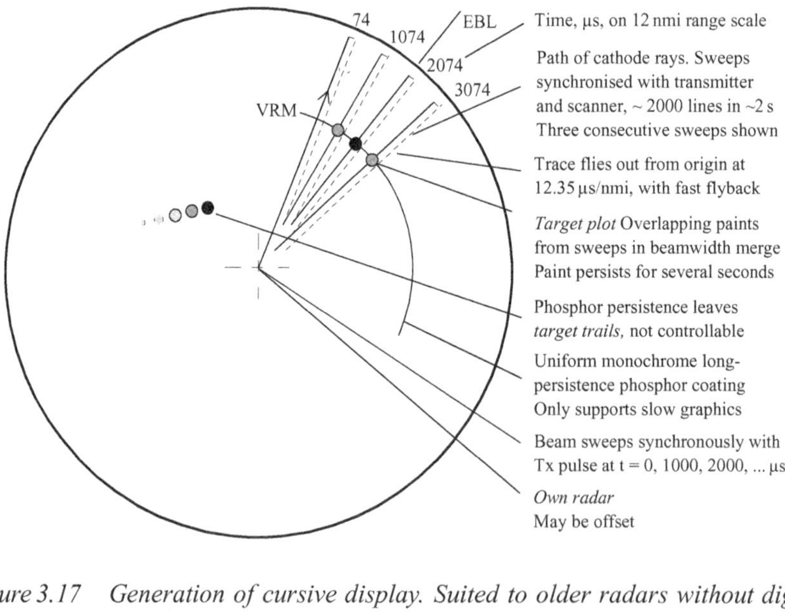

Figure 3.17 Generation of cursive display. Suited to older radars without digital signal processors

use green or orange. The phosphor fluoresces momentarily when struck by the beam, after which it continues to phosphoresce with an afterglow for several seconds, not necessarily in the same colour. Cascade screens combine long- and short-persistence phosphor layers with sealed-in optical colour filters.

In these older radars, each time the transmitter fires, the beam sweeps radially outward from own radar's position near the centre at a rate scaled to suit the transmission in a spoke-like or polar cursive manner, returns painting at scale range. Operation is wholly analog. There is no scan conversion and the screen is uniformly coated and not divided into discrete pixels. The sweeps rotate synchronously with the scanner, so the surveillance area is covered by a couple of thousand sweeps once per scan. The pelorus is engraved on a collar surrounding the display and is subject to parallax error, especially near maximum range, where curvature of the tube face is the greatest. Trackforming is generally performed manually using a reflection plotter, and any non-linearities of display scaling affect the trigonometry used in the calculation of CPA, etc.

The trace on later sets is made less dim on short-range scales by addition of a store from which the trace is repeatedly painted. The small ($\sim$7 dB) brightness range of long persistence phosphors is augmented by a soft limiter in the video amplifier to better indicate relative signal strength, helping show targets in precipitation clutter and noise.

Positive-going analog video bright-up pulses are applied direct to the CRT grid. Each pulse makes the phosphor at the scale location of the scatterer glow dimly, fading away after 20 s or so. If the bright-up is a clutter or noise spike, it is unlikely that the same point will be further brightened on succeeding sweeps or scans. But if the scatterer is a target, the same or an adjacent phosphor point is likely to be brightened on these sweeps or scans. The more hits, the brighter the phosphorescence; the phosphor integrates the returns sweep to sweep and scan to scan. Noise and clutter are not integrated and appear as dim speckles. The eye's excellent pattern recognition skill

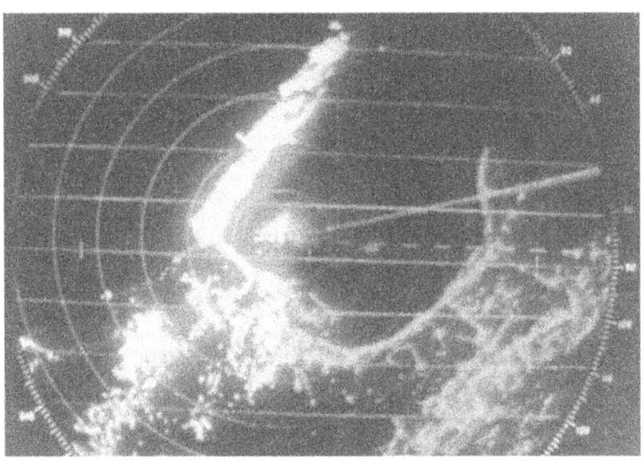

Figure 3.18 Cursive display. Brightness variation as strobe sweeps display can be mesmeric. Decca 9 GHz radar, 1.5 mile range scale, Inverness. Original monochrome orange. Author, 1983

then readily sorts echoes from noise, unless severe. Brightness is a measure of echo strength, within the limits set by the dynamic range of the phosphor, augmented by previous soft limiting applied to the video.

Thus, the tube cut-off grid voltage performs the threshold function and the phosphor integration characteristic provides pulse to pulse and scan to scan integration. The phosphor persistence maintains the display between scans and automatically provides afterglow trails of target track history, whether this information is wanted or not. Spot brilliance to some degree indicates echo strength. Minimum spot size is ~0.5 mm, which may slightly degrade the display resolution when short pulses are used with a big scanner on a long-range scale.

Radar sensitivity is optimised by biassing the grid just to extinguish the beam in absence of signal, adjusting the receiver gain to give faint background noise speckling, following the principle of setting false alarm rate to the maximum tolerable by the decision processor, here the operator's brain. The operator's mental pattern recognition capability is then used to sort the frequent noise or clutter spikes from faint but more persistent echoes. Long-persistence cursive displays well suited the earlier radars before modern computer technology was developed.

3.10.2 Cursive display problems

The scan causes a rotating strobe of background brilliance variation, all too soporific when no targets seem in sight. Moving targets automatically leave trails of diminishing brilliance, useful unless they mask other targets, but of course not under the control of the operator. The display smears badly when own ship yaws or manoeuvres, unless set to North-up presentation, and also takes half a minute or so to clear after change of range scale or off-centring. Display of tracks is rather unsatisfactory because minor scan to scan track variations due to noise cause smearing. A phosphor point is illuminated once only per scan, yielding insufficient brilliance for daylight viewing, necessitating either blackout curtains or a viewing hood which precludes two or more persons viewing together and disaccommodates the eye. Attempts at more brilliance lead to an objectionably bright 'initial flash'. Radars of the cursive era take feeds from the ship's log and compass, but otherwise tend to stand in proud isolation from other bridge equipment. Indeed, the earliest bulky equipments like the BTH RMS-1 came in their own special cabin, the operator telephoning voice-told plots to the bridge in the tradition of the crow's nest lookout. Later, a junior officer acted as radar observer at the back of the wheelhouse. The dim display and its daylight screening preclude one-man bridge operation.

Tracks are drawn by hand on an anti-parallax reflection plotter over the screen. The plotter is a semi-silvered transparent screen of opposite curvature to the tube face. Labour-intensive and permitting only a few targets to be followed, at least this chore focusses the operator's attention on nearby shipping movements. Display of alphanumeric data is difficult. One manufacturer (AEI) went so far as to provide a separate 400 mm short-persistence green alpha-numeric display, superimposed visually on an equally large orange long persistence radar tube, the pair being combined by a half-silvered mirror, hardly a compact arrangement. It was soon withdrawn.

Continued observation of a cursive system in strong clutter demands close concentration and is tiring. The eye takes time to accommodate to the dim display. Gain and brilliance controls must be kept in careful adjustment for good results. It is all too easy to take the lazy option – operate at high brilliance, low gain, fine for strong echoes, eliminating noise and clutter – unfortunately setting the grid bias threshold so high that small targets are eliminated as well.

3.10.3 Detection performance

Under ideal conditions, raw radar detection performance is little inferior to that of modern equipment and the same detectability calculations can be employed, with an operator correction to account for: (a) inappropriate control settings and (b) non-perception of targets displayed, perhaps weakly, on the screen. These human factors are difficult to quantify. They may be allowed for by assuming that the probability of detection threshold is high, or by addition of an operator loss term, L_{op}, in the radar range equation. The following has been adapted from a suggestion of Skolnik [3] but, as with all psychological effects, precision is impossible:

$$L_{op} \sim 10 \log \left(\frac{1.4}{P_D} \right) \text{ dB}. \tag{3.5}$$

For example, if desired probability of detection is 0.9 or 0.5, $L_{op} \sim 2.0$ or 4.5 dB, respectively.

3.11 Plots on the screen

Figure 3.19 depicts the main classes of target discussed in later chapters and encountered by marine and VTS radars, with an indication of the appearance of their plots. A short-sighted person using a searchlight on a clear dark night sees these targets in much better detail than can radar, which displays only those scatterers which reflect significant energy towards the scanner. No radar has remotely as good bearing discrimination as the eye and only frequency diversity is in any way analogous to colour vision. Small scanners, which cannot resolve to better than about 4°, display small targets as radial arcs several degrees wide, T on the figure. The best VTS scanners have about 0.4° resolution and reproduce small targets as crisp points.

However, unlike the eye, radar does accurately measure range. Most radars have similar range discrimination, between about 15 and 150 m according to pulselength. Poor angular resolution and detection cell size of marine radar, together with uncertainty as to which parts of the target are currently reflecting, usually preclude reliable indication of target aspect, the direction of vessel centreline, which tide streams, etc., may force off the radar track or course made good. This is unfortunate, for the Collision Regulations are written round aspect, conventionally determined by viewing side and masthead lights. Some VTS systems do have sufficient resolution to display aspect at short range, useful when the operator wishes to see a ship come off its berth, for example.

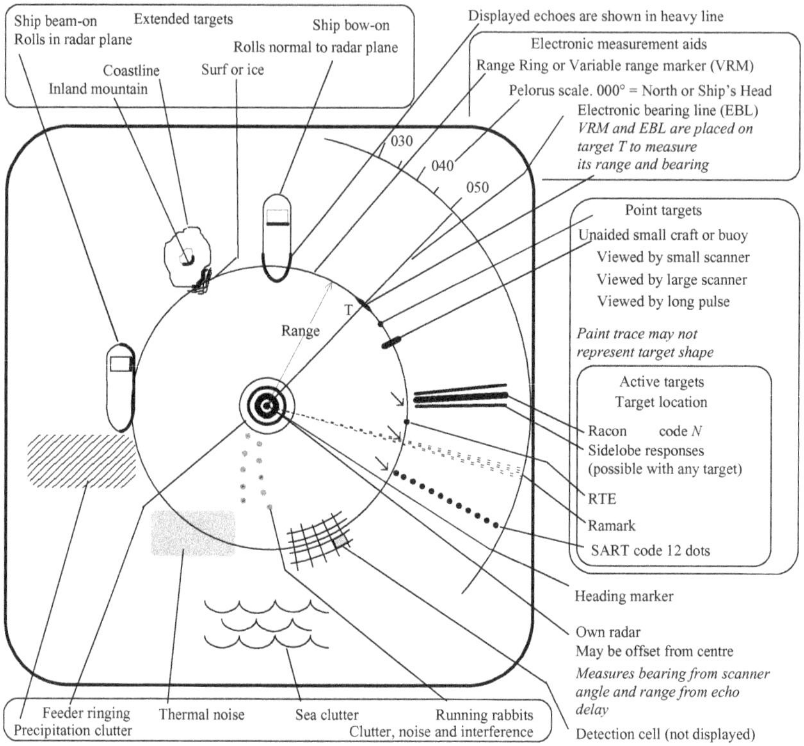

Figure 3.19 Radar display content, showing the different classes of target plot, clutter and noise. Detectability calculations differ for the passive and active targets shown. The heading marker shows the fore-and-aft line, not the course made good

3.12 Radars for special purposes

3.12.1 High speed craft

Fast ships find encounters with other vessels develop so rapidly – at several km/minute – that bearing precision is secondary to the need for low display information time lag or latency. HSC (high speed craft) radars are obliged to comply with specification IEC 60945-2. They use near-standard 3 or 9 GHz marine radars with lightweight scanners rotated fast – IMO minimum 40 rpm – so targets are updated as often as possible. Aperture is 2.7–3.6 m (9–12 ft) at 3 GHz or 1.2–1.8 m (4–6 ft) at 9 GHz; elevation beamwidths are 20°–25° and gains are around 25–28 dB on each band. The relatively low mounting height helps reduce sea clutter, aiding the essential task of detecting small targets at moderate range quickly. The low mounting is detrimental to long horizon range, which to these manoeuvrable craft is of minor importance. The relatively small scanner apertures counter the high rotation rate; number of pulses per beamwidth remains fairly conventional and so does the signal/(noise + clutter) ratio. The stringent requirement for range discrimination of 35 m laid down in the HSC Code necessitates short pulselength and many range cells.

3.12.2 Warships

The main surveillance radars carried by warships are dedicated to military tasks and not ordinarily used for navigation. Navigation radar is also carried. Although particularly robustly constructed to withstand combat shock and having some special features, particularly in the display area, it is in essence a civil marine radar, with similar transmission characteristics. Not only is this format best for the job and familiar to civilian pilots who may be taken from time to time; but it gives an identical signature to a merchant ship, enabling the warship to mingle unobtrusively into the marine environment without breaking military-radar silence or disclosing its identity to hostile SIGINT (signals intelligence). Warships on exercises in the Mediterranean are understood to have moved about undetected by 'hostile' forces for several consecutive days. Surfaced submarines manoeuvring in port areas may deploy on the sail a small portable 9 GHz radar, again with a civil marine radar emission signature.

3.13 Calibration

Radar failure may be catastrophic but is more often gradual, as the magnetron wears out after ~5000 h life, as the mixer diodes become less efficient after constant bombardment from mismatch reflections, as internal circuits drift out of tune or other adjustment and if the feeder suffers damage allowing water to seep in. Loss of performance can be too slow to be noticed by the operator, or new operators joining ship may be unaware of the radar's true potential. Besides arranging for regular servicing by qualified technicians, operators should make routine performance checks, note results and compare with previous records. Wylie [4] strongly advocates maintenance of a formal radar log in which operating experience is recorded, devoting a whole chapter to the topic. Some simple checks should be made regularly.

- Unless the sea is calm, some sea clutter should surround own ship when the radar is operated on full gain, long pulse, without differentiation.
- Any visible shipping or coastlines should be checked for echoes.
- If two radars are carried, loss of small close targets on one alone is a sure indicator of trouble.
- Built-in checking facilities should be used. These vary in complexity and the following summary is not exhaustive.

To confirm whether the transmitter is radiating full power and that receiver sensitivity is normal, a metallic chamber called an echo box was sometimes mounted near the scanner, usually just below. A colleague of the author attracted much ribald comment by trying copper float balls from water cisterns as echo boxes. Unfortunately, the equatorial joints proved poor conductors and loss was excessive. A small antenna such as a horn picked up a sample of radiated transmissions when the scanner came through echo box bearing, placed in a blind arc or astern. A small motor-driven paddle often broke up any frequency-dependent resonances to make the feather length independent of exact magnetron frequency. The box formed a multimode resonant cavity of fairly high Q factor. The microwave pulse rattled round within it, reflecting back and forth as in a mismatched feeder. The horn radiated the 'echo' for several microseconds as the reflections decayed. Some was picked up by the scanner and displayed as a performance feather or plume on box bearing. Length was a kilometre or so and gave a measure of system overall performance, including scanner and feed.

Nowadays the echo box is usually lightly coupled to the scanner feed and has preset electromechanical tuning to magnetron frequency. Its echo displays on all bearings as a sunburst around own ship; radius is a measure of the system excluding scanner and feed. For this, a small neon lamp is mounted on the static housing of the scanner drive. On each scan, near field spillover radiation should ionise the neon, causing a circuit to impress a plume at neon bearing, length indicating radiated power and acting as a transmitter monitor. The sunburst is referred to as a receiver monitor, although of course it also partially monitors the transmitter. Although these arrangements do not pretend to be precise, they have two great merits: they check the whole radar system including the scanner, and they are independent of the system. However, they cannot monitor the environment and loss of long-range performance due to adverse refraction, ducting or precipitation attenuation is not indicated.

Receiver sensitivity can be precisely calibrated by introducing wideband noise of known amplitude, the power needed to double existing noise power equalling the system noise power. Although included in some VTS systems, marine radars do not normally include such complexity.

VTS and coastal surveillance operators should identify a few weak static reference targets at moderate range and clear of blocks of clutter. Examples might include electricity pylons or lighthouses. Bearing, range and subjective echo strength should be noted as a routine. Range or bearing changes clearly indicate display error in the radar. Some strength variation inevitably occurs with weather conditions, tide height, etc., but persistently poor echoes may well indicate a radar malfunction. Buoys should not be used, particularly when subject to currents strong enough to set them over and change their RCS. Very distant targets are too much affected by atmospheric attenuation and refraction to make reliable calibration aids. It is harder for ships to make such checks, but liner trades might profitably make a point of listing a few calibration marks in each regular port of call.

3.14 References

1 MIROS, A. S.: 'Norway', reported in *Speed at Sea*, February 2002, p. 11

2 LEE, R. G.: 'Future bridge navigation', *Seaways, The International Journal of the Nautical Institute*, 2002, p. 9

3 SKOLNIK: 'Introduction to radar systems', Eq. (2.57)

4 WYLIE, F. J.: 'The use of radar at sea' (Hollis & Carter for the Royal Institute of Navigation, 5th edn.), chapter 13

Chapter 4

Echo strength in free space

'A little knowledge is a dangerous thing.'

Alexander Pope, An Essay on Criticism

4.1 Introduction

This chapter considers transmission between radar and target in an unbounded vacuum. In this hypothetical free space, environmental effects are ignored; there is no atmosphere, weather, Earth surface or other tiresome practicality. We shall derive basic forms of equations, the radar range equations, describing the energy reaching the target on the transmit leg and echoing back to the radar on the receive leg of its journey, on which later chapters will build when examining practical conditions including the environment. Application of free space equations directly to the real world without allowance for environmental effects often leads to gross errors, especially when range is long so rays traverse a lot of atmosphere. Throughout this chapter we assume the target remains of constant electrical size and is physically small enough to behave like a geometrical point.

In free space, energy radiated from a source such as a lamp or radar transmitter propagates in straight lines at the speed of light without loss; as explained in Chapter 2, Section 2.7. Doubling range, R, doubles the height and width of the beam and reduces the energy density to one quarter its original value, so energy density $\propto 1/R^2$. This inverse square law, which applies to light and sound as well as radar waves, is the reason why distant targets are illuminated weakly. Echoes from ordinary passive objects again suffer the inverse square law, making echo power reaching the scanner follow an inverse fourth power law, echo $\propto 1/R^4$. Doubling range, by an octave, reduces echo power to $\frac{1}{16}$ (-12 dB) its former value.

We now analyse transmission strengths more formally, at first ignoring radar losses and using numerical quantities, later changing to decibels.

4.2 Radiated power density

Chapter 2, Section 2.7.1, Eq. (2.6) showed that when a source radiates power P W isotropically at wavelength λ, power density, d, measured at range R m is $d = P/(4\pi R^2)$ W/m^2 of transverse cross section. If the isotropic radiator is replaced with a 'perfect' lossless directional antenna of beamwidths θ, ϕ, effective mouth or aperture area A m^2 and gain G (as we shall only be considering conditions on beam axis, we write G for G_{max} in this chapter), the energy is concentrated to produce a stronger power flux density over a smaller solid angle $\Omega = \theta\phi$ steradian. The basic antenna theory in Section 2.7.2 shows that on beam axis $G = 4\pi A/\lambda^2$ (Eq. (2.7e)). Including a transmitter loss L_t, and substituting in Eq. (2.6), power density on the transmit leg falls with range according to the inverse square law:

$$d = \frac{PG}{4\pi R^2 L_t} \text{ W/m}^2 \tag{4.1}$$

4.3 Passive reflector; radar cross section, radar range equation

4.3.1 Radar cross section

How well a passive object reflects energy back towards the radar depends on its size, shape, aspect angle to the illuminating radar and material (e.g. whether conductive metal or insulating plastic), as well as the radar wavelength and polarisation. A sphere has the unique property of looking the same from any aspect. Unlike almost all other shapes, its radar cross section (RCS), radar echoing area (REA) or [radar] cross-sectional area (CSA) is uniform, not fluctuating with the angle of aspect. Unusually, the proportion of incident energy reflected does not depend on wavelength, provided the sphere is in the 'optical region' where it is at least several wavelengths in circumference. These properties suit the sphere for use as a reference reflector and lead to the definition of RCS of an object: 'RCS is the proportion of incident energy reflected back in the direction of the source, relative to the proportion reflected by a replacement perfectly conducting sphere of cross-sectional area 1 m^2'.

Spheres reflect isotropically (in all directions). RCS, σ, equals the cross-sectional (silhouette) area:

$$\sigma = \pi r^2 \text{ m}^2. \tag{4.2a}$$

From the mechanism causing a sphere to reflect, detailed in Chapter 7, RCS can also be defined as

$$\sigma = \frac{4\pi \times \begin{array}{c}\text{(power per unit solid angle scattered in a specified direction}\\ \text{and polarisation)}\end{array}}{\begin{array}{c}\text{(power per unit area incident on the scatterer from the}\\ \text{specified direction and polarisation)}\end{array}}. \tag{4.2b}$$

RCS of practical targets often fluctuate, for example as they roll in sea waves. Fluctuation is considered in Chapter 12. Meanwhile, unless stated otherwise, we assume RCS and echo strength are the average values presented to the radar.

More complicated shapes have RCS only very roughly similar to their silhouette. For example, RCS of a flat conductive plate falls rapidly when turned oblique to the radar. Much time and computing power is needed to calculate RCS of practical targets of even moderately complex shape. RCS of point targets is discussed in Chapter 7, with ships and other extended targets in Chapter 10.

4.3.2 Two-way free space radar range equation

On the receive leg, total power reflected towards the source is

$$P_{\mathrm{refl}} = \sigma d = \frac{\sigma P G}{4\pi R^2 L_{\mathrm{t}}}\ \mathrm{W}. \tag{4.3}$$

The power density, d', reaching the scanner mouth is calculated by considering the target as a transmitter of power P_{refl}. Applying Eq. (4.1):

$$d' = \frac{P_{\mathrm{refl}}}{4\pi R^2} = \frac{\{\sigma P G/(4\pi R^2 L_{\mathrm{t}})\}}{4\pi R^2} = \frac{P G \sigma}{(4\pi)^2 R^4 L_{\mathrm{t}}}\ \mathrm{W\,m^{-2}}. \tag{4.4}$$

We denote the echo power delivered to the receiver by the scanner as $S_{\mathrm{e(FS12)}}$, subscript (FS) indicating free-space conditions and (12) that both transmit and receive legs are included. The signal fed to the receiver is aperture × density

$$S_{\mathrm{e(FS12)}} = A d'. \tag{4.5}$$

Substituting in Eq. (2.7e) for A, in Eq. (4.4) for d' and adding receiver losses L_{r}

$$S_{\mathrm{e(FS12)}} = \frac{G\lambda^2}{4\pi}\frac{P G \sigma}{(4\pi)^2 R^4 L_{\mathrm{t}} L_{\mathrm{r}}} = P G^2 \sigma \lambda^2 (4\pi)^{-3} R^{-4} [L_{\mathrm{t}} L_{\mathrm{r}}]^{-1}\ \mathrm{W}. \tag{4.6a}$$

This form of the radar range equation connects the radar parameters (such as transmitter power, scanner gain, received signal strength) with target and external factors (such as RCS and range). There are numerous variants. For example, the equation can be written to give R in terms of receiver bandwidth, noise figure, probability of detection, etc. To be exact, Eq. (4.6a) is one form of the two-way radar range equation for a passive target in free space. As expected from considerations of conservation of energy, it confirms Section 4.1 that echo power follows an inverse fourth power (R^{-4}) law. Note that for a given signal strength $S_{\mathrm{e(FS12)}}$, operation at longer range demands an increase of radar parameters P or G, or of target size σ:

- doubling P or σ increases R by $2^{1/4} = 19$ per cent
- doubling G (e.g. by doubling scanner area A, say by doubling aperture width) increases R by $2^{1/2} = 41$ per cent, because the benefits of high gain accrue on both transmit and receive legs.

Eq. (4.6a) can be written in terms of aperture area, A, rather than gain:

$$S_{\mathrm{e(FS12)}} = \sigma A^2 P (4\pi)^{-1} \lambda^{-2} R^{-4} [L_{\mathrm{t}} L_{\mathrm{r}}]^{-1}. \tag{4.6b}$$

This highlights that for constant aperture, doubling wavelength λ reduces R by 41 per cent for a given signal strength. In practice, σ tends to fall at long wavelength and multipath further also affects performance.

The range equation predicts that halving scanner gain, G, reduces echo strength to one quarter. However, low G means more solid angle is illuminated, in general raising azimuth beamwidth, giving more echoes per scan. Chapter 3, Section 3.6, indicated that this improves signal to clutter ratio, partly recouping the reduction in echo strength. We shall encounter several similar secondary effects through the book. The range equation merely indicates echo strength, not echo detectability – which also involves competition from noise and clutter – and is just the first step of the road leading to calculation of whether a given target will be seen by a particular radar under particular circumstances.

In basic radar theory, detectability is ultimately dependent on the mean power transmitted, P_m, independent of the form of modulation. The reasoning is as follows. We rewrite Eq. (4.6a) as

$$R^4 = \frac{PG^2\sigma\lambda^2(4\pi)^{-3}[L_t L_r]^{-1}}{S_{e(FS12)}}. \tag{4.6c}$$

We replace $S_{e(FS12)}$ by S_1, the minimum single pulse which can be detected and put $S_1 = nkTBs$ with the meanings of Chapter 3, Section 3.3.2, Eq. (3.2a); s = minimum SNR for detection. If there are N pulses in the packet and these are integrated within a coherent system, the minimum packet power which can be detected, S_N, is inversely dependent on N:

$$S_N = \frac{S_1}{N} = \frac{nkTBs}{N}.$$

If pulselength $= \tau$ and pulse repetition frequency $= F$, and assuming a matched filter so $B = 1/\tau$,

$$P_m = P\tau F, \qquad \text{so } P = \frac{P_m B}{F}.$$

Substituting in Eq. (4.6c) for maximum detectable range R_{max}

$$R^4_{max} = \left[\frac{P_m \boldsymbol{B}}{\boldsymbol{F}}\right] \boldsymbol{G^2\sigma\lambda^2(4\pi)^{-3}[L_t L_r]^1} \left[\frac{\boldsymbol{N}}{\boldsymbol{nkTBs}}\right].$$

The terms in bold type are independent of the form of modulation, and $N \propto F$, so the B, F and N terms cancel and

$$R^4_{max} \propto \frac{P_m G^2\sigma\lambda^2(4\pi)^{-3}[L_t L_r]^{-1}}{nkTs}, \tag{4.6d}$$

which is dependent on the mean power but does not contain terms in P, F, τ or B which depend on the form of modulation. Whether P_m (for marine radars

$P_M \sim 10\,\text{W}$) is radiated continuously as in a broadcast radio transmitter or is concentrated in short pulses as in marine radar is basically a matter of practical convenience. It is possible to make continuous-wave radars, primarily measuring radial velocity by Doppler frequency shift, rather than range by time delay as in marine radar. Range is then determined by the phase shift of a video-frequency amplitude or phase modulation superimposed on the transmission. Such radars are ill-suited to marine use and will not be discussed until Chapter 16. Nevertheless, it is well to be aware that the current marine/VTS radar format is neither essential nor some unique arrangement ordained by heaven. The format could be replaced should need be, for example, as pressure intensifies on the electromagnetic spectrum. Indeed, telecommunications users might like to consign pulse radars with their extreme EIRP (equivalent isotropic radiated power, Section 4.5.2) to a place far from heaven.

4.4 Active target

Active targets receive signals from the radar, process them and retransmit using some electronic device, Chapter 8. They have a receiving antenna, aperture A_r and gain G_r. Published parameters of active targets include their internal losses, so these do not feature separately in the range equation. Using similar arguments to those leading to Eq. (4.6b), the one-way radar range equation between a transmitter and receiver in free space is:

$$S_{\text{tgt(FS1)}} = A_r d = \left\{\frac{\lambda^2}{4\pi}\right\} \frac{PG}{4\pi R^2 L_t} = \frac{PGG_r\lambda^2}{(4\pi R)^2 L_t}\ \text{W}. \tag{4.7a}$$

This equation follows the inverse square law. We can adapt it for the response leg, from the target of transmitter power P_t, transmitting antenna gain G_t and no losses. We include radar receiver loss, to give the signal strength at the radar receiver from the response transmission. It also follows the inverse square law:

$$S_{\text{r(FS2)}} = \frac{P_t G_t G \lambda^2}{(4\pi R)^2 L_r}\ \text{W}. \tag{4.7b}$$

4.5 Range equations in practical form

4.5.1 *Extensions for practical environment*

The next chapters will examine environmental effects, showing the atmosphere introduces a loss, L_A, dependent on range and that multipath interference modifies the free space signal by a multipath factor M in a range-dependent manner. These effects apply to each of the radar-to-target and return legs. In free space L_A and $M = 1$ numerically (0 dB, no loss). It is convenient to introduce the terms here although we cannot yet evaluate them. We drop the free space (FS) subscript as the real environment is intended and convert to the decibel forms frequently used in later chapters.

4.5.2 Full radar range equation, dB

When the terms are all expressed in dB or dBW, Eqs (4.6) and (4.7) become:

Practical two-way. Echo received at radar from passive target:

$$\begin{aligned} S_e &= S_{e(FS12)} - 2L_A + 2M \\ &= \mathbf{P + 2G + 20\log\lambda - 30\log(4\pi)} + \sigma \\ &\quad - 40\log R - L_t - L_r - 2L_A + 2M \text{ dBW.} \end{aligned} \quad (4.8)$$

EIRP, P_{EIR}, is the transmitter power needed to give the same flux density if the directional scanner were replaced by an isotropic radiator of unity (0 dB) gain and is a measure of the radar plus scanner's ability to direct power to the target.

$$P_{EIR} = P + G - L_t \text{ dBW.} \quad (4.9)$$

Eq. (4.6a) becomes:

$$S_e = \mathbf{P_{EIR} + G + 20\log\lambda - 30\log(4\pi)} + \sigma - 40\log R - L_r - 2L_A + 2M \text{ dBW.} \quad (4.10)$$

Practical one-way. Interrogation received at an active target (Eq. (4.7a)) is

$$\begin{aligned} S_{tgt} = S_{tgt(FS1)} - L_A + M &= \mathbf{P + G} - L_t + \mathbf{20\log\lambda - 20\log(4\pi)} + G_r \\ &\quad - 20\log R - L_A + M \text{ dBW} \end{aligned} \quad (4.11a)$$

or

$$S_{tgt} = \mathbf{P_{EIR} + 20\log\lambda - 20\log(4\pi)} + G_r - 20\log R - L_A + M \text{ dBW.} \quad (4.11b)$$

Practical one-way. Response received at radar from an active target (Eq. (4.7b)) is

$$\begin{aligned} S_r &= S_{r(FS1)} - L_A + M \\ &= P_t + \mathbf{G + 20\log\lambda - 20\log(4\pi)} + G_t - 20\log R - L_r - L_A + M \text{ dBW} \end{aligned} \quad (4.11c)$$

or, where P_{tEIR} is the target EIRP ($P_{tEIR} = P_t + G_t$ dBW):

$$S_r = P_{tEIR} + \mathbf{G + 20\log\lambda - 20\log(4\pi)} - 20\log R - L_r - L_A + M \text{ dBW.} \quad (4.11d)$$

4.5.3 Reduced equations

The radar terms shown bold in the above equations can be lumped together as a figure of merit, F. Denoting range in km units, R_{km}, to avoid inconveniently large numbers, F_{12} is the free-space echo at the radar from a target with unit RCS ($\sigma = 1\ \text{m}^2$) at 1 km

($R_{km} = 1.0$). F_{12} is a figure of merit for the radar over transmitting and receiving legs. The 120 term reflects use of km, $1000^{-4} : 1 = -120$ dB:

$$F_{12} = P + 2G + 20\log\lambda - 30\log(4\pi) - 120 - L_t - L_r$$
$$= P + 2G + 20\log\lambda - 153 - L_t - L_r \text{ dBW}. \tag{4.12}$$

Using figure of merit, Eq. (4.10) becomes

$$S_e = F_{12} + \sigma - 40\log R_{km} - 2L_A + 2M \text{ dBW}. \tag{4.13}$$

Similarly, for one-way transmission, we get a reference equation for the interrogation reaching the target, where F_1 is a radar transmitter figure of merit. The −60 term reflects conversion to km.

$$S_{tgt} = F_1 + G_r - 20\log R_{km} - 60 - L_A + M \text{ dBW}, \tag{4.14}$$

$$F_1 = P + G + 20\log\lambda - 20\log(4\pi) - 60 - L_t = P + G$$
$$+ 20\log\lambda - 82 - L_t \text{ dBW}. \tag{4.15}$$

Where F_2 is the radar receiver figure of merit, the response at the radar is:

$$S_r = F_2 + P_t + G_t - 20\log R_{km} - L_A + M \text{ dBW}. \tag{4.16}$$

$$F_2 = G + 20\log\lambda - 20\log(4\pi) - 60 - L_r = G + 20\log\lambda - 82 - L_r \text{ dBW}. \tag{4.17}$$

The radar range equation includes the following values worth noting:

$$20\log\lambda \sim -20 \text{ 'dB'} (3\text{ GHz band}, \lambda \sim 0.1\text{ m}),$$

or

$$20\log\lambda \sim -30 \text{ 'dB'} (9\text{ GHz band}, \lambda \sim 0.032\text{ m})$$

$$10\log(4\pi) \sim 11.0 \text{ 'dB'}.$$

4.6 Calculations and graphs

The range equation is so important that it is worthwhile illustrating how it is used for calculation of echo strength and graphical depiction of variation of echo strength with range.

4.6.1 Fixed range example

Suppose we wish to find the free-space echo from a navigation buoy of RCS 10 m^2 at 10 km range received by a typical deep-sea ship's 9 GHz radar having: $f = 9400$ MHz giving $\lambda = 3.19$ cm, 25 kW transmitter pulse power, scanner gain 1260, total transmitter loss 4 dB, total receiver loss 5 dB.

For comparison we set out a range budget, with linear and decibel alternatives, Table 4.1. Splitting the decibels into positive and negative columns helps avoid minus sign blunders, so easy with hand calculation. This method is well suited to initial 'order of magnitude' tests to check feasibility of a proposed system, or as a cross-check for gross errors within more detailed calculations – computers are very accurate but work on the garbage in, garbage out principle, never questioning inputting blunders.

For this radar $F_{12} = -85.9$ dBW. Multiplying and dividing numerical values, $S_{e(FS12)} = 2.572 \times 10^{-12}$ W. Adding and subtracting decibel equivalents (expressed to nearest 0.1 dB), $S_{e(FS12)} = -115.9$ dBW, which converts by the methods of Chapter 2, Section 2.1.7 to 25.7×10^{-12} W. The difference is a trivial rounding error of 0.1 per cent.

The calculated echo power of less than a hundred-thousandth of a microwatt would probably be detectable in free-space conditions, but environmental effects such as Earth curvature might prevent detection in practice.

4.6.2 Graphs

Calculations as Table 4.1 for a spot range do not reveal the full story. It is worthwhile preparing a graph of signal strength versus range to indicate whether signal is changing rapidly and so might be unduly sensitive to parameter variations.

The 'obvious' way to do this is as follows.

1. Draw linear range (abscissa, X-axis) and power (ordinate, Y-axis) scales on a sheet of ordinary linear graph paper.
2. Calculate (as Table 4.1) echo strength at several spot ranges.
3. Plot the echoes as a series of points at the proper ranges and powers.
4. Insert the minimum detectable signal curve.
5. Draw a smooth curve through the points, freehand or by French curve.

The result is Figure 4.1. The echo curve falls sharply at short range and less sharply as range increases. We know from Eq. (4.6a) that it has the form power $\propto R^{-4}$, but this is by no means obvious from viewing the graph.

The echo power curve in itself is only part of the story, which is why a curve of the radar's minimum detectable signal (MDS), the minimum echo strength needed for detection, has been added. For simplicity, MDS is here assumed 0.5×10^{-12} W irrespective of range; in practice it would vary. At short range, the echo exceeds MDS and the target is detected. At long range, the echo falls below MDS and is undetectable. The intersect gives maximum free-space detectable range or first detection range, which can be scaled off the graph as 15.1 km.

For most purposes this presentation is tiresome; unless the range bracket of interest is less than about 2 : 1 the curve clings close to the axes and is difficult to read. Changing to a linear decibel power scale – equivalent to logarithmically in watts –in

Table 4.1 Range budget by linear and decibel methods

Free space				
Quantity	Numerical value	dB (pos)	dB (neg)	Notes
Dimensionless except as stated				
Equation	*Eq. (4.6a)*	*Eq. (4.8)*	*Eq. (4.8)*	
P	25 000 W	44 dBW		
G^2	1 587 600	62 dBi		$1260^2 = 1\,587\,600$; $2 \times 31 = 62$
λ^2	0.00102 m^2		−29.9	$(0.0319)^2 = 0.00102$
Conversion	$(1/4\pi)^3/10^{12}$		−153.0	Negative dB as < 1
$1/L_t$	1/2.51		−4.0	
$1/L_r$	1/3.16		−5.0	
Operation	*Multiply*	*Add*	*Add*	Risk of decimal point error when using numerical values
		106	−191.9	*Add dB (+ve) and dB (−ve) columns*
F_{12}	**2.572×10^{-9}**	↘	**−85.9**	-85.9 dB $= 2.5704 \times 10^{-9}$ There is a small rounding error
σ	10 m^2	10 dB m^2		
$1/R^4$ (10 km)	$1/10^4$		−40	
		10	−125.9	*Add dB (+ve) and dB (−ve) columns*
	Multiply	↘		
$S_{r(FS12)}$	**2.572×10^{-12} W**	**−115.9 dBW**		$= -120 + 4.1$ dBW $= 10^{-12} \times 2.570$ $= 2.57 \times 10^{-12}$ W

Figure 4.2 opens things up a bit, but some disadvantages remain:

- the curve is tedious to calculate and draw
- the law relating echo strength to R is not revealed
- the curve becomes very cramped at short range
- it is not easy to change a parameter, for instance target RCS or transmitter power.

Figure 4.3 overcomes the problems. As in Figure 4.2, power is scaled linearly in dBW. But range is now also to a logarithmic scale, so each cm now represents a certain range ratio. This linearises the echo strength line, which falls −40 dB (power ratio $1:10^{-4}$) when range is increased by a factor of 10, immediately revealing the R^{-4} law ($S \propto R^{-4}$) followed by the echo.

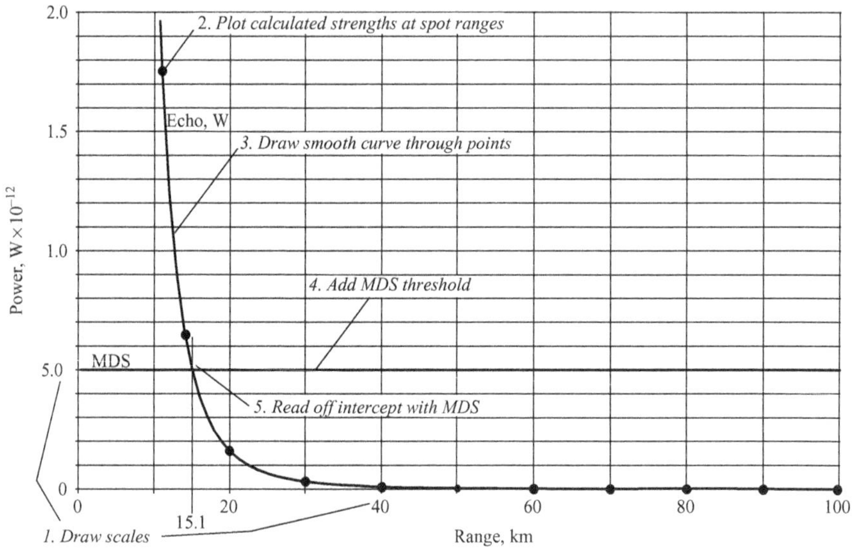

Figure 4.1 *Echo strength versus range, linear power and range scales. Echo strength has to be plotted from a series of values calculated at judicious ranges. Note how rapidly power falls at short range. Law is not obvious and short ranges are cramped. The maximum detectable range intersect at 15.1 km is difficult to determine accurately*

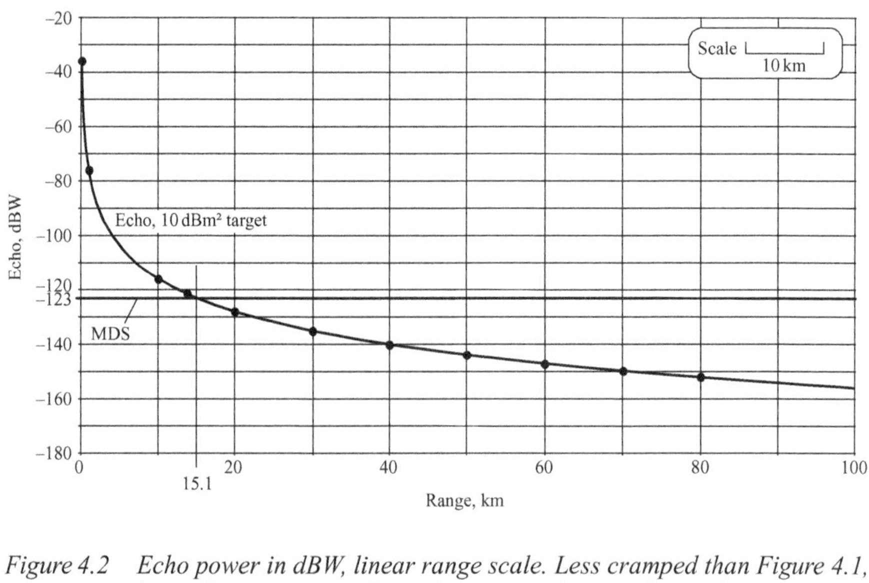

Figure 4.2 *Echo power in dBW, linear range scale. Less cramped than Figure 4.1, but still inconvenient. Draughting procedure similar to Figure 4.1.*

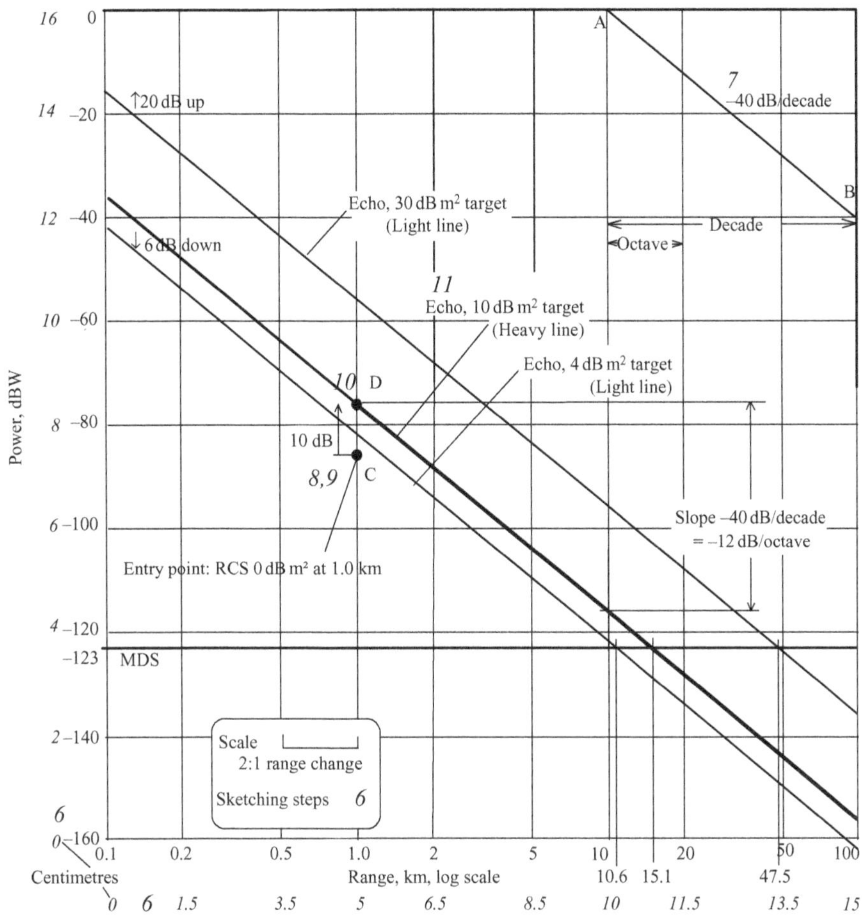

Figure 4.3 Logarithmic range scale, otherwise as Figure 4.2. Simple to draw and instructive to use. Linearises echo/range law, revealing that echo is changing −40 dB/decade and so following R^{-4} law. Easy to assess changes to parameters, for example changed RCS, light lines

The procedure is easy, accurate and quick, with only a single calculation.

6. If log graph-paper is not at hand, it is easy enough to mark off plain paper with a range axis. If interested in ranges between 0.1 km and 100 km, which is three decades, with graph width 15 cm to give 5 cm/decade, mark 0.1 km at the start, 0.2 km at 1.5 cm ($5 \log 2 \sim 1.5$), 0.5 km at 3.5 cm ($5 \log 5 \sim 3.5$), 1.0 km at 5 cm, and so on to 100 km at 15 cm as shown in italics.

As with Figure 4.2, mark the vertical axis as echo strength linearly in decibels to a convenient scale, say 1 cm per 10 dB between say −160 dBW at the foot of the page and 0 dBW near the top. Few radars will detect signals weaker than −130 dBW, but it is convenient to take the graph down further to aid construction of the curve.

7. Draw construction line AB in a corner at slope −40 dB/decade (R^{-4} law).
8. Calculate F_{12} per Eq. (4.12) or Table 4.1 with $R = 1000$ m and $\sigma = 0$ dB m^2. This gives the reference echo from a 0 dB m^2 (1 m^2) target at 1 km.
9. Enter on graph (point C).
10. Enter point D, σ dB above C (here 10 dB), to represent the echo at 1 km from the 10 dB m^2 target.
11. Draw a straight line parallel to AB (a navigator's parallel rule is handy) through D, produced to the sheet borders. The curve, of the form $y = x^z$, becomes linearised with slope z, so $\log y = z \log x$. This is the echo from a 10 dB m^2 target in free space.

The graph is no longer cramped. To examine the effect of changing a parameter, we merely slide the curve vertically up or down by the appropriate number of dB. For example, an echo of a ship of RCS 30 dB m^2 (1000 m^2) lies a further 20 dB up and a dinghy of RCS 4 dB m^2 (2.5 m^2) lies 6 dB down from our 10 dB m^2 buoy (light lines). The intercepts with MDS indicate the ship has maximum free-space detectable range 47.5 km and the dinghy 10.6 km.

It is equally easy to judge the effect on detection range of changing MDS, e.g. when using another pulselength/receiver bandwidth combination. One merely slides the MDS line up or down by the appropriate number of dB.

4.6.3 *Computer spreadsheet and charting*

Using a personal computer (PC), life is easy. All we do is compile a simple spreadsheet with cells for each of the terms of whichever of Eqs (4.8)–(4.11) suits the system, using either numerical or dB forms. A column of a hundred or so ranges, incremented from the lowest to the highest of interest, is followed by a column containing the chosen equation. The PC can then automatically 'chart' the relationship. Most Windows computers have EXCEL spreadsheet facility, which permits computation of logarithms, allowing quick numerical/dB swapping. And the chart can scale itself logarithmically, so if we produce a Figure 4.1 lookalike which turns out to be cramped, a Figure 4.3 version can quickly be generated. To try the effect of changing a parameter, we merely enter the revised value in the appropriate cell. The signal strength graphs in this book were generated in this way. The IEE website (www.iee.org) includes comprehensive spreadsheets which include environmental effects. Chapter 14 gives full details. It is all too easy to enter an incorrect value in one of the many cells necessary to quantify all the environmental, radar and target variables, and it is good practice to make rough sketches similar to Figure 4.3 as a check for gross error.

4.7 Limitations of free space formulae

The free space forms of the radar range equation are beguilingly easy to apply. They ignore the environment – factors such as the horizon, precipitation attenuation, multipath and clutter, which are often imperfectly known. The free space equations are a step to development of more realistic but necessarily more complicated expressions which include these factors. Later chapters show the free space range equations can sometimes understate performance. More importantly, they often very seriously overstate system performance because of environmental effects inseparable from practical marine conditions, particularly at long range. The next chapter starts to consider the way the environment affects echoes.

Chapter 5

Environmental effects on propagation

'The Earth, the air, the sea, the snow; the rain, the fog, the winds that blow.'

Anon

5.1 Scope of chapter

The previous chapter looked at a radar and target in unbounded free space. In reality, the space traversed is bounded by the sea surface, which acts as an imperfect mirror. Oblique incident energy is forward-reflected specularly (Chapter 2, Section 2.1.1, Figure 2.1) with an efficiency which depends on surface smoothness. Phase is also shifted at the point of reflection, which is called the grazing point. Arriving via this longer route some nanoseconds later than the direct ray, the indirect rays are delayed in phase as well as being reduced in amplitude. Constructive or destructive interference results from the system geometry and modifies the signal strength at the target.

The sea surface follows the curvature of the Earth, introducing a horizon, whose range depends on the heights of the scanner and target, and on the refractive index variation with height of the atmosphere. At long range, multipath interference is replaced by a diffraction mechanism which carries some energy to targets beyond the horizon. Heavy seas may physically obscure or screen the target from time to time, reducing the maximum available probability of detection. This effect is discussed in Chapter 12.

The atmosphere is lossy, particularly during precipitation. Near the end of this chapter, we develop the atmospheric loss term L_A to account for the attenuation of air, water vapour and precipitation in the radar–target path.

The processes recur as the echo returns to the radar. The overall result is usually to reduce detection range so much that performance calculated under free space conditions is hopelessly optimistic.

This chapter equips us to develop multipath factors M to describe the differences from free space propagation other than atmospheric loss, in Chapter 6 for point targets and in Chapter 9 for extended targets. Insertion of atmospheric loss and multipath factor convert the radar range equation from free space to real conditions, allowing echo strength at all ranges to be calculated as accurately as our knowledge of the environmental conditions permits.

Precipitation and the sea surface also scatter energy. That returning in the direction of the scanner as unwanted clutter is the subject of Chapter 11. The environmental effects are all dynamic. Movement of hydrometeors in the atmosphere and of sea-waves cause random signal fluctuations, discussed in Chapter 12. In this chapter we consider mean values.

Although it is easy to see when the sea is rough or there is precipitation, refraction is normally invisible and its insidious effects are too often ignored. We therefore make no apology for looking at refraction and its effects in close detail in the following two sections.

5.2 Atmospheric refraction

5.2.1 The problem

The dielectric constant, ε, of a vacuum, such as free space, is defined as 1.0. (This is the relative dielectric constant, relative to the absolute dielectric constant, properly called the electric constant, ε_0, $\sim 8.854 \times 10^{-12}$ F/m. (Farads per metre)) The air molecules and water vapour in the atmosphere very slightly increase the dielectric constant, increasing the refractive index, n ($n = \sqrt{\varepsilon}$) to a value a few hundredths of 1 per cent above 1.0. This very slightly reduces the velocity of propagation from the free-space velocity of light, c, to c/n, reducing the wavelength. If the refractive index were uniform, radar rays would continue to travel in straight lines, just as they do in free space, and the phenomenon would be of trivial importance. However, the rays between marine scanner and target traverse the bottom few tens of metres of the atmosphere. Here temperature and moisture content vary with height, making refractive index also change with height, curving the rays in the vertical plane by an amount dependent on meteorological factors and on range, as indicated in Figure 5.1(a). Refraction therefore affects the ranges at which events such as the horizon occur. At very short range refraction and the associated curvature can usually be ignored, just as Earth curvature is ignored when using plane trigonometry to solve short-range navigational problems. At long range, the relative phasing of direct and indirect rays is very significantly affected, and detection range of distant targets can vary by a factor of two or more as the weather changes from hour to hour. Propagation through the atmosphere is important to radio reception, particularly at the lower frequency broadcast bands and has been closely studied since the 1920s, well pre-dating radar. Air masses are sometimes identified by the thermal and moisture properties of the source region: Antarctic (AA), arctic (A), polar (P) and tropical (T); continental (c) relatively dry, maritime (m) relatively moist.

We assume the refractive index, n, of a lamina sheet of atmosphere of constant height h_a above sea level remains constant through the whole scanner/target range bracket. (This usually reasonably represents reality although local inhomogeneities can arise, causing reflection in the form of distributed angel echoes, a form of clutter.) Laminae of differing heights, h_a, have differing refractive index.

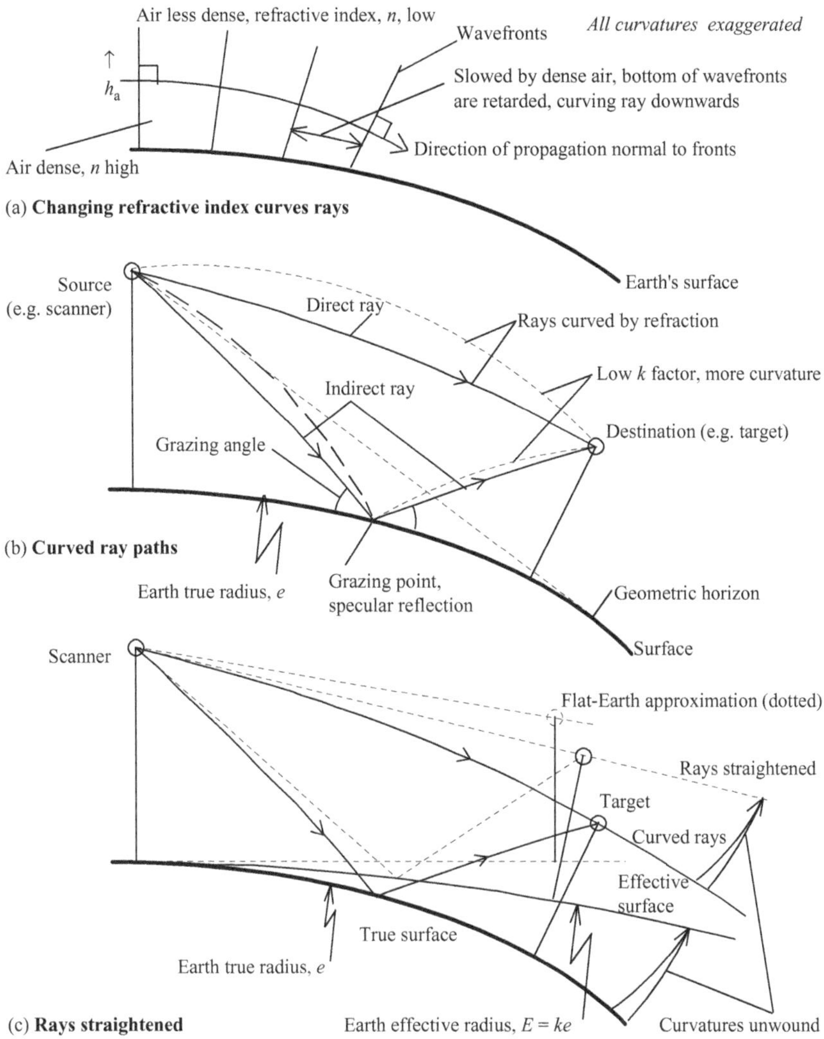

Figure 5.1 *Refraction and ray paths. Dense, moist air near the surface slows the lower portions of the rays, curving them downwards. Path calculations assume straight direct and indirect rays above an Earth whose radius is assumed to be k times actual (c)*

As mentioned in Chapter 2, Section 2.4.1, Eq. (2.3), wavelength, λ, varies with frequency, f, velocity of propagation, c, and refractive index, n:

$$\lambda = \frac{c}{fn} \text{ m.}$$

For a given frequency, raising n from 1 by inserting the atmosphere reduces λ, the distance travelled per cycle, hence reducing the velocity of propagation below the free space value. A trivial range error ~300 mm/km arises, much less than other sources of range error such as detection cell quantising, and readily corrected by a small adjustment to the display scaling factor.

Much more important than the absolute value of n is its rate of change with height, the refractive gradient, caused by changes of density and moisture content with height. The refractive gradient curves the rays by the same order of magnitude as the curvature of the Earth. At high altitude and over land masses, the refractive gradient is reasonably uniform and unvarying. Unfortunately, within the first few tens of metres above the sea surface, meteorological conditions strongly affect the gradient, in turn varying ray curvature. Neither refractive index nor its gradient is easy to measure.

Refraction of course also affects the indirect ray between source and destination and therefore significantly influences the interference between direct and indirect rays.

5.2.2 *Equivalent geometries*

We now quantify these statements. To accord with meteorological usage, some expressions use km rather than metres. In some textbooks, definitions of certain terms differ between km and metres, introducing powers of 10^3. We let:

h_a = height of lamina of atmosphere in question, km

n = refractive index of this lamina

dn/dh_a = refractive gradient (rate of change of refractive index with respect to height, km)

ρ_r = ray curvature, rad/km

e = true Earth radius, 6.371×10^6 m

e_{km} = e expressed in kilometres, 6371 km

E = effective Earth radius

k = effective Earth radius factor

N = radio refractivity

p = barometric pressure, hPa (hectopascal, numerically identical to and superseding the millibar (mbar). 1 hPa = 0.7473 mm of mercury)

T = absolute temperature, K

w = partial pressure of water vapour, hPa (usually written e, but we prefer w to avoid confusion with dielectric constant and Earth radius).

The algebra describing the curved ray paths of Figure 5.1(b) is difficult. It is usual to shorten the odds by an approximation which assumes the rays travel in straight lines above an Earth whose hypothetical radius, E, differs from the true radius, e, by an effective Earth radius factor, Earth radius correction factor or (atmospheric) refraction factor, k:

$$k = \frac{E}{e}. \tag{5.1}$$

This approximation is shown in Figure 5.1(c) (for $k \sim 2$). The geometry underlying Eq. (5.1) is valid because the ranges and heights used in marine radar are very much less than the Earth's radius. The process is akin to unwinding the curvatures until the ray path is linear. Factor k is a pure number and is independent of radar frequency. However k differs at optical frequencies, which are not affected by moisture.

A simpler approximation, discussed more fully in Section 5.6, assumes the Earth is flat and the rays travel in straight lines. This flat-Earth approximation (indicated in Figure 5.1(c)) is tantamount to putting $E = \infty$, with $k = \infty$ also. It is valid under some circumstances and we shall use it from time to time, but is inaccurate at long range.

5.2.3 *Calculation of refraction factor from meteorological parameters*

The following treatment is based on Hall *et al.* [1] and on Skolnik [2]. It enables factor k to be calculated in the rather unlikely event that the meteorological parameters in the atmospheric volume traversed by the rays are accurately known. Meteorological measurements taken at a shore station may not reliably apply above the water surface off a coast. Especially near a coast, k may also change along the scanner/target path, vitiating our calculations.

As already noted refractive index depends on the dielectric constant:

$$n = \sqrt{\varepsilon}. \tag{5.2a}$$

For the lower troposphere (the atmosphere up to 11 km), $n \sim 1.0003$. To magnify the small changes of n which are of interest, we use a radio refractivity, N:

$$N = (n - 1) \times 10^6. \tag{5.2b}$$

When, say, $n = 1.000315$, $N = 315$, a typical mid-latitude value. Refractivity at the sea surface is sometimes written N_s. The rate of change is also magnified so

$$\frac{dN}{dh_a} = 10^6 \frac{dn}{dh_a}. \tag{5.2c}$$

The value of N depends on the partial pressure of the atmospheric water vapour, w; absolute temperature, T, and barometric pressure, p, which are meteorologically inter-related (Reed and Russell [3]):

$$N = 0.373 \times 10^6 \left(\frac{w}{T^2}\right) + 77.6 \left(\frac{p}{T}\right). \tag{5.3}$$

The contribution of water molecules gives the first, 'wet', term, which is usually dominant. Nitrogen and oxygen molecules give the second, 'dry', term. The constants were empirically derived from observations. Except in very calm weather, either the wind or waves make the air turbulent, moisture plucked from the sea surface being stirred into the bottom few metres of the air column, moistening it. Usually, the following conditions apply.

- The air gets drier at height.
- The rate of change of p with h_a is negative, p falling exponentially to 35 per cent of the surface value at a scale height, H, around 8 km and not varying much with the weather.
- The rate of change of T with h_a is negative, except in fog, typically falling at 1°/100 m, the dry adiabatic lapse rate. Lapse rate is the magnitude of the decrease of an atmospheric parameter with height.
- The rate of change with h_a of the wet term is usually, but not always, negative. This governs the gradient dN/dh_a, which in turn governs ray curvature, ρ_r.

Up to $h_a > 0.1$ km, N usually decreases exponentially; $n = n_{\text{surface}} \exp(-h/H)$ and near the surface $dn/dh_a \sim -39.2 \times 10^{-6}\,\text{km}^{-1}$. The lapse rate $-dN/dh_a$ often differs from its standard value of 40 'N units'/km (12 'N units'/1000 ft).

The lower parts of wavefronts encounter higher n and propagate more slowly, curving the rays downward as shown in Figure 5.1(a). Alternatively, we could say that a ray transmitted obliquely upwards encounters decreasing refractive index, making it curve downwards in conformance with Snell's law in optics. The curvature is approximated by

$$\rho_r \sim \left[-\frac{1}{n}\frac{dn}{dh_a}\right]^{-1}. \tag{5.4a}$$

As $n \sim 1$,

$$\rho_r \sim \frac{-10^6}{dN/dh_a}\ \text{rad/km}. \tag{5.4b}$$

From the geometry,

$$k = \frac{\rho_r}{\rho_r - e_{km}}. \tag{5.5a}$$

Substituting Eqs (5.4) into Eq. (5.5a) and putting $e_{km} = 6371$ km

$$k = \frac{1}{(1 + 6371 dn/dh_a)} = \frac{1}{(1 + 6.371 \times 10^{-3} dN/dh_a)}. \tag{5.5b}$$

When $dN/dh_a = -39.2$ (sometimes taken as -40), which is approximately the standard lapse rate, $k = 4/3$. This k value has been adopted for general radio/radar use, giving $E = 8495$ km. By assuming this fictitious radius, E, for the Earth, the rays may be regarded as travelling in straight lines in a standard atmosphere, simplifying the ray geometry.

Some textbooks use a variant approach. If the Earth is assumed flat, $k = \infty$, a fictitious ray curvature can be adopted to preserve the actual curvature relative to the surface, using fictitious air having a modified refractivity, M as follows:

$$M = N + \frac{h_a}{e} \times 10^3 = N + 157h_a\text{'N units'}. \tag{5.6a}$$

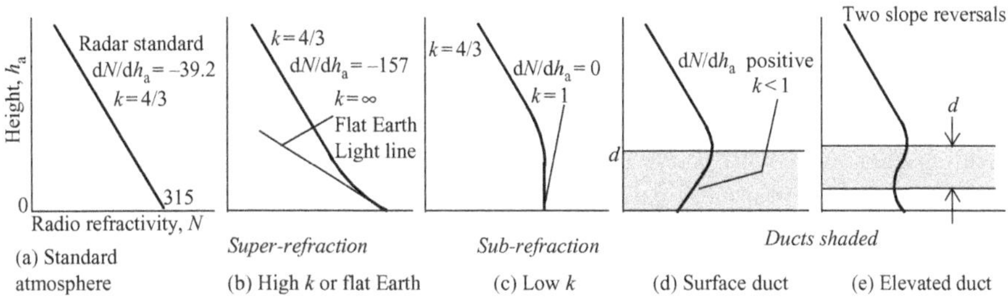

Figure 5.2 Radio refractivity profiles. Idealised, differing weather conditions. Refractive gradient is the reciprocal of the slope of the curves. Not to scale

The 10^3 term arises from the definition of h_a in kilometres. The rate of change with height is the modified refractivity gradient. For the standard atmosphere $\mathrm{d}M/\mathrm{d}h_a = 117$.

$$\frac{\mathrm{d}M}{\mathrm{d}h_a} = \frac{\mathrm{d}N}{\mathrm{d}h_a} + 157. \tag{5.6b}$$

5.2.4 *Standard atmosphere; four-thirds Earth approximation*

The radar standard atmosphere when the lapse rate gives 39.2 'N units'/km and $k = 4/3$ is also called the four-thirds Earth condition. The vulgar fraction seems better than writing 1.333, which confers a quite spurious sense of highly refined scientific precision. At optical frequencies, which are unaffected by the wet term of Eq. (5.3), the equivalent k value is 1.163 ($E = 7407.4$ km). Horizon range is proportional to $\sqrt{k}$, Section 5.5.7, so when $k = 4/3$, the radar horizon lies about 7 per cent beyond the visual horizon. The refraction profile of Figure 5.2(a) depicts the radar standard atmosphere with $k = 4/3$ and $\mathrm{d}N/\mathrm{d}h_a = -39.2$ 'N units'/km. Figure 5.2 is plotted to a base of N, so high $\mathrm{d}N/\mathrm{d}h_a$ represents a shallow slope with low $-\mathrm{d}N/\mathrm{d}h_a$.

The 'four-thirds Earth' or US Central Radio Propagation Laboratory convention is appropriate to airborne radar. It represents the mean condition at 700 m over the continental United States and often approximates the atmosphere at high altitude over land. It does not adequately represent conditions just above a sea surface and is too often automatically and unthinkingly chosen for marine radar purposes. Putting $k \sim 2$ might be a better all-round choice.

5.2.5 *Anaprop*

Anomalous propagation, anaprop, occurs when the local weather substantially changes the refraction gradient from the standard value, lapse rate and k significantly differing from 39.2 and 4/3, respectively. Sub-refraction occurs when $k \ll 4/3$ (say $k < 1.0$) and $\mathrm{d}M/\mathrm{d}h_a > 117$; super-refraction is when $k \gg 4/3$ (say $k > 2.0$) and $0 < \mathrm{d}M/\mathrm{d}h_a < 117$. Sub-refraction occurs when the coastal surface loses heat by radiation on a clear, still night and inversion (temperature and/or moisture increasing with height) causes w to rise with height. In the extreme, radiative fog may occur. On the other hand, if the air is dry, as in North Africa and the Middle East, the inversion causes super-refraction and ducting. The inversion is rapidly destroyed by solar heating after sunrise. Effective Earth radius E and k rise as lapse rate rises. Anaprop chiefly occurs in relatively calm conditions, when there is little turbulence to stir the atmosphere, and is commonly caused by a considerable difference between the wind and sea temperatures near land masses, although some anaprop is almost always present at low altitude under marine conditions. As lapse rate and k are difficult to determine in practice, anaprop has been the all-too-convenient scapegoat for many an unexpected performance shortfall over the years. For example, when a Canadian Coast Guard buoy tender observed a racon on the St Lawrence River to have half its expected maximum range during very calm autumn weather, the author had to call anaprop in aid. There was nothing obviously at fault in the radar, and the racon operated to full specification in subsequent laboratory tests. Maritime weather forecasts currently do not warn of anaprop.

In a typical atmosphere where $T = 293\,\mathrm{K}\,(20°\mathrm{C})$, $P = 1000\,\mathrm{hPa}$ and $w = 15\,\mathrm{hPa}$, the variation in N, δN, is

$$\delta N = 0.26\delta p + 4.3\delta w - 1.4\delta T.$$

Within a quite small height increment, δT may change by several degrees and δw by several Pascals, causing $\mathrm{d}N/\mathrm{d}h_\mathrm{a}$ to change by several 'N units', forming a waveguide called a duct, the waveguide walls being the points of inflexion or the sea surface. Evaporation ducts are commonest and result from δw, change of partial pressure of water vapour. A surface duct is bounded below by the surface. An elevated duct is bounded above and below by changes of $\mathrm{d}N/\mathrm{d}h_\mathrm{a}$. Within ducts, $\mathrm{d}M/\mathrm{d}h_\mathrm{a}$ is negative. Evaporative surface duct heights, according to Hall *et al.* [1, Section 6.3.2], who go into considerable detail, are typically: 5–6 m in the North Sea; 13–14 m in the Mediterranean and $>$20 m in the Arabian Gulf.

Advection ducts occur in coastal regions of seas such as the North Sea, and when the sea is enclosed by hot, dry land. Examples are again the Mediterranean and the Gulf. A warm, dry, air mass flows off the land onto the cooler sea. When a summer cyclone exists over northern Europe, warm dry air may be carried by advection to the North Sea, where it overlies cooler and moister air, giving high humidity lapse rate and temperature inversion, leading to thick (to 200 m) surface ducts which may persist near coasts for several days and are most pronounced in early evening.

The following Sections 5.2.6–5.2.9 are all examples of anaprop. Rays having obliquity $<1°$ (nearly all marine radar rays) can rarely cross duct boundaries.

5.2.6 *Super-refraction; high k; super-standard surface layer*

When, at low altitude, $10^{-6}e_\mathrm{km} \times \mathrm{d}N/\mathrm{d}h_\mathrm{a} = 1$, lapse rate $=$ 157 N units/km (48 N units/1000 ft), $k = \infty$ and the Earth appears flat; $E = \infty$ (Figure 5.2(b)). Here the rays curve down just enough to match the curvature of the Earth and the horizon recedes to infinity. Absence of horizon effects can cause targets to be detected at unusually long range, often beyond the nominal horizon. Performance may equal or even exceed the free space value. For example, in the Gulf, radar operating on an 89 km (48 nmi) range scale at 1000 pps (maximum unambiguous range 150 km) often disconcertingly displayed spurious islands in the sea at 75 km. These were 'second time around' echoes of inland mountains at 225 km, the echoes not arriving until the next display timebase sweep. PRF stagger would have decorrelated and broken up these false echoes. Wylie points out that in anticyclonic weather there is usually a subsidence of warm dry air which, settling over the surface of the sea, may produce super-refraction conditions. In areas like the Red Sea, super-refraction conditions may be present more often than not. A temperature difference of 3°C–6°C, often well exceeded, may suffice.

5.2.7 Negative k

If lapse rate exceeds 157 or M is negative, the rays would curve downwards, making horizon range fall again. The equivalent Earth would become concave with negative k. The author is not aware of any reported cases of this phenomenon.

5.2.8 Sub-refraction; low k; sub-standard surface layer

Particularly in fog, at night or early forenoon and in winter, a body of cold surface air may move under a body of moisture-laden air. According to Wylie [4], sub-refraction also occurs when cool air blows over a relatively warm sea, as in the Polar regions and near very cold land masses. The refraction gradient at the surface is depressed and k is less than 4/3, shown in Figure 5.2(c); N is negative and k can fall to 0.8 or less. The rays curve upwards. This condition is potentially hazardous; the effective Earth radius is reduced, the radar horizon is nearer and maximum detection ranges may be much less than expected under just those conditions when radar is needed most.

5.2.9 Ducts

The following description of ducting mainly follows Kerr [5] and Barton and Leonov [6, p. 151]. When warm dry air passes over a cool ocean surface, the water cools and moistens the bottom air layer. Temperature increases unusually rapidly with height. The resulting stability prevents mixing and the water vapour content decreases rapidly with height. As shown in Figure 5.2(d), lapse rate may be negative in the first tens of metres above the surface before turning positive. This change forms a surface or evaporation duct, in which rays can be trapped and propagate long distances, alternately reflected from the sea surface and the layer rather like waveguide transmission, giving anomalously long detection ranges.

In calm weather or thunderstorms, a meteorological inversion layer often forms some metres above the sea surface, modifying the N profile to the form shown in Figure 5.2(e). Figure 5.3(a) depicts ray paths when the scanner lies within the duct. Elevated ducts are often found seasonally in the tradewind region between the mid-ocean high pressure cells and the equatorial doldrums. The most famous tradewind

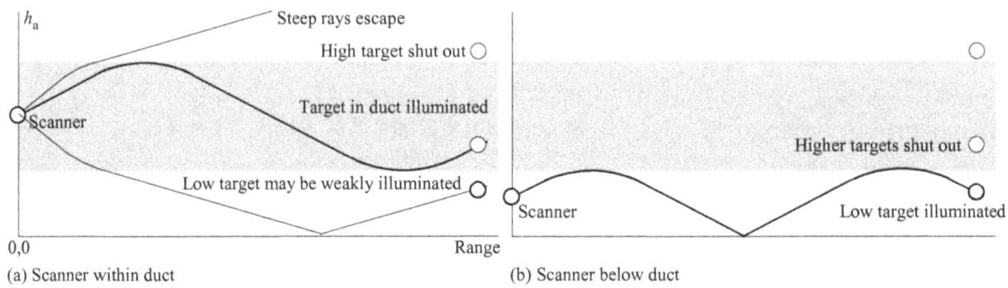

Figure 5.3 Rays in elevated duct. Refractivity profile as Figure 5.2(e). Target illumination is strongly dependent on scanner, duct and target heights. Not to scale

areas lie between Brazil and the Ascension Islands, and between southern California and Hawaii.

Some ducting is likely to be present over all water surfaces. When ducts are strong and the scanner is above the layer, rays may be deflected upwards; low targets are shut out and may be detected weakly if at all. Rays may also become trapped in elevated ducts to the detriment of targets beneath. Maximum wavelength which can be trapped varies with duct height d metres and is:

$$\lambda_{\max} = 0.0077\sqrt{d} \tag{5.7}$$

- if d is less than 17.3 m, neither 3 nor 9 GHz band can be trapped;
- when d is between 17.3 and 165 m, the 9 but not 3 GHz band can be trapped;
- when d is higher still, both bands can be trapped.

Rays within a duct, Figure 5.3(a), or below an elevated duct, Figure 5.3(b), are constrained and partially focussed in elevation and the usual R^{-4} echo strength law changes to R^{-2} because the beam spreads out only in the horizontal plane. In principle and disregarding any 'waveguide loss', the echo strength would reduce 6 dB when range doubles, rather than the free space 12 dB, explaining the exceptionally long detection ranges sometimes observed. If both the scanner and target are above the duct, the illumination intensity is more or less its normal value, but targets across a boundary from the scanner are at best weakly illuminated.

Wylie states that surface ducts are more significant to marine radar than elevated ducts, and ducting is usually associated with a sharp decrease of moisture content which is often accompanied by a temperature inversion, that is, a sudden increase in temperature.

5.2.10 Conditions causing anaprop

In calm tropical waters in daytime k can rise to 10 or more, falling at night. In temperate latitudes k usually lies between 1 and 2, although extremes are wider, perhaps 0.5–10. When trialling shipborne defence radars in UK waters in daylight, fair weather,

Chaplin[1] found $k \sim 2$ often matched observed echo strengths. Note that k tends to be high during:

- warm airflow over cool ocean
- evening
- summer
- settled weather.

It is perhaps more important to know when k is likely to be low:

- fog
- night/forenoon
- winter
- storms
- windward of ice sheets.

When calculating the reliable first detectable range of any proposed radar/target combination in poor weather, it is prudent to assume k is less than $4/3$, perhaps 0.8 or even lower. The method of performance calculation we shall develop handles k but cannot adequately allow for ducting and so may tend to misrepresent maximum detectable range.

5.3 Measurement of refraction factor

Meteorological stations worldwide launch radiosonde balloons twice daily to sample the troposphere, primarily to capture data for weather forecasting and frequency optimisation of short-wave long range radio transmissions. The balloon tracks seldom cover sea areas of interest to us. The p, T and w data gathered in the lowest 100 m is too coarse and scanty to be of much assistance in determination of effective k-factor for marine radar although minisondes are better in this respect. Refractive index can be measured by high-precision refractometer, a waveguide cavity whose resonant frequency is a function of the n of air within. Height profiles are obtained by flying the refractometer in a helicopter or aircraft.

It is often difficult to position shipboard meteorological instruments to take the necessary delicate meteorological measurements from which to calculate k per Eqs (5.3) and (5.5b). An attempt by a team including the author to obtain n by measurement of p, T and w at different heights and thus deduce dN/dh_a and n in conjunction with the experiment described in Section 5.3.3 failed utterly, perhaps because the platform, the 1600 ton *Pharos*, disturbed the local atmosphere.

The following procedure allows k to be deduced from radar observations, using a ship equipped with an ordinary navigational radar, preferably 9 GHz, and a station, such as a lighthouse, equipped with a racon mounted high. Racons are point targets which respond at constant known power. No special modification is needed to radar

[1] Chaplin M. D., formerly Design Authority, Radar Type 909, GEC-Marconi, personal communication.

or racon. The method uses the fact that multipath nulls are sharp and dependent on k-factor, as well as on the readily ascertainable radar frequency and heights of scanner and racon antenna. The procedure is as follows.

1. By noting qualitatively the variation of observed displayed response strength as the ship approaches the racon, actual ranges of nulls can be estimated with some precision. Preferably the radar is operated on a constant range scale at constant gain and always observed by the same person, who should have no other task. The observer scores signal strength subjectively on a rising scale: 0 = no hits observed, 1 point = single weak racon response (hit) per racon 'on' period, rising to say 10 points when a full set of strong responses is seen with sidelobe interference. The score is then plotted versus target range to give a curve of the general form of Figure 5.4 from which null numbers and exact null ranges can be accurately determined.
2. Using a PC and the IEE spreadsheet described in Chapter 14, Section 14.1, sets of theoretical calculations of response strength versus range are plotted for the radar and racon heights above sea level including any tide correction, for several successive k values, as shown in Figure 5.4. The value best fitting the observed null ranges gives the average k for the radar/racon path at the time of observation.
3. Ideally, k is confirmed by a simultaneous parallel exercise in the other frequency band.

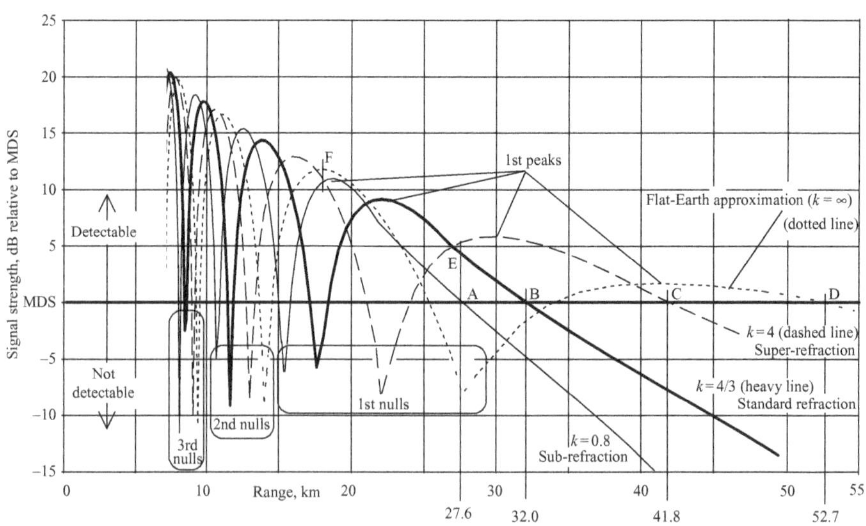

Figure 5.4 Dependence of point target response on refraction. Peak and null ranges vary with refraction factor k. Flat-Earth approximations give ranges which are somewhat too high. Computed for 9 GHz band, sea state 1; scanner height 15 m; one-way response from constant-power racon target, height 30 m

It will be seen that response strength varies rather slowly at maximum detectable range, so the latter parameter is heavily influenced by minor changes of radar sensitivity and atmospheric attenuation. Positions of multipath peaks are also ill-defined, making both these parameters less suitable criteria than the sharply defined null ranges.

Figure 5.5 summarises hitherto unpublished k-factors deduced from observed racon nulls during a particular voyage of the Lighthouse Tender *Pharos*; Figure 5.6. The weather was changeable and k varied between 0.8 and at least 4.0 – possibly as high as 8.5. Although individual results are subject to experimental error of maybe 10–20 per cent, 3 and 9 GHz band measurements agreed, and k certainly varied over a wide range. By the law of cussedness, low k with its reduction of maximum detection range occurred in fog and at night; just when the radar was most needed. Also as predicted, high k was observed in fine weather in excellent visibility when the radar was less essential for safe navigation.

Wide variations of refraction factor are sufficient to swamp quite large variations in radar sensitivity or target size. For example, suppose for a particular racon mounted at 30 m, the Figure 5.4 trial observed maximum detectable range 27.6 km. Unless k had been ascertained, one could not be sure whether this was a full-strength racon observed with sub-refraction ($k = 0.8$, point A on Figure 5.4) or perhaps a 4 dB weaker racon with standard refraction, point E, or some other combination. Such uncertainty can utterly vitiate comparative trials between competing products. Unless undertaken at the same place and time, trials results may say more about the atmosphere than the devices on test. Ideally the trial range should be chosen so response strength is nearly independent of k at a multipath peak, point F on the figure.

5.4 Ray geometry; geometrical optics

5.4.1 Introduction

We now turn to the geometrical relationship between the direct and indirect rays between a point source and a point target, using the ray tracing methods of geometrical optics. The procedure accurately represents reality when range is well inside the horizon, an area called the interference region. Chapter 6 will consider the boundary range of this region, and what happens at longer ranges where diffraction becomes significant.

We assume straight rays above a spherical effective Earth, as shown dashed in Figure 5.1(c). We defer consideration of atmospheric attenuation. The direct ray between source and destination does not touch the sea surface. The indirect ray path is a few centimetres longer, travelling via a reflection from the sea surface at the intermediate grazing point. This small but significant path-length difference increases as range decreases. The resultant signal reaching the destination is the vector sum of the direct and indirect ray voltages. When the phases coincide there is signal reinforcement; when they oppose, there is complete or partial cancellation. This interaction, called multipath interference, causes the violent echo fluctuations depicted in Figure 5.4.

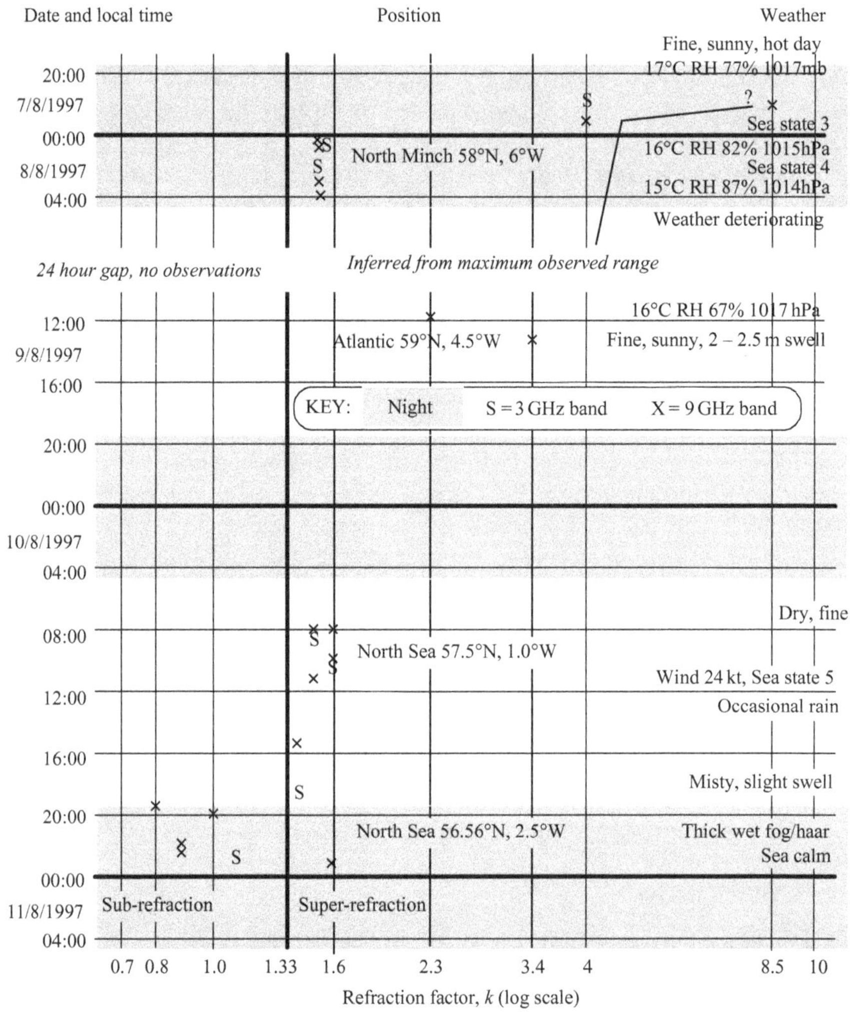

Figure 5.5 *Observed values of k-factor. Summer, temperate zone. Shows variation between day and night, and dependence on weather. Derived by observing racon null ranges on voyage of Northern Lighthouse Board tender Pharos from Oban to Leith, Scotland. Refraction varied widely and did not always approximate the 'standard' $k = 4/3$ value. Haar is a local form of advection fog. Equipment parameters as Figure 5.4*

Analysis of the geometry yields the path length difference (independent of wavelength) and so the phase angle (wavelength dependent) between the component rays. It also gives the geometrical angle of incidence, α, at the grazing point, which governs the strength and phase of the forward reflection for a given sea state, affecting the voltage and phase of the indirect component at the destination. The grazing angle

Figure 5.6 *Lighthouse Tender Pharos. Used for atmospheric refraction measurements of Section 5.3 and PPI photographs of Figures 2.6 and 8.2(a). Twin Atlas radars: 9 GHz scanner bracketed from foremast and 3 GHz scanner from main-mast. Lying at NLB Depot, Stromness, Orkney Islands. Author, reproduced by permission of the Northern Lighthouse Board*

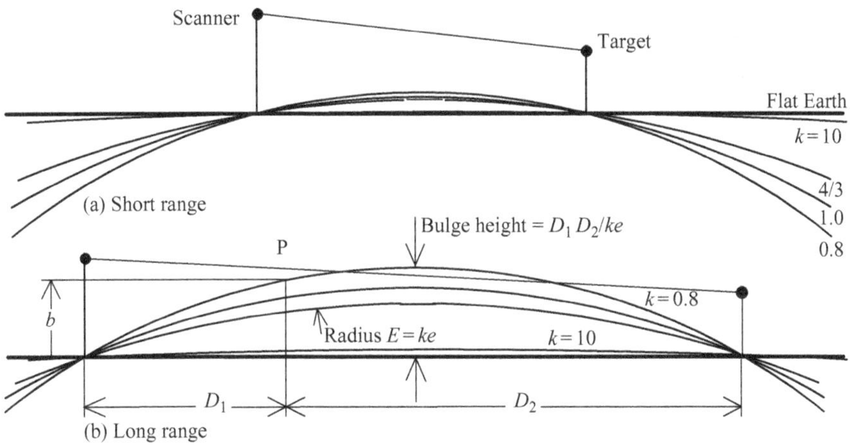

Figure 5.7 *Dependence of k-factor on ray geometry. At short range geometry is almost independent of k-factor and the flat-Earth approximation can be used. Doubling range much increases the significance of k (indeed, in this example the target falls below the horizon when k = 0.8) and the flat-Earth approximation loses validity*

forms a useful parameter linking range and multipath. Initially the source is the scanner and the destination is the target; on the return leg the roles reverse but the geometry is unaltered.

5.4.2 *Importance of k depends on range*

Figure 5.7, which is not to scale, places a scanner and target on several effective Earths of differing k-factor. At short range, the surface height between them is hardly affected when meteorological changes affect k and we could assume the Earth to be flat without introducing much error. But when range is doubled the surface bulges up into the region traversed by rays between scanner and target, particularly when k is low. This is why the flat-Earth approximation, which sets k to infinity, is reasonably valid at short range but introduces significant error at long range. We next analyse the ray geometry assuming a curved Earth, followed by an analysis using the flat-Earth approximation.

5.5 Geometrical analysis, curved Earth

5.5.1 Ray paths

We assume the scanner vertical beamwidth is so wide that the sea surface and the target are both fully illuminated by the main beam, and the radar cross section of the target is independent of the angle of view – the target is an isotropic reflector. The sea surface acts as a mirror, the grazing angle of incidence, α rad, equalling the angle of reflection. We assume that a sea surface extends from scanner to target. If land intervenes at the grazing point, scanner and target heights should be referred to grazing point height as zero. The reflection coefficient of land is generally low, making multipath less severe. Allowance must be made for any tidal height variations when considering Earth-fast VTS scanners, or targets like racons on lighthouses. All quantities, including range and Earth radius, must be expressed in metres, with angles in radians. We first examine the relative phasing and angles in the vertical plane of rays reaching the target, assuming complete forward reflection at the grazing point for the moment. Actual grazing point behaviour is deferred to Section 5.7.

We make the following assumptions, all fully justified by the geometry of marine radar systems.

- All rays travel in straight lines, their refraction curvature being transferred to a fictitious curvature of the Earth's surface, radius $E = ke$.
- Ranges of interest are much greater than scanner and target heights and much less than E.
- Negligible error occurs if R is measured along the surface rather than along straight line FG.
- Taking the scanner and target masts to be parallel causes negligible error.
- Path length timing difference is much less than transmitter pulselength so direct and indirect pulses fully overlap in time at the destination.

Figure 5.8 shows a scanner S mast-mounted at height H above the local surface at F. This figure is by no means drawn to scale. Angles are grossly exaggerated, for example $\angle$SOT, shown 20°, rarely exceeds 0.5° in reality, representing 74 km if $k = 4/3$. Direct rays travel to point reflector target T at range R. The reflector is at height h, the foot of its non-reflecting mast meeting the local surface at G. The indirect ray is specularly reflected at the grazing point Z, the tilt of the surface at Z relative to SQ increasing with grazing point range, D_1. The surface through Z intercepts the masts at A and B.

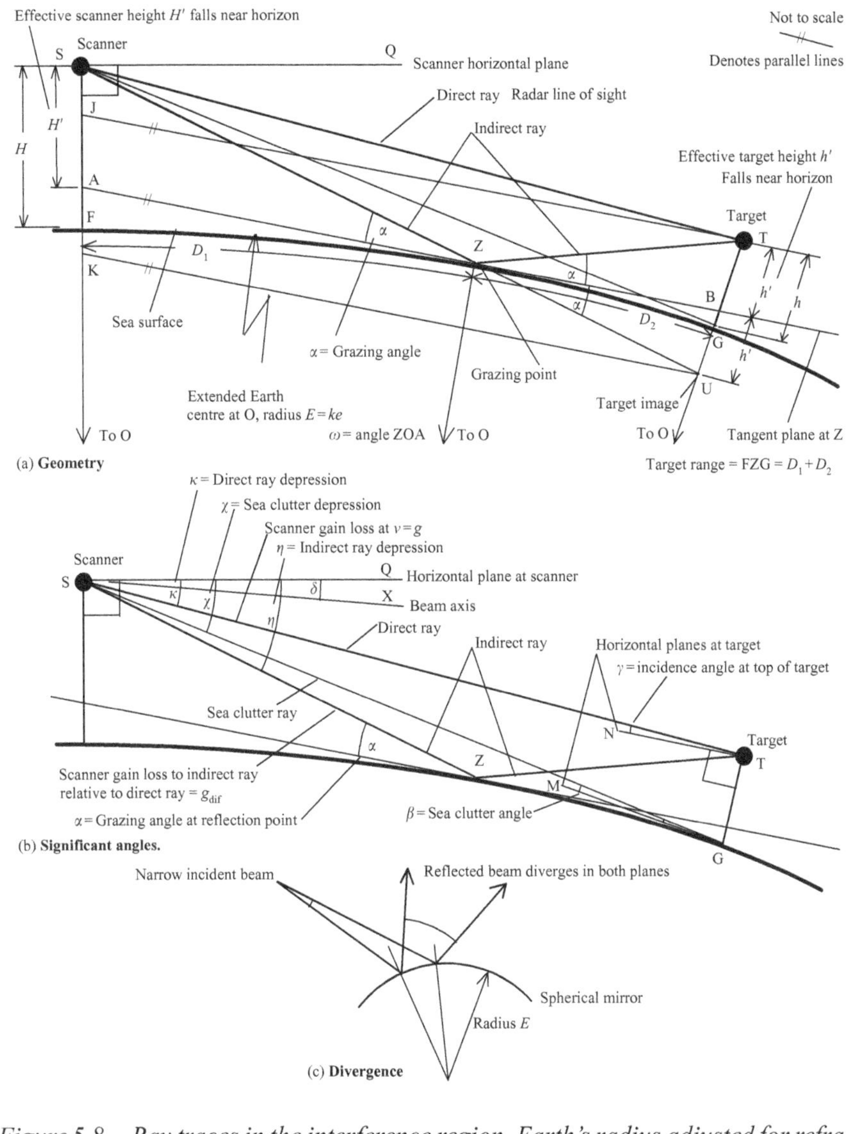

Figure 5.8 Ray traces in the interference region. Earth's radius adjusted for refraction to make rays travel in straight lines. Scanner-to-target and return echoes follow the same paths. Angles in (b) are all shown positive

This is the geometry of straight rays above plane reflecting sheet AZB. Effective heights of radar (SA $= H'$) and target (TB $= h'$) fall as range increases. Phase shift between direct and indirect rays is proportional to the path length difference, Δ, between direct ray ST and indirect ray SZT. The path difference is only a tiny proportion of path length; for example a 3 GHz radar at 10 km range contains 100 000 cycles in the direct path and perhaps 100 001 in the indirect path. It is difficult to express Δ directly in terms of range. A rigorous treatment of the geometry introduces several second-order terms. Luckily our assumptions enable many simplifications without significant error. We use the grazing angle α as a parameter linking signal strength with range. (Actual behaviour near the horizon is severely modified by diffraction, discussed in Chapter 6.) Kerr's [5] rigorous treatment has 81 pages of mathematics. The present author is unsure of the origins of our simplified treatment; it was in use by the late Dr K W Milne within the British Thomson-Houston Co prior to 1961. Other treatments are possible, see, for example, Rohan [7].

5.5.2 Range

Triangle OZA has a right angle at Z, so $\mathrm{OZ}^2 + \mathrm{ZA}^2 = \mathrm{OA}^2$, that is

$$E^2 + D_1^2 = [E + (H - H')]^2 = E^2 + 2E(H - H') + (H - H')^2.$$

Dropping the last and negligibly small second-order term and re-arranging,

$$H' = H - \frac{D_1^2}{2E}. \tag{5.8}$$

In triangle SAZ, $D_1 \ll E$ so $\angle$SAZ approximates a right angle, so

$$\alpha = \arctan\left(\frac{H'}{D_1}\right). \tag{5.9a}$$

Substituting for H' in Eq. (5.8)

$$\frac{D_1^2}{2E} = H - D_1 \tan\alpha \tag{5.9b}$$

$$\therefore \quad \tan\alpha = \frac{H}{D_1} - \frac{D_1}{2E}. \tag{5.9c}$$

The equations for h' are of similar form. By symmetry with Eq. (5.9b),

$$\frac{D_2^2}{2E} = h - D_2 \tan\alpha. \tag{5.9d}$$

From the geometry,

$$D_2 = R - D_1. \tag{5.10}$$

Substituting for $\tan\alpha$ and D_2 in Equation (5.9d), multiplying by $2E$ and collecting terms

$$(R - D_1)^2 = 2E\left[h - (R - D_1)\frac{H}{D_1} - \frac{D_1}{2E}\right]$$

so

$$2D_1^3 - 3RD_1^2 + (R^2 - 2Eh - 2EH)D_1 + 4EHR = 0. \tag{5.11}$$

The roots are difficult to extract. However, expressions for range, etc., in terms of α as parameter, are readily obtained by the following procedure. Substituting for

H' per Eq. (5.9a) in Eq. (5.8) and extracting the positive root gives the grazing point range:

$$D_1 = \sqrt{E^2 \tan^2 \alpha + 2EH} - E \tan \alpha. \tag{5.12a}$$

Similarly

$$D_2 = \sqrt{E^2 \tan^2 \alpha + 2Eh} - E \tan \alpha. \tag{5.12b}$$

Addition gives range:

$$R = D_1 + D_2 = \sqrt{E^2 \tan^2 \alpha + 2EH} + \sqrt{E^2 \tan^2 \alpha + 2Eh} - 2E \tan \alpha. \tag{5.13}$$

For given scanner and target heights, k-factor and range, Eq. (5.13) defines α in an inaccessible form. For a given α, values of H', h' and R depend on E and therefore on k and the weather. We are concerned with small grazing angles where $\tan \alpha \sim \alpha$.

If the target height is much smaller than scanner height, $R \sim D_1$ and, from Eq. (5.9b),

$$R \tan \alpha = H - \frac{R^2}{2E}.$$

or

$$\alpha \sim \frac{H}{R} - \frac{R}{2E}. \tag{5.14a}$$

Similarly, if the scanner is much lower than the target

$$\alpha \sim \frac{h}{R} - \frac{R}{2E}. \tag{5.14b}$$

When H and h are both significant, α is obtained by summing the above:

$$\alpha \sim \frac{H + h}{R} - \frac{R}{E}. \tag{5.14c}$$

Eqs (5.14a)–(5.14c) become hopelessly inaccurate near the horizon and should be used with discretion. Together with Eq. (5.10), they can simplify determination of α and hence the other geometrical parameters at a spot short range. Otherwise it is best to construct a spreadsheet, the subject of Chapter 14, having a set of α values sufficiently wide to bracket the range of interest. This method facilitates construction of such graphs as Figure 5.4, showing the variation of parameter values with range and clearly indicating how ranges of interest relate to peaks and nulls.

When constructing spreadsheets, Eq. (5.14a) is used to find α_{min} at the minimum range of interest R_{min}, and horizon range R_{H} is inserted using Eq. (5.23). Eq. (5.14d) then provides an approximate value for α at a given nominal range R_{nom}. Using this α, an accurate range value can be calculated from Eq. (5.13).

$$\alpha \sim \alpha_{\text{min}} R_{\text{H}} R_{\text{min}} \frac{(1/R_{\text{nom}}) - (1/R_{\text{H}})}{R_{\text{H}} - R_{\text{min}}} \text{ rad.} \tag{5.14d}$$

5.5.3 Path difference of indirect ray

The indirect ray additional path length Δ is obtained by evaluating similar triangles SAZ, TBZ. Drawing a line JT through T parallel to AB, the angle at J approximates a right angle. Direct path length ST is obtained from Pythagoras's theorem on triangle STJ:

$$\mathrm{ST}^2 \sim \mathrm{JS}^2 + \mathrm{TJ}^2 = (H' - h')^2 + R^2 = H'^2 + h'^2 - 2H'h' + R^2. \tag{5.15a}$$

The indirect ray reaches the target from the scanner via specular reflection at Z, forming a virtual or mirror image at U, below the target. BU = TB. Again by Pythagoras's theorem,

$$\mathrm{SU}^2 = (H' + h')^2 + R^2 = H'^2 + h'^2 + 2H'h' + R^2. \tag{5.15b}$$

Indirect path length SZ + ZT = SU and exceeds direct path length ST by Δ:

$$\Delta = \mathrm{SU} - \mathrm{ST}. \tag{5.15c}$$

Subtracting Eq. (5.15a) from Eq. (5.15b) and substituting for Δ per Eq. (5.15c)

$$\mathrm{SU}^2 - \mathrm{ST}^2 = 2\Delta\mathrm{ST} + \Delta^2 = 4H'h'.$$

Putting ST = R and ignoring the second-order term Δ^2,

$$\Delta = \frac{2H'h'}{R}. \tag{5.16a}$$

The indirect ray contains an extra Δ/λ cycles of alternating wave. Thus path length difference causes the indirect component of voltage to lag the direct component by a phase angle Φ, where

$$\Phi = \frac{2\pi\Delta}{\lambda} = \frac{4\pi H'h'}{R\lambda}\ \text{rad}. \tag{5.16b}$$

If the reflection at Z introduces phase shift ψ rad (often $\psi \sim \pi$; discussed in Section 5.8), total phase shift ϕ is given by

$$\phi = \Phi + \psi. \tag{5.16c}$$

Angle Φ is small at long range but rises as R decreases and triangle SZT gets fatter. At short range, its value may be many radians. This causes the indirect ray to swing between constructive and destructive interference with the direct ray, the resultant oscillating in magnitude. As Φ is proportional to frequency, these events occur nearer the horizon in the J and 9 GHz bands than at 3 GHz.

If scanner height H is raised by an increment δH, from the geometry H' also rises by δH; Φ rises by an increment $\delta\Phi$. Substituting in Eq. (5.16b),

$$\Phi + \delta\Phi = \frac{4\pi h'(H' + \delta H)}{R\lambda}$$

and the rate of change of phase shift with scanner height is

$$\frac{\mathrm{d}\Phi}{\mathrm{d}H} = \frac{\delta\Phi}{\delta H} = \frac{4\pi h'}{R\lambda}\ \text{rad/m}. \tag{5.16d}$$

By symmetry, the rate of change with target height is

$$\frac{\mathrm{d}\Phi}{\mathrm{d}h} = \frac{4\pi H'}{R\lambda} \text{ rad/m.} \tag{5.16e}$$

The rate of change with frequency, f, is obtained by differentiating Eq. (5.16c) (as usual c is the velocity of propagation):

$$\frac{\mathrm{d}\Phi}{\mathrm{d}f} = \frac{4\pi H'h'}{cR} \text{ rad/Hz.} \tag{5.16f}$$

5.5.4 *Useful angles*

Several angles beside α, which feature in detectability calculations, are depicted in Figure 5.8(*b*) (where the angles shown are positive) and listed in Table 5.1.

- *Beam axis angle of depression, δ.* During system installation, the beam axis of the scanner may be inclined a degree or so below the horizontal to improve illumination of the horizon. If mounted on a ship, roll or pitch may also incline the axis. Angle of depression, δ rad, positive downwards, accounts for these inclinations.
- *Direct ray depression, κ.* The direct ray is usually depressed below the scanner horizontal plane. Depending on δ, significant loss of gain may occur with some VTS scanners whose vertical beamwidth is narrow. The angle of depression of the top of the target is found by drawing line JT through the target top T and parallel to the tangent plane AB through Z. SJ $= H' - h'$. JT is parallel to AZ so $\angle$TJS $= \angle$ZAS $= \pi/2 + \omega$ and $\angle$STJ $= (H' - h')/R$. The required angle is

Table 5.1 Useful angles

Angle	Symbol	Figure 5.8	Equation
Angles relative to local horizontal			
Grazing angle	α	SZA = TZB	5.13
Sea clutter angle	β	SGM	5.21a
Incidence angle at top of target	γ	STN	5.19
Angles relative to horizontal at scanner			
Beam axis depression	δ	QSX	–
Direct ray depression	κ	QST	5.17
Indirect ray depression	η	QSZ	5.18
Sea clutter depression	χ	QSG	5.20
Scanner angles (design parameters)			
Azimuth beamwidth	θ	–	–
Elevation beamwidth	ϕ	–	–

given by

$$\kappa = \angle\text{QST} = \angle\text{STJ} + \angle\text{ZAS} - \frac{\pi}{2} = \omega + \frac{H' - h'}{R}.$$

Substituting for ω:

$$\kappa = \frac{D_1}{E} + \frac{H' - h'}{R}. \tag{5.17}$$

- *Indirect ray depression, η.* Indirect ray equivalent of κ.

$$\omega = \angle\text{ZOA} = \arctan\frac{D_1}{E} \sim \frac{D_1}{E}.$$

$$\angle\text{OZA} = \frac{\pi}{2}, \qquad \text{so } \angle\text{ZAO} = \frac{\pi}{2} - \omega \quad \text{and} \quad \angle\text{ZAS} = \frac{\pi}{2} + \omega.$$

$$\therefore\ \angle\text{ZSA} = \pi - \left(\frac{\pi}{2} + \omega\right) - \alpha = \frac{\pi}{2} - \omega - \alpha.$$

$\angle\text{OSQ} = \pi/2$, so the angle of depression is

$$\eta = \angle\text{QSZ} = \frac{\pi}{2} - \left(\frac{\pi}{2} - \omega - \alpha\right) = \omega + \alpha = \frac{D_1}{E} + \alpha. \tag{5.18}$$

- *Incidence angle at top of target, γ.* Used when target RCS varies in its elevation plane. The horizontal plane through the top of the target, NT, is inclined to the horizontal plane at the scanner, SQ, by angle $\angle\text{SOT} = R/E$.

$$\therefore\quad \gamma = \kappa - \frac{R}{E}. \tag{5.19}$$

- *Sea clutter depression angle, χ.* Affects effective scanner gain to sea clutter.

$$\chi = \angle\text{QSG} = \angle\text{QST} + \angle\text{TSG} = \kappa + \frac{h'}{R}. \tag{5.20}$$

- *Sea clutter angle, β.* This is the value of χ at target range R. It is the (positive) grazing angle of sea clutter at that range and affects the clutter return. MG is the local horizontal at the foot of the target, G. It is parallel to plane NT so must be inclined to the horizontal SQ at the scanner by $\angle R/E$; it cannot be negative:

$$\text{If } \left(\chi - \frac{R}{E}\right) > 0, \qquad \beta = \chi - \frac{R}{E}, \quad \text{otherwise } \beta = 0. \tag{5.21a}$$

When $\chi < R/E$, $\beta < 0$, the foot of the target is obscured by the horizon and sea clutter at the target ceases to be illuminated or observed. The target horizon exceeds the sea clutter horizon because the target has height. At short range, from Figure 5.17(*a*)

$$\beta \sim \arctan\left(\frac{H}{R}\right). \tag{5.21b}$$

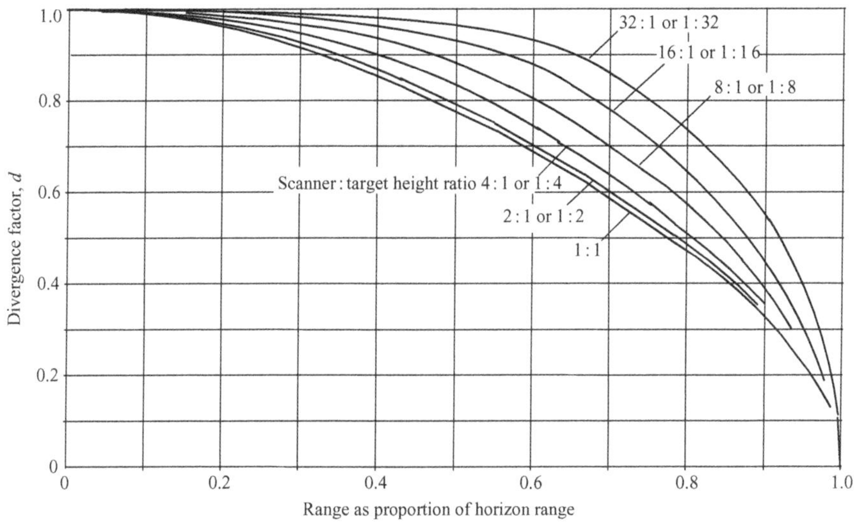

Figure 5.9 Divergence factor. Falls near the horizon. Curves approximate

5.5.5 Divergence factor

Reflection from a spherical convex mirror broadens the beam in both azimuth and elevation planes; Figure 5.8(c). The on-axis field strength after reflection is reduced by a divergence factor, d. Rohan [7, Section 6.6] gives detailed formulae for this curvature effect and Barton and Leonov [6] give the following approximation

$$d = \frac{1}{\sqrt{1 + 2D_1 D_2/(ER \sin \alpha)}} \text{ numerical.} \tag{5.22}$$

The value of d is independent of wavelength but of course depends on range through Eqs (5.12) and (5.13). It is nearly 1.0 at short range where $D_1 D_2/R \ll E$, falling to zero at the horizon, where $\alpha = 0$. Figure 5.9 shows how divergence varies with range for several ratios of scanner and target heights, range being expressed as a proportion of horizon range R_H, which will be explained shortly.

5.5.6 Variation of geometrical parameters with range

The values of α, H', h', Δ and d all fall as range increases, becoming zero at the radar horizon. Figures 5.10(a) and (b) plot these parameters for typical long and short range systems respectively for $k = 4/3$. Increasing k increases horizon range, in effect stretching all the curves to longer kilometric ranges without altering their general shape. The curves solely depend on geometry, so are independent of radar wavelength. However, a given Δ of course causes larger phase shift when wavelength is small, giving the 9 GHz band larger phase shift than the 3 GHz band.

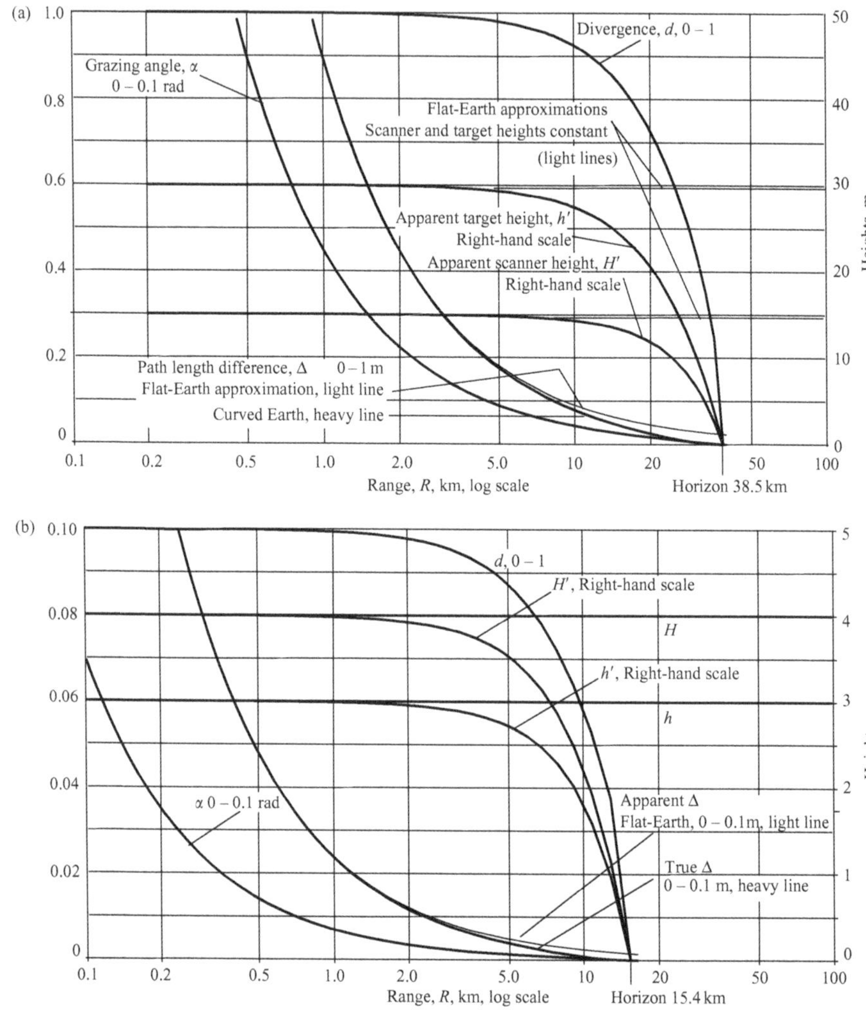

Figure 5.10 Geometrical parameter values. (a) Long-range system. As range increases, the apparent scanner and target heights fall and the flat-Earth approximation becomes inaccurate for calculation of path length difference Δ and hence multipath phase shift. Grazing angle α falls to zero at the horizon. Near the horizon divergence d falls, reducing the strength of the indirect ray. Heights use right-hand scale; other parameters use left-hand scale. $H = 15\,m$, $h = 30\,m$, $k = 4/3$, as Figure 5.4. (b) Short-range system. $H = 4\,m$, $h = 3\,m$, $k = 4/3$. The flat-Earth approximation introduces little error in short-range systems.

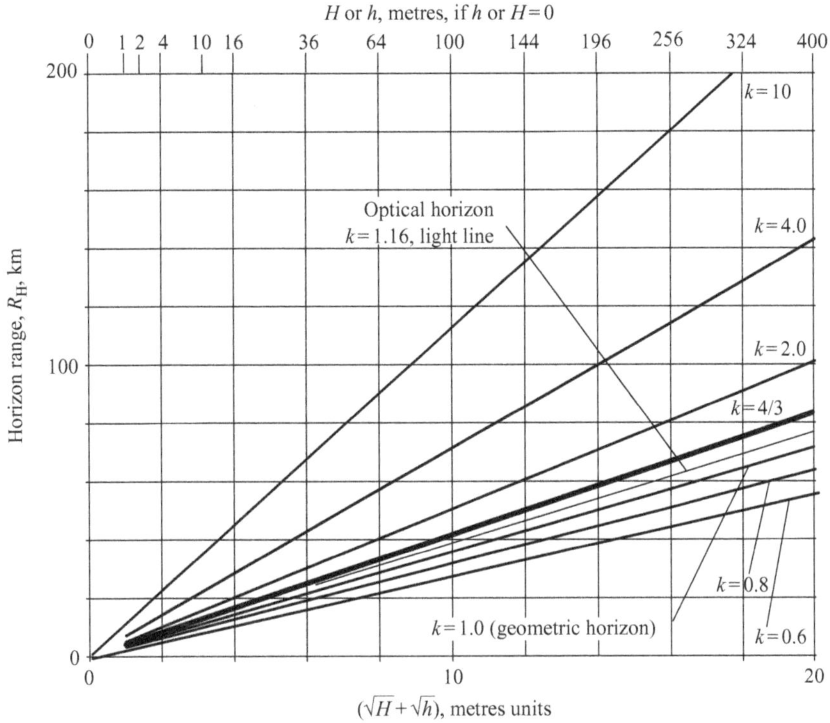

Figure 5.11 Horizon range. Plotted to a base of the sum of the square roots of the scanner and target heights, for various k-factors. Putting $h = 0$ gives the sea clutter horizon

5.5.7 Horizon

The target is at the scanner's radar horizon range, R_H, when the direct ray just grazes the sea surface at Z, so $\alpha = 0$. Here the direct and indirect path lengths become equal; $\Delta = 0$ and $\Phi = 0$. Note that, as with the geographical range of a light, horizon range of a point target is calculated to the reflecting element, not its foot. As with a light, the target may also be too feeble to be detected at the horizon. When $\alpha = 0$, Eq. (5.13) simplifies to:

$$R = R_H = \sqrt{2ke}\left(\sqrt{H} + \sqrt{h}\right) \text{m}. \tag{5.23a}$$

Putting $e = 6371$ km,

$$R_{Hkm} = 3.570\sqrt{k}\left(\sqrt{H} + \sqrt{h}\right) \text{km}. \tag{5.23b}$$

This equation is plotted in Figure 5.11 for a family of k values. Putting $h = 0$ gives the horizon range, seen by the scanner; $H = 0$ gives the horizon range seen by the tip of the target, summation gives R_H.

When $k = 4/3$,

$$R_{\mathrm{Hkm}} = 4.122(\sqrt{H} + \sqrt{h})\ \mathrm{km}. \tag{5.23c}$$

When $k = 4/3$ and H and h are in feet

$$R_{\mathrm{Hnmi}} = 1.23(\sqrt{H_{\mathrm{ft}}} + \sqrt{h_{\mathrm{ft}}})\ \mathrm{nmi}. \tag{5.23d}$$

Terms H and h have equal weight in Eqs (5.23) and may be interchanged, so R_{H} is the same for say $H = 30$ m, $h = 15$ m as for $H = 15$ m, $h = 30$ m. Even when one term is much larger than the other, doubling the height of the larger term cannot increase R_{H} by more than $\sqrt{2}$ or 41 per cent. When the flat-Earth approximation is used, the horizon recedes to infinite range. R_{H} is unaffected by wavelength.

Horizon is less decisive at radar than at optical wavelengths because diffraction (Chapter 6, Section 6.5) allows some signal to propagate beyond. So certain radars can detect big targets some distance over the horizon. On the other hand, diffraction starts to decrease signal strength below the value calculated from geometrical optics well inside horizon range. Nevertheless R_{H} remains a useful point of reference. The horizon range of the sea surface, R_{s}, governs the maximum range of sea clutter and is found by putting $h = 0$.

5.5.8 *Multipath peak and null ranges*

Multipath peaks or lobes occur when the direct and indirect rays are in phase and $\cos(\Phi + \psi) = 1$:

$$\Phi + \psi = 2\pi, 4\pi, \ldots, 2n\pi\ \mathrm{rad} \qquad [n \text{ is a positive integer}]. \tag{5.24a}$$

It is usually permissible to take $\psi = \pi$, see Section 5.8.3. Peaks then occur when

$$\Phi = \pi\,(\text{first peak, nearest horizon}),\ 3\pi, \ldots, (2n-1)\pi\ \mathrm{rad}. \tag{5.24b}$$

Similarly, there are nulls when $\cos(\Phi + \psi) = -1$, i.e. when

$$\Phi + \psi = 3\pi, 5\pi, \ldots, (2n+1)\pi\ \mathrm{rad} \tag{5.25a}$$

so if $\psi = \pi$,

$$\Phi = 2\pi\,(\text{first null, nearest horizon}),\ 4\pi, \ldots, 2n\pi\ \mathrm{rad}. \tag{5.25b}$$

Because Φ depends on H' etc., it also depends on k and hence on weather, as expected.

For simplicity of analysis we have assumed the grazing point to be a horizontal surface at sea-level datum height. Except in calm weather, the sea is not horizontal; the surface oscillates up and down as waves wash past and can locally incline by several degrees. These factors make the grazing point dither in height and range. The peak and null ranges dither likewise, so all calculated ranges must be taken as means, with a random variation which increases with wave height. The next chapter considers multipath in more detail.

5.5.9 *Effect on detection range*

Maximum detection range depends strongly on the k-factor. For example, Figure 5.4 of Section 5.3 was calculated for the trial ship's 9 GHz radar and the appropriate scanner and racon heights. Only when the response exceeded the radar's minimum detectable signal (MDS) was it detected. Calculated ranges were as follows.

- Standard refraction (point B, $k = 4/3$): maximum calculated detection range 32.0 km (100 per cent datum), horizon 38.5 km.
- Sub-refraction (point A, $k = 0.8$): detection 27.6 km (86 per cent), horizon 29.8 km.
- Super-refraction (C, $k = 4$): detection 41.8 km (131 per cent), horizon 66.7 km.
- Flat-Earth approximation (D, $k = \infty$): detection 52.7 km, horizon at infinity.

On the trial, maximum observed detection ranges were generally consistent with the above. Ranges of the shorter-range nulls are less dependent on k because ray curvature is proportional to range.

5.6 Flat-Earth approximation

5.6.1 *Geometrical analysis*

The geometry is a simplification of the curved Earth case considered earlier. Figure 5.12(a) shows rays travelling from a scanner to a point target in straight lines above a flat Earth. Scanner and target effective heights remain H and h, respectively, at all ranges. With this proviso, the analysis is similar to that of Section 5.5.3. Triangles SFZ and TGZ are similar. The direct path length ST is obtained by Pythagoras's theorem for triangle STJ:

$$\mathrm{ST}^2 = \mathrm{JS}^2 + \mathrm{TJ}^2 = (H - h)^2 + R^2 = H^2 + h^2 - 2Hh + R^2. \tag{5.26a}$$

The indirect ray reaches the target via specular reflection at Z, with a target mirror image at U. GU = TG. Again by Pythagoras's theorem

$$\mathrm{SU}^2 = (H + h)^2 + R^2 = H^2 + h^2 + 2Hh + R^2. \tag{5.26b}$$

$$\Delta = \mathrm{SU} - \mathrm{ST} \tag{5.26c}$$

$$\mathrm{SU}^2 - \mathrm{ST}^2 = 2\Delta\mathrm{ST} + \Delta^2 = 4Hh.$$

Putting ST $= R$ and ignoring the second-order term Δ^2,

$$\Delta = \frac{2Hh}{R} \tag{5.27a}$$

$$\Phi = \frac{2\pi\Delta}{\lambda} = \frac{4\pi Hh}{R\lambda}\ \text{rad}. \tag{5.27b}$$

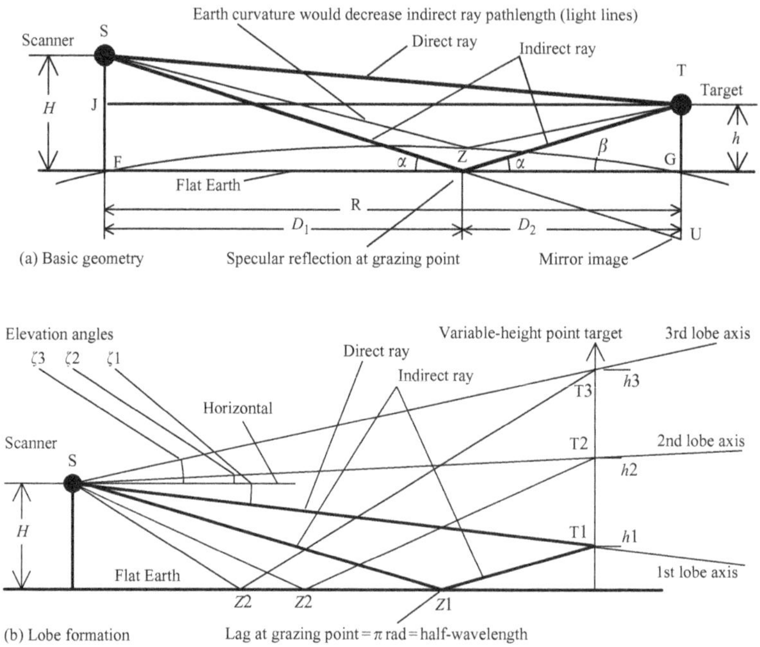

Figure 5.12 Ray traces, flat-Earth approximation

This expression overstates Φ under real curved-Earth conditions where $H > H'$ and $h > h'$. The grazing angle is

$$\alpha = \arctan\left[\frac{H+h}{R}\right]. \tag{5.28}$$

5.6.2 Approximate multipath ranges

The flat-Earth method gives the following approximations. The nth multipath peak is when $\phi = 2n\pi$. If the phase shift at the surface is π, $\phi = (2n-1)\pi$ and $\sin\phi = 1$, its highest possible value, substituting in Eq. (5.27) gives

$$\frac{2\pi Hh}{R\lambda} = (2n-1)\frac{\pi}{2}, \qquad \text{so } R = \frac{4Hh}{(2n-1)\lambda}. \tag{5.29a}$$

Flat-Earth peak ranges P_1, P_2, etc. are:

$$P_1 = \frac{4Hh}{\lambda} \text{ m (first peak, longest range)}$$

$$P_2 = \frac{4}{3}\frac{Hh}{\lambda} = \frac{P_1}{3} \text{ m}$$

$$P_3 = \frac{4}{5}\frac{Hh}{\lambda} = \frac{P_1}{5} \text{ m}$$

and so on. The nth peak is at:

$$P_n = \frac{P_1}{2n-1}\ \text{m}. \tag{5.29b}$$

Flat-Earth nulls occur when $\sin(2\pi Hh/R\lambda) = -1$, its lowest possible value. Null ranges are:

$$\begin{aligned}
N_1 &= \frac{2Hh}{\lambda} = \frac{P_1}{2}\ \text{m (first null)} \\
N_2 &= \frac{1}{2}\frac{2Hh}{\lambda} = \frac{P_1}{4}\ \text{m} \\
N_3 &= \frac{1}{3}\frac{2Hh}{\lambda} = \frac{P_1}{6}\ \text{m} \\
N_n &= \frac{1}{n}\frac{2Hh}{\lambda} = \frac{P_1}{2n}\ \text{m}.
\end{aligned} \tag{5.30}$$

The method is accurate at short range but gives somewhat high values at long ranges, particularly for the first peak and null. Ranges calculated by Eqs (5.29) and (5.30) may be regarded as the maximum likely to be observed in practice. Peak and null ranges for the example system are indicated on Figure 5.4, which indicates how their errors increase with range.

5.6.3 Tailoring null ranges

It is sometimes desirable to avoid nulls at ranges critical to the VTS traffic scenario, or to site racons to avoid ships being in nulls when within the mandatory channels of traffic separation schemes (TSS) as far as possible within the variable ship scanner and tide height constraints. When planning such systems, the rate of change of null range with scanner height is of interest and is obtained by differentiation of Eq. (5.30) with respect to H;

$$\frac{\mathrm{d}R}{\mathrm{d}H} = \frac{2h}{\lambda}\ \text{m range change per metre of scanner height change.} \tag{5.31}$$

For example, a fishing boat with radar reflector at $h = 4$ m will be in the first null of a 9 GHz band VTS scanner at $H = 15$ m when $R \sim 2 \times 15 \times 4/0.032 = 3750$ m (Eq. (5.30)). Eq. (5.31) predicts that when a falling tide raises the effective scanner height, null range increases by $2 \times 4/0.032 = 250$ m per metre fall of tide.

5.6.4 Vertical lobe structure

Path length difference Δ increases with target height. The resulting path phase difference Φ will increase and there will be alternate heights of destructive and constructive interference, exact values depending on frequency, sketched in Chapter 6, Section 6.3.1, Figure 6.1. This is equivalent to the scanner radiating a fan of lobes in the vertical plane, instead of reflecting at the grazing point. These lobes are independent of, usually narrower than, and additional to, the scanner's inherent elevation polar diagram, which for current purposes is assumed wide and uniform.

Figure 5.12(*b*) shows a flat-Earth system with targets T1, T2, T3 at heights h_1, h_2, h_3, with phase reversal at grazing points $Z1$, $Z2$, $Z3$, respectively. When indirect ray path S, Z1, T1 exceeds direct ray path ST by $\lambda/2$, total phase difference is λ. Constructive interference forms a resultant lobe at elevation angle $\zeta 1$. Similarly higher target positions T2, T3,. . ., form lobes at elevation angles $\zeta 2$, $\zeta 3$ and so on, with destructive interference between the lobes. The angles are found as follows, using the flat-Earth approximation. Eq. (5.27a) relating Δ to target height is $\Delta = 2Hh/R$. With the usual phase reversal at the surface ($\psi = \pi$ rad, equivalent to $\lambda/2$), the path length difference is n wavelengths, where

$$n = \frac{1}{2} + \frac{\Delta}{\lambda} = \frac{1}{2} + \frac{2Hh}{R\lambda}. \tag{5.32a}$$

When $(n + \frac{1}{2})$ is an integer, the indirect ray is in phase opposition with the direct ray, giving inter-lobe nulls. Peaks are when n is an integer. The first (lowest) peak is when $n = 1$. Substituting in Eq. (5.32a)

$$h_1 = \frac{R\lambda}{4H}. \tag{5.32b}$$

The second peak is when $h_2 = 3R\lambda/(4H)$ and the mth when $h_m = (2m - 1)\times R\lambda/4H$. Bearing in mind that the resultant beam emanates from the scanner at height H, the lobe is inclined upwards at angle $\zeta 1$, where

$$\zeta 1 = \arctan\left(\frac{h1 - H}{R}\right) = \arctan\left[\frac{\lambda}{4H} - \frac{H}{R}\right]. \tag{5.32c}$$

The mth lobe inclination is

$$\zeta m = \arctan\left[\frac{(2m - 1)\lambda}{4H} - \frac{H}{R}\right]. \tag{5.32d}$$

Figure 5.13 indicates approximately how the first three lobe elevations vary with scanner height in the 3 and 9 GHz bands, assuming flat Earth and an arbitrary 10 km range. An aircraft target approaching at low altitude would encounter the lobes in turn and the vertical lobe structure may be of some interest to marine users making rendezvous with helicopters, for example, when embarking a pilot. As an aside, early in the Second World War, Chain Home radar operators were able to deduce heights of enemy aircraft from their null ranges, having estimated atmospheric refraction from observed nulls of calibrating auto-gyro flights at known altitude. Luckily, the enemy did not rumble the poor low-level cover of these low frequency (~30 MHz) radars and obligingly came in at 15 000 ft.

Calculation of performance of marine radar systems from the lobe structure is usually unfruitful, but the concept of, say, a lighthouse racon or other point target ‘flying in’ through the lobes as own ship approaches may help to visualise the effects of multipath interference. At extreme range, the target is below the lowest lobe, as illustrated in Chapter 6, Figure 6.1. In military and jouralistic parlance, it is ‘under the radar beam’, making sea-skimming missiles so deadly to warships lacking airborne radar cover. Note that increasing scanner height depresses the lobe, assisting detection

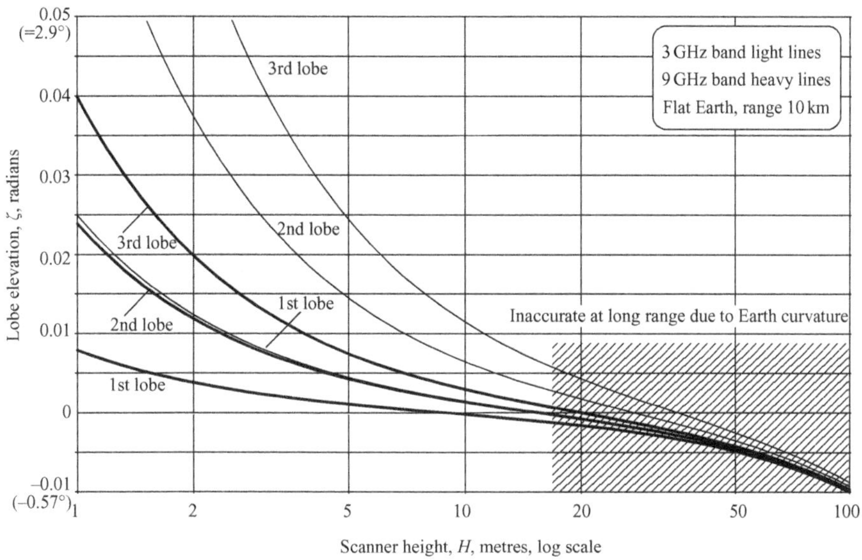

Figure 5.13 Lobe elevations. Approximate elevation angles of lowest three lobes at 3 and 9 GHz bands. The first 3 GHz lobe almost coincides with the second 9 GHz lobe

of low targets at long range. Note also that when H is low enough for the flat-Earth assumption to be valid, lobe elevations are more or less tripled at 3 GHz relative to 9 GHz, so low targets remain 'below the beam' to shorter range.

5.6.5 Dispersion

The split of echoes between direct and indirect paths is dispersive, for the indirect echo arrives later. In the extreme, the indirect component of echo pulse would cease to overlap the direct echo. The shortest pulselength in common use is 50 ns, which would cease to overlap when $\Delta = 15\,\text{m}$. From Eq. (5.27a), this occurs at range $R = 2Hh/\Delta$. Substituting high values of 100 and 30 m for H and h, respectively, the echo components cease to overlap at ranges below 400 m, where $\alpha > 18°$, but here reflection from the surface is very low (Section 5.7.5) weakening the indirect ray. Dispersion is therefore rarely significant in practice.

5.7 The sea

We here remove the approximation we have made of perfect reflection from the sea surface at the grazing point. The amplitude and phase of the forward reflection in fact depend on the roughness of the surface, the electrical properties of the water, the plane of polarisation and the spherical figure of the Earth. But first we need to look at the properties of sea waves. Overfalls occur when currents flow over shallows,

Figure 5.14 *Capillary waves and heavy traffic. Sunlight emphasises capillary waves in fair weather with a light breeze. VTS systems need extensive target tracking capacity in such busy waters as Tokyo Bay, with many ships under way and at anchor. Author*

or two currents meet. Tsunami, also called seismic sea-waves or tidal waves, are triggered by underwater geological events such as earthquakes or mud slides. With these exceptions, waves are generated by wind (Figure 5.16).

5.7.1 Capillary and gravity waves

Roughness is described by a sea state number (see Section 5.7.5). A breeze (wind speed <3 m/s, 6 knots, Beaufort wind scale 0–1) playing on calm water (sea state 0–1) causes capillary waves with amplitude of a few millimetres and wavelength of a few centimetres; although formally restricted by oceanographers to <25 mm, we may admit rather longer wavelengths. Capillaries are a surface effect, containing little momentum, and often appear as intersecting diagonal ripples, scallops or facets. Capillaries do not propagate beyond the breeze area and immediately cease when the breeze drops. Figure 5.14 shows sunlight glinting off capillary waves in fair weather with a light breeze.

When a wind arises, it transfers energy to the sea surface, dragging the water into movement, rotating the water molecules in vertical circles. At the surface, ordinary sea-waves start to form, called gravity waves by oceanographers but not to be confused with the gravity waves of theoretical physics which relate to fluctuations in the gravitational field. The form is approximately cycloidal with sharp peaks at the crests. A cycloid, also called a trochoid, is the U-shape curve traced by a point on a circle rolling along a line. Its form is $x = h/2\,(t - c\sin t)$, $y = h/2\,[(c\cos t) - 1]$, where parameter t increases from 0 to 2π during the cycle and h is the peak–peak height. Shape is set by parameter c: for a sea, $c = 1$, falling to around 0.25 for a swell.

Wind pushes on the wave flanks, distorting the cycloid by steepening the lee slope. The vertical rotation couples energy to the water at depth, so a very great mass is affected, accumulating huge momentum. While the wind persists, the waves are termed a sea. After some hours of steady wind, wave height ceases to rise further and reaches an equilibrium value for the wind speed and water depth. This ‘fully developed sea’ is the only time wave height can be directly related to present wind speed. Here the power input from the wind is balanced by power propagated outwards

Figure 5.15 *The fine structure of gravity waves persists in a seaway. The reflecting facets are tilted by the large gravity waves to increase sea clutter reflection while reducing forward reflection and multipath interference. From MV Pharos. Author, reproduced by permission of the Northern Lighthouse Board*

to generate a swell in areas beyond the wind. The wave crests progress at one quarter to one half wind speed with standard deviation around one eighth wind speed, according to Barton [8]. Discrete sea spikes of water are thrown up at irregular intervals and may persist for up to 10 s. When the wind ceases, momentum causes a swell to persist for some hours. Its waveshape is more sinusoidal than the U-shape waves of a sea. Unless the local wind drops completely to leave an 'oily' swell, capillary waves are always superimposed on the bigger gravity waves or swell, and, when tilted by the latter, are responsible for most of the retro-reflection which causes sea clutter. Figure 5.15 shows the fine structure persisting in a seaway.

Wind speed exceeding about 10 m/s, or less if the sea has not reached equilibrium, causes whitecaps whose spume is akin to severe localised rain.

Wavelength of gravity waves depends on water depth, the distance (fetch) over which the wind is active and other factors and is about 10–125 times wave height. Wave slope rarely exceeds $\frac{1}{10}$ (6°) in open water. Height and wavelength vary wave to wave; open-ocean storm wavelengths can exceed 200 m. Breakers on a shoal or beach are caused by a modification of waveform as the waves reach shallow water; the reduced depth forces wavelength to decrease and height to increase until the waves become unstable and topple, breaking when the height exceeds $\frac{1}{7}$ the length. Heavy rain tends to suppress wave height somewhat – dare we say it damps the waves? Oil or thin ice on the surface, or an obstruction to windward, also deaden the sea. In shallow water, the sea bed contours also influence wave height.

The energy in waves striking a shoreline is partly absorbed as spray and partly reflected. When reflected and incident wave trains meet, strong peaks develop locally. The process is analogous to the generation of standing waves in a mismatched microwave feeder. In enclosed areas such as the North Sea or the Great Lakes, reflected waves augment the roughness and the sea tends to become confused. An L-shaped coastline such as SW France/N Spain acts as a corner reflector; wind over the long N Atlantic fetch raises waves which bounce twice and return on a reciprocal course, giving the Bay of Biscay its reputation for *mal de mer*. Wind shifts cause the sea raised by the new wind direction to interact with swell persisting from the old, again giving a confused sea and occasionally producing abnormal waves of great height, see Chapter 11, Section 11.6.6.

Retro-reflection from each wave causes sea clutter, discussed in Chapter 11, Section 11.6. The statistical distribution of clutter spike amplitude from confused seas differs from those of a steady sea or swell, with higher probability of spikes greatly exceeding mean clutter power, necessitating particularly high signal to mean clutter power ratio for high probability of detection, discussed in detail in Chapter 11, Sections 11.7.2–11.7.4, and Chapter 12, Sections 12.4.2 and 12.4.3.

Chapter 7 will examine the micro and macro structures of man-made reflectors. Capillary waves may be regarded as the micro-structure of the sea surface and gravity waves as its macro-structure.

5.7.2 Radar reflection, capillaries alone

Figure 5.16(a) shows a flat-calm sea with no capillary waves. The sea surface is a good electrical conductor, so rays at an acute grazing angle are specularly reflected, as from a glossy optical mirror, with little loss. There is near-perfect forward reflection, so

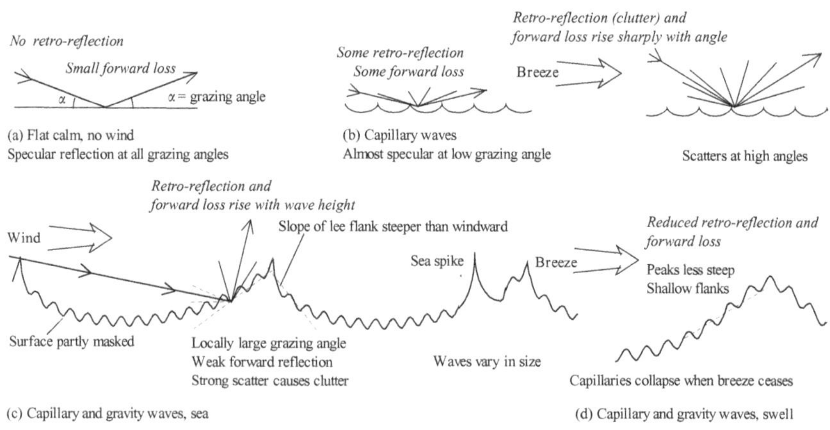

Figure 5.16 Scattering at the sea surface. Capillary waves scatter the rays; gravity waves tilt the surface and increase the return (clutter) component at expense of forward (multipath reflection) component. Wave tips may screen small targets

multipath interference is severe; conversely retro-reflection toward the radar is minute and sea clutter is small.

When capillaries are present but not gravity waves, Figure 5.16(*b*), a scattering process is introduced. An optical analogy would be a sheet of semi-gloss paper. At acute grazing angles reflection is almost specular. As the angle rises, the forward reflection component falls (reducing multipath interference) and there is more scattering over a wide solid angle. The component returned towards the radar as sea clutter rises significantly. Sea clutter therefore is most severe when the angle of incidence is large (several degrees), that is, when range is short and the scanner is high.

If the capillary wavelength subtends half a radar wavelength in the plane of incident radar waves, Bragg resonance occurs, increasing the effective roughness of the capillaries. This effect is unpredictable but more likely to occur in the J (κu) and 9 GHz bands than at 3 GHz, whose wavelength exceeds that of most capillaries. Resonance introduces a few decibels uncertainty into the clutter return strength for a given wave height. Radar detection cells are always much larger than a capillary wavelength, so individual capillaries cannot be resolved on the display and the weak clutter returns are almost random from scan to scan with near-Gaussian (normal) distribution.

5.7.3 Radar reflection, gravity waves

A sea with capillaries superimposed on gravity waves appears mottled, indicated in Figure 5.16(*c*). As the sea gets rougher, gravity waves tilt the local surface so that the radar sees the leading flanks of the waves at a steeper angle. Trailing faces are partially masked, but the overall effect is to raise the mean effective grazing angle to capillaries. So rough seas are poor forward reflectors (of interest in this chapter) but strong clutter sources (Chapter 11).

Above about sea state 5, wave front inclination ceases to steepen further and the scattering is not significantly increased by further rise in sea state – clutter ceases to increase and forward reflection remains weak. When the radar is looking into the wind, viewing the steep lee slopes, clutter is about 5 dB higher and forward reflection less than when looking down-wind on the shallower windward slopes. As well as reducing gravity wave height, rain tends to break up the capillaries and makes reflection more specular, these effects together being equivalent to a reduction in effective sea state.

Wavelength of gravity waves may be many metres, precluding Bragg resonances. On the other hand, crests may be far enough apart to exceed detection cell width. Waves are then resolved as individual objects, giving a stippled effect on the radar display. Individual wave paints can be quite strong and often persist for several successive scans – the clutter return has ceased to be random and takes on much of the character of a conventional target. Chapter 12 will show that this makes detection of small targets particularly difficult in heavy seas.

Swell has shallower flanks for a given wave height, shown by Figure 5.16(d), and scatters in the same way as a sea of lower height. The reflective properties of a given sea state probably also vary with location, between open ocean and enclosed seas with wave-reflection from coasts. Of course, when the wind speed is changing the sea behaves intermediately between a fully developed sea and a pure swell. Importantly for radar, capillaries remain present, superimposed on the big gravity waves, unless the local wind drops completely.

It is not surprising that published data for reflection, obtained at differing times and places with differing equipment and experimental techniques, sometimes appears inconsistent, further bedevilled by a multiplicity of measures of roughness.

5.7.4 Wave height

No wind is uniform in strength and direction from instant to instant, so individual waves vary in height in a quasi-random manner. Except as stated below, 'height' is measured peak-trough. There are many ways of expressing wave height. Unfortunately not all writers define which they are using. The commonest measures include the following.

1. Significant wave height. Standard deviation h_s of height distribution, (sometimes denoted σ_s or σ_h), expressed in metres. For a cycle divided into n equal time increments of heights $h_1, h_2, \ldots, h_n$, average h_a, standard deviation $= 1/n \times \{\sum[(h_1-h_a)^2+(h_2-h_a)^2+\cdots+(h_n-h_a)^2]\}^{1/2}$. The mean square surface height is h_s^2 and is the statistical variance about its mean. Because reflection phenomena are more fundamentally linked to this statistical representation of events than any other measure, it is perhaps the most satisfactory measure for calculation of reflection properties. The distribution of wave heights is not Gaussian.

   For a steady train of uniform oscillations of peak-trough height h

   $$h_s \sim Ah \tag{5.33}$$

   Factor A depends on the wave shape. For a train of equal cycloids, $A = 0.27$. For a swell approximating a sine wave, $A = 1/2\sqrt{2} = 0.35$.
2. $H_{1/3}$, the peak-trough height exceeded by $\frac{1}{3}$ of the waves, also called the significant wave height H_s. High waves carry disproportionately higher energy, so this description of wave height is biassed toward high waves by defining the significant wave height as the peak-trough amplitude exceeded by 1 in 3 rather than 1 in 2 of the waves observed. The relationship between $H_{1/3}$ and sea state is included in Table 5.3. $H_{1/3}$ is connected with σ_s by Burling's relationship:

   $$H_{1/3} = 4h_s. \tag{5.34}$$

   For a steady train of true cycloids, peak-trough height $h = H_{1/3} = h_s/0.27 = 3.7h_s$, but Eq. (5.34) uses $4.0h_s$ to account for the typical height distribution in a quasi-random train of actual sea-waves.
3. Average peak-trough wave height $= \dfrac{1}{1.6}H_{1/3} = 0.625H_{1/3} = 2.5h_s$. (5.35)
4. Peak-trough height exceeded by one wave in ten, h_w.

   $$h_w = 2.04 \times \text{average} = 1.275H_{1/3} = 5.1h_s. \tag{5.36}$$

Table 5.2 Wind and wave height

Beaufort wind scale (number)	Mean wind speed (knots)	Wind description	Average wave height (m)	Probable max. wave height $H_{1/3}$ (m)	h_s per Eq. (5.37) (m)
0	< 1	Calm	0	0	0
1	1–3	Light air	0.1	0.1	<0.01
2	4–6	Light breeze	0.2	0.3	0.03
3	7–10	Gentle breeze	0.6	1	0.09
4	11–16	Moderate breeze	1	1.5	0.21
5	17–21	Fresh breeze	2	2.5	0.42
6	22–27	Strong breeze	3	4	0.72
7	28–33	Near gale	4	5.5	1.14
8	34–40	Gale	5.5	7.5	1.71
9	41–47	Strong gale	7	10	2.43
10	48–55	Storm	9	12.5	3.33
11	56–63	Violent storm	11.5	16	4.44
12	>64	Hurricane	14 or over	–	5.76

5. Beaufort wind speed number, K_B, is said by Kerr [5, Section 1.4] to be related to h_s for a fully developed sea by the following equation. However, Kerr indicates this formula may understate wave height, so we prefer not to use it:

$$h_s = \frac{K_B^3}{300}\ \text{m}. \tag{5.37}$$

Table 5.2 gives the relationship of Beaufort number, wind speed and wave height for a fully developed sea in deep water. Table 5.3 contains part of the World Meteorological Office (WMO) table connecting sea state and wave height ($1\ \text{kt} = 0.5144\ \text{m/s} = 1.852\ \text{km/h}$).

5.7.5 Sea state

Although wave gauges are maintained at coastal and some offshore sites, wave height at sea generally has to be estimated by eye. The mariner's practical unit is the sea state number. Estimation is subjective – among other factors, the sea always appears more formidable from a small craft! Experienced observers often differ by 1 unit. Any tendency of observers to over-estimate sea state in strong wind might partly balance the additional effect of wind-induced flank slope. IMO and WMO sea state names and their relationship with wave height brackets are given in Table 5.3, based on Skolnik [9]. The data relating reflection and sea conditions is so imprecise that little or no accuracy is lost by expressing sea states as whole numbers. We shall always assume the highest wave height in the bracket for a given sea state, for example, $h_s 0.3\ \text{m}$ for SS3. It will be noted that there are gaps in the h_s column, which we have closed by the Modified h_s column. In spreadsheet calculations, Chapter 14, we

Table 5.3 Sea state and wave height

Sea state	Description		Wave height (m)			
	IMO	WMO	$H_{1/3}$	h_s	Modified h_s	Eq. (5.38a)
0	Calm	Calm, glassy	0	0	0–0.01	0.010
1	Smooth	Calm, rippled	0–0.1	0.01–0.025	0.01–0.025	0.0282
2	Slight	Smooth, wavelets	0.1–0.5	0.025–0.125	0.025–0.125	0.1282
3	Moderate	Slight	0.6–1.2	0.15–0.3	0.125–0.30	0.2937
4	Rough	Moderate	1.2–2.4	0.3–0.6	0.30–0.60	0.6008
5	Very rough	Rough	2.4–4.0	0.6–1.0	0.60–1.0	1.0274
6 upwards			> 4	> 1	> 1	

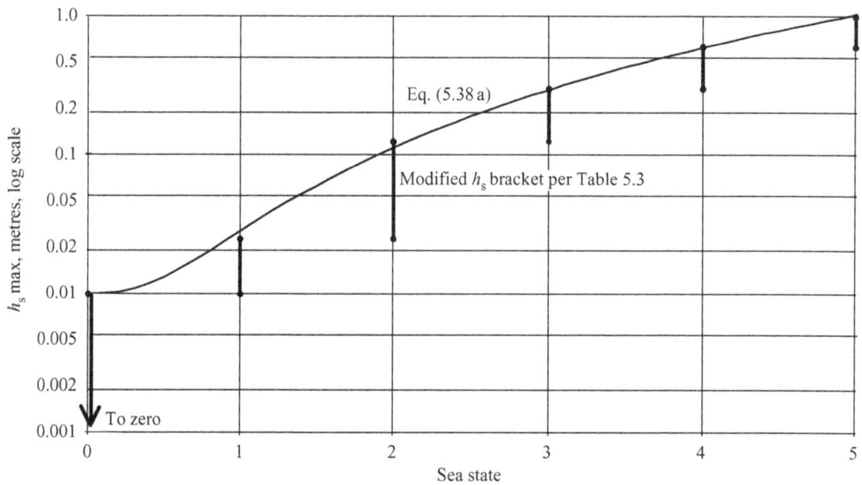

Figure 5.17 Sea state and wave height. Shows that Eq. (5.38a) is a good approximation

shall express sea roughness by modified h_s and convert to sea state by Table 5.3 to find the sea state, from which to derive sea clutter by the methods of Chapter 11, Section 11.6.2. Sea state (SS) and h_s are connected approximately by the following empirical equation (metres units), graphed in Figure 5.17:

$$\text{Modified } h_s \text{ max} \sim 0.01 + 0.0182\text{SS}^{2.5} \text{ m} \tag{5.38a}$$

from which

$$\text{SS} \sim 4.966\ (\text{modified } h_s \text{ max} - 0.01)^{0.40}. \tag{5.38b}$$

5.8 Forward reflection from the grazing point

5.8.1 Reflection coefficient amplitude

Having considered the geometry of the indirect ray, we now examine how it is affected by the forward reflection at the grazing point both in amplitude and phase, prior to calculation of multipath coefficients in Chapter 6. Instead of the geometrically perfect mirror assumed so far, we have to cope with the vagaries of the real sea surface.

The forward reflection coefficient ρ of the surface at the grazing point is the (scalar) voltage specularly reflected towards the target per volt of incident signal. It is the product of three terms:

- ρ_0, reflection from the surface of a plane smooth air-reflector (e.g. water) interface, see Section 5.8.2;
- d, divergence, describing the effect of Earth curvature, a geometrical effect already discussed in Section 5.5.5;
- ρ_s, reflection coefficient of surface roughness, accounting for waves, see Section 5.8.4.

$$\rho = \rho_0 d \rho_s. \tag{5.39}$$

5.8.2 Reflection coefficient, ρ_0, of smooth plane surface

Although there may be land at the grazing point, water is the reflector of most interest. The dielectric constant, ε, is high and almost the same whether fresh or salt. The conductivity of pure water is low, but sea water, S, is raised by the conductive dissolved salt ions. Land at the grazing point has low dielectric constant and lowish conductivity, values varying widely according to texture and wetness. Although the following treatment examines the reflection process in detail, for a water surface it often remains sufficient to assume perfect forward reflection with 180° phase reversal, as earlier in the chapter.

When an oblique ray strikes the plane surface of a thick block of lossy dielectric, such as calm water, a small proportion of the energy passes into the material and is lost as heat. The remainder is specularly reflected from the surface. This forward-reflected component of incident voltage ρ_0', is a vector quantity which may be resolved into a scalar reflection coefficient ρ_0 at phase angle ψ. The pair is sometimes called the square of the complex index of refraction. The terms depend on plane of polarisation, grazing angle α, the dielectric constant and conductivity of the reflector, and to a lesser degree on radar frequency.

The following relationships, given by Terman [10], are derived from the laws of optics and apply to non-magnetic reflectors.

Horizontal polarisation:

$$\rho_0' = \frac{\sin\alpha - \sqrt{\varepsilon' - \cos^2\alpha}}{\sin\alpha + \sqrt{\varepsilon' - \cos^2\alpha}}. \tag{5.40a}$$

Table 5.4 Surface coefficients

Reflector	Dielectric constant, ε	Conductivity, S (S/m)
Sea water	81	3–5
Fresh water	80	0.001–0.01
Wet earth	5–20	0.001–0.1
Dry earth	2–5	10^{-5}–10^{-4}

Vertical polarisation:

$$\rho_0' = \frac{\varepsilon' \sin\alpha - \sqrt{\varepsilon' - \cos^2\alpha}}{\varepsilon' \sin\alpha + \sqrt{\varepsilon' - \cos^2\alpha}} \tag{5.40b}$$

where $\rho_0' = \rho_0\angle\psi$ (ψ in radians), α is the grazing angle (rad) and ε the dielectric constant of reflector (relative permittivity, vacuum taken as unity, scalar number).

$$\varepsilon' = \varepsilon - \mathrm{j}\left(0.18 \times 10^9 \frac{S}{f}\right), \qquad \text{a complex number} \tag{5.40c}$$

where S is the bulk conductivity of reflector, S/m, Siemens per metre, scalar, f the frequency (Hz, scalar) and $\mathrm{j} = \sqrt{-1}$.

The factor 0.18 is sometimes replaced by $1/2\pi$ or $\frac{1}{6}$; the difference in outcome is small. The factor 10^9 becomes 10^{11} if S is expressed as the reciprocal of bulk resistivity in ohm cm units rather than our SI units of ohm m; f is sometimes replaced by $300 \times 10^6/\lambda$.

Meikle [11] gives typical values for reflection coefficients, reproduced in Table 5.4.

Eqs (5.40a) and (5.40b) are evaluated as follows, coefficients A–K being scalar and sets of real or imaginary terms being grouped in square brackets. We first evaluate $\sqrt{\varepsilon' - \cos^2\alpha}$.

$$\sqrt{\varepsilon' - \cos^2\alpha} = \sqrt{[\varepsilon - \cos^2\alpha] - \mathrm{j}\left[10^9 \frac{S}{2\pi f}\right]}$$

$$= \sqrt{A - \mathrm{j}B} = \frac{A - \mathrm{j}B}{(A^2 + B^2)^{1/4}} = C - \mathrm{j}D, \tag{5.40d}$$

where $A = [\varepsilon - \cos^2\alpha]$, $B = [0.18 \times 10^{12} S/f]$, where f is in Hz, $C = A/[A^2 + B^2]^{1/4}$ and $D = B/[A^2 + B^2]^{1/4} = BA/C$.

Eqs (5.40a) and (5.40b) can be combined as

$$\rho_0' = \frac{\varepsilon' \sin\alpha - \sqrt{\varepsilon' - \cos^2\alpha}}{\varepsilon' \sin\alpha + \sqrt{\varepsilon' - \cos^2\alpha}} = \frac{E - \mathrm{j}F - (C - \mathrm{j}D)}{E - \mathrm{j}F + (C - \mathrm{j}D)} \tag{5.40e}$$

where the following hold.

Horizontal polarisation:

$$E = \sin\alpha \quad F = 0.$$

Vertical polarisation:

$$E = \varepsilon \sin\alpha \quad F = B \sin\alpha.$$

Collecting terms and multiplying through by the complex conjugate of the denominator, $[E + C] + \mathrm{j}[D + F]$, to turn the denominator into a real number:

$$\rho_0' = \frac{[(E+C)(E-C)+(F-D)(F+D)] - \mathrm{j}[2(CF-DE)]}{(E+C)^2+(D+F)^2} = \frac{G}{K} - \mathrm{j}\frac{H}{K}$$

where $G = (E + C)(E - C) + (F + D)(F - D) = E^2 + F^2 - (C^2 + D^2)$, $H = 2(CF - DE)$, $K = (E + C)^2 + (D + F)^2$.

Converting,

$$\rho_0 = \frac{\sqrt{G^2 + H^2}}{K} \text{ numerically} \tag{5.40f}$$

and

$$\psi = \arccos\left[\frac{G}{K\rho_0}\right] \text{ rad.} \tag{5.40g}$$

5.8.3 *Reflection coefficient variation*

Although Eqs (5.40a) and (5.40b) are compact, Eqs (5.40d) and (5.40e) show they mask complicated relationships between reflection coefficients and grazing angle, depicted for some reflector materials in Figures 5.18 and 5.19. The figures concentrate on the 3 GHz band, since 9 GHz marine radars are confined to horizontal polarisation. They plot ρ_0 and ψ to a base of α, logarithmic scaling giving more open depiction of the low α values of most practical importance. Terrain parameters vary so widely that the 'land' curves can be indicative only. Maritime radars invariably operate with grazing angle $\alpha < 0.35$ rad ($\sim 20°$), corresponding to 275 m range for a VTS scanner as high as 100 m above the sea surface.

With horizontal polarisation, the ($\sin\alpha$) terms of Eq. (5.40a) are always smaller than the $\sqrt{(\varepsilon' - \cos^2\alpha)}$ terms. When α is small, $\rho_0' \sim -1.0 + \mathrm{j}0$, equivalent to $1.0\angle\pi$, meaning perfect reflection with phase reversal. Specular reflection at low grazing angle is efficient because little energy is refracted into the calm water. If the polarisation is horizontal, there can be no nett field at the conducting surface, induced current loops setting up an equal and opposite field to the incident field, equivalent to reversing the phase of the reflected ray.

As grazing angle rises, ρ_0 falls only slightly, to a degree dependent on conductivity, and ψ always remains almost exactly π rad. Figures 5.18 and 5.19 confirm Eq. (5.40a) is much ado about nothing and for almost all horizontally polarised

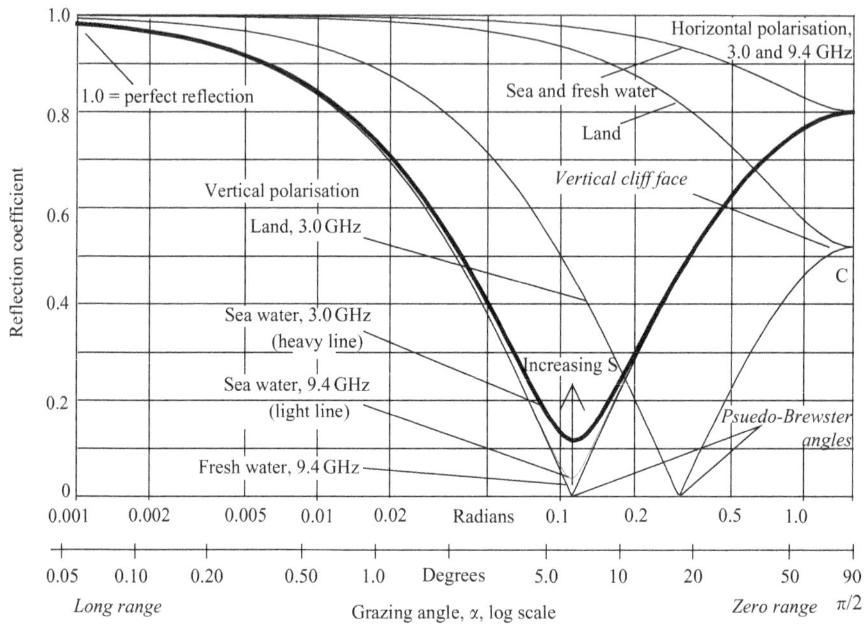

Figure 5.18 Forward reflection coefficient amplitude. Smooth plane surface term. Sea water $\varepsilon = 81$, $S = 4$. Fresh water, $\varepsilon = 80$, $S = 0.003$. Land, $\varepsilon = 10$, $S = 0.01$. C describes conditions at a wet vertical cliff face

marine purposes a plane smooth surface may be regarded as a perfect phase-reversing mirror, $\rho_0' \sim 1\angle\pi$ or $-1\angle 0$ which is the same thing, as previously stated.

When $\alpha = \pi/2$ (normal incidence, where horizontal and vertical polarisations become merely a matter of surface rotation), Eqs (5.40a) and (5.40b) both simplify to the same expression:

$$\rho_0 = \frac{1 - \sqrt{\varepsilon}}{1 + \sqrt{\varepsilon}}. \tag{5.40h}$$

Circularly polarised rays contain equal horizontally and vertically polarised components. At low α the forward reflection coefficient to CP is therefore the average of the horizontal and vertical polarisation values.

Vertical polarisation causes much different behaviour because the $(\varepsilon' \sin\alpha)$ terms of Eq. (5.40b) become significant; ρ_0 falls as α rises from zero, reaching a very low value at a grazing angle called (by analogy with optics) the psuedo-Brewster angle, α_B, before recovering at higher α. At vertical incidence where $\alpha = \pi/2$, ρ_0 has to be identical to the HP value. As α goes through α_B, ψ falls from near π to near zero, very abruptly when conductivity is low. At α_B, conductance is zero and ρ_0 would be zero if the reflector had zero conductivity. The residual ρ_0 is also dependent on wavelength and dielectric constant. Unfortunately the psuedo-Brewster angle is

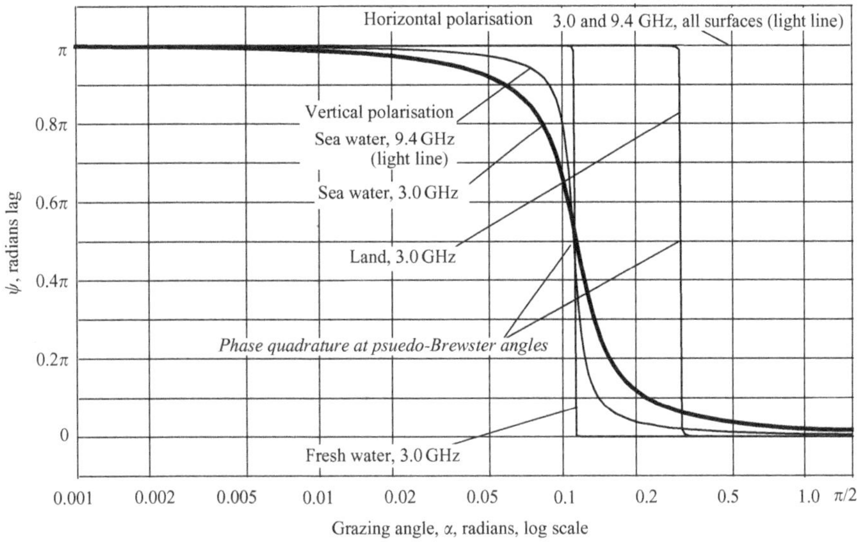

Figure 5.19 Forward reflection coefficient phase. Smooth plane surface term. With vertical polarisation, phase reverses at the psuedo-Brewster angle

less than the maximum grazing angle reached in some short-range applications. Here $(E^2+F^2) = (C^2+D^2)$, $G = 0$, ρ_0 is minimum and $\psi = \pi/2$. The psuedo-Brewster angle, α_B, is approximated at radar frequencies by

$$\sin \alpha_B \sim \frac{1}{\sqrt{\varepsilon + 1}}. \tag{5.40j}$$

For example, when $\varepsilon = 81$ (sea water), $\alpha_B = 0.11$ rad $= 6.3°$. All other reflectors have lower ε and hence higher α_B; for example, the 'land' of Figure 5.18 has $\alpha_B = 17.5°$.

It is not usually necessary to make much use of ψ, which merely modifies the range at which multipath nulls occur at very short-range. The short-range nulls (at ranges where $\alpha > \alpha_B$) of a horizontal/vertical polarisation pair of co-sited radars interleave, one benefiting from constructive interference at the ranges at which the other suffers destructive interference. Use of such a pair in diversity mode would improve cover in this area. But the ranges in question are so short that the complexity is seldom, if ever, justified.

5.8.4 Reflection coefficient, ρ_s, of surface roughness

This scalar term, sometimes called the Fresnel reflection coefficient, is the proportion of incident voltage specularly reflected in a forward direction from a fully conducting plane rough surface. It is independent of radar polarisation but depends on grazing angle, radar wavelength and roughness of the sea-waves; the latter depending on their height, period, steepness, and direction of wavefront, themselves affected by the present and past winds, fetch, water depth, topography, etc. Various empirical

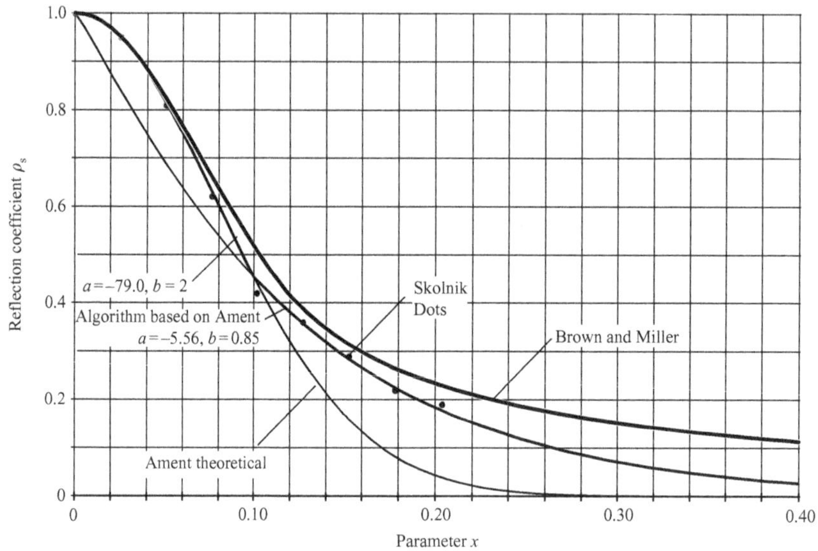

Figure 5.20 *Sea surface forward reflection coefficient, ρ_S, plotted to base of $x = (h_S \sin\alpha)/\lambda$. This expression accounts for the effects of wave height, grazing angle and radar wavelength. ρ_S may differ considerably from values shown. Ament's theoretical values when $x < 0.1$, an algorithm fits experimental data at higher x. Brown and Miller's modification gives higher ρ_S at high x. High ρ_S gives high multipath and occurs when x is low: calm sea (low h_S), long range (low α), or high λ (3 GHz band)*

formulae exist for ρ_S, not always in good agreement, probably because of differing measurement techniques and the differing qualities of waves at inshore, oceanic, etc., test locations.

A formula due to Ament, quoted in Blake [12], excludes the effect of shadowing at small grazing angles. In our terminology

$$\rho_S = \exp(-Y) \tag{5.41a}$$

where $Y = 8\pi^2 x^2$ and $x = (h_S/\lambda)\sin\alpha$.

The trials data had wide scatter and this expression is acknowledged to understate ρ_S when $x > 0.1$. Here Ament gives an empirical curve. We have devised a modification of Eq. (5.41a) which also fits the empirical portion and is valid for all values of x:

$$\rho_S \sim \exp[a(x)^b] \tag{5.41b}$$

if $x < 0.1$, $a = -8\pi^2 = -79.0$, $b = 2$ (to match Eq. (5.41a)), otherwise $a = -5.56$, $b = 0.85$ (to fit Ament's empirical curve).

The algorithm and source data have been plotted in Figure 5.20(a) and show good agreement. The inflexion at $x = 0.1$ is presumably smoothed out in actuality.

Skolnik [13] gives similar data, shown as dots in Figure 5.20. Brown and Miller in Blake [12] add an extra term to Eq. (5.41a) to raise ρ_0 when x is high. As shown in Figure 5.18, this also avoids the inflexion at $x = 0.1$. Being more recent than Ament, we shall use this interpretation to describe surface roughness:

$$\rho_s = \exp(-Y) \times I_0(Y) \text{ numerically.} \tag{5.41c}$$

$I_0(Y)$ is the Bessel function of order zero with imaginary argument, Y. When n is an integer, $n!$ means factorial $n = 1 \times 2 \times 3 \cdots \times n$, and $\Sigma_{n=0}^{\infty}$ is the sum of the terms taking $n = 0, 1, 2, \ldots, \infty$

$$I_0(Y) = \sum_{n=0}^{\infty} \frac{Y^{2n}}{2^{2n} n! n!}. \tag{5.41d}$$

This expression is difficult to evaluate, but is approximated by

$$\text{if } Y < 1, \qquad I_0(Y) \sim 0.220Y^2 + 1, \tag{5.41e}$$

otherwise

$$I_0(Y) \sim \left[1 + \frac{1}{8Y}\right] \times (2\pi Y)^{-0.5} \exp Y. \tag{5.41f}$$

Figure 5.21 plots Eq. (5.41d) and the error in the approximation of Eq. (5.41e), which is always less than 5 per cent.

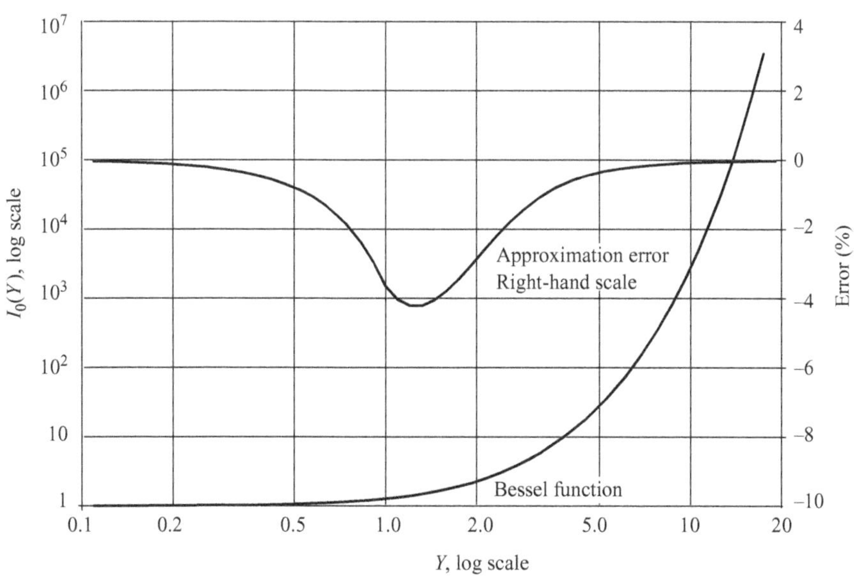

Figure 5.21 Bessel function. Eq. (5.41d) and error of approximation Eq. (5.41e)

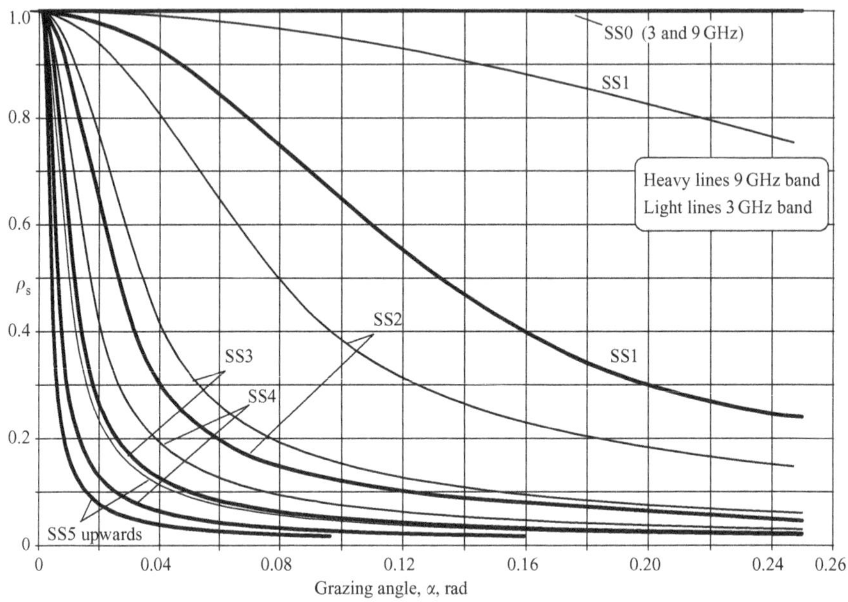

Figure 5.22 Forward reflection coefficient of surface roughness, ρ_S. Falls as sea state, grazing angle and frequency increase

5.8.5 Values of ρ_S

Figure 5.22 plots Eq. (5.41c) to show how ρ_S varies with sea state and grazing angle in the 3 and 9 GHz bands, falling from 1.0 towards 0 as the quantities increase. Of course, there are so many imprecisions behind the algorithm that it cannot pretend to accurately predict ρ_S, especially when $(h_S \sin\alpha)/\lambda > 0.21$, where the algorithm of Eq. (5.41b) has been extrapolated beyond its experimental data.

5.8.6 Values of ρ

The overall reflection (Eq. (5.39)) is plotted in Figure 5.23. It rises from a low value to near unity before gradually being driven down by the divergence factor to zero at the horizon. Peak values would occur nearer the horizon at 9 GHz than in the 3 GHz band shown. When range is less than about 10 per cent of horizon range and sea state is low, vertical polarisation delivers markedly lower forward reflection. Polarisation has less effect at long range where the grazing angle is much below the psuedo-Brewster angle.

5.9 Atmospheric and precipitation losses

5.9.1 Causes of loss

Hydrometeors (precipitation and fog particles) in the atmosphere absorb, scatter and attenuate signals, so less energy reaches the target and less still returns as an echo. The

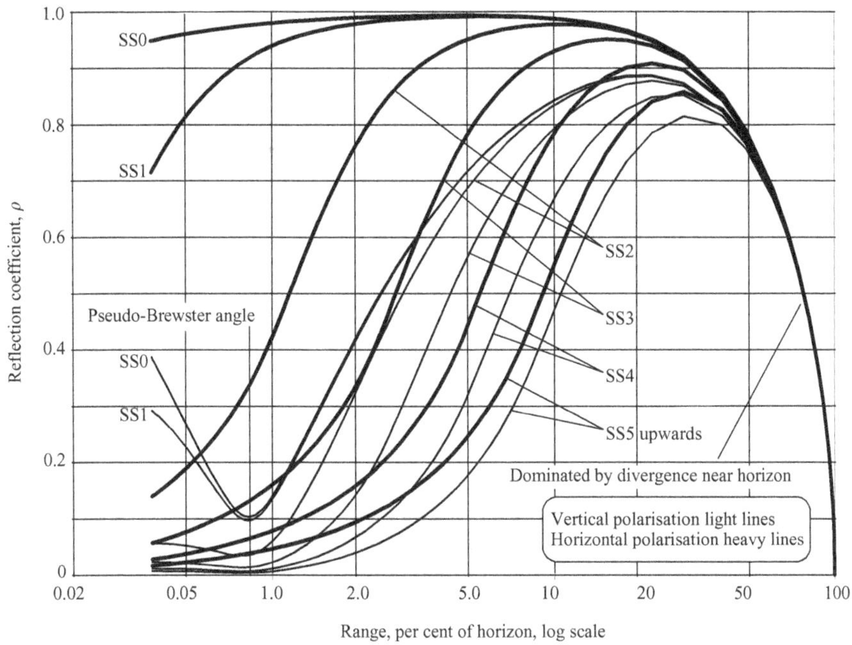

Figure 5.23 *Overall forward reflection coefficient, ρ, variation with range and sea state (SS). Shape of curves somewhat dependent on scanner and target heights. Drawn for these heights each 15 m, $k = 4/3$. Horizon range 31.9 km. At fairly long range, low sea states and horizontal polarization, ρ tends to be high*

clutter formed by energy scattered on radar bearing is discussed in Chapter 11. The attenuation is dependent on precipitation rate, type and extent, wavelength and, for rain in particular, form of polarisation. Even clear air is not quite lossless; its oxygen and water vapour both absorb a little energy. Each of these losses can be expressed in decibels per kilometre of range bracket through which the phenomenon occurs; the whole path or the range bracket occupied by a rain squall, for example. Atmospheric loss of course reduces clutter as well as target echoes and recurs on the response path.

Precipitation rates along the radar/target path are not always uniform or easy to determine, and rates measured at a coastal station may be higher than out at sea, precluding precise assessment of attenuation on any given occasion. The higher precipitation rates tend to occur in squalls occupying only part of the path, introducing less total attenuation. High precipitation rates of course impair optical as well as radar visibility. Although perhaps infrequent, they may pose a disproportionately high navigational hazard, so system performance calculations should be biassed towards the worst case of high whole-path precipitation rates.

Time-distributions of rainfall rates for different areas of the world are available from the ITU.[2] Although intended for radio propagation purposes, they may help indicate the likely severity of precipitation loss and clutter on a radar service at a specific location. For example, in the United Kingdom, there is about 0.01 per cent likelihood (~1 h/year) of precipitation exceeding 25 mm/h.

5.9.2 Rain

RCS, σ, of a single small conducting sphere such as a raindrop, having radius $a < 0.1\lambda$ (in the Rayleigh region, circumference $\ll$ wavelength), varies approximately as:

$$\sigma \sim 9 \times \frac{16\pi^5 a^6}{\lambda^4} \sim \frac{44\,067 a^6}{\lambda^4}\ \mathrm{m}^2. \tag{5.42}$$

Mass, of course, depends on a^3, so scattering for a given mass of raindrops per cubic metre is very dependent on the statistical distribution of droplet diameters. Both absorption and scattering attenuate signals traversing the path. As rain rate increases, average drop size, number of drops per cubic metre and the vertical velocity of the drops all increase in a manner dependent on the type of rainfall, for example, thunderstorm or drizzle. This multiple dependence causes attenuation per kilometre to increase disproportionately fast. Kingsley and Quegan [14] give values for rain attenuation, l_{R} dB/km, versus rain rate r mm/h at 18°C, shown as points in Figure 5.24. The figure also gives meteorologists' designations ranging from drizzle to excessive, sometimes called tropical; the terms differ somewhat from lay usage. We have interpolated values given for other wavelengths to find attenuation at J (κu) band, $\lambda \sim 2$ cm. Attenuation is one to two orders of magnitude worse at 9 GHz band than 3 GHz, and J-band is attenuated even more, unsuiting it for long range service. From the data points, we have devised reasonably well-fitting algorithms, shown as lines in the figure:

$$l_{\mathrm{R}} = 10^{[c+d\log(r+0.00001)]}\ \mathrm{dB/km\ one\text{-}way}. \tag{5.43}$$

At 18°C, values of constants c and d are:

3 GHz band	$c = -3.523$,	$d = 0.884$
9 GHz band	$c = -2.062$,	$d = 1.247$
J-band (~15 GHz)	$c = -1.403$,	$d = 1.182$.

The factor 0.000001 in the equation prevents log 0 problems when $r = 0$. For example, a J-band radar in heavy rain ($r = 16$ mm/h) at 18°C suffers attenuation of $10^{-1.40+1.18\log 16} = 10^{0.02086} = 1.0492 \sim 1.05$ dB/km one-way. This may not look too bad, but over a 20 km path, two-way attenuation is $2 \times 20 \times 1.05 = 42$ dB, rendering this band inoperable.

According to Skolnik [2], attenuations are generally about 15 per cent less dB/km at 0 and 40°C. Rivers [15] however, while giving 18°C attenuations similar to Skolnik, states attenuations should be multiplied by 2 at 2°C and by 0.7 near 30°C; perhaps

[2] ITU-R Recommendation P.837-1, Characteristics of precipitation for propagation modelling, 1994.

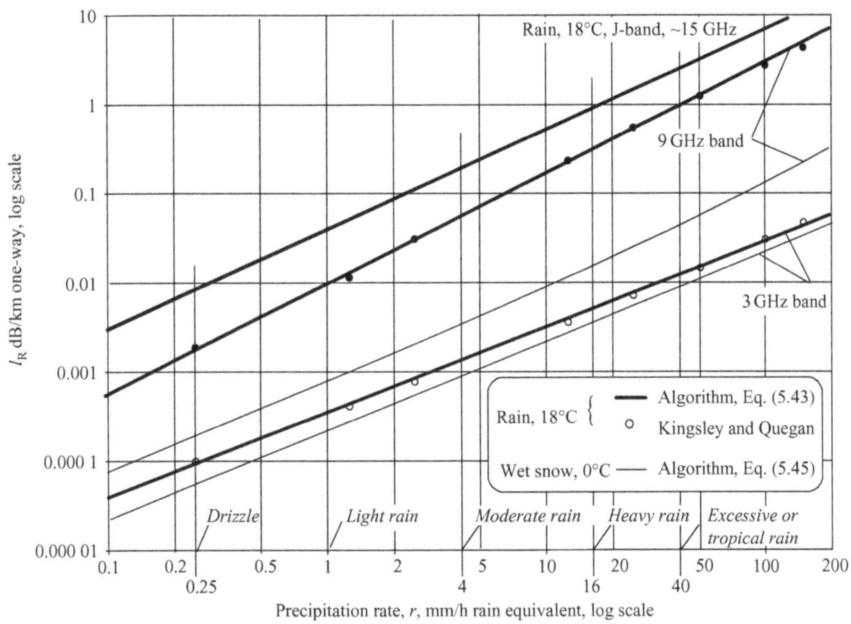

Figure 5.24 Precipitation attenuation. Rain shown for 18°C, attenuation about 15 per cent less at 0°C and 40°C

another example of the variability of natural phenomena. Because falling raindrops are not quite spherical, horizontal polarisation gives l_R dB values roughly 10 per cent higher than vertical.

Surprising as it may seem, formulae exist to predict rain-storm diameters. Meikle [11, p.198] notes that CCIR's reference model has a diameter about one-third of the following expression for diameter quoted in Nathanson [16]:

$$\text{storm diameter} = 41.595 - 23.608 \log r \text{ km}. \tag{5.44a}$$

The equation predicts diameters of about: 56 km (drizzle), 42 km (light rain), 27 km (moderate rain), 13 km (heavy rain) and 3.8 km (excessive rain). One would raise an eyebrow at the number of significant figures in the equation until rumbling that it probably merely links observations that diameter falls from about 30 nmi in drizzle to about 2 nmi in excessive rain, not the only time that change of units masks the spirit of a statement. Personal opinion and experience must decide whether to confine rain attenuation to these restricted range brackets or to take the conservative course of assuming the rain to occupy the whole radar-target path. Despite the equation, tropical rain exceeding 16 mm/h frequently persists through many kilometres together. The assumption used ought always to be stated, as the effect on long-range detection can be considerable. Examples in this book assume any precipitation to be whole-path unless stated otherwise. Figure 5.25 indicates the total storm diameter and attenuations from

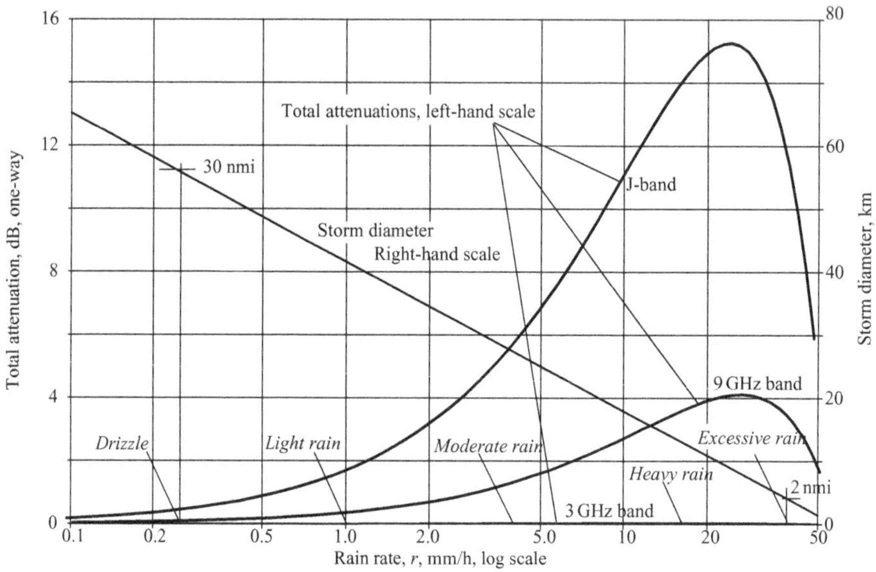

Figure 5.25 Rain storm attenuation. Eq. (5.44a)

Eqs (5.43) and (5.44a), peaking at 0.05 dB (3 GHz band), 4.1 dB (9 GHz band) and 15.3 dB in the J-band.

Goddard [17] gives a more refined formula for radio link integrity purposes in which the equivalent precipitation path length (eppl) rises with target range:

$$\text{eppl} = \frac{R}{1 + R/\{35 \exp(-0.015 r_{0.01})\}} \text{ km.} \tag{5.44b}$$

R is the target range, km and $r_{0.01}$ is the rainfall rate exceeded for 0.01 per cent of the time, about 50 min/year. For the United Kingdom, $r_{0.01} \sim 28$ mm/h and Eq. (5.44b) reduces to: eppl $= R/(1 + 0.043R)$, so when $R = 5$ km, eppl $= 4.1$ km, approximately 80 per cent of the path, but when $R = 50$ km, eppl $= 15.9$ km or 32 per cent.

5.9.3 *Snow and hail*

Relative to rain of the same water content, dry snow and hail cause less attenuation (unless, according to Wylie, the hailstones reach about 19 mm diameter), but wet snow rather more. Note that r is the equivalent rain rate, obtained by melting a collected sample, not the depth of snow accumulating per hour, which is about $5r$–$10r$. In general, snowstorms have lower r than rainstorms except perhaps in Polar regions, and the wide variability of snowflake form and size affect the attenuation. The following, suggested tentatively in Blake [12], has been included in Figure 5.24:

$$l_S = 34.9 \times 10^{-12}\frac{r^{1.6}}{\lambda^4} + 22 \times 10^{-6}\frac{r}{\lambda} \text{ dB/km one-way.} \tag{5.45}$$

5.9.4 Fog, low cloud and sandstorms

Fog is an aerosol suspension of minute water droplets in air and is not a gas. Occasionally fog seems to cause disproportionate loss of performance, probably because the quiet conditions causing the fog also introduce ducting or reduce refraction k-factor. There are three sorts of fog.

- Hill fog arises when low stratus cloud intercepts high land. It does not concern us.
- Radiation fog occurs at night under clear skies and light wind. Heat is radiated from the land to outer space, cooling the surface and the air mass just above it, whose moisture condenses to fog. Although basically a land phenomenon, radiation fog may spill offshore.
- Advection fog is the usual sort of sea-fog. It occurs when relatively warm, moist, air encounters a cold (e.g. sea) surface and cools until it can no longer support all its moisture.

These fogs have different drop size statistical distributions so, for a given suspended mass of water, have differing optical visibilities and radar attenuations. Untrained observers tend to underestimate visibility in fog. Dense fog causes about as much attenuation as moderate rain. The light lines of Figure 5.26 (based on Barton and Leonov [6, p. 151]) relate optical visibility, V km, in the horizontal plane with

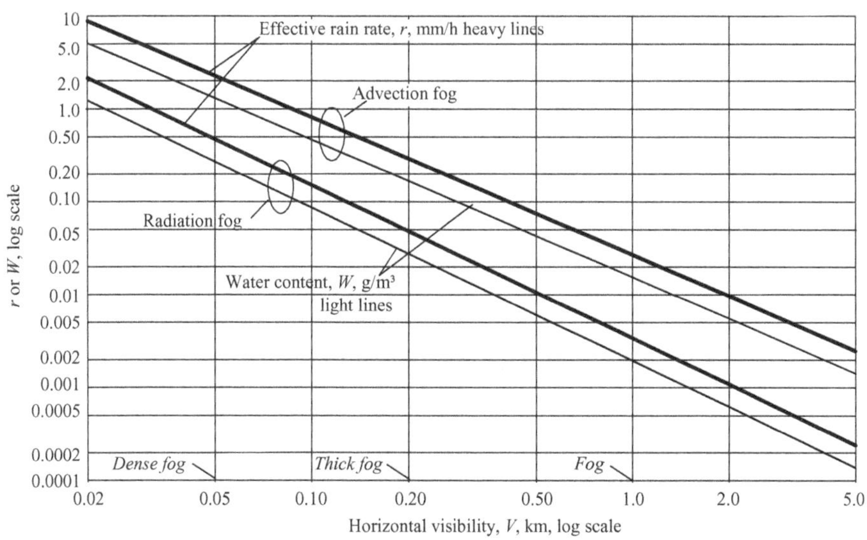

Figure 5.26 Water content and effective rain rate of fog. Use of log/log scaling gives linear curves

atmospheric water content, W g/m^3, for advection and radiation fogs. These curves are of the form

$$\log W = g - h \log V \tag{5.46a}$$

where $g = -1.81$, $h = 1.48$ (advection fog), $g = -2.71$, $h = 1.65$ (radiation fog).

Fog causes approximately the same attenuation as rain of the same water content, but generates negligible clutter. Although not strictly accurate, it is sufficient for us to assume equivalent rain rate to be proportional to water content, based on rain rate r of 4 mm/h having water content $W = 2.3$ g/m^3:

$$W \sim 0.575r \text{ g/m}^3, \quad \text{so } \log W \sim \log r - 0.240. \tag{5.46b}$$

Substituting in Eq. (5.46a),

$$\log r = g + 0.24 - h \log V \tag{5.46c}$$

$$\therefore \quad r = \frac{10^{g+0.24}}{V^h} \text{ mm/h.} \tag{5.46d}$$

Substituting: advection fog, $r = 0.00269/V^{1.48}$; radiation fog, $r = 0.00339/V^{1.65}$.

These relationships, shown full-line in Figure 5.26, convert visibility to equivalent rain rate, enabling attenuation to be calculated by Eq. (5.43), as shown in Figure 5.27. As an example of the calculations, in thick advection fog, $V = 0.2$ km, from Eq. (5.46d), $r = 0.0269/0.2^{1.48} = 0.2913$ mm/h, roughly equivalent to drizzle.

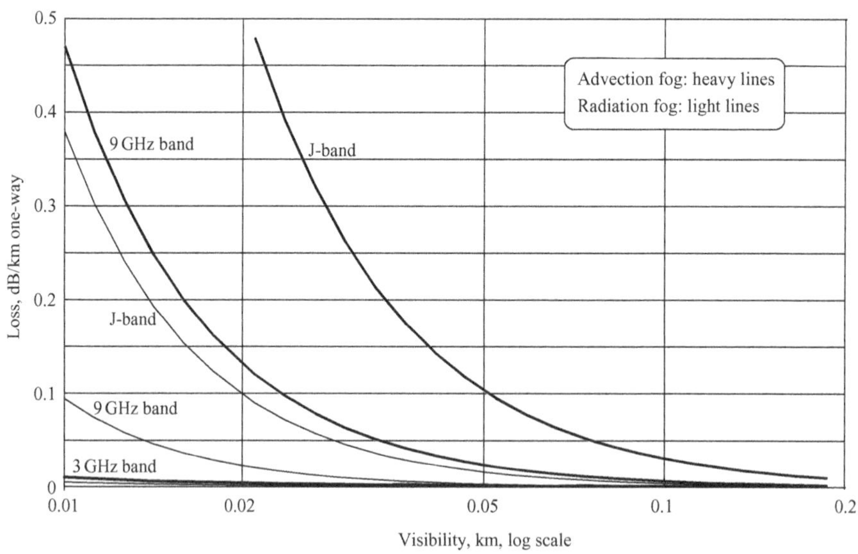

Figure 5.27 Fog attenuation. Attenuation is worst in dense advection fog at high frequency. Attenuation is always trivial in the 3 GHz band

Substituting in Eq. (5.43b) for a 9 GHz band radar, $l_R = 10^{-2.06+1.25 \log 0.2913001} = 10^{-2.729} = 0.00186$ dB/km one-way.

Other sources quote somewhat different fog attenuations, perhaps reflecting the variability of natural phenomena.

- Wylie [4, Figure 54] confirms that fog attenuation is inversely related to temperature.
- Goldstein [18] gives an alternative equation connecting attenuation with moisture content and frequency at 18°C; values are approximately doubled at 2°C and halved at 30°C;

$$l_F = 48.7 \times 10^{-3} W f^2 \text{ dB/km} \qquad [f \text{ in GHz}]. \tag{5.47}$$

  For $W = 2.3$ g/m^3 and $f = 9.4$ GHz, by Eq. (5.47) $l_F = 0.0099$ dB/km one-way. By Eq. (5.46b), $l_F = 0.049$ dB/km one-way.
- Skolnik [13, Table 24-12] quotes the fog or cloud attenuation l_F for 30/90/300 m visibility at 0°C as

$$\begin{aligned} &\text{9 GHz band} \quad 0.20/0.04/0.007 \text{ dB/km;} \\ &\text{3 GHz band} \quad 0.02/0.004/0.001 \text{ dB/km one-way, respectively.} \end{aligned} \tag{5.48a}$$

- Kerr [5, Eq. (1.18)] gives J-band (15 GHz) attenuation 0.9 dB/km one-way for 2.3 g/m^3 atmospheric water content. At temperature T°C, within the range 0–40°C, attenuation is

$$(1 - 0.024T) \times \text{ the 0°C value.} \tag{5.48b}$$

Cloud, a form of fog, behaves similarly to true fog unless there is precipitation within the cloud which perhaps evaporates before reaching the surface.

Wylie [4, p. 93] reports that sandstorms in the Gulf seem to give as much or slightly more attenuation than fog for the same visibility.

5.9.5 Clear air attenuation

Even in clear air without precipitation or fog, oxygen and water vapour molecules introduces some attenuation by molecular absorption. The effects are wavelength dependent and tabulated in Table 5.5 (interpolated from Kerr). The oxygen component is small and unvarying. The uncondensed water vapour component rises sharply with

Table 5.5 Molecular absorption, air at sea level (dB/km one-way)

	Factor	3 GHz	9 GHz	15 GHz
Oxygen	p	0.0066	0.0070	0.0080
Water, g/m^3	q	0.000025	0.00034	0.0016

Table 5.6 Mass of water vapour per cubic metre saturated air

Temperature, °C	0	5	10	15	20	25	30	35
g/m^3	4.84	6.76	9.33	12.71	17.12	22.84	30.04	39.18

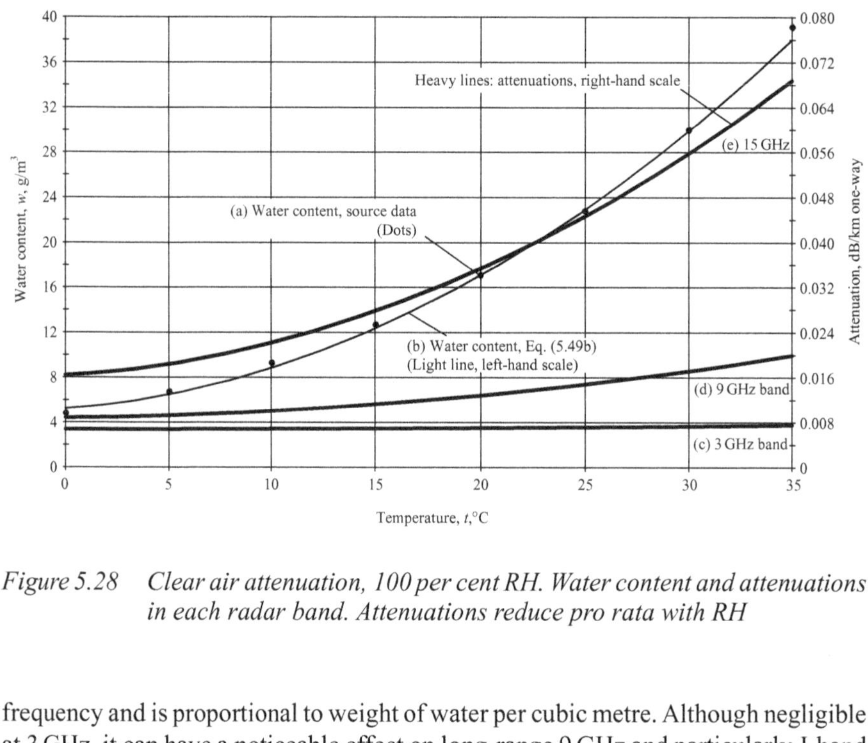

Figure 5.28 Clear air attenuation, 100 per cent RH. Water content and attenuations in each radar band. Attenuations reduce pro rata with RH

frequency and is proportional to weight of water per cubic metre. Although negligible at 3 GHz, it can have a noticeable effect on long-range 9 GHz and particularly J-band systems in clear weather.

Warm air will carry much more moisture than cool without saturating into fog. Relative humidity (RH), H, expressed as a percentage, is the proportion of actual to saturated moisture content for the temperature in question. At a given temperature, water content, W, is proportional to H. Table 5.6 relates saturation water content ($H = 100$) with temperature, t°C. The empirical quadratic expression best fitting the table is

$$W \sim 5.25 + 0.13t + 0.023t^2 \text{ g/m}^3. \tag{5.49a}$$

Figure 5.28(a) plots Table 5.6. Figure 5.28(*b*) plots Eq. (5.49a). The agreement is sufficient for our purpose.

Clear air attenuation, l_C, is the sum of the oxygen (factor p of Table 5.5) and water absorption component, factor q:

$$l_C = p + \frac{qWH}{100} \text{ dB/km one-way.} \tag{5.49b}$$

Figure 5.28(c–e) relates water content and attenuation with temperature for saturated air ($H = 100$ per cent). Attenuation rises moderately with temperature and more strongly with frequency. At 50 km range and 35°C, saturated air introduces two-way attenuations of 0.75, 2.0 and 6.9 dB at frequencies of 3, 9 and 15 GHz, respectively. Attenuations are reduced at lower RH and lower temperature, so clear air attenuation is never very severe; attenuations are roughly the same as advection fog of visibility 60 m, or 1 mm/h light rain. Clear air generates no clutter.

Example: at 20°C, 60 per cent RH, 9 GHz band: $l_D = 0.0070 + (0.6 \times 0.00034 \times (5.25 + 0.13 \times 20) + 0.023 \times 400) = 0.0105$ dB/km one-way. Equivalents for the 3 GHz and J-bands are 0.0069 and 0.024 dB/km, respectively.

5.9.6 *Spray*

Rescue helicopters have reported[3] 75 ft (23 m) waves in winter storms in the open North Atlantic throwing spray to 150 ft (45 m). Most rays of practical marine radar/target systems traverse this height band. One guesses the mean droplet size of spray thrown up by breaking heavy seas would exceed that of rain of the same, considerable, mass per cubic metre, giving substantial attenuation.

5.9.7 *Total atmospheric attenuation*

We handle the above attenuations by insertion of an atmospheric loss term, L_A, in the range equation. Assuming precipitation along the whole path, if R_{km} is range in kilometres:

$$L_A = R_{km}(l_P + l_S + l_F + l_C) \text{ dB one-way.} \tag{5.50}$$

5.9.8 *Foliage*

Foliage may sometimes intrude into the beam of VTS scanners. Foliage attenuation may be of the order of:

3 GHz band	0.4 dB/m
9 GHz band	0.9 dB/m
J-band	2 dB/m

Signal loss is very severe when foliage exceeds a few metres thickness. Loss is probably seasonal, particularly for deciduous trees, and dependent on their surface wetness. Loss fluctuates as foliage is blown by the wind, modulating strength of

[3] *Daily Telegraph*, London, 5 January 1998, reporting an air–sea rescue.

echoes traversing it and affecting probabilities of detection. Foliage clutter may overload the receiver if the range is very short, receiver paralysis perhaps then affecting detectability of targets lying just beyond.

5.10 References

1 HALL, M. P., BARCLAY, L. W., and HEWITT, M. T. (Eds): 'Propagation of radiowaves' (The IEE, 1996, 1st edn.), Chapter 6
2 SKOLNIK, M. I.: 'Introduction to radar systems' (McGraw-Hill, New York, 1983), Chapter 12
3 REED, H. R. and RUSSELL, C. M.: 'Ultra high frequency propagation' (Boston Technical Publishers, 1964)
4 WYLIE, F. J. (Ed.): 'The use of radar at sea' (Hollis and Carter for the Royal Institute of Navigation, 1978, 5th edn.), Chapter 5
5 KERR, D. E. (Ed.): 'Propagation of short radio waves, MIT Radiation Laboratory' (Re-published IEE Publications, 1987), Section 1.4
6 BARTON, D. K. and LEONOV, S. A. (Eds): 'Radio technology encyclopedia' (Artech House, London, 1996)
7 ROHAN, P.: 'Surveillance radar performance prediction' (The IEE, 1983), Sections 5.3 and 5.4
8 BARTON, D. K.: 'Modern radar systems analysis' (Artech House, London, 1988)
9 SKOLNIK, M. I. (Ed.): 'Introduction to radar systems, International Student Edition' (McGraw-Hill, 1983, 2nd edn.), Table 13-1
10 TERMAN, F. E. (Ed.): 'Radio engineers' handbook' (McGraw-Hill, 1943), Section 10.5, Eq. (27)
11 MEIKLE, H.: 'Modern radar systems' (Artech House Radar Library, 2001), Table 6.10
12 BLAKE, L. V.: 'A guide to pulse radar maximum range calculations' US NRL Report, 23 December 1969
13 SKOLNIK, M. I. (Ed.): 'Radar handbook' (McGraw-Hill, New York, 1970), Chapter 2, Eq. (56) and Figure 20
14 KINGSLEY, S. P. and QUEGAN, S.: 'Understanding radar systems' (McGraw-Hill, New York, 1992), Table 9-1
15 RIVERS, W. K.: in BLAKE, L. V. (Ed.): 'Radar range performance analysis' (Artech House, London, 1986)
16 NATHANSON, F. E.: 'Radar design principles' (McGraw-Hill, New York, 1991)
17 GODDARD, J. W. F.: in 'Propagation of radiowaves', Section 9.2.2
18 GOLDSTEIN, H.: in BLAKE L. V. (Ed.): 'Radar range performance analysis' (Artech House, London, 1986)

Chapter 6
Multipath of point targets

'The shortest distance between two points is a straight line.'

Euclid

6.1 Introduction

6.1.1 The problem

Chapter 5 showed that near the Earth's surface, beside the direct scanner–target ray, we must also consider the horizon, and the indirect ray reflected from the sea surface. Chapter 4 introduced the concept of a 'multipath factor', M, accounting for these effects and converting one-way transmission from free space to Earth's surface conditions, except for atmospheric attenuation, which is accounted separately as loss factor L_A. These terms feature in the radar range equation, Eq. (4.6). M is a power ratio. Other authors sometimes use a voltage ratio, F, called a pattern propagation factor; $M = F^2$ numerically.

At shorter ranges an interference region exists where the indirect ray can be nearly as strong as the direct ray. Because the relative phasing changes with range it is impermissible simply to add the ray powers. The indirect ray may either reinforce or partially cancel the direct ray, the resultant being bigger or smaller than either ray alone, the numerical value of M varying between almost 4 and 0. Working in decibels, the multipath factor is added to the direct ray dBW and lies between almost 6 and $-\infty$ dB.

At long range the direct and indirect rays merge. At broadcast-band radio frequencies, considerable signal diffracts round the Earth, permitting reception well beyond the horizon. The effect is negligible at optical frequencies. Radar occupies an intermediate position, diffraction sometimes permitting detection out to around 1¼ times horizon range. Diffraction would not be expected from the method of geometrical ray tracing used in Chapter 5. Between the interference and diffraction regions lies an intermediate transition region where both interference and diffraction are in play.

The ionosphere does not reflect centimetric waves, precluding the extreme-range operation sometimes possible with short-wave radio.

Multipath and diffraction, which are both reciprocal and work on both transmission and reception legs, have great practical effect on signal strength and target maximum detectable range. Ordinarily M has the same value on both transmit and receive legs, giving an M^2 term in the range equation for passive echoes. A few active targets, discussed in Chapter 8, have offset response frequency or vertically spaced antenna pairs which cause interrogation and response multipath factors to differ slightly; here the overall factor is the product of the leg factors, $M_1 M_2$ replacing $M^2 (M_1 + M_2$ replacing $2M$ in decibel terms).

Atmospheric attenuation is irrelevant to this chapter since it affects direct and indirect paths equally and is ignored except where stated. The present chapter is confined to point passive targets, but in Chapter 8 will use the one-way multipath factors when calculating active target performance. Extended targets, such as ships and coastlines, whose reflecting elements extend upwards from the waterline, behave somewhat differently and their multipath will be considered in Chapter 9.

6.1.2 Definition of multipath factor

We define the one-way multipath factor, M, as:

> The ratio of the signal power reaching the destination to the power which would do so if the Earth's surface were not present.

The 'destination' may be the target illuminated by a radar, or the scanner receiving echoes from a target. M is the ratio between the signal power actually received and that which would exist if the sea surface, the horizon and atmospheric refraction were absent.

6.1.3 Correction for scanner elevation beamwidth

Elevation beamwidths of racon and RTE antennas and most scanners are usually so wide that, unless there is severe roll, both direct and indirect rays are sufficiently near the nose of the radiation pattern to suffer negligible off-axis attenuation. At short range this may cease to be the case for some high VTS scanners with narrow beams. Here attenuation may differ slightly between direct and indirect rays, the latter usually being further offset from beam axis and encountering lower gain. A correction term becomes necessary for difference in gain between the two paths. It is convenient to include it within the multipath factor.

6.1.4 Chapter layout

After disposing of the beamwidth correction term we study addition of point to point direct and indirect rays in the short-range interference region, building on Chapter 5. Next follows diffraction, followed by the transition region, leading to fairly rigorous but lengthy expressions for multipath factor of a point target at any range.

The chapter concludes with a simplified approach to the near diffraction region and simple graphical methods for initial estimations and cross-checks on detailed computations.

6.2 Effective scanner gain

So far, we have assumed scanner gain to be independent of the inclination of the direct and indirect rays leaving or reaching it. This approximation is invariably adequate at long range, but scanner vertical beamwidth ought sometimes to be allowed for at short range. Chapter 2, Section 2.8.1, Eq. (2.10c), indicated how scanner gain is reduced by g_{el} dB when the ray is offset in elevation from beam axis. Chapter 5, Section 5.5.4 and Table 5.1, identified the angles associated with ray geometry and Figure 5.8(*b*) depicted them. As we only have to consider elevation angles in this chapter, subscript el will be omitted.

We identify losses related to the various angles by subscript, thus g_ν is the (power) loss factor to a ray at angle ν (g, >1 numerically, positive dB), relative to G_{max}.

$$\text{Effective gain off-axis} = G = \frac{G_{max}}{g}. \tag{2.10c}$$

In free space, when the scanner axis is depressed δ rad and the target is κ rad below the scanner horizontal and so is offset $\kappa - \delta$, gain loss of the direct ray, $g_{\kappa-\delta}$, is found by putting $\nu = \kappa - \delta$ in scanner elevation gain equation (Eq. (2.10c)). Effective scanner gain $G = G_{max} - g_{\kappa-\delta}$ dB and is used as the gain term in the free-space range equations (Eqs (4.6a), (4.7a) or (4.8a)). Similarly, the indirect ray, inclined at angle η, has loss $g_{\eta-\delta}$.

Relative to the direct ray, the differential additional off-axis loss, g_{dif}, almost always approximates 1 and so can be neglected. Its value is

$$g_{dif} = \frac{g_{\kappa-\delta}}{g_{\eta-\delta}} \text{ numerically.} \tag{6.1b}$$

6.3 Multipath regions

6.3.1 Regions

The interference-to-transition and transition-to-diffraction region boundaries are indistinct and rather arbitrary. Figure 6.1 illustrates the vertical lobe structure introduced in Chapter 5, Section 5.6.4, and indicates the regions, which are as follows.

1. *Interference region.* At short ranges, up to a transition range R_A, the signal reaching the destination is modified by interference between the direct and indirect rays. Some authors refer to the transition range as the critical range, but we shall differentiate between these quantities in Section 6.8.1. Ray tracing by geometrical optics as in Chapter 5, Section 5.5, accurately describes operation in the interference region, where we denote the one-way multipath factor of a point target as m_p.

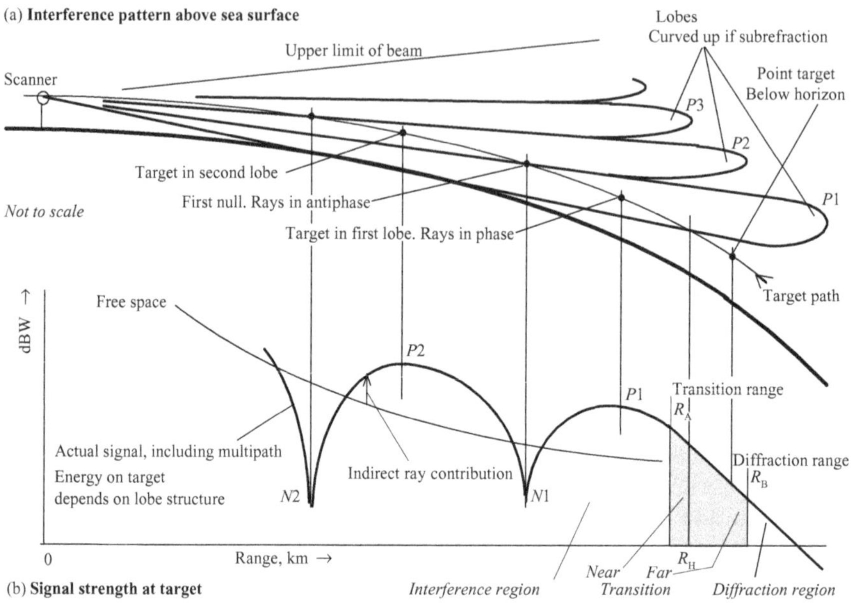

Figure 6.1 *Multipath lobes. As range closes, a point target comes from the diffraction region, through the transition region to the interference region, where it traverses a series of multipath lobes causing signal strength to vary, as shown in (b). Multipath breaks up the relatively wide elevation scanner beam into a number of narrow vertical lobes. Raising scanner height or reducing wavelength increases the number of lobes and depresses the lowest, increasing maximum detection range*

2. *Diffraction region.* When the edge of an object intrudes into the ray path, geometrical optics cannot fully predict what happens. Diffraction of electromagnetic rays around an obstacle can only be accurately described by the mathematics of wave mechanics, beyond the scope of this book. Figure 6.2(a) shows rays striking a conducting obstacle, inducing circulating currents on its surface. Acting as an antenna, the currents re-radiate weakly over a wide arc. Some of the radiation enters the shadow beyond the obstacle, while some modifies the field shortly in advance of it.

Alternatively, a plane wavefront can be considered as a system of parallel co-phased wavelets. The obstacle blocks wavelets in its path but oblique radiation from nearby unblocked wavelets penetrates the shadow area, Figure 6.2(*b*), which is based on Hall *et al.* [1].

The Earth's surface may be thought of either as a collection of obstacles, which together cause some energy to propagate beyond the horizon, or as a moderately conducting sphere on which are induced surface currents extending beyond the horizon, causing local radiation. In practice, unquantifiable anaprop and ducting, described

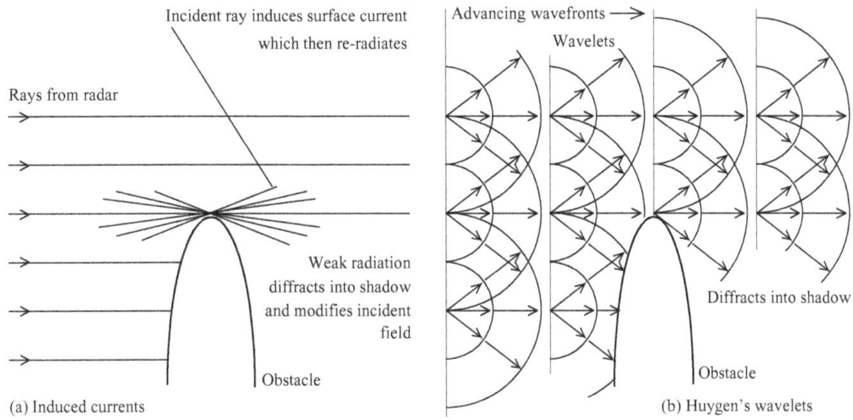

Figure 6.2 Diffraction. Alternative elementary mechanisms of diffraction at an obstacle

in Chapter 5, Section 5.2, may mask diffraction effects. When the target or scanner is very low, the diffraction region commences at quite short range, well within the horizon. Surface targets (of zero height) are within the diffraction region at all ranges. With these exceptions, few targets are big enough to remain detectable far into the diffraction region.

Some authors [2] recognise only two regions, scattering and diffraction, with their boundary at the radar horizon. On the other hand, it is sometimes convenient to identify near and far sub-regions, etc. as shown in Figure 6.3, with loose boundaries.

Figure 6.3 shows variation of multipath factor with range and connects the region boundaries with indirect ray phase shift and attenuation respectively. Drawing to a base of percent horizon range partly generalises the figure for differing scanner and target heights and refraction k factor. Beyond the diffraction boundary range R_B, the direct and indirect rays fuse and diffraction is dominant. We denote multipath in this region m_d. Although variation of strength in the diffraction region is not truly a multipath effect, there being only one path, it seems convenient to retain the term. Diffraction region calculation is a necessary preliminary for accurate calculation of the following.

- *Transition region.* As range increases from R_A to R_B, interference gradually gives way to diffraction. We denote multipath factor in this transition region as m_t. No theoretical derivation concisely describes this region. Performance was formerly often defined by drawing, freehand or by French curve, a rather arbitrary smooth curve between the interference and diffraction region graphs. We shall computerise this process.

The multipath factor to be used, M, depends on the region occupied by the target: $M = m_p$, m_t or m_d, depending on range.

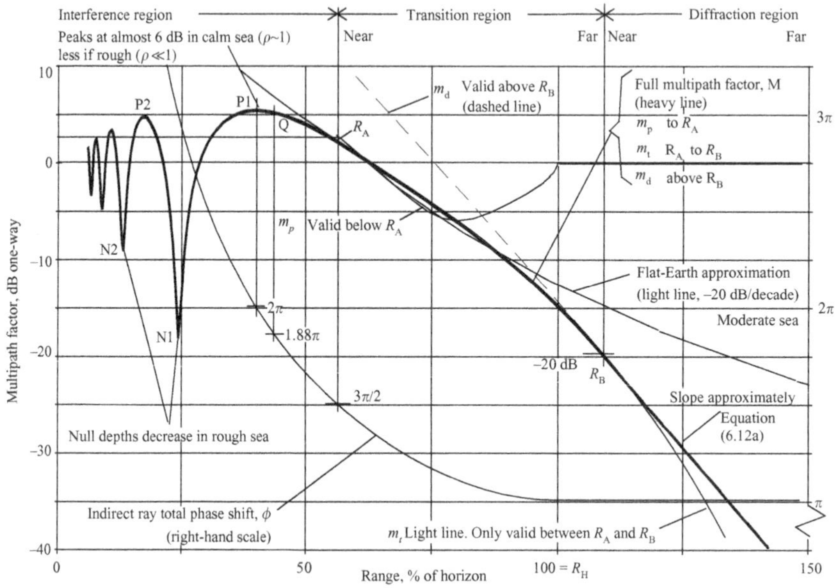

Figure 6.3 Interference, transition and diffraction regions. Showing multipath amplitude and phase variations, one-way. Full multipath factor M (heavy line) is the overall multipath factor, changing from m_p through m_t to m_d as range increases. Note how m_p becomes invalid at long range and m_d at short range. Path length phase shift ϕ governs events in the interference region; falling to π at the horizon, where $\phi = \pi$. Drawn for: scanner and point target heights 35 and 4 m, or vice versa; sea state 3; $k = 4/3$; horizon range 32.7 km; horizontal polarisation. Per cent horizon range of the events depends somewhat on scanner and target heights, and on radar band, here 9 GHz; at 3 GHz the events occur at shorter range. The flat-Earth approximation (Section 6.7.2) is indicated. Point Q relates to Figure 6.7

6.3.2 Boundaries

Following Blaise [3], who based his treatment on Kerr [4], we define transition range R_A as the range (within the horizon) where indirect ray pathlength total phase shift, ϕ, is $3\pi/2$ rad. Usually, the phase shift at the surface, ψ approximates π (Chapter 5, Section 5.8.3, Figure 5.19) so at R_A, path length phase shift $\Phi \sim \pi/2$, since $\phi = \Phi + \psi$ (Section 5.5.3, Eq. (5.16(*c*)) bringing the direct and indirect rays to phase quadrature. We will justify this definition of R_A in Sectin 6.8.3.

The lower boundary of the diffraction region is much less distinct. Again following Blaise, we rather arbitrarily define diffraction range, R_B, as the range where the one-way diffraction loss $m_d = -20$ dB. Kerr in effect uses $\Phi = \pi/2$ and -18 dB, while both Meikle [2], and Barton and Leonov [6] prefer $3\pi/4$ and the horizon. Choice of criteria can alter m_t by a couple of decibels, ours being the most conservative

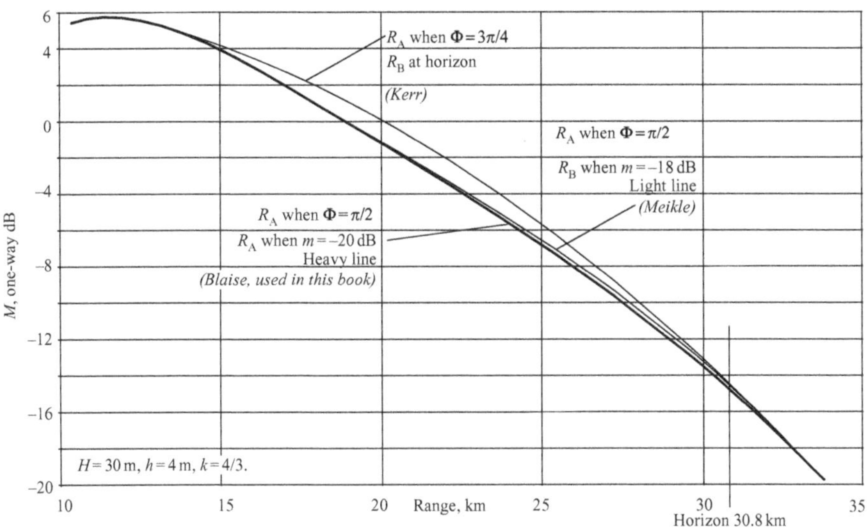

Figure 6.4 *Effect of boundary changes on m_t. For Blaise, Meikle and Kerr boundaries. In this example, boundary differences cause less than 2 dB variation in transition region multipath factor*

(lowest m_t), as shown in Figure 6.4. So performance beyond the interference region is subject to some uncertainty on the score of boundary criteria, any doubts as to how well theory reflects actuality, and of course by the possibility of ducting.

For a given radar/target combination, if no range of interest much exceeds R_A, it is sufficient to ignore diffraction and assume M falls in a square-law fashion beyond R_A, Section 6.8.3.

6.3.3 *Transition and diffraction boundary ranges*

It is often useful to form a quick rough estimate of the boundaries without embarking on detailed calculation. From the flat-Earth approximation, Chapter 5, Section 5.6.1, Eq. (5.27b), when $R = R_A$

$$\Phi \sim \frac{\pi}{2} \sim \frac{4\pi Hh}{R_A \lambda}$$

$$\therefore \quad R_A \sim \frac{8Hh}{\lambda}. \tag{6.2a}$$

Doubling scanner or point target height therefore approximately doubles R_A. For given heights, R_A is three times longer at 9 GHz than at 3 GHz. Some textbooks fail to warn of the error introduced by the flat-Earth approximation within this equation, which under some circumstances predicts R_A will exceed horizon range (Chapter 5, Section 5.5.7, Eq. (5.23)), an obvious nonsense. True R_A is always less than the flat-Earth approximation. Figures 6.5(a) (3 GHz) and (b) (9 GHz) plot families of an

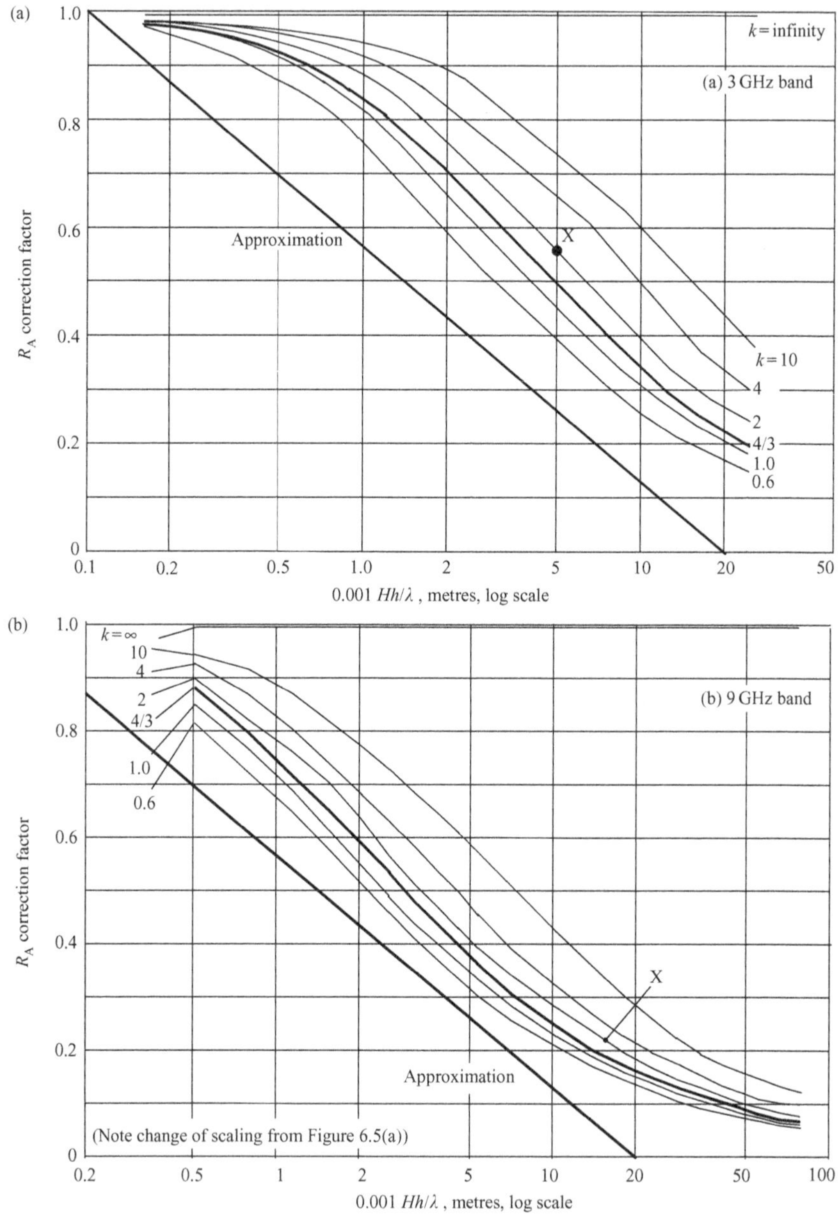

Figure 6.5 *Correction for flat-Earth approximation of transition range. Actual (curved-Earth) transition range R_A is often much less than the range calculated from Eq. (6.2) for a flat Earth. These correction curves remain approximately valid for all ratios of H to h*

R_A correction factor to a base of $0.001Hh/\lambda$ for differing values of refraction factor k, allowing the Eq. (6.2) flat-Earth value of R_A to be approximately corrected for Earth curvature. When k is low and Hh/λ is high (exceeding about 500 m) the error becomes gross, true R_A falling to a less than a quarter of the Eq. (6.2) value. For example, in a 3 GHz system ($\lambda = 0.1$ m), scanner height $H = 50$ m, point target height $h = 10$ m, Eq. (6.2) gives uncorrected $R_A = 40\,000$ m and $0.001Hh/\lambda = 5.0$. If refraction factor $k = 2$, the intercept X in Figure 6.5(a) gives correction factor $= 0.56$ so $R_A \sim 0.56 \times 40\,000\,\text{m} = 22.4$ km. For the same H, h and k, in the 9 GHz band ($\lambda = 0.032$ m), uncorrected $R_A = 125$ km but corrected $R_A \sim 27.9$ km.

Entry of low scanner and target heights in Eq. (6.2a) shows that R_A becomes very low; for example, putting $H = h = 1$ m in the 3 GHz band gives $R_A = 0.08$ km, all ranges of practical interest lying in the transition or diffraction regions where multipath peaks and nulls do not occur. Our treatment of multipath demands calculation within the interference region as a preliminary to examination of the other regions, so we have to admit it is not well suited to these extreme conditions. Multiplication of the right-hand side of Eq. (6.2a) by a linear approximation $[1.87 - 0.435 \log(Hh/\lambda)]$, shown in Figure 6.5, gives an expression which tends to understate R_A:

$$R_A \sim \left[15.0 - 3.48 \log \frac{Hh}{\lambda}\right] \times \frac{Hh}{\lambda}\ \text{m}. \tag{6.2b}$$

Figure 6.6 plots the variation of R_A, R_B and horizon range R_H with target height for two representative scanner heights at 3 and 9 GHz, the three panels relating to different k factors. R_A typically lies between 0.1 and 0.85 of R_H. The slope lines indicate that $R_A \propto h$ when h is low, migrating towards $R_A \propto h^4$ when h is high; similarly for H. For given H, h and k, R_B is of course always higher than R_A. When H and h are moderately high, R_B is a few per cent above R_H.

6.4 Interference region

6.4.1 Value of multipath factor

Reflection of the indirect ray at the sea surface was discussed in Chapter 5, Sectin 5.8. At the grazing point the ray suffers *voltage* reduction by the reflection coefficient $\rho(\leqslant 1)$ of the sea surface, where it is phase shifted by $\psi \sim \pi$ rad. The path length difference gives an additional range-dependent phase shift Φ. In the following we assume the indirect ray is displaced from the nose of the scanner elevation pattern, so the transmitted indirect ray *power* is reduced by g_{dif}, ($\leqslant 1$) as discussed in Chapter 2, Section 2.8.1, Eq. (2.10b). The indirect ray voltage component reaching the target, V_{ind}, is less than the direct ray component V_{dir}:

$$V_{ind} = \rho\sqrt{g_{dif}}\,V_{dir}.$$

V_{ind} is also subject to total phase shift

$$\phi = \Phi + \psi. \tag{5.16c}$$

Figure 6.7 is a vector diagram of direct, OP, and indirect, PQ, ray voltages. As phase shift rises and reflection coefficient falls with range, the locus of the indirect ray forms a spiral. This causes the resultant to oscillate in amplitude above and below the value of the direct ray voltage, multipath factor oscillating about 0 dB as in Figure 6.3, the oscillation amplitude dying away as range shortens.

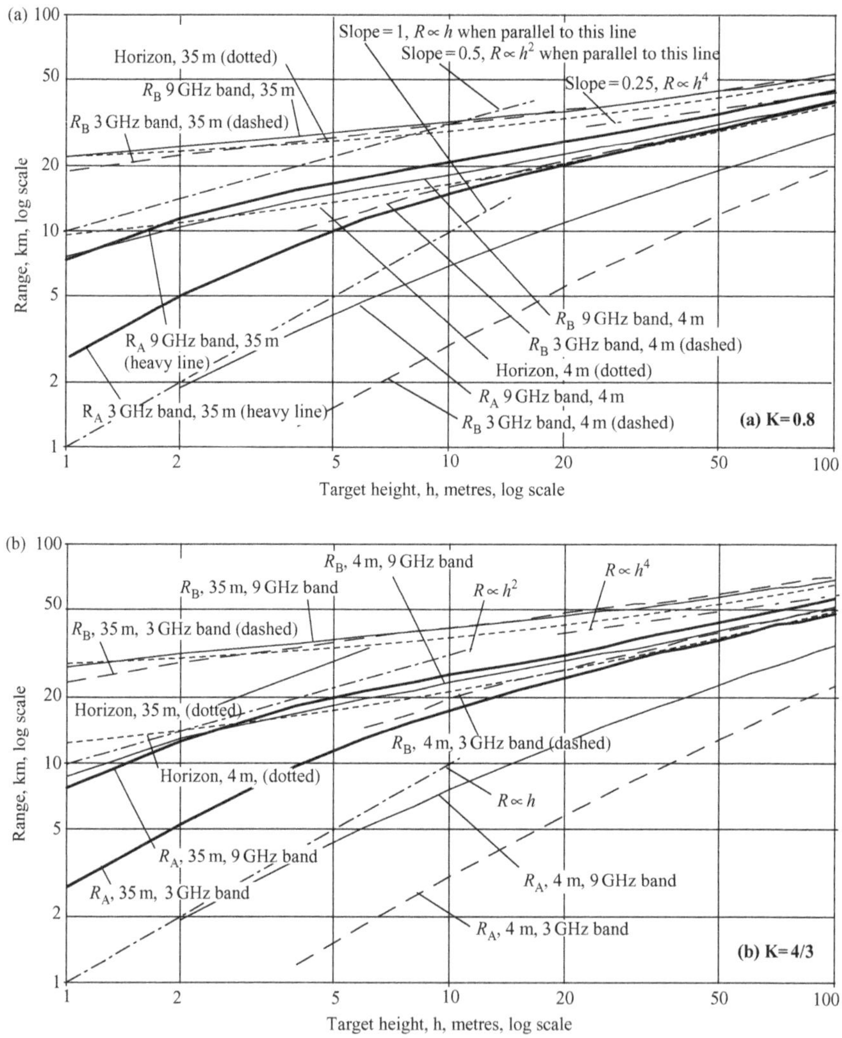

Figure 6.6 Range versus target height. Horizon (R_H), transition (R_A) and diffraction (R_B) true ranges plotted against target height, h, for representative scanner heights, 3 and 9 GHz bands and three refraction (k) values. Asymptotes chain-dotted, horizon dotted

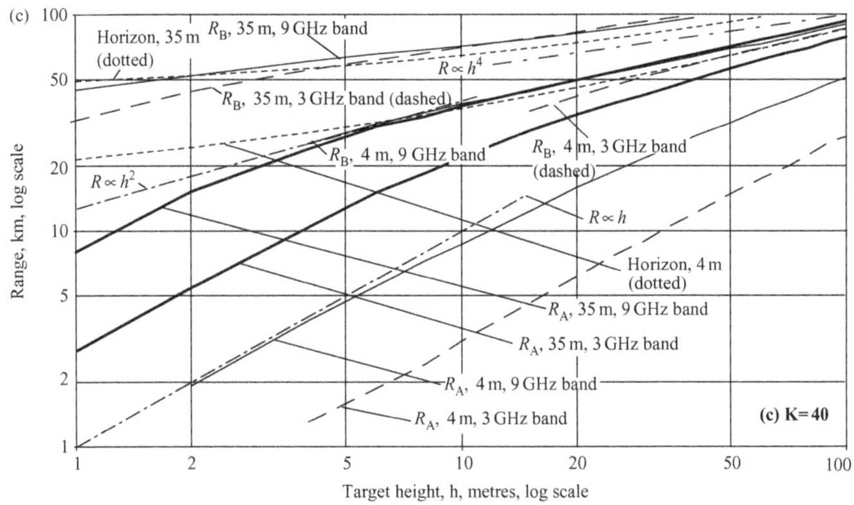

Figure 6.6 Continued

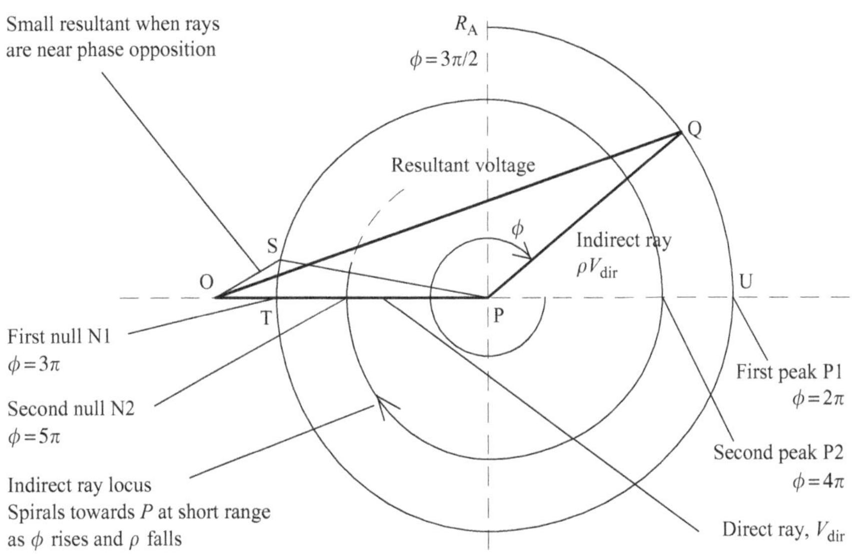

Figure 6.7 Vector addition of rays. The resultant voltage OQ varies between maxima when the direct and indirect rays are in phase with $\phi = (\Phi + \psi) = 2\pi$ (peak P1, point OU), $4\pi, \ldots$, and minima of OT etc. when they are in phase opposition with $\phi = 3\pi, 5\pi, \ldots$. The indirect ray is shown with $\rho = 0.92$, $\phi = 5.9$ rad, where $V_{res}/V_{dir} = 1.8$ and $m_p = 20 \log 1.8 = 5.1$ dB, point Q on Figure 6.3

The resultant voltage of the direct and indirect components is found by vector addition. In Figure 6.7

$$\frac{\mathrm{PQ}}{\mathrm{OP}} = \rho\sqrt{g_{\mathrm{dif}}}. \tag{6.3}$$

The numerical multipath factor is the ratio of resultant direct plus indirect ray power, to direct ray power, so

$$m_{\mathrm{p}} = \left(\frac{V_{\mathrm{res}}}{V_{\mathrm{dir}}}\right)^2 = \left(\frac{\mathrm{OQ}}{\mathrm{OP}}\right)^2 = 1 + \left(\frac{\mathrm{PQ}}{\mathrm{OP}}\right)^2 + 2\frac{\mathrm{PQ}}{\mathrm{OP}}\cos\phi.$$

Substituting and rearranging,

$$m_{\mathrm{p}} = 1 + \frac{\rho^2}{g_{\mathrm{dif}}} + 2\rho\sqrt{g_{\mathrm{dif}}}\cos\phi \text{ numerically, one-way transmission.} \tag{6.4}$$

Of course, ρ, g_{dif} and ϕ are each linked to the grazing angle, α. In Figure 6.7, ϕ is shown in a quadrant where the resultant OQ is larger than the direct ray. If the indirect ray is near phase opposition to the direct ray, shown by OS, where ϕ is near 3π, 5π, etc. radians, the resultant is smaller. In the limiting case of $g_{\mathrm{dif}} = 1$ and a perfectly reflecting sea surface, $\rho\sqrt{g_{\mathrm{dif}}} = 1$, $V_{\mathrm{res}}/V_{\mathrm{dir}}$ varies between 2 and 0 and m_{p} varies between 4 (6 dB) and zero ($-\infty$ dB), the cosine term of Eq. (6.4) making m_{p} oscillate with peak–peak amplitude dependent on $\rho\sqrt{g_{\mathrm{dif}}}$. The process is repeated on the return leg from a passive target, so when $\rho\sqrt{g_{\mathrm{dif}}} \sim 1$, the echo varies between a maximum of 16 $\times$ free-space echo (12 dB) with nulls between the peaks which are deep when $\rho\sqrt{g_{\mathrm{dif}}}$ is high, as occurs with calm sea at low grazing angle.

At R_{A}, the upper boundary of the region, $\phi = 3\pi/2$ and $\cos\phi = 0$, $m_{\mathrm{p}} = 1 + \rho^2 g_{\mathrm{dif}}$, lying between 1 (0 dB) when $\rho^2 g_{\mathrm{dif}} = 0$, and 2 (3 dB) in the more usual case where $\rho^2 g_{\mathrm{dif}} \sim 1$. Beyond R_{A}, divergence factor d (Chapter 5, Section 5.5.5, Eq. (5.22), Figure 5.9) falls towards zero at the horizon, apparently forcing ρ to zero per Section 5.8.1, Eq. (5.39). This would suppress the indirect ray and force m_{p} upwards to unity, so at or beyond the horizon echoes would take their free-space values. This flies flat in the face of reality, where signal strength is found to fall steeply. The reason is diffraction, which causes the multipath factor to fall steadily at longer range, Section 6.5.3.

6.4.2 *Average value of multipath factor*

When $\psi = \pi$ and $\rho\sqrt{g_{\mathrm{dif}}} = 1$ (e.g. calm sea), Eq. (6.4) approximates to

$$m_{\mathrm{p}} = (2 - 2\cos\phi) = 4\sin^2\frac{\phi}{2} \text{ numerically.} \tag{6.5a}$$

In dB terms,

$$m_{\mathrm{p}} = 10\log\left(\sin^2\frac{\phi}{2}\right) + 6\,\mathrm{dB}. \tag{6.5b}$$

(Writing $20 \log (\sin \phi/2) + 6\,\text{dB}$ would invite trouble when the sine term is negative.) Averaged over a range bracket containing a complete multipath cycle (ϕ increasing by 2π radians), the indirect ray makes a positive nett contribution to the resultant because it delivers more signal power near peaks than is lost through destructive interference near nulls; Figure 6.3 shows the peaks (above 0 dB) are wider than the troughs. The average indirect ray contribution is calculated by integrating the multipath factor with respect to ϕ, using a standard mathematical integral: $\int \sin^2 ax\,\mathrm{d}x = x/2 - (\sin 2ax)/(4a) + C$. As we are interested in a definite interval, we ignore C, the constant of integration.

$$\begin{aligned}\text{Average } m_{\text{p}} &= \frac{2}{\pi}\int_0^{\pi/2} 4\sin^2\frac{\phi}{2}\,\mathrm{d}\phi \\ &= \frac{4(\pi-1)}{\pi} \sim 2.727 \text{ numerically } (4.36\,\text{dB}). \qquad (6.6)\end{aligned}$$

As the range shortens, ρ falls towards zero. The weak peaks and nulls crowd together because ϕ increases rapidly, going through many radians of phase shift for a small range change. When $\rho\sqrt{g_{\text{dif}}} \to 0$ (rough sea, short range) the indirect ray power $\to 0$, there is no multipath contribution and $m_{\text{p}} \to 1$ (0 dB). Operation then approximates the free space condition (apart from atmospheric attenuation) and echo strength follows an R^{-4} law, falling at 40 dB/decade.

6.4.3 *Narrow pulses*

The bandwidth occupied by very short pulses spreads through an appreciable wavelength bracket and hence a wide multipath phase bracket, partially filling deep nulls, especially high-order nulls. For example, the fifth multipath null is when $\phi = 11\pi$. If centre frequency is 3000 MHz ($\lambda = 0.10\,\text{m}$), $\rho = 0.9$, $\psi = \pi$ (calm sea) and $g_{\text{dif}} = 1$, then $\Phi = 10\pi$ and path length difference $\Delta = 0.50\,\text{m}$. Eq. (6.4) gives $m_{\text{p}} = 0.01$, so the two-way resultant echo is $-20 \log 0.01 = -40\,\text{dB}$ on the direct component. However, a 0.05 μs pulse occupies 20 MHz, with wavelengths between 0.09967 and 0.10033 m, so ϕ occupies the bracket $11\pi \pm 0.114$ rad. At band edges $2m_{\text{p}} = -34.1\,\text{dB}$, weighted average $\sim -38\,\text{dB}$. The example shows the spread of frequency is usually too small to be significant. A possible exception is frequency offset racons, discussed in Chapter 8.

6.4.4 *Diversity*

Frequency diversity can be used to reduce the severity of multipath nulls by employment of differing frequencies. Alternatively twin scanners at differing heights but similar or identical frequencies can be used in a height diversity system. Diversity is fully discussed in Chapter 12, Section 12.8. Here we merely remark that although on ships in particular there are usually various constraints on available height, it is well worthwhile when planning twin systems to select the height spacing to minimise simultaneous nulls to point targets such as the reflectors on buoys and fishing vessels (often mounted with $h \sim 4$ m) at the navigationally important ranges around 6 km for k factors between say 0.8 and 2.

6.5 Diffraction region

6.5.1 *The nature of diffraction*

Diffraction makes calculation much more complex and is most significant at long wavelength – 3 GHz band – and in conditions favouring free-space detection far beyond horizon range; that is, when scanner and target heights are low, the radar is powerful and the RCS large. Even then, ducting may be of more practical importance. Direct and indirect rays coincide in the diffraction region so there is no scanner differential gain; $g_{\text{dif}} = 1$, allowing this factor to be dropped from our expressions. Detection far into the true diffraction region is unusual but many radars can operate into the transition region, where some diffraction effect is in play and accurate evaluation unfortunately demands prior diffraction-region calculation.

6.5.2 *Calculation of diffraction*

We base our treatment on Blaise [3], which in turn is based on a very long mathematical development in Kerr [4] who derived a pattern propagation factor, F, by which the free-space voltage is reduced in the diffraction region. Because our (one-way) multipath factor is a power ratio, F^2 is equivalent to m_{d}. Expressing Blaise and Kerr in our terminology:

$$F = \sqrt{m_{\text{d}}} = 2\sqrt{\frac{\pi R}{L}}\, f(Z) f(z) \times \exp \frac{-2.02R}{L} \quad \text{numerically (one-way transmission).} \tag{6.7a}$$

The value -2.02 applies for a standard refraction/height profile without ducts (Chapter 5, Section 5.2.9, Figure 5.3(*a*)). Squaring to give power, and writing in decibel terms,

$$m_{\text{d}} = 10.99 + 10\log\frac{R}{L} - 17.545\frac{R}{L} + 20\log[f(Z)] + 20\log[f(z)]\ \text{dB} \tag{6.7b}$$

where R is the actual range (m) and

$$L = \text{a 'natural range unit'} = \left(\frac{E^2\lambda}{\pi}\right)^{1/3} \tag{6.8}$$

E is the effective Earth radius, ($E = ke$, thus introducing atmospheric refraction; Earth radius, $e = 6.371 \times 10^6$ m)

$$U = \text{a 'natural height unit'} = 0.5\left(\frac{E\lambda^2}{\pi^2}\right)^{1/3} = \frac{L^2}{2E} \tag{6.9}$$

$F(Z)$ or $f(z)$ is the 'height-gain' function of scanner or target height, respectively, when expressed in natural units of:

$$Z = \frac{H}{U} \quad \text{and} \quad z = \frac{h}{U} \tag{6.10}$$

H, h are the scanner and target true heights (m).
To give a feel for the quantities, when $k = 4/3$:

$E = 8.49 \times 10^6$ m,

$L \sim 13\,190$ in the 3 GHz band or 9022 in the 9 GHz band.

$U \sim 10.24$ or 4.8, respectively.

Kerr connects Z and z with functions $f(Z)$ and $f(z)$ by a graph, Figure 6.8(a). Kerr's Fig. 2-20 was calculated during the Second World War, before availability of computers, for a series solution and an asymptotic expansion, which he does not detail. Particularly as we shall be using a spreadsheet method of calculation, reference to a graph is inconvenient, so we have found an algorithm to represent empirically the height-gain function, graphed in Figure 6.8(*b*), with error in Figure 6.8(c):

$$\log[f(Z \text{ or } z)] = \log(Z \text{ or } z) + 0.001401[2 + \log(Z \text{ or } z)]^6. \tag{6.11}$$

If Z or $z < 3$, the second term in Eq. (6.11) becomes negligible and $\log[f(Z \text{ or } z)] = \log(Z \text{ or } z)$; Figure 6.8(d). That is, when H and h are low (specifically, when H or

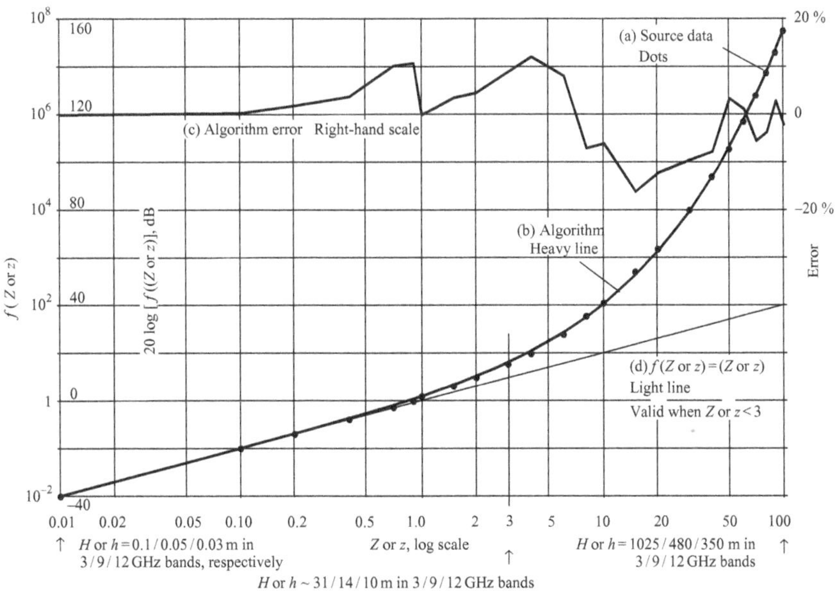

Figure 6.8 Height-gain function. $f(Z \text{ or } z)$ *rises steeply when* Z *or* $z > 3$. *The error curve shows that Eq. (6.11) reasonably matches the curve given in Kerr*

$h < 31$ m or <14 m in the 3 or 9 GHz bands, respectively), $f(Z \text{ or } z) \sim (Z \text{ or } z)$, as shown on Figure 6.8.

Equation (6.7b) connects m_d with R by a bell-shaped curve. when $k = 10$, minimum useable $(H + h)$ values at 3.0, 9.4 and 14.0 GHz approximate 3.2, 1.6 and 1.2 m respectively. To accommodate lower height pairs, the transition range boundary criterion must be reduced, say from −20 to −30 dB. The impact on accuracy is unknown.

6.5.3 *Change of multipath factor with range*

Once well into the diffraction region, the $\exp(-2.02R/L)$ term dominates Eq. (6.7) and m_d falls at a rate determined as follows.

Eq. (6.7b) can be expressed in the form

$$m_d = (\text{constants in a given system}) + 10\log\frac{R}{L} - 17.55\frac{R}{L}\ \text{dB}.$$

At long range $R \gg L$ so $R/L \gg \log R/L$, permitting the simplification:

$$m_d = \text{constant} - 17.55\frac{R}{L}\ \text{dB}.$$

Slope = rate of change of m_d with $R = \mathrm{d}m_d/\mathrm{d}R$. The rule for differentiation of $y = ax^n$ is $\mathrm{d}y/\mathrm{d}x = anx^{(n-1)}$. Rate of change of m_d with $R = \mathrm{d}m_d/\mathrm{d}R = -17.55/L$ dB/m.

Substituting for L and putting $E = ke$,

$$\frac{\mathrm{d}m_d}{\mathrm{d}R} = -0.742 \times 10^{-3} \times k^{-2/3} \times \lambda^{-1/3} = -0.742k^{-2/3}\lambda^{-1/3}\ \text{dB/km}. \tag{6.12a}$$

Unusually, this expression is in terms of dB/km rather than dB/decade. Figure 6.9 depicts variation of $\mathrm{d}m_d/\mathrm{d}R$ with k. Eq. (6.12a) is sometimes expressed as

$$\frac{\mathrm{d}m_d}{\mathrm{d}R} = -25\,700(\lambda k^2 e^2)^{-1/3}\ \text{dB/km}. \tag{6.12b}$$

Note the following.

- In the far diffraction region, multipath factor falls at a constant number of dB/km, nearly independent of scanner or target heights (H or h).
- When H and h are high, the far diffraction region is more distant.
- When k is low, m_d falls most steeply, restricting performance in the diffraction region as well as at short range.
- The wavelength dependence favours the 3 GHz band.
- As indicated in Figure 6.3, slopes in the transition and near diffraction regions are somewhat shallower than predicted by Eq. (6.12b) and Figure 6.9.

The total rate of change of echo strength at longest ranges has three major components.

- The free space inverse fourth power relationship of −12 dB/octave (−40 dB/decade). At long range, this gives quite a small rate; for example −0.35 dB/km at 50 km.

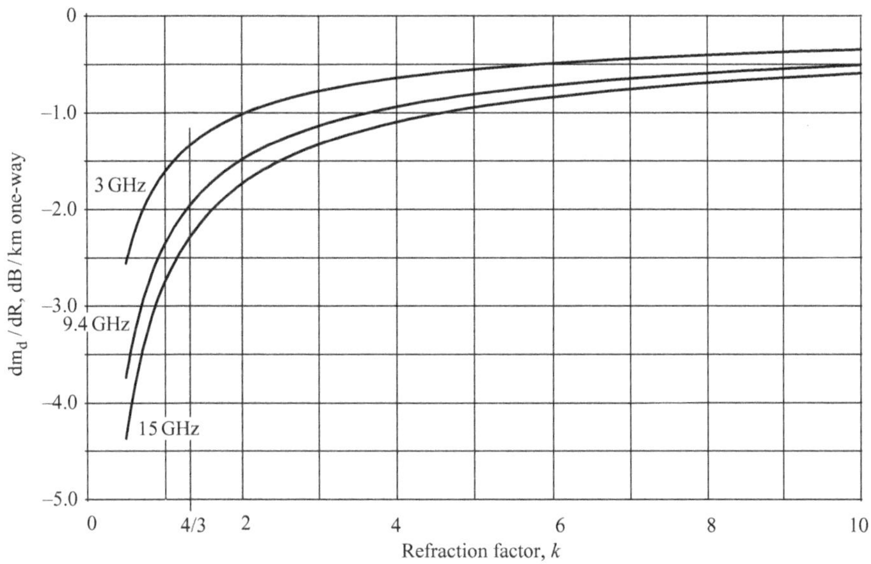

Figure 6.9 Slope of multipath factor in far diffraction region

- Atmospheric attenuation (Chapter 5, Section 5.9), which only exceeds 0.2 dB/km two-way at short wavelength and in heavy precipitation.
- $2\,\mathrm{d}m_{\mathrm{d}}/\mathrm{d}R$ (Figure 6.9), for example −3.9 dB/km two-way at 9 GHz when $k = 4/3$. This last term usually dominates at the longest ranges of practical interest.

6.5.4 Effect of height

At any fixed range, where R/L is low, Eq. (6.7a) simplifies to $\sqrt{m_{\mathrm{d}}} \propto f(Z)f(z)$ numerically. If H or $h < 31$ m (3 GHz) or <14 m (9 GHz), $f(Z \text{ or } z) < 3$ and we can put

$$m_{\mathrm{d}} \propto Z^2 z^2 \propto H^2 h^2, \text{ numerically.} \tag{6.13}$$

As echo strength after two-way propagation is proportional to m_{d}^2, it is proportional to H^4 and h^4. Doubling scanner height H or target height h therefore increases echo strength in the diffraction region by a factor of 16 (12 dB). This relationship remains reasonably accurate even when the heights exceed 31 or 14 m, and into the transition region. It is helpful when deciding necessary scanner height, particularly in VTS systems.

6.6 Transition region

6.6.1 Approach

The transition region lies between the transition and diffraction boundaries (R_{A} and R_{B}). Diffraction and multipath interference each have some effect and no simple equation describes signal strength. We first offer a formal detailed method of calculation, necessitating prior calculation of the diffraction region, despite the fact that few radar/target systems penetrate such long range, followed (Sections 6.7.2 and 6.8) by simplified methods which are generally satisfactory in the near transition region, to which many radar/target systems are confined. Again we assume $g_{\mathrm{dif}} = 1$ and omit this term from the equations. As range increases, m_{t} falls and rate of change of m_{t} with range approximates $\mathrm{d}m_{\mathrm{d}}/\mathrm{d}R$.

6.6.2 Solution of multipath equation

The simplest way to link the known interference and diffraction region multipath factors would be to assume transition region multipath factor, m_{t}, falls linearly with range, from the *values* of m_{p} at R_{A} to m_{d} at R_{B}, but this would give artificial kinks at R_{A} and R_{B}. Before the days of computers we would have sketched m_{t} so it also blended smoothly with the *slopes* of m_{p} at R_{A} and m_{d} at R_{B}. Today we can generate a 'French curve' by computer. Either procedure merely blends from interference to diffraction regimes in the simplest manner yielding a smooth curve, without attempting to conform to the underlying physical process.

The curve must meet the following conditions.

$$\begin{aligned} &\text{At } R_{\mathrm{A}} : m_{\mathrm{t}} = m_{\mathrm{p}} \quad \text{and} \quad \text{slope } S_{\mathrm{A}} = \frac{\mathrm{d}m_{\mathrm{t}}}{\mathrm{d}R} = \frac{\mathrm{d}m_{\mathrm{p}}}{\mathrm{d}R} \\ &\text{At } R_{\mathrm{B}} : m_{\mathrm{t}} = m_{\mathrm{d}} \quad \text{and} \quad \text{slope } S_{\mathrm{B}} = \frac{\mathrm{d}m_{\mathrm{t}}}{\mathrm{d}R} = \frac{\mathrm{d}m_{\mathrm{d}}}{\mathrm{d}R}. \end{aligned} \tag{6.14}$$

These four degrees of freedom suggest a cubic equation:

$$m_{\mathrm{t}} = p + qR + rR^2 + sR^3 \text{ dB}. \tag{6.15}$$

To solve Eq. (6.15) we must evaluate constants p, q, r and s. When calculating performance in Chapter 14 we shall use a spreadsheet method which throws up R values from a set of discrete grazing angles α. We here denote quantities in the spreadsheet range column next below and above the nominal by single ($'$) and double ($''$) primes, respectively. The algebra looks daunting but all we have to do in practice is to embed formulae for p, q, r and s (derived below) in the spreadsheet and let the computer take the strain.

From the spreadsheet we look up the following.

- Range R'_{A} next below true R_{A} at which $\phi = 3\pi/2$. Here $m = m'_{\mathrm{A}}$ (the value of m_{p} at R'_{A}).
- $S_{\mathrm{A}} = (m'_{\mathrm{p}} \text{ at } R'_{\mathrm{A}} - m''_{\mathrm{p}} \text{ at } R''_{\mathrm{A}})/(R'_{\mathrm{A}} - R''_{\mathrm{A}})$.
- Range R''_{B} (next above R_{B} at which $m_{\mathrm{d}} = -20$ dB, and $m = m''_{\mathrm{B}}$ (the value of m_{d} at R''_{B}).
- $S_{\mathrm{B}} = (m'_{\mathrm{d}} \text{ at } R'_{\mathrm{B}} - m''_{\mathrm{d}} \text{ at } R''_{\mathrm{B}})/(R'_{\mathrm{B}} - R''_{\mathrm{B}})$.

For brevity we put: $c = R_{\mathrm{B}} - R_{\mathrm{A}}$; $d = (R_{\mathrm{B}}^2 - R_{\mathrm{A}}^2)$; $e = (R_{\mathrm{B}}^3 - R_{\mathrm{A}}^3)$.

Differentiating Eq. (6.15) gives the slope of the m_t curve:

$$\frac{dm_t}{dR} = q + 2Rr + 3R^2 s.$$

At R_A, $dm_p/dR = dm_t/dR$, so $q + 2R_A r + 3R_A^2 s = S_A$

$$\therefore \quad q = S_A - 2R_A r - 3R_A^2. \tag{6.16}$$

Similarly, at R_B,

$$\frac{dm_p}{dR} = \frac{dm_t}{dR}, \qquad \text{so } q + 2R_B r + 3R_B^2 s = S_B. \tag{6.17}$$

Eq. (6.17) – Eq. (6.16) gives:

$$2cr + 3ds = S_B - S_A \qquad \text{so } r = \frac{S_B - S_A - 3ds}{2c}. \tag{6.18}$$

Also at R_A, $m_p = m_t$, so by substitution in Eq. (6.15)

$$p + R_A q + R_A^2 r + R_A^3 s = m_A. \tag{6.19}$$

Similarly at R_B, $m_d = m_t$ so $p + R_B q + R_B^2 r + R_B^3 s = m_B$

$$\therefore \quad p = m_B - R_B q - R_B^2 r - R_B^3 s. \tag{6.20}$$

Eq. (6.20) – Eq. (6.19) gives

$$cq = m_B - m_A - dr - es. \tag{6.21}$$

Substituting for q from Eq. (6.16):

$$c(S_A - 2R_A r - 3R_A^2 s) = m_B - m_A - dr - es.$$

Collecting terms, substituting for r from Eq. (6.18) and dividing out:

$$s = \frac{m_B - m_A - cS_A - (d - 2cR_A)(S_B - S_A)/(2c)}{(2cR_A - d)(3d)/(2c) - (3cR_A^2 - e)}. \tag{6.22}$$

Substituting for s in Eq. (6.18) gives r, substituting r and s in Eq. (6.21) gives q and substituting q, r and s in Eq. (6.20) gives p, solving Eq. (6.15) and giving values of m_t for spreadsheet calculation.

6.7 Overall multipath factor

6.7.1 Full method

When range is less than R_A, interference region multipath factor m_p applies. When range exceeds R_B the diffraction region multipath factor m_d is used; intermediately we use the transition region multipath factor, m_t. This gives a more detailed version of Eq. (6.9), valid at all ranges:

$$\begin{aligned}&\text{If } R < R_{\mathrm{A}}(\phi > 3\pi/2, \text{ Eq. (5.16c)}), \quad M = m_{\mathrm{p}} \text{ (Eq. (6.4))};\\ &\text{if } R > R_{\mathrm{B}}(m_{\mathrm{d}} < -20\,\mathrm{dB}), \quad M = m_{\mathrm{d}} \text{ (Eq. (6.7b))};\\ &\text{otherwise } M = m_{\mathrm{t}} \text{ (Eq. (6.15))}.\end{aligned} \tag{6.23}$$

6.7.2 Flat-Earth approximation

A flat-Earth approximation is sometimes used, as follows, but introduces significant error so is not recommended. Turning to Chapter 5, Section 5.6.1, Figure 5.12(a), direct ray path length ST is given by solution by Pythagoras's theorem of the right-angle triangle SJT:

$$\mathrm{ST} = \sqrt{\mathrm{FG}^2 + \mathrm{SJ}^2} = \sqrt{R^2 + (H - h)^2}. \tag{6.24a}$$

The indirect ray path length SZT = SZU is

$$\mathrm{SZT} = \sqrt{\mathrm{FG}^2 + (\mathrm{FS} + \mathrm{GU})^2} = \sqrt{R^2 + (H + h)^2}. \tag{6.24b}$$

SZT − ST is the path length difference Δ. Path length phase shift, Φ, is $2\pi\Delta/\lambda$. Assuming perfect phase-inverting reflection at the surface ($\Psi = \pi$), although flat-Earth multipath factor is not available from a simple equation, it can easily be plotted against range from a straightforward spreadsheet program, giving Figure 6.10 for

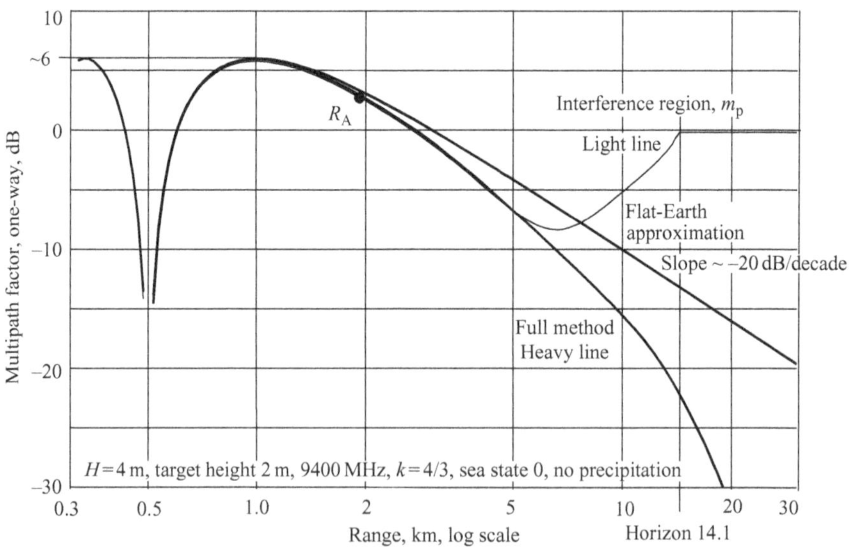

Figure 6.10 Multipath factor. Low scanner and target, no clutter. Even under these favourable conditions the flat-Earth approximation overstates performance at long range

a low scanner with a low point target. Accuracy is excellent to R_A, somewhat beyond the first peak at 1 km but at longer range the flat-Earth approximation gives signal strengths several decibel too high, the overstatement being 8.6 dB each way at the horizon. When the scanner is high, say 40 m, the flat-Earth approximation is excellent at short range but overstates the first peak range by about 1 km or 10 per cent and signal strength at the horizon is again overstated by ~8.5 dB each way. The flat-Earth method settles to a one-way multipath slope of −20 dB/decade at long range, equivalent to echo strength falling 80 dB/decade, 24 dB/octave or R^{-8}.

Even at long range, the flat-Earth approximation is much better than use of the interference region multipath factor alone, primarily because the latter's divergence term eliminates the indirect ray at the horizon and forces m_p to 0 dB. Divergence does not occur when there is no surface curvature.

6.8 Two-zone method

6.8.1 General form of multipath/range relationship

Few systems have sufficient sensitivity to operate into the far transition or diffraction regions. We have seen that at short range, multipath factor fluctuates in value with an amplitude dependent on sea surface reflection coefficient, ρ, whereas in the near transition region multipath factor falls at about 20 dB/decade. The boundary between these near and far zones is called the transition or critical range, R_C, lying near the transition range, R_A. The two-zone approximation to be described enables reasonably quick and accurate calculation of multipath factor to be made into the near transition region but becomes inaccurate at longer range.

At short range, average echo strength through a complete multipath cycle follows an R^{-4} law. Beyond critical range there is an additional R^{-2} multipath factor each way. The critical range is therefore the range at which the echo power/range law changes from inverse fourth to inverse eighth power. As we shall see in Chapter 9, when targets are extended in height short range multipath fluctuations are muted, and the critical range concept becomes especially useful. We start by examining the rate of change of multipath factor at transition range R_A.

6.8.2 Rate of change of multipath factor at R_A, calm sea

From Eq. (5.16b), $\Phi = 4\pi H'h'/R\lambda$. Substituting in Eq. (6.5a):

$$m_p = 4\sin^2\frac{2\pi H'h'}{R\lambda}. \tag{6.25a}$$

The chain rule for differentiation states that if $h(x) = f(g(x))$, then $h''(x) = f''(g(x))g''(x)$, where double primes denote the first differential of the functions $f(x)$, $g(x)$ and $h(x)$. Hence, the rate of change of m_p with R is

$$\frac{dm_p}{dR} = \frac{2\pi H'h'}{R^2\lambda}\sin\frac{2\pi H'h'}{R\lambda}. \tag{6.25b}$$

At R_A, $2\pi H'h'/R\lambda = \pi/2$ so the sine term is unity and

$$\frac{dm_p}{dR} \text{ at } R_A = \frac{2\pi H'h'}{R^2\lambda}. \tag{6.25c}$$

6.8.3 Approximation for multipath factor in near transition region

We first assume a calm sea, then a rough sea and finally a moderate sea.

6.8.3.1 Calm sea

If there is perfect reflection ($\rho = 1$, approximated by calm sea, divergence d assumed unity), by analogy with Eq. (6.25a) the flat-Earth approximation for multipath factor m_{fe} is

$$m_{fe} = 10 \log \left\{ 4 \sin^2 \left[\frac{2\pi Hh}{R\lambda} \right] \right\} \text{ dB}. \tag{6.26}$$

When R is small, the expression represents the multipath peak and null structure. From Eq. (6.6), taking the interference region average value of m_p ($\rho = 1$) to be $m_{p\,av} = 4(\pi - 1)/\pi \sim 2.726 = 4.36$ dB and ignoring the peaks and nulls gives line AX of Figure 6.11.

The relationship changes when the sine term falls to unity at the critical range, so

$$R_C = \frac{2\pi Hh}{\lambda}. \tag{6.27a}$$

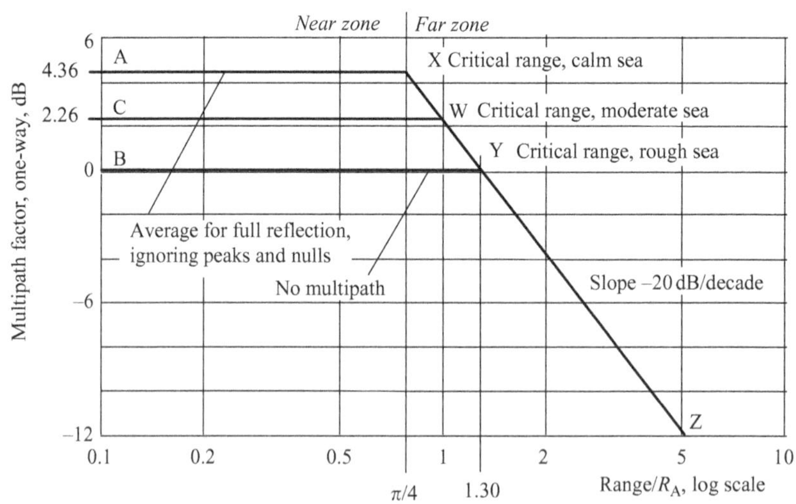

Figure 6.11 Multipath approximations. Interference and near transition regions

Substituting in Eq. (6.2a) ($R_A \sim 8Hh/\lambda$) for the total reflection condition gives point X:

$$R_C = \frac{\pi}{4} R_A. \tag{6.27b}$$

From this, $R_A = 4/\pi R_C$. This flat-Earth value may be corrected using Figure 6.5. At R_C, $m_t = m_{pav} = 4.36$ dB. Above R_C, $2\pi Hh/(R\lambda) < 1$ so $\sin[2\pi Hh/(R\lambda)] \sim 2\pi Hh/R\lambda$ and here

$$m_{fe} = 4.36 - 20 \log \frac{R}{R_C} \text{ dB} \tag{6.27c}$$

represented by line XZ. Adding the free space −20 dB/decade term, it follows that when scanner and target effective heights are assumed not to fall as the horizon is approached, echo strength above critical range falls at 80 dB/decade two-way, an R^{-8} law shown as XZ in Figure 6.12.

When the inverse range law is steep, a moderate change in system sensitivity makes little difference to detection range. Rates of change of range with system sensitivity are:

R^{-4} law, 5.9 per cent/dB; R^{-8} law, 2.9 per cent/dB;

R^{-12} law, 1.9 per cent/dB.

6.8.3.2 Rough sea

If there is no surface reflection ($\rho = 0$, approximated by rough sea), $m_{fe} \sim 0$ dB at short range. Multipath factor can be represented on Figure 6.11 as horizontal line BY

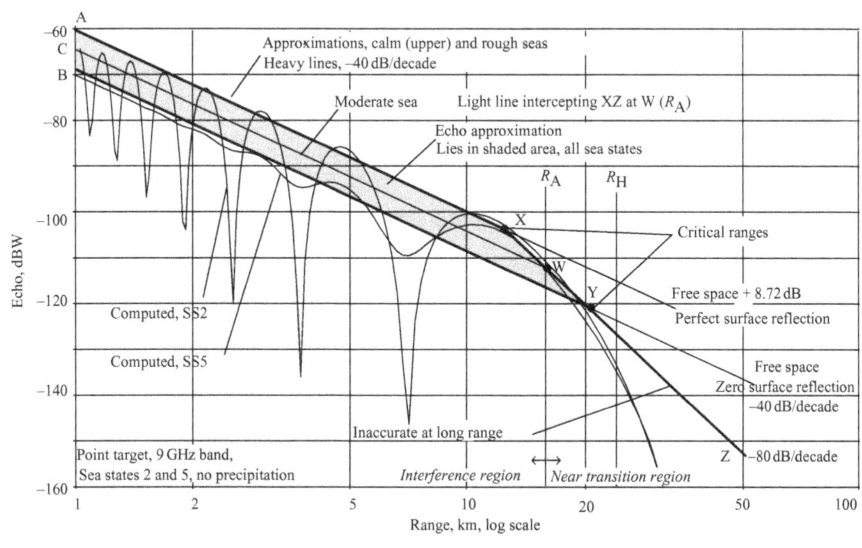

Figure 6.12 Echo strengths corresponding to Figure 6.11

at 0 dB, intercepting XZ at Y. Critical range is now $10^{4.36/20} = 1.65\times$ Eq. (6.27b) critical range:

$$R_C = 1.65 \times \frac{\pi}{4} R_A = 1.30 R_A. \tag{6.28}$$

Therefore, for a rough sea:

$$\text{If } R < R_C \text{ (per Eq. (6.28))}, m_{fe} = 0,$$

$$\text{otherwise } m_{fe} = 4.36 - 20 \log \frac{R}{R_C} \text{ dB}. \tag{6.29a}$$

6.8.3.3 Moderate sea

Eq. (6.28) shows that critical range rises with sea state and always lies between the limits 0.785 and $1.30 \times R_A$. As $1.0 \times R_A$ lies close to their geometric mean, this therefore forms a reasonable average critical range with moderate sea state, justifying the choice of $\phi = 3\pi/2$ as the R_A criterion in Section 6.3.2, line CWZ representing an average condition. In other words, the transition region commences at the average critical range. Figure 6.12 uses the same notation and shows that CWZ provides a reasonable guide to echo in the interference and near transition regions. The value of W at range R_A is $4.36 - 20 \log 4/\pi = 2.26$ dB. Therefore, CWZ is represented for average conditions by the following.

$$\text{If } R < R_A, \quad m_{fe} = 2.26, \qquad \text{otherwise } m_{fe} = 2.26 - 20 \log \frac{R}{R_A} \text{ dB}. \tag{6.29b}$$

m_{fe} can be substituted for M in the radar range equation of Chapter 4, Eq. (4.6b) to give Eq. (6.30), which is reasonably accurate if the following hold.

- H and h are each less than about 8 m (not the case in Figure 6.3), or the maximum range of interest does not exceed 15–20 km. Higher, the expression overstates long range echo strength, particularly when k is low and frequency is high (e.g. in the 9 GHz band).
- The sea is rough; it does not reproduce the peaks and nulls characteristic of calm sea.

$$S_e = S_{e(FS12)} - 2L_A + 2m_{fe} \text{ dB}. \tag{6.30}$$

6.8.4 *Approximate multipath factor near horizon*

Figure 6.3 shows multipath factor falls sharply in the transition region. From Eq. (6.25b), dm_p/dR ostensibly follows an R^{-2} law, causing echo strength to fall at 40 dB/decade plus 40 dB/decade for the free-space R^{-4} component, as Eq. (6.29a). However, H' and h' also fall as the horizon is approached and the overall one-way multipath law steepens to nearly R^{-4}; that is, m_p falls at roughly 40 dB/decade. Two-way echo strength law therefore totals R^{-12} approximately, falling very sharply by as much as 36 dB/octave, indicated in Figure 6.12 by the divergence of the computed echo strength from line XZ near Z.

Allowing for the uncertainty, overall multipath factor can be expressed as

$$\text{If } R < R_{\mathrm{A}} \text{ (Eq. (6.2a))}, \quad M = m_{\mathrm{p}} \text{ (Eq. (6.4))};$$

$$\text{otherwise } M \sim \left(\frac{R_{\mathrm{A}}}{R}\right)^{(2 \text{ to } 4)} \text{ numerically.} \tag{6.31}$$

6.8.5 *Very low scanner or target*

If the scanner or target lies on the waterline, R_{A} falls to zero; as already remarked, such systems remain in the transition or diffraction regions down to zero range. In the limit where H or h tend to zero, Chapter 5, Section 5.5.7, Eq. (5.23a) gives $R_{\mathrm{H}} = \sqrt{2ke} \times \sqrt{\text{height of other element}}$, so R_{H} remains substantial. On the other hand, transition range $R_{\mathrm{A}} \sim 8Hh/\lambda$ tends to zero. This shows that the method of ray tracing breaks down, which is not surprising, since diffraction is in play. There is no indirect ray and the concept of grazing angle becomes meaningless. We can get round the difficulty by putting the surface element at a nominal low height, say H or $h = 1$ m, then apply the usual calculations.

6.9 Sketching echo strength

6.9.1 *Use of sketches*

You want to be able to detect a particular buoy at 20 km under conditions of adverse refraction ($k = 0.8$) and excessive rain ($r = 40$ mm/h). Your 9 GHz VTS radar's minimum detectable signal is −140 dBW. (For simplicity MDS is assumed constant but in reality would suffer in the severe clutter caused by excessive rain.) A pushy young salesperson comes along to sell you a super-reflector, RCS 20 dBm^2. His laptop takes a little time to digest all the impressively voluminous data he keys in about your system, including all radar parameters, latitude and longitude, sea and air temperatures, the sign of the zodiac and your secretary's telephone number. This is your opportunity for a few quick calculations and a rough sketch in preparation for his super-accurate (but undisclosed) program's result – detection range 23.352959 km. Splendid, program gives results to the millimetre, so it must be right! But do you believe it?

Your sketch must account for multipath and precipitation attenuation, as well as the free-space conditions sketched in Chapter 4, Section 4.6.2, Figure 4.3. The method of Figure 6.12 (Section 6.8.3) is rather too simplistic. Written out in full, Eq. (6.23) is extremely lengthy and almost impossible to calculate by hand, although readily handled by a spreadsheet. It can however be sketched to fair accuracy. Sketches for one- or two-way operation have several other uses.

- As a preliminary assessment of whether a proposed system looks sufficiently promising to justify detailed calculation.

- To illustrate concepts during discussions.
- Using round numbers, as cross-checks on detailed calculation to confirm absence of gross blunders; we have all managed to forget the odd 2π or factors of 1000 for watts/kilowatts, metres/kilometres and so forth!

The following procedure may initially look long-winded, but after a little practice it should be possible to complete a curve in a couple of minutes. The procedure for sketching echo strength versus range may easily be adapted to produce curves of multipath factor alone. The following steps are indicated on Figures 6.13(a) and (b).

6.9.2 Scales

1. It is usually best to use a logarithmic range axis, as in Figure 6.13(a), to avoid cramping at short range and to linearise the inverse square and inverse fourth power laws. If log graph-paper is not to hand, it is easy enough to mark off plain paper as described in Chapter 4, Section 4.6, as shown in italics.
2. Mark the vertical axis as echo strength linearly in decibels to a convenient scale, say 1 cm per 5 dB between −160 dBW at the foot of the page and −60 dBW near the top.
3. Draw −40 and −80 dB/decade (R^{-4} and R^{-8} laws) construction lines AB, AC, respectively. Alternatively, draw lines falling 12 and 24 dB, respectively, for an octave, say between 50 and 100 km.

6.9.3 Sketching echo, fair weather

The procedure assuming calm sea, no precipitation is as follows, salient points being indicated on Figure 6.13(a).

4. Calculate free space reference echo for target RCS 0 dB m^2, F_{12}, at reference range $R = 1.0$ km by the method of Chapter 4, Eq. (4.6b). Enter point D (here −89.9 dBW).
5. Step up by target RCS, here 20 dB, to −69.9 dBW, point E. Although steps 4 and 5 may be combined, separating them facilitates re-running the procedure for differing target RCS values.
6. Draw straight line EF, slope −40 dB/decade, parallel to AB, to represent the echo in free space with no atmospheric attenuation.
7. Find R_A by Eq. (6.2), corrected by interpolating in Figure 6.4, using the appropriate k factor and radar band. If $H = 30$ m, $h = 4$ m, 9 GHz band, $R_A \sim 16$ km. Enter point J on line EF at 16 km (log 16 = 1.20 so point lies at 12.0 cm along the axis).
8. In this area signal strength approximates an R^{-8} law (Section 6.8.1). Therefore draw line GH (parallel to AC) at −80 dB/decade through J.
9. Draw line KL 10 dB above EF to represent the locus of multipath peaks. With total reflection at the sea surface, the locus would be 12 dB above the free space line. Using 10 dB allows for imperfect reflection and remains valid within a couple of decibels up to at least sea state 2.

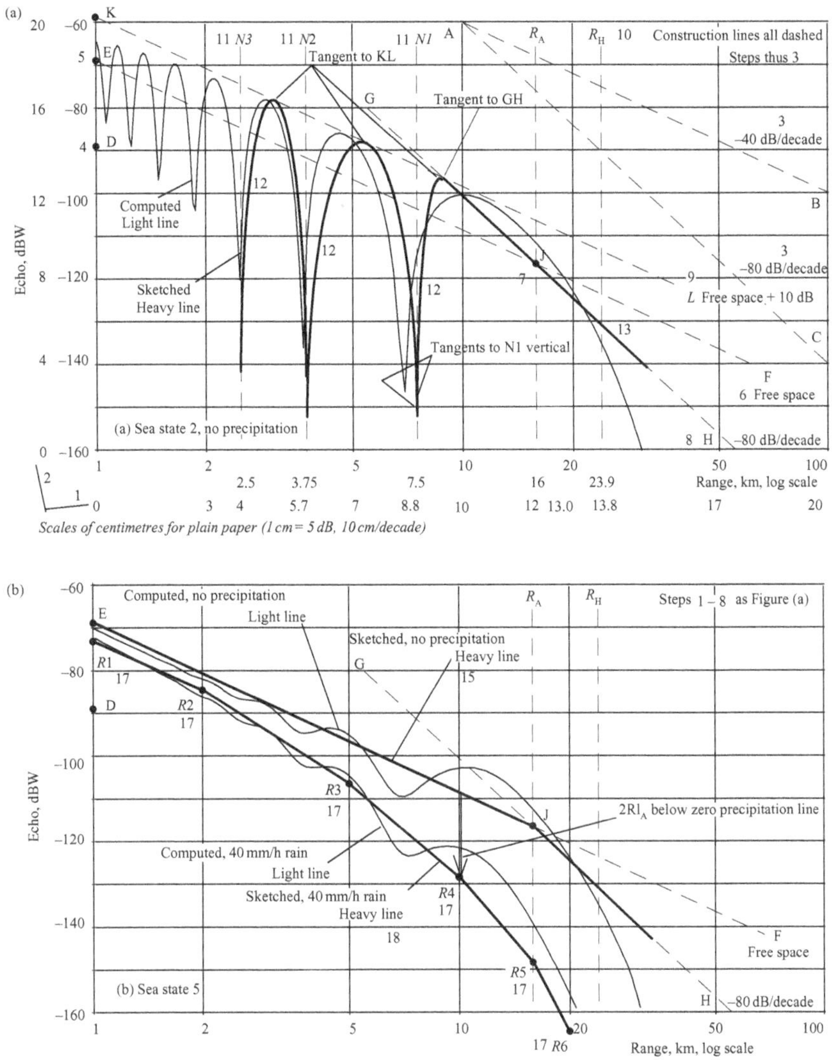

Figure 6.13 *Sketching echo strength. 9410 MHz radar, mounted at $H = 30$ m, viewing 100 m^2 (20 dB m^2) point target at $h = 4$ m. (a) Fair weather (SS2, no precipitation). (b) Foul weather (SS5, zero and excessive rain). The suggested scales just fit A4 paper*

10. If of interest, calculate horizon range from Chapter 5, Section 5.5.7, Eq. (5.23), and mark on graph: $R_H = 3.570\sqrt{k}(\sqrt{H}+\sqrt{h}) = 3.570 \times \sqrt{0.8}(\sqrt{30}+\sqrt{4}) =$ 23.9 km (at $10 \log 23.9 = 13.8$ cm).
11. Calculate null ranges $N1, N2, \ldots$ from Eq. (5.30). Being on a flat-Earth basis, these ranges will be somewhat too high. $N1 = 7.5$ km, $N2 = 3.75$ km,

$N3 = 2.5$ km. Take logarithms and mark these ranges on the graph sheet at 8.8, 5.7 and 4.0 cm, respectively.

12. Draw ∩ curves (freehand, using French curves or stencils; the figure used ellipses) tangential to KL and to the $N1$, $N2, \ldots$ verticals. The curve to the right of $N1$ is tangent to $N1$, GH and KL.
13. Emphasise the ∩ curves to J, and the KL line to its right to represent the (echo strength)/(range) curve of the system. Null depths cannot be accurately drawn (they are too deep in the figure), but at least they are a reminder that nulls exist. Comparison with the computed curve per Eq. (6.23) shows error is generally less than 5 dB, which is perhaps consistent with the uncertainties inherent in the radar, target and environmental parameters of typical systems.

A final touch: a blunt soft pencil reminds that all radar performance calculations are imprecise, a point easily overlooked with clean computer-generated curves!

6.9.4 *Sketching echo, rough sea*

To produce a sketch of echo strength of a point target in a rough sea, with negligible reflection at the sea surface (Figure 6.13(b)) proceed as follows, initially for zero precipitation.

14. Follow steps 1–8. The horizon range is unchanged.
15. If there is negligible atmospheric attenuation, for example, no precipitation, the echo is represented by line EJH.
16. If there is significant precipitation, check the one-way total atmospheric attenuation rate, l_A, from Chapter 5, Section 5.9, Figure 5.22. Figure 6.13(b) is drawn for 40 mm/h 'excessive' rain, $l_A = 1.0$ dB/km one-way. The two-way attenuation is therefore 2.0 dB/km.
17. Drop perpendiculars from EJH at half a dozen ranges (here 1, 2, 5, 10, 13.2, 20 km), calculate total atmospheric attenuation (assumed to cover the whole path) $= 2L_A = 2l_A R$ at each range and mark points $R1$–$R6$, $2L_A$ below EJH, for example, for $R2$, 4 dB at 2 km.
18. Join $R1$–$R6$, either by straight lines (as in the figure) or by a smooth curve.

    Note

    - the sketched line is a reasonable approximation to the computed line,
    - the first peak is under-represented,
    - the severe performance loss caused by excessive rain at 9 GHz.

6.9.5 *Really rough sketch*

The minor nature of the multipath ripple in the interference region of Figure 6.13(b) suggests an even simpler procedure based on Figure 6.13 and shown in Figure 6.14.

19. Draw the free space line EF as before, marking point J at R_A.
20. Ascertain l_A.
21. Mark a point P, $2R_A l_A$ dB $= p$ dB below J, representing the echo in precipitation at range R_A.

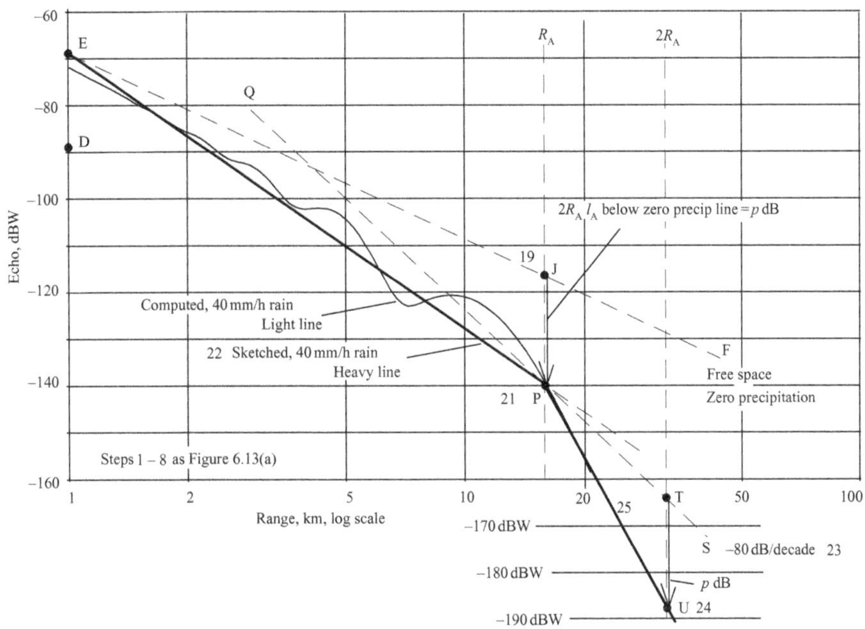

Figure 6.14 Rough sketch. Echo strength ignoring interference, foul weather. System as Figure 6.13

22. Draw a straight line through EP, extended to the right, representing the free-space line in precipitation. The error at the short range of E is negligible. There are also errors near $R_A/2$.
23. Draw line QS through P at slope −80 dB/decade.
24. At range $2R_A$ step down from QS by p dB, point U, unless this is well below signal strengths of practical significance. Figure 6.14 shows this extends below the graph sheet – have scrap paper and adhesive to hand!
25. Emphasise line EPU, which represents the echo strength.

6.9.6 Accuracy

Comparison of the sketched ranges with detailed calculations shows fair agreement except that maximum detection range is overstated for the more sensitive radar with no precipitation. This is because fall of effective scanner and target heights near the horizon causes signal to fall more steeply than −80 dB/decade. The error would have been less at 3 GHz, where the interference peaks and nulls are displaced to lower ranges. Null ranges are somewhat too high, as they use the flat-Earth approximation.

The methods of Figure 6.13(b) or the rougher Figure 6.14 agree that the range of a radar of MDS = −140 dBW in SS5 and 40 mm/h rain is 13–16 km, what an old colleague called 'definitely 14-ish'. No way can the reflector salesman's 23.3… km be right! You can now throw him out; or place an order, including, well down page 4 of your Terms and Conditions,

> Clause 13.3.2 (a). Vendor shall demonstrate operation to 23 km range on Buyer's radar in excessive rain to the satisfaction of an independent Test House.

Knowledge is power. Your choice.

6.10 References

1 HALL, M. P., BARCLAY, L. W., and HEWITT, M. T. (Eds): 'Propagation of radiowaves' (IEE Publishing, 1996, 1st edn.), Figure 4.4
2 MEIKLE, H.: 'Modern radar systems' (Artech House, London, 2001)
3 BLAISE, P.: 'Estimated range of radar beacons'. *International Association of Lighthouse Authorities XI Conference*, Brighton, UK, 1985, Paper 6.3.1
4 KERR, D. E.: 'Propagation of short radio waves, MIT Radiation Laboratory Series, vol. 5' (McGraw-Hill, 1951, re-published IEE, 1987)
5 BARTON, D. K. and LEONOV, S. A. (Eds): 'Radar technology encyclopedia' (Artech House, London, 1997)

Chapter 7

Passive point targets

'It has long been an axiom of mine, that the little things are the most important.'
Sir Arthur Conan Doyle, The Adventures of Sherlock Holmes, A Case of Identity

7.1 Introduction

This chapter examines the mechanisms causing insulators and metallic shapes to reflect and goes on to discuss the reflection to be expected of point objects. It draws on material previously published by the author [1] and by others who are acknowledged in the text.

7.1.1 Structure of RCS discussions

Radar rays striking an object are partly absorbed as heat and partly scattered or re-radiated, the scatterer acting as an antenna. The scattering mechanism differs between insulators and conductors, but the overall results are generally similar. The component of re-radiation which reaches the radar receiver is the echo. The measure of how well an object returns rays incident on it toward the source is called the radar cross section (RCS), σ, and is a measure of the mean retro-reflecting quality of the target. It describes the apparent target area as seen by the radar. RCS is of prime importance to us, for an object whose RCS is too low or is zero cannot be detected, whatever its physical size. The reflection leaving the device cannot exceed (incoming flux density) $\times$ (device area). Formal definition of RCS is within Section 7.4.2. Although the formal definition of radar echoing area differs, values are identical and the terms are used interchangeably. The following all assumes that the target lies in the far field of the reflector where range $>2l^2/\lambda$, l being target maximum projected width and λ the wavelength.

RCS is an area and may be expressed in square metres. It is the measure of the 'radar size' of the object under the specified conditions, more specifically the retro-reflecting quality of the target in the direction of the illumination. Strictly speaking, this ordinary RCS is the monostatic RCS (as seen by the source of illumination), the bistatic RCS being the proportion reflected in a specified direction when illuminated in a (generally different) specific direction. Unless stated otherwise, the monostatic value is always to be inferred.

The radar range equations (Chapter 4, Section 4.3.2, Eq. (4.3), and following), show echo power is proportional to σ. So if RCS is doubled, echo power also doubles. For this reason RCS may alternatively be expressed in decibels relative to $1\,\text{m}^2$, an RCS of say $2\,\text{m}^2$ being expressed as $10\log 2 = 3.01\,\text{dB}\,\text{m}^2$. Precipitation and sea clutter within the detection cell also have radar cross sections and the detectability of an echo often hinges on its RCS relative to the clutter RCS within the cell.

So far we have considered the target 'size' or radar cross section merely as a term in the radar range equation. Here we shall investigate the variation of RCS with viewing angle or 'aspect'. This variation can be depicted by two-dimensional polar diagrams – polar graphs of RCS versus azimuth or elevation angle of illumination – or by more general three-dimensional surface contour plots or target pattern maps (TPM) in Cartesian coordinates whose colouring, shading or contour lines represent RCS values in azimuth or elevation just as ground heights are shown on maps. This chapter deals with the RCS of objects which are sufficiently small for their direct and indirect echo rays to be treated as coming from a geometrical point; and which are passive, with no electrical augmentation of the incident energy, the response depending entirely on reflection. Active devices such as racons and RTEs substitute an oscillator or amplifier for the reflector and are covered in Chapter 8. Large targets such as ships, to which the geometry of point target rays is inapplicable, are sufficiently important for separate treatment, Chapter 10. All these discussions concentrate on the average or mean RCS. Chapter 12 includes discussion of the fluctuations about the mean as the radar or target move in a seaway, affecting probability of detection.

Chapter 4 derived forms of the one- and two-way radar range equations for free space, including atmospheric loss, L_{A} evaluated in Chapter 5, and multipath, M evaluated for point targets in Chapter 6, to account for environmental factors.

In the present chapter, we study the target RCS term, which can be inserted directly in the two-way range equation Eqs (4.6a) and (4.6b), allowing the echo S_{e} from a passive point target to be determined. Study of the reflecting properties of materials and simple basic shapes allows us to draw out some principles applicable to more complex structures. Then we shall examine some of those objects, which may be geometric shapes specifically designed to reflect, sometimes called passive signature enhancers; or practical targets such as lighthouse lanterns not specifically so designed.

Purpose-made passive reflectors rarely exist in isolation. Often they are mounted on floating platforms such as navigation buoys or small craft. Knowledge of the TPM enables performance to be predicted when the platform tilts or its azimuth aspect changes. We also have to examine the effect of a pair of reflectors, such as the buoy structure plus its reflector, for the RCS of the pair is likely to differ markedly from either of its elements. This and other concepts raised in this chapter will be applied to extended targets in Chapter 9.

7.1.2 Applications of point passive reflectors

Point reflectors are used to enhance detectability of poorly reflecting targets such as yachts by providing reasonably large, uniform and reliable radar cross section, mounted high for good horizon range. (Devices intended to improve detection of navigation marks such as buoys and lighthouses work in the same manner but are termed aids to navigation.) Beside sheer size of RCS, uniformity is important as the target heels and its heading changes. So we are interested in (a) how much of the incident energy the object reflects, expressed as RCS, (b) variations when tilted, described by the azimuth and elevation polar diagrams or by the TPM and (c) acceptance of differing polarisations.

7.1.3 Meanings

Here and in Chapter 8, we ascribe particular meanings to certain terms.

- Skin RCS, skin echo: unaided RCS of the platform surface, its echo. 'Skin' originated as military jargon for (echo) of an airborne practice target not augmented by a transponder.
- Aid: device intended to augment skin RCS.
- Passive aid: one containing no source of power to augment the incident energy; for example, a metal reflector.
- Active aid (or reflector): one augmenting the incident energy by inclusion of electronic circuits and a power source; for example, a racon.
- Point (aid): one whose volume is small enough to approximate a point when tracing rays.
- Platform: the structure carrying the aid, which may be a buoy, lighthouse, yacht, etc.
- Leg: radar-to-aid or aid-to-radar direct or indirect (multipath) path.
- Equator of aid: horizontal plane through aid when upright.
- Interrogation: pulse from a radar scanner.
- RCS: the effective radar size of an object at the current tilt and azimuth.
- Response or echo: signal returned to the radar.

Both metals and insulators can reflect. Metals are of greater importance, but we first dispose of insulators.

7.2 Reflection from insulators

7.2.1 Basic process

With the notable exception of water, insulators are always rather poor reflectors, for reasons developed below. When radar rays in air strike an insulating material the energy splits four ways depending on the angle of incidence and the material's dielectric constant or permittivity, ε, and lossiness, discussed in Chapter 5, Section 5.8.2.

- When rays strike the insulator, much of the energy is accepted and traverses the material.

- Some of that energy is dissipated as heat.
- A component of the incident energy, the amount dependent on ε and the angle of incidence, is reflected at each air/insulator interface. That at the first interface is the primary reflection.
- If the interface is flat, the reflection is specular; if rough, the reflected component is scattered through a broad solid angle.

For air, $\varepsilon \sim 1.0$ but is higher for all other insulators. Often ε falls somewhat as frequency rises, but we can ignore this dispersion. When a ray within a first dielectric having dielectric constant ε_1 strikes a second dielectric of dielectric constant ε_2 at grazing angle α, an electric field is set up in the second material. The magnetic but not the electric field is reversed and there is phase reversal. The fields cause a mirror-like reflection of a portion of the incident energy, the remainder refracting into the second material. The ratio of reflected to incident ray voltage amplitudes (the voltage reflection coefficient, ρ) depends on the plane of polarisation. Chapter 5, Section 5.8.2, Eqs (5.40a) and (5.40b), for horizontal and vertical polarisation respectively, gave ρ in terms an air to lossy dielectric interface. When the ray crosses at normal incidence ($\alpha = \pi/2$) from a vacuum (or, approximately, air, $\varepsilon = 1$) to a lossless dielectric of dielectric constant ε, ρ becomes independent of polarisation and has been encountered in Chapter 5, Section 5.8.3, Eq. (5.40):

$$\rho = \frac{1 - \sqrt{\varepsilon}}{1 + \sqrt{\varepsilon}}.$$

The quantity is negative, indicating phase reversal. The proportion of the incident power reflected is R ($R \leq 1$).

$$R = \rho^2 = \frac{\text{power reflected}}{\text{incident power}}.$$

At normal incidence

$$R = \left[\frac{\sqrt{\varepsilon} - 1}{\sqrt{\varepsilon} + 1}\right]^2. \tag{7.1a}$$

Perfectly reflecting insulators would have the same RCS, σ, as the metal reflectors to be considered next if secondary reflections are ignored. Practical imperfectly reflecting insulators, where $\varepsilon < \infty$, have the return voltage reduced by factor ρ per reflection, so if there are n reflections, for example, 2 in the case of Figure 7.3(g),

$$\text{effective RCS from the primary reflections} = R^n \sigma \text{ m}^2. \tag{7.1b}$$

It is not feasible to make practical point reflectors from insulating materials, although some metallic models are protected by insulating skins.

Although small antennas can be made from insulators, for example, the Polyrod, they are seldom encountered in marine practice. Dielectric lenses, made from low-loss plastics, are used in a few passive aids. Because of the much longer wavelength, microwave lenses often look much different from their optical cousins. Dielectric lenses are used in some of the proprietary passive reflectors mentioned in Section 7.7.

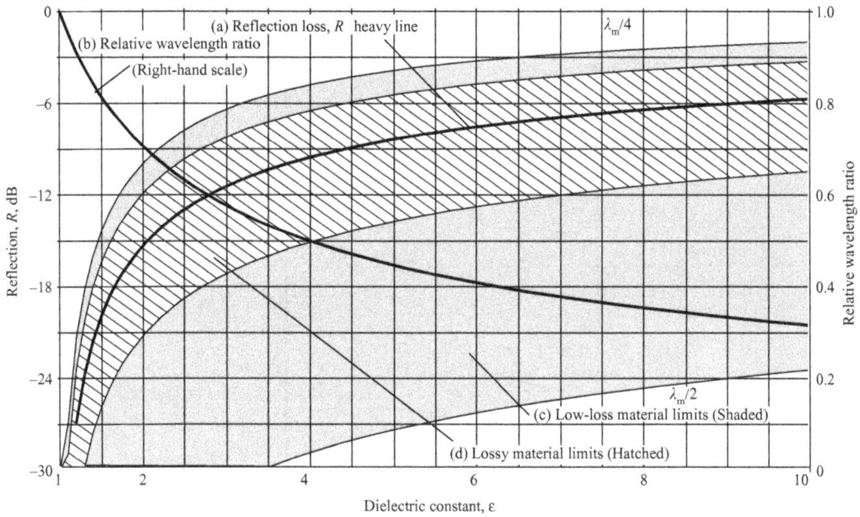

Figure 7.1 *Reflection strength from insulators. Viewed normally. Unless ε is unusually high, little of the incident energy is reflected. Wavelength falls within high-ε dielectrics*

7.2.2 Secondary reflections

The higher dielectric constant reduces velocity of propagation in the slab, reducing wavelength, λ_m. Relative to the ordinary free-space or characteristic wavelength λ_0,

$$\text{relative wavelength ratio} = \frac{\lambda_m}{\lambda_0} = \frac{1}{\sqrt{\varepsilon}}. \qquad (7.1c)$$

Eqs (7.1a) and (7.1c) are plotted in Figures 7.1(a) and (b) (the latter in dB form), while (*c*) shows the range of reflection values when the material has rather low loss ($L = 0.95 \sim 0.45$ dB per pass, curve (c)). Reflection becomes extremely dependent on slab thickness relative to wavelength. When the material is relatively lossy ($L = 0.707 = 3$ dB per pass, curve(c)), the range of reflection values is less. As loss increases further, the secondary reflections are highly attenuated and the reflection approaches the Figure 7.1(a) value, which is the primary reflection.

Figure 7.2 shows diagrammatically the component of energy not reflected entering the material at A. With a thick lossy slab, much of this energy is absorbed, at a rate governed by the material's loss tangent or tan δ (δ is the in-phase component of the current vector). The ray emerges into the air at the far side at B, accompanied by another reflection process, without phase reversal because the new interface is from high to low dielectric constant. This reflection is subject to another transmission/reflection when it reaches the front of the slab at C. The emergent component forms a secondary reflection whose phasing depends on slab thickness. The resultant of the primary reflection at A and secondary reflections at C, E, G, . . . are sensitive to phasing, which depends on slab thickness. Residual energy continues to traverse

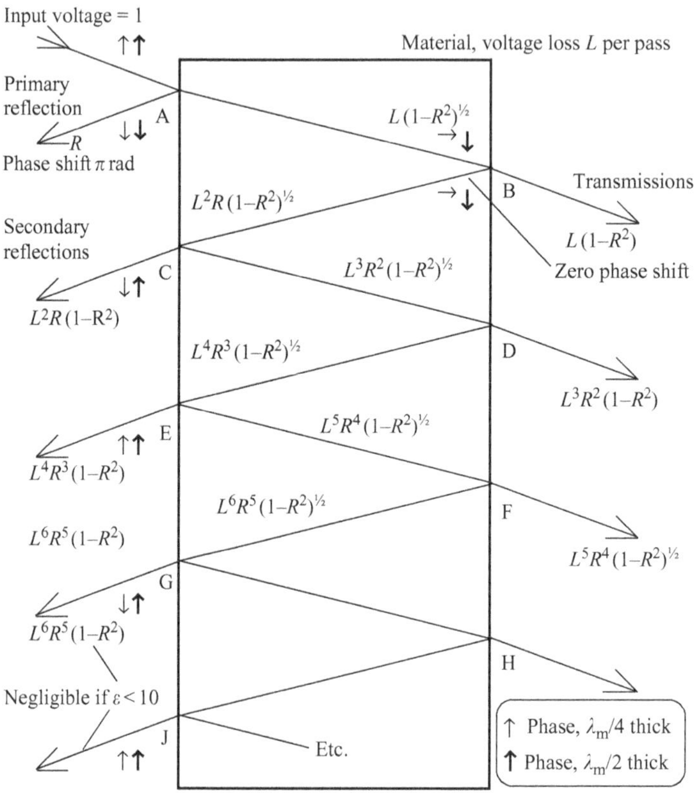

Figure 7.2 Reflections in insulating slab. Initial reflection at A is heavily modified by subsidiary re-reflections at C, E, etc., making the nett reflection dependent on dielectric constant, thickness and lossiness, and difficult to calculate

to and fro in the slab, losing energy to air at each reflection and eventually petering out, akin to the reflections in a mismatched transmission line.

In Figure 7.2, the signal from the radar has 1 power unit and 1 voltage unit. Upon striking the slab at A, the reflected voltage component is R, containing R^2 power units and the transmitted voltage is $(1 - R^2)^{1/2}$. By the time this ray reaches the far side, point B, attenuation has reduced its amplitude by a factor L. At B, the reflected ray is R times that reaching B and so on. The voltages for the first few reflections are shown in the figure. Each reflection voltage is $(LR)^2$ its predecessor.

If the material thickness is $\lambda_m/4$, $3\lambda_m/4$, etc., the secondary reflection at C is in phase with and reinforces the primary reflection at A, whereas when thickness is 0, $\lambda_m/2$, λ_m, etc., all the secondary echo components are at antiphase to A (thick arrows) and the nett reflection is minimum, the lower edges of the Figure 7.1 shaded areas. The reflections partially cancel (especially if the material has low loss), reducing the total reflection towards $-\infty$ dB and making the slab nearly transparent.

In short:

- reflection from a slab depends on radar frequency, slab projected thickness, dielectric constant and lossiness, plus of course any reflection from anything beyond the slab;
- reflections from an insulator are always imperfect;
- ε has to be high for strong reflection;
- reflection is extremely dependent on thickness and lossiness of the insulator;
- reflection may also be influenced by subsidiary reflections beyond the insulator;
- there are usually too many uncertainly known variables for accurate calculation of reflection.

7.2.3 Materials

Very broadly, dielectric constant tends to increase with density (specific gravity). Some special ceramics used in capacitors have dielectric constants of 1000 or more, but every-day non-metals typically have $\varepsilon \sim 4$, which halves the velocity of propagation and the wavelength. Black coloration may indicate presence of carbon powder in the material, making it particularly lossy. The entrapped gas within foamed material 'dilutes' the dielectric constant, which is lower than for solids. As noted in Chapter 5, water is a special case with the unusually high $\varepsilon \sim 80$ at radar frequencies.

1. GRP (glass reinforced plastic) is a rather low-loss material used in boat-building; $\varepsilon \sim 3$. From Figure 7.1 or Eq. (7.1a), $R \sim -11.4$ dB, so the nett surface reflection is likely to be at least 11 dB down on a metal surface.
2. Glazed windows are generally too thin to reflect much. Glass has $\varepsilon \sim 4$–6, low loss; $R = -9.5$ to -7.5 dB.
3. Timber has low ε, (softwood ~1.4, hardwood ~1.9; $R \sim -21.5$ to -16 dB, respectively) and is moderately lossy, particularly when damp. Indeed, tapered beech plugs a few wavelengths long make excellent laboratory waveguide loads. Wooden boats are therefore even poorer targets than GRP ones.
4. Rubberised textile sheet of rigid inflatable boats (RIBs) is lossy and generally too thin for significant reflection; $\varepsilon \sim 4$.

7.2.4 Reflecting shapes

Figure 7.3 depicts rays striking insulators. Several things can happen.

1. Thin insulator with air beyond. Boat sails, scanner windows and some active device radomes fall into this category. There is little absorption loss. The primary reflection on entry is almost cancelled by the anti-phase secondary reflection and there is negligible nett reflection. The insulator is radar-transparent, Figure 7.3(a), and R tends to $-\infty$ dB.
2. Thin insulator backed by metal. The secondary reflection is nearly perfect and the reflected energy is stronger, Figure 7.3(b). If the insulator is slightly lossy paint, the nett reflection is basically that of the underlying metal, slightly attenuated; R tends to 0 dB.

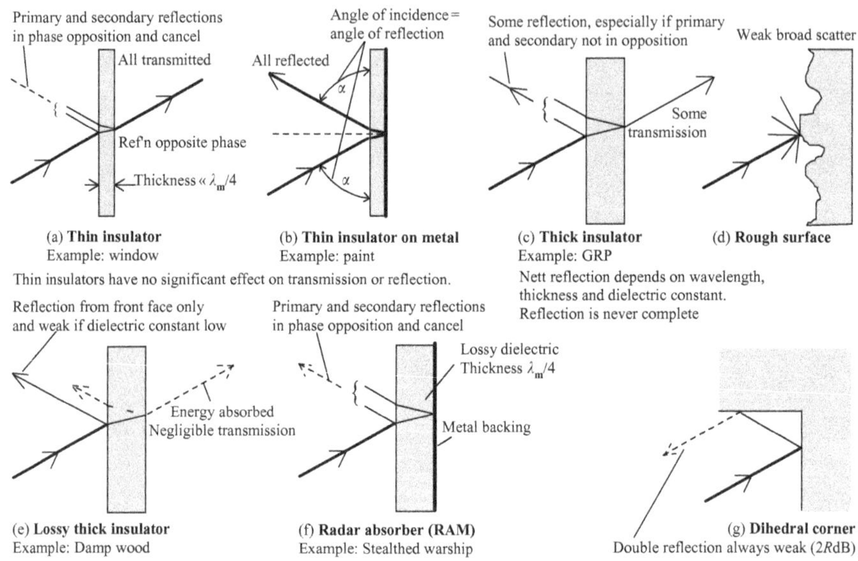

Figure 7.3 Reflecting insulator shapes

3. Insulator up to ~20 mm thick, with air beyond. Energy at the exit insulator/air surface suffers the same partial reflection, at a different phase, Figure 7.3(c). Some goes forward, particularly when thickness is chosen to give equal and opposite phasing at entry and exit. Some is reflected, to a practical maximum of say $R \sim -3$ dB. The rest is absorbed as loss. Some racon radomes are made a half-wavelength thick (~10 mm at 9 GHz), cancelling input and output reflections to maximise transmission.
4. Rough or corrugated surface. When the root mean square or standard deviation of surface roughness exceeds about $\lambda_m/4$, the reflection scatters over a broad solid angle and the monostatic component becomes very weak, Figure 7.3(d).
5. Insulator lossy ($\tan\delta$ high). Although there is a reflection from the front face dependent on ε, most of the energy is absorbed. The reflection is weak, Figures 7.1(a) and 7.3(e).
6. Radar absorbent material (RAM). These are lossy materials whose thickness, ε, and sometimes magnetic permeability, μ (using ferrite materials), have been juggled to make the weak reflection via a metal backing cancel the air/insulator front face reflection, Figure 7.3(f); R tends to $-\infty$ dB. The military stick RAM to warship masts to 'stealth' or hide them from radar. RAM also can be used to prevent unwanted operation of racons on an unwanted bearing, eg avoiding triggering from ships with radars running while tied up nearby.
7. Multiple reflections become progressively weaker, so are most unlikely to produce significant overall reflection, Figure 7.3(g). For n bounces, the reflection is nR dB. Again, R tends to $-\infty$ dB. For example the dihedral corner shown,

if made of timber with $\varepsilon = 1.7$, finally reflects 35.2 dB down, only 0.3×10^{-3} of the incident power being reflected.

7.3 Reflection from conductors

7.3.1 Principles

The mechanism of reflection from a conductive elementary scatterer which is sufficiently large relative to the wavelength to obviate resonance effects is as follows. Energy (power flux) arriving from a radar induces currents on the conductor surface. From optical theory, the incident wave is stopped as it reaches the reflector surface by creation of an equal and opposite voltage reflection vector immediately clear of the surface to give zero resultant on the surface. This is a necessary effect of the fully conductive nature of the surface, where almost no voltage is required to induce the currents on the conductive surface. Again because of full conduction, the incident energy cannot be absorbed as heat, so has all to be re-radiated from the current loops. The magnetic field set up by these loops, together with the electric field, re-radiates the incident energy, acting as a transmitter antenna. The energy forms a beam whose solid angle depends on the inverse of the aperture dimensions as for any antenna. Depending on the scatterer shape, a proportion of the scatter returns toward the radar. If the scatterer is big or there are many scatterers on an object, a pencil or fan beam is formed as from an array of antenna elements. The vector sum of the scatterer returns sets the effective 'size' of the object for radar purposes; RCS is a property of the object, not the radar. It usually varies with frequency, plane of polarisation and viewing aspect, but not with range, transmitter power or the other radar or environmental parameters. Although RCS of simple shapes can be measured and calculated with accuracy, most practical targets are too complex for accurate theoretical prediction. Most conducting targets fall into one of a few broad categories:

When circumference of a sphere of radius a is much less than a wavelength, called the Rayleigh region, RCS varies as a^6/λ^4 and cannot be predicted by ray tracing methods. The only very small targets to interest us are quasi-spherical hydrometeors, which cause clutter. The RCS of a detection cell containing raindrops is also dependent on the drop size distribution, which varies with the rate and type of rainfall, see Chapter 11.

When the circumference approximates a half-wavelength, the Mie region, the target becomes resonant and RCS is sharply dependent on frequency. As there is no significant civil marine target in this category, we shall not consider it further.

7.3.2 Target dimensions very many wavelengths

When the dimensions are many wavelengths long, as for ships and coastlines, TPM is much influenced by minor deviations from flatness in the target surfaces, that is, their micro-geometry. These targets extend over such a wide height bracket that multipath is not constant through the height range, further modifying the echo. Such targets may be big enough to overflow the detection cell. This class of target, too large to be considered as a point, is reserved to Chapter 10.

The basic shapes discussed next, and the aid to navigation reflectors later in the chapter, are several wavelengths long. They can be described by geometric ray tracing methods to the individual reflecting elements. Metals are almost invariably good conductors with small resistive loss. Almost all the incident energy is perforce reflected in some direction, but not necessarily towards the illuminating radar. Holes of diameter less than a quarter wavelength usually have little effect, so fine welded mesh reflects well.

7.4 Reflection from basic metal shapes

7.4.1 Introduction

Because the ray does not penetrate the surface and metals usually approximate 'pure' conductors having negligible resistivity, metal reflections are subject to fewer factors than the insulator reflections considered earlier. The basic conducting shapes, Figure 7.4, are the building blocks of more practical targets and illustrate several reflection principles. We use the methods of geometrical optics, which gives reasonably accurate results when all dimensions exceed about 2.5 wavelengths, resonance

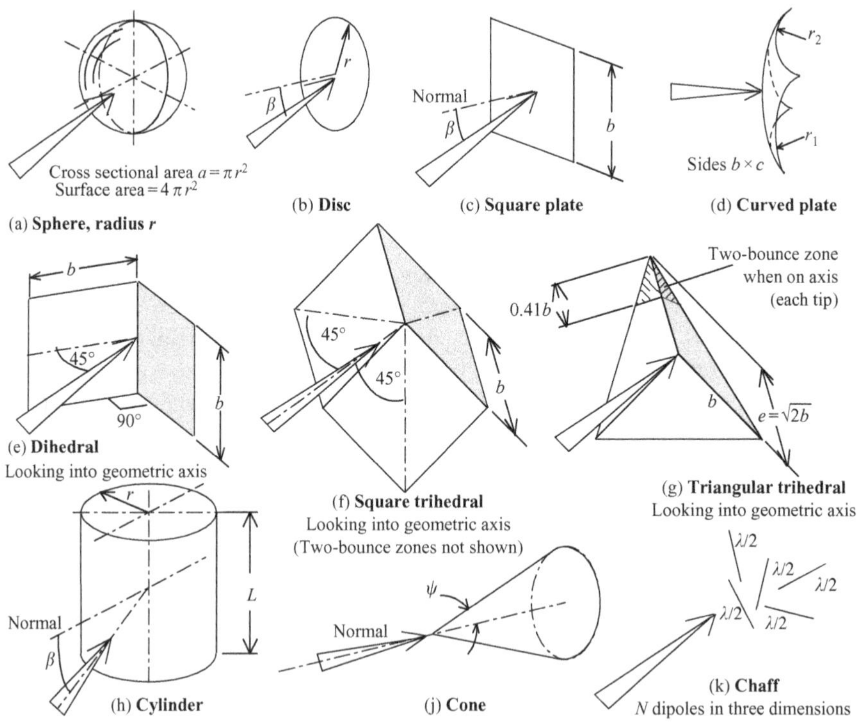

Figure 7.4 Simple metallic shapes. For RCS values see text

effects start to introduce inaccuracy in smaller devices. Our treatment is elementary but shows why devices perform as they do, without the mathematical thickets which bedevil rigorous analysis. We omit certain secondary effects of small practical significance, such as parasitic radiation at edges. Except in a few special cases, reflection is almost independent of the plane of polarisation if linear. The special case of circular polarisation is considered in Section 7.4.10. One principle is fundamental: no passive target can reflect more energy than falls upon it.

We assume the following.

- All surfaces conduct perfectly and are smooth, with surface roughness $\ll 0.1\lambda$ rms. Although skin effect increases resistance of metals at microwave frequency, resistive loss usually remains negligible and metal surfaces form good mirrors at centimetric wavelengths.
- The target is illuminated by a distant horizontal illuminator, such as a scanner.
- Targets are too small to extend beyond the beam or the range cell.
- The target is 'large', all dimensions much exceeding $\lambda/2$, so resonance effects are negligible.

We will work primarily in numerical quantities rather than decibels. Some authors include monostatic reflection within the general case of bistatic reflection. The literature reflects the interest of the military in bistatic performance – for example, a missile homing on a target illuminated by a 'lamp set' ground radar – and their need to reduce own-platform RCS, but we confine ourselves to the monostatic case, where the same radar illuminates and receives the echo. Much RCS work (including some on marine targets) unfortunately remains security classified.

7.4.2 Calculation of RCS; definitions

When a plane-wave ray of flux density d W/m^2 from a distant radar strikes a lossless conducting object (such as a flat plate normal to the beam) of projected area A m^2 the power intercepted is $P = dA$ W. (This is fundamental to most targets but applies in a modified manner with some endfire antennas such as Yagis when used as reflectors.) For conservation of energy, this incident power must all be re-radiated. Following the argument developed for antennas in Chapter 2, Section 2.7.2, the object has aperture a and acts as an antenna of axial gain $G = 4\pi A/\lambda^2$ (Eq. (2.7e)), forming a beam whose solid angle, Ω, is, from Eq. (2.7a)

$$\Omega = \frac{4\pi}{G} \text{ sr.} \tag{7.2a}$$

The output flux in the echo beam is P and the power per unit solid angle is P/Ω. Substituting,

$$\frac{P}{\Omega} = \frac{PG}{4\pi} \tag{7.2b}$$

$$= \frac{dA^2}{\lambda^2} \text{ W/sr.} \tag{7.2c}$$

RCS may be defined firstly as: '$\sigma = 4\pi \times$ (power per unit solid angle scattered in a specified direction)/(power per unit area in a plane wave incident on the scatterer from the specified direction)'. Therefore

$$\sigma = 4\pi \times \frac{dA^2}{\lambda^2 d} = \frac{4\pi A^2}{\lambda^2} \text{ m}^2. \tag{7.2d}$$

As a sphere, radius r, has perfect symmetry, it has the same cross sectional area, πr^2, at all aspects. If the conducting object is spherical, the induced currents flow uniformly round the whole surface, there being nothing to stop them doing so. The re-radiation is therefore omnidirectional through the whole 4π steradians or 'isotropic'. Substituting in Eq. (7.2a), $G = 1$, irrespective of wavelength or size, if it remains 'large'.

For a sphere of unit projected area ($A = 1\text{ m}^2$ cross sectional area), the output flux density in the echo beam is, by substitution in Eq. (7.2b),

$$\frac{P}{4\pi} = \frac{d \times (\text{area} = 1)}{4\pi} = \frac{d}{4\pi} \text{ W/sr.} \tag{7.2e}$$

When the (monostatic) radar cross section, σ, of a scatterer was introduced in Chapter 4, Section 4.3, it was defined as 'the proportion of incident energy reflected back in the direction of the source, relative to the proportion reflected by a replacement perfectly conducting sphere of cross sectional area 1 m^2'. Substituting in Eqs (7.2b) and (7.2e),

$$\sigma = \frac{dA^2/\lambda^2}{d_1/4\pi} = \frac{4\pi A^2}{\lambda^2} \text{ m}^2.$$

This is identical with Eq. (7.2d), so the two definitions of RCS are equivalent. They apply to all perfect reflectors.

RCS depends on (aperture area)^2 because the reflector in effect doubles as both a receiving and a transmitting antenna. We next examine some simple metal shapes, shown in Figure 7.4, which is based on Kingsley and Quegan [2].

7.4.3 *Sphere*

Symmetry gives the sphere the unique property that performance as a reflector is uniform in every direction. Partly for this reason the sphere is used as a reference against which other target RCS are evaluated. In the optical region where the sphere is 'large' RCS, σ m^2, is defined as the projected area:

$$\sigma = A = \pi r^2 \text{ m}^2. \tag{7.3a}$$

This can be written $\sigma = S/4$ where S is the surface area. Spheres are curved in both planes. Such figures have the property that RCS is independent of wavelength, another reason why spheres are used as reference reflectors.

$$\begin{aligned} &\text{Solid beamwidth, } \Omega = 4\pi \text{ sr} \qquad \text{[all directions].} \\ &\text{Gain, } G = 1 \qquad \text{[by definition].} \end{aligned} \tag{7.3b}$$

Although spheres work at any inclination, unfortunately the lack of focus condemns them to feeble RCS, the reflection being spread thinly in all directions. A sphere as big as 1 m diameter ($r = 0.5$ m) has RCS $0.79\,\mathrm{m}^2$ ($-1.05\,\mathrm{dB\,m}^2$) at all frequencies, which is too low for detection in moderate clutter.

In the so-called Rayleigh region where the circumference is small ($r \ll \lambda/2\pi$), RCS is

$$\sigma \sim \frac{1000\pi r^6}{\lambda^4}\,\mathrm{m}^2. \tag{7.3c}$$

In the intermediate Mie region where $r \sim \lambda/2\pi$, resonances cause RCS to fluctuate by some 5 dB about the Eq. (7.3a) value as r/λ changes.

Rohan [3] gives the RCS of an ellipsoid of dimensions a, b and c, where θ in a spherical system of coordinates is measured from the Z-axis, while ψ, in the XY plane, from the X-axis, as being

$$\sigma = \frac{\pi a^2 b^2 c^2}{(a^2 \sin^2\theta \cos^2\psi + b^2 \sin^2\theta \sin^2\psi + c^2 \cos^2\theta)^2}\,\mathrm{m}^2. \tag{7.3d}$$

For a prolate spheroid $a = b$, and for a sphere $a = b = c = r$.

7.4.4 *Disc and flat plate*

When a flat circular disc of radius r is illuminated normally (Figure 7.4(b), $\theta = 0$), the energy intercepted is proportional to the presented area, which like the sphere is $\pi r^2\,\mathrm{m}^2$. Again the energy is all re-radiated, but this time the disc focusses the reflection into the cone of a pencil beam pointing at the radar. The radar, not knowing the target shape, receives more and thinks the target is bigger – that its RCS has gone up. The disc has gain. So much of the energy is in the beam that little is available at other directions. Monostatic reflection (at illuminator bearing) has increased at the expense of bistatic (at other bearings).

If we probed the surrounding space, we would find reflected field to be maximum normal to the disc. Going off to any side the reflection weakens. As angle increases, there is in turn a first null, a sidelobe, another null and further smaller and smaller sidelobes. Not surprisingly, the pattern resembles that of a uniformly illuminated scanner, Chapter 2, Section 2.8.1, Figure 2.29. Azimuth and elevation apertures are each $2r$. A big flat disc has a narrow pencil beam and high gain.

From Eq. (7.2d), $\sigma = 4\pi A^2/\lambda^2$. $A = \pi r^2$, so for a circular disc viewed normally

$$\sigma = \frac{4\pi^3 r^4}{\lambda^2}\,\mathrm{m}^2. \tag{7.4a}$$

More generally, an elliptical disc of semi-major axes r_1, r_2 in azimuth and elevation respectively viewed normally has

$$\sigma = \frac{4\pi^3 r_1^2 r_2^2}{\lambda^2} = \frac{4\pi}{\lambda^2} \times [\mathrm{area}]^2\,\mathrm{m}^2,$$

$$G = \frac{4\pi^2 r_1 r_2}{\lambda^2},$$

$$\Omega = \frac{\lambda^2}{\pi r_1 r_2}\,\text{sr},$$

$$\text{azimuth beamwidth: } \theta = \frac{\lambda}{\sqrt{\pi} r_2}\,\text{rad},$$

$$\text{elevation beamwidth: } \phi = \frac{\lambda}{\sqrt{\pi} r_1}\,\text{rad}. \tag{7.4b}$$

For a square flat plate of side b illuminated normally ($\theta = 0$ in Figure 7.4(c)), Eq. (7.4b) holds and Eq. (7.4a) becomes

$$\sigma = \frac{4\pi b^4}{\lambda^2} = \frac{4\pi}{\lambda^2}\,[\text{area}]^2\,\text{m}^2$$

so for a rectangular flat plate, sides $b \times c$:

$$\sigma = \frac{4\pi b^2 c^2}{\lambda^2}\,\text{m}^2,$$

$$G = \frac{4\pi bc}{\lambda^2},$$

$$\Omega = \frac{\lambda^2}{bc}\,\text{sr}, \tag{7.4c}$$

$$\theta = \frac{\lambda}{c}\,\text{rad},$$

$$\phi = \frac{\lambda}{b}\,\text{rad}.$$

Note that doubling either both side lengths, or the radius, quadruples area and multiplies RCS by 16, which is 12 dB. As circular and rectangular plates both have $\sigma = 4\pi/\lambda^2 \times [\text{area}]^2\,\text{m}^2$, we conclude this formula applies generally to flat plates of any shape provided the minimum dimension well exceeds a wavelength to avoid resonances. RCS is inversely proportional to λ^2. It follows that no small flat reflector can have high RCS except at short wavelength; RCS at 3 GHz, where $\lambda = 10$ cm, is one-tenth the 9 GHz value where $\lambda = 3.2$ cm. Expressing RCS in terms of area simplifies the expressions and assists estimation of irregular shapes:

$$\text{RCS} = \frac{4\pi}{\lambda^2} \times [\text{area}]^2\,\text{m}^2. \tag{7.4d}$$

A flat plate viewed normally has:

$$\text{aperture area} = \text{actual area}$$

$$\text{gain } G = \frac{4\pi}{\lambda^2} \times \text{area} = 12.57 \times \left[\frac{\text{area}}{\lambda}\right]^2$$

$$\text{solid beamwidth } \Omega = \frac{\lambda^2}{\text{area}}\,\text{sr}. \tag{7.4e}$$

Even a small plate, area say $10\lambda^2$, has quite small solid beamwidth, here 0.1 sr.

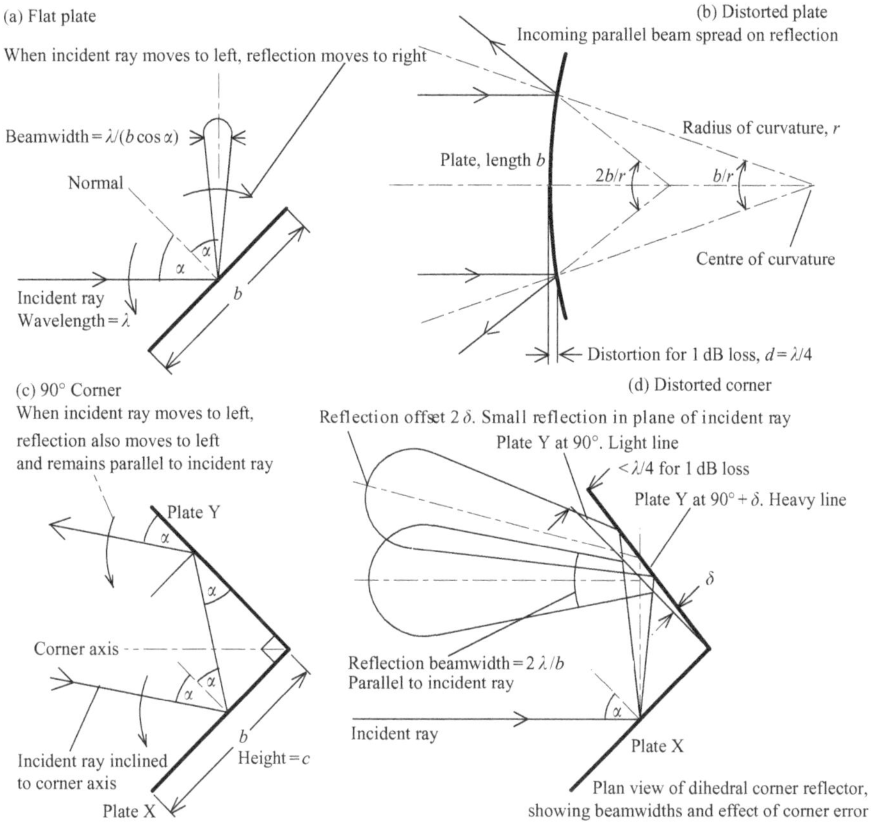

Figure 7.5 *Reflection from flat plate and corner. Plan views. A 'large' metallic truly flat plate (a) gives narrow beam and high RCS. A corner (c) works through a large azimuth angle. Both are spoilt by micro-geometry dependent flatness distortion, d (b), or corner angular error, δ (d), respectively*

Illuminating the rectangular flat plate obliquely at angle α (Figure 7.5(a)), the beam reflects like a mirror, with beamwidth at the complementary angle

$$\frac{\lambda}{b\cos\alpha}\ \text{rad.} \tag{7.5}$$

When illuminated normally ($\alpha = 0$), beamwidth between half-power points is λ/b. Therefore the reflection toward the illuminator is 3 dB down when tilted to $\alpha = \lambda/2b$. Any target develops monostatic RCS only when the radar lies within the reflected beam, so slanting the plate reduces RCS. The monostatic RCS at any obliquity, α, is obtained from antenna theory using the methods of geometrical optics.

For a square plate, side b:

$$\sigma = (\text{on-axis RCS}) \times \cos^2\alpha \times \left\{ \frac{\sin[2\pi b \sin(\alpha/\lambda)]}{2\pi b \sin(\alpha/\lambda)} \right\}^2. \tag{7.6}$$

The expression for a circular disc is similar. Where J_1 denotes a Bessel function of the first kind:

$$\sigma = (\text{on-axis RCS}) \times \cos^2\alpha \times \left\{ \frac{J_1[4\pi r \sin(\alpha/\lambda)]}{4\pi r \sin(\alpha/\lambda)} \right\}^2. \tag{7.7}$$

The {...} term confers a system of sidelobes of the general form of Figure 2.31.

Again, as the plate is flat, RCS is inversely proportional to λ^2 so is proportional to (frequency)^2. The beamwidth is governed by the $2\pi b$ or $4\pi r$ terms in these equations, falling as b or r rises. A dead-flat plate many wavelengths long therefore only reflects strongly when almost exactly normal to the beam, irrespective of shape. For example, a square plate of side 1 m viewed normally has (from Eq. (7.4d)) peak 9 GHz RCS = 12 270 m^2 = 40.9 dB m^2, and (from Eq. (7.5)) beamwidth = λ/b = 0.032 rad = 1.8°, meaning RCS is reduced by 3 dB by 0.9° tilt. At 3 GHz, RCS is 10 dB lower and beamwidth about $\sqrt{10}$ higher (not 10 because the solid angle increase is shared between both planes), giving 1256 m^2 = 31 dB m^2 and 5.7°, respectively. Raising size raises on-axis RCS but reduces beamwidth, so, viewed obliquely, average RCS through a complete sidelobe cycle remains nearly independent of plate size.

7.4.5 Macro- and micro-geometry; distorted plate

The treatment is based on J. M. Williams [4]. So far we have assumed the plate is truly flat and considered only the effects of its size, in other words its macro-geometry. We now consider the effects of irregularities, the micro-geometry. If the plate is curved solely in the plane of the b-dimension, with radius of curvature r and deviation from flatness d (Figure 7.5(b)), from elementary geometry $d = b^2/(2r)$.

Figure 7.5(b) shows the beam is reflected as a wedge of angular width $2b/r$. A very big circle surrounding the plate would be illuminated over an arc $2b/r$, the proportion of its circumference being $(2b/r)/2\pi = b/(\pi r)$, so gain relative to an isotropic (sphere) scatterer is $\pi r/b$. Similarly, if a plate, sides $b \times c$, area $A = bc$, is curved in both planes with radii r_1 and r_2 as shown in Figure 7.4(d), the gain is $\pi r_1 r_2/a$. RCS is area × gain and is $A\pi r_1 r_2/(bc)$. Term A cancels bc, so:

$$\text{for doubly curved plate, } \sigma = \pi r_1 r_2 \text{ m}^2, \tag{7.8a}$$

$$\text{effective aperture} = \frac{\lambda}{2} \times \sqrt{r_1 r_2},$$

$$G = \frac{2\pi\sqrt{r_1 r_2}}{\lambda}, \tag{7.8b}$$

$$\Omega = \frac{2\pi}{\sqrt{r_1 r_2}}.$$

Performance becomes dependent on radii of curvature but independent of size. Being curved in both planes like a sphere, RCS is also independent of wavelength. Curvature,

Eq. (7.8a), takes over from the flat plate expression, Eq. (7.4), when both expressions give the same RCS. Here

$$\pi r_1 r_2 = \frac{4\pi A^2}{\lambda^2}.$$

Therefore, RCS falls significantly below that of a truly flat plate when

$$r_1 r_2 < \frac{4(bc)^2}{\lambda^2}. \tag{7.8c}$$

From analogy with the singly curved plate, $d^2 = b^2c^2/(4r_1r_2)$ so $r_1r_2 = b^2c^2/4d^2$

$$\therefore \quad \frac{4(bc)^2}{\lambda^2} = \frac{b^2c^2}{4d^2}$$

which reduces to

$$d = \frac{\lambda}{4}. \tag{7.9a}$$

A concave surface gives the same result, and we can say that if a plate of any shape is distorted from true flatness by more than $\lambda/4$ peak–peak, RCS falls and beamwidth rises. Irregularity of ships' plating therefore generally has a profound effect on RCS. Standard deviation of distortion $\sim 1/2\sqrt{2}\times$ peak–peak. Thus, to maintain RCS,

$$\text{standard deviation of irregularity (the s.d. or rms value)} \sim \frac{1/4}{2\sqrt{2}\lambda} \sim 0.09\lambda \text{ m}. \tag{7.9b}$$

Regular undulations, spaced of the order of an exact number of half projected wavelengths (e.g. corrugated iron, walls of ISO shipping containers) enable Bragg resonance to occur, making reflection critically dependent on plane of polarisation, frequency and angle of incidence as well as corrugation orientation, shape and spacing.

7.4.6 Dihedral corner reflector

For trihedral corners see Section 7.7.1. Two vertical flat plates X and Y butted together at 90° form a dihedral corner reflector, Figures 7.4(e), 7.5(c) and (d). If illuminated from left of axis, the multiple reflection is also to the left, not to the right as in a mirror – the response follows interrogator angular movement, making the device horizontally retro-directive. But if illuminated from above, the single reflection is below axis, as for a mirror. So a vertical dihedral has high RCS through broad azimuth but narrow elevation angles. When illuminated nearly on the geometric axis, rays bounce twice: horizontal rays striking X are reflected to Y and vice versa. Reflections from both components always return toward the source, even when considerably off axis, as shown in Figure 7.5(c). Because there are two bounces, the reflected beamwidth is doubled relative to a flat plate at 45°, Figure 7.5(d). When looking on axis, $\alpha = 45°$ and $1/\cos\alpha = \sqrt{2}$. The horizontal beamwidth from illumination of plate X alone would be $2\sqrt{2}\lambda/b$, but interference from the component striking plate Y reduces overall beamwidth to $2\lambda/b$ rad.

The presented mouth area of the device, for plates individually b wide $\times$ c high, is $\sqrt{2}bc$. From the above macro-geometry a dihedral illuminated as Figure 7.4(*e*) has peak RCS:

$$\sigma = \frac{8\pi b^2 c^2}{\lambda^2} = \frac{4\pi}{\lambda^2} \times [\text{mouth area}]^2 \ \text{m}^2. \tag{7.10a}$$

This has the form of Eq. (7.4b). The object is made from flat surfaces, so RCS is inversely proportional to (wavelength)^2.

$$\begin{aligned} &\text{Effective aperture} = \sqrt{2}bc = \text{mouth area} \\ &G = \frac{4\sqrt{2}bc}{\lambda^2} \\ &\Omega = \frac{\lambda^2}{\sqrt{2}bc}. \end{aligned} \tag{7.10b}$$

If we can be sure it is mounted truly upright, is illuminated horizontally and is accurately made (requirements not readily achieved when the dimensions are large), the dihedral can form a useful reflector when traffic is confined to within 20° of axis. RCS falls by 10 dB when 30° off axis. For example, if on-axis RCS is 20 dB m^2 at 3 GHz, $bc = 0.14$ m^2. At 9 GHz, RCS rises to 30 dB m^2. To maintain performance over a reasonable elevation angle, plate height c must be low, necessitating considerable width for high RCS. A commercial reflector attempting to get good elevation performance from a cluster of dihedrals, gimballed by floating within an enclosure, proved unsuccessful, probably for mechanical reasons.

When the illuminator is offset in azimuth by exactly ±45° from device axis, one plate becomes normal to the beam and reflects strongly as an ordinary flat single plate. Such off-axis strong reflections through narrow angles are termed flashes.

7.4.7 Distorted corner

The treatment, based on J. M. Williams [5], details the theoretical and practical effects of micro-geometry distortion on dihedrals and trihedrals. If the corner of a dihedral is put out of square by δ rad, the reflection is deflected by 2δ rad from the radar; Figure 7.5(c). When $\delta = \lambda/4b$, RCS is reduced by 1 dB, so δ should not exceed about this amount. Put another way, the far edge should always be within $\lambda/4$ of its correct position, irrespective of size, representing 8 mm at 9 GHz and 25 mm at 3 GHz. For example, at 3 GHz, when $b = 0.1$ m, $\delta < 4.6°$ (corner angles between 85.4° and 94.6°) would suffice to maintain RCS within a decibel, but if $b = 1$ m, the manufacturing accuracy should be within 0.46°. This again has important consequences for RCS of ships and other large targets containing re-entrants near 90°. When $\delta = \lambda/2$, RCS falls by 3 dB and when $\delta = \lambda/b$ the reflection lies in the first null and is zero ($-\infty$ dB).

7.4.8 Practical effects of micro-geometry

We have seen that deviations from plate flatness exceeding $\lambda/4$ spoil performance. Generalising: performance falls off and the target ceases to follow the equations for a perfect reflector when irregularities exceed $\lambda/4$ peak–peak or about 0.09λ rms. A plate classed perfect at long wavelength may become irregular at shorter wavelength. For example, a dihedral corner with $b = 0.5$ m horizontally and $c = 0.28$ m vertically lies within the criterion that dimensions should exceed 2.5λ at 3 GHz ($\lambda = 0.1$ m). If the plates are distorted by 25 mm, the device is just within the flatness criterion in this band, with about 1 dB RCS loss, pulling 3 GHz RCS down from 20 to 19 dBm^2. But at 9 GHz, let alone J(ku) band, the plates are by no means flat and RCS will not rise so far as predicted by Eq. (7.10a). Depending on the shape of the distortion, RCS might not rise at all. Performance will be uncertain and disappointing.

With any passive reflector, it is always wise to measure performance at the shortest intended wavelength. If manufacture is there sufficiently accurate, RCS will fall fairly predictably at longer wavelengths until dimensions approach $\lambda/2$ when resonance effects become likely. Unfortunately, the inescapable laws of physics demand that devices intended to operate through several wavebands need to be big enough to get enough RCS at the longest wavelength and precise enough to maintain performance at the shortest. Alas, size, stiffness and accuracy entail wind resistance, weight and cost.

7.4.9 Edges and rods

Small-diameter cylinders and the edges of plates reflect only the parallel component, so their cross section depends on the polarisation. Yacht masts and near-vertical rigging reflect poorly to horizontally polarised radars, see Section 7.5.1.

7.4.10 Circular polarisation

Chapter 2, Section 2.7.13, discussed circularly polarised scanners, explaining that immediately forward of the reflector surface, optical theory demands that the reflected voltage vector has to be phased to cancel the incident vector. If, say, the incident vector is circularly polarised, rotating right-handed, the reflected vector continues to rotate clockwise but goes backwards, so in its own direction of propagation its sense is reversed to left-handed. As the circularly polarised scanner which generated the original ray will not receive reversed sense it cannot receive the reflection. Effective target RCS is reduced to zero. If the reflector is a dihedral or other structure with two reflections, the rotation is reversed twice and the scanner does receive the echo. In general, simple flat or curved surfaces and structures with an odd number of reflections (e.g. plates, trihedrals, octahedrals) are not received but those with an even number of reflections are. Deviations from truly circular polarisation, called ellipticity, reduce the cancellation (refer to Eqs (2.8)). Raindrops have an odd number of reflections and rain clutter is suppressed in principle. Raindrops are not quite spherical so cancellation is imperfect, even when the polarisation is perfectly circular.

Rods and edges are polarisation-dependent and reflect only the parallel component of polarisation, introducing loss. Practical complicated targets such as ships and aircraft exhibit roughly 3 dB RCS reduction to circular polarisation relative to linear polarisation of any plane.

7.5 Other geometric shapes

Certain other shapes are of interest, often as components of larger structures.

7.5.1 Cylinder, metal wire

Figure 7.4(h) shows a vertical cylinder, radius r and length L. As it is curved in one plane and flat in the other, RCS is inversely proportional to wavelength and the device acts as a sphere in azimuth and a flat plate in elevation. The RCS at elevation angle β is:

$$\sigma = \frac{2\pi r L^2}{\lambda} \times \cos^2\beta \times \left\{\frac{\sin[2\pi L \sin(\beta/\lambda)]}{2\pi L \sin(\beta/\lambda)}\right\}^2 \text{ m}^2. \tag{7.11a}$$

Note this is a $(\sin x)/x$ variation as for a uniformly illuminated antenna aperture, indicating the beam has sidelobes. On the broadside ($\beta = 0$), Eq. (7.11a) reduces to

$$\sigma = \frac{2\pi r L^2}{\lambda}, \tag{7.11b}$$

$$\begin{aligned}
&\text{effective aperture} = L\sqrt{\frac{r\lambda}{2}} \text{ m}^2,\\
&G = \frac{2\pi L\sqrt{2r}}{\lambda^{3/2}},\\
&\theta = \frac{\sqrt{2}\lambda^{3/2}}{4\pi L\sqrt{r}} \text{ rad}, \qquad (7.11c)\\
&\theta = 2\pi \text{ rad} \qquad [\text{omnidirectional}],\\
&\Omega = \frac{\lambda^{3/2}\sqrt{2/r}}{L} \text{ sr}.
\end{aligned}$$

As tubular metal masts are unlikely to be straight within $\lambda/4$, Eq. (7.11a) may well overstate RCS. Diameter of rigging is usually too small for the equation to apply. For a perfectly conducting wire, Chu (quoted in [3], Section 7.4) gives

$$\sigma = \frac{\pi L^2/C \sin^2\theta \sin C \cos^4\psi}{\{(\pi/2)^2 + [\ln(\lambda/(\gamma\pi a \sin\theta))]^2\}} \text{ m}^2 \tag{7.11d}$$

where: L = length of wire ($L \gg \lambda$), a = radius of wire ($a \ll \lambda$), γ = Euler's constant = $1.781, \ldots$, θ = angle between wire and direction of incidence, $C = 2\pi L/\lambda \cos\theta$ and ψ = angle of the polarisation direction relative to the plane formed by the wire and the direction of incidence. There is strong polarisation dependence: when the wire lies in the plane of polarisation, $\psi = 0$ and $\cos^4\psi = 1$.

But when perpendicular, $\psi = 90°$ and $\cos^4 \psi = 0$, so near-vertical rigging is invisible to horizontally polarised radars. When $\psi = 0$ and the wire is viewed normally with $\theta = 90°$, RCS oscillates rapidly with L. Peak values (e.g. for horizontal wire viewed normally by a HP radar) vary with radius as follows:

$$\text{when } a = 0.33\lambda,\ 0.1\lambda,\ 0.033\lambda \text{ or } 0.01\lambda;$$

$$\sigma = 1.10L^2,\ 1.12L^2,\ 0.59L^2,\ 0.29L^2\,\text{m}^2,$$

respectively.

Values for braided steel rigging might be somewhat lower.

7.5.2 Circular cone

The RCS looking on the tip (Figure 7.4(j)) is:

$$\sigma = \frac{\lambda^2}{16\pi} \tan^4 \psi. \tag{7.12}$$

RCS depends on tip half-angle, ψ, but not on size, as for a doubly curved plate. Unusually, RCS rises with wavelength. This formula indicates that the bow-on RCS of craft having sharply faired pointed bows is likely to be tiny. For $\psi = 30°$, $\sigma = -56.5\,\text{dB m}^2$ at 9 GHz and $-46.5\,\text{dB m}^2$ at 3 GHz. Normal to the cone axis, reflection is deflected and monostatic RCS is reduced.

7.5.3 Frequency effects

P. D. L. Williams [6] points out the frequency dependence on reflector curvature.

- For two infinite radii of curvature, for example, a flat plate, $\sigma \propto \lambda^{-2}$.
- For one infinite and one non-zero radius, for example, a cylinder, $\sigma \propto \lambda^{-1}$.
- For one infinite and one zero radius, for example, a wedge, $\sigma \propto \lambda^0$.
- For one finite and one non-zero radius, for example, a spheroid, $\sigma \propto \lambda^0$.
- For one finite non-zero and one zero radius, for example, a curved edge, $\sigma \propto \lambda^1$.
- For two zero radii, for example, the apex of a cone, $\sigma \propto \lambda^2$.

7.6 Requirements for practical reflectors

Most reflectors so far discussed are rarely used as they stand, but form the basis for practical devices to improve detectability of poorly reflecting targets.

7.6.1 Legal requirements, specifications

Since 2002, new vessels of tonnage less than 150 gt have been required by SOLAS Regulation V/20 to carry an active or passive radar reflector, operating in the 3 and 9 GHz bands to SOLAS V/25 para. 1.2.6 ‘if practicable’. At time of writing in 2004, an IMO NAV subcommittee draft resolution for a revised performance standard for reflectors has been tabled to IMO Maritime Safety Committee, for adoption at MSC 78th session. If adopted, reflectors for SOLAS Chapters V (conventional ships) and X (high speed craft), fitted after a provisional date of 1 July 2005, should have a 'Stated Performance Level' RCS at least 7.5 m^2 (8.75 dB m^2) in the 9 GHz band and at least 0.5 m^2 (−3.0 dB m^2) in the 3 GHz band. The Stated Performance Level should be maintained over a total of at least 280° azimuth; not remain below this level over any single angle of more than 10° (a null), and not have distances between nulls of less than 20°. For power driven vessels and sailing vessels designed to operate with little heel (catamarans/trimarans), this performance should be maintained through angles of athwartships heel ±10°; for other sailing vessels heel is ±20°. Recommended mounting height is 4 m. For small craft, weight should be 5 kg maximum with volume less than 0.05 m^3. Active reflectors (RTEs) are permitted and should also conform to ITU-R M.1176. An ISO technical standard would be produced to flesh out methods of testing.

The UK High Court has ruled it to be negligent seamanship not to hoist a reflector if carried onboard. To lawyers, 'if practicable' means 'if it can be done' rather than 'if convenient'. SOLAS applies to all ships, and to IMO a ship is almost anything which floats on salt water. Many national administrations have granted blanket exemptions relieving small leisure craft from the obligation to carry reflectors; this position may change. It is becoming a legal requirement for ships to formally plan passages, including consideration of weather likely to be encountered, hence clutter, which may influence their choice of reflector.

Reflectors fitted earlier should conform to the Annex to IMO Resolution A.384(X), which calls up reflectors to ISO 8729:1987 (E) 'Shipbuilding – Marine radar reflectors', replicated as British Standard specification BS 7380:1990. This, the only previously specified reflector, is widely regarded as too small for detection in moderate or severe clutter. Its specification is structured to accept a competently designed octahedral reflector (Section 7.7.2) of peak equatorial RCS, σ_{nom}, 10 m^2 in the 9320–9500 MHz band. The UK MCA currently requires all 'Code' ships (carrying fare-paying passengers) exceeding 15 m waterline length to carry an ISO 8729 reflector, but it is understood the length is likely to be reduced to 12 m and the reflector specification to be aligned with the revised IMO requirement, if and when issued. Vessels as small as 12 m are deemed capable of ocean passage, but Code vessels are licensed to ply only within strictly defined coastal waters, apart possibly from repositioning passages from one port to another.

The revised IMO draft specification was a difficult international compromise between the perceived need for better RCS without wide 'holes' to combat severe clutter; the desire to frame the requirement to permit passive reflectors; equipment availability (few passive and no active fully-compliant devices were available at the time) and the cost, bulk and overturning moment penalties inherent in high-RCS passive reflectors. Although a technical case can be made for RCS some 10 dB higher than specified, the revision, made by consensus after lengthy study, calculation, discussion and experiment, marks a considerable advance. From initial informal suggestion for a revision to the MSC decision took 6 years, par for the course.

7.6.2 Measurement of point aids

Point aids to navigation are small enough to be placed in anechoic chambers for far-field TPM measurement; Rayleigh distance is typically of the order of a metre. The aid is placed on a turntable and illuminated with known flux density from a small fixed microwave transmitter via one of a pair of horn antennas of accurately defined gain, testgear accuracy being traceable to international standards. Another co-located horn feeds a calibrated receiver. As the turntable slowly rotates, echo power is measured and recorded at numerous azimuth increments of say 0.5°. RCS at each angle is calculated from the radar range equation, perhaps confirmed by substitution of a known reflector such as a sphere. The measurement set is repeated at successive elevation angles to give either a set of polar diagrams or a TPM. An elevation polar diagram is also taken. Any residual echo from the chamber walls is allowed for by further measurements with the aid removed. The whole procedure is repeated at several frequencies, mid-band and band edges as a minimum. The process is automated and results should be repeatable. As part of Type Approval, the test house will certify accuracy to within a few tenths of a decibel, incomparably better than obtainable on sea trials with all the vagaries of weather and multipath.

7.6.3 Commercial reflectors

Uniform azimuth polar diagram is desirable except in a few special cases. If the target is afloat, elevation polar diagram should be broad enough to retain good RCS through all likely platform heel angles plus a degree or so to cater for depression of the radar beam.

ISO 8729 reflectors have been widely fitted to navigation buoys. Reflectors of higher RCS are used by Harbour Authorities to mark bridge piers and other fixed hazards where only a limited sector is of interest, for example, a beacon in a narrow estuary. High RCS reflectors are inevitably bulky and expensive. Many have restricted elevation beamwidth. Octahedral reflectors have rather irregular TPM. In sympathy, ISO 8729 unfortunately allows broad gaps in the TPM: in the azimuth plane RCS must exceed -6 dB relative to $10\,m^2$ (i.e. $2.5\,m^2$) over a total azimuth angle of 240°, with no gaps wider than 10°; in elevation, RCS must exceed -12 dB m^2 over a total of 240° azimuth when inclined by $\pm 15°$, with no other restriction on nulls.

We shall see in later chapters that 9 GHz shipborne radar in moderate or severe clutter needs more target RCS for detection, so compliance with ISO 8729 does not assure the platform will be seen; the revised requirement is better but still insufficient in severe clutter.

There are many reflectors on the yachting market. Within their limits, the better ones work well and have been type-approved to ISO 8729. Many are made from clusters of trihedrals, Section 7.7.3, covered by a smooth plastic skin which does not readily snag sails. RCS is generally around 10 dB m^2 peak and 4 dB m^2 through a useful solid angle, with no holes in the TPM bigger than $5° \times 5°$. These devices are capable of raising the echo above modest sea and precipitation clutter, but must not be relied on in heavy clutter. Unfortunately, the yachtsman is offered others of doubtful worth, some being too small for worthwhile RCS to be possible; others have poor

TPM with high RCS through uselessly small solid angles. Such devices lull the user into a false sense of security without much reducing collision risk.

Detection range can vary widely as the target swings between aspects exhibiting minimum and maximum RCS. All reputable reflector suppliers provide polar diagrams in azimuth and elevation planes. The latter may represent longitudes of best performance, the angle of cut being selected to mask holes in the cover. These would be revealed by the TPM, which covers all the whole solid angle through which the device is to operate. Polar diagrams may be scaled linearly in RCS, m^2, as shown in the polar diagram of a fictitious reflector, Figure 7.6(a), or in dB m^2. As zero response then would be $-\infty$, the centre as well as the periphery of the diagram have to represent specific RCS values. Polar diagram shape therefore depends on scaling as shown in Figures 7.6(b) and (c). Published diagrams of some commercial devices have delightfully vague scales. As always, *Caveat Emptor* – let the buyer beware.

Figure 7.6(b) shows peak RCS is in excess of 18 dB m^2. Sounds good. But the figure reveals that we can only rely on 14 dB m^2 on the equator. A TPM (not depicted) shows average of 14 dB m^2 with a minimum of 12 dB m^2 through $\pm 15°$ tilt angle in the elevation plane, RCS falling steeply with more tilt. Possible data-sheet descriptions might include:

- 12 dB m^2 minimum RCS within $\pm 15°$ elevation (accurate but unflattering);
- nominally 14 dB m^2 within $\pm 15°$ elevation (perhaps the fairest description, although there are holes having less RCS);
- 16 dB m^2 (the mean round the equator, saying nothing about holes or performance when tilted);
- up to 63 m^2 (very flattering, the peak on the equator; 63 m^2 sounding even bigger than 18 dB $m^2$and the 'Up to' can quietly slither away). *Caveat Emptor* again.

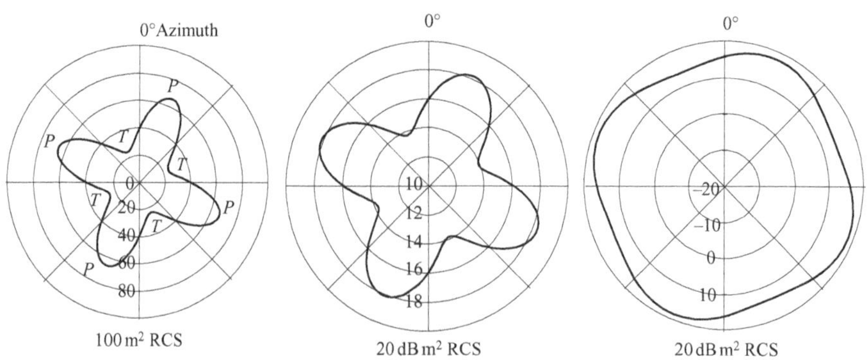

(a) Scale 0–100 m^2 (b) Scale 10–20 dB m^2 (c) Scale −20 to 20 dB m^2

Figure 7.6 Azimuth polar diagram. At first glance (c) looks a much more uniform device than (a) or (b); in fact they are all the same fictitious device plotted to differing scales

7.6.4 Problems with reflectors

Small vessels, including those granted exemption from SOLAS requirements, are encouraged to carry passive reflectors (or RTEs, Chapter 8), mounted at least 4 m above waterline to increase detection range. Several problems deter more extensive voluntary carriage of effective passive reflectors.

1. High RCS is costly and inherently necessitates size. The accompanying weight and windage are singularly unattractive to racing yachtsmen. Unfortunately, these factors deter too many yachtsmen from enhancing their radar detectability by wearing reflectors.
2. Sailing craft often heel to large angles – in excess of 20° – for sustained periods of time. But the elevation half-beamwidth of most point aids is much less. Effective RCS falls far below equatorial data sheet values when the platform heels in the radar plane. It is not often practicable to suspend the aid from halliards to keep it vertical and gimballing adds yet more weight aloft.
3. A to N reflectors cannot be coded for station identification.

Passive reflectors differ widely but share basic attributes deriving from the laws of physics. Some parameters depend on the size of the target relative to the wavelength. The full theoretical basis of reflection and RCS becomes surprisingly complex when applied to quite simple shapes.

7.7 Practical reflectors

7.7.1 Trihedral

This section is partly based on Jasik [7, Section 13-11]. Trihedral reflectors are made from three flat plates butted at right angles. As with a dihedral, single-bounce reflection occurs when any of the plates is normal to the beam, giving a high reflection flash through a narrow angle. When the beam is directed along the geometric axis, rays striking a plate near its tip bounce twice and go off at an angle oblique to the incident ray. They are lost to the illuminating radar and do not contribute to RCS. The two-bounce zones occupy about a third of the mouth area when normally illuminated, but move about and vary in size with angle of illumination.

However, triple reflections via every plate in turn track the illuminator in angle, so trihedrals work retro-directively through broad angles in both planes. After three bounces, rays return towards the source. RCS remains high to linearly but not circularly polarised rays even when the illumination is horizontally or vertically angled substantially off the normal to the mouth plane, making the device useful as a reflector to mark otherwise inconspicuous obstructions. Because on-axis RCS is high for compact dimensions, is calculable from the dimensions and almost constant near the axis, trihedral targets are sometimes used for radar calibration. Calculation of polar diagrams is surprisingly difficult because of the three-dimensional geometry. Reflection from each plate follows the general form of Eq. (7.6), before transforming to the trihedral frame of reference. As the angle of illumination is changed, one plate becomes more oblique, but its loss of reflected flux density is offset by reduction of obliquity of the other plates. The upshot is that the azimuth and elevation polar diagrams are flatter than the classical $(\sin x)/x$ pattern of a uniformly illuminated antenna aperture of Chapter 2, Section 2.8.1, Figure 2.29. Jasik [7, Figure 13-10] shows a trihedral with notched sides, giving flat RCS within ±2 dB out to ±30°, at the expense of considerable reduction of axial RCS, but such fancy shapes are rarely seen in practice.

The following equations assume dimensions are at least several wavelengths long, otherwise RCS in general falls off somewhat. As construction is from flat plates, RCS $\propto \lambda^{-2}$ (Section 7.5.3). Trihedrals take three main forms. Quadrantal or square plates can be used, the latter as in Figure 7.4(f), but we concentrate on the commoner triangular form shown in Figure 7.4(g). The peak RCS of the triangular mouth, area A, of the trihedral viewed on the axis of symmetry or geometric axis is:

$$\sigma = \frac{4\pi A^2}{\lambda^2}\ \mathrm{m}^2 \tag{7.13a}$$

The mouth, height h, is normal to the axis and has edge length e. $A = h^2/\sqrt{3} = \sqrt{3}e^2/4$. Radial length (to inner corner) $b = e/\sqrt{2}$ so

$$A = \frac{\sqrt{3}b^2}{2}\ \mathrm{m}^2 \quad \text{and} \quad \sigma = \frac{4\pi h^4}{3\lambda^2} = \frac{3\pi e^4}{4\lambda^2} = \frac{3\pi b^4}{\lambda^2}\ \mathrm{m}^2. \tag{7.13b}$$

$$\text{Effective aperture} = \frac{\sqrt{3}b^2}{2}\ \mathrm{m}^2 \quad G = \frac{2\pi\sqrt{3}b^4}{\lambda^2} = 10.88\frac{b^4}{\lambda^2}. \tag{7.13c}$$

Eq. (7.13a) resembles Eq. (7.4c) for RCS of a flat plate. Squareness and flatness tolerance requirements are again $\lambda/4$; Wylie [8] notes that a manufacturing error causing displacement of 0.62λ (20 mm in the 3 GHz band) in the three plates results in 10 dB RCS reduction. Like a dihedral, the reflection follows the illumination when swung off-axis, but now in each plane, so effective azimuth and elevation beamwidths are wide and not dependent on size if side lengths are large. Triangular trihedrals are effective over 0.75 sr (±22° × ±22° to −3 dB points), the −10 dB points being near ±30°. Robust trihedrals are easy to make and sometimes used as calibration targets of calculable RCS, but manufacturing tolerance errors in large sizes can easily degrade RCS by a couple of decibels, understating performance of devices under test.

When $b = 0.23$ m, peak $\sigma = 26.08\ \mathrm{m}^2 = 14.2\ \mathrm{dB\,m}^2$ in the 9 GHz band and $4.2\ \mathrm{dB\,m}^2$ at 3 GHz.

A square-plate trihedral has twice the mouth area, thrice the gain and nine times the RCS of a triangular trihedral, but has only about two-third the effective beamwidths, so is seldom used, particularly afloat. Quadrant plate trihedrals have intermediate performance and are sometimes used as marine reflectors.

7.7.2 Octahedral

The trihedral only operates within a restricted azimuth arc. A mechanically simple cluster of six (usually triangular) trihedrals, shown in Figure 7.7, is called an

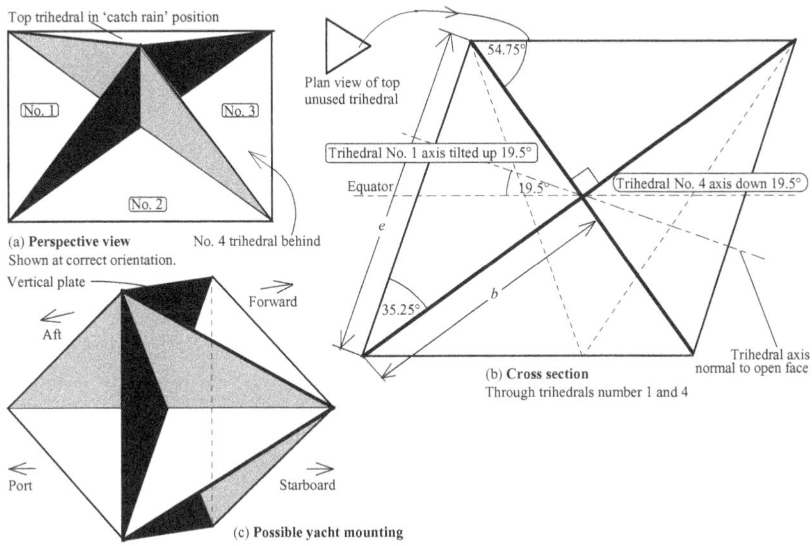

Figure 7.7 Octahedral reflector. The usual triangular pattern

Figure 7.8 Octahedral reflectors marking small craft. None is mounted in the preferred 'catch rain' position. Author

octahedral (or octohedral) and gives moderate performance throughout the full azimuth. The design was developed before the Second World War and is notable for preceding practical centimetric radar. The octahedral is widely used, Figure 7.8, and it is worthwhile examining its strengths and weaknesses. Correctly mounted with one trihedral upwards (the 'catch-rain' position) one downwards, six are active, with their geometric axes alternately offset ±19.5° in elevation. On the equator, at 60° azimuth intervals there are peaks equalling the nominal data-sheet RCS of the device, σ_{nom}, at the frequency in question.

The following relate specifically to triangular-plate octahedrals, the commonest sort. Because the equator passes well clear of the trihedral axes, σ_{nom} is about 2.6 times (4.2 dB) less than trihedral peak RCS, and is:

$$\sigma_{nom} \sim \frac{1.2\pi b^4}{\lambda^2}\ \mathrm{m}^2 \quad \text{at the equatorial peaks.} \tag{7.14a}$$

RCS is proportional to (frequency)^2 and (side length)^4; hence to (volume)$^{4/3}$. For $\sigma_{nom} = 10\,\mathrm{m}^2$ in the 9 GHz band, $b \sim 0.23$ m.

The nominal RCS of the device is usually taken as the equatorial peak value of the TPM, here 10 m^2. This octahedral fits within a sphere of 0.46 m or 18 inches nominal diameter, volume 0.4 m^3. At 3 GHz, peak RCS falls to 0 dB m^2 nominal, actually usually 1–2 dB lower because the dimensions cease to be $\gg\lambda$.

Between the peaks lie deep nulls having widths of 10° between −6 dB points, the so-called critical gap, which ought to be related to the likely yaw of a vessel mounting the reflector (shown in Chapter 10, Section 10.11, Figure 10.11). If initially on a peak, ±5° yaw drops effective RCS to σ_{nom} − 6 dB, becoming 2.5 m^2 at 9 GHz and as low as 0.25 m^2 (−6 dB m^2) at 3 GHz. Average RCS on the equator is about 5 dB below nominal peak value. There are also 'flashes' of high RCS through narrow angles from single and twin bounces and from plate edges. The author would prefer to refer to the device as '0.25/2.5', meaning effective RCS (except for holes less than 10° wide) exceeds 0.25 m^2 at 3 GHz and 2.5 m^2 at 9 GHz. The revised specification would give a '0.5/7.5' device.

At 15° heel, three trihedrals, say Nos. 1, 3 and 5, are only 4.5° off their axes and their peak RCS rises to 4 dB above σ_{nom}. Unfortunately the other three are now 34.5° off-axis and are in the sidelobe region, with negligible RCS. So the tilted polar diagram becomes even poorer, with three deep nulls each about 80° width, further reducing effective size and restricting useable angle of heel to about ±10°. Although an octahedral made about 5 dB oversize, with $b \sim 0.3$ m, would meet the stated performance level of the IMO revised reflector specification of Section 7.6.1, its wide nulls would fail the 10° or 20° tilt requirements and would by no means meet the 0.05 m^3 volume clause.

Octahedrals made from circular discs, radius r, are in effect clusters of quadrant trihedrals. At the 10° gap level, RCS falls about 10 dB, inferior to the triangular variant. Peak equatorial RCS is

$$\sigma_{nom} \sim \frac{2.8\pi r^4}{\lambda^2}\ \mathrm{m}^2. \tag{7.14b}$$

Relative to Eq. (7.14a), RCS is raised more or less proportionately to the square of the increased area of metal [$2.8/1.2 \sim (\pi/2)^2$].

Despite the danger of presenting a null to the radar for lengthy periods, octahedrals have been widely used, being cheap, light and fairly small. The shape is inherently robust and not too prone to performance deterioration through distortion in rough service. RCS rises sharply with size and is always at least 10 dB less at 3 GHz than at 9 GHz. Some of the octahedrals in service are much too small to have useful RCS. Fold-flat or inflatable 'emergency' octahedrals are unlikely to attain the quarter-wavelength accuracy needed to realise their potential RCS unless precision made and rigid, and performance can disappoint.

Unfortunately, many users mount octahedrals incorrectly, perhaps tilted to match the jaunty lines of the boat. Worse, it is all too convenient to rig with one plate horizontal and two vertical. This looks neat and will not fill up with rain but gives a dreadful pattern. Wylie [8, Figure 8.8(c)] suggests yachts might mount octahedrals as Figure 7.7(c) with one plate vertical with its top edge horizontal on the fore and aft line, orientation tending to that of Figure 7.7(a) when the yacht heels. Driving a relatively small-diameter mast through the centre does not much reduce performance – there is no particular concentration of energy at the centre.

7.7.3 Trihedral clusters

The simplest useful cluster takes five trihedrals, with the bottom surfaces inclined down at 35°, giving five lobes with intermediate nulls of about the same dB depth as those of octahedrals, but 4.2 dB higher peak RCS for the same trihedral size, the geometric axes now lying on the equator. This pentagonal cluster lends itself to mounting around buoy daymarks as shown in Figure 7.9(a). Other arrangements are possible, Figure 7.9(b), for example, in general pattern smoothness increasing with the number of trihedrals.

We have seen that the octahedral is by no means perfect: TPM is lumpy, with significant 'holes' even at small inclination. Only six of the eight available elements are used – and then inefficiently, not on their axes. The plates are exposed to the weather and may corrode, introducing ohmic loss. Several proprietary reflectors seek to improve TPM by using clusters of encapsulated trihedrals, Figures 7.10(a)–(c). Long weathering may make their outer plastic skins lossy, reducing RCS. Fixings are arranged to discourage mounting at the wrong angle. Outlines tend to be cylindrical rather than spherical, minimising wind resistance and helping the device to hang vertically from yacht halliards, so reducing performance loss when heeled.

For a given volume, many small trihedrals give lower RCS (averaged over the sphere) than a few large ones. Assuming what is not usually quite correct, that dimensions of the elements remain several wavelengths long, halving trihedral length reduces RCS to $\frac{1}{16}$ but only cuts volume to $\frac{1}{8}$. On the other hand the individual trihedral axes can be aligned more favourably and the better proprietary reflectors of the multi-trihedral type far surpass octahedrals. They have better reflection patterns in both azimuth and elevation, without broad, deep holes, see Figure 7.10(c), and occupy less volume than an octahedral of the same effective RCS. Design from first principles is difficult and many effective products have come from those people, notably the late John Firth, lucky enough to be endowed with microwave green fingers – like gardening, success at microwave design demands flair as well as scientific knowledge. As always, care needs to be taken how RCS is specified – peak or mean

Figure 7.9 Trihedral clusters. (a) Pentagonal cluster. Many buoys such as this green-painted Starboard Hand Mark rely on clusters of trihedral corner reflectors for the bulk of their RCS, the shapes and daymarks reflecting poorly. Light and solar panels above. Author. (b) Octagonal cluster. Class 1 Special Mark buoy with octagonal trihedral cluster between the daymark and the solar panels which power the light. The girder mast also reflects as an extended target. Many buoys of this importance carry racons. Shown being deployed by lighthouse tender. Reproduced by permission of Trinity House Lighthouse Service

values may be quoted, and unfortunately not all suppliers can back their performance claims with independent test certification. Buyers should remember that none of the suppliers can surmount the laws of physics and no very small device can have very high RCS.

In practice, RCS at 3 GHz is often about −11.5 dB relative to the 9 GHz band, because the device dimensions are fewer wavelengths, rather than the 9.9 dB predicted by Eqs (7.14); this is one reason why the revised IMO specification has 3 GHz RCS of −11.8 dB on the 9 GHz value.

Many reflector designs using other principles have been produced over the years, for various purposes and not always based on sound theory. Some of the better ones are briefly discussed below. The new IMO requirement is likely to stimulate innovation.

7.7.4 Luneberg lens

Luneberg lenses feature concentric shells of insulating material such as foamed plastic, dielectric constants rising towards the centre to form a convex lens. There are three main variants.

1. The conventional Luneberg focusses onto a metal reflector at the opposite side. In cylindrical form it operates over 135° azimuth.

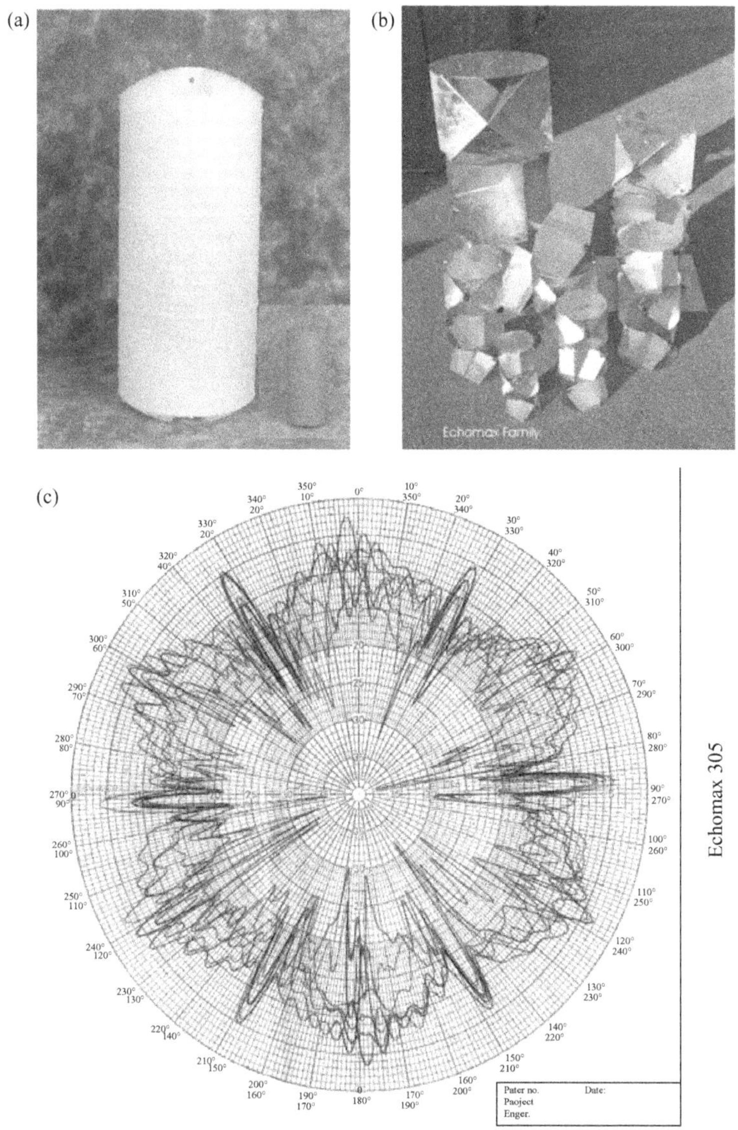

Figure 7.10 *Proprietary trihedral cluster reflectors for small craft: 9 GHz band. All reproduced by permission of Echomax Products, Dunmow, UK. (a) Large and small models for halliard or bracket mounting. Smooth outer glass-fibre weatherproof casing reduces risk of snagging sails. (b) Outer case removed, showing stacked trihedral clusters. (c) Azimuth polar diagrams. Echomax model 305. Equatorial, and with ±1°, ±2°, ±3° offset. Scaled $-20\,dB\,m^2$ at centre to $+20\,dB\,m^2$ at rim. Claimed peak RCS $63\,m^2$, with 100 per cent at $10\,m^2$, and to exceed ISO 8729, RORC and ORC requirements. Length 710 mm, weight 4.6 kg*

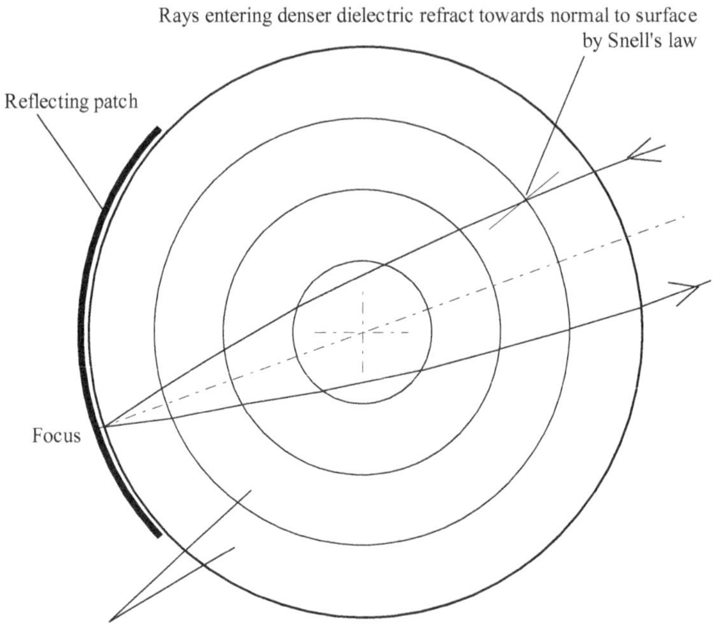

Figure 7.11 Luneberg lens. Cross section showing ray paths

2. The spherical Luneberg has nested spherical shell dielectrics assembled onion fashion, and can operate over 172° azimuth. The effect is very loosely akin to a convex lens rotated into a sphere, Figure 7.11. To focus on the opposite surface [7, Section 13-13], a lens of radius r has to have dielectric constant n at radius pr:

$$n = \sqrt{2 - p^2} \tag{7.15}$$

Elevation beamwidth between −3 dB points is $25\lambda/(\text{length})$. Although they tend to be expensive, TPM is uniform and spherical Lunebergs are sometimes used as calibration targets. Maximum RCS approaches that of a metal disc normal to the beam, Eq. (7.4a). In practice there may be a couple of dB loss and a lens of diameter $2r = 190$ mm with acceptance angles of $70° \times 70°$ has RCS of about 8 dB m^2 at 9 GHz and −2 dB m^2 at 3 GHz.
3. The Iton type, of Russian origin and described by Barton and Leonov [9] reflects from a rod at the centre but is more difficult to realise. It operates over the full 360° azimuth.

7.7.5 Helispherical reflector

This proprietary design somewhat resembles a Luneberg lens surrounded by a cage of metal strips inclined at 45°. Horizontally or vertical polarised interrogations are resolved into components normal and parallel to the strips. The former pass the strips, are focused by the shells onto the other side of the cage or onto an internal metal collar and are reflected with the original polarisation. Smooth TPM is claimed. Certain proprietary reflectors (Polaref) use similar principles. Peak RCS is quoted as

$$\sigma = \frac{r^4}{\lambda^2}\,\mathrm{m}^2. \tag{7.16}$$

7.7.6 *Lens reflectors*

The proprietary Lensref has nested dielectric shells, used as a pair of back to back lenses to focus incoming rays onto a small reflector. Each lens has a wide acceptance angle. The two azimuth gaps between the lenses are filled by trihedral reflectors. Good TPM is claimed in elevation and azimuth, without serious gaps.

7.7.7 *Chaff*

A wire placed normal to the ray axis and in the electric plane reflects strongly when a half-wavelength long, where it acts as a resonant dipole antenna. The military have long used clouds of dipoles (Figure 7.4k), called chaff, as artificial clutter to mask friendly aircraft. It might be possible to incorporate dipoles in the glass reinforced plastic used in construction of small craft to improve RCS.

The RCS of N resonant dipoles randomly dispersed in three dimensions within the detection cell is

$$\sigma \sim 0.15N\lambda^2\,\mathrm{m}^2. \tag{7.17}$$

To obtain RCS $10\,\mathrm{m}^2$ at 3 or 9 GHz, the target would need to contain about 7000 or 70 000 visible random dipoles, respectively.

7.7.8 *Phased patch array reflectors*

A trihedral cluster can be regarded as an array of small short-circuited antenna elements each reflecting through a very broad solid angle, positioned to preserve the resultant wavefront phasing through broad angles of look. Similar retro-directive arrays can be made from antenna elements interconnected as Van Atta [10] or Pon [11] systems, in effect forming phased arrays. Wavefront phasing is preserved either by a geometric arrangement of delay lines or by active phase conjugation. These arrays can be made in either conformal (Pon) or planar (Van Atta or Pon) shapes which are being developed by GCS Ltd.[1] They are not limited to any one type of radiating element and particularly lend themselves to low-loss laminated printed circuit board construction. The board, see Figure 7.12, carries an array of patches – a form of broadband resonant dipole – somewhat akin to the patch array scanners mentioned in Chapter 2, Section 2.7.8, but differently interconnected, the stripline lengths being

[1] J. Grothusen, Guidance Control Systems Ltd., Leicester UK, personal communication, January 2004.

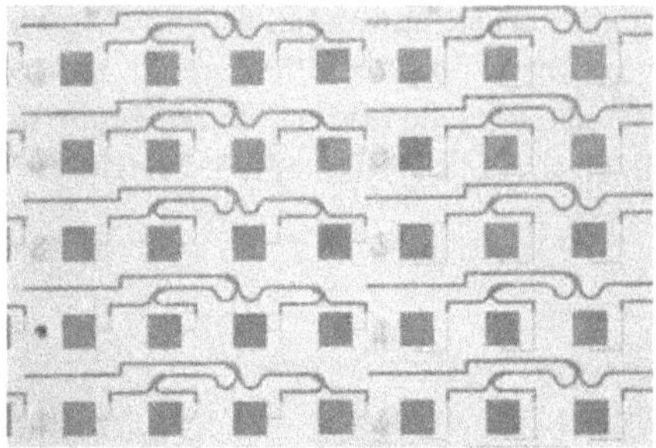

Figure 7.12 *Van Atta retro-directive reflector. Part of experimental flat array showing quads of patch radiators connected by equi-phased striplines. Reproduced by permission of Guidance Control Systems Ltd, Leicester, UK*

arranged to maintain the required phasing through broad angles at the operating frequency. Lengths are relatively short, so beamwidths are preserved through a fairly wide bandwidth. The firm reports that prototype flat panels 0.3 m × 0.3 m aperture and <1 cm thick have peak RCS of 24 m^2 (13.8 dB m^2) with azimuth and elevation −3 dB lobe widths ±27° × ±12° at 9 GHz. Except for bandwidth, this performance approaches that of corner reflectors, but within much smaller volume.

Initially development is concentrating on coded retro-directive panels intended to highlight reference points on offshore oil structures. A video-frequency code applied to switching diodes controls the phasefront, modulating the echo to give frequency-shifted sidebands. The devices remain passive because all the echo energy derives from the incoming illumination. The panels work with small specialist frequency-modulated continuous wave (FMCW) radars on offshore supply vessels. The radar receivers are tuned to the modulation and detect the small panels at up to ~1 km range amid clutter and strong adjacent structure echoes. Using a vertically spaced pair of differently coded panels obviates multipath problems. The system is designed to provide accurate range and bearing for dynamic positioning (DP) systems, remaining operative in weather which defeats conventional laser and acoustic systems.

7.8 Miscellaneous point targets

Some other targets of practical significance which approximate points do not lend themselves to detailed analysis.

7.8.1 Aircraft

Light aircraft are generally small enough to approximate point targets. They do not remain within a single detection cell long enough to be readily detected on marine or VTS radar even under ideal conditions. According to Skolnik [12], RCS of a fixed wing aircraft nose-on approximates

$$\sigma \sim 0.01 L^2\,\mathrm{m}^2. \tag{7.18}$$

L is the length in metres; for example, if $L = 10$ m, RCS $\sim$1 m^2 or 0 dB m^2. TPM is usually 'busy' with many flashes and holes.

7.8.2 Helicopters

It may occasionally be required to detect helicopters for SAR or sea-pilot rendezvous. Helicopters generally fly below 1000 ft (300 m), when they will be within the scanner beam at ranges exceeding $\sim$2 km. Helicopters are point targets, so their echoes exhibit a multipath structure, exaggerated by their great height relative to most conventional marine point targets. For example the first (furthest) null at 1000 ft altitude to a 9 GHz band scanner at 15 m is at about 80 km. The echo when on an axial course therefore fluctuates repeatedly, with nulls each kilometre or so. The author has tracked a four-seat Bolkow Bo 105s helicopter, All Up Weight 2.4 tonnes, out to beyond 20 km with good probability of detection except in the nulls, indicating its nose-on and tail-on RCS approximated 20 dB m^2 at 9 GHz. The helicopter was weakly detectable to over 45 km, indicating that RCS jittered widely, a phenomenon called 'blade flash' by the military. TPM is busier and RCS is higher than fixed-wing aircraft of the same weight, which tend to be more streamlined.

7.8.3 Buoys and lighthouses

Unless fitted with active or passive reflectors or racons, the relatively small size, height and unfavourable shape of some buoys makes them poor targets, Figures 7.9(a) and (*b*). Before the mid-twentieth century, many lighthouses were built of concrete or masonry, whose loss when more than about 20 cm thick, the usual case, masks reflection from internal metal stairs, etc. The usual coned outline further reduces monostatic RCS. Early light sources such as Argand lamps and paraffin vapour burners demanded huge catadioptric lens systems to get the long ranges which were essential in pre-radar days. These lighthouses, even when modernised, often retain the original large ($\sim$5 m diameter) lantern structures and copper domes; Figure 7.13. Although the glazing is a poor reflector, the surrounding liberal metalwork, including balconies, handrails, dome, etc. reflects well. Observation of a few such lighthouses indicates that RCS may be around 10–15 dB m^2, approximating the visual projected area including glass. The structure forms a reflector more or less at the height of the light and is sufficiently small to cause multipath peaks and nulls. Modern structures are usually much more compact, so may have lower RCS. Metallic towers form stronger targets, extended in height, their multipath factors being small for reasons developed in Chapter 9. Lighthouses are designed to be visually conspicuous, but it by no means follows that they are conspicuous on the radar display unless fitted with a racon.

Wylie [8, p. 159] gives some detection ranges for typical 9 GHz merchant ship radars in benign clutter.

High focal plane buoy	9 nmi (16.7 km)
Radar reflector buoy	7 nmi (13 km)
First-class buoy	3–4 nmi (5.6–7.4 km)
Second-class buoy	2 nmi (3.7 km); RCS $\sim$5 ft^2 (0.46 m^2).

Figure 7.13 *Cape Wrath Lighthouse. Typically massive metalwork of older lighthouses reflects well but their tapered masonry towers reflect poorly. Author, reproduced by permission of the Northern Lighthouse Board*

7.8.4 Birds

Gulls flock to fish, so fishing vessels sometimes seek their radar echoes. Individual birds cause what are sometimes called dot angel echoes and flocks may cause ring angels. Birds may fly high enough to give long horizon range. Detection ranges up to 20 nmi (37 km) have been claimed for flocks at 3 GHz, less at 9 GHz. RCS of a bird is said to approximate a water sphere of the same mass, M kg. If so, effective radius $r = 0.062M^{1/3}$. Birds are surprisingly light, the bulky plumage contributing little weight and less RCS. The upper limit of the Rayleigh region of a spherical reflector is when $r = 0.1\lambda$, corresponding to $M = 4.2$ g at 3 GHz, less at higher frequencies. For water of $\varepsilon = 80$, proportion of incident power reflected, $R = 0.8$ and assuming the radius is sufficient to lift operation out of the Rayleigh region, sphere RCS equation Eq. (7.3a) gives

$$\text{bird RCS} \sim 0.01M^{2/3}\ \text{m}^2. \tag{7.19a}$$

For example, a bird weighing 100 g would have RCS ~ -26 dB m^2.

Many birds probably lie in the Mie region where resonance effects change RCS up to a further ± 5 dB. Assuming the birds in a flock vary in mass, these resonances perhaps cancel, so when there are N birds in the detection cell volume, very roughly

$$\text{flock RCS} \sim 0.01NM^{2/3}\ \text{m}^2. \tag{7.19b}$$

7.8.5 Man

The RCS of a man (from the context, perhaps a fully accoutred soldier) is reported by Skolnik [12, Table 2.2] to be -3 to 1.1 dB m^2, enough to significantly raise RCS of small dinghies.

7.8.6 Scanners

Occasionally, one radar may illuminate the scanner of another. When not transmitting, a scanner presents a mismatch to incoming signals. To find its effective RCS, it may be treated as an RTE having less than unity amplifier gain, see Chapter 8, Section 8.11.1. The beamwidth as a reflector between -3 dB points equals the scanner -1.5 dB beamwidth, approximating 0.72 of the -3 dB beamwidth.

7.8.7 Flotsam

Ships encounter a wide variety of potentially hazardous floating objects. Ropes or nets risk fouling the propellor while more solid items may cause hull damage. Perhaps the most dangerous is a 40 ft ISO container floating awash and displacing nearly 80 tonnes. Such objects usually have low RCS and are often impossible to detect by radar. The pyramidal corner of the container looks to the radar like just another wave. Least bad detection performance would be obtained from a scanner mounted so low that it would be prone to wave impact damage – the author never regretted turning down a Scandinavian proposition to develop a solid-state phased array radar to integrate with the ship's bow for detection of flotsam in the arc ahead. HSCs are at especial risk from flotsam and are required to mount night vision equipment to augment the Mark 1 eyeball, which remains the best daylight flotsam detector.

7.9 Tilting a point target aid

7.9.1 Introduction

Here we consider what happens when the radar scanner remains upright but the target is tilted or rolls, perhaps when on a buoy set over in a current or on a yacht rolling in a seaway. Figure 7.14 shows a target rolling, viewed by radars in the roll plane (Radar A) and normal (Radar B). Of course, the radar will usually be oblique, when the angle can be resolved into roll-plane and cross-plane components.

Sometimes the radar platform (e.g. a ship) will also roll or tilt, either in or normal to the target roll plane. Effective echo strength is then further modified; in the roll plane by the scanner elevation polar diagram and in the latter case by its cross-polar characteristic.

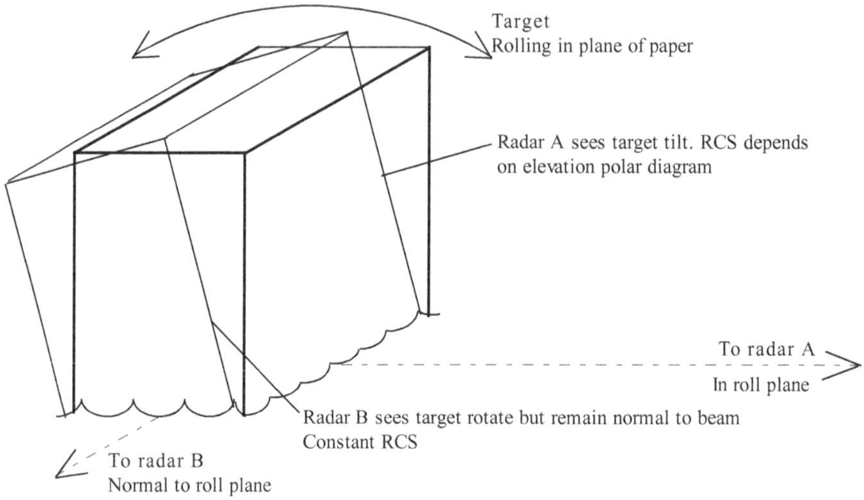

Figure 7.14 *Effect of target roll. RCS fluctuation of a rolling target depends on the angle of the radar to the roll plane*

7.9.2 Radar in roll plane

For Radar A in the roll plane, target effective RCS is determined by its TPM.

- Response strength is governed by the target's elevation polar diagram.
- The target always sees the interrogation at the same polarisation without any cross-polarisation loss.
- The scanner always sees the same response polarisation; again without cross-polarisation loss.

If the target is tilted by an angle ν, say 10° towards the radar, Radar A illuminates it at +10° in target elevation polar diagram, which often reduces RCS.

7.9.3 Radar normal to roll plane

Neglecting any scanner beam depression, the target always presents its equator to Radar B but tilts to one side relative to the plane of the scanner rays. The plane of polarisation seen by the target now inclines:

- The target elevation pattern is immaterial and full on-axis RCS is retained.
- Usually the target echo polarisation plane remains that of the interrogation irrespective of tilt and there is no polarisation loss.
- If the target is a flat-plate derivative such as an octahedral, RCS remains unaltered. If it is a Helispherical reflector or a Luneberg lens, the echo plane of polarisation may be tilted. If so, the scanner receives a rotated polarisation, incurring response loss. RCS is maximum when untilted and falls to zero ($-\infty$ dB) when 90° cross-polarised. The voltage vector depends on $\cos\nu$ so effective gain, σ_{eff}, falls by

$20 \log \cos \nu$:

$$\sigma_{\text{eff}} = \sigma_{\text{nom}} - 20 \log \cos \nu \text{ dB}. \tag{7.20}$$

Response is 3 dB down when $\nu = \pm 45°$.

7.10 Combination of point targets

7.10.1 The problem

So far when tracing ray paths to determine multipath effects we have considered a lone reflector small enough to act as a geometrical point. Most real targets are an assembly of scattering elements. Except in the simplest combinations the polar diagram or TPM of practical targets becomes very complicated and well-nigh impossible to predict in detail. Indeed possession of an accurate TPM would not much assist calculation of detectability unless aspect was known with accuracy and radar frequency and polarisation were identical to that used in the TPM determination. As a preliminary to consideration of ship targets in Chapter 10, we here examine what happens when a pair of point target elements lie within the detection cell, for example when a reflector is mounted on a small buoy. Even a simple pair can include so many variables – spacing, relative RCSs, passive or active, etc. – that we can only paint a broad outline to illustrate complexities and warn of pitfalls.

The resultant echo at a distant radar from a close-spaced pair of passive targets clearly contains a contribution from each. The energy reaching the scanner from one element will usually be phase shifted from the other, as the elements will almost always differ in range by at least a centimetre or so. The resultant is found by vectorial addition of the voltage vectors of the elements. Although the method can be extended to find the TPM of reflector clusters such as octahedrals, the calculations rapidly become very laborious.

7.10.2 Assumptions and notation

We make the following simplifying assumptions, not all of which are likely to remain fully valid in a practical system:

- free space conditions, without an indirect ray from the sea surface or atmospheric attenuation;
- each element $T1$, $T2$ is a geometrical point;
- RCS of each element remains constant at any aspect so its TPM is isotropic (that of a reflecting sphere);
- scanner is so distant that rays striking the elements are parallel;
- elements are so close that propagation loss is same to each;
- reflections from one element do not strike the other – there is no second bounce;
- elements do not mask one another;
- the radar/targets plane may be either (a) horizontal, yielding the azimuth polar diagram for two elements side by side, or (b) vertical, giving the elevation diagram for one element above the other.

We use the following notation:

σ_1, σ_2 = RCS of element $T1$, $T2$ alone, respectively, m^2,
σ_r = resultant RCS of $T1$ and $T2$ together, m^2,
L = spacing of $T2$ from $T1$, m,
n = number of wavelengths between $T1$ and $T2$,
δ = range difference between $T1$ and $T2$, m,
ν = radar bearing relative to $T1$ and T2 broadside, rad,
S_1, S_2 = echo power at radar from $T1$ or $T2$, respectively, W,
S_r = resultant echo power from$T1$ and $T2$ together, W,
V_1, V_2 = echo voltage at radar from $T1$ or $T2$, respectively, V,
V_r = resultant echo voltage, V,
ϕ = Phase shift of V_2 on V_1, rad,
C_a, C_b = constants which need not be evaluated.

7.10.3 Resultant performance of pair

Figure 7.15(a) is adapted from a paper [1] of the author. It shows reflectors $T1$ and $T2$ spaced n wavelengths apart and illuminated obliquely by a distant radar at range R. From the geometry, $T2$ range exceeds $T1$ range by δ where

$$\delta = n\lambda \sin \nu. \tag{7.21a}$$

The interrogation at $T2$ therefore lags $T1$ by δ/λ wavelengths, equivalent to phase shift $2\pi\delta/\lambda$ radians, so the response component at the radar from $T2$ lags that from $T1$ by $\phi = 4\pi\delta/\lambda$ rad. Substituting in Eq. (7.21a),

$$\phi = 4\pi n \sin \nu \text{ rad}. \tag{7.21b}$$

The free space radar range equation (from Chapter 4, Eq. (4.6b) without losses) gave the echo power at the radar from a target whose RCS is σ:

$$S_{e(FS12)} = PG^2\sigma\lambda^2(4\pi)^{-3}R^{-4} \text{ W}.$$

For a given radar working with T_1 alone at a given range, we can write this as

$$S_1 = C_a\sigma. \tag{7.22}$$

$S_1 \propto V_1^2$, so we can write Eq. (7.22) as:

$$V_1^2 = C_b\sigma_1. \tag{7.23}$$

Similarly for $T2$ alone and correcting for the changed range:

$$V_2^2 = C_b\sigma_2 \times \frac{(R+\delta)^{-4}}{R^{-4}}.$$

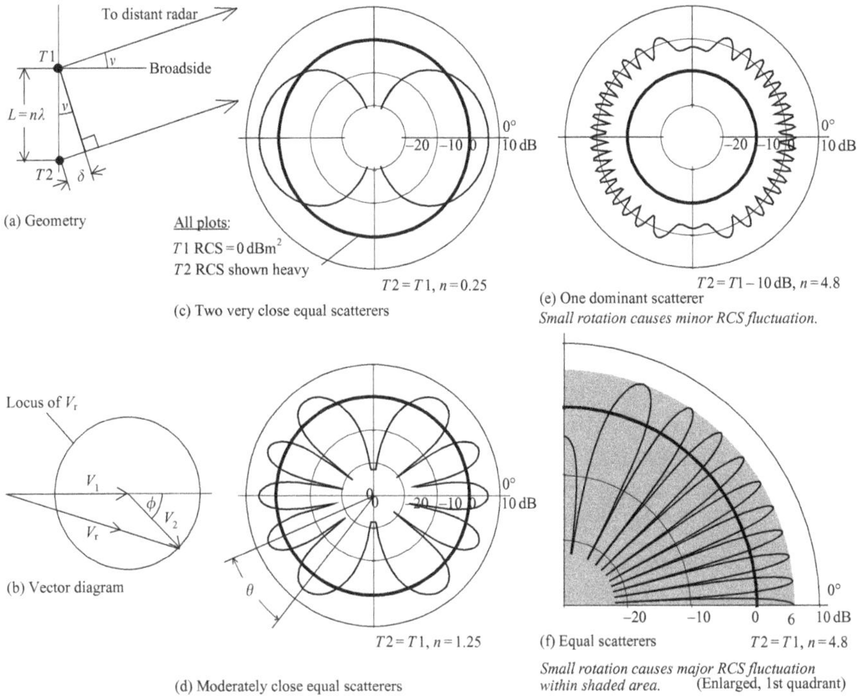

Figure 7.15 Pair of point target elements. Although each element has constant RCS and circular polar diagram, interference causes the resultant effective target polar diagram to be far from uniform. For a target pattern map see Figure 7.17

Because $R \gg \delta$, $R + \delta \sim R$,

$$V_2^2 = C_b\sigma_2. \tag{7.24}$$

The vector diagram of V_1 and V_2 is shown in Figure 7.15(b) and is evaluated as:

$$S_r = S_1\left[1 + \frac{V_2^2}{V_1^2} + 2\left(\frac{V_2}{V_1}\right)\cos\phi\right].$$

Substituting:

$$S_r = (C_a\sigma_1)\left[1 + \frac{C_b\sigma_2}{C_b\sigma_1} + 2\left[\sqrt{\frac{C_b\sigma_2}{C_b\sigma_1}}\right]\cos(4\pi n\sin\nu)\right]$$

$$= (C_a\sigma_1)\left[1 + \frac{\sigma_2}{\sigma_1} + 2\sqrt{\frac{\sigma_2}{\sigma_1}}\cos(4\pi n\sin\nu)\right]. \tag{7.25a}$$

This is equivalent to the echo from a single point target whose RCS is σ_r where

$$\frac{\sigma_r}{\sigma_1} = 1 + \frac{\sigma_2}{\sigma_1} + 2\sqrt{\frac{\sigma_2}{\sigma_1}}[\cos(4\pi n \sin \nu)] \text{ numerically.} \tag{7.25b}$$

When $\sigma_2 = \sigma_1$ as in Figures 7.15(c), (d) and (f),

$$\frac{\sigma_r}{\sigma_1} = 2[1 + \cos(4\pi n \sin \nu)] \text{ numerically.} \tag{7.25c}$$

To the radar, the resultant looks the same as the echo from a single target of fluctuating RCS. Peaks (sometimes called grating lobes after the striations observed when light shines through the narrow slits of an optical grating) occur when σ_r/σ_1 is at its maximum. Here, $\cos(4\pi n \sin \nu) = 1$ and $(4\pi n \sin \nu) = 0, 2\pi, 4\pi$, etc. This confirms that on the broadside RCS is always high because path lengths are equal.

Nulls occur when $\cos(4\pi n \sin \nu)$ is at its minimum value of -1.0. Here $(4\pi n \sin \nu) = \pi, 3\pi$, etc. and $\sin \nu = \frac{1}{4}n, \frac{3}{4}n$, etc. There are therefore $8n$ lobes in the diagram, more or less uniformly spread through the 360°, the average lobe–lobe angle θ (Figure 7.15(d)) being:

$$\theta_{av} = \frac{45}{n} = \frac{45\lambda}{L} \text{ degrees.} \tag{7.26}$$

For example, when reflecting elements are spaced at $L = 1.0$ m and wavelength $\lambda = 0.1$ m (3 GHz), $n = 10$ and $\theta_{av} = 4.5°$. A 9 GHz radar, $\lambda = 0.032$ m, would have $n \sim 31$ and $\theta_{av} \sim 1.4°$.

Peak and null RCS are given by

$$\frac{\sigma_r}{\sigma_1} = 1 + \frac{\sigma_2}{\sigma_1} \pm 2\sqrt{\frac{\sigma_2}{\sigma_1}}. \tag{7.27}$$

7.10.4 Examples

Figures 7.15(c)–(f) are polar diagrams which plot σ_r for various values of n and σ_2/σ_1. Alternatively, the plots can be read as the ratio σ_r/σ_1, calculated per Eq. (7.26) and expressed in decibels. Figure 7.15(f) shows only the first quadrant, the others being symmetrical.

1. The simplest case (Figure 7.15(c)) is when both scatterers have equal RCS and are spaced a quarter-wavelength. Viewed broadside ($\nu = 0°$ or 180°), the echoes are in phase, $V_r = 2V_1$, $S_r = 4S_1$ numerically, and $\sigma_r = \sigma_1 + 6$ dB. Adding the second scatterer has not merely doubled but quadrupled the RCS. But as ν moves round to 90° or −90°, $T2$ echo swings to phase opposition, the echoes cancel and we pay the penalty: σ_r falls to zero ($-\infty$ dB m^2). At these angles V_1 and V_2 are equal and opposite; adding the second scatterer has destroyed the echo of the first. Overall, the polar diagram has completely changed from a constant-RCS circle to a figure-8 shape.
2. In Figure 7.15(d) separation n has been increased to 1.25 wavelengths. Again there is a peak on the broadside but several deep nulls (corresponding to

$(4\pi n \sin \nu) = \pi$, 3π and 5π), separated by +6 dB peaks, are traversed before reaching the end-on condition.

3. Figure 7.15(e) is for wider separation, $n = 4.8$ wavelengths (say 0.16 m at 9 GHz or 0.48 m at 3 GHz) with one scatterer dominant, with +10 dB RCS on the other. The wider spacing has increased the number of peaks and nulls, but RCS is dominated by the larger scatterer and there is only moderate RCS fluctuation as the overall target rotates through a few degrees.
4. Figures 7.15(f) is as for (e) but the scatterers are equal size. The polar diagram now exhibits numerous severe lobes. Near 90° and 180°, the shape of the diagram is critically dependent on n. When $4n$ is an odd integer there is a null (as c and d). When $4n$ is even there is a peak. Otherwise there is a part-peak, as (f) where $4n = 19.2$. The number of lobes is $8n = 38.4$; 38 full lobes and two partly formed lobes at 90 and 180°, the latter causing wide arcs of low RCS. As little as 1 per cent change in wavelength or element spacing (~0.5 mm) would change n to 38.0 and full peaks would be developed at 90° and 180°, sharply changing the appearance of the polar diagram. If unsure of the exact direction of the centreline, all one can really say is that the RCS will lie somewhere in the shaded area and minor changes in aspect or frequency will cause significant fluctuation from the average (Figure 7.16).

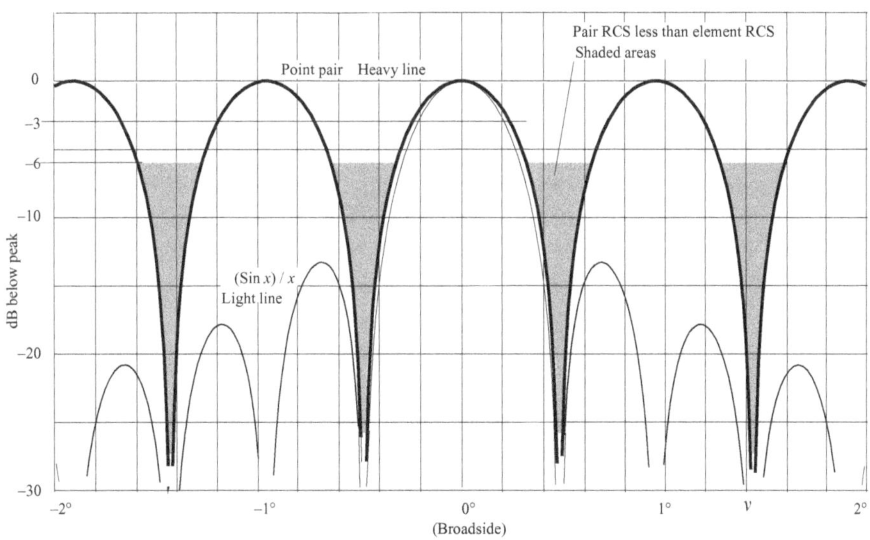

Figure 7.16 *Pair of point targets, broadside beamwidth. Radiation pattern. Elements of equal RCS, spacing 30 wavelengths (e.g. ~1.0 m at 9 GHz). Pair RCS on axis is +6 dB on single element, beamwidth 0.46° between half-RCS points. Deep nulls at ±0.48°, ±1.44°, etc. A uniformly illuminated antenna of* $(\sin x)/x$ *form, aperture (4° × 30°) and aperture length (4 m in the 9 GHz band), has a similar main beam lobe*

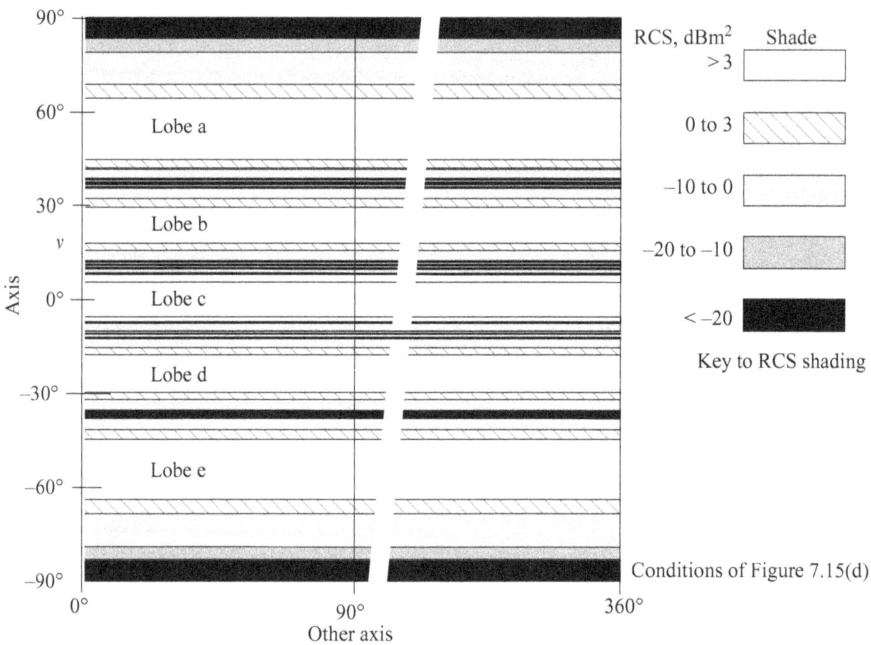

Figure 7.17 Target pattern map, two point elements vertically disposed

7.10.5 Response in other plane; TPM

The elements are symmetrical about the centreline so the response in the plane normal to the one considered is constant. When the centreline is vertical, the TPM takes the form of parallel horizontal stripes as shown in Figure 7.17, which illustrates the method of contour shading. A radar on the horizon, $\nu = 0°$, always sees constant high RCS at all azimuth angles.

7.10.6 RCS fluctuation

If one element of the pair were mounted on a platform of negligible skin RCS afloat in rough water, the distant radar would see a steady 'non-fluctuating' or 'hard' RCS of $\sigma_1 = 0\,\text{dB}\,\text{m}^2$, unvarying from scan to scan however the platform rolled in the sea. In Chapter 12 we shall see that this is termed a Swerling Case 0 target. Twin-element targets reasonably approximate Case 0 when $\sigma_2/\sigma_1 > -20\,\text{dB}$.

The target equal-pair RCS would fluctuate between $-\infty$ and $+6\,\text{dB}\,\text{m}^2$. The platform rolls or pitches with period of a few seconds, comparable to the scan period. The platform might be upright on one scan, RCS $+6\,\text{dB}\,\text{m}^2$, and tilted to a null with very low RCS the next. The precise details of the pattern would be of little interest and in practice when there are several scattering components are unlikely to be calculable. Hence it is usual to quote average RCS, which in the simple two-scatterer illustration

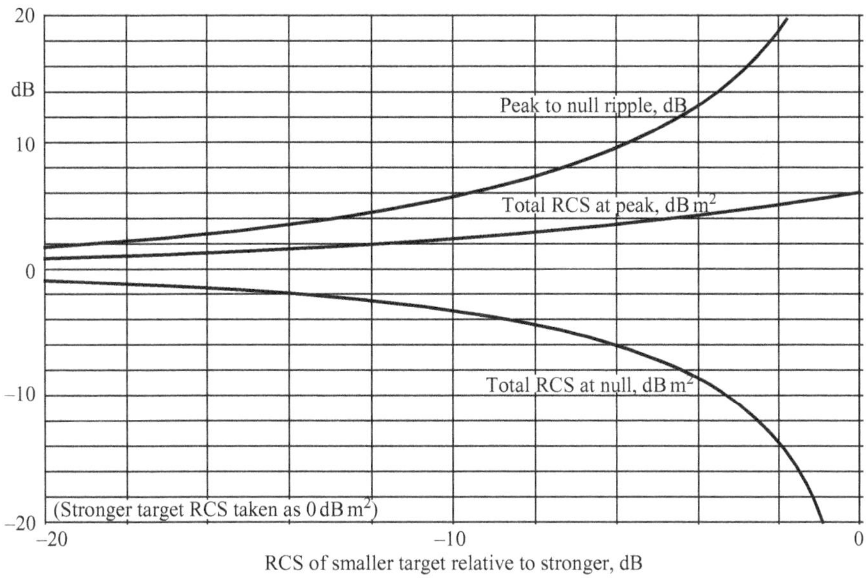

Figure 7.18 Peak and null values of effective RCS of pair. When smaller target approaches RCS of bigger element, peak RCS rises, null RCS falls and pattern ripple becomes large

is $\sigma_1 + \sigma_2$ m^2, and state that the target fluctuates. Chapter 12 will include two important broad fluctuation cases.

- Swerling Case 1 where there are many similar-size scatterers. Figure 7.15(f) is a first approximation to this case.
- Swerling Case 3 where there is one dominant scatterer and several smaller scatterers, approximated by Figure 7.15(e).

Figure 7.18, based on Eq. (7.27), shows how peak, null and ripple are related to the RCS ratio of the elements of the pair. If for example the ripple is not to exceed 6 dB peak to peak, the smaller element would have to be smaller than −9.5 dB of the larger element, or the larger 9.5 dB bigger than the smaller, almost reaching our definition of Case 3. If, say, the smaller had RCS = 3 dB m^2 (2 m^2), the larger ought to be made $3 + 9.5 = 12.5$ dB m^2 or 18 m^2 and the resultant would fluctuate between $12.5 + 2.5 = 15$ dB m^2 (32 m^2) and $12.5 - 3.5 = 9$ dB m^2 (8 m^2).

7.10.7 Tilt

Imagine the element pair to be two reflectors on the mast of a yacht of negligible platform RCS and rolling in a seaway. The TPM strips now assume a sinusoidal shape, Figure 7.19. Viewed by a radar on the horizon, RCS depends on azimuth aspect. It remains at the former (maximum) value when the radar lies in the 0°/180° plane normal to the roll plane. In the 90°/270° roll plane RCS varies with tilt. As azimuth

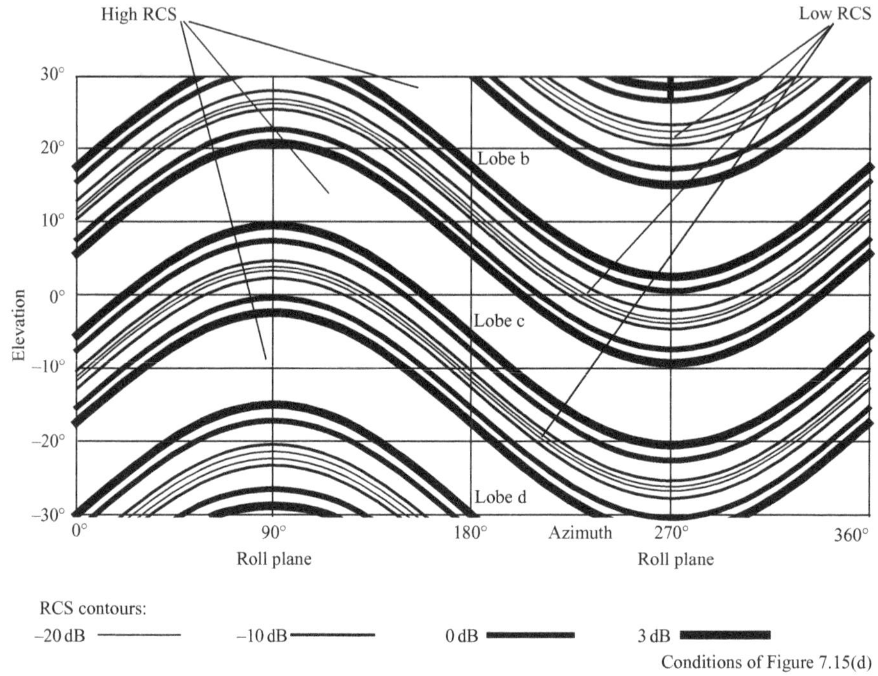

Figure 7.19 *TPM, platform tilted. Tilt 15°. RCS on horizontal (0° elevation) now varies strongly at differing azimuth bearings. TPM drawn using contour line method*

angle changes, RCS varies between the maximum (6 dB) at 0° azimuth, goes through a null (<−20 dB) at 50°, partly recovers within the next lobe, reaching 0 dB at 90°, falls back into the null at 140°, regains the peak RCS at 180° and so on. The RCS of a tilted element pair is therefore far from constant.

7.10.8 Practical performance

Turning from free space to practical conditions, we draw the following conclusions.

1. Only if one element RCS exceeds +16 dB on the other, is the RCS ripple less than 3 dB peak to trough.
2. Pattern becomes very sensitive to radar frequency and minor spacing change when the pair separation is large.
3. Two reflectors of within ±6 dB RCS give particularly poor patterns.
4. If the target is large, say a metre or more in length, its RCS fluctuates with small motions of target, scanner or grazing point. This affects detectability; a statistical approach is necessary, Chapter 12.

5. If the element RCSs at the angle of heel are nearly equal, a single composite reflector is formed, with RCS equal to the sum (in square metres) of the contributors. Its elevation and azimuth polar diagrams contain lobes dependant on the spacing of the elements. The effective mounting height is proportional to the weighted average heights of the elements.
6. In practice, the RCS of each element is not constant, its polar diagram having peaks and nulls. The resultant polar diagram can be constructed by insertion of the appropriate values for σ_1 and σ_2 at each value of ν, to give a resultant diagram with bigger peak/null ratio than shown in Figure 7.19, which was drawn for constant σ_1 and σ_2.
7. Even if only one reflector is carried, the platform's skin RCS may be sufficient to spoil the polar diagram. Although the skin RCS is distributed in height, an approximation to behaviour is obtained by considering it as a point reflector centred on the centroid of its RCS/height profile. It may pay to keep the reflector well clear of the centroid of the skin reflection to make the resultant diagram more like Figure 7.15(f) than (d), so platform movements in the seaway carry less risk of lurking in a null.
8. When two reflectors are well spaced in height, their indirect ray pathlengths will differ slightly and they will not experience quite the same multipath factors. In essence the direct ray will illuminate the composite target at a different elevation angle and phase from the indirect ray. Multipath nulls for one element will differ in range from nulls for the other element and may slightly reduce the depth of resultant response nulls observed at the radar when multipath is severe, that is, at the longer ranges in calm water when $\rho \sim 1$.

7.11 References

1 BRIGGS, J. N.: 'Specifications for reflectors and radar target enhancers to aid detection of small marine radar targets', *Journal of Navigation*, 2002, **55** (1), p. 23. (see also discussion on that paper in Forum, *Journal of Navigation*, **55** (2), p. 330)

2 KINGSLEY, S. and QUEGAN, S.: 'Understanding radar systems' (McGraw-Hill, New York, 1992), Table 2.1

3 ROHAN, P.: 'Surveillance radar performance prediction' (Peter Perigrinus for IEE, 1983), Section 7.4

4 WILLIAMS, J. M.: 'Fundamentals of scattering from large complex targets', *Microwaves and RF Conference*, London, 1996

5 WILLIAMS, J. M.: 'The radar cross section of non-orthogonal corner reflectors, symmetrically illuminated', *Military Microwaves*, 1988

6 WILLIAMS, P. D. L.: 'Civil marine radar' (2002), Table 4.2

7 JASIK, H. (Ed.): 'Antenna engineering hand book' (McGraw-Hill, New York, 1961)

8 WYLIE, F. J. (Ed.): 'The use of radar at sea' (Hollis and Carter for the Royal Institute of Navigation, 1978, 5th edn.), p. 207

9 BARTON, D. K. and LEONOV, S. A. (Eds): 'Radar technology encyclopedia' (Artech House, London, 1998), Table R-25

10 VAN ATTA, L. G.: 'Electromagnetic reflector', US Patent No. 2,908,002; 6 October 1959

11 PON, C. Y.: 'Retrodirective array using the heterodyne technique', *IEEE Transactions on Antennas and Propagation*, 1964, **12**, pp. 176–80

12 SKOLNIK, M. I. (Ed.): 'Introduction to radar systems, International Student Edition' (McGraw-Hill, New York, 1983, 2nd edn.)

Chapter 8

Active targets

'The gods help those who help themselves.'

Aesop

8.1 Introduction

8.1.1 Passive and active reflectors

Chapter 7 showed the radar cross section of a passive object to be inherently associated with its physical size, because no dumb object can reflect more power than it intercepts from the interrogating radar. A metallic reflector may be considered as a receiving antenna directly feeding the selfsame antenna in transmit mode. Insertion of an amplifier (taking power from a battery) between these antennas would clearly boost the reflection or response. An active reflector or radar target enhancer (RTE) is such an antenna/amplifier/antenna device. The poor performance associated with physically small passive reflectors or antennas is augmented by the amplifier, so RCS is no longer tied to size. It is not possible to modulate the response of an active reflector of this kind with any form of identification code. The response is merely an amplified replica of the interrogation, with the same frequency and pulselength, appearing indistinguishable from an ordinary passive echo.

Replacing the amplifier with a miniature radar transmitter, triggered to deliver a pulse or coded group of pulses when the received interrogation pulse exceeds an amplitude threshold, turns the device into a racon (radar responder beacon) or search and rescue transponder (SART), which, although operationally distinct, conceptually differ from one another only in their response identification codes. Radars can determine range and bearing of stations carrying RTEs or racons in the usual way. Range of SARTs is subject to some uncertainty because of their particular response coding format.

Discarding the receiver and its antenna, and letting the transmitter run free gives a ramark (radar mark), whose bearing the radar can measure. Lacking a receiver, ramarks cannot synchronise their transmissions with the radar, so range cannot be determined.

As point devices, all active targets are subject to multipath interference as described for passive point targets in Chapter 6.

The display traces of active devices are included in Figure 3.19 in Chapter 3, Section 3.11. Most of them respond to an interrogation pulse from primary radar and so can be classed as secondary radars. They are also occasionally classed as radar transponders, although it is doubtful whether the term is properly applicable to RTEs. Most work in-band, the response being within the interrogation frequency band, usually at radar frequency. We shall also describe a few cross-band devices where response frequency differs substantially from that of the interrogation. The interrogation is almost always the ordinary radar transmission used for detection of passive targets. Except for ramarks, the triggering process synchronises the response to the radar timebase and places the response at the correct range on the radar display. Correct bearing is assured by the narrowness of the scanner beam. Proposals have been made from time to time to mount transponders on vessels to indicate manoeuvres. For example, the Shipbuilding Research Association of Japan[1] trialled a 9 GHz band collision prevention radar transponder whose response was four single dots when turning to starboard, three dot pairs when turning to port and three dot triplets when going astern, with no response when going straight ahead. Operation was technically similar to a fast-sweep racon or SART described in this chapter. The device seems not to have been perpetuated. A few fishery protection vessels (Chapter 1, Figure 1.6) and oil rigs transiting the English Channel were marked by racons in the guise of transponders in the 1970s, navigation warnings also being broadcast; but such transponders seem not to be currently in service.

Modulated reflectors can impress information on the echo, with or without amplification. It is unlikely any are yet in marine service, but they are briefly discussed in Section 8.11.2.

8.1.2 Historical

The historical notes in Chapter 1, Section 1.3, briefly mentioned active devices. Although envisaged for marking lighthouses in the UK Post War Radar scheme about 1944, implementation of racons was deferred until marine radar was firmly established and equipment had become more reliable about 1956. The UK Admiralty Surface Weapons Establishment Civil Marine Branch, having trialled and discarded ramarks (Section 8.9) then built three prototype racons. Mariners reported favourably on ease of use and assistance to navigation, particularly one on the Bar Lightvessel off Liverpool. In the late 1960s, AEI put a production-engineered chain into regular service on British and Irish lighthouses and one or two manned lightvessels (Chapter 1, Figures 1.5(b) and (c)). These bulky units were intended for long range landfall in the days before satellite navigation, so had sensitive (−95 dBW) receivers coupled to relatively powerful (11 dBW) response transmitters at 9 GHz. Use of discrete transistors mitigated power supply problems on offshore lighthouses. No suitable solid state device then being available, a special low-power mechanically tuned magnetron

[1] Shipbuilding Research Association of Japan, Report 207, March 1994.

transmitter was used. Tuning rods in the cavities worked through a bellows seal. Between 0.12 and 0.31 mm travel of a plunger tuned frequency by 200 MHz and a fiddly cam-driven precision mechanism had to be devised, with adjusters for sweep amplitude and centre frequency. Despite much fun and games in Production Test, it proved reliable in service. Kelvin Hughes supplied similar racons which followed the ASWE design more closely, with thermionic valves.

Successful introduction and development of the racon service owes much to the commitment and long-term support of several national Lighthouse Authorities. At a time when their only experience of electronics was off-the-shelf radio installations, a few marine radars on their tenders and some MF radio beacons of conventional valved design, it required courage to buy, install and guarantee to support complex and relatively untried semiconductor equipment at remote stations like Bell Rock, accessible at half tide in fair weather by ship's boat. At the time there were cheap 'trannie radios' for the kitchen but much military electronics design still used valves. One remembers a chief lighthouse engineer touring the design laboratory: 'Ah, so that's a transistor. Very small – is it reliable?' Mercifully he was never confronted with today's microprocessors.

Advances in solid state microwave technology soon made racons small enough for deployment on buoys, Kelvin Hughes being first in the field, followed by GEC-AEI (predecessor of GEC-Marconi, Chapter 1, Figure 1.6). The shorter ranges of interest were met by receivers of −72 dBW and frequency-multiplier transmitters of −6 dBW. These swept-frequency equipments were superseded by agile models from 1980, developed by Ericsson in Sweden for rapid display refresh rate and including a 3 GHz channel for the first time. Parallel British work initially focussed on a higher performance solid-state landfall device to dissuade tankers from hugging the West coast of Scotland. A superhet receiver gave higher sensitivity (−78 dBW) and a frequency control loop cancelled severe chirp from the 0 dBW Gunn transmitter oscillator. More recent racons have tended towards lower maximum detection range because GNSS now obviates the landfall need, although there is a welcome recent trend back to higher sensitivity. Eventually AIS may take over the task of marking buoys and lighthouses and make racons redundant.

Search and rescue transponders were a Japanese initiative, introduced from the early 1980s following their earlier work on ramarks and racons.

Active reflectors arrived in the late 1980s as microwave solid state amplifiers matured. They have been used mainly in fairly small sizes at 9 GHz to mark yachts as an improvement on bulky passive reflectors. Their use may spread to small ships as more powerful models are developed. Lighthouse authorities are beginning to put them on the smaller buoys in place of expensive racons. Figure 8.1 shows one being fitted to a buoy for trial as an AtoN beacon.

8.1.3 Features of active devices

Advantages over passive devices include the following.

1. Response power (the reply leaving the device) is not limited to (incoming flux density) × area as with passive devices, breaking free from the

Figure 8.1 *Trialling an RTE. Trinity House engineer fitting a radar target enhancer to a buoy for evaluation. Buoys are visually identified by name, shape, topmark, colour and light rhythm, here Prince Consort, North Cardinal Mark, painted yellow and exhibiting a white Q or VQ flashing light. RTEs provide no identification bur racon responses are Morse-coded. Reproduced by permission of Trinity House Lighthouse Service*

size/wavelength/RCS relationship inherent to passive reflectors and giving high effective RCS in small size. This advantage is often decisive.

2. Target pattern map (TPM) is set by the antenna design and can be made excellent.
3. Devices can be arranged to report presence of interrogations or can analyse radar traffic.
4. Response pulselength, polarisation and frequency need not necessarily equal those of the interrogation, opening the door to user-selectable responses.

There are disadvantages.

1. When the interrogation is so strong that insufficient response power is available to deliver a response pulse of the expected or 'prospective' power, active devices can saturate, restricting response power at short range, which may cause problems in clutter.
2. Devices may overload, more interrogations being received than the transmitter will handle, reducing response quality or temporarily interrupting the service, see next section.
3. Devices are often restricted to a single band for commercial reasons.
4. Service is completely reliant on the battery or other power source.
5. Active devices are generally more expensive than equivalent passive devices.
6. Inappropriate polarisation can preclude reception by the radar.
7. The devices may not respond at all to interrogations just outside the marine bands, including certain VTS radars. The response power may vary somewhat within the band, dependent on the saturated power/frequency characteristic of the transmitter element (racons, SARTs and saturated RTEs), or the gain/frequency characteristic of the amplifier (unsaturated RTEs).
8. The device may generate harmonics of the microwave frequency, especially when the response transmitter is saturated (racons, SARTs and saturated RTEs). Active device harmonic EIRP is likely to be less than −10 dBW, much lower than for a radar, and not very troublesome to other spectrum users.

Active device antennas have to operate through large solid angles (360° × ~25°) and so have much lower gain than radar scanners (~6 dB instead of ~30 dB). But because the microwave transmitter has to work over only a one-way path, response

(a) (b)

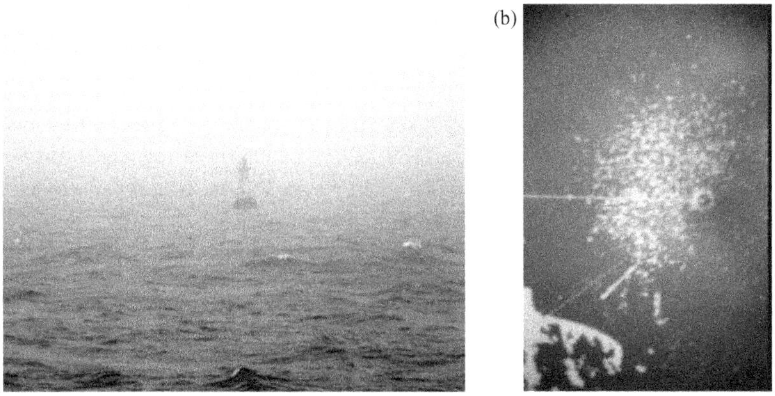

Figure 8.2 Racon in buoy service. Abertay Buoy, Estuary of River Tay, East coast of Scotland. Both: Author, reproduced by permission of the Northern Lighthouse Board. (a) Rough service. The motion was so strong on this foggy August day that a planned maintenance lift had to be postponed. (b) Racon response on PPI. Racon response, bearing 210° cutting through sea clutter, racon range ~1.5 nmi. Swept frequency 9 GHz racon, code T

power can be quite small, of the order of a watt (0 dBW). Likewise receiver sensitivity can also be modest, around −60 to −80 dBW, while radars need ~−120 dBW for two-way detection of distant passive targets.

Active devices contain relatively few components and are easily and thoroughly tested against externally calibrated test equipment in the supplier's factory before delivery. Racons in particular are made in small quantities and suppliers are usually happy for customers to witness final test, seen not as an imposition but an opportunity to foster good relations. Mean time between failures is usually several years unless mechanically damaged, a particular hazard on some mid-channel buoys. Units are sealed and have to be taken to a workshop for servicing, perhaps at three year intervals, so are at minimal risk of ad hoc technical meddling. Racons and some SARTs and RTEs include self-test facilities, although antennas are often excluded. Buoy racon service conditions are arduous, with frequent breaking seas and considerable shock loading when the anchor cable snubs, see Figure 8.2(a). At least one canny lighthouse authority equips buoys with its oldest racons, well written down in value! Racons run continuously but SARTs are at the other extreme, only energised for routine serviceability checks and in emergency. Unlike older valve equipments, such widely differing operating conditions have little impact on performance.

8.1.4 Overload

Unlike passive reflectors, active targets are sometimes unable to respond to all valid interrogations, if:

- two interrogations are received simultaneously;
- a second interrogation arrives during a previous response, or during the subsequent internal circuit recovery time or blocking period of up to 100 μs;

- too many interrogations are received per second, exceeding the rated duty of the transmitter; some are then ignored to protect against overheating. Duty is response pulse 'on' time per unit time, averaged over say 1 s.

Overload rarely degrades responses from correctly designed active targets to a noticeable degree. For example, 100 radars each triggering eight single 25 μs response pulses from a racon every 2.5 s represents 0.8 per cent duty, which is about one-tenth of rated duty.

8.1.5 Interference

When an aid is interrogated by another's radar, the response may be picked up by our radar, with an unwanted trace on our display. Such interference is rarely troublesome because of the following.

1. Most responses to other radars will be outside our receiver's passband.
2. We can only receive interference when both scanners bear on the aid. With beamwidths of say 1°, on an average, both scanners bear simultaneously only on one scan in 360, say four times per hour.
3. Other radar prf will usually differ from ours and each response for it will occur at a different quasi-random range on our display. Those that are not off-screen will appear scattered amongst several range cells, will not be accepted by sweep–sweep or scan–scan correlation circuits and will also tend to be subconsciously ignored by the operator on raw-radar displays.

The equivalent isotropic radiated power (EIRP) and receiver sensitivity of all active devices are so low that mutual device-to-device triggering interference is extremely unlikely.

8.1.6 Response law; effective RCS

Echo strength from all passive devices is proportional to interrogation strength, and in free space varies as $(\text{range})^{-4}$. However, although interrogation strength at active devices of course increases as range closes, the response may not rise proportionally, but may be generated at constant power. The signal received at the radar is then only subject to the inverse square law, varying as $(\text{range})^{-2}$, equivalent to reception from a device whose effective RCS $\propto (\text{range})^{-2}$, so a racon does not have a fixed value of equivalent RCS. Because radars are primarily designed for passive targets, active device detectability may fall at short range.

8.1.7 Specifications and legal requirements

IALA[2] Recommendation R-101r1, December 2000, 'IALA Recommendation on marine radar beacons (racons),' defines a racon as 'A transmitter receiver device

[2] ITU Radio Regulation 4.40.

associated with a fixed navigation mark which, when triggered by a radar, automatically returns a distinctive signal which can appear on the display of the triggering radar, providing range, bearing and identification information'. Racons are covered by IMO's Performance Specification for radar beacons and transponders, A.615(15) superseding A.423(XII). ITU-R Recommendation[3] M824-2 applies. IALA Recommendation R-101r1 expands on these and includes an Annex, 'IALA Guidelines on Racon Range Performance'. The only current RTE specification[4] calls for RCS to exceed 8.2 m^2 (9.14 dB m^2). IMO[5] gives guidance on SART operation. SOLAS Regulation V/20, requiring vessels of constructional cross section less than 150 gt to carry a radar reflector includes the usual clause allowing not-inferior equivalent systems and equipment, opening the door to RTEs. ITU-R M.824-2 Annex 1 recommends that the response duration should approximate 20 per cent of the maximum range requirement of the particular racon. Because radar receiver bandwidth may be as low as 1 MHz, short dashes should not be less than 1 μs.

8.1.8 Structure of chapter

First it is necessary to discuss antenna polarisation compatibility, which affects all active devices. We follow this with a description of swept frequency and agile racons, including block diagram and full performance equations, plus a briefer discussion of special forms of racon, going on to SARTs, which share many racon characteristics. Brief reference is then made to ramarks. RTEs are discussed in detail. The factors in play when an active device is tilted are then analysed. We end by considering the resultant signal from an active device combined with a passive reflector. Probability of detection of active devices is deferred to Chapter 12, Section 12.9. Terminology follows Chapter 7, Section 7.1.3.

8.1.9 Polarisation compatibility

Almost all passive targets reflect incident rays without significant change of polarisation, the chief exception being raindrops and reflectors having an odd number of bounces (e.g. octahedrals), which change the hand of circular polarised (CP) interrogations. The property is exploited by some CP VTS scanners which substantially attenuate rain clutter by switching-in a circular polariser when necessary. No radars are believed to be deliberately slant polarised, but slant is introduced by pitching or heeling.

At 9 GHz, racons and SARTS are required to operate with horizontal polarisation (HP), and ships' radars in this band are required to include a HP mode for compatibility. At 3 GHz, ships' radars are at liberty to use any polarisation (although most or all use HP), and racons in this band ought to be compatible with transmissions which are HP, vertically polarised (VP) and CP (either hand of rotation); there are no 3 GHz SARTs.

[3] ITU-R Recommendation M.824-2, Technical Parameters for Radar Beacons (Racons).

[4] ITU-R Recommendation M.1176.

[5] IMO SN/Circ.161 as revised. 'Operation of Marine Radar for SART Detection.'

RTEs are at the moment confined to 9 GHz and tend to use plus and minus 45° slant linear polarisations for receive and respond to minimise feedback from output to input antenna elements.

Considering the response antenna, the vector radiated at 45° slant can be resolved into equal horizontal and vertical orthogonal components, each $1/\sqrt{2}$ voltage and $\frac{1}{2}$ power as explained in Chapter 2, Section 2.7.1, and shown in Figure 8.3. A CP signal alternates between horizontal and vertical planes, again with half the power in each, so linearly polarised racons, RTEs and SARTs are detectable by VP and HP linearly polarised marine and VTS scanners and by CP VTS radars. As scanners and active device antennas are reciprocal devices, operation on the interrogate leg is identical. Table 8.1 summarises how well polarisations are accepted by radar scanners and by active devices. Only wrong-hand CP and wrong-slant linear polarisations are completely rejected. Some other combinations suffer 3 dB polarisation loss, often absorbed within device rated performance as declared in the data sheet.

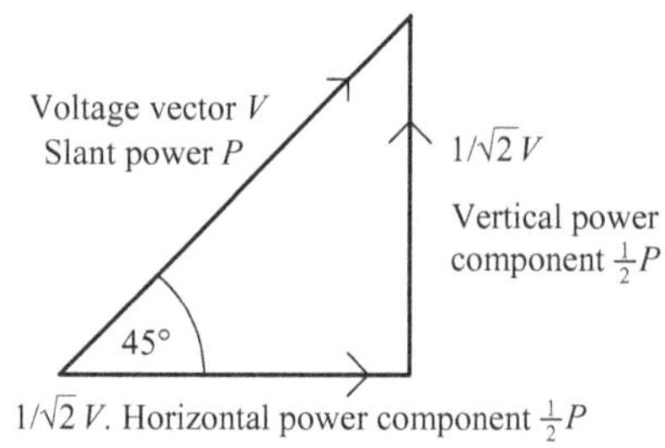

Figure 8.3 Orthogonal components of slant polarisation

Table 8.1 Polarisation compatibility

Incident signal polarisation	Antenna polarisation					
	HP	VP	LH CP	RH CP	Left slant	Right slant
HP	1	0	$\frac{1}{2}$	$\frac{1}{2}$	$\frac{1}{2}$	$\frac{1}{2}$
VP	0	1	$\frac{1}{2}$	$\frac{1}{2}$	$\frac{1}{2}$	$\frac{1}{2}$
LH CP	$\frac{1}{2}$	$\frac{1}{2}$	1	0	$\frac{1}{2}$	$\frac{1}{2}$
RH CP	$\frac{1}{2}$	$\frac{1}{2}$	0	1	$\frac{1}{2}$	$\frac{1}{2}$
Left slant	$\frac{1}{2}$	$\frac{1}{2}$	$\frac{1}{2}$	$\frac{1}{2}$	1	0
Right slant	$\frac{1}{2}$	$\frac{1}{2}$	$\frac{1}{2}$	$\frac{1}{2}$	0	1

1 = 0 dB loss; $\frac{1}{2}$ = 3 dB loss; 0 = infinity loss.

8.2 Description of conventional racons

8.2.1 Function

Racons are by international agreement reserved for the aids to navigation (A to N) services operated by Lighthouse and Harbour Authorities to mark buoys, lightvessels, lighthouses, and other points of navigational significance. It is likely that racons will be partially or totally superseded by AIS radio technology by perhaps the year 2015. Meanwhile they remain an important navigational aid. Before advent of GNSS they often assisted ships to make landfall. Their current tasks are to:

- make lighthouses and other navigational features distinctive;
- distinguish and identify an echo as being a navigation aid amid other echoes, for example, a buoy in a busy fairway;
- provide emergency marking of recent wrecks;
- mark channels under bridges;
- miscellaneous duties, such as marking difficult tows. The devices are then properly called transponders, even though perhaps electrically identical to racons.

Racon design has to reconcile conflicting requirements of small size, high reliability, robustness, wide temperature range, low power consumption and wide-band operation – the author was not alone among racon engineers in handsomely overspending the design budget. Some operating authorities call for 50‘*g*’ 11 ms shock tests and it is distinctly interesting to see one's prototype take half a dozen drops of nearly a metre, pick it up, see if it rattles and try it on a radar. Twin-band agile racons are relatively expensive, around £25 000. Some quartermasters take the command ‘steer on the buoy’ all too literally and collision damage can be severe, sometimes influencing the choice between a racon and a passive reflector. Racons are produced by firms in the AtoN industry whose primary business is lights, buoys and sound signals rather than radar manufacture. IALA plays an important role by ensuring compatibility with marine and VTS radar systems, both informally and within IMO. In the 1970s, the author was invited to outline his firm's current racon developments to a conference hosted by IALA. One recent proud development, a step-sweep racon (Section 8.6.1) produced at the request of a major lighthouse authority, drew the immediate comment from a leading radar manufacturer's engineer ‘It won't work with our latest radars with scan to scan correlation!’ Red faces, but luckily it proved possible to modify the design before anything ran aground.

Racons contain a receiver to trigger and synchronise commencement of the response transmission with the interrogation pulse. Responses are longer than radar pulses and are coded for station identification by modulating the transmitter on/off, giving a Morse-code radial line stretching outwards from the range and bearing of the platform, as shown in Chapter 3, Section 3.11, Figure 3.19, and Figure 8.2(b). Codes are shown on charts and listed in the Admiralty List of Radio Signals, yachting almanacs, etc. Table 8.2 details the Morse and SART codes. Figures 8.4, 8.6 and 8.7 show how the devices respond to radar interrogation pulses.

Table 8.2 Racon (Morse) and SART codes

Letter	Racon code	Letter	Racon code	Letter	Racon code
A	– ——	J	– —— —— ——	S	– – –
B	—— – – –	**K**	—— – ——	**T**	——
C	—— – —— –	L	– —— – –	U	– – ——
*__D__	—— – –	**M**	—— ——	V	– – – ——
E	–	**N**	—— –	W	– —— ——
F	– – —— –	**O**	—— —— ——	**X**	—— – – ——
G	—— —— –	P	– —— —— –	**Y**	—— – —— ——
H	– – – –	**Q**	—— —— – ——	**Z**	—— —— – –
I	– –	R	– —— –		

*__D__ denotes Recent Danger.
Racon codes should commence with a long (bold type).
Long = 3 × short. Space = short. Inter-letter space = 3 × short.
Letter pairs may be used, e.g. GT —— —— – ——.
SART . . . . (12 dots).

To suit the radar displays, racon pulselengths are scaled at half the speed of light, 6.67 μs response pulselength giving 1 km trace equivalent. Some agile racons adapt response pulselength to interrogation pulse length or strength in an attempt to maintain approximately constant displayed visual length, say 25 mm, as the display range scale is changed. Near change points, response pulse length may jitter, making the code difficult to read and perhaps misleading the operator into thinking the response to be some form of interference rather than a signal.

8.2.2 Swept frequency and agile types

Conventional racons fall into two distinct types, sharing many features.

1. Swept frequency racons make no attempt to measure the frequency of an interrogation, apart from a simple microwave bandpass filter to reject those well outside the band. Each in-band interrogation triggers a response. The response frequency sweeps autonomously through the band at a slow rate ($\sim$1.5 MHz s^{-1}), irrespective of interrogations. All radars within range are served in turn as the sweep traverses their operating frequencies. Although many responses are wasted, swept frequency racons, although obsolescent, require minimum circuitry and are relatively inexpensive. Some early models (Kelvin Hughes) included band-edge cavity resonators to control sweep limits, others (GEC-Marconi) relied on constancy of preset sweep voltage applied to a varactor (variable-capacitance) diode in the tuned circuit of an FET (field-effect transistor) or Gunn (transit-time negative resistance diode) oscillator.
2. Frequency agile racons (alternatively known as Interrogation Frequency Response racons) contain additional circuits to measure the frequency of each

interrogation pulse as received and quickly tune the response frequency to accord. They can quasi-simultaneously serve all radars within range. Although more complex, they have almost completely displaced swept racons. Frequency accuracy is achieved by use of a common frequency-to-voltage high speed flash convertor to sample interrogation and response. Measurement accuracy, and hence response frequency error, requires good signal to noise ratio at the receiver, easily achieved at the relatively low receiver sensitivity needed to obtain the required range.

All racon responses are generally several microseconds long, equivalent to a kilometre or more on the display. Thanks to the brain's pattern recognition ability, even weak racons are easily perceived by raw radars in clutter. If displayed on every scan, the response might overlie and mask targets lying at somewhat longer range. This danger is obviated by displaying for approximately 15 s (IALA Recommendation R101r1 Section 2.0.3 minimum recommended time) each minute or so, sufficient for the operator to spot and assimilate the response, the racon being muted and the screen clearing at other times.

In the 9 GHz band, racons are only required to cover 9300–9500 MHz and not the 9200–9300 MHz extension.

Racons generally comprise a metallic base containing the circuit modules, surmounted by a cylindrical radome containing the antenna elements, coned to deter birds from perching and worse. The assembly is watertight, under a metre high, and weighs around 12 kg. Low voltage d.c. power is often drawn from a battery, usually charged by solar cells. Great care is taken to reduce consumption to a watt or less, depending on trigger rate.

8.2.3 Traffic capacity

To respond effectively to all interrogations within the scanner beam, the racon must be:

- not muted;
- at radar frequency;
- within the scanner beam;
- synchronised with radar transmission.

In a busy sea area, a racon may have perhaps as many as 100 radars within its operating range. Each radar engages it for only say eight responses of 25 μs plus 50 μs recovery on each 2.5 s scan, or 0.002 per cent of the time, so on average 100 radars raise occupancy to 2.4 per cent and transmitter duty to 0.8 per cent, well within the rating of typically 10 per cent. Of course, on some scans interrogations from two radars overlap and responses are lost, but almost always sufficient remain from the packet to maintain detectability even when the radar employs sweep–sweep correlation. On rare occasions nearly a whole scan's responses may be lost, but the next two successive scans will usually maintain the service, perhaps with a temporary deterioration of visual appearance. Racons can therefore serve many radars without significant loss of quality.

8.2.4 *Interference*

In the early days fears were expressed that 'undesirable proliferation of racons' would overload radar displays with too much data, but one does not hear of practical problems with this so-called congestion interference, partly because racons are too expensive to be deployed in excessive quantity and partly because analysis indicates interference is unlikely to occur. When serving each of say 100 radars with eight responses per 2.5 s, the racon averages 320 responses per second. These rarely cause noticeable interference on any particular radar. Its scanner with azimuth beamwidth of say 1° would pick up on average $(320/360) \times 2.5 \sim 2.2$ unwanted responses per scan, spread randomly through the pulse repetition interval. Usually these interference responses will be rejected by the radar's pulse to pulse correlation circuit. Even if displayed, the operator would see weak single responses at random range and at racon bearing, which the brain subconsciously rejects.

8.2.5 *Detection at the racon*

Interrogations received at a (usually omnidirectional) fixed antenna feed a receiver via an overload protection device. A few racons on fixed stations use a directional antenna to serve a restricted azimuth arc. As these antennas have to serve smaller solid angles, their gain can be increased, benefiting EIRP and hence detectability in clutter and maximum detection range, since a common receive and transmit antenna is used.

The detection process differs from that within the radar receiver, where a whole packet of echoes can be integrated. The racon detection decision has to be made instantly for each single interrogation pulse, since successive interrogations are likely to be from different radars. Each signal or unwanted receiver noise pulse whose strength exceeds minimum detectable signal (MDS) crosses the detection threshold to trigger a full-power Morse coded pulse group, unless suppressed by a sidelobe suppression circuit (Section 8.2.9). The threshold is set high enough to reject almost all random noise spikes. The question of clutter does not arise on this leg. A preset receiver attenuator or threshold adjustment is included, to permit sensitivity to be reduced on stations not requiring long range, for example, in harbours. Low sensitivity reduces triggering by unwanted scanner sidelobes at wrong bearings. High sensitivity (−80 dBW, available with superhet receivers) is particularly desirable when low-power small-craft radars must receive responses at long range. The detection process was described in Chapter 3, Section 3.5.3.

8.2.6 *Swept frequency racon response*

No technical reason precludes 3 GHz or twin band swept frequency racons, but it is believed that only 9 GHz models have been produced. Slow sweep racons do not measure incoming interrogation frequency, but manage response frequency to give service to all ships' radars. Figure 8.4 depicts the signal flow when a radar interrogates the racon. (a) Represents the packets of interrogations as the scanner sweeps through

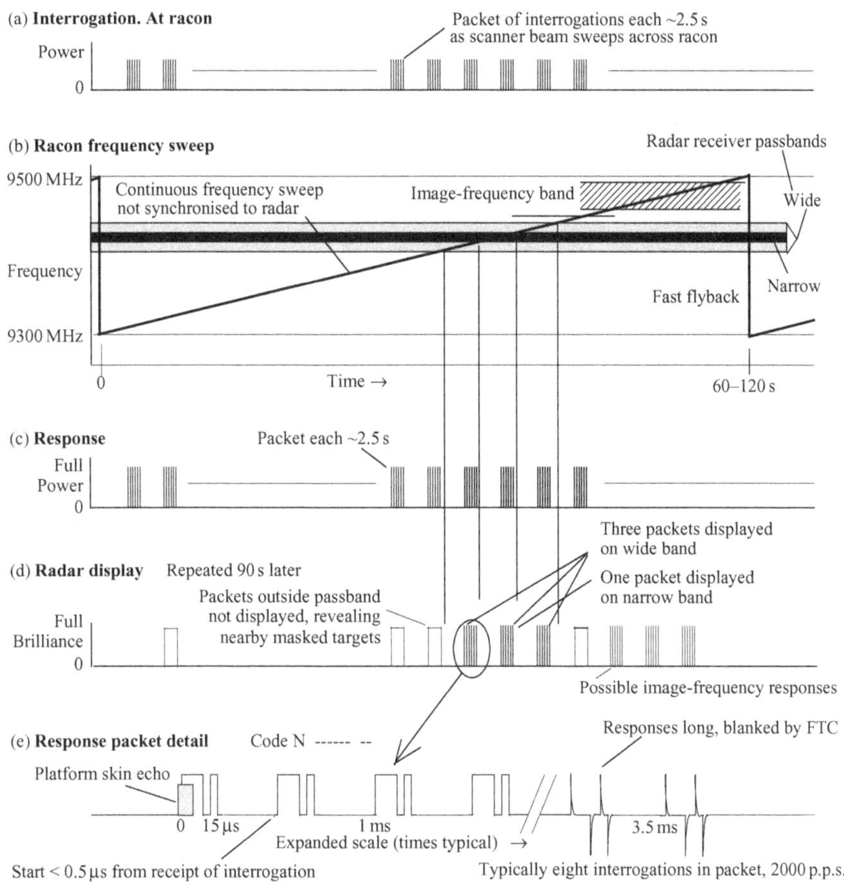

Figure 8.4 *Swept frequency racon timing. Responses are displayed only when the racon transmitter frequency lies within the radar passband. More packets are displayed when the radar is set to short pulse, where passband is wider*

station bearing, the racon receiving a packet of pulses each 2.5 s (24 rpm scanner), with several interrogations within the beamwidth. (b) Shows the response transmitter frequency. Frequency rises at around 2.2 MHz s^{-1}, independent of any interrogations, from the lower to higher band edge in typically 90 s (IALA recommendation R101-r1 limits are 60–120 s), followed by rapid flyback. On each sweep the racon frequency sooner or later traverses the radar receiver passband, which may be wider or narrower according to the radar pulselength selected by the operator.

(c) Shows the responses, one per interrogation whether or not the response frequency coincides with radar frequency. Each is followed by a dead time of $\sim$50 μs during which the transmitter recovers and no interrogations are accepted from any radar. (d) Indicates which responses are within the receiver passband and are displayed. With wide bandwidth typically three successive scans are displayed and with narrow bandwidth one only. Depending on the details of radar receiver design, packets may also be displayed intermediately at image frequency, a few tens of seconds before or after the main responses, perhaps at low sensitivity. Responses are not displayed at other times, clearing the screen to reveal nearby echoes without permanent masking. Navigators find indication of racon position once per 90 s to be generally sufficient, a minimum of 45 s being used in harbours and other close-range stations and up to 120 s in open waters. The racon runs continuously, without additional muting.

Enlargement (e) shows the response in more detail. The racon circuits introduce a response delay, during which the structure or skin echo of the racon station (perhaps a lighthouse) may paint. If the response started instantaneously when triggered, it would display at station range. Some delay is inevitable from the finite bandwidth of the racon receiver, but is kept below 0.5 μs (range error <75 m, IALA Recommendation R101-r1 maximum 0.7 μs) to minimise error on the display. Filter delay tends to diminish at high signal level, reducing range error by a few tens of metres. There is, or was, a polite fiction that the delay revealed the station structural or skin echo, which the navigator would always use for positioning rather than placing unseamanlike reliance upon an electronic device not under personal control.

The radar differentiator or FTC, if engaged, removes the bulk of the relatively long response from the display, as indicated at (e) for a couple of responses. Dots may remain at pulse edges.

Traces run outwards from just beyond racon station true position to form a radial line on racon bearing, broken into the Morse code identification. There may be a small overlap with the later part of the station echo, within which the resultant signal may fluctuate pulse to pulse as the components beat at response error frequency.

Swept racons use omnidirectional antennas typically made from a short length of reduced-height waveguide, 23 × 1.6 mm internally, radiating through three or four vertical slots on both broad faces. The patterns from each slot set join up to give 360° cover with only a couple of decibels peak–peak ripple. Taller variants with more slots increase the elevation aperture to give less elevation beamwidth for groundfast stations, with proportionately higher gain: for example, three slot pairs ±12°, 5 dB; eight pairs ± 4°, 10 dB. Putting eight slots on one face of a full-height waveguide gives ± 40° azimuth ×±4° elevation, 16 dB gain, for long-range harbour approach stations.

8.2.7 Frequency agile racon response

Most racons are of the frequency agile type, serving both the 3 and 9 GHz bands. Figure 8.5 shows a modern racon specially adapted for the flammable hazardous environment of offshore oil platforms. Although the platform structures give excellent echoes, some navigators, GPS, GLONASS and Loran-C notwithstanding, are tempted to use them as waypoints, coming hazardously close to read the platform name. A racon gives an identification at safe range. Each channel is basically separate but some sub-systems are shared. Agile racons include a frequency-measuring

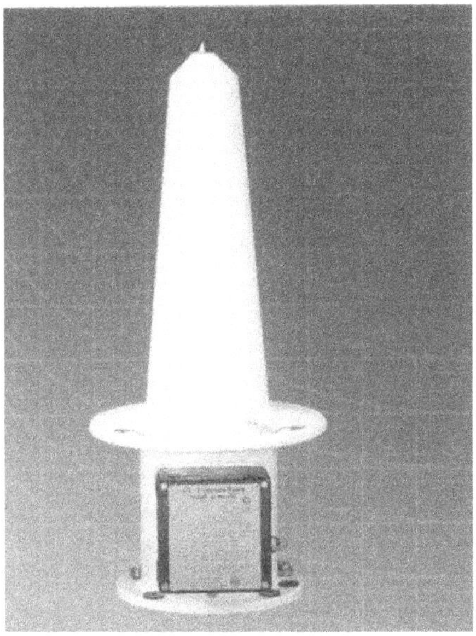

Figure 8.5 *Twin-band agile racon. This variant is for hazardous environments on offshore oil platforms. Tideland SeaBeacon 2 Ex System 6. Height 808 mm. Claimed performance: frequency matching accuracy ±2 MHz, system sensitivity (general marine use) −80 dBW, response power 1 W. Reproduced by permission of Tideland Signal Corp, Houston, Texas*

circuit which automatically tunes the response transmitter to that of the current interrogation pulse. Figure 8.6 shows the response process and assumes 'our' radar (*a*) and another (*b*) lie within range. Their interrogations are of course uncorrelated in timing, frequency or prf. The figure shows both happening to bear on the racon station together, as occurs on average one scan in 360 for scanner beamwidths of 1°, say once per quarter hour. If the interrogation frequencies are at opposite ends of the band, the response may take up to 0.7 μs to retune, increasing the range error to 105 m maximum.

The racon is activated at intervals by a muting timer (c). This prevents permanent masking of nearby echoes (as does the sweep in swept frequency racons) and conserves power during quiescent periods, an important consideration on some buoys. Unless muted, the response transmitter is tuned (d) to the frequency of each interrogation pulse within less than a microsecond. It accords with the frequency of the current interrogation, jumping back and forth when multiple radars' interrogations are interleaved, hence the term 'agile'.

Each non-muted interrogation triggers a response (e) at the appropriate frequency. Own radar displays only responses which it has triggered (f), those for the other radar usually lying outside own radar's passband. If the passbands overlap, the other radar's responses will be accepted. But as the prfs are almost sure to differ, the interfering responses will slide in apparent range from one interrogation to the next and are

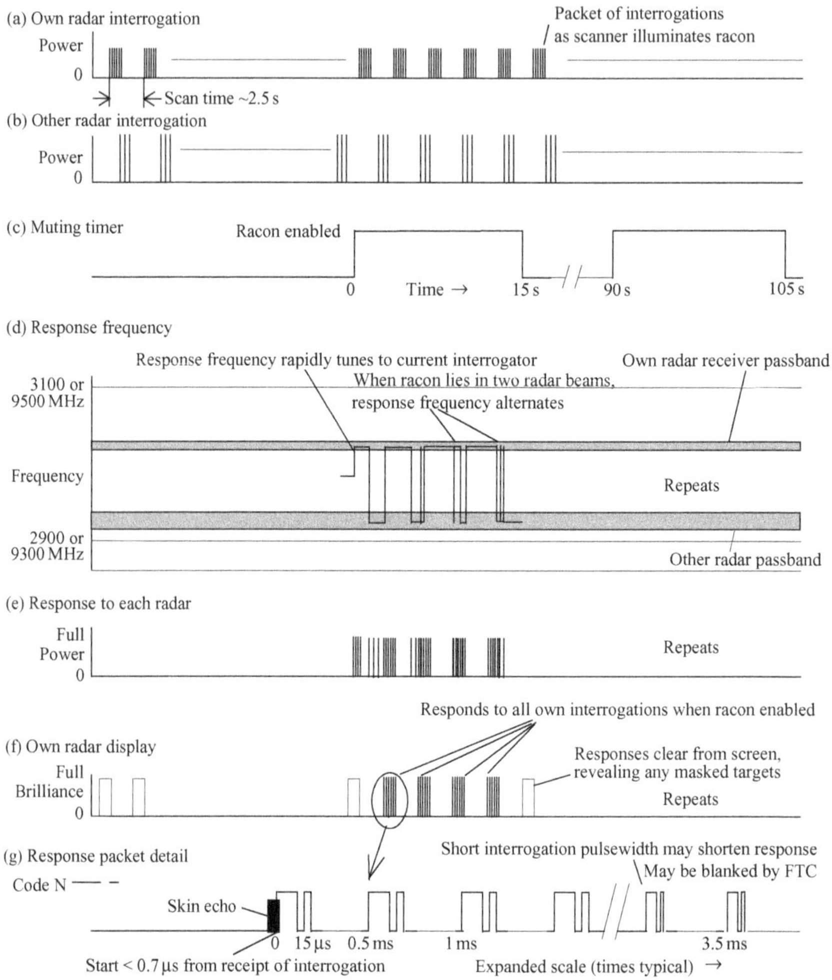

Figure 8.6 Agile racon timing. Shown responding to interrogations from two radars in same band

unlikely to register sufficient hits in any single detection cell to be accepted as valid for display. A recovery period of a few tens of microseconds follows each response. The radar displays (g) as for a swept racon. In an attempt to maintain approximately constant displayed visual length, say 20 mm, as the display range scale is changed, some agile racons adapt response pulselength through a range of ~4 : 1 as interrogation pulse length or strength changes. Near change points, response pulse length may jitter, making the code difficult to read and perhaps misleading the operator into thinking that the response is some form of interference rather than a racon.

The following sections concentrate on agile racons, mentioning the special features of the swept type where these differ.

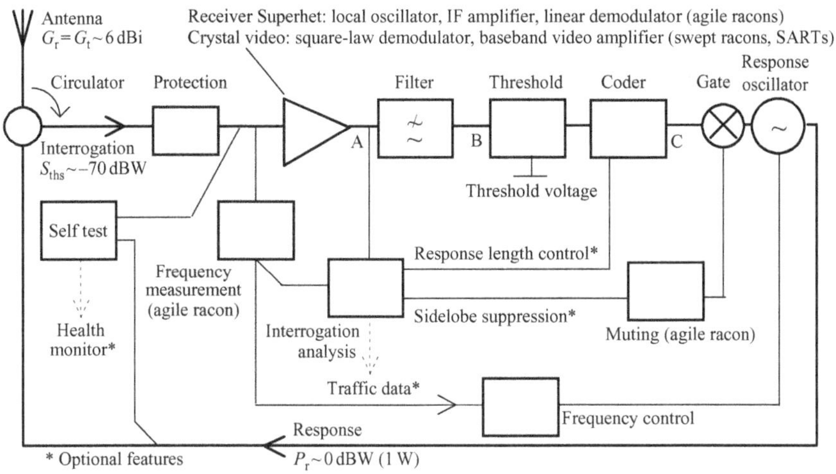

Figure 8.7 *Racon block diagram. The response transmitter fires at full power when interrogated by a radar. Nine gigahertz channel shown. SARTs are generally similar. Many of the functions are realised in digital software. The 3 GHz channel shares several functions*

8.2.8 Functional description

Figure 8.7 is a simplified block diagram of a frequency agile racon, swept types differing principally in their frequency management. The small antenna is usually omnidirectional in azimuth. To permit installation on rolling buoys, elevation beamwidth is fairly wide, so gain is necessarily low. Received interrogations traverse a protection circuit guarding against burn-out at extremely short range. Protection has to be self-powered in case the unpowered racon is swung through the buoy tender radar beam during deployment. Next follows a receiver and detector. There may be a low-pass filter, shown here at video frequency (baseband). The filter reduces sensitivity to short interrogation pulses, reducing futile responses to distant radars using short range scales. Any single interrogation pulse strong enough to cross a threshold triggers a Morse code group.

Within $\sim$0.5 μs, the group fires the transmitter, which then feeds the antenna via a circulator or similar duplexing component. The frequency control loop sweeps response frequency (swept frequency racons) or tunes it to the interrogation (agile racons). Provided the detection threshold is exceeded, full response power is generated, independent of interrogation strength.

8.2.9 Sidelobe suppression

Because of the (range)$^{-2}$ law, racons are rather susceptible to triggering from scanner sidelobes, full-power responses collected by the sidelobes painting at correct range over a broad arc centred on station bearing and masking adjacent targets. The lengthy responses make sidelobes particularly troublesome and distracting, as a considerable area is masked. Sidelobe suppression circuits are therefore often included in agile racons, but suppression is too technically difficult to incorporate in the simpler swept-frequency racons.

Two techniques are available to combat sidelobes: low pass filtering, described in the next section, and sidelobe suppression circuits, described here. Note the problem stems from the scanner sidelobes; the omnidirectional racon antenna cannot have azimuth sidelobes.

The sidelobe suppression circuit recognises when a received interrogation is so strong ($\sim$27 dB above threshold) that the scanner's radiated sidelobes are likely to be above racon detection threshold. Frequency and pulselength of these strong interrogations are stored in digital software. Weak interrogations, less than -27 dB on strong threshold but sharing the same dual radar identification tokens (those having the same radar signature) are deemed to be sidelobes and are suppressed under the control of a microprocessor. A distant radar of the same signature is not suppressed, because its interrogations are always too weak to pass the strong pulse threshold. As with any clever automatic facility, sidelobe suppression may occasionally turn out too clever and suppress low power main-beam interrogations from a distant ship sharing the same signature as a nearby strong radar. It is advisable only to employ the suppression facility when shipping is likely to approach within sidelobe range.

8.2.10 Target pattern map

Racon TPM is that of the antenna, which can be made with uniform azimuth response, ripple being less than ± 1 dB. TPM is therefore generally very smooth and predictable, without 'holes'. Elevation response approximates the classical $\sin x/x$ form of a uniformly illuminated aperture (as Chapter 2, Section 2.8.1, Figure 2.29, but with much lower gain). Nine gigahertz antennas are always horizontally polarised to match scanners. Three gigahertz band scanners are allowed horizontal or vertical polarisation so racon antennas in this band are usually slant or circularly polarised, reducing effective gain to the polarisation used by the scanner by 3 dB. Cylindrical slot antennas are often used, although the author used reduced-height slotted waveguide radiators.

Daymarks, main and reserve lights, racons, telemetry antennas and solar voltaic panels compete for space on buoys and the racon is not always king of the castle. Sometimes a lattice girder mast surrounds the racon, imposing an unquantified non-uniformity, probably several decibels peak–peak, to the TPM.

Several proprietary buoy designs are alleged to remain upright in rough water or a tideway and would permit higher antenna gain, but many lights authorities stick to the tried and tested heavy cylindrical steel buoy, which can absorb considerable punishment.

8.2.11 Low pass filter

To permit determination of interrogation frequency, agile racons use superheterodyne receivers. Superhets have adequately low noise to give high SNR in long-range operation of the one-way interrogate path without needing low noise pre-amplifiers (LNA).

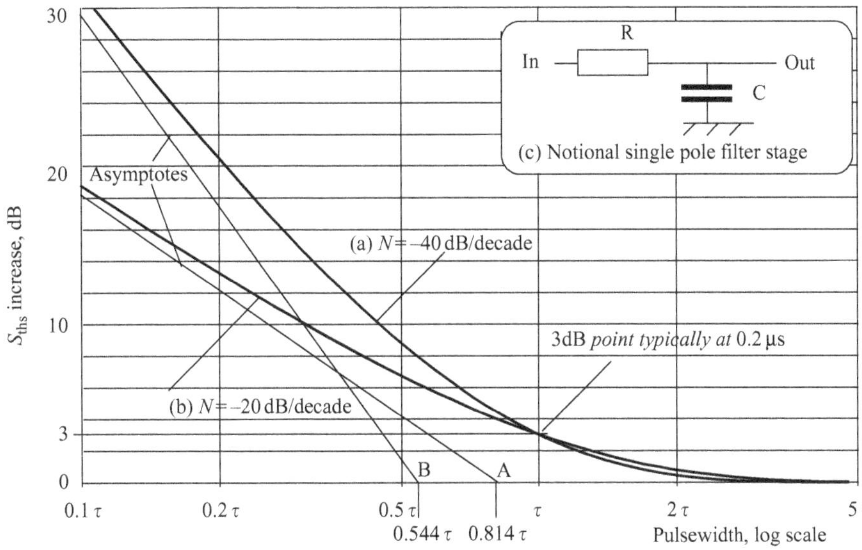

Figure 8.8 Typical racon receiver filters. Threshold rises to short interrogation pulselength, making the device less sensitive

Crystal-video receivers (swept racons) are inherently inefficient and bandwidth is minimised to reduce noise. Excessively narrow bandwidth however introduces response delay, causing the paint to commence too far beyond racon true range, navigationally unsafe. Formerly, radars transmitting short pulses were sure to be operating with the display set to a short range scale, when racon sensitivity could safely be sharply reduced, minimising futile responses which would not be displayed but would merely add to the interference received by other close-range radars. A single R–C low pass single-pole filter stage in the video amplifier sufficed, effective attenuation being squared numerically by the square-law detector characteristic, so threshold rose 40 dB/decade as pulselength fell, Figure 8.8(a).

Nowadays, the short-pulse radar may well also be driving a long-range auxiliary display or be maintaining long-range ARPA tracks, so it is desirable to retain nearly full racon sensitivity. Racon receiver bandwidth is not usually disclosed in suppliers' datasheets. It is probably reasonable to assume that the filter contains the equivalent of a single pole low-pass filter at baseband with the half-power point, $\tau = 1$ (Figure 8.9c), being near 0.2 μs, Figure 8.8(b). In summary, the filter:

- responds to strong pulses of all lengths;
- reduces unwanted responses to distant radars using short range scales and hence short pulses;
- reduces sidelobe triggering from nearby radars using short pulses;
- reduces noise bandwidth, ensuring that false responses are hardly ever triggered, despite the inherently poor noise factor of crystal-video receivers.

8.2.12 *Idling*

Supply power conservation is so important on the smaller buoys that in addition to maximising the quiescent period when the main circuits draw no power, some agile racons include an idling circuit for use in sparse traffic. When there are no interrogations, the quiescent period extends to several times normal. If a radar approaches, it wakes the racon up during one of the occasional active periods and operation reverts to normal. An incoming ship may experience a couple of kilometres loss of initial detection range or when coming out of the first multipath null, but thereafter should receive normal service.

8.2.13 *Self test*

Many agile racons contain a self-test facility, sometimes including remote monitoring and control to permit serviceability to be reported to a central control room. At hourly or other intervals service is interrupted for a quarter minute or so – half a dozen scans – while a test sequence is performed. Mean time between failures is usually several years (~50 000 h) and it is not usual to duplicate modern racons. Despite this reliability there is always a chance that a fault has occurred too recently to be promulgated, that there is no self-test system or that the test has not picked up the problem. For example, an obstruction may have been placed in front of the antenna. At one time it was said to be the practice for the garrison of a certain remote Middle Eastern island, when overdue for revictualling, to put a bucket over the racon. Some passing tanker would report the 'failure', the navigation aids service tender would come out, hand over food, remove the bucket and remind army HQ. Be that as it may, shipping should always promptly report any apparent failure.

8.3 Racon problems

Factors which may degrade racon detectability include the following.

8.3.1 *Effect of swept gain*

The swept gain facility reduces the radar receiver gain at short range, the gain law following approximately an inverse third or fourth power on the assumption that targets are passive echoes following a $(\text{range})^{-4}$ law. But as response strength follows a $(\text{range})^{-2}$ law, racon relative power falls at short range. The detection threshold may exceed the initial portion of racon responses, as sketched in Figure 8.9(a), only the later portion encountering enough gain to be displayed. The operator has to be on guard for the racon range being overstated or the response code mutilated.

8.3.2 *Tuning errors*

Figure 8.9(b) shows the radar receiver bandpass filter response curve. When switched to short pulse (SP), frequencies between x and x' are passed, bandwidth typically exceeding 10 MHz. Racon responses mistuned by at least 5 MHz are accepted without

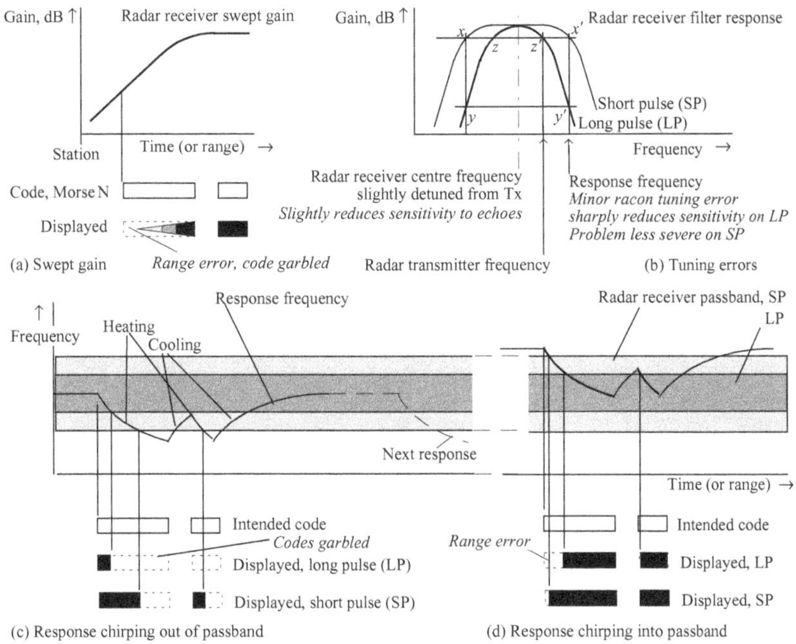

Figure 8.9 Racon coding problems. Swept gain and tuning errors can mutilate the displayed response, introducing range error and/or garbling the code. Racons are also subject to chirp

significant loss. On long pulse (LP), bandwidth is much lower, from z to z', maybe as low as 2 MHz. The racon response must now be within a megahertz of the radar interrogation for full-gain reception. A racon whose frequency was x or x' would suffer severe attenuation, points y or y'. ITU-R Recommendation M.824-2 Annex 1 requires response accuracy to be ± 3.5 MHz when pulselength $<0.2\ \mu$s and ± 1.5 MHz otherwise.

8.3.3 Chirp

The tiny semiconductor chip of the Gunn diode or field effect transistor in the response transmitter has low thermal inertia with a time-constant near 5 μs. During the response it generates ~1 W microwave out, with efficiency ~10 per cent, ~9 W going to heat the chip by maybe 100°C, with some cooling between pulses. This heating tends to pull the response frequency as the pulse proceeds, as shown in Figures 8.9(c) and (d). If uncontrolled, this frequency change or chirp (from the bird-like tones of early wireless telegraphy transmissions suffering a similar problem) throws part of the response out of the radar receiver passband, mutilating the display. If tuning was initially correct, the later part of the response gets lost, mutilating the code, Figure 8.9(c), but if the start is lost (Figure 8.9(d)), dangerous range error results, the station appearing too distant.

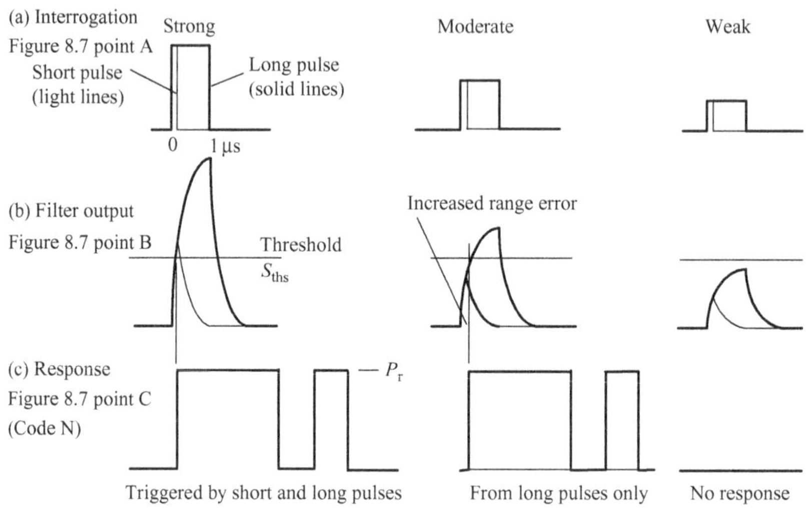

Figure 8.10 Racon waveforms. Interrogation pulse length affects sensitivity when a low-pass filter is used

To accommodate narrow passbands, chirp should preferably remain below 1 MHz. Agile racons control chirp by sampling the response as it is generated, feeding back a correction via the frequency measurement loop. Swept frequency racons inject a pre-programmed correcting waveform into the tuning voltage. It remains difficult to manufacture microwave oscillators capable of operating over a wide temperature range with enough power, tuneable over 200 MHz with low chirp, and this is a main reason for the high cost of racons.

Figure 8.10 shows waveforms at three points of Figure 8.7, for interrogations of three strengths and two pulselengths. When a pulse is applied to the filter, it slows the build-up of output voltage. Figure 8.10(a) shows that both short and long strong pulses cross the threshold; at (b) long pulses of moderate amplitude build up far enough to cross the threshold but short pulses cease before threshold voltage is reached and do not trigger; (c) shows that weak pulses of all lengths are rejected. This is an example of a receiver whose bandwidth is not matched to signal pulselength. Filtering has little effect on maximum detectable range, which in practice is so long that the radar probably will be using long pulses to which the filter has negligible attenuation. The finite passband introduces range error which luckily falls as interrogation strength increases and so is least at short range.

8.4 Racon performance analysis

8.4.1 Notation

We next analyse performance of racons, other active devices following suit in later sections. The performance equations allow for certain racons which respond at a frequency offset from the interrogation. To facilitate later analysis of RTEs having separate receive and transmit antennas, we allow each antenna to have its own height. Racon parameters are quoted nett of any internal losses. To the notation of the radar range equations of Chapter 4 are added the following.

G_r, G_t = Receive, transmit device antenna gains to signal with same polarisation as radar scanner, dB. Usually $G_r = G_t$.

h_r, h_t = Receive, transmit antenna heights, metres. Usually $h_r = h_t$.

M_1, M_2 = Interrogate, response leg multipath factors, dB. Usually $M_1 = M_2$.

P_r, P_p, P_t = Device rated, prospective and actual transmitted response powers respectively, dBW.

P_{tEIR} = Transmitted response equivalent radiated isotropic power.

$$P_{tEIR} = P_t + G_t \text{ dBW}. \tag{8.1}$$

S_{tgt} = Signal reaching device per one-way radar range equation Chapter 4, Section 4.4, Eqs (4.7a) and (4.7b), dBW.

S_{ths} = Threshold received signal power dBW, below which device does not respond.

S_{thsL} = Threshold sensitivity to long pulse, quoted in data-sheet for interrogation pulselength say 1 μs.

S_r = Response received at radar from active target per Chapter 4, Section 4.5.2, Eqs 4.8(a) and (4.8b), dBW.

S_e = Echo received at radar from passive reflector per Chapter 4, Section 4.3.2, Eqs (4.6a) and (4.6b), dBW.

σ_{sat} = Racon equivalent RCS, dB m^2. (Suffix denotes response power is fixed, σ_{sat} being used for commonality with saturated RTEs.)

$G, \lambda, L_t, L_r, L_A, P$ and R have their usual meanings.

8.4.2 Interrogation received at racon

Unless the strength of an interrogation pulse, S_{tgt}, from a radar at range R is sufficient to trigger the racon, there is no response. Chapter 4, Section 4.4, Eq. (4.7a) is the one-way radar range equation for the interrogation received at an active target, reproduced here in slightly modified form, the term $\{\cdots\}$ being radar EIRP. For M we have substituted the interrogation leg multipath factor, M_1, in case the racon includes response frequency offset.

$$S_{tgt} = \{P + G - L_t\} + G_r + 20 \log \lambda - 20 \log(4\pi) - 20 \log R - L_A + M_1 \text{ dBW}. \tag{8.2a}$$

The reduced form of Eq. (4.7b) in which F_1 is the radar transmitter figure of merit per Eq. (4.10) becomes

$$S_{tgt} = F_1 + G_r - 20 \log R_{km} - L_A + M \text{ dBW}. \tag{8.2b}$$

Being one-way, the interrogation power of course follows an inverse square law ($-20\log R$) apart from multipath M.

The low-pass receiver filter may be realised in many ways, but performance usually approximates that of one or more simple single-pole resistor-capacitor stages, shown in Figure 8.8(c). For a single stage having time-constant T seconds (e.g. $T = \mathrm{RC}$ in the figure), the input voltage, V_t, at time t to obtain unity output voltage is $V_t = 1 + \exp(-t/T)$. The 3 dB point, at time τ, is when $V_t = \sqrt{2}$ and timeconstant $T = 0.814\tau$. When $t \ll T$, the slope is -20 dB/decade. If there are $N/20$ stages so the slope is $-N$ dB/decade, each stage timeconstant has to be

$$T = -\frac{\tau}{\ln[1 - 2^{(-10/N)}]}\ \mathrm{s}.$$

The threshold is raised to

$$S_{\mathrm{ths}} = S_{\mathrm{thsL}} + N\log\left[1 - \exp\left(-\frac{t}{T}\right)\right]. \tag{8.2c}$$

Figures 8.8(a) and (b) show how S_{ths} rises at short interrogation pulselength, making the racon less sensitive, when $N = 40$ and 20, respectively. Filter stage timeconstants T are shown at A and B.

8.4.3 *Probability of detection*

Because of random receiver noise, probability of detection (P_{D}) rises quickly (within about 3 dB) from near-zero to near-unity when S_{tgt} reaches the racon data sheet threshold value, S_{ths}. The interrogation approximates the ideal non-fluctuating Swerling Case 0, see Chapter 12. Accurate calculation of P_{D} with S_{tgt} is deferred to Chapter 12, Section 12.9. In practice, little error is incurred by taking $P_{\mathrm{D}} = 0$ (no responses) if the detection threshold is not exceeded, rising abruptly to $P_{\mathrm{D}} = 1$ (response from each interrogation) otherwise:

$$\text{if } (S_{\mathrm{tgt}} < S_{\mathrm{ths}}), \qquad P_{\mathrm{t}} = \text{zero } (-\infty\ \mathrm{dB}), \quad \text{otherwise } P_{\mathrm{t}} = P_{\mathrm{r}}. \tag{8.3a}$$

Of course, initiation of a response is no guarantee that the radar receiver will be sufficiently sensitive to detect the racon; each leg has to be calculated separately.

When interrogation strength at a frequency agile racons is poor, receiver noise precludes accurate measurement of interrogation frequency. The response frequency then varies randomly either side of the interrogation and effective response power within the radar receiver passband falls. When τ is the measurement time (approximately the radar pulselength) and q is numerical SNR (defined in Chapter 12, Section 12.2.2), the frequency error δf is given by Meikle [1] as

$$\delta f = \frac{1}{\tau\sqrt{2q}}\ \mathrm{Hz\ rms}. \tag{8.3b}$$

8.4.4 *Response on axis*

Assuming the racon fires, the response one-way range equation, an adaptation of Eq. (4.8a), now gives the signal at the radar, S_{r}. The term $[\cdots]$ is the racon EIRP.

We use the response leg multipath factor M_2:

$$S_r = [P_t + G_t] + G + 20\log\lambda - 20\log(4\pi) - 20\log R - L_r - L_A + M_2 \text{ dBW}. \tag{8.4a}$$

Using the reference equations (Eq. (4.8b)), in which F_2 is the radar receiver figure of merit per Eq. (4.11):

$$S_r = F_2 + [P_t + G_t] - 20\log R_{km} - L_A + M_2 \text{ dBW}. \tag{8.4b}$$

The noise component within the response is negligible. Between responses, no power is applied to the transmitter and it therefore generates no noise.

8.4.5 *Equivalent RCS*

Eq. (8.4a) is similar to the two-way radar range Eq. (4.6a) for the echo of a target whose RCS, σ_{sat}, is as Eq. (8.5a):

$$S_e = P + 2G + 20\log\lambda - 30\log(4\pi) + \sigma_{sat} - 40\log R - L_t - L_r \\ - 2L_A + M_1 + M_2 \text{ dBW}.$$

The equivalent RCS is the quantity which makes the racon response equal an echo from a passive target of that RCS at the same range. $S_r = S_e$ when

$$\sigma_{sat} = [P_r + G_t] - \{P + G - L_t\} + 10\log(4\pi) + 20\log R + L_A - M_1 \text{ dBm}^2 \tag{8.5a}$$

$$= [\text{racon EIRP}] - \{\text{radar EIRP}\} + 10\log(4\pi) + 20\log R \\ - L_A - M_1 \text{ dBm}^2. \tag{8.5b}$$

We could also write Eq. (8.4a) as for a passive target:

$$S_r = P + 2G + 20\log\lambda - 30\log(4\pi) + \sigma_{sat} - 40\log R - L_t - L_r - 2L_A \\ + M_1 + M_2 \text{ dBW}. \tag{8.5c}$$

Eq. (8.5a) shows the effective RCS, σ_{sat}, of a racon is not constant. Beside terms $[\cdots]$ dependent on the racon, there are terms $\{\cdots\}$ dependent on the radar and also multipath and loss terms dependent on the environment. The free-space echo from a passive target follows an R^{-4} law while the racon response law is R^{-2}. It follows that, to the radar, the racon is like a passive target whose RCS follows an R^2 law, equivalent RCS rising with range:

$$\sigma_{sat} \propto 20\log R \text{ dBm}^2. \tag{8.5d}$$

In general, the concept of equivalent RCS is unfruitful in the racon context except perhaps for comparison with passive reflectors or RTEs. Racon RCS may fall to such a low value at short range that detection in clutter is impaired.

8.4.6 Sidelobes

At short range, scanner sidelobes may trigger the racon, triggering responses which may be strong enough to be displayed. Performance can be calculated by inserting the scanner sidelobe gain for G in the racon performance equations. As σ_{sat} depends inversely on G (Eq. (8.5a), G term is negative), racon effective RCS is higher to sidelobes than main beam and short range racons without sidelobe suppression facilities display sidelobes relatively strongly. The sidelobe problem is ameliorated if racon sensitivity reduces to short interrogation pulselength. Many modern racons contain sidelobe suppression systems, Section 8.2.9.

8.4.7 Example

Figure 8.11 diagrammatically compares performance of a typical racon and a passive reflector, working with a typical radar set alternatively to long (LP) and short (SP)

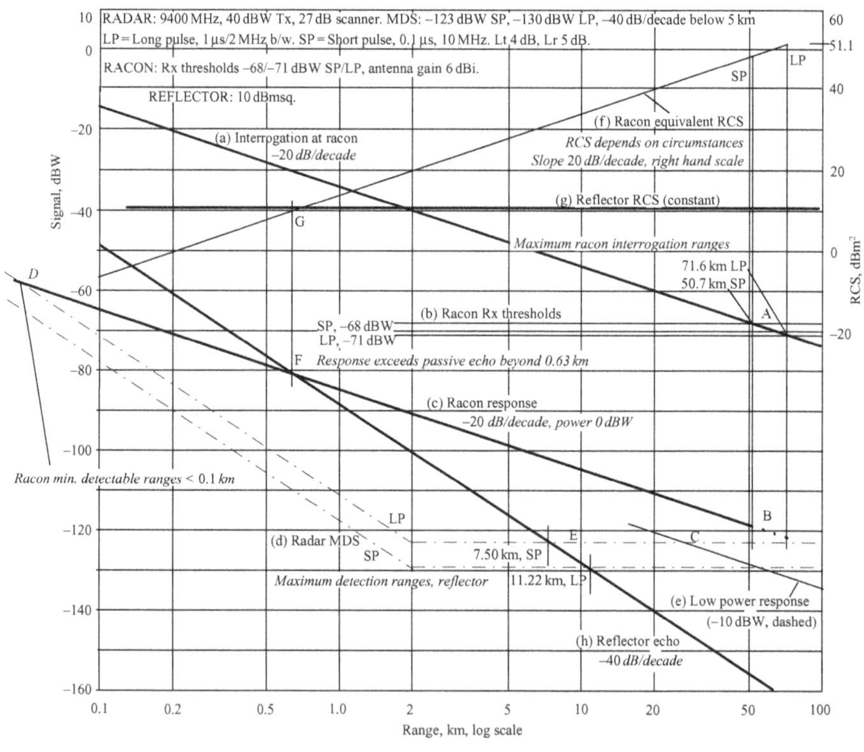

Figure 8.11 Racon and reflector main beam response comparison. Free space, no atmospheric attenuation. Racon superiority falls as range closes. Racon sidelobes are more troublesome than those of passive reflectors of the same maximum detectable range. Increasing racon receiver sensitivity would increase maximum detectable range in the system shown

pulselengths. For simplicity, free space conditions have been assumed; in practice multipath, M, and atmospheric attenuation, L_A, would take their toll of maximum ranges and mask the underlying linearities of the figure. Assumed radar parameters are indicated in the figure; minimum detectable signal falls at 40 dB/decade to 5 km range, beyond which it stabilises.

Curve (a) graphs the interrogation reaching the racon receiver, which intersects the racon threshold (b, −68 dBW SP, −71 dBW LP) at A. Curve (c) shows the prospective response power reaching the radar receiver. Its intercepts at B with the radar receiver MDSs, curves (d), are at longer range, so the interrogation path defines the racon maximum detectable ranges, 50.7 or 71.6 km, respectively. If the racon transmitter power were 10 dB lower, curve (e), the response power would be the dominant factor, with maximum ranges at points C. The racon response exceeds radar MDS down to point D, defining the minimum ranges at which the racon is detectable, <0.1 km. Curve (f) shows how the racon equivalent RCS rises with range.

Curve (g) is the reflector RCS, which of course remains constant at all ranges and curve (h) is the reflector echo, following the usual R^{-4} law. It intersects the radar MDS at point E, giving maximum detectable ranges of 7.5 and 11.2 km, respectively.

The echo and the racon response are equal at 0.63 km, the echo being stronger at shorter ranges. At f the racon equivalent RCS and reflector true RCS are of course equal, point G.

Figure 8.12 applies to the sidelobe performance. Racon sidelobes are detectable from 0.48 or 1.03 km (points J), to 2.02 or 2.85 km, points K, depending on radar pulselength, whereas the passive reflector sidelobe echo (l) is about 25 dB too weak to display sidelobes at any range, because it depends on the scanner sidelobe discrimination two-way, not one-way as for a racon.

The example illustrates how racon detection may be impaired at short range by the sloping RCS characteristic. On the other hand, racons have a long response trace whose outer portion may remain visible when the platform is buried in clutter.

IALA R101r1 Annex A contains a set of guidelines giving families of graphs relating expected maximum range with mounting height for representative radar parameters within the 3 and 9 GHz bands, derived by application of equations similar to Eqs (8.1)–(8.4) to the standard radar range equation. Figures 8.13 and 8.14 reproduce a couple of the IALA graphs. They clearly show the vertical one-way multipath lobe structure of Chapter 6, Figure 6.1, and how it is influenced by wavelength.

8.4.8 Balance between legs

To maximise response to compete with clutter, the racon EIRP should always be as high as possible. EIRP comprises (a) antenna gain, constrained by the need to cover a wide solid angle, and (b) power, which is costly.

If the racon is required to operate to long range, sufficient receiver sensitivity to long pulses should be provided fully to exploit the response transmitter's capabilities with the mix of radars transiting the sea area in question. If ships also come close to the racon, sidelobe suppression should be included or sensitivity should be reduced as in the example.

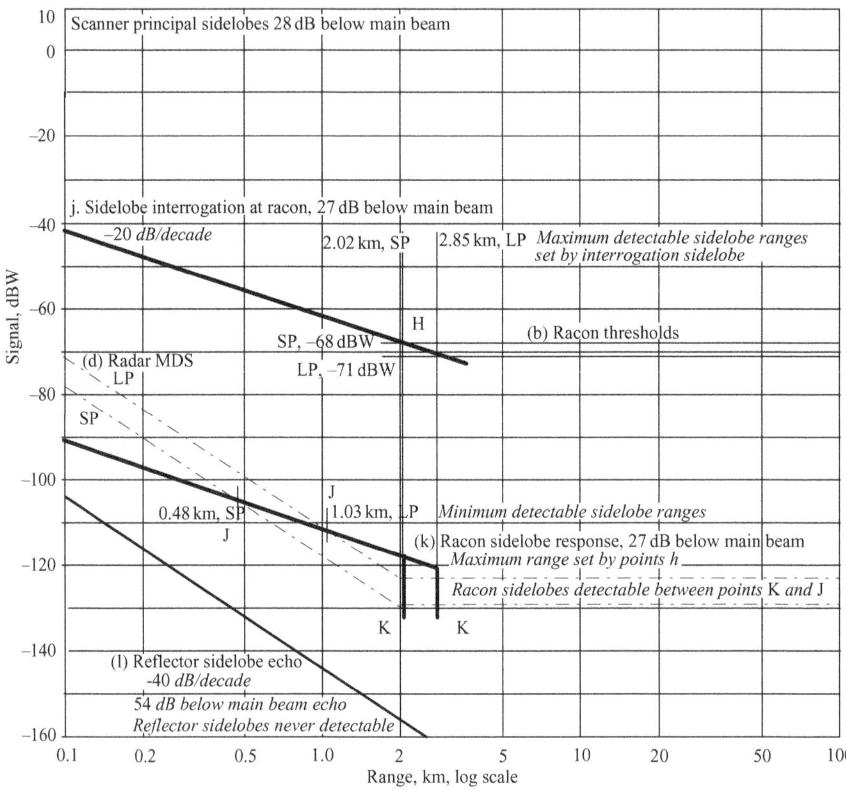

Figure 8.12 Racon and reflector sidelobe response comparison. Conditions as Figure 8.11. Racon sidelobes tend to be more troublesome than those of passive reflectors

Sometimes long range is not required – in harbours, for example. Here it is best to retain maximum response power to break through clutter, reducing receiver sensitivity to the minimum which assures triggering at the maximum range of interest from the minimum EIRP radar. This course minimises generation of sidelobe interference, so special sidelobe suppression circuits may not be required. There is also less triggering by ships beyond maximum range, reducing general interference and power consumption.

Substituting S_{ths} for S_{tgt} in Eq. (8.2a) and MDS for S_r in Eq. (8.4a), rearranging to give expressions for 20 log R and equating these, the range equation takes the form

$$20 \log R = \{P + G - L_t\} + G_r + 20 \log \lambda - 20 \log(4\pi) - S_{ths} - L_A + M_1$$
$$= [P_t + G_t] + G + 20 \log \lambda - 20 \log(4\pi) - \text{MDS} - L_r - L_A + M_2.$$

The interrogation and response legs share the same scanner, racon antenna and environmental factors such as multipath and atmospheric attenuation. If $G_t = G_r$

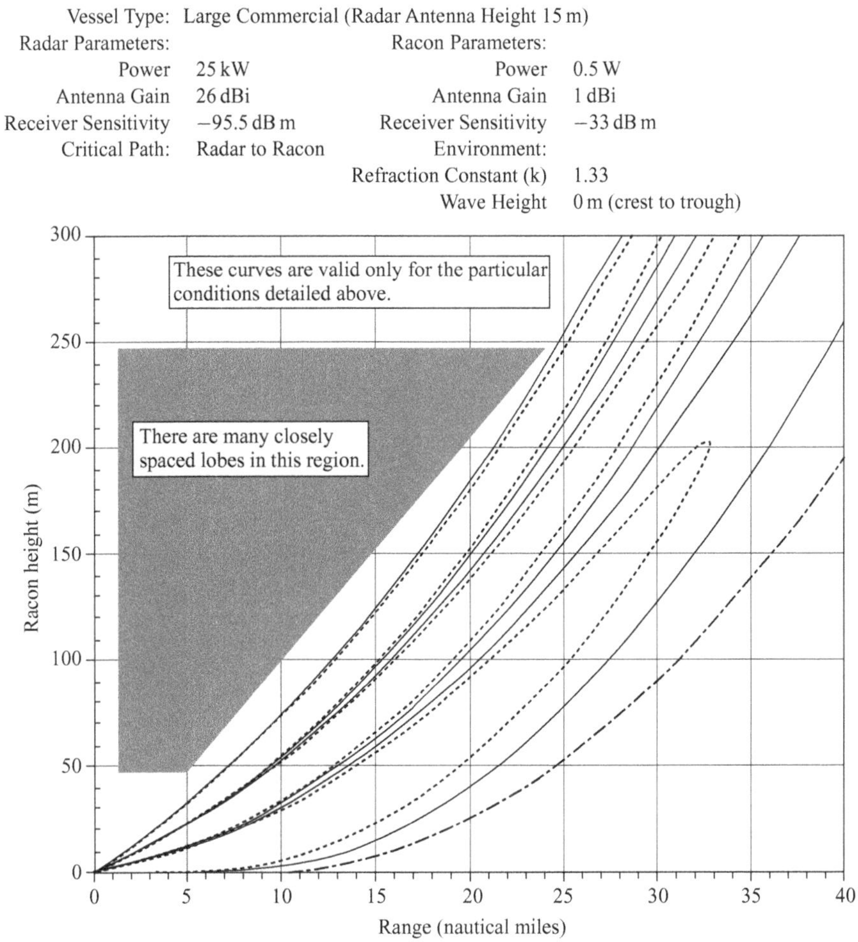

The graph shows the maximum geometric range and the first three lobes caused by multipath effects due to reflection at the sea surface, plotted as a function of racon height above sea level. Interference nulls are also present above the third lobe, but are generally only of smaller range duration. To use the graph, enter with racon height (assumed to be the same as the associated light if not specifically quoted) and read off the maximum detection range and the ranges at which signal loss due to the multipath effects will be experienced. The curves inside the main lobes are calculated for a 6 dB loss of performance and indicate the range reduction if such losses are present

Figure 8.13 Racon range diagram, 3.0 GHz. Large commercial vessel, normal propagation. The chain-dotted line represents maximum geometric range. Reproduced by permission from IALA Guidelines on Racon Range Performance, Figure 8

Vessel Type: Large Commercial (Radar Antenna Height 15 m)

Radar Parameters:		Racon Parameters:	
Power	25 kW	Power	1 W
Antenna Gain	31 dBi	Antenna Gain	4.5 dBi
Receiver Sensitivity	−95.5 dB m	Receiver Sensitivity	−40 dB m
Critical Path:	Radar to Racon	Environment:	
		Refraction Constant (k)	1.33
		Wave Height	0 m (crest to trough)

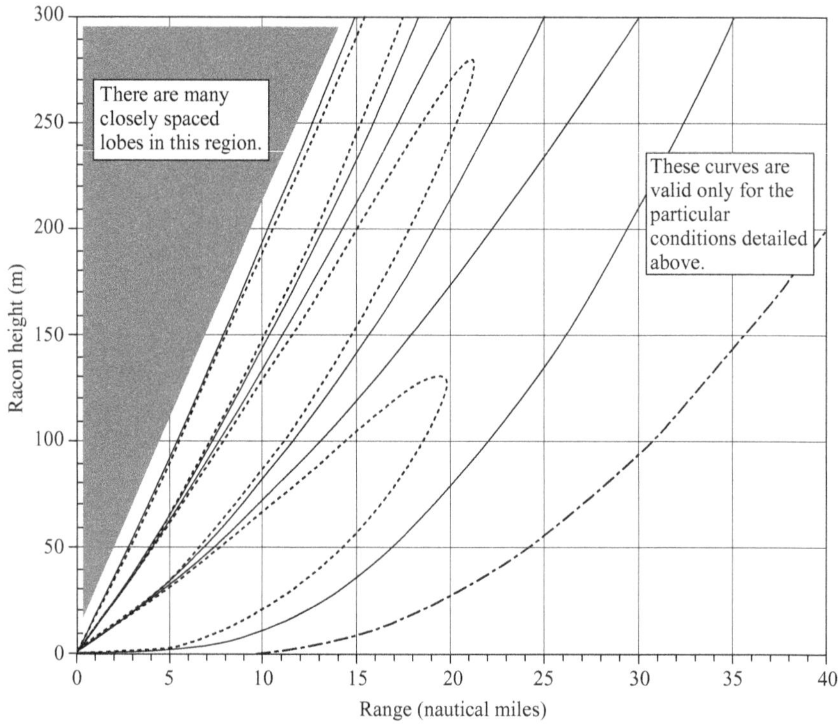

The graph shows the maximum geometric range and the first three lobes caused by multipath effects due to reflection at the sea surface, plotted as a function of racon height above sea level. Interference nulls are also present above the third lobe, but are generally only of smaller range duration. To use the graph, enter with racon height (assumed to be the same as the associated light if not specifically quoted) and read off the maximum detection range and the ranges at which signal loss due to the multipath effects will be experienced. The curves inside the main lobes are calculated for a 6 dB loss of performance and indicate the range reduction if such losses are present

Figure 8.14 Racon range diagram, 9.4 GHz. Large commercial vessel, normal propagation. The chain-dotted line represents maximum geometric range. Reproduced by permission from IALA Guidelines on Racon Range Performance, Figure 6

and $M_1 = M_2$, many terms cancel, leaving a simple expression showing the balance of radar (left-hand side) and racon (right-hand side) parameters necessary for equal leg sensitivities:

$$P + \text{MDS} - L_t + L_r = P_t + S_{ths}. \tag{8.6}$$

An extra decibel of response power, P_t, is usually more expensive to provide than a decibel of receiver sensitivity, S_{ths}, so it makes sense to raise the latter somewhat beyond that predicted from Eq. (8.6) for the highest radar transmitter power, P, likely to be encountered; radar MDS does not vary much from model to model.

8.4.9 Interaction

When two or more racons are operated in proximity, for example, on the deck of a buoy tender before deployment, repeated mutual triggering does not occur should one of them be triggered by receiver noise, another racon or a radar. This is partly because of the combination of low transmitter power, low antenna gain and modest receiver sensitivity; but more particularly because the receiver is purposely inhibited for tens of microseconds after a response transmission. Maximum range at which one racon can trigger another is given by application of a rearrangement of Eq. (8.2a), substituting racon parameters for the 'radar' transmitter and scanner:

$$20 \log R = \{P_t + G_t - L_t\} + G_r + 20 \log \lambda - 20 \log(4\pi) - S_{ths} - L_A + M_1 \text{ dBW}. \tag{8.7}$$

Inserting the parameters of the racon of Section 8.4.7, setting $L_t = L_A = 0$ and M_1 to its maximum value of 6 dB gives $R = 35$ m as the maximum range at which one such racon can trigger another, assuming no inhibition.

Within GEC-Marconi as a routine we operated new racons for 50 h, triggered at 1000 pps from an internal test-pulse generator, before retest to confirm stability of parameters. For convenience, half a dozen might be left running together on the soak-test bench, supply current ammeters indicating triggering activity. Out of curiosity or devilment, on several occasions we tried to get the racons to 'talk among themselves'. They never did.

8.5 User-selectable racons

8.5.1 The problem

The racons described so far have inconveniences.

- They can be difficult to perceive in clutter, despite the long response, especially when differentiation (anti-clutter rain) is engaged.
- Main beam and sidelobe responses continue to display when no longer needed by the operator, distracting attention and possibly masking important targets.
- Responses are not instantly available on demand but depend on the autonomous muting timer or sweep generator.

Several determined attempts have been made to provide a user-selectable racon service to participating ships while retaining an ordinary non-selectable service for all comers. The systems outlined below have all undergone technically satisfactory sea trials. To prevent inadvertent continual suppression of echoes, they all incorporate spring biassed switches or an equivalent at the radar. In no case has the benefit been deemed sufficient to offset the cost of radar modifications for general use. With the advent of AIS, it seems unlikely that a user selectable racon service will ever be widely adopted.

8.5.2 Fixed frequency and fixed offset frequency racons

User-selectable racons responding at band edge and received through a dedicated radar receiver channel were envisaged in the Post War Radar scheme and channels at 2900–2920 and 9300–9320 MHz were reserved for them. Experimental implementation was deferred until the 1970s, by which time the early hoghorn plus cheese squintless scanners had given way to slotted waveguides, whose azimuth beam squints away from boresight in a frequency-dependent manner. Successful trials aboard UK lighthouse tenders with specially modified radars showed that racon, radar and racon plus radar modes could indeed be selected. However, it was realised there would be excessive (>10 dB) scanner differential squint loss when a transmitter lying near the upper band-edge is coupled to a narrow beamwidth slotted array. Solutions took some years to find.

The advent of frequency agile racons in 1980 enabled GEC-Marconi to reduce the loss to less than 2 dB by arranging an agile racon to respond a preset 50 MHz fixed offset from interrogation frequency, e.g. response at 9400 MHz from 9450 MHz interrogation, 9350 MHz from 9400 MHz, etc. Service to unmodified radars would have been on a time-share basis. Nowadays a lower offset such as 10 MHz would suffice to separate racon and 'radar' (i.e. echo) channels, as in the later USIFAR, Section 8.5.4. Cost of the additional radar receive channel told against the scheme, which never entered service. Performance calculation for fixed offset frequency (FOF) racons is as for user-selectable interrogable frequency agile racons (USIFAR).

8.5.3 ITOFAR

Meanwhile interrogation-frequency time offset frequency agile racons (ITOFAR) were developed by L. M. Ericsson as an integral part of their frequency agile Ericon racon development programme under contract to the Swedish Board of Navigation to improve display in clutter, ice clutter being an especial problem in the Baltic. The system has operated successfully in the difficult conditions of the Swedish Archipelago and elsewhere. To ordinary radars, ITOFAR behaves as a standard frequency agile racon already described, where the response immediately follows interrogation and is displayed superimposed on nearby clutter returns. However, co-operating radars can command the racon to delay its response for a preset time, by setting pulse repetition interval to a certain precise value, typically 500.0 μs. The response then appears in the dead time between interrogations, where clutter returns are negligible. Matching delay circuits at the radar superimpose ordinary echoes in correct register on the delayed response. The operator can temporarily suspend the echoes and clutter, the racon response alone remaining at its proper range and bearing for clutter-free scrutiny. The racon trace may also be suppressed when not required, preventing masking of nearby targets.

8.5.4 USIFAR

USIFARs are a more recent and neat solution of the offset problem, developed in the late 1990s and offered by Tideland Signal Corporation, Houston, in their SeaBeacon agile racons. USIFARs also serve unmodified radars and are stated to be compatible with ITOFAR. A 3 MHz frequency-modulation is applied to the response of an otherwise standard agile racon. Its interrogation-frequency response component is claimed to fall by less than 1.5 dB, but there are sidebands of comparable power, offset 3 MHz. Ordinary radars accept the first component and operate normally, with a small performance loss. Co-operating radars take the modulation frequency via a separate sharply tuned receiver channel, whose bandwidth may be made narrow, as it only handles long responses, so retaining adequate signal to noise ratio and compensating the reduced power in the sideband. When the operator selects responses alone, they are displayed at the correct location, but with passive echoes and clutter reduced by more than 20 dB. When calculating this racon, noise figure, bandwidth, etc. appropriate to each radar channel must be used, as well as the appropriate response power component.

8.6 Miscellaneous in-band racons

The following descriptions of some other racons and transponders which have been tried and discarded may save the trouble of re-inventing these particular wheels. As far as known none is in current service.

8.6.1 Step-sweep racons

The sweep time of conventional slow-sweep racons has to be long enough for at least one, preferably three, scans to lie within the radar receiver bandwidth, limiting sweep rate to about 3 MHz/s maximum, necessitating sweep time of 45 s or more, too long for some short-range applications. Relying on there being eight or more interrogations per scanner beamwidth on the shorter range scales, the 200 MHz of the 9 GHz band is divided into four sub-bands. In absence of an interrogation, the sweep idles in one sub-band, sweeping say 9300–9350 MHz at 3 MHz/s, sweep time 16.7 s. The first two interrogations of the scan trigger responses at whatever part of the sweep has been reached, say 9320 MHz. The sweep then steps up 50 MHz to 9370 MHz and carries on in the 9350–9400 MHz sub-band. The next two responses in the scan are at 9370 MHz. There is another 50 MHz step to 9420 MHz in the third sub-band, and so on. The effect is of a transmitter for each sub-band, each responding to two interrogations in eight, paired to prevent rejection by radar receivers employing pulse to pulse correlation. There is some loss of paint density on cursive displays, but otherwise the display is almost indistinguishable from a slow-sweep racon, with the benefit of display refresh each 16.7 s. Although generally superseded by agile racons, step sweep might merit consideration where frequent displays are needed at lowest cost.

8.6.2 Fast-sweep racons

These preceded SARTs (Section 8.8) and work on similar lines. The main differences were that a standard racon coding scheme was used and that the sweep was free running, not starting from band edge when interrogated. Forward sweep rate was ~50 MHz/μs, with return at 200 MHz/μs. Response dots displayed at random points within the code envelope, a scan's worth appearing as a speckled Morse response. Fast sweep racons were discontinued because:

- sweep loss in narrow band radars was severe, for reasons outlined in 8.2.11;
- pulse to pulse correlation receivers, which rejected them, came into service;
- agile racons became available, doing the job better.

Another early fast-sweep racons operated almost exactly like a SART, with sweep as well as response triggered by the interrogation. At least one collision was blamed on confusion between the resulting dot pattern and the echo of an anchored ship.

8.6.3 High power racons

The first racons to go into regular service, already referred to, were primarily intended for long-range landfall and used 9 GHz band magnetron transmitters, duplicated for reliability. The output of about 11 dBW was complemented by receiver sensitivity as high as −95 dBW. The local oscillator of a superheterodyne receiver was tuned to mid-band. The IF response was flat from 4 to 100 MHz and both sidebands were utilised. The hole in the response between 9396 and 9404 MHz was filled by modulating the local oscillator frequency through 8 MHz peak–peak at a quasi-random 1 kHz modulation frequency. Coupled to twin antennas of 7 dB gain, EIRP was some 12 dB more than modern racons. When mounted 75 m above sea level, observed detection range was 60–90 km, extending into the diffraction region. High power racons were made obsolete by the introduction of far cheaper and power-economical solid state devices. Use of radar to make landfall has sharply diminished since introduction of GNSS, but if similar performance were needed today, it would seem better to team a high-gain omnidirectional antenna of narrow elevation beamwidth with a standard agile racon.

8.7 Cross-band racons and transponders

All the racons described above are in-band devices; the response lying within the 3 or 9 GHz IF band, and automatically painting direct on the display. System bandwidths are inherently wide enough to give good range resolution. In-band racons rely on the scanner to provide high gain as well as good bearing resolution, necessitating only low power from the response transmitter, an important practical consideration. Radar receivers are unsuited to extraction of much data from the response; more or less limited to a Morse letter, sufficient to distinguish amongst a few racons within an area but little else.

8.7.1 Radar/radio systems

Radio transmissions have to use narrow bandwidth for spectrum economy and so a radio-frequency response cannot have the sharp edge needed for ranging by pulse methods. On the other hand, walkie-talkie radios show how low (audio) bandwidth depresses noise and permits a data stream (e.g. detailed position messages) to be transmitted at low power between omnidirectional antennas. There are various ways of combining the positioning advantages of radar with the data transmission advantages of radio; radio on one leg with radar on the other, or radar both legs with a supplementary data interrogation or response by radio. Although none has been widely adopted, several have been proposed over the years for the following.

- Racon service, but advent of GNSS has lessened the need to transmit data.
- Ship to ship identification transponders. AIS is intended for this role, but has not yet won full acceptance by the marine community.
- Carry-aboard transponders reporting ships transiting a waterway, for which AIS now offers a neater solution. Examples include the carry onboard radar transponder (CORT) used at one time by Suez Canal pilots and the Canopus system proposed in the 1970s for the Manchester Ship and Kiel (Nord-Ostsee) Canals.

Detectability calculations follow the same principles as conventional racons. Sometimes similar calculations have to be made at the radio frequency, these follow conventional radio engineering practice.

8.7.2 Radar automatic identification system

As an alternative to the IMO-mandated AIS, which is based on GMDSS (the radio-frequency Global Marine Distress Signalling System), and which relies on a navigation satellite constellation such as GPS (Global Positioning System), Midar have put forward their RAIS. This system delivers the radio call-signs of radar targets of interest and is independent of satellites. All vessels and sea-marks such as lighthouses which are to respond as targets would carry an International Standard Interrogation Transponder (ISIT) containing 3 and 9 GHz omnidirectional antennas feeding receivers. There would also be a C band (~5 GHz, NATO G band) receiver–transmitter with its own omnidirectional antenna. Antenna gains, azimuth and elevation cover follow racon practice. Ships equipped to interrogate targets would have a similar transponder outfit, linked to their navigation radar(s). The system has been successfully trialled at 5.5 GHz but no operational frequency has been allocated at time of writing.

To interrogate a radar echo to find its radio callsign, the operator sets own ship's C band transmitter to fire synchronously with own ship's 3 or 9 GHz radar transmission pulse when the radar scanner bears on the target. The intended target's transponder only then receives simultaneous pulses, instructing it to respond at C band with a 32-bit digital code of the vessel's Lloyd's Register (LR/IMO) number, unique to the hull irrespective of changes of ownership, flag and callsign. Vessels at other ranges or bearings receive the C band and radar interrogations at different times so remain silent, preventing unwanted responses from interfering, called garbling.

C band responses received after a delay matching the range of the echo being interrogated are accepted by the ship. An on-board database, regularly updated, converts the LR/IMO number to the target's current radio callsign, which is displayed adjacent to the radar.

The atmospheric attenuation of C band is intermediate between the 3 and 9 GHz band values. The C band interrogation is received inherently free of clutter and the response can be made clutter-free by insertion of a time delay. The C band receiver sensitivity and bandwidth are similar to radar receivers, 40–50 dB more sensitive than racon practice. The C band sub-system uses omnidirectional antennas to avoid the expense of scanners rotating synchronously with the radar scanner. To give range comparable to racons and compatible with VHF ship–ship radio range, the C band response solid-state transmitter delivers of the order of 10 dBW.

8.8 SARTs

8.8.1 Purpose

Search and rescue transponders (SARTs) are restricted to the search and rescue (SAR) task of marking vessels and liferafts in distress and are carried by at least a quarter of each roll-on, roll-off (ro-ro) passenger ship's rafts. Electronically, but not operationally, they are akin to swept frequency racons and have similar sensitivity, response power and antenna characteristics, although the sweep regime differs. A supply battery is included within the watertight casing. SARTs are usually deployed on the liferaft mast, inside the non-metallic canopy, constraining mounting height to a metre or so above the water. Trials have established that canopies introduce little loss. Operation is confined to the 9 GHz band. It is the need to operate with SARTs which constrains 9 GHz radars to horizontal polarisation.

IMO COMSAR sub-committee's scenario is that the distressed vessel transmits a Mayday message, including position, on the GMDSS system and/or activates an EPIRB beacon whose presence and approximate location are relayed by GNSS satellite to a maritime rescue coordinating centre (MRCC) which alerts nearby shipping, the lifeboat service or SAR helicopters as appropriate. Diaza [2] has reported on operational experience. Life-rafts can be difficult to spot and may drift from the reported position. The SART is intended to mark them on the 9 GHz navigation radar (mandatory on all ships and carried by SAR aircraft) to a range of 5 nmi (~9 km). The SART includes an interrogation indicator, which hopefully prompts the raft coxswain to discharge flares at the right moment.

SARTs cannot repeatedly inter-trigger for the same reasons as racons (Section 8.4.9).

8.8.2 Sweep regime

Figure 8.15 shows the response process. Although different makes may vary somewhat, typically each interrogation pulse (a) at any frequency in the 9200 (to suit aircraft) to 9500 MHz band generates a single long response pulse (95.4 μs, b) during which the transmitter frequency, starting at band edge, sweeps rapidly (0.4 μs forward and 7.5 μs return) through the whole marine 9 GHz band 12 times (c). Provided the transmitter always sweeps through at least the full band, chirp during the pulse is not too serious; for example, the sweep might be 9200–9550 MHz at the start, chirping down 50 MHz to 9150–9500 MHz by the end of the pulse. As SARTs are merely a low power emergency aid, seldom energised, a certain amount of out of band interference is tolerable, which helps reduce cost.

8.8.3 Display on radar

When its scanner bears on the SART, the interrogating radar potentially displays (d) a dot each time the response sweeps through radar frequency. Depending on the location of the radar frequency within the band, the first strong dot may be delayed

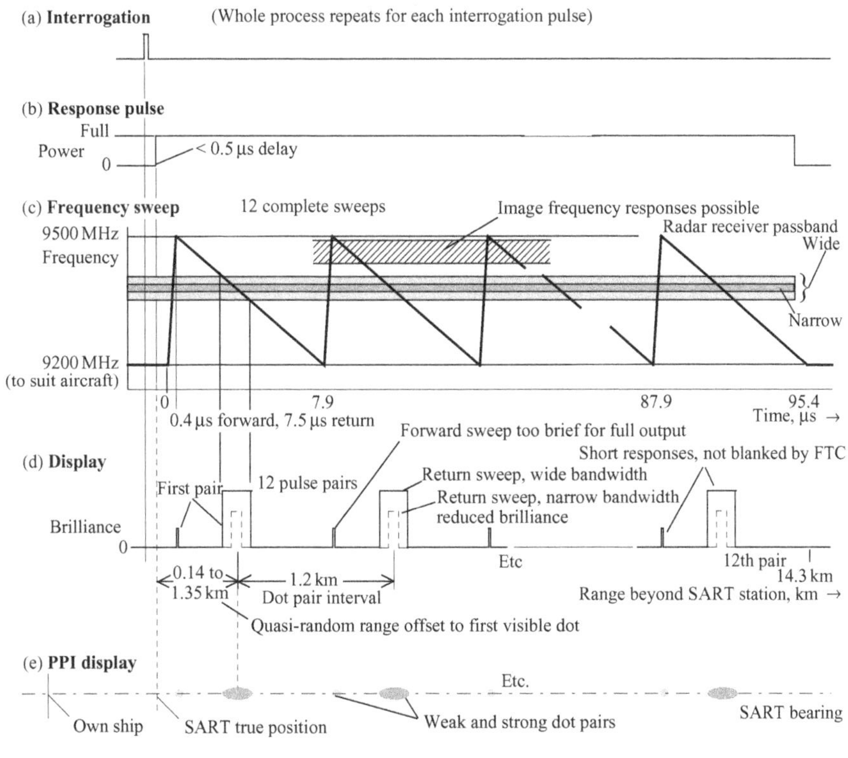

Figure 8.15 SART timing. Shows effect of radar receiver bandwidth and hence of pulselength

up to 7.5 μs, painting up to 1.35 km beyond true position (as usual, scaling is at half the speed of light). The first dot is followed by forward and return dots to a total of 12 pairs, spanning a 13 km range bracket; note this is not a measure of detection range. Each interrogation within the scanner beamwidth generates a response group with identical timing; the dots of each response coincide, brightening the dot pattern and avoiding rejection by radars using pulse to pulse correlation. As with swept racons, image frequency reception may occur, displaying an interleaved set of dots.

The dots from the quick forward sweep are usually too weak to display and the 12 return dots appear alone. All dots are so short that they are unaffected by the radar FTC facility. SART responses are harder to perceive than the solid lines of racons, especially in clutter, despite the brain's pattern recognition skills, and pose headaches to designers of radar detection systems. Some radars include a dedicated SART mode, in which receiver bandwidth is widened and features such as pulse–pulse correlation are disabled.

As with racons, the radar receiver swept gain may suppress the early dots in the response and these may anyway be delayed several microseconds, making the target appear to be at longer range. Navigators may therefore choose to make an oblique approach to reduce the risk of running the liferaft down.

8.8.4 Performance equations – sweep loss

Most swept frequency racon performance calculation equations apply. The fast sweep may cause the response to be attenuated within the radar receiver. If the latter's bandwidth is narrow, SART responses (particularly the very quick 0.4 μs forward sweep) change frequency so fast that the echo is attenuated – the response flashes through radar frequency before the receiver bandpass filter has time to accept it. An alternative viewpoint is that the pulse energy is distributed through a wide spectrum, only a small part lying within the receiver passband. The author learnt this bandwidth problem the hard way long ago when developing a fast sweep racon which worked fine on the test radar but had to be withdrawn from sale after failing miserably in service with radars of narrower bandwidth.

Circuit theory predicts that a filter of bandwidth B Hz attenuates sweeping signals by L_s unless the sweep rate Z Hz/s is much less than B^2. Filter loss L_s is somewhat dependent on the shape of the filter response curve but approximates

$$L_s \sim 10 \log \left(\frac{1+Z}{B^2} \right) \text{ dB.} \tag{8.8a}$$

In other words, the bandpass filter acts as a single-pole low pass filter, rolling off at 6 dB/octave or 20 dB/decade when the receiver bandwidth is narrow, with 3 dB loss when $\sqrt{Z} = B$. The SART forward sweep has $Z = 300$ MHz per 0.4 μs so $\sqrt{Z} = \sqrt{300 \times 10^6/(0.4 \times 10^{-6})}$ Hz $= 27.4$ MHz and $Z = 750$ MHz/μs. The return sweep takes 7.5 μs, here $\sqrt{Z} = 6.4$ MHz and $Z = 40.6$ MHz/μs.

$$\text{For 1 dB loss,} \qquad B = 1.97\sqrt{Z}. \tag{8.8b}$$

When $Z = 40.6$ MHz/μs, $B \geq 12.5$ MHz for 1 dB loss.

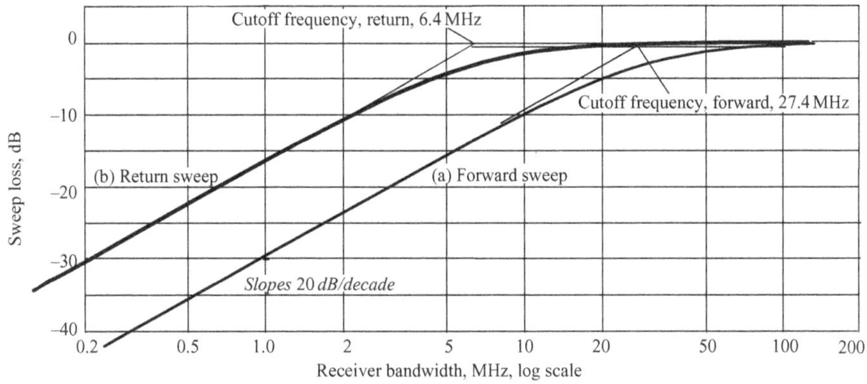

Figure 8.16 *Sweep loss. Radar receiving SART. Approximate; depends on shape of radar receiver passband. Use wide bandwidth (short range scale/short pulse) to optimise detection of SARTs*

Figure 8.16 plots forward and return sweep losses to a base of radar receiver bandwidth, B. On long range scales, many radars have $B \sim 2$ or 3 MHz, giving forward sweep loss of about 22.8 or 17.3 dB, making the dots difficult to detect. The return sweep suffers 10.5 or 7.4 dB loss respectively. Switching to a shorter range scale using say 15 MHz bandwidth reduces forward sweep loss to 6.4 dB, with return sweep loss 0.7 dB. It is therefore preferable to use a wide bandwidth (i.e. a short range scale) for SART detection, putting up with the additional thermal noise. Wide bandwidth is synonymous with narrow pulselength, which reduces clutter. Unfortunately wide bandwidth is rarely available on the fairly long range scales the operator would initially prefer to use on SAR missions. It has to be said that the sweep times mandated by IMO for SARTs seem rather too fast to accommodate the narrower bandwidths of modern radars.

Although really a radar processing loss, it is more convenient to treat L_s as a SART transmitter loss. It is not included within the EIRP. The range equation for the response at the radar is similar to Eq. (8.4a) for a racon, modified by the additional filter loss term:

$$S_r = [P_t + G_t] - L_s + G + 20 \log \lambda - 20 \log(4\pi) - 20 \log R - L_r - L_A + M_2 \text{ dBW}. \quad (8.9)$$

Relative to a racon of the same receiver sensitivity, response power and antenna gain, SARTs are likely to give inferior performance because:

- racons incur less filter loss;
- racon responses are more readily perceived;
- racons are usually mounted higher.

The racon curves of Figures 8.11 and 8.12 apply to SARTs except that performance is depressed by the sweep loss.

8.9 Ramarks

A ramark is half a racon, with transmitter but no receiver. Unlike all the other active devices, strictly speaking it is therefore neither a transponder nor a secondary radar. After trialling ramark beacons in 1949, the United Kingdom concluded that racons were preferable. Ramarks have been used (principally in Japan) as aids to navigation, marking buoys and lighthouses. A transmitter generates a steady train of microwave pulses, power ~1 W, spread across a frequency range which usually is a restricted part of the marine 9 GHz band centred on 9375 MHz as favoured by local shipping (not all international radars can participate). Swept-racon-style sweep and muting are included. The radiator is a racon-style antenna. Any nearby radar of suitable frequency accepts the component of energy lying within its receiver bandwidth. Maximum range is typically 10–20 km. The transmission prf is of necessity unrelated to radar prf so response dots are scattered radially across the display at all ranges from zero, building up a fuzzy line on ramark bearing as shown in Chapter 3, Section 3.11, Figure 3.19, accuracy being around 1°, depending on scanner beamwidth. This response feather is on station bearing and extends beyond any clutter which may mask the central portion of the display, so device power can be relatively low. The radar observer can determine position by taking cross-bearings. Many navigators would prefer three lines of position, providing a 'cocked hat' to facilitate error estimation. The LOPs can be provided by other ramarks or conspicuous passive landmarks. As with fast sweep racons, ramark responses are not rejected by fast time constant (FTC) features in the radar receiver. Neither coding nor range information can be extracted.

Most modern displays regard such unsynchronised signals as interference from another radar, to be rejected by the pulse to pulse correlation circuit, therefore ramarks have dropped out of use. Detectability can be calculated by treating them as SARTs which are always triggered. Because of the line nature of the response, relatively low single-pulse probability of detection is sufficient.

8.10 Radar target enhancers

8.10.1 Principle

An RTE, often known as an active reflector, is an electronic device consisting basically of a microwave amplifier connected between small receive and transmit omni directional antennas, intended to enhance the echo of small targets. Received radar pulses are amplified and retransmitted at increased power without significant delay, lengthening or other change. For regulatory reasons, RTE responses cannot be coded and RTEs are not generally regarded as being transponders. The radar displays an artificial echo at RTE position (see Figure 3.19), looking exactly like the echo of a passive reflector target or vessel, but of useful strength and uniform TPM. As with racons, elevation beamwidth can be traded for antenna gain. RTEs are completely inoperative outside their designed frequency band. The electrical power requirement of a few watts is trivial to SOLAS ships and well within the capacity of many yachts.

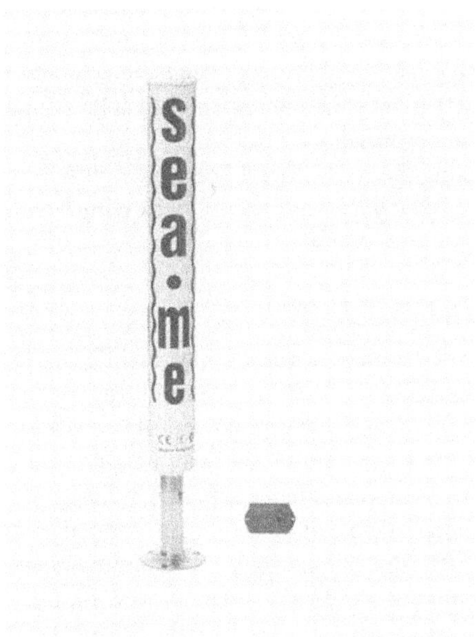

Figure 8.17 *Radar target enhancer. Sea-me 9 GHz type. Claimed maximum RCS 63 m^2, average 34 m^2, meeting ITU-R 1176 and environmental BSEN 90645:1997. Dimensions 416 mm × 50 mm diameter, weight 410 g, 12 V power consumption 0.75–3 W. Cockpit display module (right) warns of interrogations. RTEs are widely used to improve RCS of small craft and AtoN buoys. Reproduced by permission of Munro Engineering Ltd, Wincanton UK*

RTEs usually have narrow tubular construction to separate the antennas and prevent self-oscillation, see Figure 8.17. RTEs may be thick on the ground in some marinas, but the similarity of the responses to the yacht passive echoes limits congestion interference. Uptake of RTEs was at first slow, mainly small 9 GHz models for the amateur yacht market. By 2004, many more were being installed worldwide, also finding their way into some offbeat applications – marking remotely controlled jetskis towing gunnery practice targets, for example. Twin-band and 3 GHz models are in development.

RTEs are used in the same way as passive reflectors to increase RCS of vessels, offering compactness with a possible cost penalty. RTEs are cheaper than racons so are occasionally used to mark buoys. Interrogations are not 'detected' and reconstituted as with racons and SARTs. Instead, the response pulse is an amplified copy of the interrogation with negligible (nanoseconds) delay, restricting range error to a metre or so.

Some yacht-market 9 GHz band RTEs include an indicator of interrogations, warning the skipper to sharpen the lookout as a radar is nearby. This facility does not satisfy the yacht's Colregs duty to maintain an efficient lookout, and should be used cautiously, because:

- there is no guarantee that the RTE response is detectable through clutter;
- the radar range scale in use may not extend to yacht's range;
- the radar may go unobserved;
- the ship may be using the 3 GHz band or no radar at all;
- the RTE may be masked at close range by deck cargo, etc.

8.10.2 Basic description

Figure 8.18(*a*) is a block diagram of a single-channel RTE, usually 9 GHz band, with typical waveforms added (Figures 8.18(b)–(d)). Interrogations are received by a small omnidirectional antenna, usually slant polarised. After a burnout protection circuit, it feeds a microwave amplifier whose bandwidth covers one marine band. As with racons, twin-band devices are essentially separate but may share some components. The amplifier has several stages and overall unsaturated gain, G_a, is about 47 dB (9 GHz). Bandwidth may not extend to VTS radars using frequencies adjacent to the 9300–9500 MHz marine band. The final 'power' amplifier stage is capable of handling a certain rated peak output power P_r of the order of 0 dBW or 1.0 W. Suppliers sometimes quote equivalent isotropic radiated power (EIRP), which is $P_r + G_t$. In this

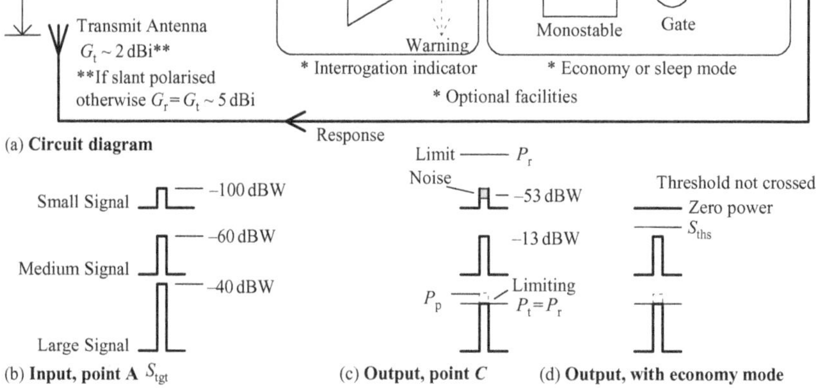

Figure 8.18 *RTE block diagram. Typical waveforms are shown. Operation is independent of pulse width. Compared with a racon, Figure 8.7, the RTE is basically a microwave amplifier without detection and regeneration of the microwave signal. Parameters shown give 10 dB m² RCS*

example, EIRP = 2 dBW or 1.6 W. A limiter prevents transmitter overload. For purposes of calculation, the overall frequency response of amplifiers plus antennas is usually assumed to remain constant across the band, falling abruptly to zero at band edges. If this is not the case, performance will vary with frequency. A second antenna similar to the receive, but if polarised of opposite slant, transmits the responses. Each antenna is designed as far as possible for omnidirectional response in the azimuth plane and fairly wide elevation beamwidth to cater for roll; ITU-R Recommendation M.1176 requires ±15° between half-power (−3 dB) points of the pair, much the same as buoy racons. Antennas are sometimes made from an array of three patches printed on a low-loss substrate. Wide beamwidth reduces sensitivity to platform heel at expense of gain and hence of effective RCS, as we shall see. The combined response of the antennas in each plane defines the TPM. As with racons, it is relatively easy to design for smooth TPM.

The short feeders connecting the antennas to the amplifiers propagate more slowly than the speed of light, and the signal takes a little time to traverse the amplifier chain. The response is therefore delayed by a few nanoseconds, typically causing less than 1 m range error on the radar display. Although navigationally insignificant, this delay affects the response of RTE/reflector combinations, Section 8.15.4.

The great problem of any high-gain amplifier device is self-oscillation, sometimes called ringaround, some of the response finding its way back into the receive antenna and starting a new spurious response – we have all been deafened by howl from public-address loudspeakers when the microphone is too close. Oscillation occurs when the gain round the loop exceeds losses. To be precise, vector nett in-phase component of gain in a continuously oscillating loop has to be unity (the Nyquist criterion). In practice the oscillation frequency adjusts itself to meet the phasing condition and the amplitude builds up until saturation reduces nett gain to unity. In a two-antenna RTE the loop is receive antenna, amplifiers, transmit antenna, radiated signal leakage to receive antenna. An oscillating RTE would display as a ramark.

The receive/detect/regenerate/inhibit mechanism of racons prevents oscillation and permits their receiver and transmitter to be coupled by a circulator to a common antenna. Even if a leakproof RTE circulator were available, antenna mismatch would reflect enough response into the receiver for oscillation, or there might be reflection from a nearby mast outside the designer's control. The usual solution includes two antennas separated by half a metre or so, slant polarised at opposite hands, with careful radome design to prevent signal creep. Twin band RTEs would almost certainly be taller to accommodate the additional channel. To the receive antenna, responses reflecting once off a nearby mast or other one-bounce object are cross polarised and rejected. A 45° slant polarised antenna will accept any radar polarisation, horizontal, vertical or circular, with loss of half the signal (Section 8.1.9, Table 8.1). Likewise, the radar scanner accepts slant polarised responses as echoes. The resulting polarisation losses of 3 dB each way are included within RTE specifications where appropriate. The difference in antenna heights above sea level makes interrogation and response leg multipath factors M_1 and M_2 differ slightly, splitting interrogation and response multipath null ranges, so RTEs responses suffer slightly less multipath than point reflector echoes.

RTEs without economy or sleep modes include no detection process, so questions of probability of detection only arise when we come to consider how the radar detects responses in Chapter 12.

8.10.3 Ancillary facilities

An optional economy circuit may comprise a detector, video-frequency amplifier, monostable with threshold and gate. When no interrogations are present, thermal noise at the video amplifier output is insufficient to cross the threshold, there is no detector output and the output stage current supply is cut off to conserve power. Any medium-size interrogation pulse from a radar transmitter is detected to give a video pulse which after amplification exceeds the threshold, flipping the monostable for a few microseconds during which the gate is opened for the power stage to respond. Detection may either use a conventional microwave diode as shown in Figure 8.18(a), or may detect the rise in supply current when the power amplifier handles a signal pulse train, cheaper but usually less sensitive.

Alternatively, power may be conserved using a sleep mode. If a train of interrogations is received, a slow-acting circuit opens the gate for some seconds so subsequent scans from the radar evoke responses. The auxiliary video amplifier bandwidth may be somewhat restricted to make the RTE less sensitive to narrow pulses to prevent futile responses to interrogations from distant radars using short range scales. As with racon filters, this also reduces sidelobe responses.

Calculation of economy and sleep mode sensitivities can be treated along the lines of racon receivers, Section 8.4. Economy circuits also provide a convenient drive to the optional interrogation indicator. Sensitivity of an RTE's optional visual or audible alarm facility is nearly as high as that of a racon receiver, and the alarm will usually be triggered by a distant radar at range well in excess of RTE detection range at the radar.

To guard against continuous oscillation, a timeout circuit may briefly interrupt the power supply should the detector indicate a response exceeding about 100 ms.

8.10.4 Specification

IMO's current revision of its specification for shipborne active or passive reflectors, see Chapter 7, Section 7.6.1, may stimulate introduction of more powerful RTEs and spread availability from the 9 GHz band to 3 GHz. Meanwhile RTEs are sometimes tested against relevant clauses of the passive reflector specification, ISO 8729. The draft ITU-R Recommendation for RTEs (ITU-R M.1176) limits maximum response power to 10 W EIRP (including antenna gains; and presumably to restrict any spurious out of band interference) and tolerates unsaturated RCS at 9 GHz as low as 9 m^2, which is insufficient for detection in severe clutter. No specification currently addresses the 3 GHz band.

8.10.5 Radar cross section

By the inverse square law, halving the range of an unsaturated RTE increases power at its receiver ×4. The amplifier delivers ×4 response power. The inverse square law again applies, to give ×16 echo at the radar. As a passive reflector would also give ×16 echo, the unsaturated RTE has a definite RCS. Eventually as range shortens the power stage reaches its rated maximum. The device saturates and no further output increase is possible as interrogation strength rises yet more; this condition occurs with racons at all ranges. RCS now falls as range closes, because response power remains constant, like a racon. Received response power continues to rise, but more slowly. Unfortunately, the term 'saturation' is also sometimes confusingly applied to traffic capacity, the number of responses per second the RTE will support.

Because RTEs saturate at short range, may not respond at all at very long range and have vertically separated antennas, no single figure adequately describes their RCS under all conditions. Variation of response at the radar is different from passive reflectors. As RTEs are now replacing reflectors for some tasks, it is interesting to compare predicted performance. To find RCS we consider how response power varies with interrogation strength.

8.10.6 RTE response on axis

RTE parameters are quoted inclusive of any internal losses. We adapt the notation and treatment already used for racons:

G_r, G_t = Receive, transmit antenna gains to signal with same polarisation as radar scanner, dB

G_a, G_o = RTE amplifier, overall gains (unsaturated), dB

$$G_o = G_r + G_a + G_t \tag{8.10}$$

h_r, h_t = Receive, transmit antenna heights, metres

M_1, M_2 = Interrogate, response leg multipath factors, dB

P_r, P_p, P_t = Device response power: rated when saturated, prospective, actually transmitted, dBW

P_{tEIR} = Transmitted response equivalent isotropic radiated power. $P_{tEIR} = P_t + G_t$, dBW

S_{tgt} = Signal reaching device per one-way radar range Eqs (4.7a) and (4.7b), dBW

S_{ths} = Threshold received signal power dBW, below which device does not respond. May be pulselength-dependent per Figure 8.10, and only applies to RTEs having economy mode

S_r, S_e = Signal received at radar from active target, echo received at radar from passive object, both per radar range Eqs (4.6a) and (4.6b), dBW

If the signal is less than the optional threshold, there is no response. If above threshold (or if there is no threshold circuit, in which case the 'threshold' voltage can be considered zero), the full prospective response power is delivered up to the rated power limit. Adapting Eq. (8.3):

$$\text{if } S_{tgt} < S_{ths}, \quad P_t = 0, \quad \text{otherwise if } P_p < P_r, \quad P_t = P_p, \quad \text{otherwise } P_t = P_r. \tag{8.11}$$

Assuming the RTE is not inhibited, its received signal, S_{tgt}, is given by radar range Eq. (4.7a) and by the racon equation Eq. (8.4a). Prospective response power P_p is raised by the amplifier gain:

$$P_p = S_{tgt} + G_a \text{ dBW}. \tag{8.12}$$

We consider first the unsaturated condition, followed by saturated.

At long range when *unsaturated* the full prospective response power is delivered: $P_t = P_p$ per Eq. (8.11), so Eq. (8.12) gives:

$$S_{tgt} = P_t - G_a \text{ dBW}. \tag{8.13}$$

Substituting for S_{tgt} in Eq. (4.7a), the unsaturated response power is:

$$P_t = P + G + G_r + 20 \log \lambda - 20 \log(4\pi) - 20 \log R - L_t - L_A + M_1 + G_a \text{ dBW}. \tag{8.14}$$

Applying the one way range Eq. (4.8a) for the response leg, the unsaturated echo at the radar is:

$$S_r = P_t + G_t + G + 20 \log \lambda - 20 \log(4\pi) - 20 \log R - L_r - L_A + M_2 \text{ dBW}. \tag{8.15a}$$

Substituting for P_t from Eq. (8.14):

$$S_r = P + 2G + G_r + G_a + G_t + 40 \log \lambda - 40 \log(4\pi) - 40 \log R - L_t - 2L_A - L_r + M_1 + M_2 \text{ dBW}. \tag{8.15b}$$

Putting $G_r + G_a + G_t = G_o$ and $P + G - L_t = P_{EIR}$ (at the radar) gives the tidiest version of the response signal at the radar:

$$S_r = P_{EIR} + G + G_o + 40 \log \lambda - 40 \log(4\pi) - 40 \log R - L_r - 2L_A + M_1 + M_2 \text{ dBW}. \tag{8.15c}$$

When the RTE is *saturated* at short range, prospective power exceeds rating and only the latter can be delivered. From Eq. (8.11), $P_t = P_r$. Signal at the radar, S_r, is obtained by substituting P_r for P_t in Eq. (8.15a) to give an expression identical to that for the racon, Eq. (8.4a):

$$S_r = P_r + G + G_t + 20 \log \lambda - 20 \log(4\pi) - 20 \log R - L_r - L_A + M_2 \text{ dBW}. \tag{8.16}$$

8.10.7 Unsaturated RCS

Making the approximation $M_1 + M_2 = 2M$ and rearranging Eq. (8.15c) gives:

$$S_r = P_{EIR} + G + \mathbf{G_o} + \mathbf{20\log\lambda} - \mathbf{10\log(4\pi)} + 20\log\lambda - 30\log(4\pi) - 40\log R - L_r - 2L_A + 2M\,\text{dBW}. \tag{8.17}$$

Substituting σ_{unsat} for the bold terms makes the equation identical to the two-way radar range equation for a passive target whose RCS $= \sigma_{unsat}$, Eq. (4.6a). The unsaturated RCS of an RTE (which applies at long range and is quoted in datasheets) is therefore the bold terms of Eq. (8.17). Except for substitution of $(M_1 + M_2)$ for $2M$, to the main beam of the radar, the unsaturated RTE seems indistinguishable from a point reflector whose RCS is σ_{unsat}:

$$\sigma_{unsat} \sim G_o + 20\log\lambda - 10\log(4\pi) = G_a + G_r + G_t + 20\log\lambda - 10.99\,\text{dB m}^2, \tag{8.18a}$$

which boils down to:

$$\begin{aligned} &3\,\text{GHz band } (\lambda \sim 0.1\,\text{m}) : \sigma_{unsat} \sim G_o - 31\,\text{dB m}^2, \\ &9\,\text{GHz band} (\lambda \sim 0.032\,\text{m}) : \sigma_{unsat} \sim G_o - 41\,\text{dB m}^2. \end{aligned} \tag{8.18b}$$

So to replicate a passive reflector having, say, 20 dB m^2 RCS at 9 GHz and 10 dB m^2 at 3 GHz, G_o must be 41 dB at 3 GHz and 61 dB at 9 GHz. If $G_r = G_t = 2$ dB, $G_a = 37$ or 57 dB respectively (Eq. (8.10)). Prevention of self-oscillation with such high gain demands great care during the detailed design phase. The RTE of Figure 8.18 has RCS 10 dB m^2.

The unsaturated RCS is a property of the RTE alone. Unlike a racon, it is independent of the radar parameters and only depends on range to the small extent that antenna offset makes M_1 differ from M_2.

8.10.8 Saturated RCS, saturation range

At saturation range, R_{sat}, the amplifier reaches its maximum available (rated) power output. Further halving of range now gives only ×4 rather than ×16 echo strength, so effective RCS falls to a quarter. This is the saturation condition, where RCS $\propto$ (range)^2 as with a racon, thus an RTE which is detectable at moderate range may become undetectable when saturated at short range. When saturated, we obtain the response at the radar by substituting P_r for P_t in Eq. (8.15a):

$$S_r = P_r + G_t + G + 20\log\lambda - 20\log(4\pi) - 20\log R - L_r - L_A + M_2\,\text{dBW}. \tag{8.19}$$

We find R_{sat} by substitution in Eq. (8.14), with $P_t = P_r$:

$$P_r = P + G + G_r + 20\log\lambda - 20\log(4\pi) - 20\log R_{sat} - L_t - L_A + M_1 + G_a.$$

Therefore

$$\begin{aligned}20\log R_{\text{sat}} &= P + G + 20\log\lambda - 20\log(4\pi) - L_{\text{t}} - L_{\text{A}} \\ &\quad + M_1 + G_{\text{a}} + G_{\text{r}} - P_{\text{r}} \\ &= [\sigma_{\text{unsat}} - G_{\text{t}} - P_{\text{r}}] + \{P + G - L_{\text{t}}\} - 10\log(4\pi) - L_{\text{A}} + M_1.\end{aligned} \tag{8.20a}$$

P_{rEIR} is the response EIRP $[P_{\text{rEIR}} = P_{\text{r}} + G_{\text{t}}]$ and P_{EIR} is the radar EIRP $\{P_{\text{EIR}} = P + G - L_{\text{t}}\}$. We can write Eq. (8.20a) as:

$$20\log R_{\text{sat}} = \sigma_{\text{unsat}} - P_{\text{rEIR}} + P_{\text{EIR}} - 10\log(4\pi) - L_{\text{A}} + M_1. \tag{8.20b}$$

Saturation range is therefore dependent on the radar and atmosphere as well as the RTE. When saturated, effective RCS is found by equating Eq. (4.6a) for a passive target with Eq. (8.14a):

$$\begin{aligned}&P + 2G + 20\log\lambda - 30\log(4\pi) + \sigma_{\text{sat}} - 40\log R \\ &\quad - L_{\text{t}} - L_{\text{r}} - 2L_{\text{A}} + M_1 + M_2 \\ &\quad\quad = P_{\text{r}} + G_{\text{t}} + G + 20\log\lambda - 20\log(4\pi) \\ &\quad\quad\quad - 20\log R - L_{\text{A}} - L_{\text{r}} + M_1 + M_2.\end{aligned}$$

Therefore

$$\sigma_{\text{sat}} = P_{\text{rEIR}} - P_{\text{EIR}} + 10\log(4\pi) + 20\log R - L_{\text{A}} + M_1 \text{ dB m}^2. \tag{8.21}$$

This expression is similar to that for a racon, Eq. (8.5b). Variation in M_1 as range changes may cause several transitions between saturated and unsaturated modes. Note that high P_{r} is beneficial, reducing the range at which the RTE saturates and increasing saturated RCS, σ_{sat}. The saturated RCS TPM depends solely on the response antenna and is independent of the RTE receive antenna, except in the skirts where it comes out of saturation.

Saturation range also increases with radar transmitter power and scanner size (high EIRP), and reduces as RTE rated EIRP increases. It is difficult to provide sufficient RCS to surmount clutter at short range. At 0.5 km, rated transmitter power might have to breach the ITU 10 W EIRP limit. It is certainly unrealistic to expect RTE responses to remain detectable in severe clutter down to 0.5 km. The problem is compounded when elevation beamwidth is wide, with low antenna gain.

8.10.9 Sidelobes

As always, it is scanner rather than active target's sidelobes that cause the problem. When the main beam fails to saturate the RTE, sidelobe performance is identical to that of a passive reflector of the same RCS. At short range, where the main-beam response is reduced by saturation, the sidelobes remain unsaturated and are not reduced. At the radar receiver, the main beam to sidelobe ratio falls. In this sense, RTEs have poorer sidelobe performance than passive reflectors. However, RTE sidelobes are always less severe than racons of the same maximum detectable range unless the racon includes a sidelobe suppression facility.

8.10.10 Target pattern map

Eq. (8.17) contains G_r and G_t terms within G_o, so the unsaturated target pattern map depends on both antennas. Their individual patterns are likely to be smooth, similar to those of racons, and without serious 'holes'. It is possible to rotate one relative to the other in azimuth so its residual peaks coincide with the other's troughs, further smoothing the azimuth pattern. But such refinement may be hardly worthwhile, since skin echo from the craft or other platform carrying the RTE nearly always spoils the overall pattern, as discussed in Chapter 7, Section 7.10.

If the -3 dB elevation beamwidth of each antenna is ϕ, the -6 dB unsaturated beamwidth of the RTE is also ϕ and the -3 dB beamwidth, which ought to be the width quoted in the data sheet, is 0.72ϕ, so $\phi = 1.39 \times$ data sheet elevation beamwidth. The values 0.72 and 1.39 are on the reasonable assumption that the antenna aperture is uniformly illuminated with a classical $(\sin x)/x$ elevation pattern. At short range, when saturated, Eq. (8.21) applies, which includes the response antenna only (term G_t); elevation -3 dB beamwidth widens to ϕ, being $1.39\times$ the quoted long-range width. One needs to be quite sure what is really being quoted in datasheets; salesman's 'specmanship' may slip in the saturated value instead of the more useful but smaller unsaturated value. When calculating ϕ from quoted values of G_r or G_t, allowance must be made for any slant polarisation loss.

8.10.11 Noise power output

Like all amplifiers, the RTE input stage generates thermal noise. Unless inhibited between responses by sleep or economy gating arrangements, this noise is amplified and transmitted continuously. If strong enough, the radar would pick it up to display a ramark-like trace. Noise power calculation uses the following notation:

P_{nd} = noise power density, W/Hz at RTE transmitter output
n = RTE receiver noise factor, giving noise figure, $N = 10\log n$
K = Boltzmann's constant, 1.38×10^{-23} J/K
T = absolute temperature (conventionally 290 K;
$-10\log(KT) = 204$ dBW/Hz)
S_{nd}, S_n = Noise power density dBW/Hz, noise power, dBW, at radar receiver.

The noise spectrum is assumed to be uniformly spread throughout the amplifier bandwidth (which may be taken to equal the radar band, as the designer will have minimised bandwidth to save cost). P_{nd} is proportional to the amplifier noise figure and power gain. Numerically

$$P_{nd} = nKTG_a \text{ W/Hz}.$$

In dB terms

$$P_{nd} = N - 204 + G_a \text{ dBW/Hz}. \tag{8.22}$$

N is rarely quoted in RTE datasheets. Low noise amplifiers with $N \sim 3$ dB are available but are unlikely to have been fitted on cost grounds, so it may be reasonable to assume $N \sim 10$ dB. Noise is never sufficient to saturate the transmitter. The noise density received at the radar is obtained by substituting P_{nd} in Eq. (8.15a):

$$S_{nd} = P_{nd} + G_t + G + 20\log\lambda - 20\log(4\pi) - 20\log R - L_r - L_A + M_2 \text{ dBW/Hz}. \tag{8.23}$$

The average received noise power is proportional to radar receiver bandwidth, B:

$$S_n = S_{nd} + 10\log B \text{ dBW}. \tag{8.24}$$

8.10.12 Example of RTE noise

The author has no personal experience of interference from RTEs and has seen no anecdotal or formal reports of noise interference – if no news is good news it is probable that the current generation of low-RCS 9 GHz RTEs do not transmit significant noise. However, high-RCS models, with higher gain amplifiers, may radiate more noise. For analysis, we take a 9 GHz RTE of rather high unsaturated RCS $\sigma_{unsat} = 17$ dB m^2 (50 m^2), effective antenna gains $G_r = G_t = 3$ dB, $N = 20$ dB (a high value for illustration), $P_r = 3$ dBW ($P_{rEIR} = 6$ dBW, as the racon example in Section 8.4.7). From Eq. (8.18a), $G_o = 58$ dB; so by Eq. (8.10), $G_a = G_o - G_r - G_t = 52$ dB. We make the free-space assumption that $L_A = M_2 = 0$ dB, and the RTE is operating with the radar of Figure 8.11, Section 8.4.7 above, having receiver beamwidths B of 2 MHz long pulse (LP) and 10 MHz short pulse (SP).

From Eq. (8.22), response noise power density as transmitted, $P_{nd} = 20 - 204 + 52 = -132$ dBW/Hz. Over the 200×10^6 Hz of a marine band, total noise power is $-132 + 10\log(200 \times 10^6) = -49$ dBW. This is 12 μW and far below saturation for any practical RTE, even allowing for occasional high spikes containing several times the average noise power.

From Eq. (8.23), noise density received at the radar at range $R = 1000$ m is given by: $S_{nd} = -132 + 3 + 27 - 29.9 - 22 - 60 - 5 - 0 + 0 = -218.9$ dBW/Hz, so from Eq. (8.24) the noise power received at 1 km is $S_n = -218.9 + 10\log(2 \times 10^6$ LP, or 10×10^6 SP$) = -155.9$ dBW LP or -148.9 dBW SP, falling at 20 dB/decade as range increases, as shown in Figure 8.19.

Suppose an RTE lies at 0.02 km from the radar operating on SP. The radar receives -114.9 dBW noise in its 10 MHz bandwidth, point X. The noise is continuous and will be 'displayable' whenever threshold sensitivity is more negative than -114.9 dBW, at range Y (1.26 km) outwards, unaffected by choice of LP or SP. At range exceeding 0.06 km, X lies below MDS and the noise is not detected. Depending on the radar

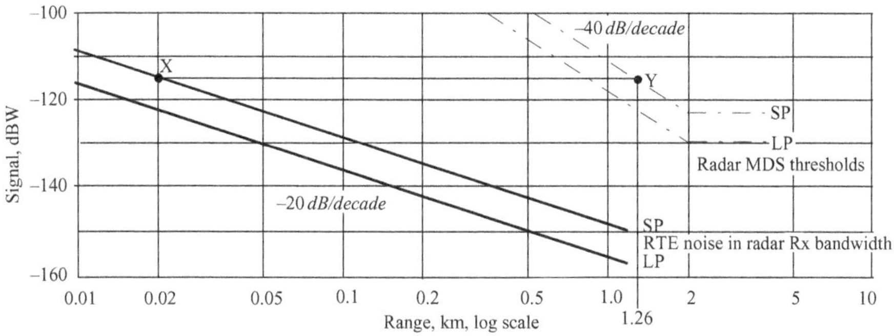

Figure 8.19 RTE noise. Radar as Figures 8.11 and 8.12

signal processor, the noise will either:

- be displayed as a radial ramark trace on RTE bearing if the radar has little processing capacity (e.g. older types with cursive display), or
- will depress receiver sensitivity at that bearing.

We have taken a rather extreme example with high noise factor to illustrate the concept. Results would be similar for a more powerful radar with $G = 34$ dB and 2 dB loss, and an RTE of 10 dB noise figure. Radar receiver thermal noise is 10 dB or so below MDS and it could be argued that RTE noise becomes troublesome when it equals radar noise. If so, RTE noise is several decibels more severe than depicted in Figure 8.19.

RTE noise is least negligible when radar scanner gain is high and radar receiver noise figure low. We conclude that noise of high-RCS RTEs should always be calculated but may only rarely pose a problem except perhaps to high performance VTS radars. Cures which could be applied to the RTE design include muting between responses and use of a low-noise input amplifier stage. If the noise proves troublesome, the radar operator should try reducing gain. Changing radar pulselength or engaging differentiation are likely to be ineffective.

8.10.13 Example of RTE performance

Performance of the same radar and RTE is shown in Figure 8.20, which may be directly compared with Figures 8.11 and 8.12 for a racon (where the racon antenna gain was assumed 6 dB, giving the same EIRP). The comparison is not quite fair; the racon will have wider elevation beamwidth. A 10 dB m^2 reflector has again been included for reference.

At (a) is shown the interrogation received at the RTE receive antenna, falling at 20 dB/decade; an auxiliary receiver threshold could be drawn in if applicable. Response transmitter power is at (b), rated power of 3 dBW being reached at R_{sat} (1.43 km). The response received by the radar is shown at (c). At short range, saturated, response falls at 20 dB/decade, changing to 40 dB/decade above R_{sat}. Dashed lines

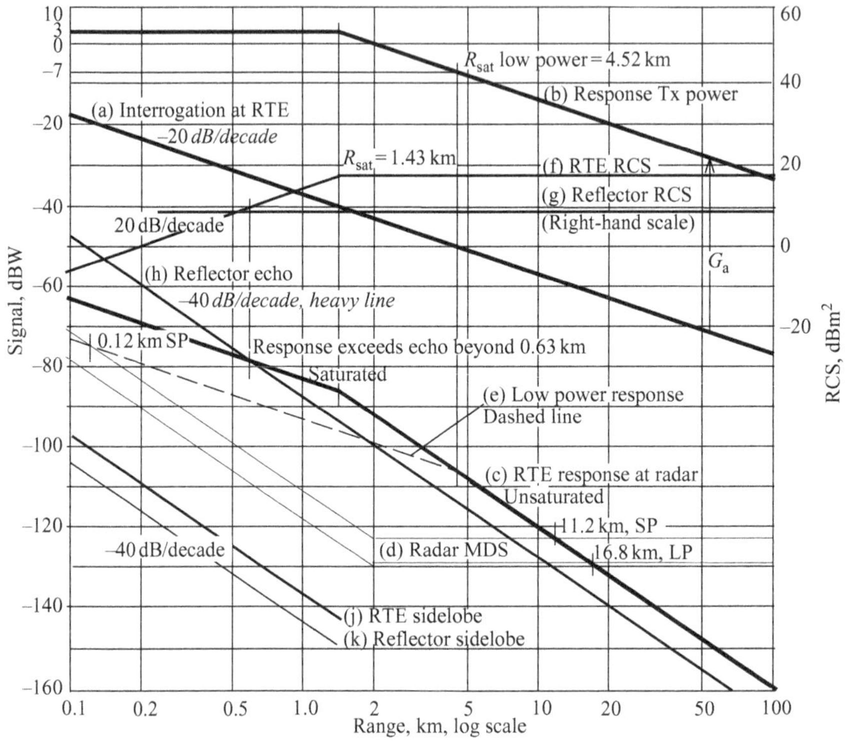

Figure 8.20 Comparison of RTE and reflector responses. Radar and reflector as in Figure 8.11

depict an RTE having 10 dB less rated power, raising R_{sat} to 4.52 km. At short range it delivers a weaker response which is more susceptible to loss in clutter or through swept gain; indeed it ceases to be detected below 0.12 km SP (and at 0.038 km LP, off the graph). At long range, both variants behave identically. The passive reflector (h) is detectable at all short ranges. A more powerful radar would raise curves (a), (b) and (c) bodily, intercepting the saturation ceiling at longer range.

In regard to slopes of the curves, at long range the RTE resembles a passive reflector, but below saturation range it becomes akin to a racon. The full-power RTE RCS (f) exceeds the 10 dB m^2 reflector beyond 0.63 km, whereas the low power variant does not do so until 1.99 km. Maximum detectable range lies in the unsaturated region and is 11.2 km SP and 16.8 km LP for the two variants.

RTE sidelobe response is shown at (j); saturation is reached at 0.06 km, off the graph. Above main beam saturation range (1.43 km) sidelobes are 56 dB below main beam response, but at shorter range the ratio falls until at 0.06 km it stabilises at 28 dB. Reflector sidelobes are shown at (k) and are always 56 dB below main beam echo. Sidelobes are many decibels below MDS and are undetectable in both cases.

8.10.14 Interaction

Can a pair of close-spaced RTEs continuously trigger one another? An expression for maximum interaction range is obtained by rearrangement of Eq. (8.15b) with $S_r = P_t - G_a$. Inserting $M = 6$ dB, its maximum possible value, gives

$$40 \log R = \sigma_{unsat} - 10 \log(4\pi) + 6 = \sigma_{unsat} - 8.97. \quad (8.25)$$

Substituting for the parameters of the above example, $R = 1.6$ m, showing interaction to be impossible at more than a few metres spacing between a pair of RTEs. This is much shorter than for a pair of racons per Eq. (8.7), because a weak interrogation evokes a weak response, rather than the full power of the racon. It is unlikely that a pair of RTEs will often come within interaction range; if they do, they will suffer increased supply power drain and may run at or near to maximum available duty. Responses to an interrogating radar may become infrequent. Both RTEs will lie within the same radar detection cell and they will be displayed as a single echo, as would a similarly spaced pair of passive reflectors.

8.10.15 Problems and opportunities

At time of writing, although two or three firms are addressing the Ports and Harbours and Buoyage Authority sectors, most RTE suppliers are concentrating on the private yacht market, which demands moderate 9 GHz performance at low cost. It is hoped that IMO performance specifications for twin-band RTEs will encourage relaxation of the current ITU-R limit of 10 W EIRP. Doing so might encourage production of twin-band RTEs of higher RCS, including models for yachts subject to very considerable angles of heel. High RCS RTEs of short saturation range are difficult to design. Amplifier gain has to be raised to a level at which ringaround is difficult to control, and rated response power becomes too high to be readily delivered by affordable solid state devices.

8.11 Miscellaneous devices

8.11.1 Scanner RCS

As mentioned in Chapter 7, Section 7.8.6, occasionally a radar may illuminate the scanner of another radar. When not transmitting, a scanner presents a moderate mismatch to incoming signals. To find the effective RCS of the mismatched scanner, it may be treated as an RTE having less than unity amplifier gain. Chapter 2, Section 2.6.2, Eq. (2.5) and Figure 2.16, connect return loss with voltage standing wave ratio (VSWR). For example, a rather poor VSWR of 2.0 has 9.54 dB return loss. On-axis RCS can be found by substitution in Eq. (8.18a), putting $G_a = -13$ dB. For a large 9 GHz ship's scanner of 34 dB gain and VSWR $= 2$, on-axis RCS is 17.5 dB m^2 or 56 m^2. This is usually insignificant; it exists through a very small solid angle (6 dB down at the nominal beam edges) and at its peak is likely to be less than the average superstructure RCS. Above their design frequency, slotted array scanners usually continue to reflect, but not at lower bands, due to waveguide cut-off; so 3 GHz scanners reflect well at 9 GHz but not vice versa.

8.11.2 Modulated reflectors

If the gain of an RTE, or the loss in the feed of a reflecting antenna, Section 8.11.1, is modulated in phase or amplitude, at say 10 MHz (perhaps by a PIN switch semiconductor element or a varactor variable-capacitance diode), it is possible to recover the modulation at a suitably adapted radar. The modulation therefore forms a response code which can, in principle, convey a limited amount of information such as station identity, or a reference point within a larger object. The modulation produces frequency sidebands and, again in principle, it is possible to operate the modulated reflector as a form of USIFAR (Section 8.5.4) to place the sidebands in a clutter-free area of the spectrum. Usually the sideband energy will be several decibels below the unmodulated echo and this modulation loss would have to be taken into account when calculating detectability. The author is not aware of any modulated reflectors currently in civil marine service, but has used a ferrite modulator and short-circuit behind a dish antenna to form an artificial Doppler-shifted 'moving' target for calibration of military radars. Of course, some energy has to be supplied to the modulator, but this energy is not available to augment response power.

Miyamotu *et al.* [3] have described how the principles underlying the phased patch array passive reflector, Chapter 7, Section 7.7.8, can be extended to introduce gain to form an active integrated retro-directive transponder. The elements of an antenna array similar to that described in Section 7.7.8 each connect with a mixer fed from a common double-frequency local oscillator and re-transmit at the conjugate phase. Using field effect transistors as mixers introduces power gain. If desired, the LO may be modulated to add a binary phase shift keyed identification modulation to the response. The author has not yet heard of such transponders entering service.

8.12 Target tilted in radar/target plane

8.12.1 General

Chapter 7, Section 7.9, showed that the effective RCS, and hence the echo strength, of passive aids depends on how they are viewed relative to the tilt plane. Response strength of active devices is subject to the same factors, plus some new ones. When viewed in the tilt plane, response strength depends on the device's elevation pattern. Viewed normal to that plane the interrogation and response may become partly cross-polarised, introducing significant attenuation at large tilt angles. In general, when the tilt plane is oblique to the scanner-target axis, the effects can be resolved into tilt- and cross-plane components.

The radar platform, if a ship, may also roll or tilt, either in or normal to the radar–target plane. Effective echo strength is then further modified by the scanner elevation polar diagram or its cross-polar characteristics respectively. We do not further consider this effect here, but assume the scanner/target direct and indirect (multipath) ray paths meet the target horizontally. Throughout the following analysis we assume uniform aperture illumination with the ordinary $(\sin x)/x$ pattern.

We let:

ν = instantaneous tilt angle

G_{sl} = antenna gain to slant polarisation (RTEs)

G_{eff} = antenna gain to the scanner's polarisation

S_{eff}, S_r = effective and on-axis response signals at radar, dBW

σ_{eff}, σ = effective and on-axis RCS, dB m^2.

When the radar lies in the target tilt plane:

- response strength and effective RCS are governed by the target's elevation polar diagram;
- planes of polarisation are unaffected without introduction of cross-polarisation losses;
- detailed performance is affected by the device type, discussed below.

8.12.2 Racons and SARTs

Gain of the antenna in receive mode is reduced when illuminated off its elevation axis, following the curve of Figure 8.21(a), reducing the maximum trigger range. Once triggered, there is no effect on generated response power and echo strength depends

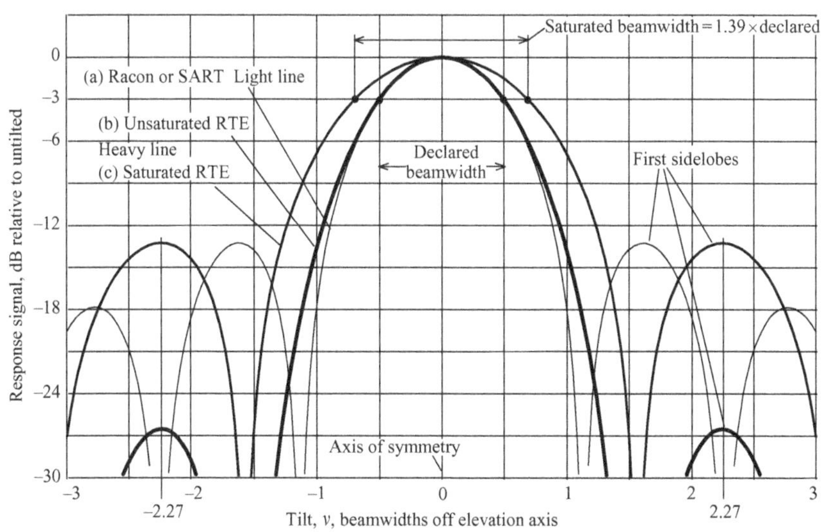

Figure 8.21 Signal variation with tilt. Radar in target tilt plane. Apart from shifts of the zero point, (a) is the elevation pattern of a racon antenna, (c) of a single RTE antenna and (b) of a pair of RTE antennas. Saturation increases RTE beamwidth

solely on transmit-mode antenna elevation pattern, Figure 8.21(a) again. For example, if the device has declared elevation beamwidth 30°, its antenna −3 dB one-way beamwidth, ϕ, is also 30° and echo strength is reduced by 3 dB at ±15° tilt. Received interrogation, response and effective RCS therefore each vary as Figure 8.21(a), which is of the form, where $x = 2.783 \times$ (elevation declared half − power beamwidth), as in Chapter 2, Section 2.8.1:

$$\text{off-axis value} = \text{on-axis value} + 20\log\left(\frac{\sin x}{x}\right)\ \text{dB}. \tag{8.26}$$

The peak value of the first sidelobe is only 13.3 dB down and occurs when $\nu = 1.69\phi$, perhaps delivering weak responses near 51° tilt. More importantly, the first nulls are when $\nu = \pm 1.1288\phi$, here at ±33.9°.

8.12.3 Unsaturated RTEs

Here response and effective RCS depend on both antenna gains (Eq. (8.18a) and are reduced by 3 dB when each antenna gain has fallen by 1.5 dB, so for a given declared beamwidth, the RTE antennas need wider beamwidth than does the racon. RTE declared beamwidth is (or should be) quoted in this unsaturated condition, so a device of 30° beamwidth again has its RCS reduced by 3 dB at ±15° tilt. Response power and effective RCS now depend on $[(\sin 0.720x)/(0.720x)]^4$ and follow Figure 8.21(b):

$$\text{off-axis value} = \text{on-axis value} + 40\log\left(\frac{\sin 0.72x}{0.72x}\right)\ \text{dB}. \tag{8.27}$$

This curve closely follows Eq. (8.26) and Figure 8.21(*a*) until tilt reaches half the declared beamwidth, then falls less sharply. Because two antennas are in play, the sidelobes are reduced in amplitude, with first sidelobe peak amplitudes of −26.6 dB, so there is little likelihood of detecting an RTE at extreme tilt via its elevation sidelobes.

8.12.4 Saturated RTEs

The device behaves rather like a racon. When tilted, reduction of effective receive antenna gain reduces saturation range (Eq. (8.21)), saturation then removing this antenna from play. The −3 dB beamwidth is 1.39× the −1.5 dB beamwidth, so effective beamwidth widens to 1.39× declared beamwidth. Response power and RCS of our 30° beamwidth device now reduce by 3 dB at $\frac{30}{2} \times 1.39 = 20.8°$ tilt. Response, RCS and single antenna elevation pattern follow Figure 8.21(c):

$$\text{off-axis value} = \text{on-axis value} + 20\log\left(\frac{\sin 0.72x}{0.72x}\right)\ \text{dB}. \tag{8.28}$$

If a racon and an RTE both have the same declared elevation beamwidth, the RTE antenna has to cover $\frac{1}{0.72}$ the solid angle, so has 0.72 (−1.42 dB) of the racon antenna gain. If the racon antenna gain was say 6 dB, the RTE antenna gain would be 4.58 dB, apart from a possible slant polarisation loss.

8.13 Target tilted normal to radar–target plane

8.13.1 General

To a radar normal to the target tilt plane, the target appears to tilt from side to side. Illumination is always equatorial but the plane of polarisation inclines to the target axis. Whether this changes effective RCS depends on whether target RCS is sensitive to polarisation. In general RCS is maintained up to at least 15° tilt. When tilted, the following apply.

1. The target elevation pattern is immaterial.
2. If the target receive antenna has the same nominal plane of polarisation as the scanner (racon, SART or some RTEs), it incurs a polarisation loss which makes it less sensitive to interrogations.
3. The response polarisation of these devices is inclined to the vertical. The scanner receives a slanted polarisation, incurring response loss.
4. If the device antenna uses slant polarisation (some RTEs), there will be a receive loss or gain depending on the direction of tilt relative to the slant, only partly cancelled by the equivalent response gain or loss at the scanner.

If the antenna is not slant polarised, the voltage vector depends on $\cos \nu$ so gain falls by $20 \log \cos \nu$ dB. Many RTE tilt properties follow from the fact that effective gain, G_{eff}, of a 45° slant polarised antenna depends on ν:

$$G_{\text{eff}} = G_{\text{sl}} + 20 \log \cos \left(\frac{\pi}{4} + \nu \right) \text{ dB}. \tag{8.29a}$$

When $\nu = 0$ (no tilt), $G_{\text{eff}} = G_{\text{sl}} - 3$ dB and G_{eff} is here the ordinary gain G_{r} so we can write

$$G_{\text{eff}} = G_{\text{r}} + 20 \log \cos \left(\frac{\pi}{4} + \nu \right) - 3 \text{ dB}. \tag{8.29b}$$

When $\nu = -\pi/4$, $G_{\text{eff}} = G_{\text{sl}}$, raising antenna gain by 3 dB; but when $\nu = \pi/4$, $G_{\text{eff}} = -\infty$ dB and the device cannot respond.

8.13.2 Horizontally polarised racons, SARTs and saturated RTEs; linearly polarised scanner

Only the response antenna function is in play. Antenna gain is maximum when untilted and fall to zero ($-\infty$ dB) when 90° cross-polarised, at 90° tilt. Response and RCS are given by:

$$\text{off-axis value} = \text{on-axis value} + 20 \log \cos \nu \text{ dB}. \tag{8.30}$$

Figure 8.22(a) refers. Response is 1 dB down at ±27°, 3 dB down at ±45° and 6 dB down at ±60°.

8.13.3 Circularly polarised 3 GHz band racons

Although 9 GHz band racons and SARTs are all horizontally polarised, it is permissible for 3 GHz band racons to use circular polarisation to respond to horizontally

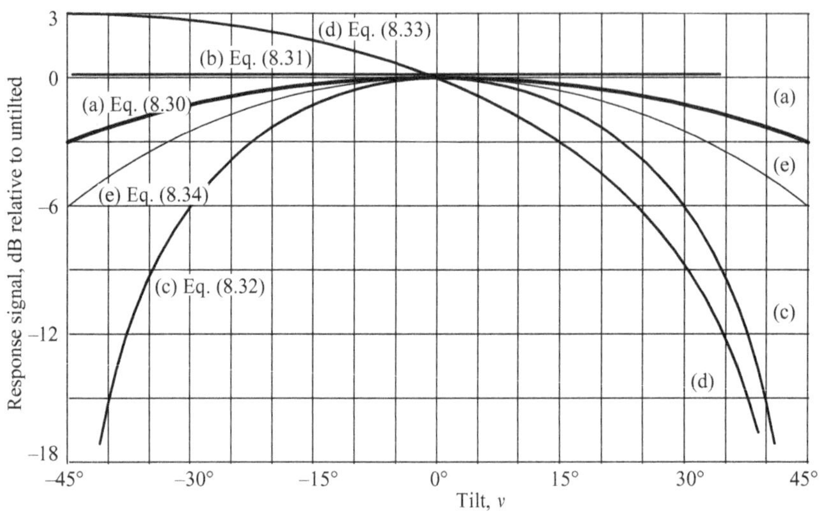

Figure 8.22 *Change of RCS with tilt. Radar normal to tilt plane. RCS change in this plane depends on cross-polarisation rather than elevation pattern. For targets following Eqs (8.30)–(8.34)*

and vertically polarised radars. These will work normally when tilted, with no performance loss, Figure 8.22(*b*).

$$\text{Off-axis value} = \text{on-axis value}. \tag{8.31}$$

8.13.4 Unsaturated RTEs, slant polarised antennas, linearly polarised scanner

When 45° slant polarisation is employed to prevent ringaround, Eq. (8.29a) shows the receiving antenna effective gain (dB) depends on $20 \log \cos(\pi/4 + \nu)$, but the scanner will be presented with polarisation at $\pi/4 - \nu$, equivalent to raising the transmit antenna gain. The on-axis value of G_o ($G_o = G_a + G_r + G_t$, Eq. (8.10)), is replaced by

$$G_o = G_a + \left[G_r + 20 \log \cos\left(\frac{\pi}{4} + \nu\right)\right] + \left[G_t + 20 \log \cos\left(\frac{\pi}{4} - \nu\right)\right] \text{ dB}.$$

Response and RCS are given by the following equation, shown in Figure 8.22(*c*). The overall effect is to give a broad symmetrical pattern having 45° between half-power points.

$$\text{Off-axis value} = \text{on-axis value} + 20 \log \cos\left(\frac{\pi}{4} + \nu\right) + 20 \log \cos\left(\frac{\pi}{4} - \nu\right) + 6 \text{ dB}. \tag{8.32}$$

The factor 6 sets to the correct on-axis value when $\nu = 0$. Loss is 3 dB when $\nu = \pm 22.5°$ and is ∞ dB at $\pm 45°$, where there can be no response.

8.13.5 Saturated RTEs, slant polarisation, linearly polarised scanner

Slant polarisation gives stronger response and RCS if tilted to bring response polarisation more into the radar's polarisation plane, but weaker when tilted the other way.

$$\text{Off-axis value} = \text{on-axis value} + 20 \log \cos\left(\frac{\pi}{4} + \nu\right) + 3\,\text{dB}. \tag{8.33}$$

Figure 8.22(d) shows that to negative tilt there is a rise, peaking at +3 dB at −45°; to positive tilt the fall is 3 dB at $\nu = 15°$.

8.13.6 Slant polarised RTEs, circularly polarised scanner

Certain VTS scanners have the facility to switch to circular polarisation to defeat rain clutter. The RTE antenna takes one plane polarised component of the interrogation, rejecting the other, irrespective of tilt. The scanner accepts any plane of response polarisation equally well and tilt does not reduce echo strength. Eq. (8.31) and Figure 8.22(b) apply.

8.13.7 Unsaturated RTE without slant polarisation

This has the $20 \log \cos \nu$ law for each antenna, so overall:

$$\text{Off-axis value} = \text{on-axis value} + 40 \log \cos \nu\ \text{dB}. \tag{8.34}$$

Figure 8.22(e) shows response and RCS are 3 dB down at ±22.5° and 6 dB down at ±30°.

8.14 Target tilted oblique to radar–target plane

Usually, the radar will be neither exactly within nor normal to the target tilt plane, but at some oblique azimuth angle ω. Effective tilts now become:

$$\begin{aligned} &\text{tilt plane component} = \nu \cos \omega, \\ &\text{normal plane component} = \nu \sin \omega. \end{aligned} \tag{8.35}$$

The total dB reduction in response or RCS is the sum of the component reductions.

Example. Suppose an RTE using slant antennas has declared (unsaturated) elevation beamwidth 25° between half-power points. It is tilted by 30° and viewed by a radar bearing 60° to the tilt plane.

The tilt plane component is $30 \cos 60° = 15°$ and the normal component is $30 \sin 60° = 26°$.

At long range, unsaturated, Eq. (8.31) and Figure 8.22(b) apply in the tilt plane. The tilt component is $\frac{15}{25} = 0.6$ beamwidths and the response is reduced by 4.4 dB.

In the normal plane Eq. (8.32) and Figure 8.22(c) apply. At $\nu = 26°$ the response is reduced by 4.2 dB. The total attenuation is $4.4 + 4.2 = 8.6$ dB.

At short range, saturated, Eq. (8.30) and Figure 8.22(a) apply; the tilt plane component of signal reduction is halved from 4.4 to 2.2 dB. The normal component is unaltered so the total reduction is $2.2 + 4.2 = 6.4$ dB.

8.15 RTE plus passive point target in free space

8.15.1 Introduction

It is sometimes thought prudent to provide a back-up passive reflector in case an RTE or its power supply should fail. RTEs respond at the same time and at the same frequency as the skin/reflector echo, and the radar receives both together. In this section, we find the resultant, generally by following the method used in Chapter 7, Section 7.10, for a pair of passive target elements, and at first assume free space conditions for simplicity. We shall find that RCS becomes very variable.

8.15.2 RTE below reflector

We start by considering the combination of perfect RTE and perfect point reflector in free space, illuminated by a distant radar. The antennas of RTEs are always mounted one above the other to maintain the required plane of polarisation. The vertical plane, with the passive reflector mounted above the RTE, is shown in Figure 8.23(a).

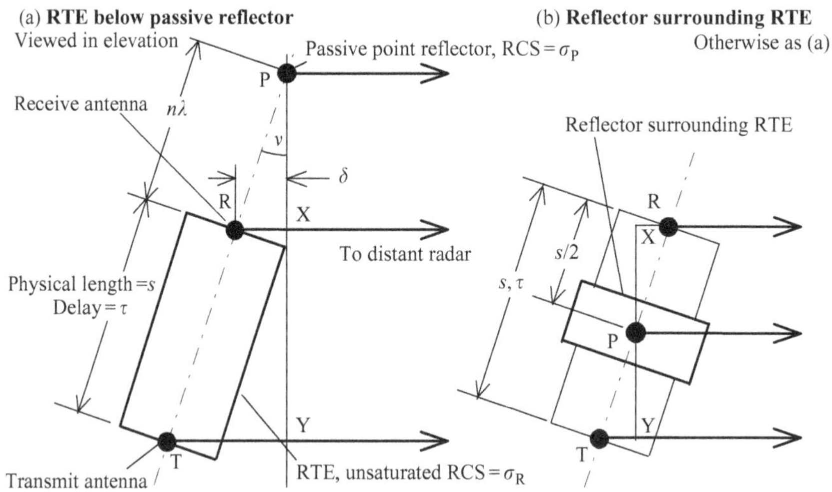

Figure 8.23 Passive reflector plus RTE. The reflector is above the RTE at (a) but surrounds it at (b), minimising pathlength differences

The following assumptions are not all likely to remain fully valid in a practical system.

- Free space conditions, with neither indirect ray from the sea surface nor atmospheric attenuation.
- Scanner so distant that rays striking elements are parallel.
- Elements so close together that scanner gain is same to each.
- Reflections from one element do not strike the other, without second bounce.
- The radar–targets plane is vertical, giving the elevation diagram for one element above the other.
- The RTE is unsaturated.

Notation:

σ_R = RCS of point reflector element P,

σ_2 = unsaturated RCS of RTE with receiving antenna R above transmitting antenna T,

σ_r = equivalent RCS of reflector and RTE together,

n = number of wavelengths spacing between P and R,

δ = range difference between P and R,

ν = radar bearing relative to reflector/RTE broadside,

V_1, V_2 = voltage vectors of responses from R and T,

ϕ = phase shift of V_2 on V_1,

s = separation of RTE antennas, m,

τ = delay between RTE reception and response, s,

f, λ = radar frequency, Hz and wavelength, m.

From Figure 8.23(a) we see that, relative to P, the interrogation at R lags by δ/λ wavelengths. Following Chapter 7, Section 7.10.3, this is equivalent to phase shift $2\pi\delta/\lambda = (2\pi/\lambda) \times n\lambda \sin \nu$ rad. There is phase shift $2\pi f \tau$ rad in the RTE. Additionally the response lags by the path length TQ $= (n\lambda + s) \sin \nu$ m, whose lag is $(2\pi/\lambda) \times (n\lambda + s) \sin \nu$ rad.

The total phase shift of the RTE response relative to the reflector echo is the sum of the above:

$$\phi = 2\pi \left(2n + \frac{\sigma}{\lambda}\right) \sin \nu + 2\pi f \tau \text{ rad.} \tag{8.36}$$

We noted that a pair of passive elements always had $\phi = 0$ where $\nu = 0$, giving an RCS peak on broadside. But the RTE internal delay remains, even when $\phi = 0$. The designer may have arranged the RTE output to be antiphase to its input at mid-band to minimise the ring-round problem, giving a null on broadside, but one cannot be sure.

When ν increases from 0 to $\pi/2$, ϕ increases by $2\pi(2n+\sigma/\lambda)$. There are therefore $2n+\sigma/\lambda$ lobes in the quadrant and the average width between nulls is

$$\text{lobe width} \sim \frac{2}{\pi} \times \left(2n + \frac{\sigma}{\lambda}\right) \text{ rad.} \tag{8.37}$$

Peak and null RCS, σ_r and σ_1, respectively, are given by Chapter 7, Section 7.10.3:

$$\frac{\sigma_r}{\sigma_1} = 1 + \left(\frac{\sigma_2}{\sigma_1}\right) \pm 2\sqrt{\frac{\sigma_2}{\sigma_1}}.$$

Except that peaks are not tied to $\nu = 0$, Figures 7.15–7.18 of Chapter 7, Section 7.10, apply. The elements are symmetrical about the centre line so the response in the azimuth plane is constant and the TPM takes the form of parallel stripes as shown in Figure 7.18. From Eq. (8.36) it is plain that the resultant RCS on any particular elevation angle ν will change rapidly with wavelength.

8.15.3 Passive reflector surrounding RTE

The rapid variation is undesirable. Can it be minimised? The best arrangement might be to surround the RTE with reflector elements, e.g. clustered trihedrals, with one RTE antenna above and the other below, as shown in Figure 8.23(b). Time τ would be minimised and adjusted to give an RCS peak at the middle of the frequency band. In practice, the designer would adjust τ to compensate for any phase shift in the reflector, whose effective centre would probably not coincide with the centreline. Bandwidth would be affected by any dispersion in the feeders and amplifiers within the RTE, and any change of τ when the RTE saturates. It turns out that the maximum value of τ (~3 ns) is the same for the 3 and 9 GHz bands. This setting might exacerbate the ringaround problem.

As angle ν increases from zero, path length RP falls but TQ rises by an equal amount, so the response remains at a peak; in other words the combination would be limited only by the antenna and trihedral elevation beamwidths, which can be made wide. In effect the reflector and RTE occupy the same point.

For example, if $\tau = 0.333$ ns, corresponding to RTE range error 0.35 m (with s approximating this length), at mid-band (say 3000 MHz), delay is 7.0 cycles and $\phi = 0$ so RTE and reflector are in phase. At band edges (2900 and 3100 MHz), the delay is 7.0±0.233 cycles with residual phase shifts $\phi = \pm(2\pi \times 0.233) = \pm 1.47$ rad; here the echoes are almost at phase quadrature and overall RCS never falls below that of the stronger device alone, normally the RTE. Unfortunately it is questionable whether there would be room for a passive reflector of worthwhile RCS.

8.15.4 Delayed RTE

Figure 8.24(*a*) is a timing diagram of a conventional reflector/RTE pair. The RTE response is delayed only slightly and the echo and response overlap almost completely. By increasing the RTE delay, τ, to say 0.25 μs, the response could be thrown clear of the echo to short pulses. Long pulses would still overlap, but even at a null would present the radar with first the skin echo with its own TPM, then the null and finally

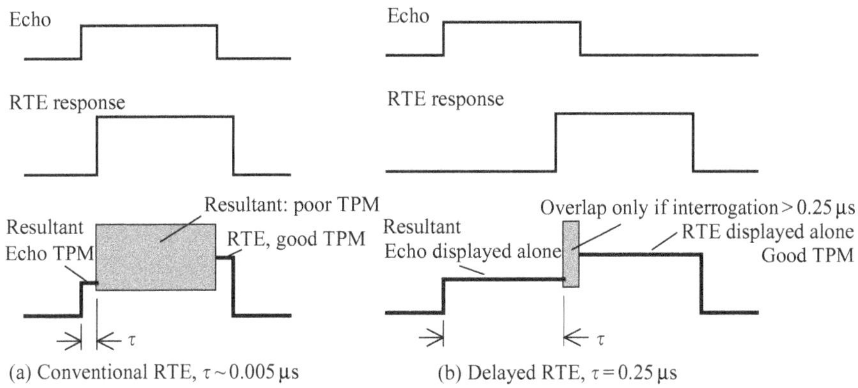

Figure 8.24 Delayed RTE response. Introduction of delay disengages the response from the echo to avoid degradation of TPM

with the RTE alone, with its smooth TPM, Figure 8.24(b). The response would lie immediately beyond the reflector echo. RTE delay could be introduced in several ways. A passive microwave delay line is conceptually simplest, but the designer might prefer to mix down, introduce analog or digital delay at IF and mix up to microwave. Whatever approach is taken, cost is added and the IMO performance specification, which currently allows only minimal delay, would require amendment (IMO currently allow racons 0.5 μs delay). Delay introduces range error, but 0.25 μs, corresponding to 38 m, might be acceptable if it increased probability of detection.

If the reflector is placed to one side of the RTE, azimuth polar diagrams take the form of Figure 7.15 and the TPM becomes more complex.

8.15.5 Practical conditions, RTE/reflector pair

All the conclusions reached for a pair of passive elements in Chapter 7, Section 7.10.8, apply. Skin echo and echoes of any passive reflectors on a platform precede the response from any racon or SART it may carry. Often the unsaturated RTE's RCS will much exceed that of the platform and will dominate performance, the skin introducing a reasonably small ripple. But as range closes and the RTE saturates, its RCS falls and the pattern ripple increases. At some short range the RCSs become equal and the nulls become infinitely deep; at some elevations if the RTE is above or below the reflector, or at some azimuth bearings if alongside it, precluding detection at these null angles.

8.15.6 Practical conditions, racons, SARTs and ramarks

The first microsecond or so of a racon response may overlie the skin echo, Figures 8.4(e) and 8.6(g). There will always be some residual response frequency error, so the resultant amplitude will beat at error frequency (which should be less than 1 MHz); one scan may be at a beat peak and the next at a null. But in general the response of all these devices occurs at a different time when there is no echo. The device polar TPM is therefore unaffected by the RCS of the target and any passive reflectors it may carry – another reason to prefer racons to RTEs for AtoN service.

8.16 References

1 MEIKLE, H.: 'Modern radar systems' (Artech House, London, 2001), Eq. (13.20)
2 DIAZA: 'Experience with search and rescue transponders or SARTs', *Journal of Navigation*, 2002, **55** (1), pp. 65–74
3 MIYAMOTU, R. Y., QIAN, Y., and ITOH, T.: 'An active integrated retrodirective transponder for remote information retrieval-on-demand', *IEEE Transactions on Microwave Theory and Techniques*, 2001, **49** (9)

Chapter 9

Multipath factor of extended targets

'Blow, bugle, blow, set the wild echoes flying.
Blow bugle: answer, echoes, dying, dying, dying.'

Tennyson

9.1 Introduction

9.1.1 The problem

Previous chapters considered point targets – so physically small that only single pairs of direct and indirect rays exist, giving definite multipath structures within the interference region. Point targets also usually have well-defined and broad (often omnidirectional) spatial target pattern maps (TPM) or radiation patterns, so their RCS does not vary much as they tilt. Although sea-waves may modulate their multipath factor, received signal is reasonably steady. In this chapter we consider how multipath affects vertically extended targets, such as ships and coastlines, whose structure extends over a wide height bracket. We shall frequently recall principles established in Chapters 5 and 6. Atmospheric and precipitation attenuations do not affect the arguments deployed in this chapter and will be ignored. We defer target roll to Chapter 10, which examines RCS values.

Vertically extended targets contain many scatterers or individual elements which combine to reflect through a height bracket extending from the waterline up to the target tip, at height j metres. This configuration introduces apparent complications which in the end turn out to allow simplification: (a) one element of the target may be in the interference region while a lower element is in the diffraction region; (b) direct and indirect rays striking low elements have different phases and amplitudes from rays hitting high elements. The element reflections combine to form the echo received at the scanner, so we cannot directly apply our theory of point targets, with its single reflection element.

Targets may also be wide enough to overflow the scanner azimuth beamwidth so that part only of their width is illuminated. Such horizontally extended targets include large ships at the shorter ranges and some coastlines at all ranges. Targets are only likely to overflow in elevation at short range with a few high gain, narrow elevation beamwidth, VTS scanners whose received echo can be treated by reducing RCS as range closes according to the geometry of the particular case. We assume target total RCS at the aspect presented to the radar is known (Chapter 10 gives values); we also assume rough micro-geometry, typical of most large targets in practice. When the surface is flat or smooth within a quarter-wavelength, the target acts as a point and is not covered by this chapter; some typical smooth shapes such as flat plates and cylinders were considered in Chapter 7. Precise calculation is so difficult we shall have to make approximations.

9.1.2 Target echo

Figure 9.1 depicts a vertically extended target approaching a scanner from long range, as did Figure 6.1 for a point target. At extreme range – beyond the horizon, point W – the target is in the diffraction region; the effective multipath coefficient is low and the echo is very weak. As range closes to X, the target tip projects into the interference region and echo strength rises. Later, at Y the target spans the first lobe. As range shortens yet further to Z, elements of the target are illuminated by several lobes. Here when one element comes to be illuminated by a lobe, another enters a null, so the total energy hitting the target as a whole is little affected by the precise lobe structure. The whole process recurs on the return leg and overall echo strength has

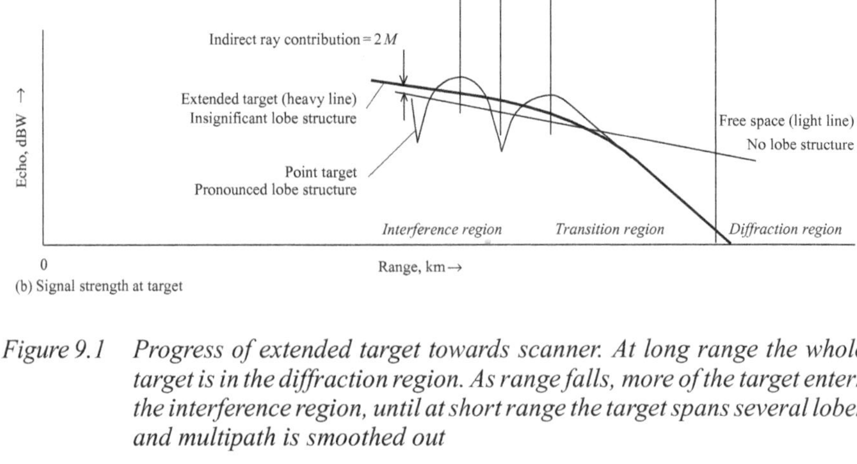

Figure 9.1 Progress of extended target towards scanner. At long range the whole target is in the diffraction region. As range falls, more of the target enters the interference region, until at short range the target spans several lobes and multipath is smoothed out

little or no nett multipath fluctuation as range changes, even in calm water. Multipath interference is therefore much less than for point targets.

The plating of extended targets such as ships is rarely if ever flat to a quarter wavelength and most coastal features are yet more irregular. Micro-structure is rough, so phase of the echo from one scattering element remains approximately independent of the phasing of elements at other heights. Such targets do not behave as a single point, but may be regarded as an assembly of independent decorrelated reflectors. The indirect rays contribute some additional energy, so the multipath factor is nearly constant and somewhat greater than unity. As always, multipath factor is unity (0 dB) in free space where there is no sea surface to give an indirect ray. We next quantify this behaviour.

9.2 Multipath of extended target, summation method

9.2.1 Summation of element echoes

We first consider a simple form of extended target, a rough vertical cylinder of uniform RCS per unit height. We develop multipath factors, enabling us to calculate the echo strength at any range of interest. From our knowledge of the multipath of point targets we develop a formula for the total echo from the cylinder, then use it as a basis for calculation of echoes from more complicated practical targets, such as ships, whose RCS per unit height changes with height.

To get a feel for the total echo, we start by dividing the cylinder into ten height elements, each small enough to approximate a point, then calculate the echo power of each element. The surface roughness causes the echoes to be randomly phased, so we can find the resultant echo power by simple addition of the element echo powers. (The equal voltage vectors of two equiphased elements add and resultant echo power $= 4 \times$ element echo. But for random phasing between 0 and 2π, the in-phase components deliver power which is dependent on the average height of a $\sin^2$ function: taken over 2π rad, $1/2\pi \int_0^{\pi} (\sin^2 x)\,\mathrm{d}x = \frac{1}{2}$, so resultant echo = sum of element echoes.) Ten elements give an approximation to true integration, which sums an infinite number of infinitesimal element echoes. Comparison of the resultant with the target's free space echo (where there is no indirect ray, so extended targets behave exactly as points) gives the two-way multipath factor.

9.2.2 Uniform RCS distribution; critical range

Figure 9.2(a) shows a cylinder having about the RCS of a coaster ship. Overall height, j, is 10 m and RCS is $1000\,\mathrm{m}^2$, divided into 10 equal elements each $100\,\mathrm{m}^2$ and 1 m high, centred on $h = 0.5, 1.5, \ldots, 9.5$ m, randomly staggered in range by a quarter wavelength or more, so the echoes have random phase.

Each element echo has been calculated as a point target by the detailed method of Chapter 6 (Sections 6.4–6.7), including the full effects of diffraction and the range-dependence of sea surface reflection coefficient. We have assumed sea state 0, giving the 'worst case' strongest multipath fluctuation as range changes. The echoes of four

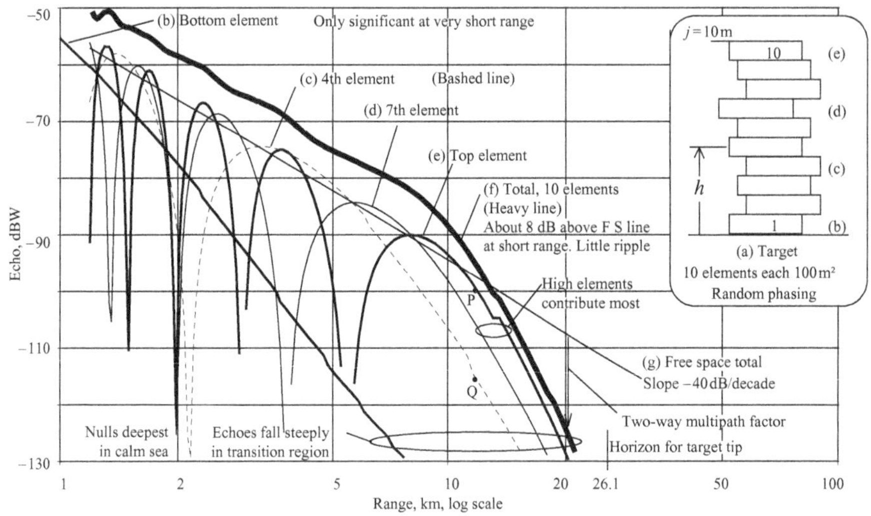

Figure 9.2 *Total echo as sum of elements. First (bottom), 4th, 7th and 10th (top) element echoes and sum of all ten. Peaks of one element's echo are balanced by nulls of another's, so overall echo has little ripple; most of that would disappear if many more elements had been taken. The indirect rays contribute enough to raise the echo a few decibels above the free space line at short range. The echo plunges steeply near the horizon. For points P and Q see Section 9.3.1. Exact shape depends on parameters, here: 9410 MHz, $H = 10$ m, $j = 10$ m, $k = 1.333$. Strong sea surface reflection; sea state 0*

representative elements are shown at (b), (c), (d) and (e). Each exhibits the familiar strong multipath pattern of peaks and nulls, tailing off beyond its interference region, through its transition region to its diffraction region beyond its own particular horizon. The lowest element has such a short horizon that it exits its interference region at very short range (transition range $R_A = 1.2$ km, diffraction range $R_B = 11.9$ km) and contributes little. At long range, the total echo comes largely from the upper elements. As one would expect, the topmost element contributes particularly strongly through the whole range bracket. In general, it can be seen that when one element's echo is at a peak, another's is near a null.

The sum of all 10 echoes is almost free of ripple and at short range lies about 8 dB (4 dB each way) above the free-space line, curve (g), which falls at −40 dB/decade in the usual way (R^{-4} law). This 8 dB represents the nett contribution of the indirect rays, would fall if high sea state reduced the sea surface reflection coefficient and is equivalent to one-way nett multipath factor 4 dB. Above a roll-off, transition or critical range (Section 9.5.1) at about 8.4 km, the resultant echo falls more sharply as the target sinks below the horizon, fewer and fewer elements making an effective contribution. Changing RCS by 1 dB here causes well under 1 km change in detectable range, indicating that imprecision in our knowledge of target RCS has less effect on maximum detection range than might be expected. Some of the residual ripple shown in the figure arises from the rather small number of elements integrated, some from the granularity of the spreadsheet used for the calculations, while some is genuine.

9.2.3 Multipath factor

Chapter 6, Section 6.4.2, Eq. (6.6) showed that, in the interference region, average one-way multipath factor for a point target above a perfectly reflecting surface was 2.727 (4.36 dB), close to our approximate 4 dB in a calm (not quite totally reflecting) sea. In the diffraction region, Section 6.5.3 (Eq. 6.12) gave the point target rate of change of multipath factor with range; the extended target has similar performance beyond the interference region.

Figure 9.3 is for the system of Figure 9.2; (a) shows the RCS distribution, while (f) repeats the summed echo from the ten elements, together with the echoes from the top six (b) and bottom six elements (d), respectively. Curve (b) lies close to (f), showing that deletion of the bottom four elements barely affects the resultant, confirming Figure 9.2. This indicates that ships' hulls do not contribute strongly to long-range RCS. On the other hand, removal of the top four elements at (d) (e.g. the superstructure) reduces the long-range echo by 10 dB. At short range RCS falls from 1000 to 600 m^2 or 2.2 dB and critical range falls. It follows that long range detection of ships and coastal features depends principally on the RCS of the upperworks.

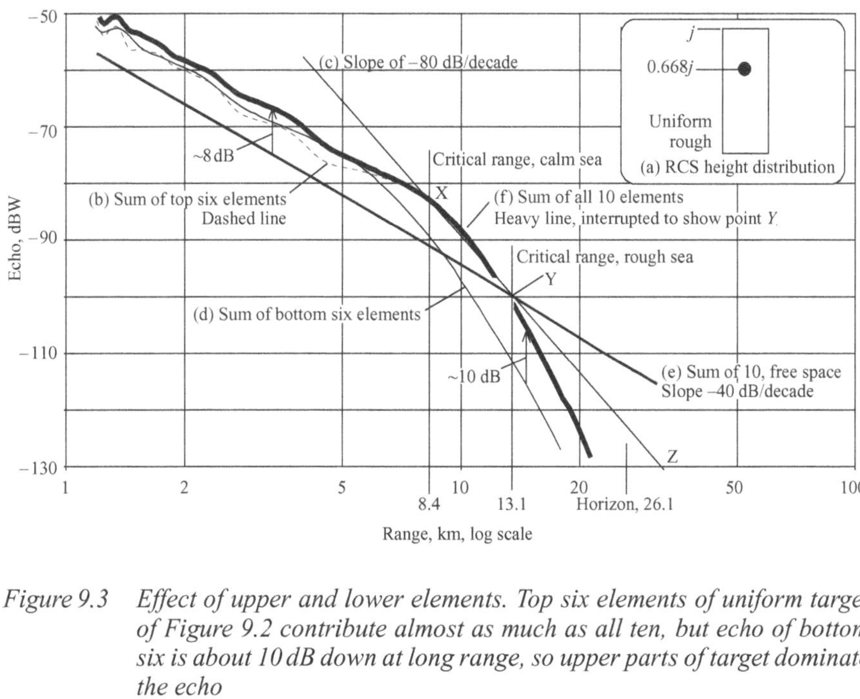

Figure 9.3 Effect of upper and lower elements. Top six elements of uniform target of Figure 9.2 contribute almost as much as all ten, but echo of bottom six is about 10 dB down at long range, so upper parts of target dominate the echo

At short range the total echo, (f), is lifted about 8 dB above the free space echo, (e), by the two-way multipath factor. At long range the total echo falls below the free space line because of diffraction effects.

9.3 Diffraction and transition regions

9.3.1 Echo variation with element height

We use numerical quantities rather than decibels in most of this section and make some approximations to reduce the number of variables. As before, we assume that the target is rough and lies wholly within the scanner beam at constant range. Constants C_1–C_5 do not require to be evaluated.

Chapter 6, Section 6.5.2, included an expression (Eq. 6.7a) for the multipath factor of a point target in the diffraction region:

$$\sqrt{m_{\mathrm{d}}} = 2\sqrt{\frac{\pi R}{L}} f(Z) f(z) \exp \frac{-2.02R}{L},$$

where $f(Z)$ and $f(z)$ are functions of the 'natural heights' of scanner and point target respectively. In a given system at fixed range. Eq. (6.7a) has the form

$$\sqrt{m_{\mathrm{d}}} = C_1 f(z). \tag{9.1}$$

If H or $h < 31$ m (3 GHz) or <14 m (9 GHz), $z < 3$, $f(z) = z$ and $z \propto h$, so we can put

$$\sqrt{m_{\mathrm{d}}} = C_2 h, \qquad \text{therefore } m_{\mathrm{d}} = C_3 h^2. \tag{9.2}$$

When target RCS $= \sigma$, the echo, S_{p}, is of the form

$$S_{\mathrm{p}} = C_4 \sigma m_{\mathrm{d}}^2, \qquad \text{therefore } S_{\mathrm{p}} = C_5 \sigma h^4. \tag{9.3}$$

This shows that, for a constant total RCS, long range echo strength depends on the fourth power of target height, so doubling height increases echo strength by 12 dB. For example, the echo of point P on the 9.5 m element of Figure 6.2 is seen to be about 16 dB above point Q on the 3.5 m element, confirmed by Eq. (9.3): $10 \log(9.5/3.5)^4 = 17.3$ dB.

9.3.2 Integration, uniform target; height factor

We can think of an extended target having total RCS σ m^2 as a stack of very many elements whose individual incremental thickness dh is small enough for each to behave as a point target at its height h metres above sea level, h varying between 0 for the lowest and j for the highest element. We at first continue to assume the target is our rough uniform cylinder, each element having the same RCS, so the RCS per unit height is constant, with random phase, Figure 9.3(a).

$$\text{Element RCS} = \frac{\sigma\,\mathrm{d}h}{j}\ \mathrm{m}^2.$$

From Eq. (9.3), its echo, S_p, is

$$S_p = C_5\left(\frac{\sigma\,dh}{j}\right)h^4 = C_5\left(\frac{\sigma}{j}\right)h^4 dh \text{ W}. \tag{9.4a}$$

Integration with respect to h over the interval $h = 0$ to $h = j$ gives the total echo, S_{tgt}:

$$S_{tgt} = \int_{h=0}^{j} C_5\left(\frac{\sigma}{j}\right)h^4\,dh = \left[C_5\left(\frac{\sigma}{5j}\right)h^5\right]_0^j = \frac{C_5}{5}\sigma j^4. \tag{9.4b}$$

Eq. (9.3) indicates the echo, S_p, from a point target at height nj (where n is the target's height factor) is

$$S_p = C_5\sigma(nj)^4. \tag{9.4c}$$

The point target is equivalent to the extended target when their echoes are the same, $S_p = S_{tgt}$. Substituting in Eqs (9.4b) and (9.4c):

$$C_5\sigma(nj)^4 = \frac{C_5\sigma j^4}{5}$$

$$\therefore \quad n = 5^{-1/4} = 0.668 \sim \frac{2}{3}. \tag{9.5a}$$

Thus, at long range a cylindrical target behaves as a point target at height $h = 0.668 \times$ (target tip height, j), provided the scanner and target 'natural heights' (Chapter 6, Section 6.5.2) are less than about 3 (31 m at 3 GHz or 14 m at 9 GHz). The relationship remains roughly correct at heights up to twice these values, see Figure 6.8, curves (b) and (d). The equivalence is valid for all values of refraction k-factor unless ducting is present. In general, effective height (sometimes called the centroid height) h is given by

$$h = nj. \tag{9.5b}$$

9.3.3 Non-uniform target

What happens if the element RCS varies with height? As an example, we take a bottom-heavy target model, Figure 9.4(a), whose element RCS falls linearly with height, so element RCS $= 2\sigma/j\,(1 - h/j)\,dh$. For this particular target, the bottom

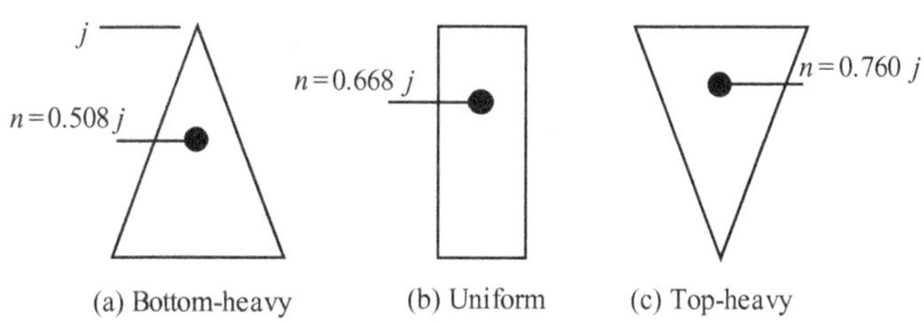

Figure 9.4 RCS distributions of model targets. Width represents RCS

10 per cent of height contributes 19 times (+12.8 dB) the RCS of the top 10 per cent. Substituting $2\sigma/j(1-h/j)\mathrm{d}h$ for the $(\sigma/j)\mathrm{d}h$ term in Eq. (9.4a) and working through to Eq. (9.5a),

$$n = 15^{-1/4} = 0.508 \sim \frac{1}{2}. \tag{9.5c}$$

Figure 9.4(b) represents the uniform target model already considered, where $n = 0.668$. Repeating with target (a) inverted so its RCS per unit height is top-heavy, rising linearly from zero at the waterline, Figure 9.4(c), with element RCS $= 2\sigma\,\mathrm{d}h/j = 2\sigma/j^2 h\,\mathrm{d}h$ gives

$$n = 3^{-1/4} = 0.760 \sim \frac{3}{4}. \tag{9.5d}$$

Chapter 6, Section 6.3.3, Eq. (6.2a) gave the transition range in the flat-Earth approximation for a point target of height h:

$$R_{\mathrm{A}} = \frac{8Hh}{\lambda}.$$

By analogy, the transition range for an extended target is at

$$R_{\mathrm{A}} = \frac{8Hnj}{\lambda}. \tag{9.6a}$$

Assuming uniform distribution with $n = 0.668$,

$$R_{\mathrm{A}} = \frac{5.34Hj}{\lambda}. \tag{9.6b}$$

The factor '5.34' varies between 4.06 (bottom-heavy model, $n = 0.508$) and 6.08 (top-heavy, $n = 0.76$).

9.3.4 *Choice of target height factor*

Usually in real life there is no firm evidence of how the target RCS is distributed with height. Luckily, the above examples show that the $-\frac{1}{4}$ exponent in Eq. (9.5) reduces the effect of RCS height distribution. If the top 10 per cent has RCS lying between ±12 dB of the bottom 10 per cent, it may be reasonable to assume that the effective centre of reflection lies somewhere between $\frac{1}{2}$ and $\frac{3}{4}$ the tip height, the latter being the highest point on the target judged to reflect significantly. Of course, Figures 9.4(a) and (c) do not represent the minimum and maximum values of n which might occur in practice. Putting $n = 0.5$ (approximating the bottom-heavy model) should give reasonably conservative detection ranges for most targets. Here

$$R_{\mathrm{A}} \sim \frac{4Hj}{\lambda}. \tag{9.6c}$$

9.4 Interference region

9.4.1 Ray geometry, cylindrical target

Numerical summation as used in Section 9.2 is tedious and not well suited to spreadsheet calculation. Neither does it yield a formula for multipath factor ready for insertion in the radar range equation. We shall derive such a formula by integration, the formal mathematical equivalent of numerical summation. Integration is the inverse of differentiation. $\int_0^h f(x)\,\mathrm{d}x$ means the definite integral of function $f(x)$ with respect to x between the limits $x = 0$ and $x = h$. It is the area under the graph of $y = f(x)$ between these limits. We shall be using definite integrals so can omit the constant of integration from the preliminary indefinite integrals. Integration takes the summation process to its limit, using an infinite number of infinitely small elements. The complete expression for target echo throughout the interference, transition and diffraction regions is too complex to be integrated with mathematical rigour. However, integration is feasible within the interference and near diffraction regions and gives a useful formula for multipath factor where it matters most.

We start the formal analysis with a uniform target of many elements extending from the sea surface to the target tip at height j, similar to Figure 9.2(a), as indicated in Figure 9.5. The radar is at range R, with scanner height H. The indirect ray grazing angle, α; reflection coefficient ρ of the indirect ray to the element, and the relative illumination of the scanner beam, g_{dif} all differ from element to element. As we know from Chapter 5, Section 5.5, the effective scanner and target heights, H' and h', respectively, also change with range, as does the indirect ray phase shift Φ.

9.4.2 Element multipath factor, flat Earth

We assume the Earth is flat ($k = \infty$), with no atmospheric loss, use the Chapter 5 notation and put:

$\psi = \pi$ (phase reversal at sea surface, accurate for horizontal polarisation and nearly so for vertical polarisation, Chapter 5, Section 5.8.3, Figure 5.19)

$H, j =$ scanner and target tip true heights

$h =$ element height. $0 < h < j$

$\mathrm{d}h =$ element thickness

$r = R\lambda/2\pi H$, a factor independent of h, used for brevity

$\sigma =$ total RCS of target, m^2

$m_{\mathrm{p}}, m_{\mathrm{e}} =$ numerical multipath factor (one-way) in interference region of point element or whole extended target, respectively

$S_{\mathrm{p}}, S_{\mathrm{tgt}} =$ echo power from element or whole target respectively, W

$y = \rho\sqrt{g_{\mathrm{dif}}}$, for brevity.

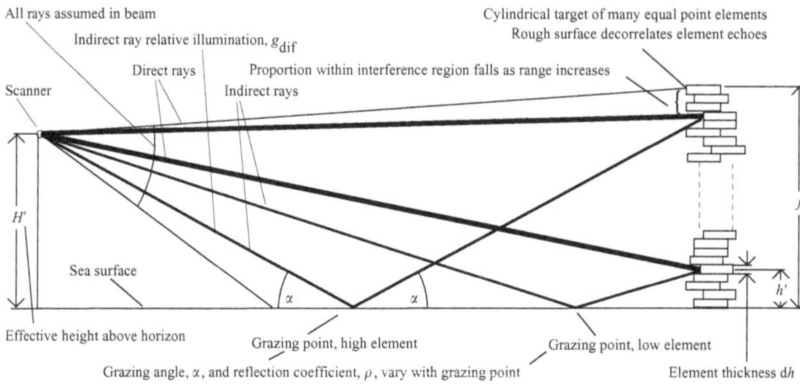

Figure 9.5 Extended target as sum of infinitesimal elements. Flat-Earth approximation, interference region. Exaggerated vertical scale

Chapter 6, Section 6.4.1, Eq. (6.4) gave the multipath factor of a point target in the interference region:

$$m_p = 1 + (\rho\sqrt{g_{dif}})^2 + 2\rho\sqrt{g_{dif}}\cos(\Phi + \psi) = 1 + y^2 + 2y\cos(\Phi + \psi) \tag{9.7a}$$

Putting $\psi = \pi$, for a point this becomes

$$m_p = 1 + y^2 - 2y\cos\Phi. \tag{9.7b}$$

From Chapter 5, Section 5.6.1, Eq. (5.27b), for a flat Earth, at target height nh, $\Phi = 4\pi Hnh/R\lambda = 2nh/r$.
Substituting and squaring, the two-way point target element multipath factor is

$$m_p^2 = 1 + 2y^2 + y^4 - 4y(1 + y^2)\cos\frac{2nh}{r} + 4y^2\cos^2\frac{2nh}{r} \text{ numerically.} \tag{9.7c}$$

9.4.3 Target multipath factor

For a uniform extended target model, the RCS per metre height is constant at σ/j m^2/m so the element RCS is:

$$\text{RCS} = \frac{\sigma}{j}\,\mathrm{d}h. \tag{9.8}$$

From the radar range equation, the element echo at range R is:

$$\begin{aligned} S_p &= (\text{free space echo of unit-area target}) \times (\text{element RCS}) \times m_p^2 \\ &= (\text{FS echo of unit-area target}) \times \frac{\sigma}{j}\,\mathrm{d}h \times m_p^2 \\ &= \frac{\text{FS echo of whole target}}{j} \times m_p^2\,\mathrm{d}h. \end{aligned} \tag{9.9}$$

The total echo at a given range is obtained by integrating the element echoes of Eq. (9.9) through the height interval $h = 0$ to $h = j$:

$$S_{\text{tgt}} = \int_0^j (\text{FS echo of target}) \times \frac{1}{j} \times m_{\text{p}}^2\, \mathrm{d}h$$

$$= \frac{\text{FS echo of target}}{j} \times \int_0^j m_{\text{p}}^2\, \mathrm{d}h. \tag{9.10a}$$

Also, by definition,

$$S_{\text{tgt}} = (\text{FS echo of target of height } nj) \times m_{\text{e}}^2 \tag{9.10b}$$

$$\therefore\ m_{\text{e}}^2 = \frac{1}{j}\int_0^j m_{\text{p}}^2\, \mathrm{d}h. \tag{9.11a}$$

Substituting per Eq. (9.7c),

$$m_{\text{e}}^2 = \frac{1}{j}\int_0^j \left\{1 + 2y^2 + y^4 - 4y(1+y^2)\cos\left(\frac{2nh}{r}\right) + 4y^2\cos^2\left(\frac{2nh}{r}\right)\right\} \mathrm{d}h.$$

Putting $\cos^2 2nh/r = \frac{1}{2}\,(1 + \cos(4nh/r))$ and separating the terms,

$$m_{\text{e}}^2 = \frac{1}{j}\left\{\int_0^j (1 + 2y^2 + y^4)\, \mathrm{d}h - \int_0^j 4y(1+y^2)\cos\left(\frac{2nh}{r}\right) \mathrm{d}h + \int_0^j 2y^2\, \mathrm{d}h + \int_0^j 2y^2 \sin\left(\frac{4nh}{r}\right) \mathrm{d}h\right\}.$$

Integrating the four terms

$$m_{\text{e}}^2 = \frac{1}{j}\left[(1 + 2y^2 + y^4)h - 2y(1+y^2)\left(\frac{r}{nh}\right)\sin\left(\frac{2nh}{r}\right) + 2y^2 h + \frac{y^2}{2}\left(\frac{r}{nh}\right)\sin\left(\frac{4nh}{r}\right)\right]_0^j. \tag{9.11b}$$

Evaluating at the limits of the definite integral and collecting terms.

When $h = 0$, $\quad m_{\text{e}}^2 = 0.$

When $h = j$,

$$m_{\text{e}}^2 = 1 + 4y^2 + y^4 - 2y(1+y^2)\left(\frac{r}{nj}\right)\sin\left(\frac{2nj}{r}\right) + \frac{y^2}{2}\left(\frac{r}{nj}\right)\sin\left(\frac{4nj}{r}\right).$$

Substituting in Eq. (9.11c)

$$m_{\text{e}}^2 = 1 + 4y^2 + y^4 - 2y(1+y^2)\left(\frac{r}{nj}\right)\sin\left(\frac{2nj}{r}\right) + \frac{y^2}{2}\left(\frac{r}{nj}\right)\sin\left(\frac{4nj}{r}\right) - 0.$$

Using decibel notation, the multipath factor for a uniform target (9.11d) above a flat Earth is therefore:

$$m_{\rm e} = 5\log\left[1+4y^2+y^4-2y(1+y^2)\left(\frac{r}{nj}\right)\sin\left(\frac{2nj}{r}\right) + \frac{y^2}{2}\left(\frac{r}{nj}\right)\sin\left(\frac{4nj}{r}\right)\right] \text{ dB one-way.} \tag{9.11c}$$

9.4.4 *Curved Earth*

At short range, curvature effects are small and the above equations are accurate. At longer ranges within the interference region, by analogy with the formulae for point targets, we allow for Earth curvature by replacing true heights by the effective heights above the horizon of scanner and target respectively, indicated by primes, using the method of Chapter 5, Section 5.5. This corrects the sine-term ripple components. We let:

H', j' = effective scanner and target tip heights at range R assuming curved Earth.
$r' = R\lambda/2\pi H'$.

For example, substituting in Eq. (9.11c), the multipath factor for a uniform target (9.11d) above a curved Earth is:

$$m_{\rm e} = 5\log\left[1+4y^2+y^4-2y(1+y^2)\left(\frac{r'}{nj'}\right)\sin\left(\frac{2nj'}{r'}\right) + \frac{y^2}{2}\left(\frac{r'}{nj'}\right)\sin\left(\frac{4nj'}{r'}\right)\right] \text{ dB one-way.} \tag{9.12a}$$

9.4.5 *Multipath factor up to critical range*

When r is small, the (r'/nj') terms tend to zero. Here

$$m_{\rm e} = \sqrt{1+4y^2+y^4} \text{ numerically.} \tag{9.12b}$$

When there is perfect surface reflection ($y = 1$),

$$m_{\rm e} = \sqrt{6} \sim 2.45 \text{ (3.89 dB one-way).} \tag{9.12c}$$

Echo strength is subject to $m_{\rm e}$ twice and is $2 \times 3.89 = 7.78$ dB above its free space value, confirming the values found less formally in Section 9.2. The underlying reason $m_{\rm e} > 1$ is because nett energy is reflected from the sea surface. If all the returns were aligned in phase instead of being random, the factor $\sqrt{6}$ would become 4, which is 6 dB one-way. Chapter 6, Section 6.4.2, Eq. (6.6) showed average $m_{\rm p}$ for a point target was closely similar, at 4.36 dB one-way.

Figure 9.6 graphs Eq. (9.12a) for various values of y between its extremes of 0 (no reflection from sea surface and no multipath interference, approximated by a rough sea) and 1.0, full multipath interference, approximated by a calm sea. The multipath ripple components are less than ±1 dB, and can usually be neglected.

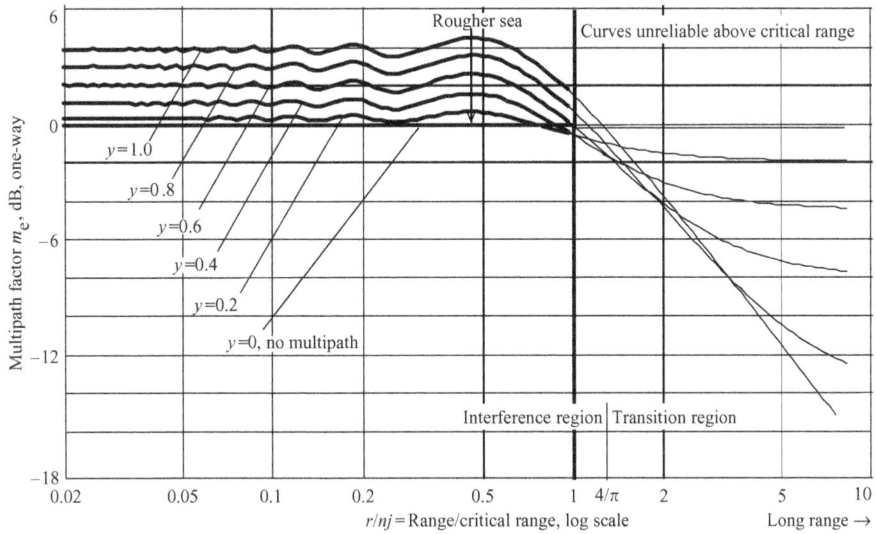

Figure 9.6 *Interference region multipath factor. Cylinder target. Factor* $r/nj = R\lambda/2\pi Hnj$. *Factor* $y = \rho g_{\mathrm{dif}}$. *With complete multipath reflection* ($y = 1$)*, when* r/nj *is less than 1, one-way multipath factor* m_{p} *approximates 3.89 dB with less than 1 dB ripple. Without multipath* ($y = 0$)*,* $m_{\mathrm{p}} = 0$ *dB. The curves roll off at* $r/nj = 1$

9.5 Approximate multipath factor into transition region

Figure 9.7 illustrates the following and may be compared with Figure 6.11 for a point target, same notation.

9.5.1 Critical range

Eq. (9.12b) remains valid until $r/nj \sim 1.0$. Here $R = 2\pi Hnj/\lambda$, at $\pi/4$ of transition range, R_{A}, where in free space by analogy with Chapter 6, Section 6.3.3, Eq. (6.2a),

$$R_{\mathrm{A}} = \frac{8Hnj}{\lambda}. \tag{9.13}$$

These expressions include n, so are not limited to the uniform target just considered. Curved Earth R_{A} can be corrected using Figure 6.5. Therefore the critical range, R_{C}, under calm sea conditions is Figure 9.7 point X:

$$R_{\mathrm{C}} = \frac{2\pi Hnj}{\lambda} = \frac{\pi}{4}R_{\mathrm{A}}. \tag{9.14a}$$

By analogy with point targets Eq. (6.28), the critical range under rough sea conditions where $y = 0$ is $(\sqrt{6})^{1/2}$ ($\sim$1.57) $\times$ calm sea R_{C}, point Y

$$R_{\mathrm{C}} = 1.57 \times \frac{\pi}{4}R_{\mathrm{A}} = 1.23R_{\mathrm{A}}. \tag{9.14b}$$

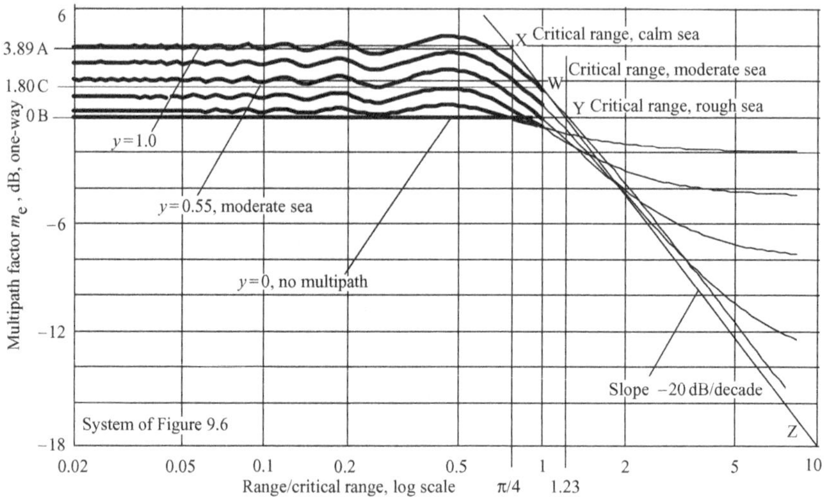

Figure 9.7 Approximate multipath factor. Moderate sea is approximated by curve CWZ. Compare with Figure 6.11

In general, if varying sea state is taken into account, critical range is not constant and is

$$R_C^1 = \frac{1.23 R_A}{(1 + 4y^2 + y^4)^{1/4}}. \tag{9.14c}$$

9.5.2 Moderate sea condition

When $R_C = R_A$, $1 + 4y + y^4 = 1.23^4$ and $y \sim 0.55$. Taking $R_C = R_A$ therefore well represents average sea conditions, as it did with point targets. Substitution in Eq. (9.12b) shows that at short range $m_e = 1.51$ (1.80 dB), Figure 9.7 line CW. Taking $R_C = R_A$ for point targets gave a rather similar value: $m = 2.26$ dB (Eq. 6.29b).

By analogy with the two-zone method of describing point target echoes (Chapter 6, Section 6.8), in the near transition region when range somewhat exceeds R_C, the multipath factor falls at −20 dB/decade, line XZ.

$$m_e \sim \frac{\sqrt{6}}{(R - R_C)^2} \text{ numerically.} \tag{9.15a}$$

In dB terms,

$$m_e = 3.89 - 20 \log \frac{R}{R_C} = 1.80 - 20 \log \frac{R}{R_A} \text{ dB.} \tag{9.15b}$$

We may therefore write an equation similar to Eq. (6.29b), valid in the interference and near transition regions in a moderate sea:

$$\text{If } R < R_A, \quad M = 1.80 \text{ dB}, \quad \text{otherwise } M = 1.80 - 20 \log \frac{R}{R_A} \text{ dB.} \tag{9.15c}$$

That is, below R_A echo strength follows an R^{-4} law and above R_A the law is R^{-8}. As range increases, targets becomes less influenced by interference and more by diffraction, until at the diffraction range R_B interference becomes negligible. Figures 6.6(a)–(c) provide approximate values for R_B for a selection of point target scenarios and it seems reasonable to retain them for extended targets, putting $h = nj$. Eq. (6.12) for a point target in the diffraction region indicates that rate of change of multipath factor with range is independent of target height. Therefore, in the diffraction region the extended target behaves as a point target of the same RCS.

9.5.3 High scanner or target

The analysis of Section 9.3 does not remain strictly valid in the transition region, which is of more practical importance than the more distant diffraction region; it also assumes scanner and target tip heights are each less than about 31 m (3 GHz) or 14 m (9 GHz) so $f(z) = z$. Formal mathematical analysis when the scanner and/or target are higher seems impossible and in Section 9.3.2 the argument leading to the expression for effective target height was confined to these limits. However, repeated numerical integration by the 10-element method indicates that extended targets continue to behave in much the same way when j exceeds the limits. Higher targets can be handled by the methods developed above without significant error, at least up to the maximum usually encountered in marine practice. This is because $f(z)$ for the effective point target is representative of the $f(z)$ values of the upper elements of the extended target, those which play the biggest part. Figure 9.8 has been calculated for a high 9 GHz band target, $j = 40$ m, nearly three times higher than the 14 m at which $f(z) = z$. Eq. (9.16) continues closely to replicate the

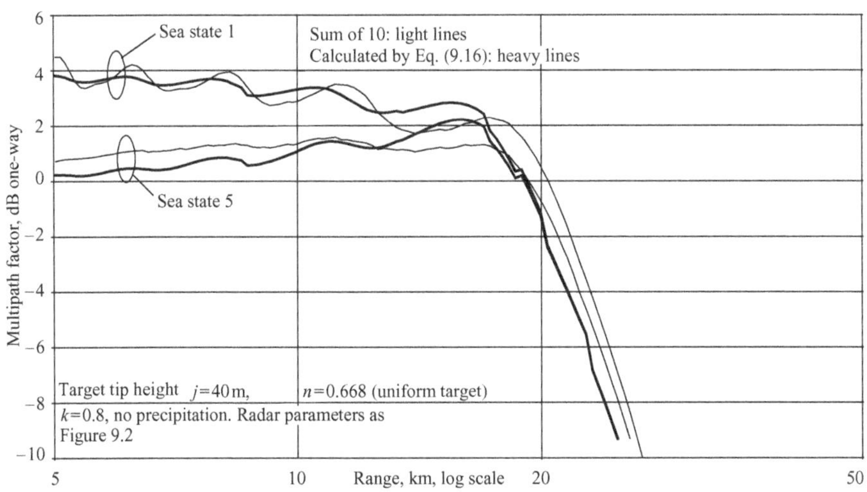

Figure 9.8 Multipath factors, high target (the computation has introduced minor irregularities).

sum-of-ten values, for the two values of sea state used. In sea state 1, M varied between 2 and 4 dB; in the rough conditions of sea state 5, M lay between 0 and 2 dB.

The arguments leading to the effective height method were founded on 'flat-Earth' considerations where atmospheric refraction k-factor is infinite. To show that the method continues to work when k is small, the figure has been computed for a low k-factor, $k = 0.8$.

Although not shown in the figures, the procedures work equally well in the 3 GHz band.

9.6 Complete multipath expression

9.6.1 Multipath factor

Adapting Eq. (6.23) (which was for a point target height h, through all three regions), including sea surface reflection phase angle ψ and putting $h = nj$ enables us to treat an extended target similarly to a point target, giving an expression for the one-way multipath factor at any range.

$$\text{If } R < R_\text{A} \text{ (Eq. (9.13) corrected by Figure 6.5)}, \quad M = m_\text{e} \text{ (Eq. 9.15b)};$$
$$\text{otherwise if } R > R_\text{B}(m_\text{d} = -20\,\text{dB}), \quad M = m_\text{d} \text{ (Eq. 6.7b)};$$
$$\text{otherwise } M = m_\text{t} \text{ (Eq. 6.15)}. \tag{9.16a}$$

That is, an extended target of overall height j can be treated as a point target of height nj provided the appropriate interference region multipath expression is used. The value of n depends on the RCS height profile and might lie between about 0.5 and 0.75 (per Eq. 9.5); putting $n = 0.5$ should usually give somewhat conservative detection range.

Eq. (9.16a) can be generalised for all passive targets.

$$\text{If } R < R_\text{A} \text{ (Eq. 6.2 or 9.13)}, \quad M = m_\text{e} \text{ (Eq. 9.15, if extended target) or}$$
$$m_\text{p} \text{ (Eq. 6.5, point target)}; \quad \text{otherwise if } R > R_\text{B}(m_\text{d} = -20\,\text{dB}),$$
$$M = m_\text{d} \text{ (Eq. 6.7B) or } m_\text{t} \text{ (Eq. 6.15)}. \tag{9.16b}$$

The multipath factor is always identical on go and return legs, to and from the target; $M_1 = M_2$.

9.6.2 Variation of echo with range

Figure 9.9 depicts echo power curves for the 1000 m^2 target of Figure 9.2. The free-space or rough sea line is (b), intersected by the $2 \times (-20)$ dB/decade roll-off line (e) for uniform target weighting at the critical range R_C, point Y. The full reflection line is (c), 2×3.89 dB above (b), intersecting the roll-off at X. The roll-off lines for top- and bottom-heavy targets are (d) and (f) respectively. The echo, (g), lies within the shaded area bounded by lines (b), (f), (c) and (d). The calculated sum-of-ten echo for

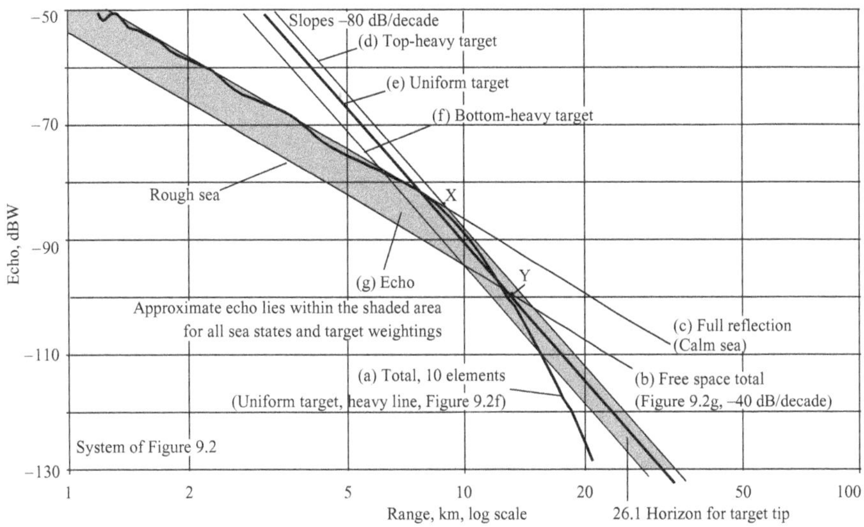

Figure 9.9 Approximate echo strength. Echo lies within shaded area under most target and sea conditions

calm conditions is superimposed at (a) for comparison; agreement is good into the early transition region, but poorer at extreme range.

Figure 9.10 compares the echo strength of an extended target, computed by the integration (a) and sum-of-ten (b) methods, which agree very closely at all ranges. The curves lie some 7 dB above the free-space curve (d) up to range R_A. Compared with a point target of the same RCS and height $0.668j$ (curve (c)), agreement is good beyond the transition range R_A but of course the curves differ at short range on account of the sharp multipath fluctuations of (c). The smoothing of the point's interference pattern by the extended target is shows very clearly, confirming that extended targets do not suffer the intermediate range weak signals at multipath nulls. Curve (e) shows the multipath phase shift of a point target at height nj. A physically small but tall target will return lower echo strength at short range than a large, low target, but its longer critical range may raise its echo above that of the low target at long range.

9.6.3 Non-uniform targets

Figure 9.11 compares the multipath factors, $\sqrt{\text{actual echo/free space echo}}$, for our extended targets with uniform, bottom- and top-heavy RCS height distribution models, calculated by the sum-of-ten and Eq. (9.16) methods. Agreement is good at all ranges. The one-way multipath factors for bottom- and top-heavy targets differ from the uniformly distributed case by about ± 2 dB, so the echo from either would lie within about ± 4 dB of that of a uniform target, equivalent to ± 10 per cent detection range change if beyond critical range.

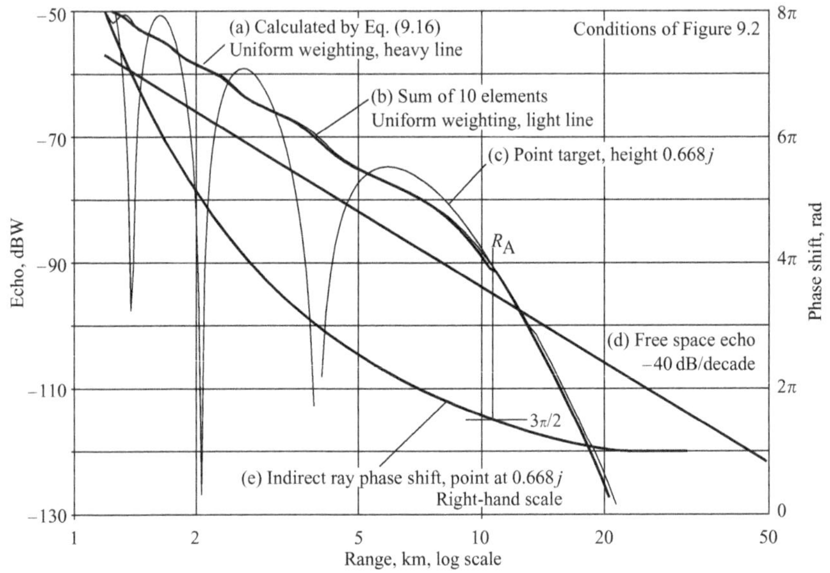

Figure 9.10 *Echo strength comparison. Extended target compared with point of same RCS, calculated by sum-of-ten and Eq. (9.16)*

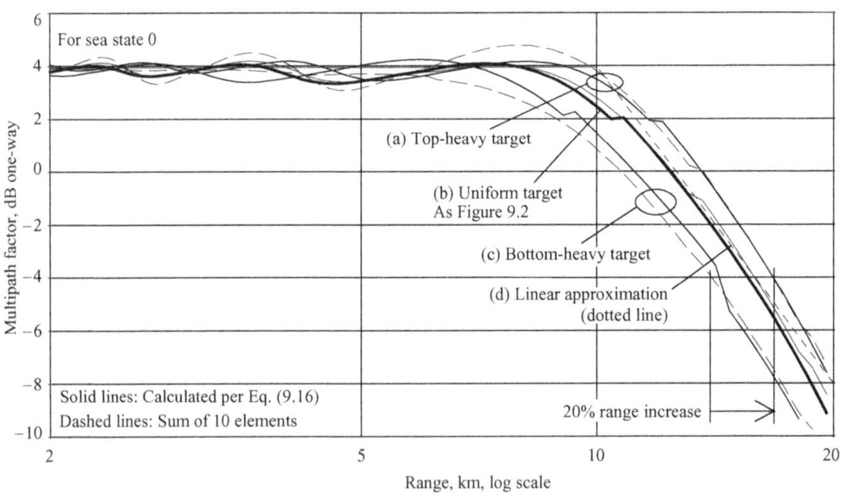

Figure 9.11 *Effect of RCS height distribution on multipath factor. Conditions as Figure 9.2 (uniform target, heavy line). Bottom- and top-heavy targets added for comparison; their multipath factors are almost identical at short range, but differ from the uniform target by about ±2 dB at long range. For given long-range m, say −4 dB, ranges differ about ±10 per cent. The curves should be smooth, the small steps being caused by granularity of the calculations*

9.6.4 Sketching echo strength

Multipath factor and echo strength can be sketched along the lines of Chapter 6, Section 6.9. Height factor n is chosen per Section 9.3.4 and R_A or R_C by Eqs (9.13) or (9.14c), respectively. The sketch then follows Figure 9.9. Alternatively, multipath factor can be sketched like Figure 9.7.

Chapter 10

Extended target reflections; ships and coasts

'Morte point is rocky and barren, sloping from its summit to low cliffs'

West Coast of England Pilot, 1960

10.1 The problem

10.1.1 Target parameters affecting detection

Detectability of extended targets depends on several of their parameters.

1. Mean radar cross section. This term, σ, in the radar range equation, affects echo strength at all ranges. RCS of practical targets such as ships is difficult to measure and more difficult to calculate with any degree of precision. RCS values of a wide range of extended targets are discussed in this chapter. In clear conditions, the echo has only to compete with receiver noise, and quite low RCS may give adequate detection range. But when clutter is severe, higher RCS is necessary.
2. Fluctuations in RCS. This chapter addresses mechanisms causing RCS fluctuations as the target aspect changes and it heels (rolls or pitches). The effect of fluctuation on detection is considered in Chapter 12.
3. Effective reflecting height (centroid height). Chapter 9 showed that the effective height of the target as a radar reflector affects the critical or transition range at which the echo changes from an R^{-4} to an R^{-8} law, affecting the multipath factor beyond, and hence long-range detectability. Multipath factor is unaffected by target height at short range, but a heavy sea which physically screens or masks low targets reduces the maximum achievable probability of detection even in favourable signal to noise plus clutter ratios.
4. Physical size relative to the detection cell. At short range, the target may exceed the height or width of the scanner beam and its axial extent may exceed the pulse length, reducing effective RCS per cell footprint.

10.1.2 Difficulty of finding RCS

When civil marine radar was introduced, the collision threats were 'rust and rivets' ships which luckily proved to be excellent reflectors. Even though their echoes were not always revealed amidst the clutter until rather short range, this was usually tolerable at the slow speeds then current, so little attention had to be paid to target RCS. But several of the diverse types of ship now at sea, including some high speed craft, have poor RCS by reason of small size, streamlined shape and sometimes non-metallic construction. Adequate reflection can no longer be taken for granted, yet today's much higher closing speeds necessitate greater detection range for safe collision avoidance. The inability of radar to detect and form tracks on small targets, often moving fast, in heavy weather has become sufficiently life-threatening for IMO to introduce a Carriage Requirement for all ships within the SOLAS Convention built after 2002, and under 150 gt, to carry a radar reflector or other means to enhance detectability to 3 and 9 GHz radars.

We saw in Chapter 7, Section 7.7 that such relatively simple shapes as octahedral reflectors have fairly complicated target pattern maps (TPMs). However, such small objects can be accurately measured at moderate cost in the controlled conditions of an anechoic chamber at any frequency of interest. Ships are by far the most important target for marine and VTS radars, yet their TPMs are hard to determine and there is no obligation on shipowners or masters to ascertain the RCS of their vessels. The echoes emanate from numerous small scattering elements. Chapter 7, Section 7.10 showed how even two simple point elements can give quite complicated TPMs, with peaks at in-phase bearings and deep interference nulls, so it is no surprise that it is difficult to calculate TPMs of such complex objects as ships – and they certainly cannot be measured within test chambers! This chapter examines reported radar cross sections of many kinds of vessel and attempts a broad theoretical justification of those values, although too many factors are in play to expect precision.

10.1.3 Factors affecting RCS seen by interrogator

Ships are so complicated in shape that the reflections of the many scattering elements, at differing phases, interfere strongly. The resulting polar diagram is 'busy' and defies precise calculation. RCS is only loosely related to the projected visual area seen from the scanner. Among vessels of the same tonnage, RCSs probably differ through variations in cargo gear, streamlining of superstructure, etc., as well as aspect. A ship's echoing characteristics may also vary with time (draught changes, deck cargo, hogging or sagging under load, in-service distortion of plating at bow, see Figure 10.1, or speed (wake, heel and squat – change of draught and trim when one end of the ship is sucked down by hydrodynamic interaction with the sea bed in shallow water). Effective RCS fluctuates sharply with minor heave, yaw, roll and pitch because of the complexity of the TPM. It also varies with inclination of the line of sight to the deck plane and with radar frequency and polarisation. We shall discuss factors affecting detectability in more detail in Section 10.6.

Figure 10.1 *Ships' bows. Bows take much punishment from the sea. Plating distortion may drastically change RCS from theoretical predictions. Plates are never flat within $\lambda/4$. The bow is no place for delicate and expensive radar! Author*

10.1.4 Estimation of effective height

In the absence of measured values, it is necessary to estimate effective height, h, of the ship's echo (Chapter 9, Section 9.3, height factor n). Physical height of the superstructure topmost significant reflector at height j, is fairly easy to determine, ignoring slender masts of small visual cross section. The (UK) Institution of Marine Engineers publishes an annual review of recent significant world-wide newbuildings including side elevations. Other sources include the handy Observers Ships [1] pocket book, containing many silhouettes, again drawn to scale. And of course all ships carry copies of builder's drawings. Section 10.3 describes experimental methods, but more usually it is necessary to guess a value for height factor n of between say 0.5 and 0.75.

These uncertainties make radar performance prediction difficult. Luckily, most merchant ships have RCS high enough to assure detection at good range. This saving grace does not apply to small craft or some high speed craft (HSC). And luckily we do not need to know the precise detail of the TPM. To enable us to predict detectability with reasonable confidence, remembering the imprecision associated with factors such as atmospheric refraction, it is enough to know the average RCS and significant deviations from it, when viewed nearly horizontally.

10.1.5 Our approach

We first explain how ship size is described, followed by the techniques and problems of RCS measurement. We next summarise reported experimental results, then offer an analytical method of assessing RCS from published data on ship size. We can do this because although all conventional ships differ, most are in many ways the same; they are made of painted conducting metal, are very many wavelengths long, have broadly similar length/breadth/height ratios, contain a fair number of quasi-corner reflectors and have large areas of not-quite-flat plating. These similarities let us make some predictions based on the basic principles of reflection of Chapter 7, Section 7.3.

We go on to examine other extended targets such as coastlines. Chapter 9 showed that multipath and diffraction affect extended targets differently from points, and we develop a calculation method to suit. All this prepares us to predict extended target detection range for the relevant radar and environmental parameters in Chapter 12.

10.2 Ship size

Ship size is usually expressed as a tonnage. Ships are put to sea to earn money, so tonnage measurements have a mercantile rather than scientific basis. In compliance with the International Tonnage Convention 1969 (ITC 69), passenger and cargo merchant ship size is measured in gross tons, gt, which we denote T. It would be wrong to ascribe a mathematical dimension such as mass to this unit; although expressed in tons, strictly speaking gt is a dimensionless quantity based on the total volume, V m^3, of all enclosed spaces. According to ITC 69:

$$T = k_1 V.$$

$$\text{Factor } k_1 = 0.2 + 0.02 \log V. \tag{10.1a}$$

From this

$$T = V(0.2 + 0.02 \log V) \text{ gt}. \tag{10.1b}$$

An approximation, Eq. (10.7) in Section 10.5.2, eases calculations. For a 1000 gt ship, Eq. (10.1b) gives a volume ~3.71 m^3 per gt, falling to ~3.21 m^3 per gt for a 100 000 tonner. Older IMO instruments using earlier tonnage measurements are taken to apply on the basis of one to one equivalence.

Other tonnage measurements which are or have been widely used include the following.

- *Nett tonnage, nt,* is based on gt, total cargo volume and other coefficients according to a formula within ITC 69.
- The *Oslo Convention* tonnage measure, superseded by gt, was based on 1 ton gross ton (tgt) = 100 ft^3 (2.83 m^3) of cargo space, with exemptions for bilges and certain other spaces.
- *Light displacement tonnage, ldt,* approximates the metalwork in the ship, indicative of residual scrap value.
- *Displacement tonnage* is the gross weight of ship and full cargo, equalling the mass of displaced water. Warships are usually quoted in displacement tons, often with poetic licence. Unlike the mercantile gt, displacement is the physical mass of the floating object, expressed in imperial long tons (2240 lbs = 1060 kg), short tons (2000 lb = 907 kg) or metric tonnes (1000 kg). Skolnik [2] gives displacement $\sim 2.5 \times 10^{-3} L^3$ tons, where L is length in metres.
- *Deadweight tonnage, dwt,* is the difference in displacement between (loadline) displacement tonnage and (empty) light displacement, representing the maximum cargo weight.
- *Suez Canal tonnage* is arrived at by a special formula for payment of Canal dues.

Figure 10.2 *One of the biggest ships afloat. Hellespont Metropolis, built 2003 and lead ship of a class of four 680 000 gt ultra large crude oil carriers (ULCC). Seen unballasted and dwarfing the powerful tug at the bow. Fully duplicated Furuno 3 and 9 GHz radars, mounted, as usual, on the accommodation block aft. Although reported to detect reflector-equipped yachts at up to 8 nmi, no way can they see a small craft near the bow. Reproduced by permission of Captain A. P. Starling Lark*

- *Compensated gross tonnage*, *cgt*, is a shipbuilding measurement indicative of constructional man-hours rather than sheer size.
- *Nett register tonnage*, *nrt*, is used to measure private motor yachts.

Tonnage is not the only measure. Container ships are sometimes quoted in TEU units, the carrying capacity (up to 9000 or more) in 20-foot equivalent units of ISO standard cargo containers, $20 \times 8 \times 8$ ft (volume $\sim 36\,m^3$) externally. Over 6000 containers may be carried, half within the hull and half above decks, stacked maybe six high. Some 'hatchless boxships' have no conventional deck. Tankers are sometimes called handy size, around 70 000 gt (handily matched to cargoes by charterers), VLCC (very large crude carriers) ~250 000 gt, or ULCC (ultra large crude carriers) up to ~650 000 gt, see Figure 10.2. Panamax ships are the largest able to transit the locks of the Panama Canal, about $228 \times 32 \times 12.5$ m draught. Capesize ships draw too much water for the Suez Canal. Tugs are characterised by bollard pull. HSC use displacement tonnage or length overall (LOA). LOA also sometimes augments tonnage for calculation of harbour dues and other charges. Ship and yacht length is sometimes quoted BP, between perpendiculars. Fishing vessels also are often characterised by length. The biggest modern cruise liners reach 150 000 gt, with air draught (height above waterline) exceeding 60 m. Many HSC are of the roll-on mixed passenger and cargo (Ropax) type; size is often given as deadweight, here meaning cargo tonnage.

None of these measurements neatly or uniformly converts to gt. Some classes of vessel such as car carriers and passenger ships may have gt as much as double their tgt values. Perhaps gt is somewhat more proportional to visible bulk above waterline, excluding deck cargo, than previous measures.

Tankers and dry bulk ships are in essence a cargo box with an accommodation/engine house stuck on the stern or occasionally the bow. The box needs only sufficient freeboard for buoyancy and to keep the hatch covers clear of the worst seas; maybe 80 per cent of the volume is below the load-lines. Ships carry half a dozen load-lines, to suit water density (Tropical Fresh, Fresh, Tropical, Summer, Winter, Winter North Atlantic), with ascending minimum freeboard and rising RCS. Unladen ships may 'ride high', not ballasted down to the load line marks, particularly near loading and discharge ports as they deballast, further raising RCS, clearly shown in Figure 10.2. Cranes and other cargo-handling gear may add significant RCS. Cruise ships are the other extreme, basically hotels offering rooms with sea view. Perhaps as little as 15 per cent of the volume lies below water and the much bulkier superstructure raises RCS. Car carriers resemble a floating multi-storey slab-sided car park, Figure 10.3. Some coasters are low-built to transit inland waterway bridges, with retractable ship's bridge and radar mast, Figure 10.4. Most other classes of ship lie between these extremes. Fresh and sea water are such good conductors that at microwave frequency negligible energy penetrates the surface and the submerged parts of hulls play no part in the reflection process.

Ships are the most important targets for all marine radars. We badly need a reasonably reliable value of RCS to insert in the range equation to enable us to determine detectability. Although it is clearly impossible to produce one simple formula

Figure 10.3 *Car carrier. Slab sides give peaky target pattern map, mitigated by minor plating irregularities. Author*

Figure 10.4 *Low-built coaster with retractable bridge and radar mast for inland waterways. MV Pandor entering Fowey, Cornwall UK. Author*

accurately relating RCS to any of the recognised methods of reporting size for all ships, it seems best to link RCS to gt.

10.3 Experimental determination of RCS and effective height

TPMs may be determined by calculation from constructional drawings, or by measurement of a physical model or the vessel itself. All these methods are difficult and costly. It is easier to infer average RCS from the tonnage, adding assumed null width and depth. Ship designs are so varied that we have to make broad approximations.

10.3.1 Military methods

The military measure small-scale physical models at correspondingly higher frequency within an anechoic test chamber, often using lasers at optical frequencies. This must be done with great care, retaining the original's minor as-built micro-geometry imperfections of plate flatness, but it is possible to determine which constructional features dominate particular RCS spikes. Expense bars the technique from the civil marine industry. The military have also obtained TPMs by flying helicopter-borne calibrated radar in a height spiral round the target, although it is difficult to obtain good accuracy. Complete three-dimensional TPMs are understood to include some merchant ships, but appear to remain classified despite the ending of the Cold War. A few test stations have specialist high-precision multi-band calibrated radars, whose comprehensive digitised recording facilities can produce accurate polar diagrams of vessels lying offshore when swung through a circle.

10.3.2 Radar measurement of typical ship RCS

A viable alternative to the above uses groundfast calibrated marine-type radar to measure echoes from ships at sea. Special ship charter is generally prohibitively expensive, so RCSs are taken from targets of opportunity. Choosing a headland location where ships change course enables RCS to be measured at several aspects, approximating half an azimuth polar diagram, but it is generally impractical to take measurements in the elevation plane to get a complete feel for the TPM. Tonnage is found from reference publications, having read ship's name by telescope. Usually aspect cannot be measured direct and is inferred from the easily measured course made good, taking care that wind or tide stream are not causing leeway.

The instantaneous RCS of a large and irregular target such as a ship varies rapidly with trivial changes of azimuth and elevation as component reflection phasing changes. It is therefore necessary to apply some smoothing to the apparent pulse to pulse RCS measurements. Too much smoothing will however mask any arcs of low RCS which might hazard detection, Section 10.11. Smoothing should be over increments of about 1° to match typical yaw. Mean RCS, but not exact peak and null bearing of pattern events, is only weakly dependent on frequency and it is sufficient to measure the target at a single spot frequency in each band.

To ensure accuracy is difficult, as with most field experiments. Best results, as usual, are obtained by a small team working slowly and methodically with full understanding of their equipment and the physical processes in play. A log of all events, however seemingly trivial, can often explain data anomalies during subsequent analysis. It is prudent to repeat the measurements on similar ships of the same class. This helps to smooth out the inevitable random measurement errors and also indicates the ship to ship variability and robustness of the method.

The radar is adapted to measure echo strength, typically using a calibrated microwave or IF attenuator, adjusted manually to the point of target paint extinction or 50 per cent blip-scan ratio, or by digital measurement of detection cell content, a more complex modification which facilitates subsequent machine analysis of gathered data. Scanner azimuth beamwidth should not be so narrow that the target overflows at short range. Likewise, long pulses should be used (for detail, see Section 10.7). Measurement is confined to ranges below the critical range for the target and scanner heights in question. Weather dependent atmospheric refraction or ducting effects are minimised by operating at short range. RCS is calculated from echo strength using the radar range equation, Chapter 4, Section 4.5.2. Below critical range, re-arrangement of Chapter 4, Section 4.5.2, Eq. (4.8), gives:

$$\sigma = S_e - P - 2G - 20\log\lambda + 30\log(4\pi) + 40\log R + L_t + L_r + 2L_A - 2M \text{ dB m}^2. \tag{10.2}$$

10.3.3 Alternative measurement strategies

Two basic calibration procedures are possible.

1. *Dead reckoning.* P, G, λ, L_t and L_r are determined by appropriate laboratory tests. At short range in clear weather L_A is small and is estimated from the data in Chapter 5, Section 5.9. Multipath factor, M, depends on sea state at the grazing point between the radar station and the target. Chapter 9 showed that short range one-way multipath factor M for extended targets lies between about 3.89 dB (calm sea, Section 9.4.5, Eq. (9.12c)) and 1.8 dB (moderate sea, Section 9.5.2, Eq. (9.15c)). A representative value is chosen to suit the experimenter's judgement of sea state at grazing-point range. Echo power S_e and range R are measured repeatedly as the ship passes, usually by setting an attenuator within the receiver to give 50 per cent probability of detection against the pre-determined noise background, R being read off the display. Values of σ are calculated by Eq. (10.2) and running averages taken as the aspect changes. If multipath or losses are ignored, calculated RCS will be several dB too high or too low respectively.
2. *Comparison with known target.* A buoy or array of buoys of known RCS are deployed at known ranges. Using the attenuator, target echo strength is equalised with a buoy forming a transfer standard; ranges are allowed for by the range equation. This procedure dispenses with the necessity to measure P, G, λ, L_t, L_r and M. The simplest transfer standard is a plastic buoy of minimal skin RCS carrying a calibrated point reflector of uniform TPM and fairly high RCS, such

as a Luneberg lens (Chapter 7, Section 7.7.4). The problem is that the multipath factor M for the extended target will have a smooth structure, quite different from the fluctuating M of the point transfer standard. This may be overcome by fitting the plastic buoy with a metal reflective lattice mast to form an extended target whose RCS is measured within an anechoic chamber. Here the problem is that heel resulting from wind or tide may tilt the mast and change the RCS shown to the radar.

Secondary problems include the following.

1. Time variations in radar sensitivity. Modern radars are usually stable once well warmed up, but care must be taken not to disturb control settings such as gain during the trial. Sensitivity should be measured before and after the trial.
2. Range variations in receiver sensitivity (swept gain). Adaptive swept gain should be disabled and the swept gain law should be laboratory-measured before and after the trial and allowed for when calculating.
3. Pitch, roll and heel presenting fluctuations of elevation aspect. Results from targets visually observed to be heeling should be regarded as suspect. Pitch and roll may cause recognisable cyclic ripples on the echo strength plot. Mean values should be taken, cautiously, retaining fluctuations of strength as aspect changes.
4. Very large targets may have Rayleigh distances of several km and may not always be in their far fields at radar range (see Chapter 7, Section 7.1.1); the practical significance of this, if any, is problematical.

10.3.4 RCS of specific vessel

When the vessel is under test station control, method may be improved. The vessel is either rotated on its axis at short range by a small workboat of negligible RCS (a rigid inflatable is ideal for small vessels) or sailed slowly in a tight circle. Questions of critical range no longer arise and the multipath remains almost constant. Heading can be accurately reported by an observer onboard, and it may even be possible to heel the vessel by adjustment of ballast to get a feel for the elevation polar diagram and TPM. The radar can be fitted with a staring antenna of accurately known gain, such as a large horn, approximating the military test stations mentioned in Section 10.3.1. An echo is received from every transmitter pulse rather than a few each couple of seconds. Integration improves SNR to the point where receiver noise becomes negligible and measurement of echo strength is more straightforward – all echoes at the appropriate range must be from the target, and can readily be measured by oscilloscope or, better, automatically recorded digitally for subsequent analysis.

10.3.5 Effective target height

Approximate critical range can be inferred from the point of inflection of an echo strength/range plot. Knowing scanner height and frequency, the target effective height, nj, can then be calculated from Chapter 9, Section 9.5.1, Eq. (9.14c). As physical height j can be found with fair accuracy, height factor n can be inferred.

A more accurate assessment of critical range should be possible if a pair of radars can be mounted one above the other at the test station. Use of two co-located radars, 1 and 2, of the same frequency (but not necessarily identical transmitter powers, scanner gains, etc.) enables a fair estimate of both target RCS and effective height, nj, to be made. Each radar is arranged as before to measure echo strength. An example illustrates the method.

1. Radar 1 is installed as high as possible, with Radar 2 directly beneath it, as near sea level as possible, say $H_1 = 20$ m and $H_2 = 4$ m. Both view the target at identical azimuth and nearly identical elevation aspect, so target TPM variations between them should be negligible. Suppose radar frequency is 9400 MHz in each case.
2. A preliminary guess is made of target effective height h from known actual height j and a guessed height factor n. If, say, $j = 16$ m and it is assumed $n = 0.5$, $h = 8$ m.
3. A value for refraction factor is estimated, to suit the prevailing atmospheric conditions, say $k = 1.333$.
4. Critical ranges R_C are determined from Chapter 9, Section 9.5.1 to be 12 km (Radar 1) and 6 km (Radar 2). The target is sailed on a tangential course or is anchored, at an intermediate range of say 10 km.
5. Echo strength at Radar 1 is measured in terms of dBW, and plotted on a graph sheet to a base of range, Figure 10.5 point A.
6. Multipath factor m_e is calculated from Eq. (9.12b) for the prevailing sea state.

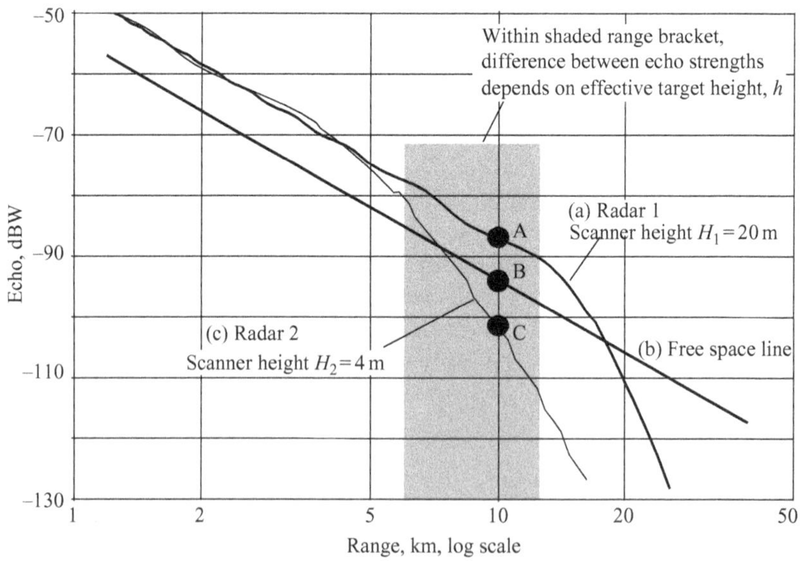

Figure 10.5 Determination of effective target height. Follows Figure 9.9

7. Deduction of $2 \times m_e$ gives the free space echo strength, point B, permitting the free space line (b) to be drawn at slope -40 dB/decade. The target RCS is now calculated from range Eq. (4.8) by insertion of the radar parameters.
8. Echo strength at Radar 2 is measured and graphed, point C.
9. A spreadsheet of echo strength is run with successive values of h. The one (c) passing through C indicates the correct h for the target.
10. The experiment should ideally be repeated at a number of ranges to smooth out at least some of the numerous secondary factors which always conspire to reduce the accuracy of radar measurements at sea.

10.4 Reported RCS values

10.4.1 Limitations

Given the difficulty of trustworthy measurement, it is not surprising that reported RCS figures are rather scanty. There are a few polar diagrams but not complete TPMs. Rarely is physical height of the ship quoted, let alone its effective reflecting height, so assessment of critical range remains a matter of guesswork. Neither is it usual, except for military purposes, to attempt to identify the contributions of the various elements – hull, superstructure, deck cargo, etc. It is difficult to guess how closely any specific ship approximates the quoted values for ships of its general type, some of which may be to detail designs no longer current. Unfortunately, published measurements do not always make clear which tonnage measurement is used, or even the radar wavelength. RCS of big ships is usually so great that detection remains possible beyond critical range where echo strength follows an R^{-8} (or steeper) law and any error in the assumed RCS causes $10^{(1/80)}$ per unit (2.9 per cent) calculated detection range change per dB error, so some RCS uncertainty is usually tolerable.

Small ships are the problem. Their RCS is so low that they may be only marginally detectable in clutter at hazardously short range, a small variation may significantly affect probability of detection, and effective height sharply affects detectable range. Here it is important to be able to calculate detectability with accuracy and it is a great pity that data on the reflection qualities of modern small ships and HSCs seems particularly sparse.

The following review of published data is not exhaustive. All reported values have been brought to dB m^2 for uniformity. Except where stated, RCS values assume the transmitter and receiver share the same plane of polarisation, usually horizontal (HH). Because some energy falls on slant edges whose re-radiation may be resolved into horizontal and vertical components, there is often also a significant cross-polarised (HV) echo.

10.4.2 Williams et al.

P. D. L. Williams *et al.* [3] in the mid-1970s used a shore based 9 GHz band HH polarised marine radar to measure targets of opportunity passing Dungeness, on the north shore of the English Channel, following the procedures of Sections 10.3.2

and 10.3.3. Basically his experienced team adopted the dead reckoning approach, but also cross-checked performance against control targets: specially prepared low-RCS ($-10\,\mathrm{dB\,m^2}, 0.1\,\mathrm{m^2}$) plastic buoys fitted with reflective lattice masts. Ballasting by long heavy poles maintained the masts upright. The radar had 0.8° azimuth beamwidth and 0.5 μs pulselength, giving a footprint about 70 m diameter at 2 nmi (3.7 km), the typical range, illuminating almost the whole target simultaneously. Aspect changed from nearly bow-on through broadside to the starboard quarter as the ships passed. Those ships have probably since been broken up. It would be good to repeat the measurements with today's traffic, and to include VLCCs, cruise ships and, in particular, small vessels and HSCs.

Range was a few kilometres (below critical range) and the radar beam was almost horizontal at the targets. High amplitude 'flashes' on the target beam were observed but discarded. Williams' Table 10 is summarised by Table 10.1 and Figure 10.6. The latter indicates each ship's observed non-flash RCS bracket by vertical lines linking the tabulated maximum and minimum. Williams [4, Chapter 4] obtained RCS generally somewhat smaller than reported by some much earlier workers on similar

Table 10.1 Summary of Williams' 1977 measurements

Ship type	Ref.	gt (tons)	Length LOA (m)	RCS min. ($\mathrm{dB\,m^2}$)	RCS max. ($\mathrm{dB\,m^2}$)	Deviation from Eq. (10.3a) (Max. dB)	Deviation from Eq. (10.3b) (Min. dB)
Merchant ships							
Inshore FV	13	5	8.5	5	10	1.7	−0.9
Small coaster	18	225	42.6	14	29	1.0	−5.8
Coaster	3	500	54.9	17	33	0.9	−5.7
Coaster	1	500	54.9	25	36	3.9	2.3
Coaster	16	500	56.7	30	41	8.9	7.3
Large coaster	2	1 000	67.1	30	37	1.3	4.8
Collier	4	1 570	73.5	25	33	−5.0	−1.8
Cargo liner	10	5 000	114.0	40.5	42	−2.0	9.4
Cargo liner	11	8 000	136.6	36	42	−4.4	3.2
Bulk carrier	6	8 200	166.4	28	41	−5.6	−4.9
Cargo	9	9 400	154.5	32	41	−6.3	−1.4
Cargo	7	10 430	166.1	28.5	42	−5.8	−5.3
Bulk carrier	12	17 500	198.1	30	45	−5.5	−5.6
Ore carrier	8	25 400	205.7	32	45	−7.4	−5.0
Container carrier (considerable deck cargo)	9	26 436	208.8	40	48	−4.6	2.9
Medium tanker	15	32 500	221	47	59	5.3	9.1
Medium tanker	17	44 700	251.2	43	62	6.7	3.9
Warship		(displacement tons, not stealthed)					
Frigate	14	2 000	103.0	38	50	10.7	10.3

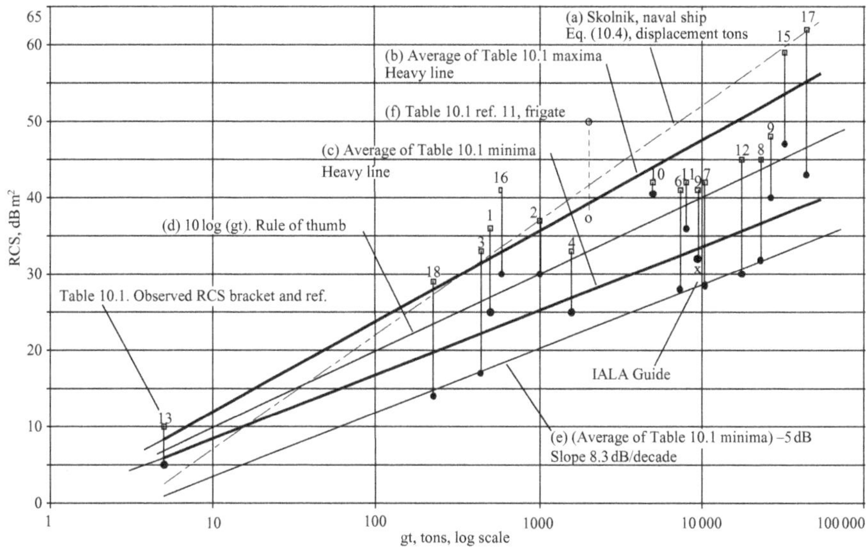

Figure 10.6 *RCS of ships. Viewed horizontally. Summarises data from Table 10.1 and other sources, 9 GHz band*

shipping, which he ascribed to their making incorrect allowance for radar losses (which can aggregate 10 dB). Minima were often when the vessel was nearly bow- or stern-on; we will show why later. Maximum and minimum values for all types of ship observed (except the warship) seem to lie within about ±5 dB of the following tonnage relationships, graphed as Figure 10.6, curves (b) and (c).

$$\text{Maximum RCS} = 11.9 \log T \ \text{dB m}^2. \tag{10.3a}$$

$$\text{Minimum RCS} = 8.4 \log T \ \text{dB m}^2. \tag{10.3b}$$

For miscellaneous targets, Williams [4, Chapter 4 and *passim*] reports the following RCS median values, dB m^2, probably in the 9 GHz band.

50 m vertical steel lattice mast	>27
Offshore gas rigs	High
Electricity pylons	High, depends on aspect
Waterlogged floating ISO container	<0
Dockside crane	60
Large lit steel navigation buoy	10
14 ft (4.3 m) dinghy, no motor	−12 to −15
Fishing boat, 28 ft (8.5 m)	3–9
Empty 5 l plastic oil container marking mooring:	∼−10

Table 10.2 IALA VTS Manual

Vessel	6 m wood	20 m wood	80 m steel	10 000 grt steel
RCS, m^2	6	25	300	1000
RCS, dB m^2	*7.8*	*14*	*24.8*	*30*
Height, m	1	3	8	15

Table 10.3 Summary of Skolnik's RCS data

Target	RCS (dB m^2)
Man (infantryman with accoutrements?)	−3 to 1.1
Small open boat	−17
Small pleasure boat	3
Pleasure boat 20–30 ft (6–9 m)	'a few m^2 at X band'
Pleasure boat 40–50 ft (12–15 m)	10
Cabin cruiser	10
Naval ship at small grazing angle	Eq. (10.4)

Rohan [5] reports the following, band unstated.

Boeing 707 aircraft	15 dB m^2	
Cessna 180 aircraft	1.8 dB m^2	
Submarine schnorkel	7 dB m^2	
Wooden craft, bow-on	$\frac{1}{3}$ × bow-on projected area	for RCS expressed in m^2.
Steel craft, bow-on	1 × bow-on projected area	for RCS expressed in m^2.

10.4.3 IALA VTS manual

The IALA VTS Manual[1] gives figures (reproduced as Table 10.2), with no indication of radar band or how RCS is distributed with height or aspect. Its 10 000 grt vessel has been shown on Figure 10.5 assuming 10 000 gt and the 9 GHz band.

10.4.4 Skolnik

Skolnik [6] includes the targets of Table 10.3 within a summary of targets 'at microwave frequencies' but not necessarily all viewed horizontally. Relative to the medians quoted, broadside flash ~13 dB, while minima are −8 dB.

[1] IALA VTS Manual, Tracking, p. 43.

For a [not stealthed] naval ship at small grazing angle, his equation 2-38 gives

$$\mathrm{RCS} = 10\log(52 f^{1/2} D^{3/2})\ \mathrm{dB\,m^2}, \tag{10.4}$$

where f is the frequency in MHz and D is the displacement tonnage in thousands of tons; Figure 10.6, curve (a).

The weak variation with frequency of Eq. (10.4) causes RCS to fall ~2.5 dB from 9 to 3 GHz. Its cause is discussed in Section 10.5. It seems reasonable to extend this frequency relationship to merchant ships.

10.4.5 *Warships*

Eq. (10.4) is said to give the average RCS for the port and starboard bow and quarter aspects of naval ships of displacement $D = 2000$–17 000 tons (presumably short tons of 907 kg), omitting the peak at broadside, at grazing incidence for bands X, S and L (9, 3 and 1 GHz). It has been plotted as Figure 10.6, curve (a), and gives higher RCS than Williams' merchant ships but intercepts his frigate RCS bracket, Figure 10.6, curve (f). It is understood that RCS of non-stealthed warships is generally higher than merchant ships of the same numerical tonnage, because (a) the former have more equipment on the superstructure; (b) displacement tonnage is used rather than gt and is small for the bulk. Warships designed or rebuilt after about 1990 are generally 'stealthed' to reduce their RCS to enemy missile radars using cleaner outlines, slanted surfaces, radar absorbent coatings and avoidance of corner reflectors. Their RCSs are classified, but at near-horizontal aspects probably at least 10 dB lower than indicated by Eq. (10.4). A modern frigate (4000 long tons displacement) has been claimed in the Press to 'have the RCS of a fishing boat'. With such precision, the journalist is unlikely to have breached the Official Secrets Acts! Williams [4] remarks that stealthing may reduce warship RCS by 20–30 dB.

10.4.6 *A rule of thumb*

A rough relationship for merchant ships, sometimes used for the 9 GHz band and graphed in Figure 10.6, curve (d), is

$$\mathrm{RCS} = T\ \mathrm{m^2} \tag{10.5a}$$

or

$$\mathrm{RCS} = 10\log T\ \mathrm{dB\,m^2}. \tag{10.5b}$$

Eq. (10.5b) gains credibility by lying midway between Eqs (10.2) and (10.3). Random broadside flashes may add ~15 dB and those at the bow and stern about 5–7 dB.

10.4.7 *Radar Technology Encyclopedia [7]*

Part of Table 14 of this encyclopaedia is reproduced as Table 10.4, but excluded from Figure 10.6 because some of the quoted sizes seem more appropriate to feet than metres.

Table 10.4 Generalised Maritime RCS values – Radar Technology Encyclopedia

Target type	Size (m)	RCS ($dB\,m^2$)	Target type	Size (m)	RCS ($dB\,m^2$)
Small open boat		−17	Freighter	610 (sic)	30
Small pleasure boat		3	Tanker		33.5
Fishing vessel	6	6	Navy picket ship	134	23–35
Cabin cruiser	12	10	Loaded tanker		28–33.3
Wooden mine sweeper	44 × 7.6	19	Loaded freighter		38.7
Patrol boat	226 (sic)	20	Navy cruiser	202	41.1
Small cargo ship		21.5	Cargo ship	85 × 12	20–60
Trawler	160 (sic)	21.8	Coast Guard, 3rd class		18.1
Surfaced submarine	93 × 8	10–23	Coast Guard, 8 × 26		25.7
Freighter	366	27	Coast Guard, 1st class		28.3

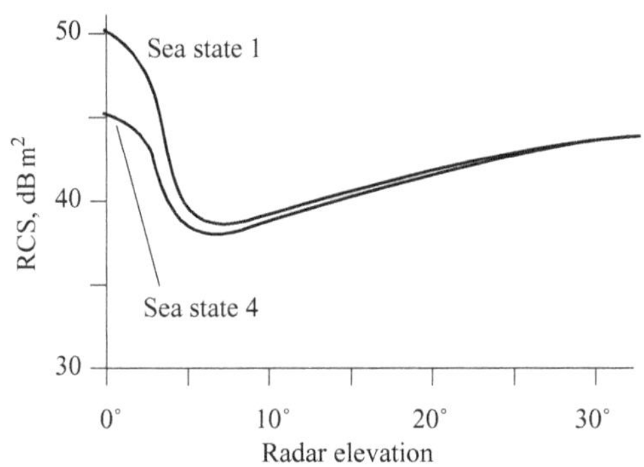

Figure 10.7 RCS versus elevation angle. Laden container ship, LOA 204 m, on beam, 9 GHz band. RCS was maximum when viewed horizontally. At 4° elevation, when measured through all azimuths, RCS varied between 32 and 42 $dB\,m^2$

10.4.8 Angle of depression

McClintock [8] states that RCS tends to vary with elevation angle of incidence of the line of sight, γ rad, approximately as Figure 10.7. (We use γ because target roll and incident beam inclinations, Chapter 5, Section 5.5.4, Eq. (5.19), probably have closely similar effects. We will not consider the effect of the differing angles of direct and indirect components of the incident signal; the added complication contributes little accuracy.) He ascribes the high RCS at low angles to reflections from vertical plating, while at high angles horizontal deck surfaces come into play. (Also the vertical shell forms a dihedral corner with the sea surface.) This relation might also describe RCS

Table 10.5 Factors X and Y

On bow	X	Y	On beam	X	Y
Minimum likely RCS	0.83	15	Average RCS	1.15	15
Average RCS	0.83	20	Maximum likely RCS	1.15	23

of a rolling or listing ship viewed on the beam. The higher RCS at low sea state was presumably multipath adding about 2 dB each way. The true RCS, shorn of any over-measurement through multipath perhaps lies a decibel or so below the Sea state 4 curve.

Inclination of the ship probably has similar effect on RCS to elevation and may occur as:

- roll or pitch in a seaway;
- trim in fore and aft plane;
- list in thwartship plane due to poorly stowed or shifted cargo, passengers crowding to one side, or leaks;
- heel in thwartship plane when turning at speed, or due to wind loading.

10.4.9 Suggested formula for merchant ships

Figure 10.6 summarises much of the experimental data. Williams' minimum RCS observed at low angle at 9 GHz (Figure 10.2(e)), could conservatively be taken as (Eq. (10.3)) -5 dB, that is, RCS $= 8.4 \log T - 5\,\mathrm{dB\,m^2}$. Inserting Skolnik's $f^{1/2}$ factor (Eq. (10.4)) and the elevation factor (Figure 10.7) gives:

$$\text{likely minimum RCS} = 10X \log T + Y - 5 \log f - 1.8\,\gamma\ \mathrm{dB\,m^2}, \qquad (10.6)$$

where T in gt 'tons', f in MHz, γ in degrees, $\gamma < 10°$. Factors X and Y are given by Table 10.5.

10.5 Theoretical basis for RCS

10.5.1 Approach

Traditional ships, as illustrated by Figure 10.8(a), with near-vertical hull and accommodation plating contain many reflecting elements – flat and curved steel panels and corners – spaced many wavelengths apart so average RCS is high. These ships have enough elements for their TPMs to resemble hedgehogs of many seemingly random peaks and nulls in both azimuth and elevation. Because panels, particularly shell panels, are never quite flat, their reflections combine to partially fill the nulls, which rarely go deeper than about 5 dB ($\frac{1}{3}$ power). Now and then a hull plate or other big plate comes normal to the radar, causing a massive RCS spike which is far

(a)

(b)

Figure 10.8 *Contrasting ferry designs. Red Funnel Line on the Solent, Southern England. Author. (a) Conventional construction. Inshore symmetrical passenger and vehicle ferry Red Eagle. Plenty of near-vertical plating and corner reflectors. High all-round RCS for the tonnage. Navigational 3 and 9 GHz radars aloft, plus small berthing radars each end, with extensive blanking arcs to avoid radiating upper-deck passengers. (b) Streamlined catamaran. Foot-passenger ferry. Absence of vertical plating forward, partially stealthing the profile and reducing RCS to approaching ships*

too intermittent for reliable detection. TPMs differ in shape between the bands, with fewer but wider nulls at 3 GHz where the ship is fewer wavelengths long.

To hide themselves from hostile radar, modern naval vessels employ several RCS-reduction 'stealth' techniques (see Lynch [9]). Techniques include:

- Slanted hull and superstructure panels to deflect reflections upward from the radar.
- Minimum number of dead flat panels to widen nulls and reduce reflection spillover towards the radar.
- No dihedrals or trihedrals – avoidance of ladders and other fussy details.
- Non-metallic superstructure.
- Metalwork within superstructure masked by lossy coatings.

This list all too well describes many modern small civilian vessels such as the ferry shown in Figure 10.8(*b*), which may also make unsatisfactory reflectors for other reasons. In all:

- gt is low;
- stealth features may further reduce RCS by several decibels;

- the few reflecting elements coarsen the TPM, with wider nulls and increased chance of deep cancellations;
- panels tend to be flatter, giving strong narrow lobes but wide nulls;
- small ships pitch and roll further, tending to slant 'vertical' panels and reduce RCS;
- on the other hand, naval architects strive to maximise carrying capacity for the gt, with more superstructure.

Most of these factors apply particularly strongly at 3 GHz.

As a counterbalance, small HSC and craft below SOLAS size may not venture into severe clutter, reducing the RCS necessary for detection.

The author is indebted to Dr J. M. Williams [10] for the ideas underlying the following presentation. We saw in Chapter 7 that RCS of flat plates and corner reflectors was affected by their overall dimensions, called the macro-geometry, and by irregularities in flatness or orthogonality, the micro-geometry. Here we shall compare the rate with which RCS is predicted to change with ship size under the macro and micro regimes and compare with the experimental results summarised in Eq. (10.6). The conceptual treatment is abstracted from the minutiae of actual targets; it is necessarily 'broad brush' – even cavalier – and indicative only. We use square metres rather than dB m^2 units and start by considering the relation between tonnage and surface area.

10.5.2 Tonnage and linear dimensions

The form of Eq. (10.1) relating T with volume is inconvenient. An approximation with error <1 per cent from 1.0 to >1 000 000 gt is

$$T = 0.205V^{1.033}. \qquad \text{That is, } T \propto V^{1.033}. \tag{10.7}$$

For a given shape where length is a linear dimension, L, volume $\propto$ length^3. Substituting in Eq. (10.7),

$$L \propto V^{1/3} \quad \text{so } L \propto T^{1/3.1} \propto T^{0.323}. \tag{10.8}$$

We next develop a baseline RCS for conceptual large targets.

10.5.3 Micro-geometric approach; baseline RCS

Chapter 7, Section 7.4 showed that large flat plates and corners whose irregularities do not exceed $\lambda/4$ reflect extremely strongly through very small angles. Plating on practical large reflectors is always irregular. This dependence on microstructure precludes large targets such as ships from behaving as predicted by elementary macro-geometrical theory. Using an approach based on flatness micro-geometry, we now show that large targets have RCS which is related to a baseline RCS for a geometrically simple target of the same size.

We start by thinking within two dimensions only. In Section 7.4.5 we saw that a strip facet (Figure 7.5(b)), length b and radius of curvature r, reflects through a subtended angle of $2b/r$. A two-dimensionally isotropic radiator of unity gain would reflect over the whole 2π radians. The facet gain is therefore $2\pi/(2b/r) = \pi r/b$. The echoing width (analogous to echoing area in three dimensions) is actual width $\times$ gain $= b \times \pi r/b = \pi r$. Over the complete circle of observation this echoing width is obtained over the arc through which the facet presents a normal, that is, through b/r radians for the convex face and a further b/r for the opposite concave face. At all other aspects the echoing width is zero. This approach is similar to that adopted for antennas, where to simplify calculation the actual rounded beamshape is replaced by a rectangle having full gain between half-power points and zero elsewhere.

Over the whole circle, the average echoing width is

$$\frac{1}{2\pi} \times \frac{2b}{r} \times \pi r = b, \quad \text{which is half the whole (front plus back) target perimeter.} \tag{10.9}$$

This expression for a curved facet is independent of wavelength, unlike the echoing length of a straight strip. Generalising: the average RCS in two dimensions is $\frac{1}{2}$ the total external perimeter. Of course, this says nothing about RCS on any particular aspect.

Turning to three dimensions, for a plate facet size $b \times c$ and radii of curvature r_1, r_2 (Figure 7.4(*d*)), similar arguments apply, except we have to consider solid angles and our circle of 2π rad is replaced by a sphere of 4π sr. Over the whole sphere, the average echoing area is

$$\frac{1}{4\pi} \times \frac{2bc}{r_1 r_2} \times \pi r_1 r_2 = \frac{bc}{4}, \tag{10.10}$$

which is one quarter the whole (front plus back) target area.

Generalising, if there are no flats or re-entrants, a closed surface has average RCS one quarter the target superficial area. The radius of curvature occurs in numerator and denominator of Eqs (10.9) and (10.10), cancelling out in the final results. So the equation holds good as the radii approach zero (a point) or infinity (a flat plate). In the latter case the RCS locally goes to infinity over a vanishingly small solid angle causing a TPM spike. Each elementary facet such as the one considered above contributes one quarter its outer area. In Sections 7.4.2 and 7.4.3, Eq. (7.3a), we have already found this relationship applies to a sphere, the simplest target shape. Facet RCS is independent of wavelength.

Eq. (10.10) forms a baseline for large targets: the average RCS in three dimensions is $\frac{1}{4}$ the total external area.

The half-perimeter and quarter-area rules may not apply to targets with re-entrants or concavities where multiple bounces can give additional returns or parts of the surface may be shadowed. For these reasons and because rust, paint and the resistivity of the metalwork introduce losses, average RCS of ships does not exactly equal $\frac{1}{4}$ the superstructure area (in essence the total 'paintable' external surface above waterline); nor does the RCS viewed on the horizon exactly average $\frac{1}{2}$ the perimeter.

10.5.4 TPM smoothness

If radius of curvature $r \gg b^2/\lambda$, adjacent reflection lobes remain distinct and the TPM is spiky, with high peaks but deep nulls. If $r \ll b^2/\lambda$, the lobes merge and the

TPM smoothens out towards that of a hemisphere. Targets composed of relatively few very flat (r large, irregularity $<\lambda/4$) facets tend to have spiky TPMs with the risk of presenting a null to the radar as in stealthed warships, but if r is small and the number of facets high, there are no deep nulls through which detection is impossible. Large vessels therefore tend to have smoother TPM than small vessels. The latter are more likely to exhibit RCS nulls, especially when wavelength is long.

10.5.5 RCS/tonnage by micro-geometry

This micro-geometrical theory indicates (Eq. (10.10)) that RCS $\propto$ area. Area $\propto L^2$, that is,

$$\text{RCS} \propto L^2.$$

Substituting in Eq. (10.8), RCS $\propto T^{2/3.1} = T^{0.645}$. This is independent of wavelength, so we could write

$$\text{RCS} \propto \frac{T^{0.645}}{\lambda^0}. \tag{10.11}$$

Eq. (10.6) indicates RCS $\propto T^X$, and Eq. (10.11) gives $X = 0.645$ which is less than the range 0.84–1.19 found experimentally (Table 10.5). Also in practice there is a mild wavelength dependence (Eq. (10.6)), so micro-geometry is unlikely to be the sole determinant of RCS. We next ignore micro-geometry and go to the other extreme, recalculating purely from macro-geometry.

10.5.6 RCS/tonnage by macro-geometry

From the macro-geometry of flat sheets and corner reflectors (Chapter 7, Section 7.4, Eqs (7.4) and (7.10)), RCS $= 4\pi/\lambda^2(\text{area})^2 \propto A^2/\lambda^2$, where A is target surface area. (On this macro-geometrical basis the RCS of the side of a VLCC riding high in light draught, say $250 \times 20\,\text{m}$, would be $108\,\text{dB m}^2$ in the 9 GHz band. The echo at a typical radar would be around 6 dBW (4 W) at 1 km, enough to damage the receiver if its protection circuit were inoperative.)

$$A \propto L^2, \quad \text{so RCS} \propto \frac{L^4}{\lambda^2}.$$

Substituting in Eq. (10.8)

$$\text{RCS} \propto \frac{T^{4/3.1}}{\lambda^2} = \frac{T^{1.29}}{\lambda^2}. \tag{10.12}$$

1.29 lies above the 0.84–1.19 range found experimentally for X and the wavelength dependence is too sharp. So macro-geometry is not solely responsible for RCS.

We next reconcile these micro and macro approaches with the experimental data.

10.5.7 Reconciliation with reported results

Eq. (10.6) and Table 10.4 show reported practical measurements give RCS $\propto T^X$ where X lies between 0.84 and 1.19. Let us include wavelength and assume that reality is approximated by the law

$$\text{RCS} \propto T^X \lambda^{-Z}. \tag{10.13a}$$

From micro-geometry, Eq. (10.11): if $X = 0.645$, $Z = 0$.

From macro-geometry, Eq. (10.12): if $X = 1.29$, $Z = 2$.

From which $Z = 3.1X - 2$. Substituting in Eq. (10.13), when $X = 0.83$, $Z = 0.57$ and

$$\text{RCS} \propto T^{0.83} \lambda^{-0.57}. \tag{10.13b}$$

RCS would fall 2.8 dB at 3 GHz relative to the 9 GHz band. (The $5 \log f$ term of Eq. (10.6) indicates 2.5 dB reduction, which is very close.) Also compare with Eq. (10.4) and the frequency dependence of simple shapes, Chapter 7, Section 7.5.3, indicating that the following configurations contribute to overall RCS:

- one infinite and one non-zero radius, for example, a cylinder, $\sigma \propto \lambda^{-1}$;
- one infinite and one zero radius, for example, a wedge, $\sigma \propto \lambda^0$;
- one finite and one non-zero radius, for example, a spheroid, $\sigma \propto \lambda^0$.

When $X = 1.19$, $Z = 1.68$ and RCS $\propto T^{1.19} \lambda^{-1.610}$. RCS would fall 10.3 dB at 3 GHz relative to 9 GHz. This is less close and we conclude that Eq. (10.13b) lies close to the truth.

Our broad-brush argument depends on many approximations, among them that (surface area above waterline) $\propto T^{2/3}$, in other words that small ships are scale models of large ones, which is questionable. Nevertheless, it indicates that

- there is some theoretical justification for the form of reported results and for Eq. (10.5);
- micro- and macro-geometry play nearly equal parts in forming RCS;
- therefore spikiness of TPM is likely to be muted;
- and the reported variation of RCS with elevation angle (Figure 10.7) is explained;
- RCS is weakly dependent on wavelength and likely to fall a few dB as we go from 15 through 9 to the 3 GHz band. Presumably at some very small size RCS starts to follow the $1/L^2$ relationship typical of many of the simple shapes discussed in Chapter 7;
- paradoxically, RCS of big vessels are biassed toward their micro-geometry and small vessels to their macro-geometry.

10.6 Features contributing to ships' RCS

10.6.1 Long-range detectability

The quarter-area rule gives a baseline RCS averaged over the visible hemisphere but does not indicate whether RCS at bearings of interest near the horizontal will be above or below that average. We shall review the factors affecting RCS in the horizontal plane to help inform our judgement of the likely RCS of specific vessels relative to their tonnage. We include a few other factors affecting long-range detection.

The maximum range at which a given radar/scanner/height combination will detect a vessel of given tonnage depends on several factors. Some are governed by the basic shape of the ship, imposed by its trade – bulk carrier, passenger, etc. Perhaps we could call these mega-geometry factors. Others are matters of design detail and styling, which affect the macro-geometry, while micro-geometry often results from the vagaries of constructional practice – flatness of plating for example.

10.6.2 Mega-geometry factors

These are as follows.

1. Maximum height of significant reflecting elements. Vessels lying low in the water have short critical and horizon ranges.
2. Targets with low freeboard, e.g. laden tankers and bulk carriers, have low surface area for the tonnage.
3. Deck cargo, notably containers, may increase target volume above that predicted by the gt formulae, Eqs (10.1) and (10.7). Where ships such as hatch-cover-less container ships have no conventional deck we mean containers or other cargo visible above the bulwarks.
4. Non-metallic construction sharply reduces RCS, as indicated in Chapter 7, Section 7.2. Currently large vessels are entirely metallic apart from windows, but small craft employ extensive glass reinforced plastic (grp), other composites and timber.

10.6.3 Macro-geometry factors

Surfaces oblique to the horizontal reflect away from a surface interrogator. RCS will be higher than baseline average if viewed normal to the slanted surface at the expense of low RCS where needed in the horizontal plane. We conclude that the following features tend to reduce RCS.

1. Bows and sometimes sterns of many faster vessels such as ferries and container ships are flared to improve sea-keeping. These 'fine lines' deflect radar beams and the hull echo of such ships viewed end-on is weak (note the low RCS of a cone, Section 7.5.2). Ships' smooth bow-wave also deflects rays and partially masks the bow. This is confirmed by P. D. L. Williams' measurements reported in Section 10.3.
2. Accommodation blocks with pronounced tumblehome.
3. Passenger vessel styling including streamlined accommodation outlines, devoid of vertical faces; a style currently also fashionable on some HSC and mega-yachts.

Concavities or re-entrants, even when not forming truly orthogonal corners, generally raise average RCS above the quarter-area baseline more than the protrusions at their edges drive it down, and can give high RCS over wide arcs. Many modern cruise ships include numerous state-room balconies, probably raising broadside RCS. Because of concavities, mean RCS (over the visible hemisphere) of ships generally exceeds the quarter-area baseline, Eq. (10.10). Macro-geometry contributions drive up local RCS on normals to large quasi-flat surfaces, so the RCS within a degree or so of horizontal, of prime interest to us, is generally several dB above baseline, see Figure 10.7.

Exactly broadside-on, most hulls and some superstructures approximate a flat plate and return a strong echo over a narrow angle. Although this 'flash' needs to be taken into account when considering radar receiver overload and dynamic range, it is too transitory for inclusion in detection calculations. Plating irregularities may spread the flash over several degrees at expense of its peak RCS.

Deck cargo is often rich in corners. ISO containers are metallic, usually corrugated, frequently stacked to 15 m or so high and can contribute high RCS through broad azimuth angles, but timber cargoes may reflect poorly.

Cargo gear such as deck cranes and tanker pipework is likely to reflect fairly omnidirectionally in elevation and azimuth except perhaps from ahead and astern where it is masked by raised bows or the accommodation block.

The accommodation of many ships is more or less a cube raised well above deck level, whose height makes it important to long range detection. The front is often raked and curved, deflecting rays and reducing RCS somewhat. Windows allow rays to reflect around the interior and re-radiate weakly over broad angles. Sides and rear of the block are often festooned with ladders, companion ways and other rich reflectors, and are less likely to be raked, giving higher RCS than the bow.

The funnel is often nearly integral with the block and often curved and tapered, reducing its RCS, although its height may increase its significance to long-range detection. The radar mast typically carries an assortment of lights and other equipment with access ladders and platforms, giving significant RCS. It is also the highest structure and may govern maximum detectable range. Wake is only a significant reflector with small fast craft (see Section 10.9.3).

Designs having relatively few facets have spiky TPMs, with low RCS between the spikes. Smaller vessels are inherently likelier to have few facets, particularly when styled with clean lines.

10.6.4 Micro-geometry factors

Dead flat surfaces, with irregularities less than $\lambda/4$ are most likely on smaller passenger craft where great attention is paid to appearance. They give narrow TPM spikes well above baseline average, with correspondingly lower RCS elsewhere, particularly in the 3 GHz band. Load-bearing hulls are likely to strain between the frames even if accurately constructed in the shipyard so are unlikely to remain geometrically flat in practice. Repeated pounding from head seas may scallop the bow area (if metallic) as the ship ages, somewhat increasing head-on RCS (Figure 10.1).

10.6.5 Stealthed vessels

We have briefly referred to stealthy targets. Although the military use a wider definition of stealth, including optical and infrared wavebands and acoustic noise, we mean targets having RCS in the horizontal plane much lower than normal for the tonnage.

These vessels only break the quarter-area rule when non-metallic or coated with radar absorbent material (RAM), but have spiky TPM with large RCS at some angles well out of the horizontal plane and correspondingly low RCS to sea skimming missiles and hostile warships.

Naval craft obtain stealth by featuring few, large, surfaces, and these slanted, dead-flat with few windows or other apertures (which would radiate over broad arcs), no 90° re-entrants or internal corners; radar absorbing material screening unavoidable masts and other potential scatterers, possibly utilising non-metals in the superstructure. Jones *et al.* [11] claim that the horizontal polar diagram of vertical plating may be reduced by 20–35 dB when masked by zigzag metal foil alternately angled at +20° and −20°. The foil is supported and concealed within foam plastic slabs. There may be other tricks of the trade less apparent from our fairly elementary analysis of the factors governing RCS. Their commanders sail in the knowledge that other shipping, reliant on radar lookout, may not detect at enough range for effective anti-collision manoeuvre. Surfaced submarines are generally devoid of vertical plating, smooth in contour and clad with acoustic tiles which are probably poor radar reflectors if not actually designed to be radar absorbent. RCS is therefore likely to be low.

Stealthed vessels may well artificially increase their RCS by carriage of a reflector or RTE in peacetime, even when well above IMO's 150 gt limit, improving their detectability for collision avoidance and concealing the true RCS from the inquisitive.

Unfortunately, for one reason or another, some civil marine craft share many of the same design characteristics and will inadvertently be stealthy. Their masters may not realise how poor targets their commands are, as their owners will certainly not have measured RCS of their vessels. If in doubt, they should carry a reflector or RTE.

10.7 Detection cell overflow

10.7.1 Azimuth overflow

Radars detect target presence by measurement of signal strength within detection cells whose width is set by the scanner azimuth beamwidth, between about 0.4° for some VTS radars and 4° for small-ship sets. A large ship, particularly at short range, may subtend a wider azimuth angle; in the example of Figure 10.9(a) two cells are occupied, halving the effective RCS per cell. From the geometry, at double the range the cell has twice the linear width (Figure 10.9(b)) and all the RCS is available within one cell for the detection process. There is thus a definite range below which effective per-cell RCS is reduced. If scanner beamwidth $= \theta$ rad, range $= R$ m, target projected width $= w$ m and the RCS is uniformly distributed:

$$\text{if } R < \frac{w}{\theta}, \qquad \text{effective RCS reduction} = 10 \log\left(\frac{R\theta}{w}\right),$$

otherwise

$$\text{reduction} = 0\,\text{dB}. \tag{10.14}$$

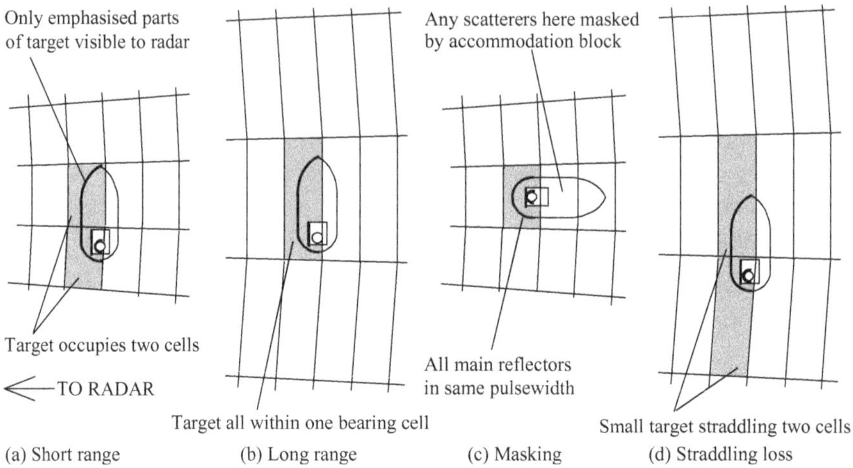

Figure 10.9 *Detection cell overflow. At short range (a) the projected target width may exceed scanner beamwidth, reducing effective RCS within the detection cell. Targets may also occupy several pulselengths in range, again reducing effective RCS (not shown). At long range (b), whole echo lies within one cell. Distant parts of targets are often masked by closer parts, fewer cells being occupied (c). Straddling loss (d) occurs when cell boundary intersects target*

The reduction may be less severe if the signal processor is arranged to integrate the contents of adjacent cells.

Ships may be up to about 300 m long. When broadside to a radar of scanner azimuth beamwidth say 1°, RCS rises with a R^1 law until it reaches its full value when all the echo fits within a single cell at 17 km. At shorter range, the usual R^{-4} echo strength law (or R^{-8} if above critical range) is changed by one order to R^{-3} or R^{-7}, respectively. It is not always clear whether allowance for detection cell overflow has been included within reported RCS figures for shipping. Luckily overflow primarily affects big ships, whose effective RCS usually remains fully adequate for detection.

10.7.2 Range overflow

Radar pulselengths vary between about 0.05 μs at short range and 1 μs, giving cell lengths between about 7.5 and 150 m, so a 300 m ship end-on occupies between 40 and 2 cells. The nearer reflecting elements on ships may mask the remainder. For example, when viewed on the stern the fo'c'sle area is usually masked by the accommodation block aft, as shown in Figure 10.9(c). Range overflow can be treated either by

- using a reduced end-on RCS, or
- using average RCS with an allowance for range cell overflow. As the reduction factor depends on pulselength, it is affected by the radar range scale selected by

the operator. When the target projects axial length L m and pulselength $= \tau$ μs, axial overflow reduction in RCS is given by:

$$\text{when } L > 150\tau, \quad \text{effective reduction of average RCS}$$

$$= 10\log\left(\frac{L}{150\tau}\right), \quad \text{otherwise reduction} = 0\,\text{dB}. \qquad (10.15)$$

10.7.3 Glint

When the echo overspills the detection cell, large close target echoes are often observed to break up into apparently distinct smaller targets, for example, the fo'c'sle and stern accommodation, with a weaker return from the intermediate deck area. Each sub-target is relatively small, with lobes in its polar diagram, roughly resembling Chapter 7, Section 7.10.3, Figure 7.15(f). Sometimes a distant radar will receive lobes from each sub-target, sometimes a pair of nulls; so the echo will fluctuate. But sometimes the stern will give a lobe and the bow a null, shifting the echo paint to the stern. Minor yaw or roll may then bring a lobe at the bow and a null at the stern, moving the paint to the bow. This apparent position dither, called glint, adds uncertainty to the detection process and makes manual or automatic track-forming more difficult. Worse, OOWs have been misled into thinking a tanker lying across own ship's bow to be a pair of fishing vessels, with a navigable gap between.

10.7.4 Straddling

Of course, even when the target is smaller than the detection cell, it may not sit neatly in the middle of a cell. Figure 10.9(d) shows a target straddling two cells (in bearing, but could equally be range), reducing effective RCS. Straddling loss is conventionally treated as a processing loss within the radar detection system rather than a reduction of effective RCS, Chapter 12, Section 12.11.

10.8 The RCS to use for ships

Eqs (10.14) and (10.15) assume the target reflects uniformly throughout its length or width but probably reasonably describe the cell-overflow variation in effective RCS occurring in practice.

Because

- bow-on is a particularly dangerous collision encounter;
- many VTS radars are so located that traffic tends to be bow on as they look down a fairway;
- RCS tends to be least near the bow;

failing specific information on the vessel in question we suggest it prudent to use Eq. (10.6) with $X = 0.83$ and $Y = 15$ as the basis for detection calculations. Results should then under- rather than over-estimate detection range for conventional non-stealthy ships. Combining Eqs (10.6) (basic RCS, terms in bold type),

(10.14) and (10.15) gives Eq. (10.16), the terms $[\cdots]$ and $\{\cdots\}$ being the azimuth and range cell overflow corrections, respectively:

$$\text{minimum likely RCS} = \mathbf{8.3\log \mathit{T} + 15 - 5\log \mathit{f} - 1.8\gamma} - \left[10\log\left(\frac{R\theta}{w}\right) \text{ when } R < \frac{w}{\theta}\right] - \left\{10\log\left(\frac{L}{150\tau}\right) \text{ when } L > 150\tau\right\} \text{ dB m}^2. \tag{10.16}$$

10.9 RCS of small craft

10.9.1 The problem

Small craft without reflectors or RTEs usually have very poor RCS. Metallic masts and stays are primarily thin vertical cylinders whose RCS is low, especially to horizontally polarised signals. Hulls are predominantly non-metallic of rather low dielectric constant and are poor reflectors. Profiles are curved and streamlined with few metallic corner reflecting elements, again reducing RCS. There are insufficient reflecting facets for their echoes to fill TPM nulls, which may remain deep through significant arcs, causing detection problems discussed in Section 10.11. Sailing vessels in particular lie low in the water and their hulls have low horizon range. RCS of sails, wet or dry, is probably negligible, but wire rigging may reflect significantly if nearly in the scanner plane of polarisation. More and more such craft are encountered far offshore. It has been suggested that the RCS of non-metallic small craft, including rigid inflatable boats, Figure 10.10, equals that of the hole made in the (conductive) water. Detection of these small targets in sea clutter remains the most difficult problem facing the radar designer. The problem is neither skill nor will, but the inherent similarity of weak echoes to sea clutter.

Figure 10.10 Rigid inflatable boat. Neither much metal, displacement, height nor wake. RCS very low, perhaps approximating that of the crew or the saucer of water displaced. Author

The only sure way to improve detection is to enhance the unaided echo with a passive or active reflector. Even then, heel may make it difficult to keep the device within its vertical acceptance angle, and Chapter 7, Section 7.10, showed that the combination of a reflector or RTE with a target of equal RCS can spoil the resultant TPM. No universally satisfactory solution is yet available. Development of powerful RTEs having wide elevation beamwidth currently seems the best way forward, although concern has been expressed by some navigators that the strong echo will cause craft so equipped to be misjudged as big ships.

A few RCS values which might apply to motor yachts were included with ship data in Section 10.4. There are so many designs of small craft that the following paragraphs considering the factors influencing their RCS can only generalise.

10.9.2 Reflecting elements

The only significant reflectors aloft on many yachts are the tubular metal mast and wire rigging. The mast is usually aluminium, a good conductor. The formula for RCS of a conducting cylinder (Chapter 7, Section 7.5.1) may give meaningful RCSs of masts if dead-straight and of rigging if taut. A straight conducting circular cylinder acts as a point reflector at mid-height, with RCS per Eqs (7.11a) and (7.11b). Broadside on ($\theta = 0$), if straight within $\lambda/4$, macro-geometry Eq. (7.11a) indicates a yacht mast having $L = 10$ m and $r = 0.1$ m, would have 9 GHz band RCS 1962 m^2 (32.9 dB m^2) with elevation beamwidth approx $0.44\lambda/L$ rad $= 0.08°$. Micro-geometry flexing will dilute these values considerably, but they show the mast cannot be relied on as a reflector over a usefully wide angle. Rigging uses stranded steel wire, a fair conductor. RCS (Eq. (7.11d)) is strongly polarisation dependent.

Sails contain no metallic material. It is sometimes asserted that wet sails reflect well, but convincing evidence is lacking. Stanchions, the galley and miscellaneous metal details may provide some reflection near the waterline.

It has been suggested that the RCS of a non-metallic sailing dinghy approximates that of the crew's bodies. From Table 10.3 this might be -3 dB m^2 per lightly clad adult, less if crouched and less still if within a lossy hull or cabin.

No data is at hand, but it seems reasonable to assume that RCS of an outboard motor may lie between -10 and 0 dB m^2 depending on size. Inboard engines may lie too low to reflect significantly. Engine and crew may be the only significant reflectors in a rigid inflatable (RIB) at low speed or with insignificant wake.

10.9.3 Displaced water

At rest, the RCS of the displaced water volume (m^3) equals the displacement mass, W tonnes, of the craft plus crew. Assuming a hemispherical 'saucer' is displaced, radius r, $W = 2\pi/3r^3$ so $r = 0.78W^{1/3}$. Taking the RCS of a hemisphere to be $\frac{1}{2}$ that of a sphere, from Eq. (7.3a):

$$\sigma = \frac{\pi r^2}{2} = 0.96W^{2/3} \sim W^{2/3}\,\text{m}^2. \qquad (10.17)$$

In dB terms, $\sigma \sim 6.6 \log W$ dB m^2.

This is very rough and ready, but indicates a dinghy plus two crew totalling 250 kg might have RCS about -4 dB m^2 (0.4 m^2), independent of radar band.

At speed, craft such as RIBs with powerful outboard engines may throw up considerable wake, which may be detectable. RCS may approximate the 'rooster tail' visual cross section, perhaps a couple of square metres bow-on, rather more on the beam. The wake of course lies behind the boat and use of its echo introduces a small positional error. The wake of fast motor-boats is sometimes found to be their biggest reflecting element, but some modern designs throw up little wake or wash in deep water, even at speed.

10.10 Fast craft

10.10.1 High speed craft

HSC are those whose service speed exceeds 30 kt (55 km/h). This is not the only definition. The International Code of Safety for HSC, 1994, using a hydrodynamics approach, defines an HSC as a craft capable of a maximum speed (metres per second) equal to or exceeding $3.7 \times (\text{displacement, tonnes})^{1/6}$, which can lead to relatively slow small craft being classed as HSC. Some of the old North Atlantic passenger liners such as RMS *Queen Mary* and SS *United States* made over 30 kt but modern cruise ships rarely exceed 24 kt (45 km/h). The current maximum speed of small fast craft is about 75 kt (140 km/h), in the types used for coastguard and gendarmerie duties. Most HSC are passenger/car ferries, running at up to 40 kt, speed and frequency of service replacing size, although cargo craft are being developed. Craft and services are evolving rapidly, displacing larger conventional ships and speeds are tending to increase. Although craft up to 200 m length exist, most are much smaller. Closing speeds between HSCs can exceed 100 kt and long-range detection is important for safety.

Little seems to have been published on RCS of HSC; data are badly needed. These fast passenger/vehicle ferries are often relatively small, without tall masts and are streamlined and devoid of vertical surfaces or dihedral/trihedral corners which contribute to monostatic RCS. Plating may have fewer irregularities than on a conventional steel ship; all in many ways akin to a stealthed warship. Although nimble enough to turn away from developing close-quarter situations and navigated with their poor visibility in mind, these craft need to be detected at long range for effective collision avoidance.

Construction is usually steel or aluminium, but there is a trend to composites, likely to have low reflectivity. Tonnage is low. For all these reasons, RCS is likely to be poor, especially on the bow. Until representative HSCs have been measured and their RCS published, it would be prudent to assume considerably less than a conventional ship of the same visual cross section. Overall height is about $\frac{1}{3}$ the length, proportionately at least as high as conventional ferries, masts and funnel excepted. Distribution of RCS with height may be rather uniform, with effective height factor $n \sim \frac{2}{3}$ (Eq. (9.5)).

10.10.2 Large motor yachts

Although usually smaller than commercial HSC, large luxury yachts share many of their reflection features. They are classed as super-yachts when over 400 tons gross, or mega-yachts when in excess of 30 m LOA. The gt measurement is not used, as they carry no cargo. Maximum speeds may be well below 30 kt, in which case they pose no special navigational problems to other traffic. The bigger yachts are usually commanded by qualified master mariners and sailed responsibly. On the other hand, some owners delight in extremes, some yachts being engined to give in excess of 75 kt sprint. Here other traffic may find timely detection difficult especially as yachts do not follow conventional merchant ship or ferry routes and may manoeuvre faster than the envelopes of the tracking algorithms within automatic tracking aids. Some racing power boats also exceed 75 kt; their specialist drivers well know they must avoid other shipping and their poor detectability has not posed problems.

10.10.3 Wing in ground (WIG) craft (ekranoplanes)

When taking off, a seaplane initially works up speed as a displacement boat. Partial vacua form on the upper surfaces of the aerofoil wings. Positive pressure on the lower surfaces is augmented by air which becomes wedged between the wing and the 'ground' (sea surface). This curiously titled wing in ground effect is also used by many seagulls. Eventually the vehicle rises clear of the water. Hydraulic drag ceases, speed rises further and the aircraft is airborne. The WIG lift drops rapidly and is negligible above a few metres altitude.

WIG effect craft, alternatively known by the Russian term Ekranoplanes, are in essence seaplanes whose short wings are shaped to maximise WIG lift. Although propelled like aircraft, they cannot truly fly but skim along a couple of metres clear of the surface. Land rarely offers a sufficiently smooth terrain and WIG craft usually operate over water, where IMO regards them as ships.

Heroic experimental military WIGs exceeding 100 m length, cruising at around 300 km/h, were developed in the former USSR on the Caspian Sea and ocean-going cargo vessels have since been proposed. Once the initial strenuous take-off effort has been made, they have the potential of high speed at much lower power than other forms of marine or air craft. Meanwhile, most current development centres on much smaller models for the leisure market, as a form of speedboat. One problem is that WIGs can only leave the surface when travelling fast. To get up to speed in displacement mode needs much more power to force the craft through the water. Possibly this problem may one day be alleviated by operating as a form of hovercraft. Nevertheless, their pilots will wish to run fast whenever possible, making early radar detection of WIGs particularly important.

At speed in displacement mode there is a large wake. In WIG mode the wedge of air generates wash but no significant wake of broken water. No RCS data is to hand on these novel craft. WIG craft tend to be streamlined and small ones may well be constructed partly from non-metals, making RCS low for the size, perhaps somewhat exceeding the RCS of a fixed-wing aircraft of the same all-up weigh because WIGs

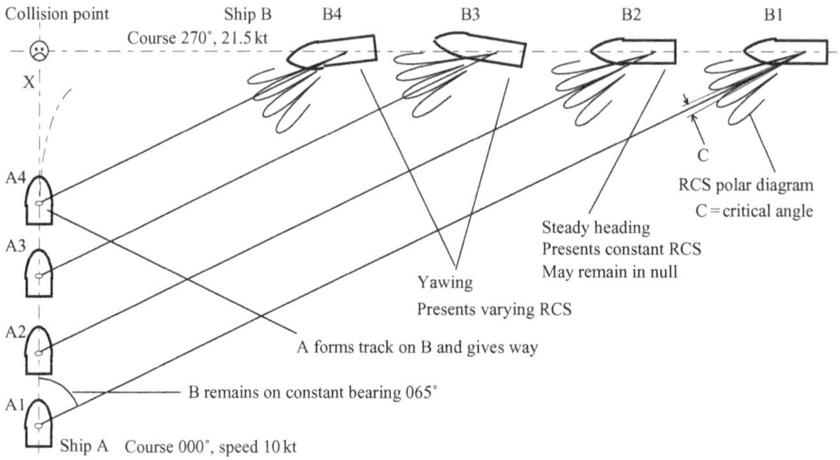

Figure 10.11 *Effect of yaw on RCS. Ships A and B are on collision courses. At instants 1 and 2, B is on a steady heading and continually presents an RCS null to A. At instants 3 and 4, B is yawing, presenting variable RCS including lobes which enable detection by A. Rolling has similar results*

may be less streamlined. RCS will vary with mode. The clearance above the surface is likely to cause the craft to appear intermediate between a point and a true vertically-extended target, with moderate multipath ripple on the echo strength/range graph. Speed may exceed 200 km/h (~110 kt).

10.11 Lobe spacing, yaw and roll

An adaptation of Figure 10.11 and some of the arguments of this section have previously been published [12]. Figure 10.11 shows two ships A and B on course for collision at X. A must form a track of B's echoes – a process requiring at least 50 per cent probability of detection (P_D) on several scans – before finding it is the give way vessel under the Collision Regulations and starting to manoeuvre.

At instant 1, B is steering on a good heading control system (automatic pilot) without yaw or roll. It happens to present an RCS null to A. Later, at instant 2, A continues to look into the null and detection remains impossible. At instants 3 and 4, B yaws by more than the lobe spacing of its polar diagram, showing A the whole range of RCS values. There are now enough strong echoes for B to be detected and tracked. Rolling introduces a similar sampling process.

If average RCS is enough for 50 per cent P_D, detection is likely when maximum lobe spacing of the target exceeds its peak–peak yaw or pitch, the so-called critical angle, C. It is the RCS within nulls only a few degrees wide, rather than average or peak values, which determines whether the target will be detected.

Yaw and pitch depend on vessel size and on weather. Luckily, the wide nulls of a small vessel may be matched by wide yaw and roll, especially in bad weather where clutter is most troublesome. In reasonable weather, many ships hold course and heading within 1° using auto pilots. It is for this reason that an averaging width of 1° was suggested for experimental determination of RCS in Section 10.3.2.

10.12 Land and shoreside features

10.12.1 Introduction

Although satellite positioning systems have reduced the former importance of marine radar as a long-range landfall aid, the mariner still uses radar to fix position relative to coastlines. Here we will think of land primarily as a target. In some circumstances land can mask other targets such as a racon marking a lighthouse and is then a form of clutter. VTS and some modern ships' radars generally portray coastline synthetically as map data. The coast echo is often blanked out from the detection process to reduce the computational load.

Coastal features are often much bigger than ships and extend over many azimuth beamwidths, but are rarely high enough to overflow the top of the elevation beam except at very short range. Coasts are so varied that it is difficult to give specific RCS values. Echo strength depends on the following.

1. RCS per square metre of terrain when viewed at normal incidence. Chapter 5, Section 5.8.3, showed that near normal incidence, voltage reflection coefficient, ρ, of the surface depends on its dielectric constant, ε, and thus on the type of rock, its porosity and wetness. Typically ε lies between 2 (dry, porous surface) and 10 (wet, porous conditions of Figure 5.18 point c). From Eq. (5.40h), power reflection coefficient ρ_0^2 (i.e. RCS) lies between 0.029 and 0.27m^2/m^2 (-15.4 to -5.7 dB) respectively.
2. Compound azimuth/elevation angle of incidence to normal.
3. Surface roughness or micro-geometry, which generally reduces the importance of angle of incidence and scatters reflection through a broad solid angle, depressing monostatic reflection coefficient. Figure 10.12 reminds us that many cliffs have very rough macro- and micro-geometries. Occasionally, tilted beds of sedimentary rock give rise to very smooth cliff faces.
4. Number of square metres in detection cell, which varies with range, scanner beamwidth and pulselength.
5. Radar band. The frequency dependence is usually slight enough to be ignored in face of the uncertainties of the data.
6. As with other targets, the usual factors of atmospheric refraction coefficient (which may differ from the k-factor out to sea), overall height, etc.

10.12.2 Coastline and rivers

The RCS depends on the following.

- *Reflection coefficient*; increases with water content within the top wavelength or so of the surface. Vegetation and clusters of buildings tend markedly to raise

Figure 10.12 *Cliffs often have rough micro- and macro-geometry and form distributed targets reflecting fairly uniformly through wide aspects. Reflectivity may depend on seasonal vegetation growth or surface wetness. Except at longest range cliffs usually overflow the scanner azimuth beamwidth. Chalk 30 m cliff, North Foreland, English Channel. Author*

the coefficient. Extrapolating from reflection coefficients cited for land viewed nearly vertically from aircraft, voltage reflection coefficient of fairly steep cliff may lie between 0.2 and 0.4, the equivalent unit RCS varying between −14 and −8 dB m^2 per square metre when viewed near normal.

$$\text{Total RCS} = \text{unit RCS, dB m}^2/\text{m}^2 + 10\log(\text{area illuminated}), \text{dB m}^2. \tag{10.18}$$

- *Coast area within detection cell.* This in turn depends on the steepness of the terrain coupled with the radar pulselength; the azimuth beamwidth and, only if the range is short or the terrain exceptionally high, the elevation beamwidth as shown in Figure 10.13.
- *Micro-geometry.* If root mean square (rms) roughness of the terrain is, unusually, less than a quarter wavelength, our study of micro-geometry tells us the surface acts as a specular mirror, with little echo unless the radar is near normal to it, Figures 10.14(a) and (b). Specular effects are absent when the rms surface roughness exceeds λ.
- *Macro-geometry.* When the surface is rugged, the incident rays are scattered through a wide solid angle and echo strength becomes much less dependent on angle of incidence, as shown in Figure 10.14(c).

Coastal echo is poorest when the land rises gently and is smooth and dry, the desert coast of Namibia being a notorious example. On the other hand, Romney Marsh near Dungeness, English Channel, has a surprisingly strong surface echo, not from the flat terrain but from numerous fishermen's huts, beached boats, motor vehicles and so forth. Waves breaking on a foreshore fronting a cliff may produce sea clutter which is only distinguishable from the land echo by careful observation, making estimation of the position of the true shoreline difficult.

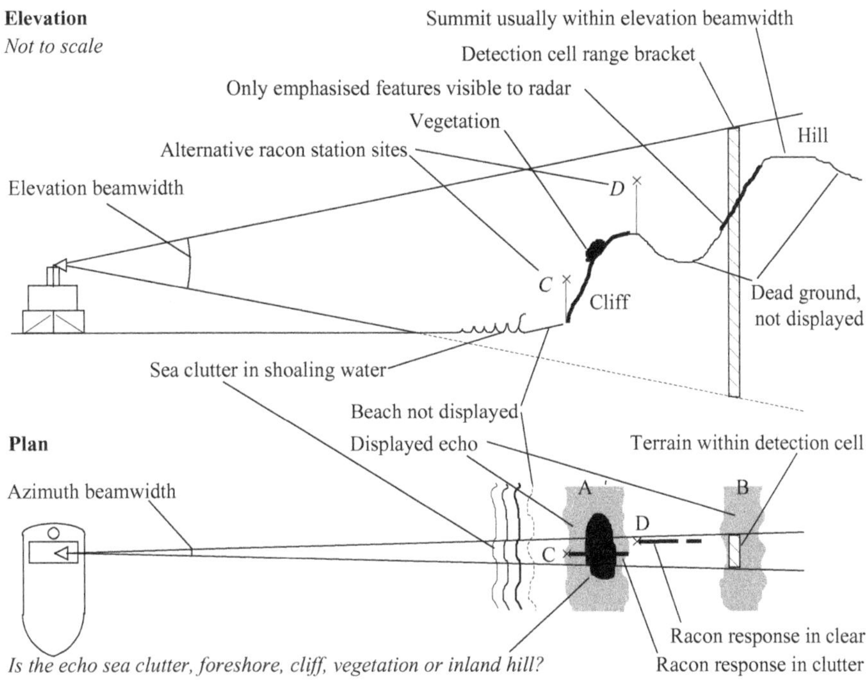

Figure 10.13 Illumination of coastline. Terrain may be partly masked. Azimuth width of coastline in detection cell is proportional to range, also to pulselength if the shore is not steep-to. At long range lower terrain A may fall below the horizon, giving a false coastline B. Beaches are poor retro-reflectors and not displayed, although surf may show strongly. The response (shown diagrammatically) of a racon at C lies over ground clutter; at D it lies over dead ground and is more easily detected

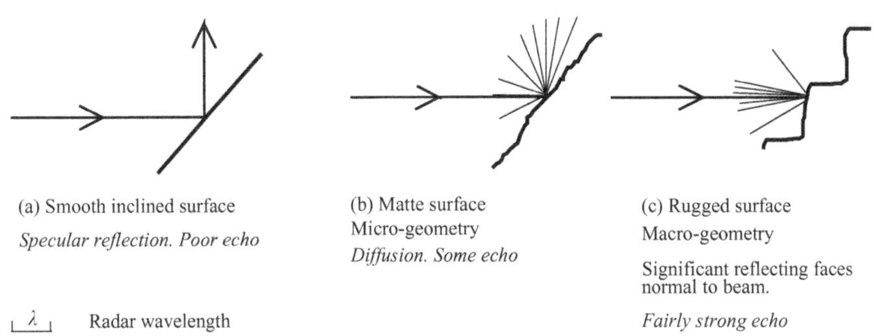

Figure 10.14 Effect of surface roughness. Except when nearly normal to the beam, a rugged surface tends to give a stronger echo, less dependent on aspect

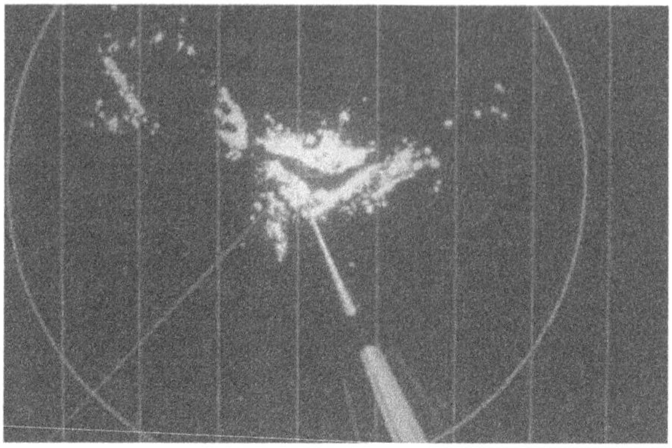

Figure 10.15 *Narrow river. Conventional radar on a short range scale is inadequate for such confined waters. Racon on test at bankside manufacturing facility. Bill Quay, River Tyne, North-east England. Author*

Figure 10.13 depicts a cliff fronting a hill, giving two distinct echo bands. At long range the cliff may fall below the horizon, causing range error when the hill echo is mistaken for it. At shorter ranges the intermediate dead ground gives an area free from land echo. Judiciously sited racons or reflectors can here display without being swamped by land clutter. Shoaling steepens waves, which may break as surf, with strong returns. Smooth beaches lie at acute angles to the rays and reflect poorly, although hillocky sand dunes often reflect fairly well. As range closes the nearer features start to be displayed, so apparent range reduces faster than true range. Position fixes from shore echoes should be treated cautiously – the echo may be from an unexpected, uncharted and visually inconspicuous feature. Brightly painted lighthouses and other sea-marks without racons may return quite inconspicuous radar echoes.

River navigation poses sufficient problems for specialist radar specifications and equipment to have been developed for the Rhine and other waterways. Clearly, long range is not received unless the ship also goes to sea, as many do. But bankside echoes are severe, particularly in industrial districts and may cause display of sidelobes or reflection echoes. Figure 10.15 shows a conventional 9 GHz display in the narrow and industrialised River Tyne, operated on the 1.5 nmi range scale. Clearly, a more open scale, say $\frac{3}{4}$ or $\frac{3}{8}$ nmi ($\sim$0.7 km) is desirable.

10.12.3 Shoals

Shoals and half-tide rocks usually have poor echoes. The echo of exposed rocks may often be augmented by breaking waves (beware the calm day and high tide!). Here sea 'clutter' becomes our target and the radar should be set up to maximise its echo; long pulse, anti-clutter sea turned off.

10.12.4 Bridges

Bridges over navigable waterways, whether metal, masonry or ferro-concrete, almost always have very large RCS, returning strong echoes. Unfortunately, their piers are

(a)

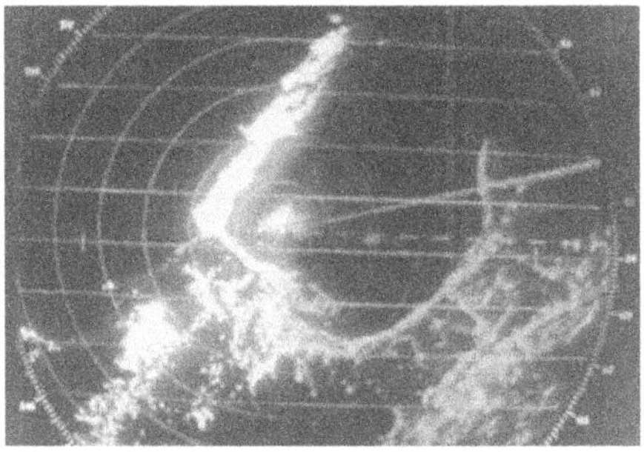

(b)

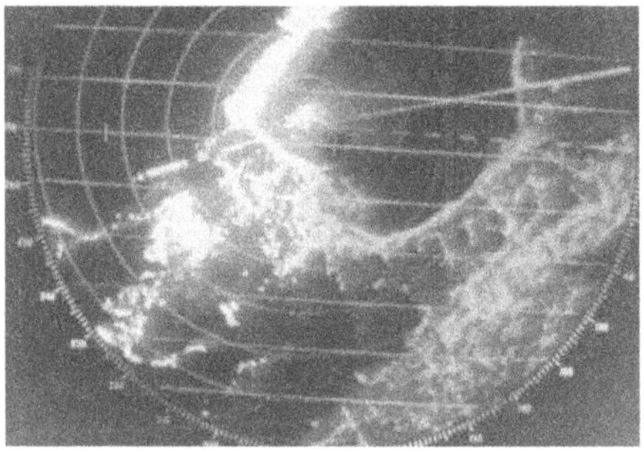

Figure 10.16 *Channels under bridges. 1983 trial to determine whether channel centrelines under bridges may benefit from marking by racon. Kessock Bridge, a cable-stayed suspension bridge with piers in the water, at the head of Moray Firth, North-east Scotland; 9 GHz radar, 0.05 μs pulselength, 0.7° scanner beamwidth, PPI cursive display, 1.5 nmi range scale, offset, range ring intervals 0.25 nmi. On MV Pharos. Note excellent resolution available from the older generation of radars, but brilliance fluctuation as the strobe sweeps the screen. Strong land returns from buildings in the City of Inverness. Author, reproduced by permission of the Northern Lighthouse Board. (a) No racon, channel centreline not obvious. (b) With racon, code K, marking channel at 235°.*

generally indistinguishable from the general echo, see Figure 10.16(a), and it is of course dangerous to assume piers are symmetrical either to the bridge echo, the banks or the deep channel centreline. Particularly when lying near the navigable channel, it may be prudent either to mark piers with reflectors placed well in front, or with an overhead racon at mid-channel, Figure 10.16(b). Effective RCS of these aids needs to be large, since the ship is likely to be using a short range scale with the gain turned down to prevent display of sidelobes from the structure or other nearby features. Racon receiver sensitivity should be low, to prevent overload of its sidelobe suppression circuit as the ship passes, but response power needs to be high to counter the low radar receiver gain; a high-gain directional antenna may be beneficial, with figure-8 azimuth response, which can be constructed from conventional full-height vertical waveguide radiating from slot arrays on each broad face.

Bridges do not generally interfere with echoes of targets seen through the span if not physically masked, with one important exception. At short range the bridge echo may be strong enough to overload the radar receiver, which is then desensitised for a period of the order of a microsecond. During that time, its operation is unreliable; false echoes may display, or, worse, true echoes may be suppressed. The remedy for all false echo problems is to reduce gain as far as consistent with detection of small targets at safe range.

10.12.5 Overhead obstructions

Although a line of distant pylons may be clearly displayed, the actual cables rarely show even at short range, so radar often fails to warn of this hazard to navigation. Marine and VTS radars have no height-finding capability and cannot warn of insufficient air-draught. Vessel traffic services needing to protect bridge spans and overhead cables use calibrated closed circuit visual and infrared television cameras with long-focus lenses to check height of approaching ships. Accuracy may be limited by optical refraction effects analogous to k-factor. The TV picture may be associated with VTS radar plots to determine target range and assist the height computation.

10.12.6 False echoes

As indicated in Figure 10.17, large metallic vertical surfaces such as nearby ships, warehouses, cranes or bridges may reflect as mirrors, causing other targets to display at false positions as well as their real ones, an effect akin to sidelobe interference. A straight line being the shortest distance between two points, false echoes are always at longer range than the true echo. They may appear to move at double speed or in a wrong direction.

10.12.7 Fluctuation characteristics

This section is partly based on Sekine and Mao [13, Section 2.1]. Many landforms contain a weakly reflecting and rather uniform background plus a scatter of physically small strong reflectors, for example the cliff shown in Figure 10.14(c) or a suburb on a hillside sloping gently up from a beach. The sloping open spaces are viewed at a fairly low grazing angle, perhaps a few degrees, but the sides of buildings have nearly normal vertical incidence and reflect more strongly. When they also chance to lie normal in azimuth, these quasi-point scatterers reflect very strongly. Echoes from such landforms therefore have an almost noiselike component which retains near-constant average amplitude, plus a scattering of spikes which vary quasi-randomly as the radar sweeps in range and scans in azimuth. Of course, a radar on a fixed platform, staring at a single small area of terrain, would collect an unvarying echo, not fluctuating in the time domain and therefore distinguishable from thermal noise, or precipitation or sea clutter, so the fluctuation is spatial but not temporal.

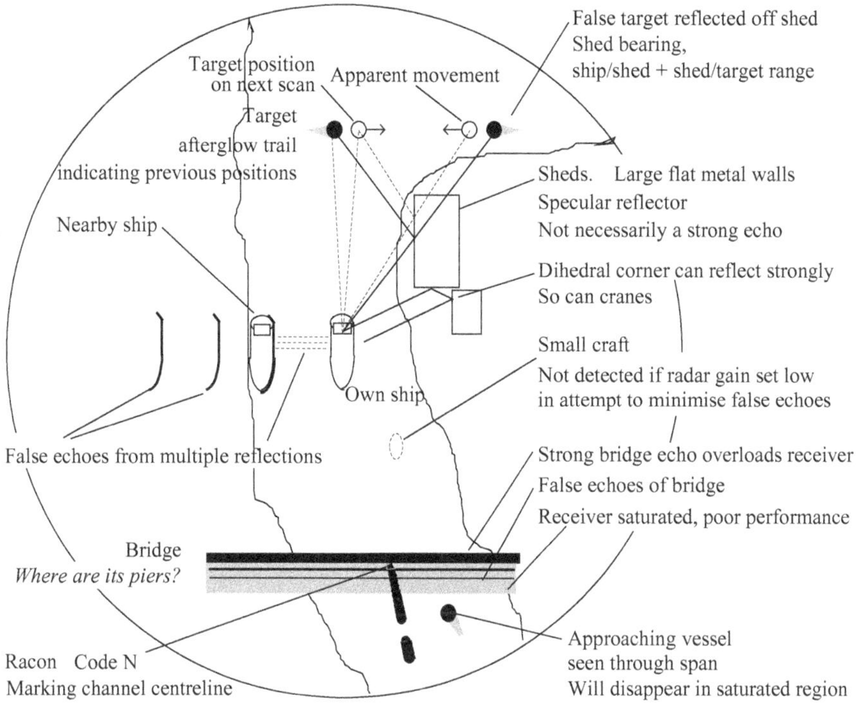

Figure 10.17 *False echoes and saturation. Radar on short range scale. False echoes via reflections from ship, shore structure and bridge span. Strong bridge echoes may overload the receiver and desensitise nearby targets. False echoes are alleviated by gain reduction*

The size of the radar detection cell affects the spikiness of the echo at the detector. A low-resolution radar, having a small scanner at 3 GHz working long pulses, will always illuminate numerous quasi-point scatterers, and the signal presented to the detector, although fluctuating, will contain few extreme events as the sweep or scan progresses; scatterers will join and leave the signal at almost the same rate, and the echo will fluctuate in a similar way to wideband noise. However, if the detection cell is small, with low azimuth beamwidth and short pulse, there will be pauses between one scatterer joining and another leaving, their reflections fluctuating sharply with time. The over-many extreme events make the amplitude distribution of the echo less noise-like. Statistical amplitude distributions are discussed in Section 11.3 and here we merely note that landforms can be represented by Weibull distribution, with varying shape parameter, c, as well as varying average reflectivity. Low value of shape parameter represents a spiky echo. These parameters may vary seasonally as foliage grows and dies back, as well as with grazing angle, radar frequency and radar resolution.

The spikes enable the landform to be displayed at long range with low probability of detection, but mean echo must be very large to ensure high P_D. If the operator

Table 10.6 Landform parameters

Sekine and Mao ref.	Terrain	Conditions	Radar	Reflectivity (dB m^2)	Shape parameter c	Depression
p. 7	Mountainous		3 GHz, 1.5°, 2 μs		0.512	
p. 7	Rural[a]		1 GHz, 1.7°, 3 μs		0.625	<0.5°
Figure 2.2	Cultivated[b]	March	9 GHz? 1.4°, 0.17 μs	−54	0.262	1.25°
Figure 2.2	Cultivated	August	9 GHz? 1.4°, 0.17 μs	−39	0.353	1.25°
Figure 2.3	Cultivated	March	9 GHz? 1.4°, 0.17 μs	−51	0.315	2.5°
Figure 2.3	Cultivated	August	9 GHz? 1.4°, 0.17 μs	−30	0.552	2.5°
Table 2.1	Wooded hills	Sweden	1 GHz, 1.17°, 3 μs		0.314	0.5°
Table 2.1	Rocky mountains	Sweden	3 GHz, 1.5°, 2 μs		0.256	?
Tables 2.1 and 2.2	Forest	Summer, Sweden	9 GHz, 1.4°, 0.7 μs		0.253	0.7°
Tables 2.1 and 2.2	Forest	November	9 GHz, 1.4°, 0.7 μs		0.266	0.7°
Table 2.2	Cultivated land	April	9 GHz, 1.4°, 0.7 μs		0.303	1.25°
Table 2.2	Cultivated land	May	9 GHz, 1.4°, 0.7 μs		0.352	1.25°
Table 2.2	Cultivated land	April	9 GHz, 1.4°, 0.7 μs		0.573	2.5°
Table 2.2	Cultivated land	May	9 GHz, 1.4°, 0.7 μs		0.429	2.5°
Table 2.2	Cultivated land	April	9 GHz, 1.4°, 0.7 μs		0.909	5°
Table 2.2	Cultivated land	May	9 GHz, 1.4°, 0.7 μs		1.0	5°

[a]: Low rolling wooded hills, grassland, wooded mountains and some man-made structures.
[b]: Northern hemisphere. Maximum and minimum of monthly data. Reflectivity taken as 0.5 P_D.

wishes to detect some other target within the landform area, such as a reflecting beacon or a racon, the landform return becomes a form of clutter. The beacon RCS will need to be much higher than the mean landform echo to ensure high probability of detection with low probability of false alarming on landform spikes. These problems will be discussed further in Chapter 12, Section 12.4.3.

Table 10.6, although not as representative of coastal conditions as one might wish, presents some typical landform parameters. It is extracted from data in Sekine and Mao [12], who give further particulars of test conditions and full source references. The seasonal changes of shape parameters should be noted. They also report (p. 13) that the standard deviation of the sample to sample spread when scanning clutter in the Cotswold Hills, England was 8.4 dB spatial and 6.3 dB temporal, while for the nearby city of Worcester, population 85 000, the spreads were 13.1 and 3.5 dB, respectively.

(a)

(b)

Figure 10.18 *Ice. Oslofjord, January. Author. (a) Young ice. Flat sheets cause little or no clutter and strong multipath interference. Leads may not be detected. (b) Broken ice. More clutter, leads shown by absence of clutter. Lower multipath interference. Difficult to distinguish the several islets, visually or by radar*

10.13 Ice

10.13.1 Introduction

The description of ice and its backscatter draws in part on Sekine and Mao [12, Section 2.3]. Lewis *et al.* [14] have produced a major textbook on the subject. Ice formed in the sea contains brine and therefore conducts. Icebergs, on the other hand, are basically fresh water, and pure frozen distilled water does not conduct. Unlike the insulating materials considered in Chapter 7, water has an extremely high dielectric constant, $\varepsilon = 80$. As expected from the discussion of reflection at the water surface in Chapter 5, Section 5.8, radar rays striking a plane of pure ice undergo nearly complete refraction at the air–ice boundary. This is a mismatch effect akin to but more severe than the light we see reflected back from a shop window. The mechanism differs but the effect is nearly as complete as replacing the ice by a metal sheet. The plane would be easy to detect at normal incidence, but hard at other angles. The backscatter from ice mainly depends on its surface wetness and physical contours, the latter formed by its type, age, history and thickness.

The cold air at the surface of an ice-field causes temperature inversion and curves radar rays upwards. Refraction k-factor falls locally to a low level, so horizon and maximum detectable ranges of the ice itself and nearby targets fall sharply, especially when the ice-field is to windward.

As with land, we may wish to see ice, perhaps to avoid collision with it, and then regard it as a radar target. Alternatively, ice may surround and mask the echo of some other feature of interest, such as a pillar lighthouse; ice is then really a form of clutter, but it is conveniently dealt with here. We shall now try to differentiate the radar characteristics of the several types of ice.

10.13.2 Ice formed in the water

Ice crystals form when the open sea cools below 0°C. On further cooling, the crystals congeal to a continuous sheet of surface ice, called young ice, which in its first year can reach 1 m thickness, Figure 10.18(a). Thin ice is less than 300 mm, other ice being classed thick. Ice which has undergone several melt-and-refreeze cycles and exceeds 2 m thick is called multi-year ice.This surface melting followed by re-freezing reduces the salinity and hence the conductance, reducing the backscatter of old ice at steep grazing angles. Freshwater lake or river ice has even lower backscatter. Onstott *et al.* [15] indicate that at low grazing angles, vertical polarisation and 1.5 GHz, lake, thin first-year and thick first-year ice all have $\sigma_0 \sim -32\,\text{dB}\,\text{m}^2$, roughly equal to the backscatter of sea state 4 or 5. At 30° grazing angle, the values are about −32, −28.5 and $-28\,\text{dB}\,\text{m}^2$, respectively. Sekine and Mao note there is a lack of data at low grazing angles, but include indications that sea ice clutter amplitudes may have Weibull distribution (discussed in the context of land clutter in Section 10.12.7) with shape parameter $c \sim 1.195$ (for Gaussian (normal) distribution $c = 1$), meaning there are fewer extreme RCS events than for normal statistical distribution. They also report measurements made by Ogawa *et al.* [16] off Hokkaido, Japan: with grazing angles less than 0.6°, Weibull shape factor was found to vary between 0.501 and 1.215. A single outlier at 1.649 indicated an area where the ice had particularly uniform irregularities or, more likely, was smooth.

In open water, small pieces of brash ice may generate clutter analogous to sea clutter, Figure 10.18(b). Clutter rises and multipath reduces. Brash ice may extend over the surface to form floes, which also wash down estuaries from rivers. The surface of floes is flat and sea-waves are damped out. Like a calm water surface without capillary waves, virtually all incident energy is reflected forward, so multipath is severe. Conversely, almost nothing is retro-reflected and the ice cannot be detected either as a target to be avoided or as clutter masking other targets. This is particularly so when the scanner is low and the angle of incidence is extremely acute.

10.13.3 Bergs and growlers

Most Northern-hemisphere icebergs, and some Southern ones, are formed as glacier snouts break up. As formed, the bergs are up to 100 m or so in each dimension, weighing around a million tons, and of random shape. They are detectable at around 10 km range.

Bergs get smaller as they age and fracture into growlers. Their shapes remain random but washed smooth by wave action. Unfortunately, the sides generally slope markedly from vertical, so monostatic RCS is much less than for a ship of the same size. Even a small growler of side 12 m and only projecting a metre or so above the water weighs a thousand tons, quite enough for severe collision damage. The small projected area, low height and stealthed shape combine to reduce RCS to the order of a square metre or even less. Even in favourable conditions it is difficult to detect growlers on ships' radar beyond a couple of miles. Lisa McDonald, skipper of the yacht Amer Sports Too participating in the Volvo Ocean Race, graphically reported[2] the ice peril off Cape Horn: 'The sheer amount of ice has made sailing through it more like driving fast through a busy car park. You expected to hit something at any moment. Crew member Miranda Merron and I watched the radar 24 h a day. . . . Counted the proper icebergs we'd sighted . . . 18, but it's the small ones that are really scary. They are about the size of a wave and in the rough seas you can't really see them in the clutter. . . . For three days I thought of nothing except the radar and getting holed. . ..'

Sea- or precipitation-clutter may mask growlers completely at all ranges. Ice echoes remains constant for much longer than the few seconds typical of waves, so scan to scan correlation at the radar receiver assists their detection; conversely scan–scan correlation no longer helps pick out small targets in ice clutter.

10.13.4 Pack and fast ice

Freely floating ice is called pack ice, that attached to the shore or sea-bed being fast ice. Unless rafted or squeezed into pressure ridges by wind or waves, pack ice formed by freezing of the water surface is low and rather featureless, with low RCS per unit area. In some cases ice patches may be detected at short range as changes in the appearance of sea clutter.

Fast ice is of altered appearance and it is difficult to relate the radar shore-plus-ice echo to the charted shoreline. RCS per square metre of the ice approximates snow-covered ground, making differentiation between them difficult. A pillar lighthouse or islet may be detectable as the head of a 'lead' or 'wake' of open water as the ice sheet drifts past. The edge of the pack is usually detectable from several kilometres to seaward.

10.13.5 Icebergs calved from glaciers

Breaking from the ice shelves which are particularly common in the Antarctic and typically rise 30 m above sea level (draught 200 m), tabular bergs may be over 100 km long, weighing many billion tonnes. The sides are sheer and RCS per unit width is so high that the bergs are easily detected at long range. But by the time they have drifted into waters frequented by shipping, they have generally partially melted, broken up,

[2] Daily Telegraph, London, 13 February 2002.

listed or capsized, and are little different from glacier-formed bergs and growlers, with the same detectability problems.

10.13.6 Optimum radar bands

It is usually easier to see other ships or an icebreaker in pack ice when using 3 GHz radar, whereas the track cut by the icebreaker may be clearer at 9 GHz. In a proposed new Polar Code, IMO would require all ships in high latitudes to carry both 3 and 9 GHz radar.

10.14 Echo strength from extended targets; sketches

The strength of the echo at the radar from a ship target follows the ordinary radar range equation:

$$S_e = P + 2G + 20\log\lambda - 30\log(4\pi) + \sigma - 40\log R - L_t - L_r - 2L_A + 2M \text{ dBW}. \qquad (4.6a)$$

L_A is given by Chapter 5, Section 5.9.7, Eq. (5.50), M by Chapter 9, Section 9.6.1, Eq. (9.16), and the RCS, σ, by Eq. (10.16). All these terms are range-dependent, so the simple-looking range equation hides considerable complexity, and accurate computation of the variation of echo strength versus range requires help of a computer

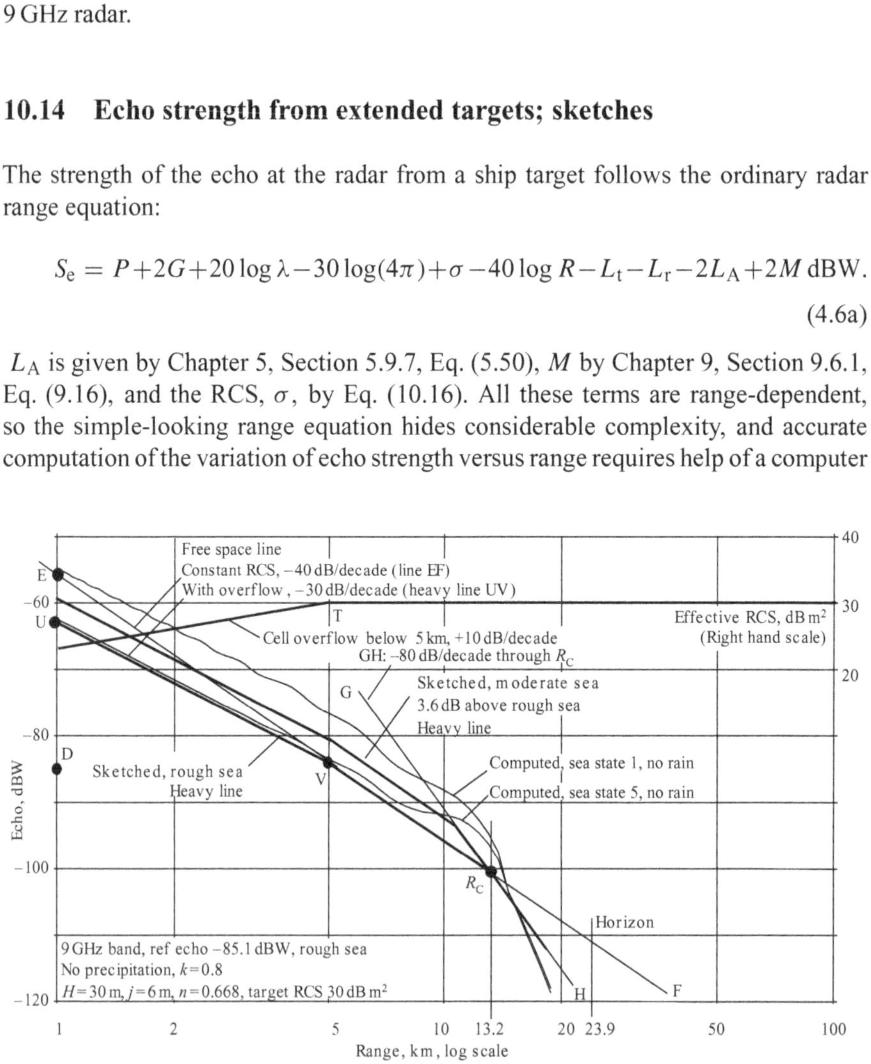

Figure 10.19 Sketching extended target echo. Echo follows R^{-3} range law in overflow region, then R^{-4} law to critical range R_C, then approximates an R^{-8} law until near the horizon

spreadsheet, which will follow in Chapter 14. However, it is fairly easy to sketch graphs of echo strength with range, the procedure closely following that for point targets, Chapter 6, Section 6.9, Figures 6.13 and 6.14. Figure 10.19 uses the notation of the earlier figures to illustrate the procedure, which is as follows.

1. The axes are set up as Chapter 6, Section 6.9. It is also convenient to add a scale of target effective RCS, shown at the right of Figure 10.19.
2. Unless target RCS is known, calculate from Eq. (10.6). For example, a crossing coaster of 1100 gt, even keel, viewed on a 9410 MHz radar has $X = 1.15$, $Y = 15$, $\gamma = 0$, giving likely RCS $= 10 \log 1100^{1.15} + 15 - 5 \log 9410 - 0 = 30.11$ dB m^2, say 30 dB m^2.
3. Visually assess RCS height factor n (Chapter 9, Section 9.3.5). The example target was thought to be uniformly distributed, so $n = 0.668$. Overall height of the highest significant reflector was judged to be the funnel top at 6 m; so $h = nj = 0.668 \times 6 \sim 4.0$ m.
4. Calculate horizon range from Eq. (5.23) and mark on graph. For $k = 0.8$ and scanner height 30 m, say:

$$R_{\mathrm{H}} = 3.569\sqrt{k}(\sqrt{H} + \sqrt{h}) = 3.569 \times \sqrt{0.8}(\sqrt{30} + \sqrt{4}) = 23.9\,\text{km}.$$

5. Next calculate overflow range from Eq. (10.14): $R = 0.0573w/\theta$ km. If scanner has beamwidth $\theta = 0.9°$ and target projected length $w = 78$ m, $R = 4.97$ km, say 5.0 km.
6. Sketch effective RCS as a straight horizontal line at 30 dB m^2 above 5.0 km (point T), sloping at 10 dB/decade below 5 km (i.e. RCS falls at short range). Effective RCS at any range can be read off; for example, 23 dB m^2 at 1 km.
7. Calculate free space reference echo for target RCS 0 dB m^2, F_{12}, at range $R = 1.0$ km by the method of Chapter 4, Eq. (4.6a). Enter point D (here −85.1 dBW).
8. Step up by target long-range RCS, here 30 dB, to −55.1 dBW, point E.
9. Draw straight line EF, slope −40 dB/decade to represent the echo in free space with no atmospheric attenuation or overflow reduction.
10. Draw line UV at −30 dB/decade to meet EF at 5.0 km to account for overflow. (If the target is a coastline, overflow might extend to longer range.)
11. Calculate free space R_{C} from Eqs. (9.13, 9.14b) using $h = 4$ m. Correct for Earth curvature by Figure 6.5, using the appropriate k factor and radar band. If $H = 30$ m, $h = 4$ m, 9 GHz band $R_{\mathrm{C}} = 13.2$ km.
12. Beyond R_{C} signal strength approximates an R^{-8} law. Draw line GH at −80 dB/decade through R_{C}.
13. Emphasise line UVR_{C}H, representing the echo under rough sea conditions with no reflection from the sea surface. Stop well short of the horizon.
14. Draw a line parallel to UVR_{C} and 3.6 dB above it, (1.8 dB each way, Eq. (9.16)) out to its intersection with GH, thereafter follow GH, to represent the echo in calm water.
15. Allow for any precipitation as in Figure 6.13 (no precipitation is shown in Figure 10.19).

Comparison with computed values in the figure shows the sketched echo strengths reasonably depict actuals within a few decibels.

10.15 References

1 DODMAN, F.: 'Observers ships' (Bloomsbury Books, 1973) and later revisions

2 SKOLNIK, M. I. (Ed.): 'Radar handbook' (McGraw-Hill, New York, 1970)

3 WILLIAMS, P. D. L., CRAMP, H. D., and CURTIS, K.: 'Experimental study of the radar cross section of maritime targets', *IEE Electronic Circuits and Systems*, 1978, **2** (4)

4 WILLIAMS, P. D. L.: 'Civil marine radar' (Privately published, 2002)

5 ROHAN, P.: 'Surveillance radar performance prediction' (Peter Perigrinus for The IEE, 1983), Section 7.10.3

6 SKOLNIK, M. I. (Ed.): 'Introduction to radar systems. International student edition' (McGraw-Hill, New York, 1983), Table 2-2

7 BARTON, D. K. and LEONOV, S. A. (Eds): 'Radar technology encyclopedia' (Artech House, London, 1997)

8 McCLINTOCK, W. J.: 'Radar signal measurement of complex targets', *Military Microwaves* **810**

9 LYNCH, D.: 'Introduction to RF Stealth' (The IEE, Stevenage, 2004)

10 WILLIAMS, J. M.: 'RCS of non-orthogonal reflectors symmetrically illuminated', *Military Microwaves*, 1988, pp. 405–10; and personal communication

11 JONES, M. R., STAFFORD, P., and ANDERSON, W. C.: 'Tenix panels: a solution to the problem of RCS reduction for ships in service' (IEEE, 2002), 0-7803–7/03

12 BRIGGS, J. N.: 'Specifications for reflectors and radar target enhancers to aid detection of small marine radar targets', *Journal of Navigation*, 2002, **55**, pp. 23–38

13 SEKINE, M. and MAO, Y.: 'Weibull radar clutter' (Peter Peregrinus for The IEE, 1990)

14 LEWIS, E. O., CURRIE, B. W., and HAYKIN, S. S.: 'Detection and classification of ice' (Research Studies Press/Wiley, New York, 1987)

15 ONSTOTT, R. G., MOORE, R. K., GOGINENI, S., and DELKER, C.: 'Four years of low-altitude sea ice broad band backscatter measurements', *IEEE Journal*, 1982, **OE-7**, pp. 44–50 (Some data is reproduced as Sekine and Mao Figure 2.28)

16 OGAWA, H., SEKINE, M., MUSHA, T., AOTA, M., OHI, M., and FUKUSHI, H.: 'Weibull-distributed radar clutter reflected from sea ice', *Trans. IEICE*, 1987, **E70**, pp. 116–120 (The measurements were at 44° 21′ N, 143° 22′ E)

Chapter 11

Noise, clutter and interference

'Confusion now hath made his masterpiece'

Shakespeare, 'Macbeth'

11.1 The importance of noise and clutter to detection of targets

Preceding chapters enable the echo strength reaching the radar from passive or active targets to be calculated in presence of precipitation and in the various sea conditions. In this chapter we examine the noise and clutter which reach the detection system, competing with and perhaps masking the wanted target echoes. The competitors comprise thermal noise generated in the receiver and elsewhere, back-scatter from precipitation, sea-waves, ice or land, and possibly short-range internal feeder mismatch reflections. Their quantification prepares us to examine the signal to noise and clutter ratios necessary for calculation of target detectability in Chapter 12. Noise and clutter, by competing with wanted echoes at the radar, set a limit to the smallest detectable echo. Some aspects of precipitation and the sea were described in Chapter 5. The special characteristics of land and ice clutter were discussed in Chapter 10, Sections 10.12 and 10.13, respectively.

As far as possible radars are designed to minimise noise and reject clutters, but their natures are too signal-like for total elimination. Modern radars go far to removal of distracting clutter paints from the display. Nevertheless, if precipitation is falling around the target, or it is in a rough sea, the resulting clutter always impairs performance, even when hidden from display by a clever data extraction system. Enquiring the prevailing sea state in the approaches to a major Mediterranean port, the author was shaken to be assured by the naval officer conducting a VTS procurement: 'Quite immaterial, modern radars are completely immune to clutter'. Beware of salesmen bearing glossy literature!

In the detection process, incoming signals are assigned to the detection cell corresponding to current scanner bearing and the range matching the elapsed time from the instant of pulse transmission. Each cell is examined to see whether it contains sufficient energy to warrant declaration of a target at its range and bearing. If the cell energy far exceeds each of its neighbours', plainly a target is present. But if severe clutter or thermal noise is providing substantial but fluctuating energy to the cells, it is much harder to be sure that a somewhat higher level within a particular cell signifies a target. We are therefore primarily interested in the clutter and noise immediately surrounding the target.

Many military radars defeat clutter by the moving target indication (MTI) technique. MTI relies on target movement being much faster than that of the surrounding waves or hydrometeors. It is not currently available to marine radars, where target and clutter speeds may be similar. Wave crests travel at about one-third the wind speed; so do some ships. And of course the target may be hove-to in the teeth of the storm. But MTI may come, see Chapter 16.

Noise and clutter each have a mean value and fluctuations about the mean. We will examine first the means, then the fluctuations. It is convenient to start with noise, including a section on the basic properties of random fluctuation. We go on to precipitation clutter, whose fluctuations are noise-like, and then sea clutter, whose fluctuations are rather different, ending with the problem of feeder mismatch reflection. 'Mean' clutter and noise powers discussed in this chapter are averaged through a time long enough to average out the inevitable instant-to-instant fluctuations. We reference power levels to the input of the radar receiver, as we did for echoes in previous chapters. This approach enables signal to clutter and signal to noise ratios to be calculated on a uniform basis. Of course, input power levels are unmeasurably small and practical measurements would be made further down the receiver chain, necessitating allowance for any bandwidth reduction or signal compression en route.

11.2 Mean noise

11.2.1 Noise power

We discussed thermal noise and its effects in general terms in Chapter 3, Section 3.3. The thermal noise generated within the radar sets such an important and fundamental performance limit that it demands further examination. Noise cannot readily be avoided – the laws of physics dictate that temperature always stirs the electrons in any resistive medium. Although tiny, thermal noise is the factor setting the limit to detection of weak targets in absence of clutter. As echo strength sinks toward the 'floor' formed by noise, it becomes harder to decide what is a valid echo and what is a noise spike – the probability of detection falls and the probability of false alarm rises until it becomes impossible to make a meaningful detection decision. But except in calm weather clutter is often dominant, making it rarely worthwhile to take extreme steps to minimise noise.

Noise is completely random, sometimes called Gaussian after the celebrated German mathematician – previous history completely fails to predict whether the next time increment will see an increase or decrease of instantaneous noise power, so the noise is decorrelated pulse to pulse as well as scan to scan. All that can be said is that, averaged over a long time, the noise will have a certain mean power level. The equivalent mean noise voltage is its root-mean-square (rms) value, found by taking the square root of the sum of the squares of many instantaneous voltage values, or the square root of the mean power. Statisticians use the standard deviation (s.d.), σ, which is equivalent to the concept of rms voltage. The mean power is the heating power, and is the square of the rms voltage into the system input resistance, or notional unit resistance. Statisticians also use variance, equivalent to the mean power, so variance $= (\text{s.d.})^2$.

The particles constituting the atmosphere are thermally agitated and introduce further noise. Passive components such as the scanner, and more particularly a long feeder if there is one, have resistive loss, so generate some noise. Importantly, these lossy components also attenuate the competing echoes. Within the transceiver, described in Chapter 3, there is some form of duplexer and a TR cell or other protection device, whose loss may decrease with time from instant of pulse transmission. On older radars, the first major receiver component is the mixer, which introduces noise of its own. Modern radars precede the mixer with a microwave low noise amplifier (LNA), designed to contribute the minimum possible additional noise. The LNA raises the signal (and the noise collected thus far) to such an extent that mixer noise, and noise contributions from the subsequent IF amplifier, are negligible in comparison and do not significantly further degrade the signal to noise ratio (SNR) which may be expressed as a numerical power ratio or in decibels.

For example, if the echo input power is 20×10^{-12} W and the noise at the LNA input is 10^{-12} W, the input SNR is $20 = 13$ dB. If the LNA has power gain 10, and the mixer following it introduces 3×10^{-12} W, after the mixer the echo power is 200×10^{-12} W and the noise is $(10 \times 10^{-12} + 3 \times 10^{-12}) = 13 \times 10^{-12}$ W, with SNR degraded to $\frac{200}{13} = 15.38 = 11.9$ dB, a drop of 1.1 dB. Without the LNA, SNR would be $\frac{20}{3} = 6.67 = 8.2$ dB, a drop of 4.8 dB. It is vital to introduce gain as early as possible; the more initial loss has to be compensated by additional gain, the more will noise be amplified. Ideally, the first amplifier stage immediately follows the scanner. Pre-amplification feeder loss cannot be compensated merely by boosting the gain of an amplifier further down the chain, since this would amplify the noise as well as the echo.

Although every active stage in the receiver introduces noise, only the input stages are significant, because their noise is amplified more than later-stage noise. The noise floor remains at the same power level relative to the receiver input however much the IF gain is increased, so increasing gain does not improve detection in noise. At short and medium ranges, noise is often swamped by clutter, which then becomes the factor limiting detection. But in absence of clutter, detection of weak targets depends almost entirely on achievement of the best available SNR.

11.2.2 Receiver input stage noise contribution

We first consider the receiver itself, without any noise contribution from the atmosphere, scanner or feeder. Recalling Chapter 3, Section 3.3, mean noise power referenced to the radar input is the product of the following.

- n, a noise factor describing the shortfall of the practical receiver from theoretical perfection. This term should be reserved for n expressed numerically. Noise figure, often written N, denotes the dB equivalent $N = 10 \log n$. Like many engineers, we confess to sometimes using the terms interchangeably.

- K, Boltzmann's constant, 13.8×10^{-24} W/degree/Hz, one of the fundamental physical constants like the speed of light.
- T_0, the absolute temperature of the conductor (conventionally taken as 290 K ~ 17°C. If no other noise-related temperatures are in play we shall often write T for T_0). The universal use of K for Boltzmann's constant and K for Kelvin is unfortunate in our context.
- B, the bandwidth, Hz, within which the noise is observed.

Noise power level fluctuates from microsecond to microsecond. Averaged over a time much longer than the reciprocal of the bandwidth, the mean available noise power, P_n, is given by Chapter 3, Section 3.3.2, Eq. (3.2a). ('Available' means usable by an external load, impedance-matched for maximum power transfer. Open-circuit, the noise emf is doubled but no power is transferred. Some authors denote mean values by a superscript bar.)

$$P_n = nKT_0B \text{ W}. \tag{3.2a}$$

Noise may alternatively be expressed as the noise power density, $P_n/B = nKT_0$ W/Hz, which is independent of bandwidth as well as the chosen IF centre frequency.

Eq. (3.2a) is often expressed in decibel form:

$$P_n = N + 10 \log(KT_0) + 10 \log B \text{ dBW}.$$

When $T_0 = 290$ K and B is expressed in MHz:

$$P_n = N - 144.0 + 10 \log B_{MHz} \text{ dBW}. \qquad \text{(3.2b modified)}$$

For example, if $N = 3.5$ dB and $B = 10$ MHz, $P_n = -130.5$ dBW, an eleventh of a micro-microwatt. We now examine these terms to suggest ways of reducing noise power.

11.2.3 Noise factor

Until the 1970s the first receiver stages were generally the duplexer, protection and mixer, with $N \sim 9$ dB at best, more or less equally divided between mixer loss and component noise; the first amplification stage followed at IF. Modern radars precede the mixer with a low-noise microwave transistor pre-amplifier operating at ordinary temperature, the LNA. Its noise figure is typically $N \sim 2.5$ dB ($n \sim 1.78$), so advent of a theoretically perfect device would give less than 3 dB overall improvement (sky noise, Section 11.2.6, would further limit the advantage gained). The noise contributions of the duplexer, protection, mixer and succeeding IF amplifier adds a little, so the overall receiver noise figure $N_r \sim 3.5$ dB ($n_r \sim 2.24$). If radar data sheets quote LNA noise figure, it is reasonable to assume receiver noise figure to be about 1 dB higher.

Logarithmic amplifiers generate constant noise of $a\pi/\sqrt{24} = 0.641a$, where a is the constant for the amplifier input/output function (Chapter 3, Section 3.5.7). They are invariably preceded by linear amplification stages, reducing their noise contribution.

11.2.4 Noise temperature

Noise is sometimes expressed as receiver noise temperature, T_r, the temperature at which a perfect ($n_r = 1$) receiver would have to operate to produce the same noise power output.

$$T_r = T_0(n_r - 1)\,\text{K}. \tag{11.1a}$$

This approach is particularly useful when introducing relatively small noise contributions, which are considered as adding to the overall or system noise temperature. For example when $N_r = 3.5\,\text{dB}$, $n_r = 2.24$ and $T_r = 360\,\text{K}$.

Eq. (11.2a) shows P_n is proportional to T_0. In principle, noise can be avoided by using reactive rather than resistive components in a parametric amplifier (paramp) or by cooling the amplifier to absolute zero temperature. Parametric amplifiers introduce other problems, including poor dynamic range. Cryogenic amplifiers need extreme refrigeration – even a modest improvement of say 2 dB would mean running the LNA at ~202 K (−71 °C), around the temperature of 'dry ice' (solid CO_2). Neither technique has been found worth the hassle in marine or VTS practice.

11.2.5 Bandwidth

The only control over noise available to the operator is by bandwidth selection, achieved indirectly through the range scale selector and the long/short pulse switch, if provided. The majority of radars automatically lengthen the pulse and reduce bandwidth in two or three steps as longer range scales are selected. Reducing B clearly reduces P_n. Unfortunately if taken too far, the high frequency components in echo pulses, then the fundamental itself, are stripped away, reducing echo size and preventing any further improvement of the all-important SNR; the baby has gone out with the bathwater. Bandwidth reduction also introduces lag or latency, leading to ranging error, discussed in Chapter 13. We noted in Chapter 3, Section 3.5.2, that the bandwidth necessary to handle the echo is inversely related to transmitter pulselength, τ. The SNR-optimum or 'necessary' bandwidth is given almost exactly by:

$$B_n = \frac{1}{\tau}. \tag{3.4}$$

That is, a pulselength of 1 μs requires 1 MHz IF bandwidth, 0.1 μs needs 10 MHz and so on. Video bandwidths (after demodulation) have to be at least half the IF value. To minimise noise, the pulselength would be switched to the longest available, at the same time switching IF bandwidth to conform with Eq. (3.4). This can be helpful but may introduce several problems.

1. The receiver response becomes sluggish, delaying the pulse output and introducing tens of metres of range error or latency, particularly unacceptable when track-forming.
2. The long pulse necessitates additional detection cell size in the range dimension, admitting more precipitation- and sea-clutter.
3. Any slight inaccuracy in local oscillator tuning throws the IF frequency out of the passband, reducing the echo but not the noise, spoiling SNR.

4. The receiver may serve multiple displays, whose operators need differing range scales. For example, while making rendezvous with a pilot cutter, the officer of the watch may be monitoring the cutter on a short scale demanding the good range resolution of short pulses, a colleague meanwhile keeping more distant potential collision threats under surveillance, making do with inferior clutter performance.

Therefore, bandwidth sometimes has to exceed the necessary value, reducing SNR. For example, a modern radar having $N_r = 3.5\,\text{dB}$, operated on long pulse, $\tau = 1.0\,\mu\text{s}$ might have bandwidth between say 2 and 10 MHz, giving P_n between -137.5 and -130.5 dBW, several decibels worse than the necessary value of -140.5 dBW predicted by Eq. (3.4).

11.2.6 Environmental noise sources

1. The ohmic losses in the atmosphere, scanner, feeder (ohmic loss component; feeder mismatch loss diverts rather than absorbs power so introduces no noise) and any other component preceding the receiver generate noise, which must be added to that of the receiver. It is convenient to do this using noise temperature.
2. 'Atmospherics' from thunderstorms are negligible at radar frequency (above 50 MHz).
3. Man-made interference is also negligible (above the UHF band, ~1 GHz) with the exception of other radar transmissions, which may display 'running rabbits' which are countered by prf jitter and pulse–pulse correlation but defy generalised numerical analysis.
4. Sky noise collected by the scanner has noise temperature up to 150 K at 3 and 9 GHz when boresight is horizontal, falling as the boresight is raised, for example, if the ship rolls.
5. Solar noise is received when the Sun lies in the scanner beam near sunrise or sunset and is then about $-140\,\text{dBW/m}^2\text{/MHz}$ at 3 or 9 GHz, rising by 20 dB at sunspot maximum. Marine radar scanner apertures seldom exceed $1\,\text{m}^2$ (ships) or $10\,\text{m}^2$ (VTS), so solar noise rarely exceeds ordinary thermal noise, and then only through a small azimuth arc.
6. Telecommunications services sharing radar bands in future may introduce quasi-random noise sufficient to degrade the noise figure by about 1 dB.

11.2.7 Atmosphere and line attenuation noise

Speaking generally, when a noise source at temperature T_{source} passes via a transmission line (scanner and feeder) having ohmic loss to a receiver, the source noise is attenuated by the line, but a fresh noise component is introduced by the line lossiness. If the power attenuation ratio of the line is L, where $L = P_{\text{in}}/P_{\text{out}}$, $(L > 1)$:

$$T_{\text{out}} = \frac{T_{\text{source}}}{L} + T_{\text{line}}\left(1 - \frac{1}{L}\right). \tag{11.1b}$$

The effective input noise temperature, T_i, is defined as the source temperature at the input which would give the same noise at the output if the line were noiseless with

$T_{\text{line}} = 0$: $T_{\text{i}} = T_{\text{out}}L$. Substituting,

$$T_{\text{i}} = T_{\text{source}} + T_{\text{line}}(L - 1). \tag{11.1c}$$

When the source is the atmosphere, if the atmospheric and line temperatures are each T_{a} and, as is usually sufficiently accurate, we take $T_{\text{a}} \sim T_0$:

$$T_{\text{i}} = T_0 + T_0(L - 1) = LT_0. \tag{11.1d}$$

If, say, the scanner plus feeder have 3 dB loss, $L = 2$ and $T_{\text{i}} \sim 580\,\text{K}$.

Scanner loss is not always easy to ascertain as it is buried within the published overall gain, but is typically ~1 dB, including the rotating joint. For example, signals within a conventional end-fed slotted array on average traverse rather more than a m of waveguide, having about 0.14 dB/m loss at 9 GHz (aggregating say 0.5 dB for aperture $a = 3$ m), beside other loss elements. Noise arising from line loss also modulates target echoes on both interrogate and return paths, contributing to target fluctuation, discussed in Chapter 12, Section 12.5.

11.2.8 System noise

When several noise contributors (e.g. the atmosphere-cum-feeder, the receiver low noise amplifier and succeeding amplifier stages) operate in series and the amplifier stages have numerical gains $G_1, G_2, \ldots$, noise factors $n_1, n_2, \ldots$, and noise temperatures $T_1, T_2, \ldots$, they combine to give a system noise factor, n_{s}, and system noise temperature, T_{s}. These quantities characterise the radar system as a whole, rather than the transceiver alone, giving the same noise output as if the first stage contributed all the noise.

$$T_{\text{s}} = T_0L + T_1 + \frac{T_2}{G_1} + \frac{T_3}{G_1G_2} + \cdots = LT_0 + T_{\text{r}} \tag{11.2a}$$

Substituting in Eq. (11.1a),

$$n_{\text{s}} = \frac{T_{\text{s}}}{T_0} + 1 = \frac{LT_0 + (n_{\text{r}} - 1)T_0}{T_0} + 1 = L + n_{\text{r}} \tag{11.2b}$$

Working in decibels,

$$N_{\text{s}} = 10\log[10^{(L_{\text{dB}}/10)} + 10^{(N_{\text{r}}/10)}]\,\text{dB}. \tag{11.2c}$$

Thus if, say, $L = 2$ and $n_{\text{r}} = 2.24$ ($N_{\text{r}} = 3.5$ dB), $n_{\text{s}} = 4.24$ and $N_{\text{s}} = 6.3$ dB. As here, LNA noise is often the most significant contributor. System noise factor is not always quoted in data sheets, for it includes the installation-dependent feeder as well as the receiver. System noise is continually present, irrespective of the transmitter operating or any targets being within range, although the receiver swept gain or the data extraction system may prevent its noise speckles from being displayed.

11.3 Noise fluctuation

Before tackling random noise fluctuations, we note some general statistical properties of random events, a few of which have already been tacitly assumed.

11.3.1 Prediction of random events

The likelihood that an unbiased tossed coin falls heads is 1 in 2, or 0.5. The coin cannot remember whether the previous half dozen tosses were all heads, all tails or a mixture and the probability of the next toss being heads always remains 0.5. In such a series, each random event is independent of its predecessors; they are uncorrelated. The next event is unknowable; even after a run of 20 wrong calls, the cricket captain is cheered by remembering there remains 50 per cent prospect of getting it right next time. A powerful computer can do no better. However, some things about a series of random events are knowable in an imprecise sense. For example, about half of a long series will fall heads, and the probability of a run of 20 heads is very low but not zero. It is 0.5^{20}, about one in a million or 10^{-6}. Statements of this sort rely on probability theory, some of whose tenets are as follows.

11.3.2 Individual noise contributors do not cancel

Each electron in the radar receiver input circuit is in random thermal motion and develops an immeasurably small random micro-voltage in the circuit resistance. Just as the random whispers from individual leaves of trees in a gale combine to produce not silence but a roar, the individual micro-voltages do not cancel to zero, but produce a significant random resultant noise voltage at the receiver input. This is because each electron contributes positive power.

11.3.3 Amplitude distribution of noise

The way in which individual results cluster about the mean is described by their Distribution, discussed in detail in Section 11.3.8. Whenever summation of many uncorrelated events of roughly the same magnitude causes random variation (e.g. electron motions causing receiver input noise) the overall amplitude variation tends toward what is called normal or Gaussian distribution. This central limit theorem remains true even when the events themselves are not entirely random, such as the echoes from individual hydrometeors within rain blown by the wind, so precipitation clutter can be treated statistically as a form of Gaussian noise. Turning back to Chapter 3, Section 3.3.2, Figure 3.4(*a*) is a typical sample of Gaussian noise voltage observed through a relatively wide system bandwidth.

11.3.4 Noise bandwidth

Noise uniformly distributed among all frequencies is called white noise by analogy with light. It is of course impossible to depict true white noise, whose bandwidth extends from zero to infinity. The frequency distribution of slow-moving echoes is that of the transmitter pulse, so depends on transmitter pulselength and rise-time. To obtain the best available SNR, the receiver bandpass filter bandwidth and response curve should be matched to the echo pulse-shape and hence to the transmitted pulse shape. This ideal or optimum case is called a matched filter. Depending on the shape (frequency extent) of the IF filter skirts, some excess signal-free noise may be collected, causing noise bandwidth slightly to exceed signal bandwidth. The filter is then sometimes called a matching filter, noise rejection being inferior by a decibel or so, but for practical purposes noise and signal bandwidths can be considered equal. Optimum in this context means the best that can be done in the conflicting circumstances and matched has a different meaning from impedance matching for maximum power transfer. Figure 3.4(b) depicts how restricting bandwidth using a low-pass filter (giving a less 'busy' appearance) reduces the noise amplitude.

In some ways noise at the output of a narrow bandpass filter is akin to a sinusoid at the filter centre frequency whose amplitude and phase wander at random – a sine wave made of jelly.

11.3.5 Amplification

Assuming the receiver has a linear transfer characteristic and enough dynamic range to avoid clipping amplitude peaks or otherwise to distort the voltage waveform, amplification does not change the distribution. Figure 3.4(c) shows the narrow-band noise sample after it has been amplified back to the original power level. The proportion of the time that amplitude exceeds say 1 V is the same as the original wideband waveform Figure 3.4(*a*) – distribution is unaltered by linear amplification or attenuation. Non-linear amplification, or a non-linear transfer function in general, gives more or less emphasis to extreme events and necessarily changes the amplitude distribution.

11.3.6 Event rate

The basic theory of information, Nyquist's theorem, states that if a function contains no frequency components higher than f Hz, it is completely determined by stating its ordinates at intervals of $1/f$ s. That is, the number of uncorrelated class intervals or noise events per unit time is proportional to system bandwidth, B. Thus there are 10^6 events/s when $B = 1$ MHz:

$$\text{Event rate} = B \text{ events/s.} \tag{11.3}$$

Two bandpass-filtered samples of wideband random noise are decorrelated when spaced at least $1/B$ s apart.

11.3.7 Amplitude and power conventions

Following electro-statistics convention we shall often normalise expressions by assuming the noise to be $\sigma = 1$ V rms and the circuit impedance 1 ohm resistive, so noise power is 1 W. In practice, the power is generally much less and the impedance much higher, but the convention enables us to suppress resistivity and similar terms from the equations. The simplification is legitimate since we shall finish up with expressions in terms of signal to noise ratio in which voltage and impedance do not appear. We shall need to be careful about whether formulae relate to voltage, or to power and RCS which are both proportional to $(\text{voltage})^2$.

11.3.8 Distribution and probability density, unmodulated white noise

The following definitions of distribution, Gaussian distribution, probability and probability density function (PDF) have been paraphrased from Clapham [1]. The distribution of a random variable such as noise is concerned with the way in which the probability varies of its taking a certain value, or a value within a certain interval. The probability of an event such as the noise spike X of the sample of noise in Figure 3.4 is denoted $P(x)$. Here and elsewhere in this chapter, operator P signifies probability rather than power. $P(x)$ must always lie between 0 (never occurs) and 1 (always occurs). If an experiment or series of n events is repeated n times and the event x occurs m times, then in the limit when n approaches infinity, m/n tends to $P(x)$. Thus, P (a coin tossed m times falling heads n times) $\rightarrow 0.5$.

For any continuous random variable x, distribution is described by the PDF, written $p(x)$, such that probability of x lying between a and b is the area under the PDF curve between $x = a$ and $x = b$. It is the probability that some particular observed value lies within an infinitesimal width $\mathrm{d}x$ centred on x. Put mathematically, where PDF is $p(x)$,

$$P(a < x < b) = \int_a^b p(x)\,\mathrm{d}x. \tag{11.4a}$$

The long-term average noise voltage, V_n, applied to load resistance r is such that the available mean noise power calculated from Eq. (3.2a) is $V_\mathrm{n}^2/r = P_\mathrm{n} = nkTB$. V_n is the rms noise voltage – that is, the voltage which defines power, just as 230 V is the rms voltage of the UK domestic mains waveform.

When Gaussian noise having instantaneous voltage v and rms value V_n is superimposed on a noise-free d.c. component μ, the PDF is given by the following. The factor $\sqrt{\mathbf{2}}$ (shown bold in Eq. (11.4c)) arises from comparison of instantaneous voltage with rms rather than with sinusoidal peak voltage. We shall encounter several PDFs of the general form

$$P(x) = \frac{f_1(x)}{\exp[f_2(x)]}$$

where $f_1(x)$ and $f_2(x)$ are simple functions of the quantity x. This equation is usually written

$$P(x) = f_1(x)\exp[-f_2(x)]. \tag{11.4b}$$

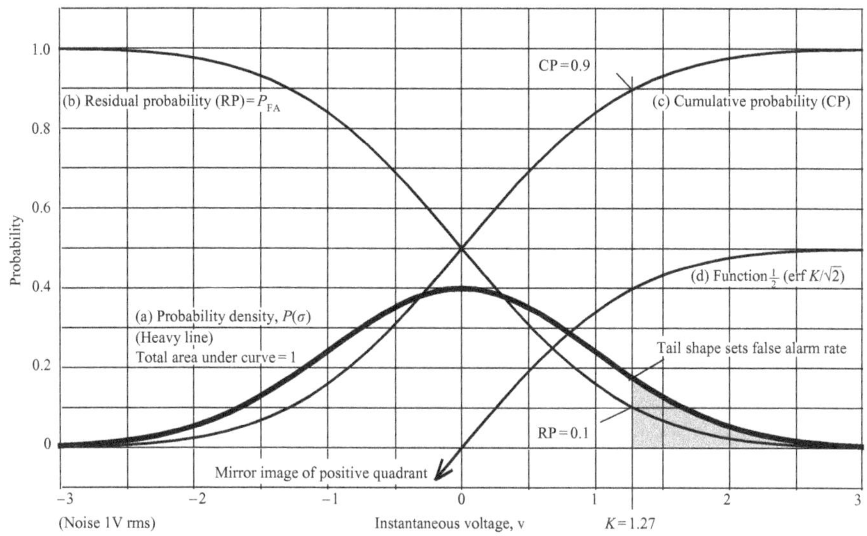

Figure 11.1 *Probability functions, Gaussian unmodulated noise. The shaded area represents the residual probability that $\sigma > K$. When $K = 1.27$, $RP = 0.10$. Curve A follows Figure 3.5. If the rms value is σ, the ordinate extends from -3σ to $+3\sigma$*

In the present case,

$$P(v) = \frac{1}{\sqrt{2\pi}\,V_n} \exp\left[-\frac{(v-\mu)^2}{(\sqrt{2}V_n)^2}\right] = \frac{1}{\sqrt{2\pi}\,V_n} \exp\left[-\frac{(v-\mu)^2}{2V_n^2}\right]. \tag{11.4c}$$

Variance, a measure of the dispersion of the events, is V_n^2. Standard deviation, the square root of variance, is V_n.

Normalising Eq. (11.4c) by putting $V_n = 1$, so variance $V_n^2 = 1$, and removing the d.c. component by putting $\mu = 0$ gives the PDF for standard Gaussian distribution:

$$p(v) = \frac{1}{\sqrt{2\pi}} \exp\left[-\frac{1}{2}v^2\right]. \tag{11.4d}$$

This bell-shaped function is shown in Figure 11.1(a) for values of σ between ± 3. Long low-amplitude tails extend beyond these limits, as shown in Figure 11.2, in theory to infinity, meaning that occasionally there will be a very high positive or negative noise-voltage spike. This is the reason why occasional noise spikes break through, even when echoes exceed noise by the substantial margin needed to ensure high probability of detection. Figure 11.1 indicates that the probability of an event exceeding what would be the peak voltage of the sinusoidal waveform of the same power, namely $P(\sqrt{2}V)$, is 0.147 or 14.7 per cent. Maximum probability density is when $\sigma = 0$, where $P(0) = 1/\sqrt{2\pi} \sim 0.40$. The curve may be thought of as the limiting

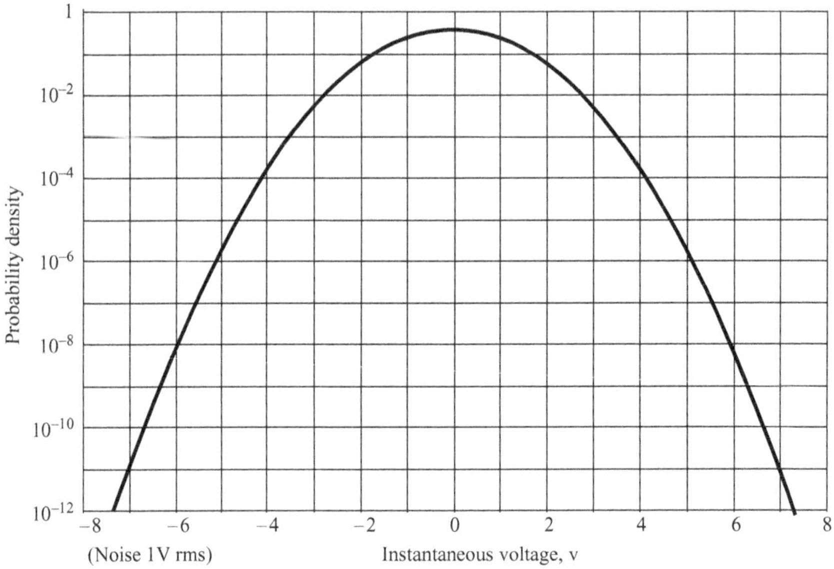

Figure 11.2 Probability density function. Gaussian distribution. Figure 11.1(a) redrawn to logarithmic scale (turning the curve to a parabola) and extended to extreme values. The tails are important to false alarm rate

case of a histogram of many measurements taken at narrowly spaced class intervals. False alarm probabilities of practical importance are so low (10^{-4} or less) that they are governed by the tail rather than the body shape of the probability density curve, so the conformance of extreme events to the nominal distribution, Gaussian or otherwise, is important. We shall return to this point when discussing sea, land and ice clutter.

The whole area under the PDF curve (Figure 11.1(a)) between $\sigma = \pm\infty$ must have unit area, since the probability of the event having a finite value must be unity. The area under the curve to the left of some threshold value K represents the cumulative probability (CP) that the random variable is less than K. The CP curve is S-shaped, Figure 11.1(c). For example when $K = 0$ the CP is 0.5, this being the probability that the variable is negative, as one would expect. CP rises with K. When $K = 1.27$, CP $= 0.90$, meaning 90 per cent of noise events are less than $1.27 \times$ (rms noise voltage).

The probability that the random variable σ exceeds K is $1 -$ CP, which we call the residual probability, RP, a reverse S curve, Figure 11.1(b). For example, when $K = 1.27$, the probability of an event exceeding 1.27 times rms is: RP $= (1-0.90) = 0.10$. This is the shaded area under the PDF to the right of K and is of interest to false-alarm calculations. The PDF is symmetrical and the probability of an event lying outside $\pm 2\sigma$ is $2 \times 0.023 \sim 0.05$ or 5 per cent. This is the two sigma error, often quoted for position-finding systems.

Expressions for CP and RP are found as follows. The probability area between events value $R = a$ and $R = b$ is the integral $\int_a^b P_n(R)\,dR$. The area under the PD curve to the right of K is RP and is the integral

$$P(R > K) = rp = \int_K^\infty \frac{1}{\sqrt{2\pi}} \exp\left[-\frac{1}{2}R^2\right] \mathrm{d}R \tag{11.5a}$$

$$= \int_0^\infty \boldsymbol{\frac{1}{\sqrt{2\pi}}\exp\left[-\frac{1}{2}R^2\right]\mathrm{d}R} - \int_0^K \frac{1}{\sqrt{2\pi}} \exp\left[-\frac{1}{2}R^2\right] \mathrm{d}R. \tag{11.5b}$$

The value of the integral shown in bold represents $\frac{1}{2}$ the unity total area under the PDF. So the residual probability that a continuously random variable exceeds R is

$$RP = \frac{1}{2} - \frac{1}{\sqrt{2\pi}} \int_0^K \exp\left[-\frac{1}{2}R^2\right] \mathrm{d}R \tag{11.5c}$$

$$= \frac{1}{2}\left[1 - \operatorname{erf}\left(\frac{R}{\sqrt{2}}\right)\right]. \tag{11.5d}$$

When

$$R = K, \qquad P = \frac{1}{2}\left[1 - \operatorname{erf}\left(\frac{K}{\sqrt{2}}\right)\right]. \tag{11.5e}$$

Erf(t) is the commonly used statisticians' error function. We use the following definition in which u is a positive parameter. Beware of alternatives.

$$\operatorname{erf}(t) = \frac{2}{\sqrt{\pi}} \int_{u=0}^{t} \exp(-u^2)\, \mathrm{d}u. \tag{11.6}$$

The function is symmetrical, $\operatorname{erf}(-t) = -\operatorname{erf}(t)$. Unfortunately error functions cannot be directly calculated, so a table has to be looked up; either as a printed sheet or a tabulation within a PC spreadsheet, compiled by summation of successive increments δu. Figure 11.1(d) graphs $\operatorname{erf}(K/\sqrt{2})$ to a base of K. When $R = 0$, $\operatorname{erf}(R/\sqrt{2}) = 0$ and $RP = \frac{1}{2}$. The 50 per cent points are at $(2\ln 2)^{0.5}\sigma = 1.1774\sigma$.

11.3.9 Effect of atmospheric and feeder noise on signals

Because the hydrometeors which cause atmospheric attenuation are constantly entering and leaving the beam in a random manner, and the electrons within a lossy feeder are likewise agitated, the atmospheric and feeder attenuations fluctuate, modulating noise onto the signal. A passive target is illuminated by a signal made noisy by passage through the feeder and atmosphere on the outward leg, and then again on its return, but a saturated RTE or a racon is only subjected to one leg's worth of this form of noise, which is best thought of as adding to signal fluctuations from other causes and is unlikely to significantly affect the target fluctuation characteristics discussed in Chapter 12, Section 12.3.4.

11.4 Mean precipitation clutter

11.4.1 Clutter mechanism

Rain, hail and snow near the target reflect some of the radar transmission back to the radar receiver as precipitation clutter, sometimes called weather clutter, illustrated in Chapter 2, Figure 2.6. The precipitation column within the detection cell forms a special sort of very high extended target, each hydrometeor (e.g. raindrop) forming a reflecting element. All are continually moved quasi-randomly by gravity and the wind, so their echoes are at random uncorrelated phase one to another and combine to appear as a return whose strength varies randomly from pulse to pulse, the mean value depending on precipitation rate and type, and radar wavelength and polarisation. Viewing successive scans, precipitation clutter always looks different in fine detail but general intensity changes only slowly – the micro-geometry continually varies while the macro-geometry generally changes slowly, over time intervals of the order of several scans. Because there are many scatterers, the central limit theorem tells us the overall distribution is Gaussian, apart from secondary factors such as wind which affect all equally so that particular component of the individual fluctuations is correlated. The RCS (or clutter power) fluctuations are therefore almost Gaussian or noise-like and there is almost perfect decorrelation from pulse to pulse and from scan to scan, see Section 11.5. Precipitation is usually assumed to extend through the whole column height within the detection cell. The noise caused by precipitation (and other atmospheric loss components) was discussed in Section 11.2.6.

Data relating precipitation rate with hours per year exceeded are available for many coastal localities; offshore rates tend to be lower. Collision or stranding risk may become high enough in severe precipitation to outweigh the small number of occurrence hours, biassing procurement towards radar with good performance in clutter – wide aperture scanner (small detection cell) in particular.

11.4.2 Mean reflectivity

We use the following notation:

a = mean raindrop radius
λ = radar wavelength
K = complex refractive index
r = precipitation rate, mm/h of water or meltwater equivalent (this practical unit is universally preferred to the SI unit, which would be m/s or kg/m^2/s)
σ_{P0} = normalised clutter reflectivity, RCS m^2 per m^3 of atmosphere within the detection cell
b, z = constants depending on precipitation type per Table 11.1.

Each raindrop is much smaller than a wavelength (it is in the Rayleigh region) where the internal scattering mechanism causes raindrop RCS to be proportional to a^6/λ^4. Radius, a, and the number of raindrops per unit volume both increase with precipitation rate, the exact manner depending on the precipitation type – stratiform or thunder rain, snow, etc. To all linear polarisations, the mean value of clutter reflectivity

Table 11.1 Precipitation constants

Precipitation type	$z \times 10^{-15}$	b	K^2
Stratiform rain	0.2	1.6	0.93
Orographic rain	0.031	1.71	0.93
Thunderstorm	0.486	1.37	0.93
Ice crystals	0.10	1.66	?
Wet snow, > 0°C	2.0	2.0	0.2
Dry snow, < 0°C	0.5	1.6	? 0.2

(RCS normalised to a per-cubic-metre basis) is, according to Morchin [2]:

$$\sigma_{P0} = \left(\frac{\pi^5}{\lambda^4}\right) K^2 z r^b \text{ m}^2/\text{m}^3. \tag{11.7a}$$

Expressed in decibels,

$$\sigma_{P0} = 50 \log \pi - 40 \log \lambda + 20 \log K + 10 \log z + 10b \log r \text{ dB m}^2/\text{m}^3. \tag{11.7b}$$

Most marine rain is stratiform. Orographic rain, resulting from moist air being forced upwards by mountains, may occur in some coastal districts. Constants K, z and b are found from Table 11.1, which is based mainly on Meikle[1]. Other sources give somewhat different values, often taking $K = 1$. Morchin gives 'snow' reflectivity as $\sigma_{P0} = (\pi^5/\lambda^4)K^2 z r^2 \text{m}^2/\text{m}^3$, where $K^2 = 0.2$ and $z = 2 \times 10^{-15}$, in agreement with the 'wet snow' in Table 11.1. Eq. (11.7b) shows that σ_{P0} rises $10b$ dB/decade of r increase.

Morchin gives no value for K^2 for ice crystals, and offers only a tentative value for dry snow. There is probably little error in either case by taking $K^2 = 0.2$. For example, from Table 11.1, stratiform rain has $z = 0.2 \times 10^{-15}$, $b = 1.6$, $K^2 = 0.93$ so $\sigma_{P0} = 56.9 \times 10^{-15} r^{1.6}/\lambda^4 \text{ m}^2/\text{m}^3$. At 3 GHz and 'excessive' rain ($r = 40$ mm/h), $\sigma_{P0} = 0.2082 \times 10^{-6} \text{ m}^2/\text{m}^3$ (-66.8 dB m^2/m^3) rising at 9.4 GHz to $20 \times 10^{-6} \text{ m}^2/\text{m}^3$, 96 times (19.8 dB) more. Wylie [3, p. 95] states that hail clutter is much less than rain clutter at the same precipitation rate unless hailstone diameter exceeds some 6 mm, and that hail rates are generally lower than rain.

Note an important practical result: because of the strong inverse wavelength dependence, rain clutter is much more important in the higher frequency 9 GHz and J bands than at 3 GHz.

Equivalent meltwater content, r, of snowstorms tends to be less than in rainstorms. Dry snow is relatively innocuous but, as shown in Figure 11.3, heavy squalls of wet snow cause 10–12 dB worse clutter than stratiform rain of the same r. It is for this reason that Scandinavian shipping has traditionally favoured the 3 GHz band. Hail

[1] H. Meikle, Modern radar systems, Artech House, 2002, Section 6.6.6, which quotes $z \times 10^{-16}$.

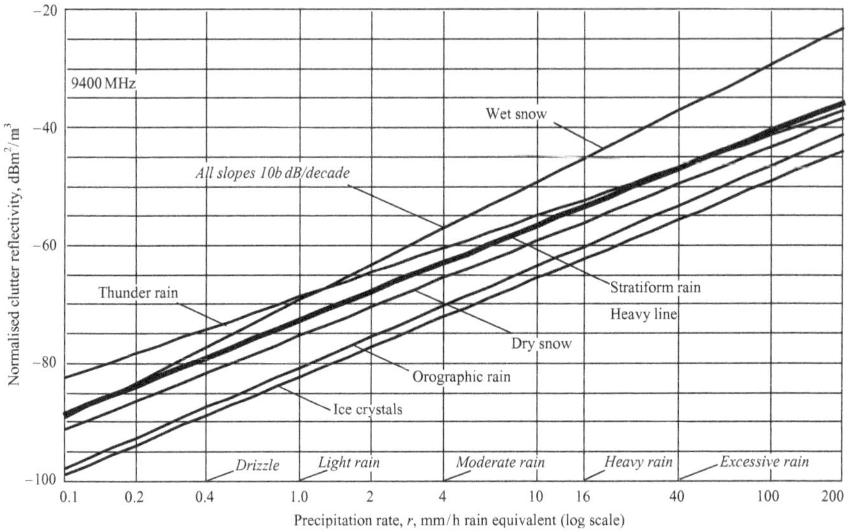

Figure 11.3 Precipitation clutter reflection coefficient, mean σ_{P0}. Precipitation rates of ice, hail and snow are expressed as mm/h of meltwater. Meteorological terms for rain rates in italics: 9 GHz band. At 3 GHz, reflectivities are 19.8 dB less, and at 15 GHz are 8 dB more

seems to be less reflective than rain, but freshly melted hail produces particularly large raindrops, which are very reflective.

It is usual to assume the ground-level type of precipitation fills the cell volume, the cloud base being above the beam. If the cloud base is illuminated (e.g. when roll tilts the scanner beam upwards), a high-RCS bright band may be seen, particularly by horizontal polarisation, centred in height on the 0°C isotherm, where predominantly horizontal melting snowflakes or hail having low nett rate of descent are concentrated. The precipitation clutter then becomes stronger than predicted from Eq. (11.7).

Figure 11.5 shows how σ_{P0} varies for each type of precipitation. It is rarely possible to measure instantaneous precipitation rate at the target and it will usually fluctuate from minute to minute. As always with natural phenomena, wind and other effects impose some variability in the constants of Table 11.1, so precipitation clutter cannot be determined with great precision from Eq. (11.1) or Figure 11.3.

11.4.3 Polarisation

As mentioned in Chapter 2, Section 2.7.11, raindrops wobble in shape but tend to reflect vertical polarisation somewhat better than horizontal. A fraction of the return is cross-polarised; horizontal transmissions giving an element of vertical polarisation in the return. This is not usually significant, because data on rain reflectivity assumes the receiver has the same plane of polarisation as the transmitter. Snow returns do not much depend on polarisation.

Internal reflections within perfect spheres of water would return circularly polarised incident signals at the opposite hand after three internal bounces, which perfect circularly polarised scanners would reject. In practice, raindrops are not perfectly spherical (particularly in heavy rain); neither are circular polarisers perfect. (Scanner imperfection is expressed as cross polarisation, the amount by which, say, a left-hand polarised antenna accepts right-hand polarised signals, Chapter 2, Section 2.7.13, Eqs (2.8).) Nevertheless circularly polarised scanners depress rain returns by 20–30 dB, which reduces rain clutter very significantly. Snowflakes have the wrong shape for snow clutter to be much reduced. Luckily snowstorm equivalent precipitation rates, r, expressed in mm/h of meltwater, are generally lower than in rainstorms.

Circular polarisation unfortunately also somewhat reduces the RCS of wanted targets, so the operator must have the facility to disengage in clear weather.

11.4.4 Mean received clutter power

At any instant the scanner illuminates a detection volume cell bounded by width $R\theta$, height $R\varphi/2$ (above the surface) and range increment $c\tau/2$, where as usual θ, ϕ are the scanner horizontal and vertical beamwidths (rad), c the velocity of propagation (299.7 m/μs) and τ the transmitter pulselength (μs). Assuming total reflection at the sea surface, the volume is illuminated twice; by the direct and indirect rays. From the geometry,

$$\text{Effective cell volume} = R\theta \times 2\left(\frac{R\phi}{2}\right) \times \left(\frac{c\tau}{2}\right) = \frac{R^2\theta\phi c\tau}{2}\ \text{m}^3. \qquad (11.8)$$

Note that doubling range increases cell volume by a factor of 4 and doubling pulse-length doubles volume. If rough sea or short range reduces the sea surface reflection coefficient, ρ, at the grazing point, precipitation clutter will be overstated by less than 3 dB. In effect we are incorporating a two-way multipath factor $M^2 = 2$ so $M = 1.4$ numerically (1.5 dB one-way). Chapter 9, Section 9.4.5, showed that a distributed target in the interference region had one-way multipath values varying between 0 dB (rough sea, high grazing angle) and 3.89 dB (calm sea, low grazing angle, Eq. (9.12c)). The precipitation column forms a vertically extended target, spanning a relatively wide grazing angle bracket. The multipath factor of the lower part of the column falls rapidly when its transition range is exceeded. Low rain-filled clouds may lie within the upper part of the scanner beam at long range and can produce rain clutter even when there is no rainfall at the surface.

Clutter RCS, σ_P is σ_{P0} (Eq. (11.7a)) × (cell volume, Eq. (11.8)):

$$\sigma_P = \sigma_{P0} \times \text{Effective cell volume} = \sigma_{P0}\frac{R^2\theta\phi c\tau}{2}\ \text{m}^2/\text{m}^3. \qquad (11.9)$$

For example, stratigraphic rain of 4 mm/h has $\sigma_{P0} = -62.9\,\text{dB}\,\text{m}^2/\text{m}^3 = 0.51 \times 10^{-6}\,\text{m}^2/\text{m}^3$. Viewed by a scanner of 1° × 25° beamwidth, 1 μs pulse, at 10 km, the effective cell volume is 114×10^6 m^3 and $\sigma_P = 58.3$ m^2 $= 17.6$ dB m^2. For detection, the echo in the cell must have comparable RCS.

This 'target' return is received as clutter power C_P. Being an extended reflection source, the echo has neither peaks nor nulls as the precipitation cell changes range. Atmospheric attenuation factor L_A (Chapter 5, Section 5.9.7, Eq. (5.50)) applies each way. C_P is obtained by substitution of σ_P for σ and M in the range equation of Chapter 4, Section 4.3.2, Eq. (4.6d), in which F_{12} is the free-space echo at the radar from a target with unit RCS at 1 km range; a 120 term converts from kilometres to metres:

$$S_e = F_{12} + \sigma - 40 \log R_{km} - 2L_A + 2M \text{ dBW} \tag{4.6d}$$

$$C_P = F_{12} + \sigma_P - (40 \log R - 120) - 2L_A \tag{11.10a}$$

$$= F_{12} + \sigma_{P0} + 10 \log \left[\frac{R^2 \theta \phi c \tau}{2} \right] - 40 \log R + 120 - 2L_A + 0 \tag{11.10b}$$

$$= F_{12} + 120 + \sigma_{P0} - 20 \log R + 10 \log \tau + 10 \log \left[\frac{\theta \phi c}{2} \right] - 2L_A \text{ dBW}. \tag{11.10c}$$

The expanded form, Eq. (11.10c), brings out the following.

1. Precipitation clutter follows a nett R^{-2} law ($-20 \log R$, -20 dB/decade) because the divergent scanner beam increases the cell volume as R^2, partly offsetting the usual point-target R^{-4} law.
2. Wide-beam scanners illuminate more clutter. Beamwidths θ and ϕ should be minimised. Sometimes easier said than done; narrow widths mean large aperture. VTS scanners can be given narrow elevation beamwidth, at a price; ϕ is constrained to $\sim 20°$ on ships' radars by the need to illuminate the horizon when rolling.
3. Pulselength τ should be minimised to reduce the axial depth of the detection cell, but this raises necessary bandwidth, increasing noise.
4. C_P is independent of scanner height H.
5. Relative to the 9 GHz band, for constant scanner beamwidths θ and ϕ, the λ^{-4} term within σ_{P0} (Eq. (11.7)) reduces C_P by 19.8 dB at 3 GHz; the 15 GHz band being 8 dB worse. Although true, this statement may mislead; for constant scanner aperture area, C_P is only 9.9 dB better at 3 GHz and 4 dB worse at 15 GHz. Nevertheless, 3 GHz radar has superior detection performance in precipitation against ship targets. The improvement is somewhat offset by a typical reduction of ship RCS of about 2.5 dB from 9 to 3 GHz, see Chapter 10, Section 10.4.9, Eq. (10.6). RCS of most passive point targets has yet more frequency dependence. Poor performance in precipitation is a main reason why the 15 GHz band is never used, except perhaps for very short-range VTS service.
6. Atmospheric loss L_A affects the target echo and clutter equally.

11.5 Precipitation clutter fluctuation

Although individually much too small to be detectable, individual hydrometeor returns combine randomly to form the 'target', which therefore fluctuates randomly just as individual electron motions combine as thermal noise. Precipitation clutter usually closely approximates Gaussian distribution. When, for example, a short-range target lies in a rain-storm, precipitation clutter in the detection cell may far exceed thermal noise so detection requires good signal to clutter ratio. At longer range or in lighter clutter, noise becomes dominant, so in general it can be seen that detection depends on signal to (clutter plus noise) power ratio, often loosely called SNR.

Precipitation clutter sometimes displays a woolly structure as scuds or showers form and reform in the wind. Here more extreme events enter the detection cell than predicted by Gaussian distribution, increasing tail amplitude. Sekine and Mao [4, Section 2.4] note that stormy and windy weather can bias the distribution from Gaussian towards Weibull (a distribution discussed in Section 11.7.4). They observed Weibull RCS shape coefficient, c, to vary between 1.0, equivalent to Gaussian distribution, and 0.825, with an average 0.945, indicative of somewhat higher tail amplitude and increased false alarm rate.

11.6 Mean sea clutter

11.6.1 Reflection mechanism

Some of the radar transmission is reflected back by rough water near the target as sea clutter, illustrated in Chapter 8, Figure 8.2(b). The return varies with wave height and a number of other factors and is much harder to predict than precipitation clutter. In many regions of the World, sea clutter is a severe problem, particularly to 9 GHz band radar. After reviewing the behaviour of sea waves, we go on to attempt a statement of RCS per square metre of surface, which enables calculation the clutter picked up by the radar, following the general method followed for precipitation.

Sea-waves were described in Chapter 5, Section 5.7. No single theory of sea clutter seems fully to agree with all observations. The author's qualitative interpretation of the clutter mechanism is as follows. Rays obliquely striking a dead-calm surface are specularly reflected as from a spherical mirror, almost all the energy going forward as a multipath indirect ray, a little being absorbed within the water mass and almost none retro-reflecting towards the source as clutter. When the surface roughens, less goes forward (Chapter 5, Sections 5.7.2 and 5.7.3) and more comes back. At the higher sea states spume probably introduces a component of droplet clutter akin to localised but severe rain clutter.

Capillary waves from local wind form saucer-shaped facets which scallop the surface and seem to be responsible for much of the sea clutter. The normalised reflectivity or backscatter coefficient (RCS per square metre of sea surface), σ_{S0}, increases with the sea clutter angle, β, at which they are viewed (Chapter 5, Section 5.5.4; Eq. (5.21a), Table 5.1, Figure 5.8(*b*) refer; β is sometimes confusingly called the grazing angle, the complement of angle of incidence). Reducing range or raising the scanner increases β. Gravity waves tilt parts of the sea surface but retain the capillaries, increasing the viewing angle, in turn increasing σ_{S0}. At low β, σ_{S0} also rises when the scanner looks up-wind into the steeper lee side of the waves, further increasing the effective viewing angle. Portions of the surface are masked and do not retro-reflect, this only partly offsetting the strong reflections from the wave flanks illuminated by the scanner. By about sea state 5 these effects are fully developed and additional increase in sea state does not increase σ_{S0} further. Swell has lower σ_{S0} than a sea of the same height, perhaps because it lacks capillaries and has lower maximum slope for a given sea state number. As the surface has no elevation, it is always beyond its multipath transition range and there is no multipath lobe structure. At the horizon range of the sea surface (less than target horizon), β falls to zero. Beyond, the surface is below the horizon and is assumed not to retro-reflect, diffraction returning negligible clutter.

Individual capillary patterns may persist for about 30 ms, so are correlated pulse–pulse but decorrelated scan–scan. They approximate Gaussian distribution when there are no gravity waves, but with a tendency to a deficiency of high-amplitude spikes – visual observation of capillaries confirms the infrequency or absence of high amplitude events.

Large individual gravity and swell waves may persist for many seconds, requiring several scans to achieve decorrelation. Particularly with short pulses ($\sim$0.1 μs), the radar may resolve and display waves as discrete targets. When wavetrains collide, the wave flanks of the resultant confused sea become unusually steep, further increasing RCS. This happens when an oceanic swell meets local storm waves at a continental shelf, for example, off the SE coast of South Africa. The clutter power distribution then has unusually high probability of a peak greatly exceeding the mean. Here and in confused seas generally, sea clutter echo is much spikier than true Gaussian distribution. This spikiness makes it more difficult for signal processing circuits to reject all clutter events – the clutter punches above its weight. Visually, the choppy waves of confined waters such as the North Sea look different from the rolling waves of oceans and σ_{S0} of a given sea state probably likewise varies with location. Statistically the height distribution becomes less Gaussian, and approximates log-normal (Section 11.7.3 and Appendix A2, Section A2.1) or Weibull distribution (Section 11.7.4) with high tails. Conversely low sea states have lower tails than predicted by Gaussian distribution. Shoals, a rough bottom or wind over tide may raise sea state locally.

In high seas at low viewing angle, some waves are masked by closer waves. Clutter is found to be somewhat dependent on the ray polarisation, perhaps because of interference patterns in the illumination. Passage of a ship may flatten sea clutter in its wake. Wylie [3, p. 109] reports that tide rips, overfalls, position of shoals and lines of demarcation between currents may be more distinctly painted than when viewed by eye.

With so many variables, it is not surprising that published data for σ_{S0} obtained at differing times and places with differing equipment and experimental techniques sometimes appear inconsistent. We will suggest an empirical algorithm to describe σ_{S0}. But one should always bear in mind that no mathematical model can fully represent the actual clutter at any specific time or place. This is a pity, given the importance of sea clutter in limiting detection of small targets. However, the algorithm gives at least some indication of likely clutter performance.

11.6.2 Clutter per unit area, σ_{S0}

The value of σ_{S0} depends on wind direction; radar polarisation; wavelength; sea clutter viewing angle β; wave height $H_{1/3}$ (or significant wave height, h_s, Chapter 5, Section 5.7.4, connected with sea state by Section 5.7.5, Table 5.3 and Eq. (5.14)), and the sea form – a fully developed sea or a swell. As the lee slope of waves is steeper than the windward slope, it reflects more strongly and the clutter extends furthest to windward, forming an oval about the viewing ship. Nathanson [5] tabulates values of σ_{S0} for 3 and 9 GHz band horizontal polarization, probably for US oceanic seas, summarised here in Table 11.2 and Figures 11.4 and 11.5. Nathanson's normalised mean sea backscatter $= \sigma_{S0}$ and his grazing angle = sea clutter angle, β. It can be seen that there is considerable scatter within the results. He states vertical polarisation gives 5 dB more clutter than tabulated. It seems reasonable to assume circular polarisation would have 2.5 dB more clutter, but would reduce clutter from whitecap droplets. In more confused coastal waters containing sea spikes, vertical

Table 11.2 Normalised mean sea backscatter (Nathanson)

Grazing angle		Sea state					
Degrees	Radians	0	1	2	3	4	5 upwards
		σ_0 (dB m^2/m^2) horizontal polarisation					
3 GHz band							
0.1	0.0018	−90	−80	−75	−68	−58	−53
0.3	0.005	−83	−74	−66	−58	−50	−44
1.0	0.018	−73	−65	−55	−48	−42	−42
3	0.053	−68	−59	−53	−46	−41	−37
10	0.18			−51	−46		−37
30	0.52			−40	−38	−37	−32
60	1.05	−32	−25				−17
9 GHz band							
0.1	0.0018		−71	−61	−53	−48	−42
0.3	0.005	−74	−66	−56	−46	−42	−39
1.0	0.018	−66	−51	−46	−40	−36	−33
3	0.053	−58	−48	−42	−39	−35	−32
10	0.18	−56	−51	−46	−37	−34	−33
30	0.52			−44	−34	−33	−26
60	1.05	−34	−26	−23	−26		−14

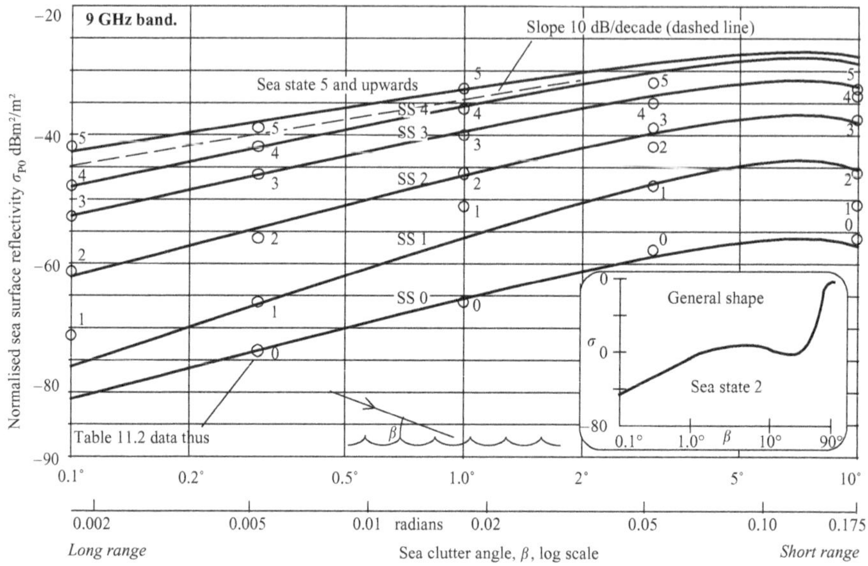

Figure 11.4 *Sea clutter mean reflection coefficient, 9 GHz band, per square metre of surface versus grazing angle at target for various sea states. Horizontal polarisation, add 5 dB for vertical*

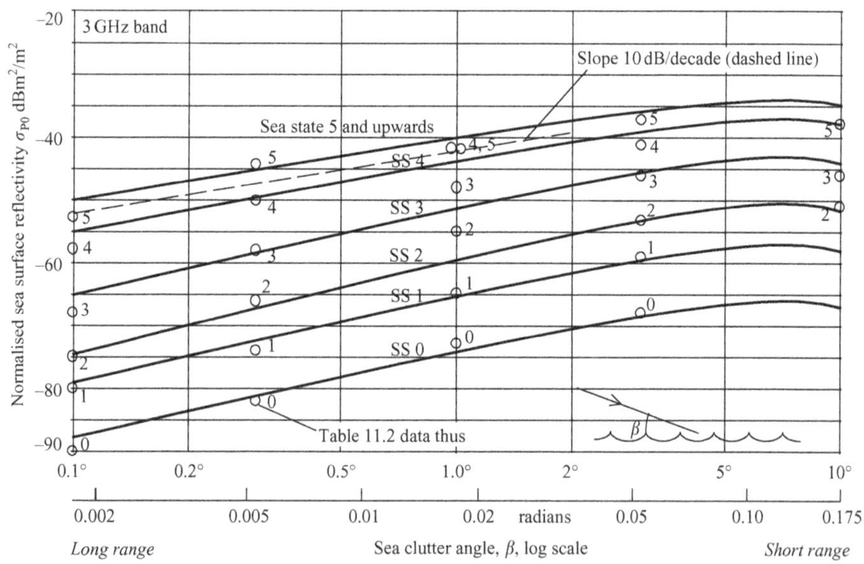

Figure 11.5 *Sea clutter mean reflection coefficient, 3 GHz band. Otherwise as Figure 11.4. Horizontal polarisation, add 5 dB for vertical*

polarisation has been found beneficial and some VTS reflector scanner polarisers can be switched among HP (for long-range detection of incoming shipping in open water), VP (for short-ranges in confused seas) and CP (rain). The reduced spikiness of circular and vertical polarisation may enable detection at slightly lower signal to clutter ratio.

Other authors give differently expressed σ_{S0} values which are difficult to align with Nathanson, perhaps reflecting the natural variability of clutter. Estimation is subjective and one wonders whether observers are also influenced by shape factors or even the size of their vessel. Anecdotal indications are that landlocked sub-tropical seas such as the Arabian Gulf having calm water with extensive capillary waves, sea state 0 or 1, exhibit clutter appropriate to sea state 2 when viewed at 9 GHz. This perhaps is because patches of capillary waves have crests approximately a multiple of $\lambda/2$ apart, where reflections combine in phase to give Bragg resonance rather than partially cancelling as they do when the phasing is random.

When sea clutter angle, β, is low, σ_{S0} is also low. It rises until $\beta \sim 0.12$ rad, where it pauses before rising further to nearly total reflection when looking vertically down on the sea (Figure 11.4 inset: $\sigma_{S0} \sim 0\,\text{dB}\,\text{m}^2/\text{m}^2$ when $\beta = \pi/2$ rad); high scanners give higher β and worse sea clutter. It is perhaps no coincidence that the maximum sustainable wave slope without toppling is 1 in 7 (0.14 rad), close to the 0.12 rad local maximum. To get a value for σ_{S0} at any value of β (hence range in a given system) we have devised as an algorithm a quadratic expression peaking at 0.125 rad (7.2°) connecting σ_{S0} with sea state and β:

$$\sigma_{S0} = A + B\log(\beta - 4\beta^2) + C\,\text{dB}\,\text{m}^2/\text{m}^2. \tag{11.11a}$$

For calculation of β see Chapter 5, Section 5.5.4, Eq. (5.21b). Wave heights, $H_{1/3}$, are from Skolnik [6]. Table 11.3 tabulates constants A, B and C.

Figures 11.4 and 11.5 plot the algorithm for sea states 0–5 in the 9 and 3 GHz bands, respectively, up to $\beta = 10°$ (0.175 rad), showing fair agreement with Nathanson. Steeper angles tend to fall outside scanner elevation beams, also they relate to ranges too short to be of much interest.

Note the following.

- σ_{S0} rises with increasing β at roughly 10 dB per decade at the higher sea states and longer ranges where $\beta < 2°$ (0.035 rad), so $\sigma_{S0} \propto \beta$ numerically.
- As β is here approximately proportional to 1/range, σ_{S0} falls with increasing range at a rate of ~10 dB per decade at the longer ranges.
- As a very rough first approximation, sea clutter rises ~5 dB per sea state number and is ~10 dB worse in the 9 GHz band than at 3 GHz, that is, $\sigma_{S0} \propto 1/\lambda^2$ approximately (the same relationship applies to a flat conductive plate reflector, Chapter 7, Section 7.4.4, Eq. (7.4a)). Lacking specific experimental data for higher (J band) or intermediate (C band) frequencies, 9 GHz band Eq. (11.11a) results could be used with addition of a term

$$20\log\left(\frac{f_{\text{GHz}}}{9.4}\right)\,\text{dB} \tag{11.11b}$$

Table 11.3 Constants for sea clutter algorithm

Sea State	Description	Wave height $H_{1/3}$ (m)	3 GHz band		9 GHz band	
			A	*B*	*A*	*B*
0	Calm	0	−49.1	14	−36.7	16.05
1	Smooth	0–0.1	−40.0	14.13	−19.1	20.6
2	Slight	0.1–0.5	−31.85	15.45	−17.3	16.2
3	Moderate	0.6–1.2	−26.0	14.13	−15.1	13.57
4	Rough	1.2–2.4	−23	11.6	−12.4	12.9
5 and upward	Very rough	2.4–4.0 and upward	−21.6	10.28	−15	10

Factor C (dB) is very roughly the sum of $C1$ and $C2$.

Horizontal polarisation	$C1 = 0$	Upwind	$C2 = 2.5$
Circular polarisation	$C1 = 2.5$	Crosswind	$C2 = -2.5$
Vertical polarisation	$C1 = 5$	Downwind	$C2 = 0$
	(Open sea)	Swell: $C2$ several dB lower	

where f_{GHz} is the frequency in GHz. Some error would doubtless arise, but it is better than nothing.

11.6.3 Wave height relation to wind speed

Barton [7] gives

$$\sigma_{S0} = \gamma \sin\beta, \qquad \text{where } 10\log\gamma = 6K_B - 10\log\lambda - 64\,\text{dB}. \tag{11.12}$$

K_B is the Beaufort wind scale number (which implicitly assumes a fully developed sea). But Barton indicates the formula is erroneous in high seas, so we shall not employ it.

11.6.4 Sea clutter mean power

The approach is similar to that for precipitation clutter, Eqs (11.8) and (11.9). The scanner illuminates, at angle β, a footprint area effectively equal to the range-dependent projected size of the detection cell, having width $R\theta$ and range increment $c\tau/2$.

$$\text{Projected cell area} = R\theta \times \left(\frac{c\tau/2}{\cos\beta}\right)\,\text{m}^2. \tag{11.13}$$

The practical maximum value of β is about 0.175 rad (10°), representing a minimum range of <600 m from a high scanner at 100 m. Here $1/(\cos\beta) = 1.015 = 0.07$ dB. The $\cos\beta$ term is therefore always close to unity. The sea forms a target of zero height ($h = 0$), so the clutter cell is always in the transition region. It is customary to work on the assumption that multipath factor $m_t = 1$ numerically (0 dB), because reported values of σ_{S0} underlying Eq. (11.11), Table 11.3 and Figures 11.4 and 11.5 will have made the same assumption.

The sea clutter horizon range, where $\beta = 0$, is found by putting $h = 0$ in Chapter 5, Section 5.5.7, Eq. (5.23):

$$\text{sea clutter horizon} = \sqrt{2keH}. \tag{11.14}$$

Sea clutter horizon range typically varies between about 5.5 km ($k = 0.6$, $H = 4$ m) and 71 km ($k = 4$, $H = 100$ m, high VTS scanner).

The clutter RCS, σ_S, is $\sigma_{S0} \times$ (projected cell area):

$$\sigma_S = \sigma_{S0} R\theta \frac{1}{2} \frac{c\tau}{\cos\beta} \text{ m}^2. \tag{11.15}$$

This 'target' is received as clutter power C_S. Being within the transition region, the return has neither peaks nor nulls and falls monotonically as the clutter cell increases in range. Atmospheric attenuation factor L_A of course applies each way. The following equation, based on the radar range equation Eq. (4.6a), gives C_S in terms of dBW.

$$C_S = F_{12} + \sigma_S(-40 \log R + 120) - 2L_A = F_{12} + \sigma_0 + 10 \log \left[\frac{R\theta c\tau}{2\cos\beta} \right] - 2L_A \tag{11.16a}$$

$$= F_{12} + 120 + \sigma_{S0} - 30 \log R + 10 \log \tau + 10 \log \theta$$
$$+ \frac{\log c}{2\cos\beta} - 2L_A \text{ dBW}. \tag{11.16b}$$

The expanded form, Eq. (11.16b) brings out the following, other things being equal.

1. Sea clutter ostensibly follows a nett R^{-3} law (−30 dB/decade) because the divergent scanner beam increases the cell area as R^1, partially offsetting the usual point-target free space R^{-4} law. But Figures 11.4 and 11.5 show that σ_{S0} falls with β when range increases, steepening the overall fall of sea clutter with range, often roughly to R^{-4} (−40 dB/decade). At long range, as the clutter horizon is approached where σ_{S0} becomes infinitely small, the law steepens yet more, so sea clutter is a relatively short-range problem.
2. Wide-beam scanners illuminate more clutter. Azimuth beamwidth, θ, should be minimised, unfortunately necessitating wide aperture. Elevation beamwidth is immaterial.
3. Pulselength, τ, should be minimised to reduce the axial length of the detection cell. Changing from 1 μs to 0.1 μs reduces clutter by 10 dB, at the expense of additional noise.
4. Although σ_{S0} is ~9 dB lower at 3 GHz than at 9 GHz, a scanner of given aperture has 5 dB more beamwidth, leaving a nett improvement of very roughly 4 dB in favour of 3 GHz. But RCS of ship targets tends to be 2.5 dB lower at 3 GHz, so the important signal to clutter ratio is only slightly (1.5 dB) more favourable.

RCS of passive point targets such as octahedral clusters usually falls 10 dB at 3 GHz, and here a 9 GHz system is some 6 dB better.

5. Earth curvature k-factor only weakly influences sea clutter, by modifying its horizon and by marginally affecting β.

11.6.5 Effect of scanner height

When the scanner is high, clutter return power C_S rises because, for a given range, β is approximately proportional to scanner height, H; also the sea clutter horizon increases (Eq. (11.14)). It is important to site the scanner as low as possible to minimise sea clutter. Unfortunately this also reduces target horizon range, and practical difficulties arise on ships – optimum height is a matter of informed case-by-case judgement unless dictated by the need to clear obstructions such as deck cargo.

Figure 11.6 is a family of plots of C_S versus range, R, for a number of scanner heights, H, for a typical 9 GHz marine radar with sea state 3, assuming long pulse operation with 3 MHz bandwidth. Sea clutter follows a law near R^{-4} at medium range, steepening near the clutter horizon. The clutter only becomes significant when it equals the receiver noise power, so for $k = 4$, about the highest atmospheric refraction commonly encountered, maximum clutter ranges (at differing scanner heights) are approximately: 0.65 km (5 m), 2.2 km (10 m), 6.5 km (30 m) and 15 km (100 m).

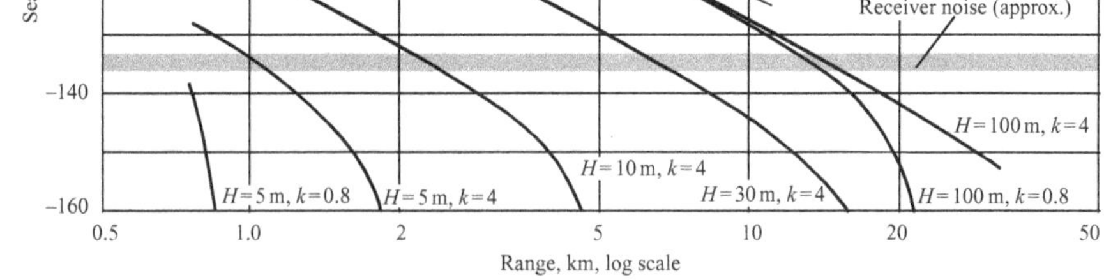

Figure 11.6 Variation of sea clutter with scanner height and range. C_S would be about 4 dB less at 3 GHz for the same scanner aperture. Sea clutter is strongly dependent on H. The slope line indicates the actual clutter/range law is about −40 dB/decade at medium range, steepening near the clutter horizon. Atmospheric refraction (k variation) is only significant near the horizon

Using short pulse with say 13 MHz bandwidth would reduce clutter powers by $10\log(13/3) = 6.4$ dB and raise noise by another 6.4 dB, approximately halving the above maximum ranges.

Under constant sea conditions, clutter falls monotonically with range as shown in Figure 11.6, but shoals, a rough bottom or wind over tide may raise distant clutter power.

11.6.6 Abnormal waves

We remarked in Section 5.6.2 that intersecting wave trains throw up some unusually high waves. Sometimes called freak or, better, abnormal waves, they can be dangerous to shipping. Individual wave heights of up to 34.7 m have been reported by Faulkner [8] and it is thought that there is 1 per cent probability of a single wave within a storm having height 2.5 times significant wave height H_S. So storm waves may occasionally well exceed 20 m. Making the very crude assumption of a rectangular pyramidal wave, height 20 m, with sides inclined at the maximum sustainable slope of 1 in 7 (0.14 rad), the face area is 2800 m^2. If the pyramid surface approximates SS2 locally, Figure 11.4 indicates the local reflectivity is -37 dB m^2/m^2 in the 9 GHz band, giving the single abnormal wave RCS of $10(\log 2800) - 37 = -2.5$ dBm^2, equal to the return from 1100 m^2 of ordinary sea surface at sea state 5 and $\beta = 1°$.

11.7 Sea clutter fluctuation

11.7.1 Sea clutter, low sea state

Chapter 5, Section 5.7 and Figure 5.16 described how energy is scattered by the capillary waves which ride on the flanks of the larger gravity waves. At low sea state, where gravity waves are small, the individual capillaries are to some degree organised or correlated so their RCS distribution is not quite Gaussian random, with disproportionately few high events relative to the rms. Their distribution has low tails. However there are, by some accounts, enough of them within the detection cell for their RCS or echo power to approximate Gaussian distribution in accordance with the central limit theorem (Section 11.3.3). At low sea states, sea clutter is often therefore treated as ordinary noise (a refinement using Weibull distribution is suggested in Section 11.7.3), sea clutter power being added to precipitation clutter and noise powers (watts not dB) to give the total noise + clutter return power.

11.7.2 Sea clutter, high sea state

Quite apart from abnormal waves, local collisions between individual waves in confused seas frequently throw up transient sea spikes, a few metres high, a few metres wide, lasting a few seconds and moving at nearly wind velocity, as mentioned in Chapter 5, Section 5.7.1. Spikes rise disproportionately in amplitude and frequency with increasing sea state, causing more RCS high amplitude events than predicted from Gaussian distribution. Wave height distributions are therefore not Gaussian.

The returns from individual large waves are very much bigger than from hydrometeors of precipitation clutter and their number is far fewer. Their sea clutter returns have a spikier appearance than Gaussian noise of the same average power, especially in the higher sea states. The effect is analogous to the intense sparkles of sunlight on the sea which cause automatic cameras to under-expose seascapes. The distribution depends on factors which include:

- whether the waves are a fully developed sea or a swell;
- the angle of observation relative to the directions of the wave train and the wind;
- the shortness of the waves (dependent in part on water depth);
- possibly the sea clutter angle;
- the radar detection cell footprint size, which is range-dependent;
- scanner polarisation.

According to Williams [9], spiky sea clutter is reduced in amplitude on vertical polarisation; some VTS scanners include a vertical polarisation option. Particularly when short pulses are transmitted and detection cells are small, the display may resolve individual spikes. Waves persist for many seconds and spikes for several, so their clutter returns are not fully random. Looking at one scan, the appearance of the next scan's clutter is predictable with some semblance of accuracy – clutter decorrelation time is several scans in length. At high sea state each wave or spike takes the form of a definite object which persists for several scans together, all too easily mistaken by the signal processor for genuine targets. The spikes necessitate setting threshold particularly high to get low P_{FA}, reducing P_D more than would precipitation clutter of the same mean power. False plots or ARPA tracks may overload the processor enough for it to dump genuine targets. Severe sea clutter is the most important problem facing marine radar designers and dictates choice of several major radar parameters, such as pulse length and scanner aperture. Adding to this, we have seen that sea clutter is vexingly difficult to quantify.

Purely statistical treatments do not always fully represent reality and many radar manufacturers prefer to test new designs against their collection of recordings of real sea clutter. It is no light task to compile such a library and it is definitely locked away as valuable 'Company Confidential' intellectual property. Decca Radar, for example, maintained research stations on the tidal River Thames and at Dungeness on the English Channel for many years, making extensive recordings for future analysis, often after a long and frustrating wait for the right weather to arrive. Designs under development could then be tested and refined at will under laboratory conditions against real target and clutter recordings. This attention to detail undoubtedly contributed to the high reputation of their radars. (Decca, now a division of Northropp Grumman Sperry Marine, manufacture the well-known BridgeMaster radars, Chapter 2, Figure 2.2).

Clutter at shore stations is not always representative of the open sea, so the author's firm (AEI) preferred to install developmental radars and recorders on a sea-going motor yacht, taking the radar to the clutter. Losing all contact with the trials team after a storm in the Western Approaches, the firm feared the worst. But bad pennies always turn up and they were snug in the recesses of Swansea Docks, writing up their notes, they said: asleep, we suspected.

Datasheets rarely quote specific performance because of the difficulties of effective simulation and the impossibility of turning on a rough sea to order for acceptance tests. Likewise, Type Approval authorities have to take their local sea area as they find it, so cannot test for target detection in clutter, with the result that the IMO radar specifications have had more to say on control knobs than clutter performance, although it is believed a more specific clutter requirement may be introduced soon.

In rough seas, the relatively small number of spikes superimposed on the Gaussian element of surface roughness reflect disproportionately large echoes and the tails of the probability distribution become significantly larger than Gaussian. Spikes travel at the speed of the prevailing wind and occupy only a few square metres. If the cell footprint is made smaller, by use of a narrow beam scanner, short pulses or reduced range, the amplitude of the Gaussian component of clutter falls proportionately, the distribution remaining Gaussian. However, all practical radars have detection cells much larger than spike footprints, and the spatial and temporal separation of spikes are so high that there may or may not be a spike within the cell at a given instant. So as cell size is reduced, spike amplitude remains relatively constant and only spike frequency falls. This behaviour, more pronounced with horizontal polarisation than vertical, differs from noise-like precipitation clutter and changes the overall distribution to a 'super-Gaussian' form having higher tails, which, particularly in constant false alarm rate (CFAR) systems, raises the detection threshold and reduces sensitivity. We need a statistical model which enables distribution to be fitted to observed results as sea state changes. Candidates include log-normal and Weibull distributions.

11.7.3 Log-normal distribution

Spikes make the distribution of rough sea roughly approximate log-normal, drawn by re-scaling the abscissa of an ordinary Gaussian (normal) distribution logarithmically. This distribution may raise tail amplitude too far, is not easy to evaluate and does not provide a ready transition which slides seamlessly from the sub-Gaussian distribution describing low sea states as roughness increases. Although we shall not use log-normal distribution, it is detailed in Appendix A2, Section A2.1.

11.7.4 Weibull distribution

This distribution has sufficient parameters to allow tailoring to differing tail strengths above or below Gaussian, and is reasonably easy to compute. Since the 1980s, it has been found to approximate observed RCS sea clutter amplitude distributions with considerable accuracy. It can also be used for land and ice clutter, which also have few major scatterers per unit area. The following is based on the full and readable treatment in the earlier part of Sekine and Mao [4]. See Appendix A2, Section A2.5 for additional details.

A 1 V rms (normalised) noise signal of Gaussian distribution, after modulation with the local oscillator to become an IF signal, has its negative components folded over into positive components, changing the distribution to Rayleigh, discussed in more detail in Chapter 12:

$$p(R) = R\exp(-\tfrac{1}{2}R^2). \tag{11.17a}$$

The Weibull distribution of RCS, σ, proportional to received clutter power, valid when $\bar{\sigma} > 0$, $c > 0$, and zero otherwise, is

$$p(\sigma) = \frac{2c}{\bar{\sigma}}\sigma^{(2c-1)}\exp\left(-\frac{\sigma^{2c}}{\bar{\sigma}}\right) \tag{11.17b}$$

where σ is the instantaneous clutter RCS or power, $\bar{\sigma}$ the scale parameter, here average clutter RCS or power and c the shape parameter. $c = 1$ for Rayleigh distribution, and is a measure of the degree of organisation of the clutter.

Although we shall not do so, because the Eq. (11.17b) form will be more convenient when considering Weibull clutter voltage in Chapter 12, Weibull distribution is often defined with shape parameter η or $C = 2c$, so $\eta = 2$ for Rayleigh distribution:

$$p(\sigma) = \frac{C}{\bar{\sigma}}\sigma^{(\eta-1)}\exp\left(-\frac{\sigma^{\eta}}{\bar{\sigma}}\right). \tag{11.17c}$$

Putting $\bar{\sigma} = 1$ in Eq. (11.17b) gives Weibull normalised RCS or power distribution, Figure 11.7:

$$p(\sigma) = 2c\sigma^{(2c-1)}\exp(-\sigma^{2c}). \tag{11.17d}$$

Integration gives the cumulative probability, Figure 11.8.

$$P(\sigma) = \text{CP} = \int_0^{2\sigma} p(\sigma)\,\mathrm{d}\sigma = 1 - \exp(-\sigma^{2c}) \tag{11.18a}$$

from which

$$\sigma = \left\{\ln\left[\frac{1}{1-\text{CP}}\right]\right\}^{1/2c}. \tag{11.18b}$$

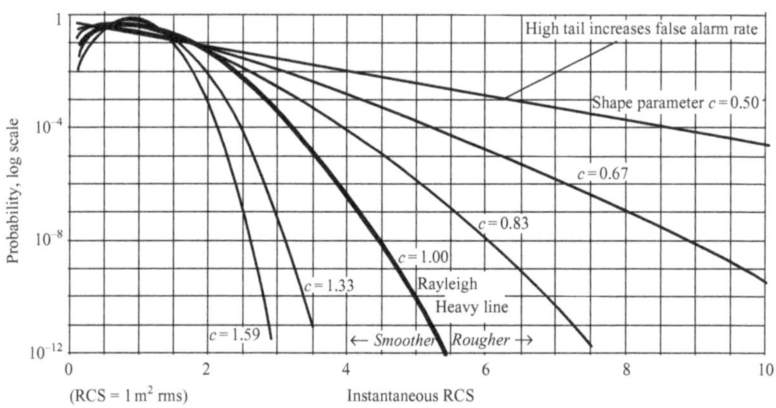

Figure 11.7 Weibull distribution

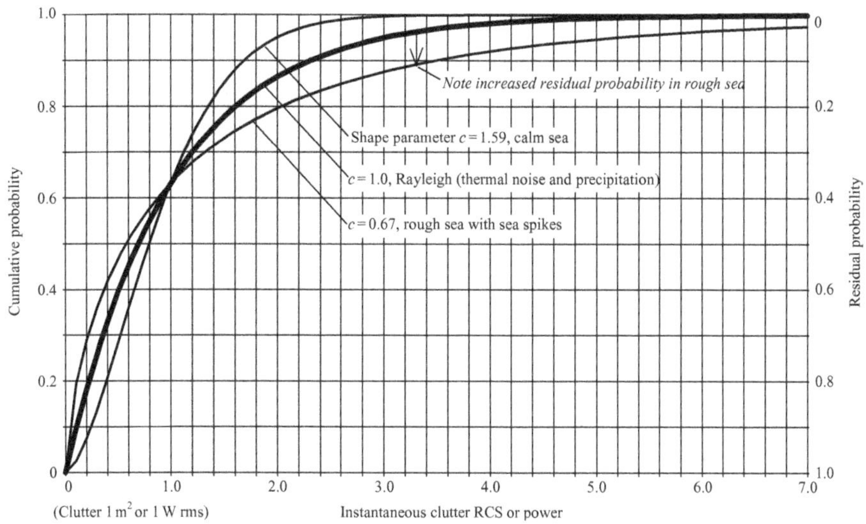

Figure 11.8 Cumulative probability, Weibull distribution

Figure 11.7 shows RCS probabilities for representative values of c. When c is low there is a long tail indicating relatively high probability of the instantaneous RCS (or received clutter power) much exceeding the rms value. The following puts Eq. (11.17a) into Weibull form. It is again normalised with rms $\sigma = 1$ and is restated in terms of instantaneous voltage, v, where $\sigma = v^2/2$.

$$\text{Voltage cumulative probability, CP} = 1 - \exp\left(-\frac{v^2}{2}\right)^c.$$
$$\text{Residual probability, RP} = 1 - \text{CP} = \exp\left(-\frac{v^2}{2}\right)^c. \tag{11.18c}$$

Table 11.4 summarises sea clutter parameters quoted by Sekine and Mao from several sources. They report that shape parameter c varies between 0.67 in 'rough sea' and 1.59 in 'smooth sea' where there are many small waves within the detection cell. Assuming that with relatively high definition marine radars at small grazing angles c rises linearly from 0.67 in sea state (SS) 5 to 1.59 at SS0, we suggest that

$$c \sim 1.59 - 0.184 \times \text{SS}. \tag{11.19}$$

It is stressed this is the author's supposition only, and it must be remembered that c values also depend in an unquantified manner on detection cell footprint (tending to 1 when footprint area is large), angle of incidence, whether sea or swell, and perhaps on plane of polarisation. Eq. (11.19) indicates that SS3 approximates Rayleigh distribution ($c = 1$). Lower SS have $c > 1$, meaning the energy reflected fluctuates less than expected for true noise, with low tail amplitude, presumably caused by wave-to-wave uniformity. For example, a surface resembling a sheet of corrugated

Table 11.4 Sea clutter parameters

Wind or sea state	Radar	Sekine and Mao ref.	Grazing angle, β	Shape parameter, c	Median RCS/m^2
Wind 10–15 kt	9 GHz	Fig. 2.13b		1.24	
Wind 30–40 kt	9 GHz	Fig. 2.14		0.67	
SS3, into sea	K, HP, 0.1 μs pulse	Fig. 2.12	1°	1.16	
SS3, into sea	K, HP, 0.1 μs pulse	Fig. 2.12	5°	1.65	
SS3, into sea	K, HP, 0.1 μs pulse	Fig. 2.12	30°	1.78	
SS3, into wind	L, HP, low res.	Fig. 2.15	0.5–0.72°	1.585	
SS2	9 GHz, VP, 40 ns pulse. Cell 31.6 m^2	Table 2.4		0.622	−21.4 dB m^2
SS5	9 GHz, VP, 40 ns pulse. Cell 31.6 m^2	Table 2.4		0.495	−16.2 dB m^2

Notes: L band ~1 GHz, cell 3 μs × 1.23°. K band ~15 GHz. HP = horizontal polarisation, VP = vertical polarisation.

iron with no high-amplitude events would cause little or no variation of the return as the uniform corrugations slid toward the radar in the breeze and c would be very high. A method of accommodating Weibull clutter distribution will be suggested in Chapter 12, Sections 12.4.2 and 12.4.3.

11.8 Short-range ringing clutter

11.8.1 Feeder ringing

Chapter 2, Section 2.6.2, explained how mismatched feeders reflect part of the transmitter pulse back to the receiver before reaching the scanner, perhaps spoiling short-range performance by competing with echoes. When lengthy feeders are employed, ringing can be sufficiently severe to mask quite strong short-range targets and can reduce system performance below IMO minimum requirements.

In the following, losses are in decibels. Figure 11.9(a) shows a transceiver connected to a scanner by a feeder, length x metres, ohmic loss F dB/m, so one-way ohmic loss $A = Fx$. Typical attenuation rates were included in Chapter 2, Section 2.6.2, Table 2.2. For simplicity, the transmitter and scanner are assumed to be equally mismatched; it is straightforward but tedious to extend the argument to differing mismatches or to intermediate mismatches such as a kinked waveguide within the feeder run. As shown diagrammatically at Figure 11.9(*b*), transmissions reach the scanner

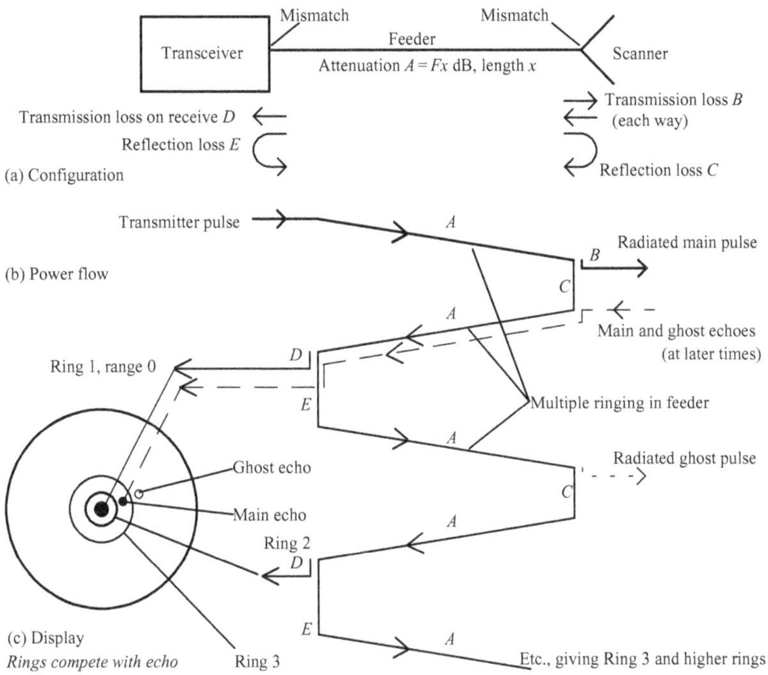

Figure 11.9 Feeder reflections. Mismatches reflect some transmitter power, which travels to and fro several times before petering out, delivering a ring of false echoes at each pass

through feeder attenuation A + transmission loss B. Target echoes are then subject to feeder attenuation A + transmission losses $(B + D)$:

$$\text{loss to echo} = 2A + 2B + D\ \text{dB}. \tag{11.20a}$$

The first reflection false-echo ring into the receiver is subject to two-way feeder attenuation, reflection loss C at the scanner mismatch and transmission loss B:

$$\text{loss to first reflection} = 2A + C + D\ \text{dB}. \tag{11.20b}$$

As it reflects at the transmitter mismatch E, again trundles forward along the feeder, is reflected at the scanner and returns into the receiver, each subsequent reflection suffers further loss:

$$\text{loss per ring} = E + 2A + C\ \text{dB}. \tag{11.20c}$$

$$\text{Total, second ring (Eq. (11.20b) + Eq. (11.20c))} = 4A + 2C + D + E\ \text{dB}. \tag{11.20d}$$

$$\text{Total loss, } n\text{th ring} = 2nA + nC + D + (n-1)E\ \text{dB}. \tag{11.20e}$$

Scanner voltage standing wave ratio is usually published and transceiver VSWR (≥ 1) may be winkled out of the supplier if not included in the data sheet. As explained in Chapter 2, Section 2.6.2, losses are found by finding the reflection coefficient of the VSWR from Eq. (2.5a): $\rho = (\text{VSWR} - 1)/(\text{VSWR} + 1)$, then substituting in Eqs (2.5b) and (2.5c) to get the mismatch transmission loss (B or D; $-10 \log (1 - \rho^2)$ dB) and mismatch reflection loss (C or E; $-20 \log \rho$ dB), respectively, for VSWR of the port in question.

Because system timing causes a target at zero range to arrive simultaneously with the first ring, that ring appears as a harmless paint at the display origin (Figure 11.9(*c*)). Although important to receiver burn-out protection, the first ring therefore does not conflict with targets. The nth ring is displayed as the $(n - 1)$th display circle:

$$\text{apparent range, } n\text{th ring} = \frac{(n-1)xc}{\text{GV}}\ m \tag{11.21}$$

where c is the velocity of light and GV is the group velocity of the feeder. For coaxial cable, GV $\sim \frac{2}{3}c$, $\sim$200 m/μs. For waveguide (Chapter 2, Section 2.6.1, Eq. (2.4b)), $\text{GV} = c \times [1 - (\lambda/2a)^2]$, a being the broad or H-plane dimension.

The equivalent echo RCS of the ring, σ_{eq}, is found by equating it with the echo from a target of the same RCS at the ringing range in question (Eq. (11.21)), using the radar range equation (Chapter 4, Section 4.5.2, Eq. (4.8)). It is sufficient to use the free space form at the short-ranges in question and to ignore all but the feeder-related losses of Eq. (11.20a).

$$\begin{aligned}\text{Target echo } S_e = {} & P + 2G + 20 \log \lambda - 30 \log(4\pi) + \sigma \\ & - 40 \log R - (2A + 2B + D)\ \text{dBW}.\end{aligned}$$

$$\text{Ring 'echo' power} = P - (\text{total loss per Eqs (11.20b) to (11.20e)})\ \text{dBW}.$$

Equating and re-arranging,

$$\begin{aligned}\sigma_{\text{eq}} = {} & 30 \log(4\pi) + 40 \log R + 2A + 2B + D \\ & - (\text{total loss per Eqs (11.20b) to (11.20e)}) - 2G - 20 \log \lambda\ \text{dBm}^2.\end{aligned} \tag{11.22}$$

Results will be imprecise because no account has been taken of mismatch phasing or the power and time dependence of losses D and E as the receiver protection TR or other device relaxes after the transmitter pulse. Ringing returns are correlated pulse to pulse, so the echo to ring ratio is not improved by pulse to pulse correlation. Echoes coinciding with ring clutter must exceed the latter by fully 10 dB for detection.

11.8.2 Example

Suppose a 9 GHz band radar is connected to its scanner via a lengthy waveguide feeder, $x = 25$ m, $F = 0.18$ dB/m (waveguide WG16, $a = 22.86$ mm). At each port VSWR $= 1.5$ (representing a moderate mismatch). For the radar: $P = 10\,\text{kW}$ (40 dBW), $G = 30\,\text{dBi}$, $\lambda = 0.032\,\text{m}$ (9365 MHz). The calculations are as follows.

Group velocity (Eq. (2.4b)) $= 0.714\,c = 214\,\text{m}/\mu\text{s}$.
From Eq. (2.5a) $\rho = (1.5 - 1)/(1.5 + 1) = 0.20$.
Feeder ohmic loss: $A = xF = 25 \times 0.18 = 4.5$ dB one-way.
Transmission losses (Eq. (2.5b)), B and $D = -10\log(1 - 0.20^2) = 0.18$ dB each.
Reflection loss $C = E = -20\log 0.2 = 13.98$ dB each.
Total loss to echo (Eq. (11.20e)) $= (2 \times 4.5 + 2 \times 0.18 + 0.18) = 9.54$ dB.
Second (first visible) ring occurs (Eq. (11.20)) at $R = 25/0.714 = 35$ m range.
Power of 2nd ring, $n = 2$ (using Eq. (11.20c)) $= P - [2nA + nC + D + (n - 1)E]\,\text{dBW} = 40 - [4 \times 4.5 + 2 \times 14 + 0.18 + 14] = -11.18$ dBW (0.762 W).
Equivalent 2nd ring RCS (Eq. (11.22)), $\sigma_{eq} = 33.0 + 40\log 35 + (2 \times 4.5 + 2 \times 0.18 + 0.18) - (4 \times 4.5 + 2 \times 13.98 + 0.18 + 13.98) - 2 \times 30 + 29.90 = 14.09\,\text{dB}\,\text{m}^2$ ($25.6\,\text{m}^2$).
Using the same method, the third ring at 70 m has $\sigma_{eq} = -10.87\,\text{dB}\,\text{m}^2$ ($0.082\,\text{m}^2$).

Because of the assumptions made, these results are not likely to be very accurate. Note how sharply high-order ring RCSs fall in amplitude. The fall would be less if VSWRs were worse. In Chapter 12 we will see that minimum detectable target RCSs need to exceed these ring clutter RCSs by about 10 dB, so only targets exceeding $\sim 250\,\text{m}^2$ at 35 m or $\sim 0.82\,\text{m}^2$ at 70 m are likely to be detected.

11.8.3 Ghost axial echoes

The figure shows that after traversing the feeder three times, a second weak 'ghost' pulse is transmitted. This of course causes a ghost second echo, delayed by the propagation delay in the feeder and causing a ghost paint on target bearing at a distance beyond it of the ring-ring separation. These ghost echoes are not usually noticeable, because only large targets are likely to have sufficient RCS to make them detectable, and the axial extent of such targets may well overlap the ghost. Any further ghosts from the secondary feeder reflections will be weaker still.

11.8.4 Receiver oscillation

The strong short-range ring signals may induce damped oscillation within the receiver circuits, giving additional rings of interference on the display. This form of clutter is a matter of detail design and is difficult to quantify.

11.9 Man-made interference

11.9.1 Other radars

Usually the most troublesome man-made interference received at marine and VTS radars is from other radars – pulsed surveillance radars used for air traffic control and military purposes, but predominantly those of the VTS and civil marine navigation services, the radars which are the subject of this book. Considering interference from typical shipborne civil marine radar received at a similar radar, pulses received at our radar from the other may be displayed as 'running rabbits' – trains of dots often looking like a ramark responder radial line or like the spokes of a slowly rotating wheel, as indicated in Chapter 3, Section 3.11, Figure 3.19. The other radar may be on own platform (ship or VTS site) or be distant. Its transmissions are of course unsynchronised to ours in prf or scan rate, so its interference paints continuously change in apparent range and bearing and are colloquially called running rabbits. Although not likely to be mistaken for genuine echo plots by an operator or by a plot extractor, the interference is a distracting annoyance and may overload trackformers or ARPA.

As an example, suppose each radar has scanner gain 30 dBi, gain to primary sidelobes 2 dBi (28 dB down on main beam) and gain to far sidelobes – 10 dBi through the 360° azimuth. In Chapter 8 we noted that, because only one path leg is in play, a racon having response EIRP of around 6 dBW may be detected to well in excess of 10 km range through the main beam of own radar, the sum of response power and the two antenna main-beam gains, called here the 'leg signal', being ~36 dBW.

We first assume both radars operate at the same microwave frequency (or, sometimes, at receiver image frequency). The interfering radar has EIRP typically 70 dBW (e.g. P_t 40 dBW, G 30 dBi), enough to interfere to horizon range or beyond at the comparatively rare occasions when the main beams are on reciprocal bearings with leg signal 100 dBW, say one scan in 180 for 1° beamwidths or eight times per hour, giving severe ramark-like interference on interfering radar's bearing. When one main beam is on a reciprocal bearing to a primary sidelobe, leg signal is 72 dB. When the primary sidelobes are on reciprocal bearings, leg signal is 44 dBW and 'ramark' interference may be experienced to >20 km range. Leg signal is 60 dBW when the main beam of either radar scans through secondary sidelobes of the other, giving spoking interference on every scan even when the interfering radar is very distant.

When the radars have differing frequencies within the same band, leg signals are reduced by the IF bandpass filter attenuation and interference ranges are reduced. Transmissions in the 3 GHz band are unlikely to be accepted by 9 GHz scanners of the slotted waveguide or VTS reflector scanners fed by waveguide; the 3 GHz frequency will be below waveguide cut-off with evanescent mode attenuation at least several tens of dB/m. But 9 GHz interference can in general be received relatively efficiently by 3 GHz scanners and feeders—depending on detail design—perhaps sufficiently powerfully to cause rectification effects in the receiver, the resulting hiccoughs displaying as echo- or noise-like disturbances.

When the interfering radar is on own platform, the scanners must be arranged well outside one another's main beams to prevent receiver burn-out, typically being sited more or less vertically in line. Nevertheless, the scanners couple through far-out elevation sidelobes and through unwanted reflections from nearby objects, the leg signal then being outside the installation designer's control. Transmission should be inhibited within such blind arcs. It is also a straightforward matter to connect transmitter pre-pulses, occurring a few microseconds before transmission, to blank off the other's receiver, resulting in a pattern of minor blank patches on the display – running rabbit-holes as it were, which are unobtrusive and seldom suppress a third-party echo. If the interference still proves severe, microwave bandstop filters tuned to reject the other radar may be inserted in the feeders, and/or the two radar transmissions may be synchronised.

None of these palliatives is available to a distant interferer. The usual, and effective, cure for interference is to accept as candidate detections only pairs of echoes occurring within the same range cell, two out of two, a special case of M out of N integration, see Chapter 12, Section 12.6.5.

11.9.2 Own ship

Electrical interference sources near the radar may impress unsynchronised noise-like signals onto the receiver, despite its screening and cable filtering. Similar signals may be generated by poor contacts, 'dry joints' caused by faulty soldering, dirty and worn rotating joint sliprings or poor earthing (grounding). The receiver noise factor is degraded, reducing detectability of small echoes, particularly when there is little or no clutter to mask the noise.

11.10 References

1 CLAPHAM, C.: 'The concise Oxford dictionary of mathematics' (Oxford University Press, Oxford, 1996, 2nd edn.)
2 MORCHIN, W.: 'Radar engineers' sourcebook' (Artech House, London, 1993), (cited in MEIKLE, H.: 'Modern radar systems' (Artech House, 2002))
3 WYLIE, F. J.: 'The use of radar at sea' (Hollis & Carter for the Royal Institute of Navigation, 1978, 5th edn.)
4 SEKINE, M. and MAO, Y.: 'Weibull radar clutter' (Peter Peregrinus for the IEE, 1990)
5 NATHANSON, F. E.: 'Radar design principles' (McGraw-Hill, New York, 1969), Tables 7-2–7-8
6 World Meteorological Organisation table, in SKOLNIK, M. I.: 'Introduction to radar systems. International student edition' (McGraw-Hill, New York, 1983, 2nd edn.), Figure 13.4
7 BARTON, D. K.: 'Modern radar systems analysis' (Artech House, London, 1988)
8 FAULKNER, D.: 'Freak waves and survival design', *Seaways, The International Journal of the Nautical Institute*, 2002
9 WILLIAMS, P. D. L.: 'Civil marine radar – a fresh look at transmitter spectral control and diversity operation', *The Journal of Navigation*, 2002, **55**, pp. 405–18

Chapter 12
Detection

'There are three kinds of lies – lies, damned lies, and statistics.'

Benjamin Disraeli

12.1 Outline

Signal processing leading to target detection was discussed in general terms in Chapter 3, Section 3.6. We now quantify the detection process. Appendix A2 expands on some matters of detail. This chapter, which owes much to expert guidance from Professor E. D. R. Shearman, enables P_D to be predicted. Although long, we will only scratch the surface of detection theory. For a taster of the extensive mathematics necessary for rigorous treatment, see for example, Rohan [1], Chapter 3. Theoretical work is ongoing, driven by the needs of telecommunications, and is leading to introduction of refinements in radar data handling software. The next chapter will discuss the errors in measurement and calculation arising from the uncertainties surrounding radar operation.

12.1.1 What we mean by detection

Historically, detection was the machine process of extracting radio signals from the carrier, originally using a coherer on the incoming Morse-modulated RF carrier and later by diode rectification or 'demodulation' of the IF. The operator then mentally extracted the Morse or speech data from the residual noise and interference, although machine Morse inkers were soon developed. The rectification process came to be called detection. Radar and radio receivers are similar in some ways, but we use detection in a wider sense, to embrace the whole task of extracting the presence of a target from the surrounding noise and clutter. Demodulation of the IF is one stage of the process and is still sometimes itself confusingly called detection, although we have been careful not to do so.

The radar receiver output contains thermal noise, mostly internally generated, and may contain precipitation- and sea-clutter returns, short-range ringing clutter and echoes from passive or active targets. Detection seeks to maximise echoes and minimise noise and clutter. These latter (except ringing clutter) fluctuate and so do most echoes. Occasionally noise or clutter spikes will be wrongly declared as targets, generating false alarms; correct detections of echoes being irrationally called detections, not true alarms. A decision, right or wrong, that a target is present is a declaration. Sometimes echoes will suffer fades (e.g. from multipath), becoming too weak to be detected.

Formerly, most of the detection workload was shared between the integrating phosphor of the long-persistence cursive PPI display and the operator, who decided which paints to accept as echoes and which to ignore as clutter or noise. Nowadays, much of the work is performed by digital software, although the operator still plays a part, optimising control settings and mentally filtering the data presented on the raster display.

Detection performance can be predicted by statistical mathematical models of noise, clutter and target fluctuation, idealised to reduce the complexities of the real world to a manageably small number of alternative scenarios or cases. Several differing modelling suites are available. We describe only one – the Swerling Cases, whose predictions have been found to give reasonable accuracy when tried on practical radar/target/environment systems and have proved an essential tool for design of today's radars, with their impressively good detection performance in adverse clutter conditions. The models also give insight to the inherent limits of radar performance and the best control settings for given scenarios. But the models are just that – models – and do not always exactly match reality.

12.1.2 Echo fluctuations

Chapters 4–10 developed procedures for calculation of the mean signal strength at the radar receiver from point or extended passive target echoes, and from active device responses (all called 'echoes' in this chapter), applicable to all environmental conditions except ducting. Those chapters explained that echo strength at a given range was often likely to fluctuate, a theme we must now develop. Causes of echo fluctuation include:

- target roll, pitch and yaw, which change the RCS presented on radar aspect;
- roll, pitch and yaw of the ship mounting the radar, which change effective scanner gain;
- variation of multipath factor from these movements;
- variation of multipath factor by movement of the sea surface at the grazing point;
- variation of precipitation attenuation as hydrometeors in the radar/target path randomly fluctuate;
- noise generated by Feeder ohmic loss.

12.1.3 Noise and clutter fluctuations

Chapter 11 discussed the mean strength and fluctuations of competing unwanted returns from the ever-present receiver noise and from precipitation, sea and the (non-fluctuating) short range ringing clutter components. Fluctuations arise from:

- random receiver, feeder and path noise components;
- the random number of hydrometeors in the detection cell, which randomises the instantaneous precipitation RCS;
- the random number and height of waves in the detection cell, which randomise instantaneous sea clutter.

We shall express target and noise/clutter fluctuations primarily as amplitude changes, which may occur at various rates:

- fast, fluctuating between one pulse and the next (each pulse's return decorrelated from its neighbours);
- slow, fluctuation being insignificant from pulse to pulse but significant from scan to scan (pulse to pulse correlation but scan to scan decorrelation);
- very slow, with fluctuation observable only through the period of several scans, with scan to scan as well as pulse to pulse correlation.

12.1.4 Detection in random noise or clutter

At given range and multipath, a change in RCS causes a proportional change of echo power, enabling our discussion to interchange freely between RCS and echo power. Echoes have to be detected in presence of noise, precipitation clutter or sea clutter, perhaps all three. To detect is to answer the question 'Does the return within this detection cell contain a target?' Because of the unpredictable fluctuations, an honest observer can only reply between 'Most unlikely', through 'Undecided, need to look for longer' to 'Almost certainly'. Only when the question is qualified to 'Does the return contains a target in excess of an agreed probability of detection while the likelihood that this declaration will be wrong (a false alarm) is less than another agreed value?' can the observer declare a simple and definite 'Yes' or 'No', and then preferably after comparison of the strength of the return in question with its neighbours and with time. 'Yes' might then mean 'more than 50 per cent target likelihood and probability of wrongly declaring Yes with no target present is less than 1 observation event in 10^6'. In engineering terms, single-pulse probability of detection (P_D) > 0.5 (sometimes called 50 per cent), with probability of false alarm (P_{FA}) < 10^{-6} (sometimes called 1 in 10^6, which is really the false alarm rate).

There is a definite theoretical limit to the P_D available for given P_{FA}, signal to noise-and-clutter ratio, fluctuation type and decision timeframe. A skilled and alert observer viewing an optimally adjusted cursive raw-radar display can sometimes come respectably close to this limit. The observer's decision process can be replicated in a more formal manner by electronic circuits and the powerful digital signal processors of modern radar come very close indeed to the limit, for they can rapidly and untiringly analyse the mass of data generated by the incoming returns and rationalise the comparison process. But the limit is always present.

For commercial reasons associated with intellectual property rights, manufacturers are reticent about their detection strategy. This in no way infers they neglect clutter performance when designing radars. On the contrary, designers acknowledge detection of weak targets in clutter as perhaps their most difficult technical challenge. Living by repeat orders, they must keep up with their competitors to stay in business. Detection performance of radars in service is remarkably good, falling little short of that theoretically possible from published parameters such as scanner beamwidth and receiver bandwidth for any given radar/target/environment system. The inevitable practical shortfall is allowed for by introduction of a processing loss. This term also often includes any shortfall introduced by the operator, who may have quite properly optimised the radar controls for another target lying in a different patch of clutter at different range.

Detection, then, is to determine whether a total return event contains a valid signal, using strategies which depend on the properties of noise and other random events described by probability theory, a branch of statistics. Two sorts of error may exist in a declaration: Type I is a false alarm, Type II is missing a target. Both usually have equal importance in data telecommunications work. Because radar final declarations are made only after examination of several sets of data (after several sweeps or scans) and there are relatively far fewer targets than noise events present, relatively more Type I errors are permissible.

12.1.5 Assumptions

The relatively slow range-dependent change of returns can be straightforwardly allowed for by re-calculation of the radar range equation and other range-dependent equations developed earlier and can be ignored in the present chapter. If the system noise passes through any form of limiter which clips or reduces high-amplitude events, such as an amplifier lacking sufficient dynamic range, the distribution is broadened, raising P_{FA} or reducing effective SNR for a constant P_{FA}. Likewise, high-amplitude echoes are reduced, further reducing effective SNR.

Therefore in this chapter, unless specifically stated:

- range and multipath effects are excluded;
- 'echoes' refers to responses from active targets, as well as true echoes from passive targets;
- the sum of signal plus clutter power is called the total return;
- all amplitudes are referenced to power levels at the input of the radar receiver;
- echo pulses are assumed rectangular, and where modulated on a carrier, the modulation is assumed sinusoidal with an integral number of cycles within the pulselength, sidestepping tiresome and unrewarding questions of pulse start and finish phasing;
- we infer actual performance from consideration of what is possible in principle;
- we normally assume that automatic gain control (AGC) and/or the operator gain control are optimised for the target in question to avoid saturation or cut-off;
- we shall not consider interference from other radars, which may cause 'running rabbits' on the display;
- we initially assume that waves do not obscure the target;
- 'noise' usually means system noise as analysed in Chapter 11, Section 11.2.8, including the scanner and feeder loss contributions as well as receiver noise;
- we remember that power $\propto$ voltage^2.

12.1.6 The detection problem

Figure 12.1 illustrates the problem of detection in clutter. It shows the digitised returns from a single transmitter pulse on a single bearing, partially filling 500 range cells whose individual capacity is 16 units of power. In (a) the random clutter is modest, the cell counts varying between 1 and 9, average 2.3. This might approximate the areas of light clutter in Chapter 2, Figure 2.6. It is pretty obvious to an observer that the high counts at Q and S represent echo pulses, but opinions might differ how to declare R and T, which in fact are not genuine echoes but particularly high clutter spikes. The next few transmissions within the pulse packet would probably continue to confirm Q and S, not R or T, and might reveal the missed echo at P; but fresh spikes would perhaps cast doubt on some of the other cells.

In Figure 12.1(b) the clutter is worse (approximating the heavy clutter areas of Figure 2.6), averaging about 4.4 counts, reducing the signal to clutter ratio.

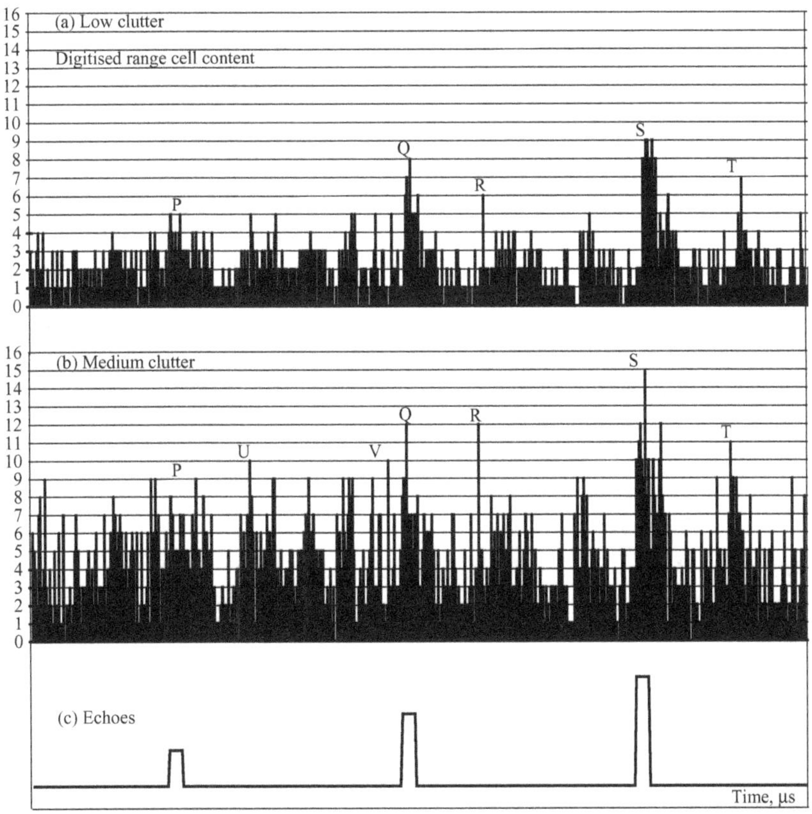

Figure 12.1 Signal in clutter: (a) shows that weak echoes can be detected with reasonably high probability in low clutter, but stronger clutter (b) reduces the probability of detection unless a higher probability of false alarms is accepted

Again, the observer can feel sure S is an echo, but may wonder about Q and R, doubt T and have grave reservations on the weak echo P. If Q is accepted, what about clutter spikes U and V? True echo T is much lower than these and some other clutter spikes. In other words the stronger clutter has reduced the operator's probability of detection of weak pulses and increased the probability of false alarms. The words in italics indicate that detection is never certain. Electronic decision making is subject to similar uncertainties – neither the cleverest person nor the best software can impose certainty when the data is so random. But although an alert operator can make almost optimal detection decisions, electronics does not get bored, drift off into thoughts of that next cup of coffee, break concentration to sign off the Garbage Log, become distracted by radio distress traffic or suffer toothache.

Target fluctuations may add to the problems posed by the inherently random nature of clutter and noise. Detection becomes yet more difficult if:

- the echo fluctuates;
- the clutter is so severe that all the cells are almost completely filled or the preceding receiver is saturated, unable to handle any stronger signal (in the example, count trying to exceed the cell capacity of 16), leaving little headroom for echoes;
- to exclude clutter, gain has been reduced until the echo barely registers, so that most of the 16 available cells are never used and are wasted.

The total return is a stream of events, most of which represent noise or clutter, perhaps including one or more echoes. Rather than burdening the digital signal processor with many returns so weak they are almost certain to be rejected, a preliminary sorting is usual, based on event amplitude. Only those events having greater amplitude than a certain threshold are accepted as single-pulse detection declaration candidate targets, the remainder being discarded. The signal processor then compares the current single-pulse declaration with the history of activity in that range cell and adjacent cells to deliver the final target declaration to the display. Singly detected pulses are sometimes called blips. The blip/scan ratio is proportional to the number of blips within a scan, so the ratio equals the single-scan probability of detection, and is unity when SNR is so high that all sweeps within the scan deliver a blip.

The process is as follows.

1. Make initial decision on each pulse as received, discarding all events which are almost certainly too low to be targets by thresholding, allowing a relatively large number of false alarms in the hope of discarding very few echoes at this stage.
2. Integrate all returns within the scan packet for the range/bearing detection cell in question, correlated echoes accumulating more rapidly than decorrelated noise spikes.
3. Possibly eliminate candidates which, though strong on one scan, are weak on the other (scan to scan correlation).
4. Possibly adapt the decision threshold to the clutter prevailing in the vicinity of the cell location (clutter mapping).
5. Declare all event groups passing this adaptive threshold as plots, to be displayed on the screen.

6. Use ARPA or ATA to form tracks from candidate plot sequences, obtained over numerous scans, which meet acceptance criteria such as maximum rate of turn and acceleration, using the method of plot association.

12.1.7 Rigour

Rigorous analysis of fluctuating echoes competing with fluctuating noise or clutter is well beyond the scope of this book, not to mention the author's understanding of statistics. Neither would such analysis yield convenient formulae accurately linking probabilities of detection and of false alarm with signal to noise-plus-clutter ratio (SNR). Beside the primary factors of fluctuation characteristic, SNR and P_{FA}, P_{D} depends to a lesser degree on secondary parameters not published by radar manufacturers, such as the frequency response curve of the radar receiver filter (strictly, its match to transmitted pulse shape). Some practical demodulators have IF-volts-in to baseband-volts-out transfer functions or characteristic curves which change from square law towards linear as signal strength rises. Square and linear approximations yield equations of markedly different form to describe probability of detection; the differences often prove to be more apparent than real, required echoes being within a few tenths of a decibel. Any refinement to P_{D} obtained by inclusion of these factors would be swamped by the practical uncertainties surrounding estimation of target RCS and of clutter intensity. And of course real targets do not always fluctuate in ways which neatly fit the mathematical models used to describe them. Comparison of different textbooks' formulae is bedevilled by differences in definition of signal to noise ratio and some other parameters.

12.1.8 Effect of receiver type

After the second demodulator of a superhet receiver, the total video return contains perhaps external noise-like clutter and certainly receiver noise whose bandwidth has been limited by bandpass filters in the receiver IF section. There may be no echo in the return. If the detection system is coherent, the true modulus and phase angle of the original microwave signal are preserved (Chapter 2, Sections 2.2.4 and 2.2.5). The unidirectional baseband pulse and noise then look exactly as though the incoming return had merely been low-pass filtered and linearly rectified immediately on receipt, without any form of distortion (but amplified), in particular the amplitude probability distributions remain unaltered.

The video of an ordinary non-coherent receiver also of course contains a demodulated unidirectional echo pulse and noise. In this case the process of mixing down to IF and subsequent demodulation, called envelope detection, discards phase information and changes the probability distribution.

12.1.9 Chapter layout

We first consider how a single unidirectional pulse superimposed on a noise background is detected, including some additions to the statistics theory of Chapter 11,

Section 11.3, with formulae for probabilities of false alarms and single-pulse probability of detection; this section is applicable to coherent reception. Non-coherent reception is considered in Section 12.3, catering for the mixing and demodulation processes within most current marine radars where the local oscillator is not phase-locked with the transmitter magnetron. We then account for the fluctuation characteristics of the different cases of target, followed by the improvement obtained by integrating the pulses within a received pulse packet. The chapter ends with several secondary aspects of detection. Appendix A2 contains further remarks on the statistics.

12.2 Direct detection of single pulse in noise

This section assumes unmodulated unidirectional signal pulses and Gaussian noise at the point of detection, applicable to detection systems where either (a) the band-limited total return is translated down to baseband using a mixer, provided coherence with the signal is maintained, or (b) is directly rectified at microwave frequency. These systems are characterised by an unchanged amplitude distribution of equivalent instantaneous noise power at the antenna and at the point of decision. The section also provides an introduction to Section 12.3, dealing with conventional non-coherent radars.

12.2.1 Detection threshold, unmodulated noise

Suppose the event stream at the point of decision contains only Gaussian noise (discussed in Chapter 11, Section 11.3.8) of long-term voltage V_n rms, instantaneous voltage RV_n, fluctuating about a mean of 0 V d.c. According to whether or not the instantaneous noise voltage exceeds a d.c. threshold preset to KV_n volts, a noise event may be either wrongly regarded as significant – a false alarm – or correctly discounted as a random spike. The basic circuit is inset in Figure 12.2; in practice

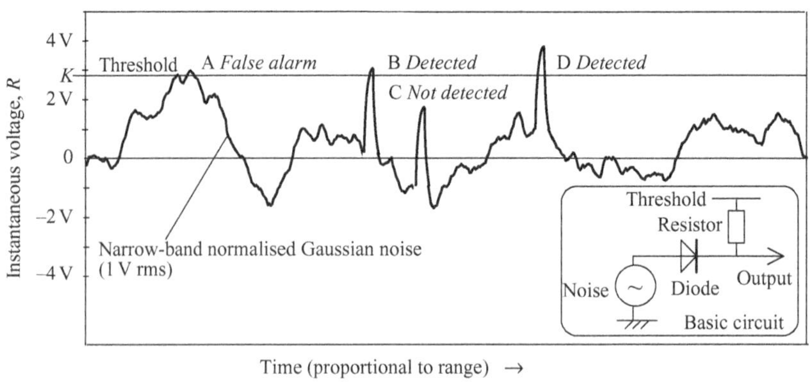

Figure 12.2 Noise and threshold. Instantaneous noise amplitude affects detectability of equal superimposed signals B–D

a fast comparator integrated circuit would be preferred. As in Chapter 11, the discussion is simplified by normalising: setting V_n to unity (1 V rms) and assuming circuit impedance notionally $1+j0\,\Omega$ (1 ohm resistive), the threshold becomes K volts, 2.8 V in the figure, which depicts a narrow-band noise specimen similar to Figure 3.4(c). The threshold is crossed when $R > K$. Such an event, point A, must be a false alarm, for we have said no signal is present; K is called the noise margin, which is really a voltage ratio, referenced to $V_n = 1$ V rms. It may also be implicitly referenced to mean noise power and expressed as $20 \log K$ dB, here ~9 dB. It is usually easier in practice to reduce the amplifier gain instead of increasing the threshold voltage to adjust noise margin. The effects are identical and are to be understood to apply throughout this chapter.

The probability that R exceeds K is the probability of false alarm, also the residual probability, RP (RP = 1−CP as in Chapter 11, Section 11.3.8). Extending Eq. (11.5e):

$$P_{FA} = P(R > K) = \mathrm{RP} = \frac{1}{2}\left[1 - \mathrm{erf}\left(\frac{K}{\sqrt{2}}\right)\right]. \tag{12.1}$$

Practical false alarm rates are so small, of the order of 10^{-4}–10^{-12}, that it is convenient to describe them by F, the exponent of P_{FA}; for example, when $P_{FA} = 10^{-4}$, $F = -4$:

$$P_{FA} = 10^F, \quad \text{so } F = \log P_{FA}. \tag{12.2a}$$

The positive quantity $-F$ is sometimes called the false alarm number. The false alarm rate (FAR, number of false alarms per second) is a function of the noise bandwidth, B_n, and the noise margin, as shown in Figure 12.3:

$$\mathrm{FAR} = B_n \exp\left(-\tfrac{1}{2}K^2\right) \text{ alarms/s}. \tag{12.2b}$$

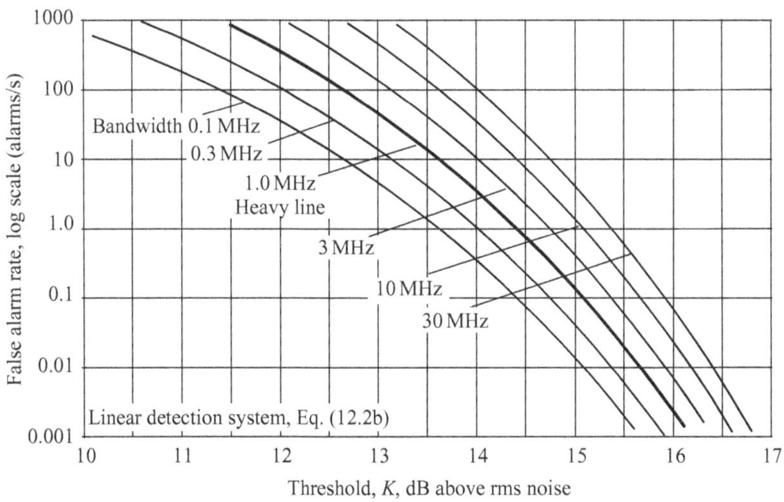

Figure 12.3 Variation of false alarm rate with detection threshold

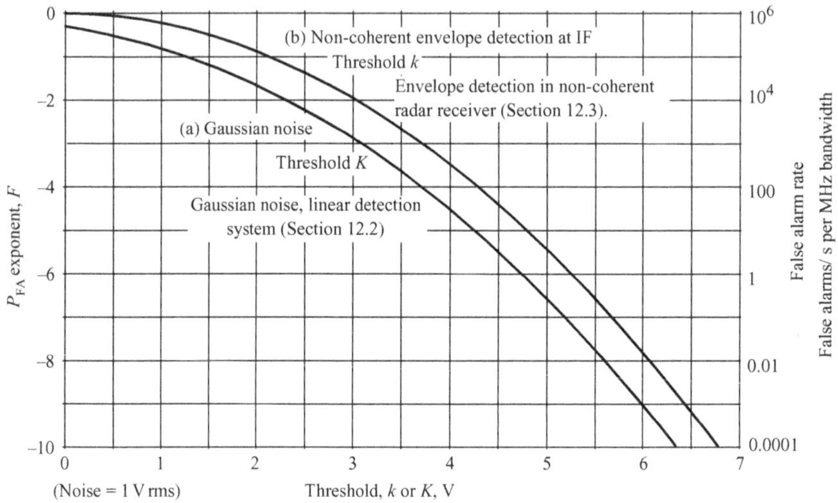

Figure 12.4 Variation of P_{FA} exponent and false alarm rate with threshold voltage

Alternatively,

$$\text{FAR} = \exp\left(-\tfrac{1}{2}K^2\right) \text{ alarms/s/Hz bandwidth.} \tag{12.2c}$$

The tolerable FAR is a function of the subsequent machine processing, or of the display tube phosphor and the operator if a simple cursive system is used; its dependence on K is shown in Figure 12.4(a). Having decided that, for example, 1 false alarm per second is permissible and system bandwidth is 1 MHz, Eq. (11.3) indicates that $P_{FA} = 1 \times 10^{-6}$, so $F = -6$.

Figure 12.4(a) shows how sharply F varies with K. For $F = -6$, K must be set to 4.75 (4.75 V d.c. for 1 V rms noise, noise margin $= 20 \log 4.75 = 13.5$ dB). In Figure 11.2, $K = 1.27$ (noise margin 2.1 dB) gave $F = -1$ so increasing K by a factor of less than 4 (noise margin change 11.5 dB) reduces P_{FA} by as much as 5 orders of magnitude. Increasing noise margin by another 1 dB from 13.5 dB changes F from -6 to -7.32, reducing P_{FA} by well over another order of magnitude.

12.2.2 Detection of sinusoidal signal

We shall assume the mean noise power remains constant and the signal strength is varied. The bandwidth-limited IF noise envelope is represented by the sum of a pair of mutually independent sine and cosine quadrature carriers whose instantaneous amplitudes are Gaussian about zero mean. An echo event into a coherent demodulator has the form of a burst of sinusoidal IF frequency superimposed on continuously-present noise. Figure 12.5, curves (a)–(c), respectively, show the IF noise and signal components and their combination. The noise is normalised Gaussian. The signal at (b) has voltage $A = 1$ V peak. The instantaneous voltage of the total return is R, shown at (c). The sinewave rms voltage is $A/\sqrt{2}$ and the power (into the notional 1 ohm load) is $A^2/2$, which we call q times the rms noise power (which is unity).

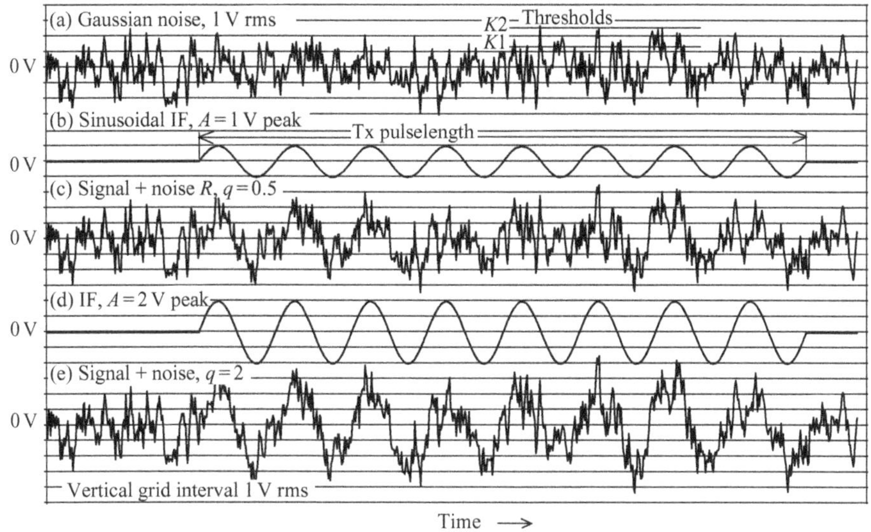

Figure 12.5 IF noisy echo pulse. In practice the noise bandwidth would be narrower

Factor q is important and sometimes called the visibility factor or single-pulse SNR. That is

$$\text{SNR} = q = \frac{A^2}{2} \text{ numerically,} \tag{12.3}$$

so $A = \sqrt{2q}$.

In general, RP is the probability that the threshold is crossed. As a signal is present, crossing now is the probability that the signal is detected, obtained by substituting $K - A = K - \sqrt{2q}$ for R in Eq. (11.4e):

$$\text{RP} = P_\text{D} = P(R) = \frac{1}{\sqrt{2\pi}} \exp\left[-\frac{1}{2}\left(K - \sqrt{2q}\right)^2\right]. \tag{12.4a}$$

This expression when graphed has the same form as the probability curve of Figure 11.1(b), unchanged in shape but shifted bodily to the right by $A = \sqrt{2q}$, equivalent to shifting the threshold to the left by A volts. Figure 12.6, curves (a) and (b) are as Figure 11.1(a) and (b), showing PDF and RP for noise alone. Performance is indicated in the table. Curves (c) and (d) are for a weak signal corresponding to Figure 12.5 curves (a)–(c), while curves (e) and (f) are for a stronger signal, as Figure 12.5 curves (a), (d) and (e). Threshold $K1$, giving P_FA 0.1 (intercept $Z1$) intersects weak signal RP at W1. The strong signal intersect is at S1. Doubling threshold voltage sharply reduces P_FA, but drastically cuts P_Ds. As so often, no gain without pain.

12.2.3 Variation of P_D with SNR

Eqs (12.3) and (12.4a) can be used to generate curves of P_D versus signal strength for a family of F values using the method of Chapter 11, Section 11.3.8, Eq. (11.5d),

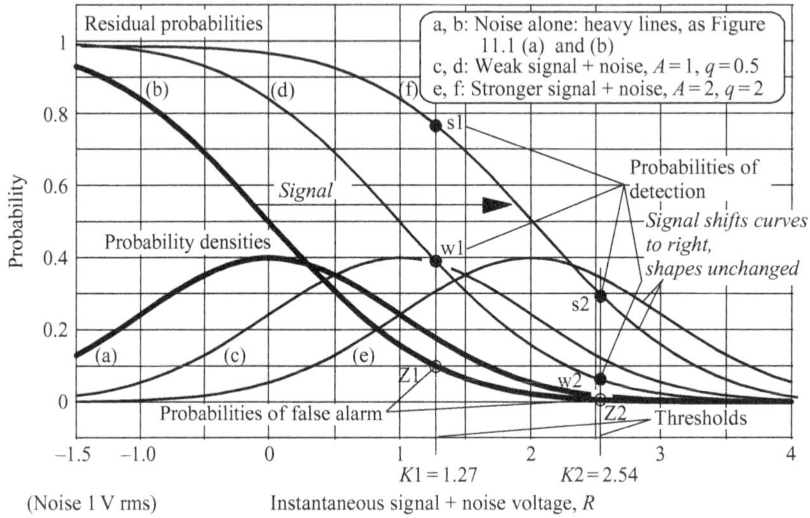

Figure 12.6 *Probabilities, signal in noise. Gaussian distribution. Probability density and residual probability curves for noise (as Figure 11.1), noise + weak signal ($q = 0.5$) and noise plus stronger signal ($q = 2$). Signal shifts PDF and RP to the right without alteration of shape. Low ($K1$) and high ($K2$) thresholds of detection. Performance as follows*

Threshold (Noise 1 V rms)	*PFA*	*PD*	
		Weak pulse, $q = 0.5$ (−3 dB)	*Stronger pulse, $q = 2$ (+3 dB)*
$K1 = 1.27\ V$	$Z1 = 0.10$	$W1 = 0.39$	$S1 = 0.77$
$K2 = 2.54\ V$	$Z2 = 0.0054$	$W2 = 0.061$	$S2 = 0.29$

giving Figure 12.7. This has been drawn to a primary baseline of signal voltage A, but the noise margin, q dB, is also indicated. P_D is almost zero when SNR $= 1$ and varies only slowly when P_D is near zero or unity, reflecting the wide tails of the probability density function (PDF). When P_D lies between about 0.2 and 0.8, P_D varies almost linearly with signal voltage change at a rate of 0.8 P_D units/V peak (0.56 units/V rms). When $A = K$, $P_D = 0.5$. Here P_D would jump abruptly from 0 to 1 if there were no noise. This approximates conditions when swept gain reduces sensitivity at short range.

Figure 12.8 presents the curves in a widely used quasi-logarithmic form which better reveals the extreme values of P_D. The ordinate is scaled linearly in terms of D:

$$D = \log\left[\frac{P_D}{1 - P_D}\right] = \log P_D - \log(1 - P_D). \tag{12.4b}$$

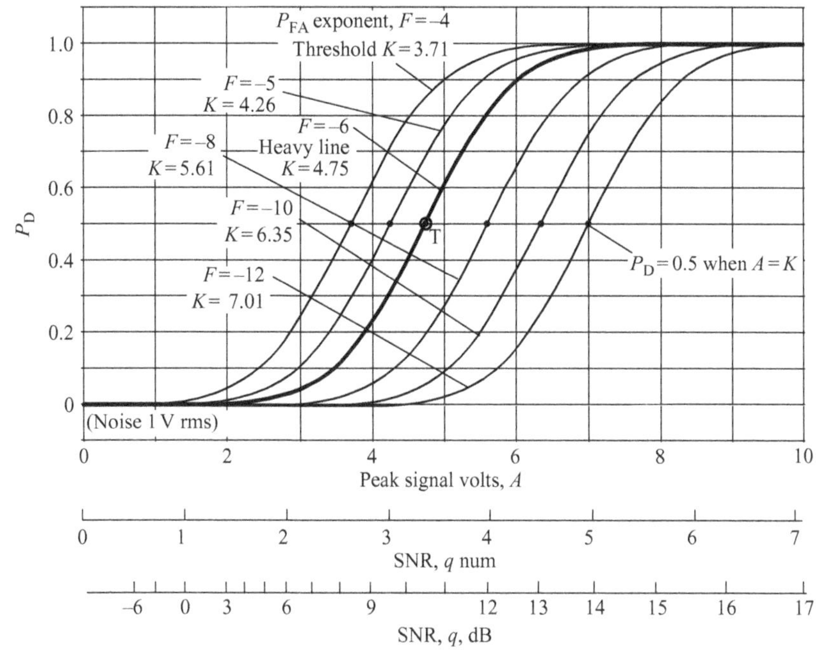

Figure 12.7 Variation of P_D with signal strength and P_{FA} exponent, linear scaling. Non-fluctuating sinusoidal signal in Gaussian noise. For point R see Section 12.9.2

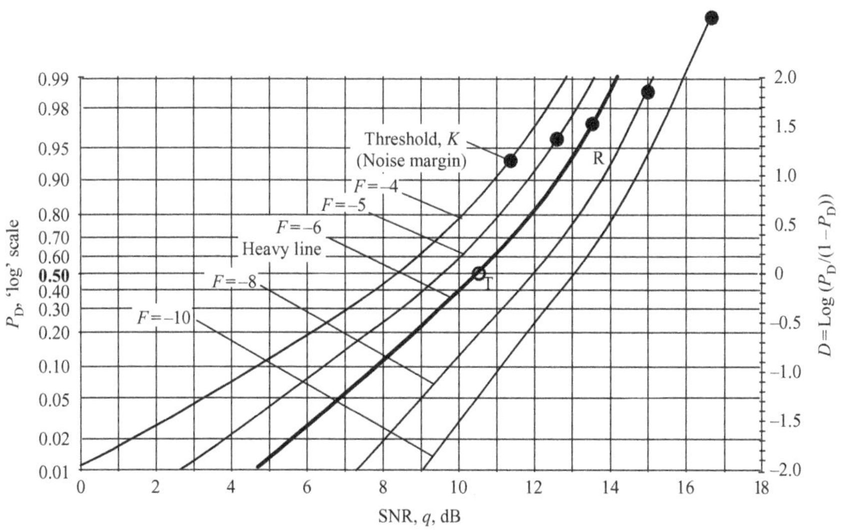

Figure 12.8 Figure 12.7 redrawn to logarithmic scales

The abscissa is q dB. The thresholds (expressed in threshold to noise ratio, $20 \log K$ dB) for the various P_{FA} have been spotted in. They lie at quite high P_Ds, because when $q = K^2$, $A = \sqrt{2K}$ and most signal peaks cross the threshold. $P_D = 0.5$ at $P_{FA} = 10^{-6}$ are typical values needed for effective detection, and require $A = 4.7$ V, corresponding to voltage strength of $3.33 \times$ noise rms voltage and SNR = 10.4 dB, point T on the figures.

Returning to Figure 12.2, equal-sized echo pulses, amplitude K, are shown at B, C and D. Echo B happens to coincide with an instant of zero noise and is just detected. Echo C, at an unfavourable noise instant, is undetectable, while D, at a favourable instant, is readily detected. Several important points can be deduced from the figure and will be confirmed analytically later in the chapter.

- Low P_{FA}, needing high threshold, requires high signal for detection.
- In noise, low P_D can be achieved with weak signals (event D).
- In noise, high P_D requires strong signals (event C would have to be increased to ensure B, C and D were all detected).

12.3 Envelope detection of echo pulse in noise

12.3.1 Detection in non-coherent receiver

Ordinary non-coherent receivers (Chapter 2, Section 2.2.1, Figure 2.10(*a*)) discard the echo phase information component. The receiver input is a microwave echo superimposed on wideband Gaussian noise. The much-simplified block diagram of Figure 12.9 shows how the receiver mixes the noisy signal down to intermediate frequency, where it is amplified as a pulse of sine waves (at say 50 MHz). A filter restricts bandwidth to minimise noise as far as consistent with acceptance of the major frequency components of the signal pulses. Point S is the filter output and is followed by the demodulator, whose output is point T. As the voltage envelope of the IF signal is preserved, this form of demodulation is called envelope detection. A wide-band video amplifier brings thhe output base-band signal to a convenient level for digitisation and subsequent processing, without affecting the mathematics. A voltage comparator (point U) then makes the initial detection decision whenever the instantaneous baseband voltage exceeds the d.c. threshold, again using the basic circuit of the inset in Figure 12.2. Some of these processes upset the conditions we assumed in Section 12.2 and affect the interplay of P_{FA}, P_D and SNR. As before, the signal processor then compares the present and previous single-pulse declarations with the history of activity in that and adjacent cells to deliver the final target declaration to the display. The function of an envelope detector is therefore to extract the modulation amplitude while rejecting the carrier, using a demodulating rectifier and a low-pass filter.

Modulation by the LO causes the envelope of the bandwidth-limited noise signal at S to be converted from Gaussian to Rayleigh amplitude distribution, see Section 12.3.3. The IF carrier supporting this fluctuating envelope has a fluctuating phase, any phase being equally probable, giving uniform phase distribution.

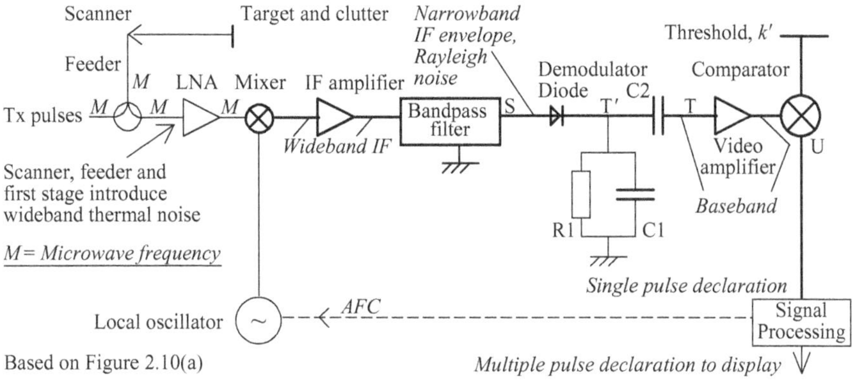

Figure 12.9 Detection block diagram, superheterodyne receiver

The diode and shunt resistor–capacitor circuit (R1, C1) rectify the IF signal and filter out the IF component to yield at T′ a unidirectional low-frequency video signal at baseband. It is common practice in marine radar receivers to insert a blocking capacitor C2 after T′, so that the waveform at T has zero mean value as shown at (*d*) in Figure 12.10. We use lower case k for threshold voltage to distinguish from the coherent detection system of Section 12.2. Provided the demodulator diode is 'perfect' (resistance zero when anode voltage is positive to cathode, infinite otherwise), the distribution at its output remains Rayleigh except for a shift in baseline voltage.

Figure 12.10 shows a time or range bracket shortly after transmission of a radar pulse. Non-fluctuating echoes $e1$–$e9$ from different targets are depicted diagrammatically at (a), $e1$–$e7$ being equally weak with $e8$ and $e9$ stronger. Of course, no noise-free waveform of this sort exists; detection would be a doddle if it did! Rayleigh-distributed noise (b) combines with the echoes to give a total return as at point S of Figure 12.9. This is shown at (c) after amplification and filtering, with the carrier oscillation at ~50 MHz shaded. The envelope outline is rounded by the narrow filter, as expected from Figure 3.4. The demodulator output is the baseband (video) waveform (d).

The radar, or the operator viewing a raw-radar display, decides a high event to be a valid candidate echo if it exceeds the threshold voltage. (For raw radar cursive displays, Figure 12.10(d) can be interpreted as brightness, the threshold being the lowest brightness discernible by the operator.) Alternative thresholds $k'1$ and $k'2$ are shown applied to the video of Figure 12.10(d) to give alternative comparator outputs (Figure 12.9 point U). In practice, the threshold (or IF gain) would be preset to suit the prevailing noise or clutter. For high threshold $k'2$, Figure 12.10(e) shows only three noise returns are wrongly accepted to form false alarms (f), so P_{FA} is low. Even so, echoes $e1$ and $e5$ are missed (m). On this very limited sample of seven weak echoes, $P_D = \frac{5}{7} \sim 0.7$.

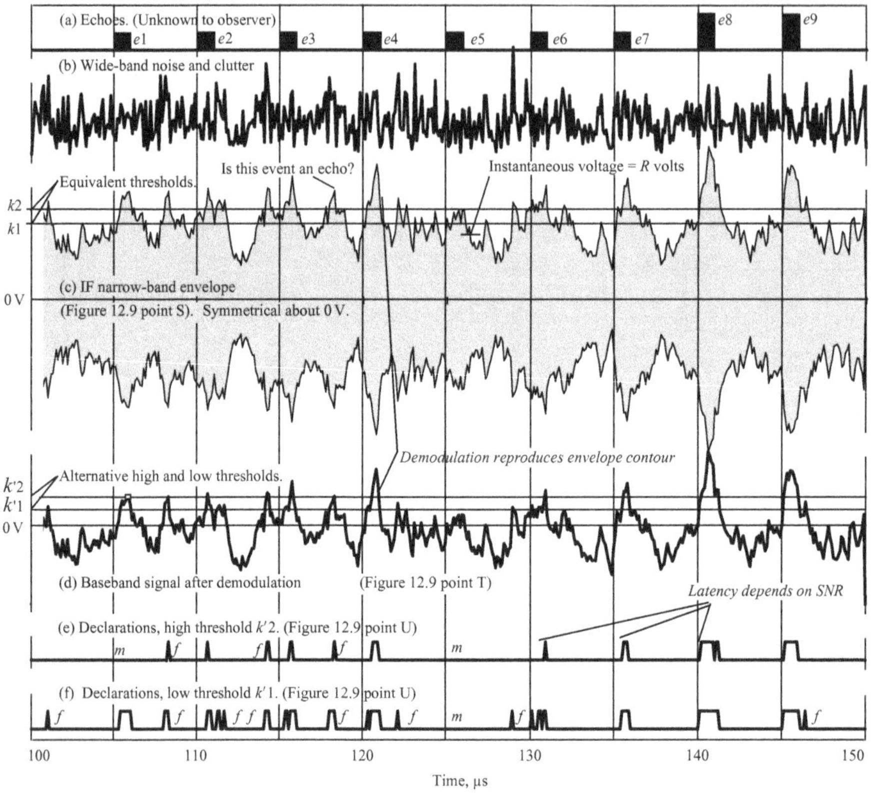

Figure 12.10 *Detection of echoes in noise. Time domain. (d) is the baseband output from the demodulator of a non-coherent receiver. When the comparator threshold is set high ($k'2$), some echoes are missed (m) but false alarms (f) are few. Reducing threshold to $k'1$ increases P_D at the expense of poor P_{FA}. Substituting equivalent envelope thresholds $k2$, $k1$ simplifies analysis*

In an attempt to improve P_D, the threshold might be lowered to $k'1$. Sure enough, curve (f) shows echo $e1$ is now detected, raising P_D. But numerous false alarms creep in where there is a noise peak but no echo, raising P_{FA}. Raw radar would display more noise speckles than echoes. The operator or the machine should compromise by setting the threshold somewhere between $k'1$ and $k'2$, to let through as many false alarms as can be tolerated. The threshold therefore forms a noise floor, the lowest input signal power level that will produce a detectable output signal in the presence of system noise and absence of clutter.

As an aside, comparison of (a) and (d) or (e) shows a small but fluctuating time delay or latency between arrival of echoes at the radar and their declaration, caused by the limited system bandwidth. Latency introduces range uncertainty which can cause errors in the historic track of the target and in track prediction, see Chapter 13.

Track-formers therefore require wide bandwidth equating to good SNR (high P_D) to produce reliable predictions, see Chapter 13, Section 13.5.

12.3.2 Equivalent envelope detector

Inspection of Figure 12.10 suggests identical detection performance would be achieved by declaring detection when the IF envelope exceeds high or low limits $k2$ or $k1$, shown in Figure 12.10(c). This concept simplifies the mathematics and gives accurate results provided the system satisfies certain conditions, more or less met by practical radars.

- The initial noise is Gaussian.
- The demodulator and video amplifier reject the carrier frequency but preserve the IF modulation envelope.
- Video bandwidth well exceeds half IF bandwidth, so the IF filter defines overall system bandwidth.
- The circuits have linear transfer characteristics (in practice the demodulator characteristics may permissibly introduce minor non-linearity).

12.3.3 Noise distribution

When Gaussian white noise is modulated onto a carrier and passed through an IF filter, the noise output voltage envelope (which cannot be negative and whose instantaneous amplitude is R, Figure 12.10(c)) takes on Rayleigh probability density function. See also Appendix A2, Section A2.2. The negative components are 'turned over', becoming positive. In general, when the rms value of variable ν is σ, so that σ^2 is the variance of ν, the Rayleigh PDF is:

$$p(\nu) = \frac{2\nu}{\sigma^2} \exp\left(-\frac{\nu^2}{\sigma^2}\right) \qquad [\nu \geqslant 0]. \tag{12.5a}$$

When ν is a voltage, σ is the power in the notional 1 ohm load. Appendix A2, Section A2.2, contains further comments on Rayleigh distribution.

In the present instance, if R were a steady peak value of the sinusoidal signal $= \sqrt{2}$ V, the rms value, σ, would be 1 V, so

$$p(R) = \frac{R}{\sigma^2} \exp\left[-\frac{1}{2}\frac{R^2}{\sigma^2}\right]. \tag{12.5b}$$

The voltage distribution of normalised ($\sigma = 1$ V rms) Rayleigh-distributed noise is

$$p(R) = R \exp\left[-\tfrac{1}{2}R^2\right]. \tag{12.5c}$$

Eq. (12.5c) is shown in Figure 12.11(a). The whole area under the PDF curve between $0 < R < \infty$ has unit area as usual. The area under the curve to the right of threshold k represents the residual probability (RP $= 1 -$ CP) that the normalised random variable exceeds k. Residual probability is shown at Figure 12.11(b).

Any event crossing the threshold is assumed to be an echo. At the moment we are assuming no signal, so crossings must be false alarms. Setting k high reduces

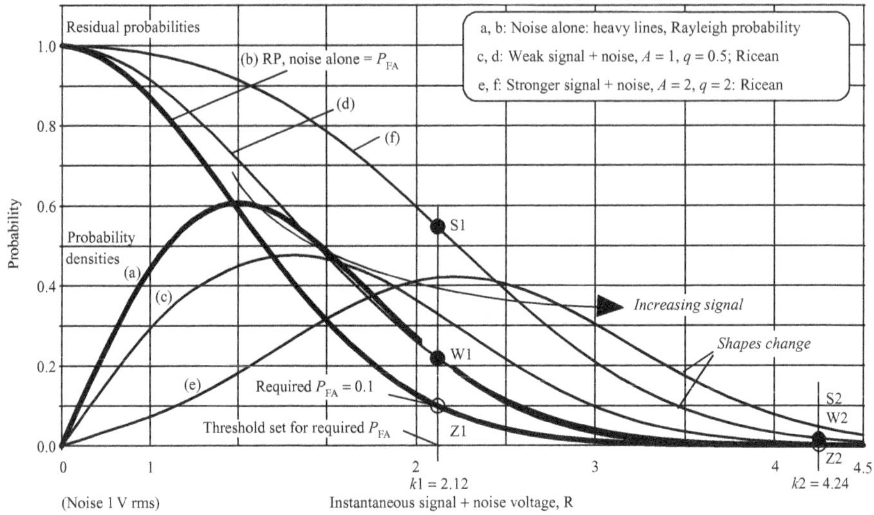

Figure 12.11 *Ricean probability functions. Noise alone, and signal in noise. At IF frequency in non-coherent receiver. Values of k1 and k2 differ from those used in Figure 12.10. Performance as follows (compare with Figure 12.6)*

Threshold (Noise 1 V rms)	*PFA*	*PD*	
		Weak pulse, $q = 0.5$ *(−3 dB)*	*Stronger pulse,* $q = 2$ *(+3 dB)*
$K1 = 2.15\,V$	Z1 = 0.10	W1 = 0.22	S1 = 0.55
$K2 = 4.30\,V$	Z2 = 0.0032	W2 = 0.0013	S2 = 0.018

false alarm rate but prevents detection of weak echoes. As before, the threshold is therefore set as low as possible, to pass the maximum tolerable number of false alarms. P_{FA} is the probability that the noise exceeds k:

$$P_{\mathrm{FA}} = P(R > k) = \mathrm{RP} = (1 - \mathrm{CP}) = \int_k^{\infty} R \exp\left(-\frac{1}{2}R^2\right) \mathrm{d}R$$

$$= \exp\left(-\frac{1}{2}k^2\right). \tag{12.6a}$$

Therefore

$$F = \log P_{\mathrm{FA}} = \log\left[\exp\left(-\tfrac{1}{2}k^2\right)\right] = -0.217k^2, \tag{12.6b}$$

so

$$k = 2.146\sqrt{-F}. \tag{12.6c}$$

Say desired P_{FA} is 10^{-6}. $F = -6$ so k is set at $2.146\sqrt{6} = 5.256 \times$ rms noise voltage. Noise margin is $20 \log 5.256 = 14.4$ dB. Figure 12.4(b) plots F to a base of k. The curve closely parallels curve (a) for coherent detection, despite the difference in form between Eqs (12.6c) and (12.1). Again F falls sharply as threshold is raised. Raising noise margin by 1 dB so that $k = 5.898$ shifts F to -7.55 and reduces P_{FA} by 1.5 orders of magnitude. For a given F, the threshold has to be raised by about half a volt, representing a loss of performance relative to coherent operation.

The false alarm rate is a function of threshold and noise bandwidth, follows Figure 12.3 and is similar in form to Eq. (12.2b):

$$\text{False alarm rate} = B_n \exp\left(-\tfrac{1}{2}k^2\right) \text{ alarms/s.} \tag{12.6d}$$

12.3.4 *Noisy signal distribution*

Addition of a signal of amplitude A volts to the normalised noise is accommodated by addition of a term to the Eq. (12.5) Rayleigh PDF, converting the distribution to one introduced by Rice [2] and discussed in Appendix A2, Section A2.3. Ricean distribution is a more generalised development of Rayleigh. PDF of the modulated noisy signal voltage is

$$p(R) = RI_0(RA) \exp\left[-\tfrac{1}{2}(R^2 + A^2)\right]. \tag{12.7}$$

$I_0\,(RA)$ is the modified Bessel function of zero order (already discussed in the context of sea forward reflection in Chapter 5, Section 5.8.4, with approximation Eq. (5.41e) and Figure 5.21). When $A = 0$, $I_0\,(RA) = 1$ and the expression reverts to Rayleigh, Eq. (12.5c). Figure 12.11, curves (c) and (e) show the PDF curve departs from the Rayleigh curve (a) and changes shape as signal strength increases. The left skirt remains tied to the origin while the peak slumps downward to the right, the area under the curve of course remaining unity. CP and RP change shape in sympathy with the PDF. The RP curves (d) and (f) are reversed-S shaped from 1 at the origin.

P_D is the probability that R will exceed k and is the area under the PDF to the right of k. The threshold for $P_{FA} = 0.1$ is 2.12, higher than for unmodulated noise ($k = 1.27$). The weak signal now has $P_D = 0.22$ but the strong signal P_D is 0.55, which is acceptable for many conditions. Doubling threshold to 4.24 reduces P_D more than for Gaussian probability: weak signal P_D becomes very low at 0.0013, strong signal $P_D = 0.017$, far too small for reliable detection. For the same P_{FA}, Ricean P_Ds are lower than Gaussian because the phase component of information has been discarded.

P_D is found by substituting from Eq. (12.7) and then Eq. (5.41e):

$$\begin{aligned} P_D &= \int_k^\infty P(R)\,\mathrm{d}R = \int_k^\infty RI_0(RA) \exp\left[-\frac{1}{2}(R^2 + A^2)\right] \mathrm{d}R \\ &= \int_k^\infty \frac{R(1 + 1/(8RA))}{\sqrt{2\pi RA}} \exp(RA) \exp\left[-\frac{1}{2}(R^2 + A^2)\right] \mathrm{d}R. \end{aligned} \tag{12.8}$$

This expression is difficult to evaluate. Approximate solutions are included in the next section and Appendix A2, Section A2.4.

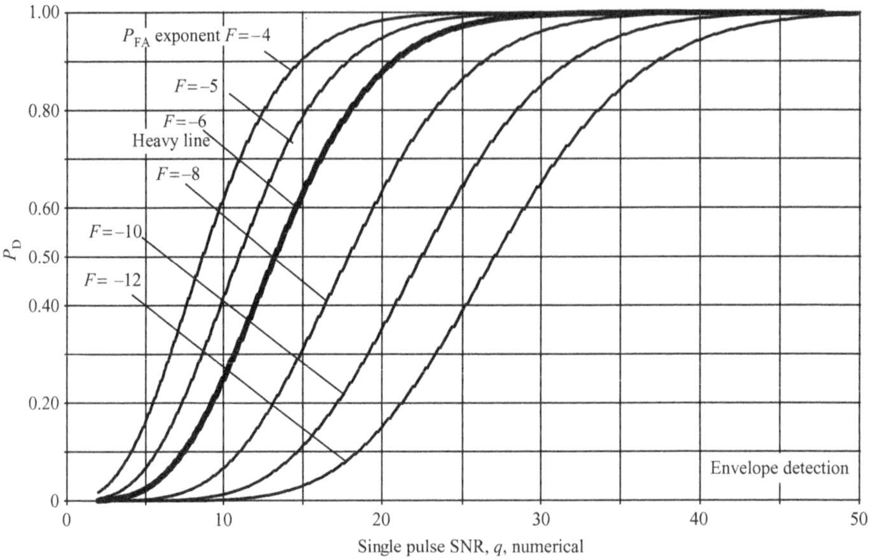

Figure 12.12 Single pulse probability of detection, non-fluctuating signal. Linear scales, differing from Figure 12.7

Figure 12.12 graphs single-scan P_D to a base of q for representative values of F. The curves are rescaled logarithmically in Figure 12.13. Comparing with Figures 12.7 and 12.8, the families of curves are similar despite the apparent differences in the underlying equations, Bessel function and all. Non-coherence introduces some loss. Within the important region where F lies between -4 and -8 with P_D between 0.1 and 0.9, this necessitates an increase in q of about 0.8 dB, which of course is included within Eq. (12.8). The thresholds again lie at high P_Ds, in excess of 0.95. When rms signal = threshold, peak signal voltage is $\sqrt{2}$ or 3 dB above threshold and only occasionally does an unfavourable noise spike in phase opposition depress the resultant enough to prevent detection.

It was noted in Chapter 11, Section 11.3.9, that atmospheric and feeder noise make the echo noisier. The effect is small and may be treated as an increase in the receiver noise background rather than as an echo component.

12.3.5 Approximations for P_D calculation

Eq. (12.8) is unfriendly, but help is at hand. Levanon [3] offers a remarkably simple approximation:

$$q = a + 0.12ab + 1.7b \text{ numerical}$$

where:

$$a = \ln\left(\frac{0.62}{P_{FA}}\right) \quad \text{and} \quad b = \ln\left[\frac{P_D}{1 - P_D}\right]. \tag{12.9}$$

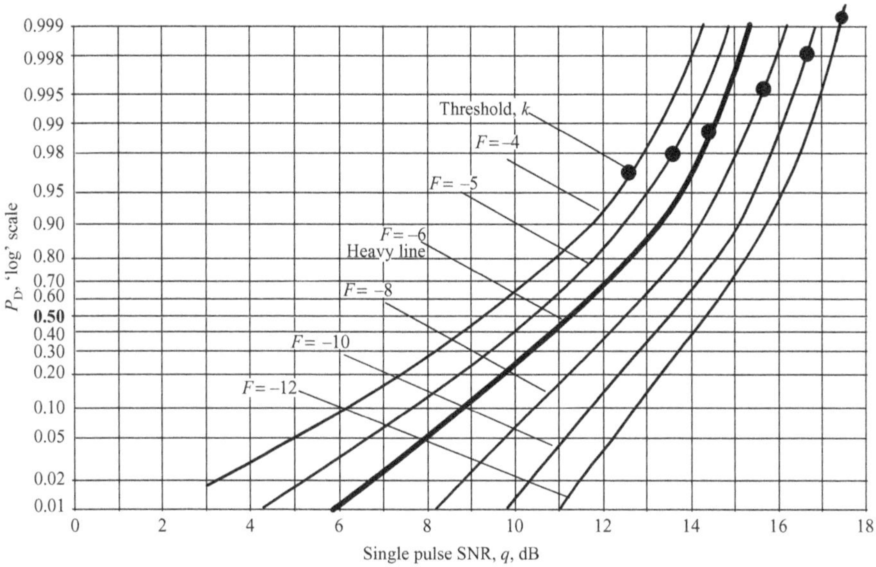

Figure 12.13 *Figure 12.12 redrawn to logarithmic scales, which better depict extreme values of P_D. Threshold indicated for each P_{FA}*

Converting to common logarithms and rearranging gives forms directly connecting single pulse SNR (q, numerical), P_{FA} exponent F, and P_D in the form $D = \log[P_D/(1 - P_D)]$, as Eq. (12.4b):

$$q = (3.783 - 0.636F)D - 0.4780 - 2.303F \text{ numerical.} \tag{12.10a}$$

To be meaningful, q must be positive, setting an F-dependent bound to lowest usable P_D:

$$P_D > 0.07\ (F = -3), > 0.04\ (F = -4), > 0.019\ (F = -6),$$
$$> 0.010\ (F = -8), > 0.006\ (F = -10), > 0.005\ (F = -12).$$

Rearranging,

$$D = \frac{q + 0.4780 + 2.303F}{3.783 - 0.636F} \tag{12.10b}$$

and

$$P_D = \frac{10^D}{1 + 10^D}. \tag{12.10c}$$

Figure 12.14 compares Levanon's approximation with P_Ds calculated per Eq. (12.8) for $F = -3, -6$ and -12. Accuracy is good to at least $P_D = 0.99$, with no unpleasant surprises at intermediate Fs. Unfortunately, P_D is overstated at low q; error exceeds 1 dB at P_Ds of 0.46, 0.19 and 0.14 respectively.

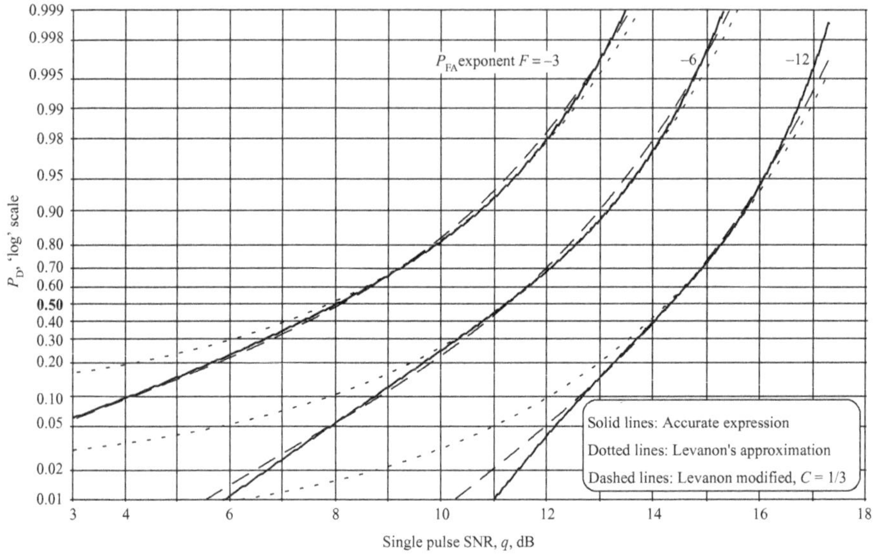

Figure 12.14 Approximations for P_D of single pulse. Non-fluctuating target. Levanon's and modified Levanon's approximations compared with Figure 12.13

The understatement is almost eliminated by addition of a correction term to Eq. (12.10b), shown bold in the following expression:

$$D = \frac{(q + 0.4780 + 2.303F) \times (\mathbf{1} - \boldsymbol{CF/q})}{3.783 - 0.636F}. \tag{12.11a}$$

An alternative correction term $[1 + F^2/(3q^2)]$ would give even less error but determination of q for given P_D and F would embroil us in solution of a cubic equation. Choosing constant C relatively low gives only partial correction at low P_D but minimises error at high P_D. Taking $C = \frac{1}{3}$ seems to offer a good compromise and is compared with Levanon and the accurate expression of Eq. (12.8) in Figure 12.14, error being shown in Figure 12.15. In these figures, q is shown in decibel form.

Given F and P_D, q can be found by extracting the positive root of a quadratic equation

$$q^2 + sq + t = 0,$$

so

$$q = \tfrac{1}{2}\left[-s + \sqrt{s^2 - 4t}\right] \text{ numerical,}$$

where

$$\begin{aligned} s &= D(0.636F - 3.783) + F(C + 2.303) + 0.478, \\ t &= CF(2.303F + 0.478). \end{aligned} \tag{12.11b}$$

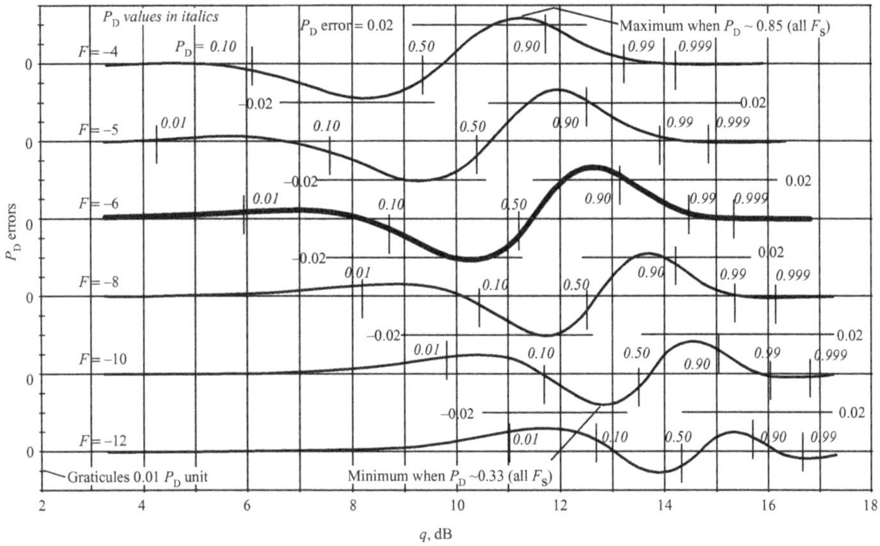

Figure 12.15 *Accuracy of modified Levanon approximation. The approximation (Eq. (12.11a),* $C = \frac{1}{3}$*) differs from the accurate expression of Eq. (12.8) by less than* ± 0.025*, errors being worst when* $P_D \sim 0.33$ *or 0.85*

A real solution requires $s^2 > 4t$, for which P_D must exceed the following values, which although higher than for the Levanon approximation, are below P_Ds commonly used for main-beam detection but not for sidelobe detection, restricting the usefulness of the modified Levanon approximation:

$$P_D > 0.24\ (F = -4); \quad > 0.18\ (F = -6); \quad > 0.15\ (F = -8);$$
$$> 0.13\ (F = -10); \quad > 0.12\ (F = -12).$$

When q and P_D (hence C) are given, P_{FA} exponent, F, is found in a similar manner by taking the positive root:

$$F = \frac{-s' + \sqrt{s'^2 - 4r't'}}{2r'}$$

where

$$\begin{aligned} r' &= 2.303\frac{C}{q} \\ s' &= \left(\frac{0.478}{q} + 1\right)C + 0.636D + 2.303 \\ t' &= 0.478 + q - 3.783D. \end{aligned} \tag{12.11c}$$

12.3.6 Accuracy

Figure 12.15 indicates the error in determination of P_D from the modified Levanon approximation Eq. (12.11a) rather than from Eq. (12.8). P_D numerical error has been plotted to a base of q for several representative P_{FA} exponents, F. The error in P_D is always less than ± 0.025. When P_D per Eq. (12.11a) $\sim$0.33 the error is maximum negative and actual $P_D \sim 0.35$. When $P_D \sim 0.85$, error is maximum positive and actual $P_D \sim 0.83$. Error is negligible when P_D is very low, very high or near 0.55.

Other sources of error include the following.

- Noise bandwidth may differ from signal bandwidth.
- The radar does not actually perform detection at IF but at video.
- Second-order effects in the radar are not reflected in the mathematical modelling.
- Demodulator transfer characteristic. Square-law demodulation is rather better in weak signals but linear is better when the signal is strong. Luckily, the nominally linear demodulators commonly used become somewhat square-law to weak signals, so our assumption of envelope demodulation of the IF waveform usually introduces less than 0.2 dB error.

12.4 Single pulse detection in clutter

12.4.1 Noise and precipitation clutter

In non-coherent systems, calculation of detection in precipitation clutter is exactly as for detection in noise because they share the same distribution, initially Gaussian. Clutter power (W, not dBW) is merely added to the noise power in the equations. Signal to (noise plus clutter) power ratio can directly replace numerical signal to noise ratio in the equations for detection in noise, explaining why signal to noise-plus-clutter ratio is so often abbreviated to SNR. To retain the desired P_{FA} in clutter, threshold k has to be raised to prevent declaration of clutter returns as targets. This reduces P_D and explains why clutter inevitably impairs target detectability, even when a clever signal processing system keeps the display screen clear. The PDF is as Figure 12.11 and Eq. (12.5b) of Section 12.3.3, when the receiver is non-coherent, otherwise Figure 12.6 and Eq. (12.4a) of Section 12.3.3 apply.

12.4.2 Clutter with Weibull distribution

Chapter 11, Section 11.7.4, showed that sea clutter may approximate Weibull distribution with shape parameter c. Land and ice clutters also approximate this distribution. The following puts Eq. (12.5a) into Weibull form to align with Eq. (11.18). It is normalised by putting $\bar{\sigma} = 1$ and restated in terms of instantaneous voltage, v, where $\sigma = v^2/2$.

$$\text{Voltage CP} = 1 - \exp\left(-\frac{v^2}{2}\right)^c.$$

$$\text{Residual probability, RP} = 1 - \text{CP} = \exp\left(-\frac{v^2}{2}\right)^c.$$

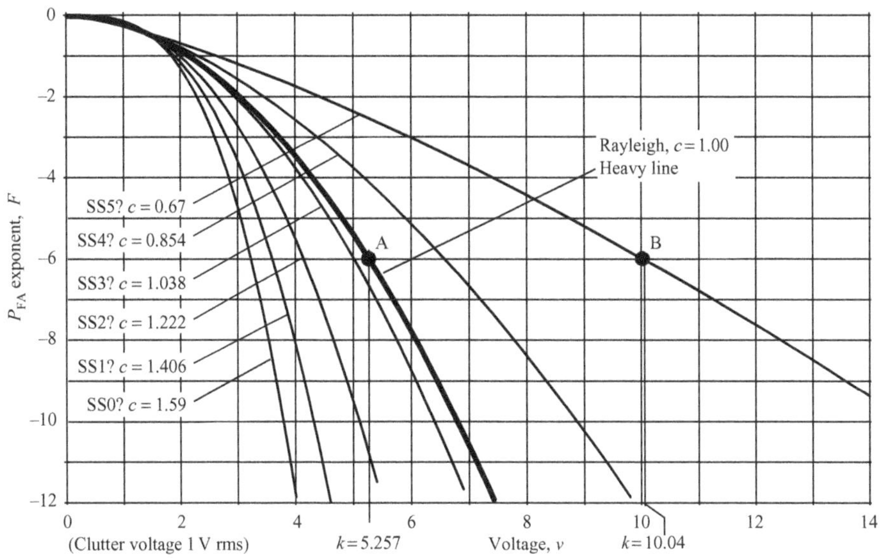

Figure 12.16 *Variation of P_{FA} exponent with threshold voltage for a range of Weibull shape parameters. Sea states tentative. The Rayleigh curve for noise or precipitation clutter follows Figure 12.4(a), whose axes had different scales*

Using subscript c to indicate conditions at shape parameter c and setting threshold k_c at the value of v which gives RP equal to the desired P_{FA} having exponent F_c:

$$10^F = \exp\left[-\frac{k_c^{2c}}{2^c}\right],$$

so

$$F_c = \log\exp\left[-\frac{k_c^{2c}}{2^c}\right] = -0.4343\left(\frac{1}{2}k_c^2\right)^c \tag{12.12a}$$

and

$$k_c = -\sqrt{2}\times(2.303F_c)^{1/(2c)}. \tag{12.12b}$$

Substitution of Eq. (11.19) (which tentatively links c with sea state) into Eq. (12.12a) links F with v and sea state as shown in Figure 12.16. Figure 12.17 shows demodulated noise waveforms for different shape parameters.

12.4.3 Equivalent sea, land and ice clutter

Detection circuits can be devised to assess shape parameter and optimise detection tactics to suit, say, a rise in sea state but it is doubtful whether they are used in the civil marine field. Usually, it suffices to assume that the threshold of a conventional detector is merely adjusted to maintain the desired false alarm rate, the threshold

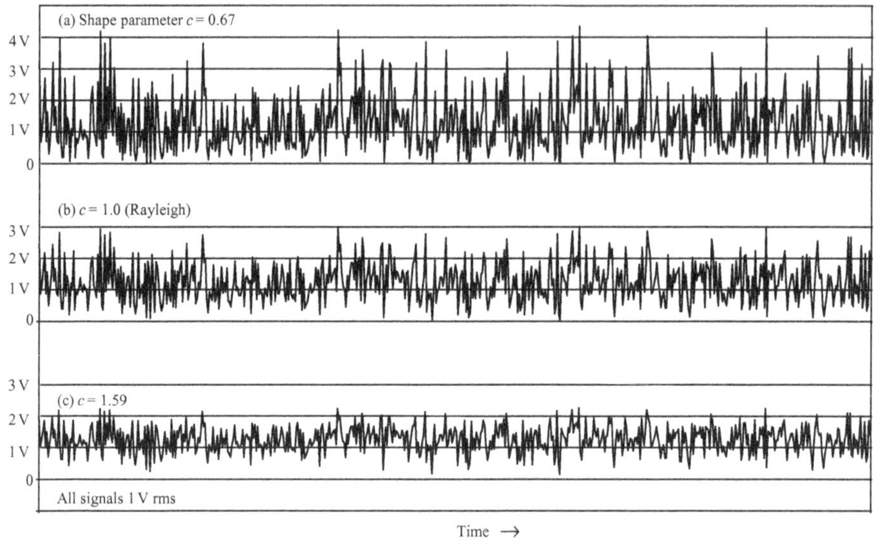

Figure 12.17 *Demodulated noise waveforms. Weibull shape parameters $c = 0.67$ (rough sea), $c = 1.0$ (Rayleigh) and $c = 1.59$ (smooth sea). Signal is demodulated so voltage is unidirectional. Higher c describes less spiky clutter*

rising as the shape parameter falls, also of course rising to accommodate the higher rms voltage. We tentatively suggest accounting for non-Rayleigh shape parameters caused by sea spikes etc. (Chapter 11, Section 11.7.2) by re-scaling the clutter voltage using a clutter weighting factor W (power by factor W^2) to give the same effect as Rayleigh distribution.

Suppose Rayleigh normalised noise or clutter is received and the desired P_{FA} is 10^{-6}, $F = -6$. In accordance with Eq. (12.6c), the appropriate threshold is $k = 5.257$ V, point A on Figure 12.16. If now the clutter is thought for argument's sake to remain 1 V rms but changes to sea state 5, $c = 0.67$, Eq. (12.12b) shows that k has to be increased to 10.04 V, point B on the figure, increasing by a voltage factor $W = 10.04/5.257 = 1.91$, preventing the increased number of high spikes from generating extra false alarms. Reduction of the rms clutter voltage to $1/1.91 = 0.52$ V rms (by 5.68 dB, point X on Figure 12.18) would restore F to -6 at $k = 5.257$. In other words, sea clutter of voltage v and shape parameter c can be treated as Rayleigh clutter, $c = 1$, of voltage Wv or of power $(Wv)^2$.

The value of W depends on F and c and is found as follows, where subscripts c and 1 denote Weibull and Rayleigh conditions, respectively. For Weibull noise, from Eq. (12.12a)

$$-F_c = 0.4343 \left(\frac{k_c^2}{2}\right)^c \qquad (12.13a)$$

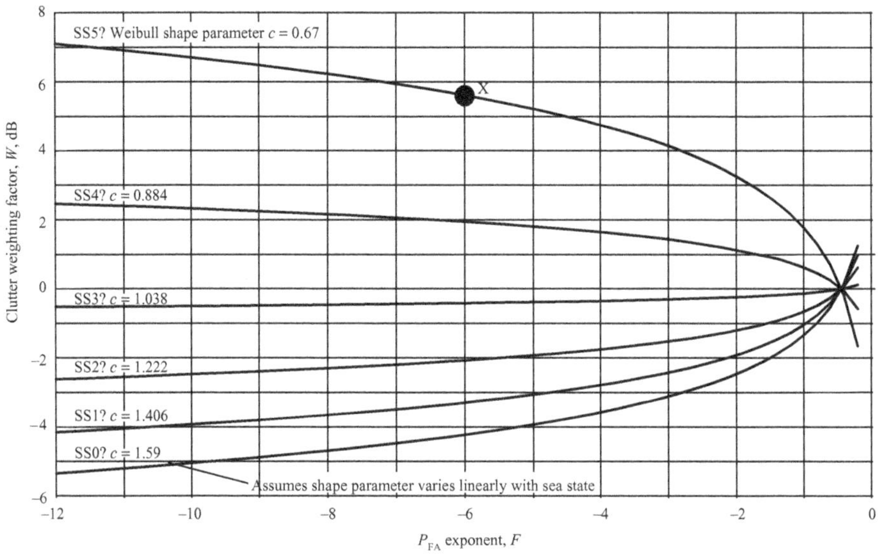

Figure 12.18 Variation of Weibull clutter weighting factor with P_{FA} and sea state. Applicable to small detection cells

and for Rayleigh noise, $c = 1$

$$-F_1 = 0.4343\left(\frac{k_1^2}{2}\right).$$

If $F_1 = F_c$ and $k_c = Wk_1$

$$0.4343\left(\tfrac{1}{2}k_1^2\right)^c W^{2c} = 0.4343\left(\tfrac{1}{2}k_1^2\right).$$

Therefore

$$\left(\tfrac{1}{2}k_1^2\right)^{c-1} W^{2c} = 1 \quad \text{and} \quad W = \left(\tfrac{1}{2}k_1^2\right)^{[(c-1)/2c]}.$$

Substituting for k_1

$$W = (-2.303F_1)^{[(1-c)/2c]} \text{ numerically}$$

or

$$W = 20\left[\frac{1-c}{2c}\right]\log(-2.303F_1) = 10\left[\frac{1-c}{c}\right]\log(-2.303F_1)\text{ dB.} \tag{12.13b}$$

Figure 12.18 shows the variation of W as P_{FA} exponent F is varied, for the family of shape factors tentatively linked to sea states. We see that SS 5 clutter boxes about 5.6 dB above its weight when $F = -6$, whereas calm-sea clutter is several dB more

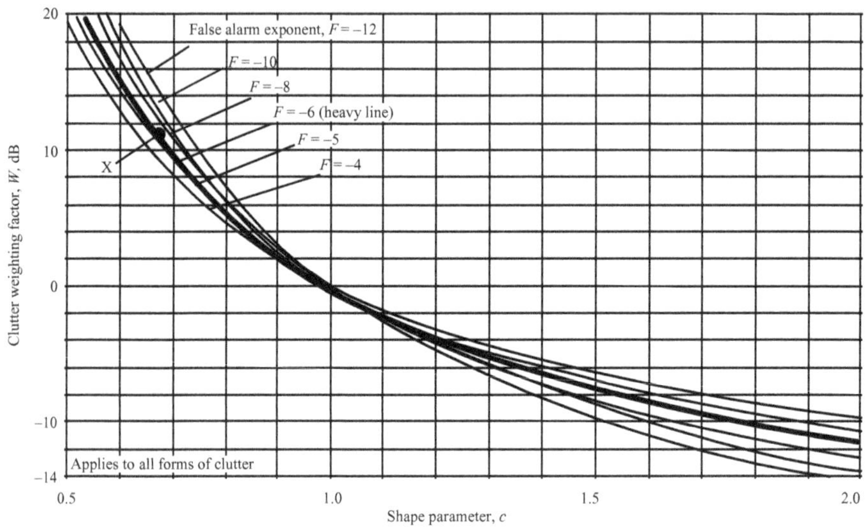

Figure 12.19 Variation of Weibull clutter weighting factor with shape coefficient. Applies to all forms of clutter

benign than expected from its RCS. W approaches 0 dB at high P_{FA} and always remains nearly 0 dB at SS3. When the sea area within the detection cell is large (at long range, long pulses, wide azimuth beamwidth), W approaches 0 dB. As no allowance has been made for this unquantified dependence, Eq. (12.13b) exaggerates W under these conditions.

Figure 12.19 is redrawn to plot W to a base of c for a family of typical F values and may also be used to weight land and ice clutter having Weibull distribution. It shows the false alarm exponent only weakly affects W within the usual working range where $-4 > F > -12$.

When there are two or more clutter/noise sources, perhaps noise and sea clutter, having different shape parameters, it is probably sufficient to apply the appropriate W factor to each source and then add their powers (W, not dB) to give the resultant noise/clutter.

12.5 Target fluctuation

12.5.1 The problem

So far this chapter has concentrated on a single echo pulse, assumed the target has definite fixed RCS so its echo is non-fluctuating with fixed power, and explained probabilities of detection based on the echo power relative to the mean of the fluctuating noise plus clutter. Without noise or clutter, non-fluctuating echoes marginally below threshold would never be detected but echoes marginally above threshold would have

$P_D = 1.0$. The target RCS or echo power probability density function (PDF) would be a narrow spike of great (theoretically infinite) height enclosing unit area, coinciding with a cumulative probability (CP) step function from 0 to 1.0. Only noise and clutter would introduce uncertainty to spoil the picture.

At very short range, noise and clutter may be small and the threshold is then set via the swept gain function to reject trivially small echoes from birds and so forth. The CP here does approximate a step function. And a few real-life longer-range 'hard' targets, such as reflectors and racons, discussed later, do indeed have nearly constant echo strength. But, as indicated in Chapter 7, Section 7.10, for point target pairs and Chapter 10 for extended targets, echo strength of most targets fluctuates significantly above and below their means. We avoid the term Fades, which might infer reductions only, fade margin then being the echo in hand above that required to achieve the required P_D.

It is not essential here for us to dissect which of the fluctuations listed in Section 12.1.2 change target RCS and which modulate the path loss of the direct/indirect ray combination at the radar receiver. It is enough to regard fluctuations interchangeably as variations either of echo power or of target RCS, whichever is convenient, about the mean values considered in previous chapters.

Even without noise or clutter, it is impossible to be sure whether a single observation of a fluctuating target shows it above or below its mean strength. Even when mean echo is below threshold, there remains some probability that a particularly strong pulse will exceed threshold and be detected; or a weak pulse from a generally strong target may fail to cross the threshold.

Fluctuation therefore broadens the echo PDF from a spike to a hump. The curve of probability that return exceeds threshold, $P\ (R > k)$, changes from a step to a rounded curve, the area under the curve of course remaining unity. Any receiver noise (or clutter) fluctuations compound the detection uncertainty. An above-average echo may sit on a low noise (or clutter) event to preclude detection, or a below-average echo may coincide with a high noise event to enhance detectability. Noise and clutter are of course unaffected by target fluctuation so the threshold remains as before, Eqs (12.6c) or (12.12b).

12.5.2 Swerling fluctuation cases

Target fluctuations differ so widely that before attempting calculations of P_D some simplifications or abstractions have to be made, by allocating targets to a few broad representative classes or cases which can be mathematically modelled. We choose to use one of the oldest models, proposed by Swerling [4], which has stood the test of time and is still widely used in detection studies. It was briefly mentioned in Chapter 7, Section 7.10.6. Swerling's famous cases include three distributions of observed RCS (or echo) values.

- *Case 0.* RCS does not fluctuate.
- *Cases 1 and 2.* Targets with many (theoretically an infinite number, but in practice exceeding 4 or 5) independent scatterers of approximately equal strength,

no one scatterer predominating; targets whose width or height are very many wavelengths. RCS has Rayleigh probability distribution.

- *Cases 3 and 4.* Targets with one dominant plus several subsidiary scatterers, or a single non-uniform scatterer subject to modest changes in viewing angle.

Cases 1 and 3 are distinguished from Cases 2 and 4 by the rate of change of RCS.

- *Cases 1 and 3.* RCS fluctuates on a time scale slower than the scan interval (~2 s), so returns on adjacent scans are uncorrelated and provide independent samples of instantaneous RCS. However, returns within a single scan *are* correlated so may not average the long-term mean; for example, the returns in a scan may be all low or all high.

Case 3 is subdivided.

- *Case 3a.* Case 3 target viewed by single radar in the ordinary way.
- *Case 3b.* Case 1 target viewed by a decorrelated pair of dual-diversity radars, adequately separated in frequency or position.

- *Cases 2 and 4.* RCS effectively uncorrelated between adjacent pulses (~1 ms apart) and therefore also uncorrelated scan to scan. For a single pulse, performance approximates Case 1 but when 10 or more pulses are integrated (Section 12.6.4), detectability approaches that of Case 3a. Case 2 applies to frequency agile radars, possibly occasionally used in VTS service, and perhaps to helicopter and small hovercraft blade flash echoes. Case 4 targets are rare in marine practice and will not be discussed further.
- *All cases.* RCS remains constant for the pulse duration of ~1 μs.

The cases are at best only approximations to reality, so detectabilities calculated with their aid will include error, adding to imprecisions in knowledge of average target RCS, actual radar parameters, environmental conditions on the day and so on. That said, the Swerling Cases offer a good footing for comparison of likely detectability of differing radar/target/environment systems. Table 12.1 summarises them and Figure 12.20 shows their probability distributions. If a target's case is unknown it is prudent to be pessimistic and assume Case 1, whose instantaneous RCS lies below the average oftener than above, necessitating higher average RCS than Case 0 or Case 3 targets to get high P_D.

12.5.3 Case 0 (Case 5) non-fluctuating target

Also called Swerling Case 5, the Marcum Case or a hard target. All echoes are identical in strength, equalling the mean. We have partially analysed this case in the previous sections. Echoes have full or perfect pulse-to-pulse and scan-to-scan correlation, meaning each sample size is related (here equal) to its neighbours rather than being random. There is unity probability that the echo power is the average value. The PDF is therefore a spike or delta function centred on average RCS, enclosing unit area, at the point where CP steps from 0 to 1, see Figure 12.20. Figure 12.13 indicates that, even in noise, quite a small RCS or echo strength increase raises P_D from 0.10 to 0.90: for example, 4.2 dB when $P_{FA} = 10^{-6}$.

Table 12.1 Target fluctuation; Swerling Cases

Swerling Case	0 (or 5)	1	2	3a	3b	4
Pulse to pulse	Correlated	Correlated	Uncorrelated	Correlated	Correlated	Uncorrelated
Scan to scan	Correlated	Uncorrelated	Uncorrelated	Uncorrelated	Uncorrelated	Uncorrelated
Fluctuation	Non-fluctuating	Slow	Fast	Slow	Slow	Fast
Target characteristics	Single reflector with large solid angle	Many independent scatterers	Many independent scatterers	One dominant plus other smaller scatterers	Dual diversity radars	As Case 3a
Typical marine targets	Good point reflector, racon or RTE alone	Ships and all physically large targets	Freq agile radar, helo blade flash	Buoy, small yacht, buoy + racon? RTE + yacht?	Case 1 target	None
Target alone: probability distribution	Line	Ricean, Eq. (12.14a)	Eq. (12.16a)	Eq. (12.17a)	Eq. (12.21a)	
Target alone: cumulative probability	0 or 1	Eq. (12.14b)	Eq. (12.16b)	Eq. (12.17b)	Eq. (12.21b)	
Target in noise: probability of detection	Eqs (12.7a), (12.8), (12.11a)	Eq. (12.15c)	Eq. (12.16c)	Eq. (12.18c)	Eq. (12.22f)	
Single-pulse SNR, P_D 0.1, P_{FA} 10^{-6}	8.8 dB	7.0 dB	~7.0 dB	7.5 dB	5.9 dB	
Single-pulse SNR, P_D 0.9, P_{FA} 10^{-6}	13.0 dB	21.2 dB	~21.2 dB	17.3 dB	15.7 dB	
Change, P_D 0.1–0.9	4.2 dB	14.2 dB	~14.2 dB	9.8 dB	9.8 dB	

Note: Quoted equations and SNRs apply to detection by non-coherent radar receiver.

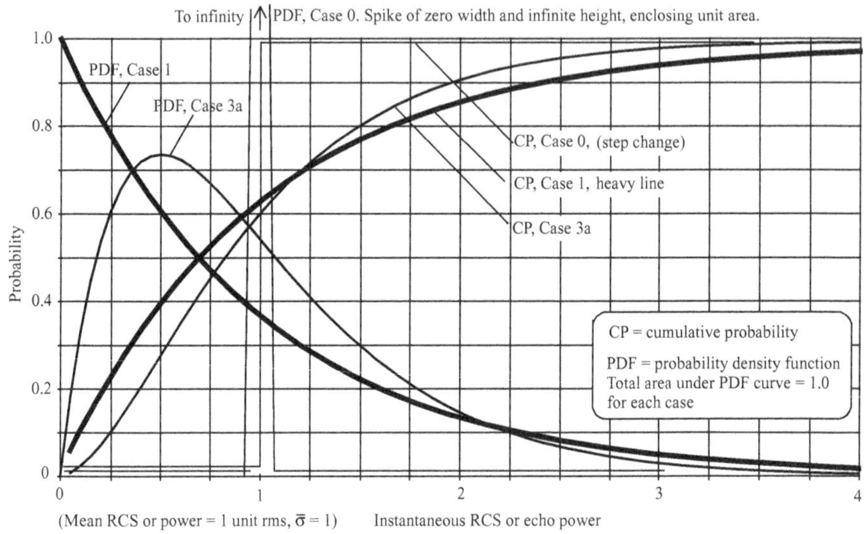

Figure 12.20 Probability distributions, Swerling Cases. Normalised. Case 0 non-fluctuating targets are represented by a spike. When the signal is strong, Case 1 has lowest cumulative probability, the integral of probability density. Case 3 is intermediate between Cases 0 and 1

Targets approximating Swerling Case 0 include the following.

- Terrain echoes without moving foliage, particular when seen by a groundfast radar.
- Radar interrogations at racons and racon responses at radars, Section 12.9.
- Platforms (particularly when groundfast) carrying radar reflectors or RTEs whose RCS exceeds skin RCS by 20 dB or more.
- Targets such as octahedral reflectors having target pattern maps whose RCS changes slowly with angle are best treated as non-fluctuating with RCS equal to the device mean, or, more prudently, the minimum RCS sustained through an angle of 10° (IMO's 'stated performance level' for reflectors, see Chapter 7, Section 7.6.1).

12.5.4 Fluctuating targets

Figure 12.21, based on Figure 12.11 for Ricean distribution, illustrates diagrammatically how the uncertainty of signal strength creates uncertainty of residual probability (equivalent to P_D), which may lie between strong and weak instantaneous RCS points S and W.

If the fluctuation is slow, as for most marine targets (Cases 1 and 3), the echo might remain above average for a whole scan, whereas quick fluctuation (Cases 2 and 4) would enable a reasonable estimate of P_D to be formed from the packet of

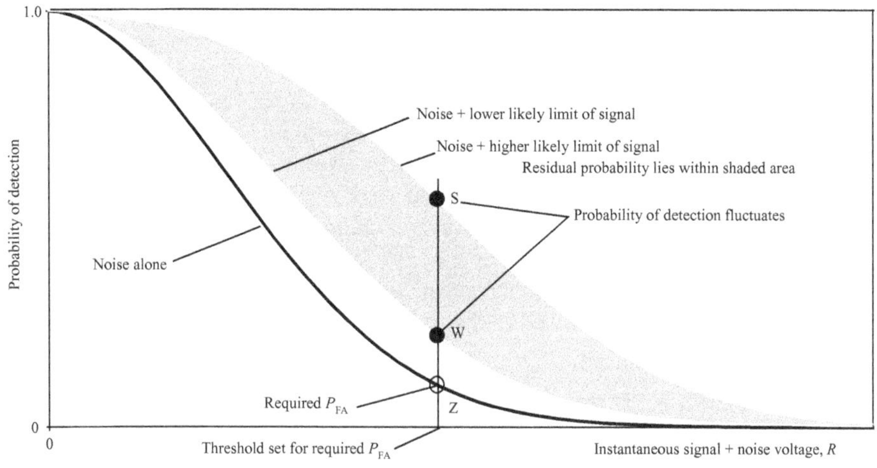

Figure 12.21 *Probability of detection, fluctuating signal. Signal fluctuation adds uncertainty to the curves. P_D of a pulse may fluctuate between points S and W*

echoes within the scan. So both the spread of instantaneous echo strengths about the mean, and fluctuation rate, must be considered, forming the reason for differentiating between Cases 1, 2 and 3. No two targets fluctuate in quite the same way, and the fluctuations of a given target may change with sea conditions, for example, increasing roll or yaw. Noise associated with the two-way attenuation of atmospheric, scanner or feeder loss modulates the echo and introduces a component of fluctuation, biasing the Swerling Case from Case 0, etc., part-way towards Case 2. This effect, partially range-dependent, is customarily ignored or lumped with the other uncertainties surrounding detectability.

12.5.5 Swerling Case 1

Chapter 7, Section 7.10 indicated why even fairly small ship targets have many scatterers, no one of which usually predominates, and large enough dimensions for the TPM to become busy, with numerous close-packed lobes and nulls in the TPM, as Figure 7.15(f). As the target moves in a seaway, the fluctuation is generally reckoned to have Rayleigh distribution, although some authors suggest it may be better described by log-normal distribution (Chapter 11, Section 11.7.3). Fluctuation rate is linked to roll and pitch periods and is slow relative to the pulse repetition interval. Swerling Case 1 approximates this very important class of target. The model assumes RCS occasionally extends towards infinity, at which point probability is infinitely low. Practical targets come close to this extreme on the rare occasions a large flat hull plate comes exactly normal to the sight-line and reflects a high-amplitude 'flash' (sometimes 30 dB above mean RCS). Writing mean RCS $= \bar{\sigma}$, and instantaneous RCS $= \sigma$, the PDF is:

$$p(\sigma) = \frac{1}{\bar{\sigma}} \exp\left(-\frac{\sigma}{\bar{\sigma}}\right) \qquad [\sigma \geqslant 0]. \tag{12.14a}$$

As depicted in Figure 12.20, which is normalised by putting $\bar{\sigma} = 1$, this PDF falls smoothly, from 1 towards 0 when σ rises from 0 towards ∞. When $\sigma = \bar{\sigma}$, $p(\sigma) = 0.37$, so on 37 per cent of occasions RCS exceeds average. More to the point, the remaining 63 per cent have RCS below average. A threshold set to detect average RCS will give $P_\mathrm{D} = 0.37$, too low for the operator to be reasonably sure of perceiving the raw target on the display, or for effective track formation, manually or by ARPA/ATA.

As usual, cumulative probability that the signal is less than the threshold, k, is found by integration:

$$\mathrm{CP} = \int_0^k \exp\left(-\frac{\sigma}{\bar{\sigma}}\right) \mathrm{d}\sigma. \tag{12.14b}$$

Figure 12.20 shows this to be a curve from the origin which is more or less a mirror image of the PDF. Putting SNR $= q$ (numerical), the PDF of signal plus noise is approximately as follows (the expression is exact for square-law detection).

$$p(n) \sim \frac{1}{1+q} \exp\left[-\frac{1}{2}\frac{k^2}{1+q}\right] \tag{12.15}$$

and

$$P_\mathrm{D} = \exp\left[-\frac{1}{2}\frac{k^2}{1+q}\right]. \tag{12.16a}$$

Substituting $k = 2.146\sqrt{-F}$ per Eq. (12.6c) gives an expression linking P_D with SNR and P_FA:

$$P_\mathrm{D} = \exp\left[\frac{2.303F}{1+q}\right] = 10^{F/(1+q)} \tag{12.16b}$$

and

$$\log P_\mathrm{D} = \frac{F}{1+q}. \tag{12.16c}$$

Therefore

$$q = \left(\frac{F}{\log P_\mathrm{D}}\right) - 1 \text{ numerically.} \tag{12.16d}$$

Figure 12.22 shows how P_D varies with q (shown in dB) for a family of P_FA exponents F. Comparison with Figure 12.13 (Case 0) confirms that somewhat lower q gives the lower values of P_D but considerably higher q is needed for high P_D. When $P_\mathrm{D} \sim 0.3$, q is independent of Case. If Figure 12.2 is taken to depict mean echo voltage, on several scans together all echoes B, C and D may be less than average, reducing P_D, while on another set all might exceed average, raising P_D and confirming that the rate of change of single-sweep P_D with q must be lower than for a non-fluctuating Case 0 target.

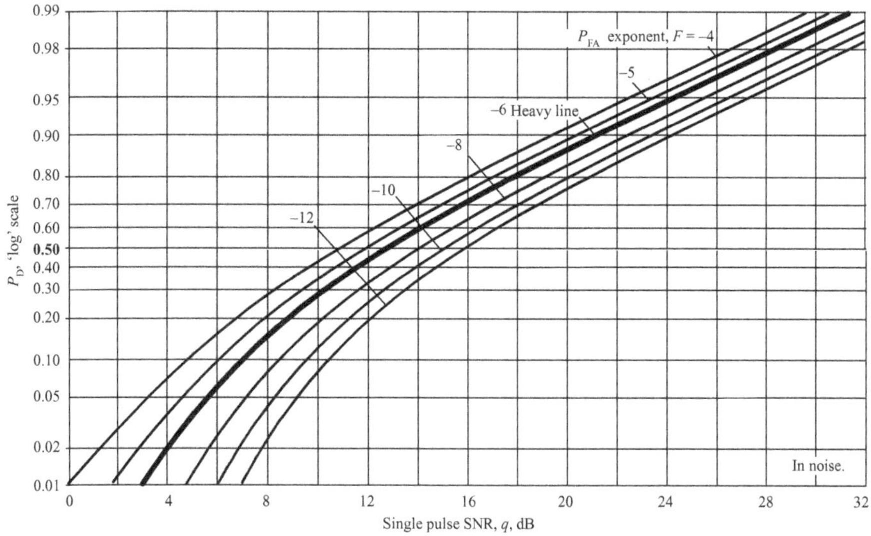

Figure 12.22 Probability of detection, Swerling Case 1 target

12.5.6 Swerling Case 2

Here each echo is a sample from a Rayleigh distribution, fully decorrelated pulse to pulse. Examples are frequency agile radar observing a Case 1 target, or conventional fixed-frequency radar observing a helicopter echo dominated by blade flash. For agility to be sufficient to change a target from Case 1 to Case 2, the return from the nearer half of the target must differ in phase by at least π rad from that of the remoter half. For target axial length L, phase difference $= 2\pi L/\lambda$, where c is the velocity of light:

$$\text{wavelength difference } \delta\lambda = \frac{2L}{c}\ \text{m} \tag{12.17a}$$

so

$$\text{frequency difference } \delta f = \frac{1}{2}\frac{c}{L}\ \text{Hz.} \tag{12.17b}$$

For example, if $L = 5$ m, the radar's frequency range must cover at least $3 \times 10^8/10 =$ 30 MHz, independent of radar band.

When N pulses having signal/noise ratio q with threshold k are integrated non-coherently, Meikle [5] states PDF is

$$p(Y) = \left[\frac{(1+q/2)^{-N}k^{(N-1)}}{(N-1)!}\right]\exp\left[-\frac{k}{1+q/2}\right] \tag{12.17c}$$

and

$$P_D = 1 - \gamma\left[N, \frac{k}{1+q/2}\right],$$

where the gamma function

$$\gamma(N,x) = \int_0^N t^{N-1}\exp(-t)\,\mathrm{d}t. \tag{12.17d}$$

When $N = 1$, single pulse, Case 2 approximates Case 1. For $N > 1$ see Section 12.6.4.

12.5.7 Swerling Case 3a

Consider a boat target, rolling with a period around 10 s, carrying a reflector whose RCS exceeds that of the boat's skin by several decibels. The boat is small, so the rather open composite TPM has only a few lobes and nulls, similar to Chapter 7, Section 7.10.2, Figure 7.14(*d*). In the couple of seconds between scans, the boat rolls several degrees, enough for a substantial change in RCS as different parts of the target pattern map are illuminated. The echo therefore fluctuates quasi-randomly from scan to scan. But in the 10 ms or so scan dwell time containing the scan's pulse packet, there is time only for a tenth of a degree roll, much less than the lobe width, and the echo barely alters – the echo is correlated pulse to pulse but uncorrelated scan to scan, with slow fluctuation. PDF is a chi-squared distribution with four degrees of freedom:

$$P(\sigma) = \left(\frac{4\sigma}{\bar{\sigma}^2}\right)\exp\left(-\frac{2\sigma}{\bar{\sigma}}\right) \qquad [\sigma \geqslant 0]. \tag{12.18a}$$

Standard deviation is $1/\sqrt{2}$. The normalised distribution, included in Figure 12.20, humps up from the origin to a maximum of 0.73 when $\sigma = 0.5$, then decays towards 0 when σ increases towards infinity. When RCS is at the mean value, PDF is 0.53, so RCS is more or less balanced about the mean; setting the threshold for mean RCS yields 53 per cent P_D. As usual, the total area under the PDF = 1.0.

$$\mathrm{CP} = \int_0^k \left(\frac{4\sigma}{\bar{\sigma}^2}\right)\exp\left(-\frac{2\sigma}{\bar{\sigma}}\right)\mathrm{d}\sigma. \tag{12.18b}$$

Putting numerical SNR $= q$, the PDF of signal plus noise approximates the following, being exact for a square law detector.

$$p(n) = \frac{1 + k^2/(4/q+2)}{(q/2+1)^2}\exp\left[-\frac{k^2}{q+2}\right] \tag{12.18c}$$

and

$$P_D = \frac{q}{q+2}\times\left[1 + \frac{k^2}{q+2} + \frac{2}{q}\right]\exp\left[-\frac{k^2}{q+2}\right]. \tag{12.18d}$$

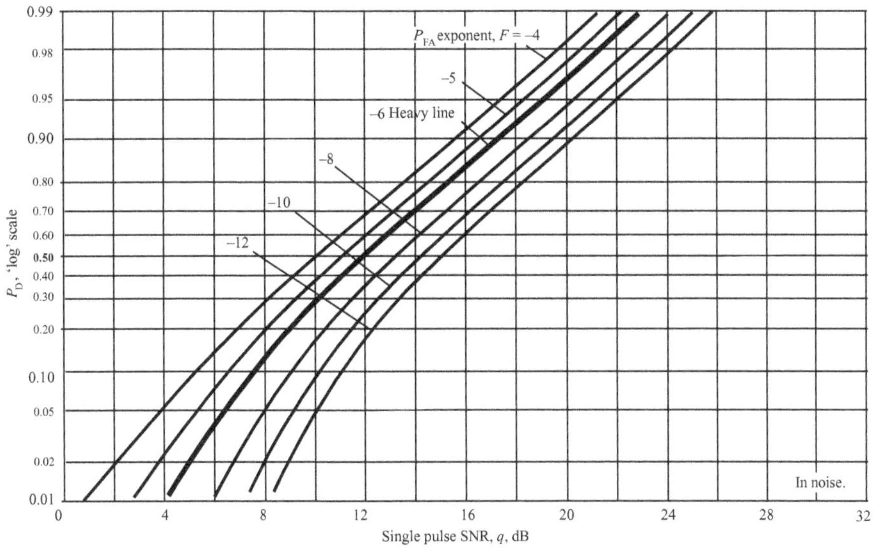

Figure 12.23 Probability of detection, Swerling Case 3a target

Substituting $k = 2.146\sqrt{-F}$ by Eq. (12.6c),

$$P_D = \frac{q}{q+2} \times \left[1 - 4.605\frac{F}{q+2} + \frac{2}{q}\right] \exp\left[4.605\frac{F}{q+2}\right]. \tag{12.18e}$$

This expression readily finds P_D knowing q and F and is illustrated in Figure 12.23, but resists re-arrangement to yield q given P_D and F. The following algorithm gives a fair approximation. It is derived from the quadratic best matching the $F = -6$ curve of Figure 12.23 in the range $0.01 < P_D < 0.99$, taking q as ordinate. Figure 12.24 shows error to be less than ± 1 dB except at extreme values:

$$q_{dB} \sim 17.2 + \frac{29.7}{F} + 4.86D + 0.38D^2 \text{ dB}. \tag{12.18f}$$

$D = \log[P_D/(1 - P_D)]$ as before. Rearrangement gives an expression for F when q_{dB} and P_D are known:

$$F \sim \frac{29.7}{q_{dB} - 17.2 - 4.86D - 0.38D^2}. \tag{12.18g}$$

12.5.8 Comparison of fluctuation cases

Figure 12.25 compares detectability of Case 0, 1 and 3a targets for two representative values of false alarm exponent, F. Relative to Case 0, although Case 1 requires a couple of decibels less signal for low P_D, it needs substantially more when P_D is high, for example, 8 dB more for $P_D = 0.9$. Put another way in Figure 12.26, introduction of Case 1 fluctuation is enough to reduce P_D from 0.9 to ~0.5. Case 3

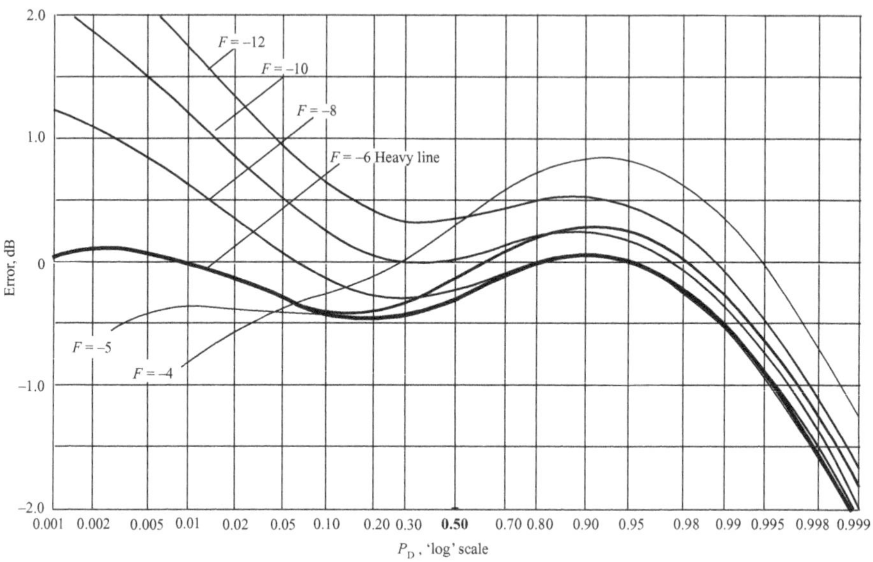

Figure 12.24 Accuracy of formula for q knowing P_D. Case 3a, Eq. (12.18f)

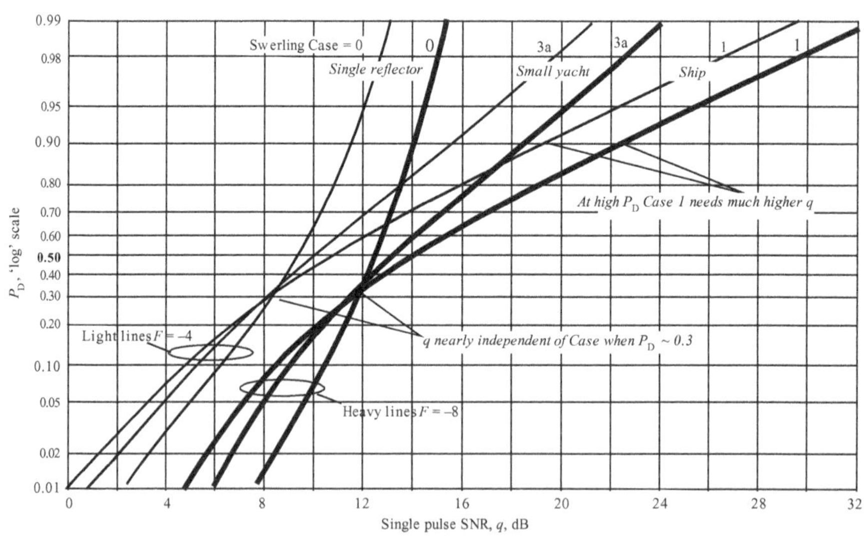

Figure 12.25 Comparison of probabilities of detection. Swerling Cases 0, 1 and 3a compared for $F = -4$ and -8. Swerling Cases 2 and 4 approximate Case 1 when few pulses are integrated

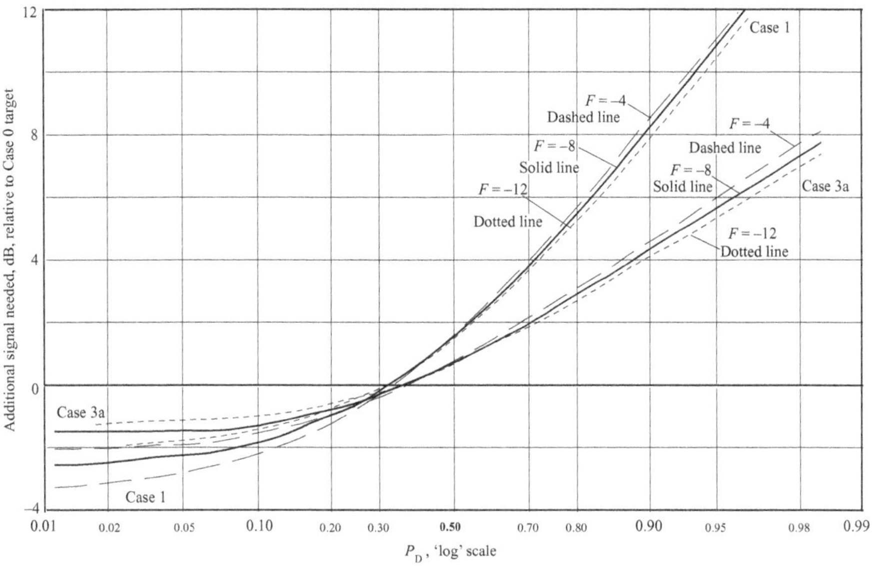

Figure 12.26 Additional signal necessary to match Case 0 P_D. Figure 12.25 re-based to emphasise the cost of target fluctuation

fluctuation is about half as bad. Target fluctuation is therefore too important to be ignored. We repeat: if the fluctuation characteristic is unknown, it is prudent to assume Case 1.

An easy simplification which usually gives conservative results is to assume the target to be non-fluctuating, Case 0, but to have artificially low RCS. Figure 12.26 shows the appropriate decrease is almost independent of F. If maximum P_D of interest is say 0.9, nominal RCS of Case 3a targets should be decreased by 4.5 dB, and Case 1 by 8.5 dB, almost irrespective of P_{FA}.

12.6 Multiple observations

12.6.1 Addition of returns

We have so far concentrated on a single pulse. Radar receivers (but not racons, Section 12.9) make their final detection decision only after receipt of at least one scan packet of several pulses, affecting detectability in two main ways:

- successive echoes always occur at target range but noise spike ranges are random, improving effective SNR;
- echo strength may fluctuate from pulse to pulse and/or scan to scan, depending on the Swerling Case of the target, affecting the observation time necessary to obtain a truly representative sample.

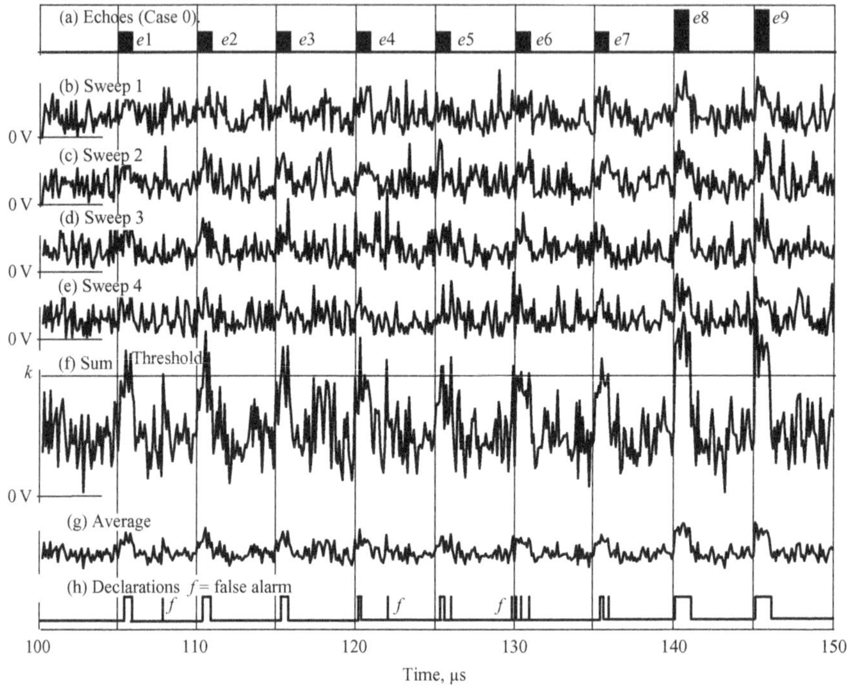

Figure 12.27 Summation of pulses in packet. Summation (here of four sweeps) increases correlated signals more than uncorrelated noise, so improving effective SNR

Figure 12.27(a) depicts the echoes of the targets of Figure 12.10(a), here considered constant from sweep to sweep (Case 0). Figure 12.27(b)–(e) are the noisy narrow-band IF returns from N successive sweeps within a single scan. Although N has been taken as 4 for illustration, in practice the packet more likely contains 6–40 returns depending on prf (hence on range scale) and on scanner azimuth beamwidth and rotation rate. All the sweep returns are the same in general form, random noise making them differ in detail. The P_D of each would be generally as Figure 12.10(e) or (f), with low P_D or high P_{FA} according to the selected threshold. Waveform (f) shows the sum of the four returns. At (g) this waveform is scaled down by a factor of 4 to give the average return. Comparison shows its echoes are much more apparent than those within any single return, with less noise – addition or integration has improved SNR.

Waveform (f) also shows a threshold, set to include three noise spikes (like the high threshold of Figure 12.10, which allowed detection of only six of the eight echoes). Figure 12.27(h) shows no echo now goes undetected. Inspection confirms the improvement stems from noise humps appearing at random ranges and accumulating relatively slowly, while echoes always appear at the same range so accumulate more rapidly – noise goes up, but signal goes up more. The radar has studied the scene for longer, rather than jumping to a declaration on the evidence of a single sweep, rather as we have to stare at an indistinct scene. In a sense the improved SNR results from increased look time, equivalent to reduced system detection bandwidth, just as narrow-band receivers give better SNR. Fundamentally, we have put more energy (watt seconds) on target.

Making the detection decision on a packet thus yields an integration improvement factor, otherwise called integration gain. This summation of a packet of already-demodulated returns before making the detection declaration in an envelope detector is called post-detection integration and is usual on marine radars. 'Post-demodulation integration' would suit our terminology better.

12.6.2 Coherent and non-coherent integration

Post-detection integration uses echo amplitudes only. The phase information is lost, because the local oscillator and transmitter are each free-running devices whose relative phases are uncontrolled and unknown. Post-detection integration is therefore also known as non-coherent integration and was described in Chapter 2, Section 2.2.5. An alternative, coherent integration, utilises phase as well as amplitude to enhance performance, Section 2.2.7. Conventional magnetron transmitters are unsuitable, currently confining the technique to certain VTS radars where a few decibels of detection improvement may justify the considerable additional cost. Use of a phase sensor in the IF system to control the LO, called a phase-locked loop, or coherent-on-receive system, Section 2.2.6, has intermediate performance.

Coherent integration adds signal and noise *voltages*. The signal voltages add coherently $\propto N^2$ while the noise voltages add incoherently $\propto N$, so coherently integrated SNR (power) improvement $= N^2/N = N$ and the voltage sum is compared in the threshold. Non-coherent integration adds signal *powers*, so speaking broadly and as modified below, the integration improvement approximates $\sqrt{N}$. The single-pulse probability curve is followed, but with all the energy concentrated in it. Often the amplitudes of the echoes to be summed are weighted to improve signal processing.

12.6.3 Integration gain or loss

The detectability improvement from integration can be treated in alternative fashions. Either a multipath gain can be calculated and inserted in the right hand side of the two-way radar range equation (e.g. Chapter 4, Section 4.3.2, Eq. (4.6a)) to give an effective echo for detectability purposes, greater than the true echo. But this effective echo neither anywhere actually exists, nor is it applicable to calculations of signal handling capacity, dynamic range, etc. The alternative approach we prefer is to leave the range equation in its single-pulse form, while reducing the effective SNR required for specific detection conditions by an integration improvement (or gain) factor.

The integration gain, usually expressed in decibels, yields the same given P_D and P_{FA} relative to single-pulse detection, but at lower echo strength. It raises the actual single pulse SNR, q, to a higher effective value, Q, enabling q to be reduced. Practical radars do not achieve the theoretically possible integration gain, because when echo returns jitter between a pair of adjacent range cells they are not fully counted, because of processing imperfections and because the system may not be of fully coherent form, the phase energy being discarded in non-coherent integration. These factors give rise to an integration loss, L_N, expressed numerically or in dB. Widespread use of the term 'loss' is unfortunate; integration always confers some gain in performance, non-coherent systems merely losing some of the prospective improvement available from a theoretically perfect coherent integrator. Subscript (dB) identifies quantities expressed in decibels rather than numerically in the following analysis.

Either one can obtain the effective SNR value by addition of the achieved dB integration gain to the single-pulse SNR, q_{dB}, or one can assume perfect coherent integration for the number of sweeps integrated, then subtract the integration loss from this ideal state. Swerling Case 2 targets behave in a special way, Section 12.6.4, but other cases all respond to integration in the following manner.

$$\text{In perfect coherent systems, numerical integration gain} = N. \tag{12.19a}$$

$$\text{In dB terms, coherent system integration gain} = 10 \log N \text{ dB}. \tag{12.19b}$$

That is, integration of say ten pulses gives 10 dB detectability improvement, so echoes 10 dB weaker can be detected with the same P_D and P_{FA} relative to single-pulse detection. Figures 12.28 and 12.29 follow Skolnik [6, Figures 2.8(*a*) and (*b*)]. Figure 12.28 shows how gain rises in response to the number of pulses integrated, Eq. (12.19b) being depicted as a straight line. The function is continuous because N need not be an integer. In passing, we note that the eye's excellent pattern recognition capability can just about recognise the sinusoidal component of Figure 12.5 waveform (c) amid the noise where $q_{dB} = -3$ dB and eight cycles are available for comparison in phase and amplitude, giving $Q_{dB} = -3 + 10 \log 8 = 6$ dB, whereas raising q by 6 dB ($\times 4$) to give waveform (e), the same certainty comes after two cycles, $Q_{dB} = 3 + 10 \log 2$, again 6 dB, tending to confirm Eq. (12.19b).

Given perfect integration and ignoring clutter, effective SNR and detectability become the product of radiated power (EIRP), pulse length (1/bandwidth) and number of pulses integrated per second, proportional to mean power on target, as predicted in Chapter 4, Section 4.3.2. This general result is used in the concept of continuous wave (CW) operation discussed in Chapter 16.

Integration gain of a non-coherent system is somewhat lower than for coherence and is mildly dependent on q, the equivalent single pulse SNR giving the required P_D and P_{FA} (or F) combination, Eqs (12.19c) and (12.19d). The expressions become somewhat inaccurate for very high q_{dB}, exceeding ~30 dB. Figure 12.28 shows the relationship for a family of q_{dB} values.

- Non-coherent system

$$\begin{aligned} \text{integration gain} &= 6 \log N + 0.342 q_{dB}[1 - \exp(-0.65 \log N)] \\ &= 6 \log N - 0.342 q_{dB}(N^{-0.282} - 1) \text{ dB}. \end{aligned} \tag{12.19c}$$

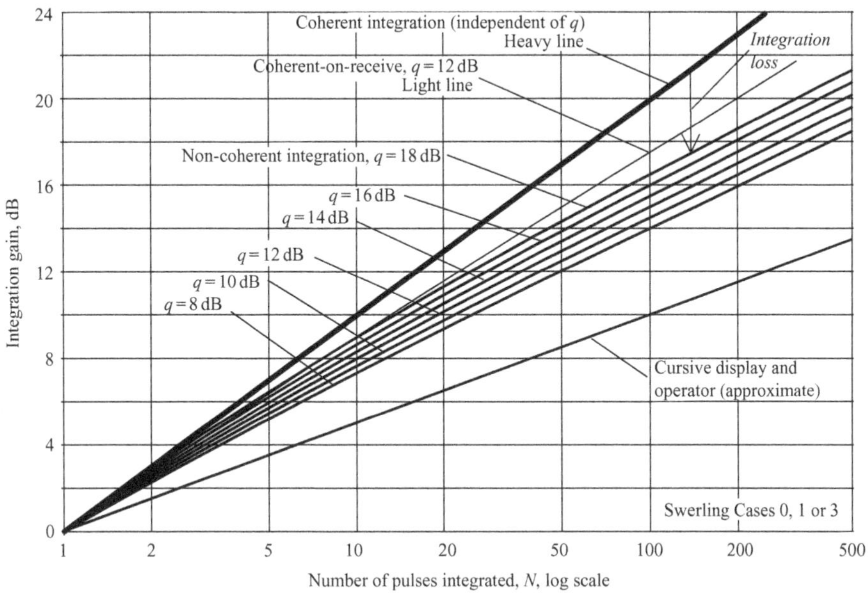

Figure 12.28 *Integration gain. Coherent systems (used in some VTS radars) utilise phase as well as amplitude information to deliver the full benefits of integration. Ships' non-coherent receivers have somewhat lower integration gain. Coherent-on-receive systems have intermediate performance. The older cursive displays have yet lower gain. Integration loss is the shortfall from a perfect coherent integrator and is shown to larger scale in Figure 12.29. Equipment imperfections somewhat diminish gain below the values shown*

By subtraction from Eq. (12.19b), integration loss, shown in Figure 12.29, is:

$$\text{integration loss} = 4\log N + 0.342q_{\text{dB}}(N^{-0.282} - 1)\ \text{dB}. \qquad (12.19\text{d})$$

- Coherent-on-receive systems have about half this loss:

$$\text{integration loss} \sim 2\log N + 0.17q_{\text{dB}}(N^{-0.282} - 1)\ \text{dB}. \qquad (12.19\text{e})$$

So by subtraction from Eq. (12.19b)

$$\text{integration gain} \sim 8\log N - 0.17q_{\text{dB}}(N^{-0.282} - 1)\ \text{dB}. \qquad (12.19\text{f})$$

Performance calculations usually yield q and N, from which P_{D} has to be determined for a given P_{FA} (exponent F). The following relationships apply.

$$Q_{\text{dB}} = q_{\text{dB}} + \text{integration gain, dB}. \qquad (12.19\text{g})$$

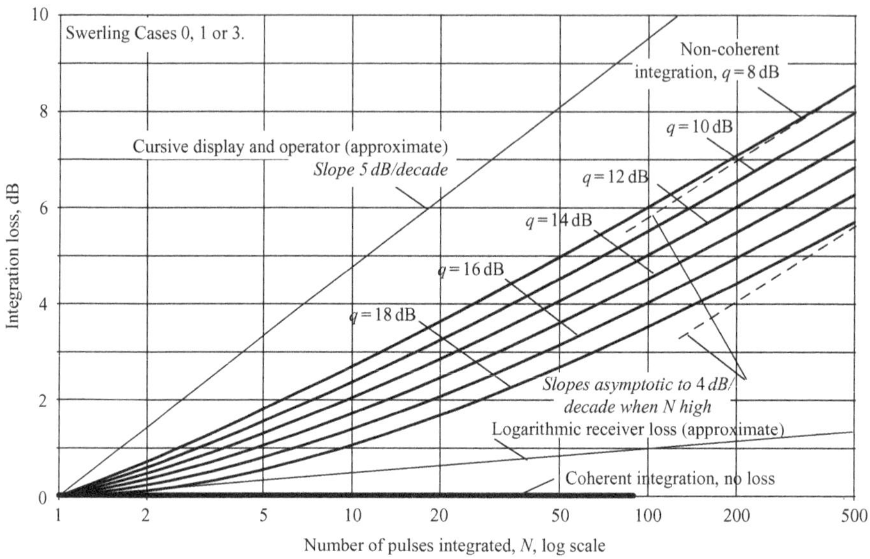

Figure 12.29 Integration loss. Also shows additional loss in logarithmic receivers

Substitution gives Q_{dB} in terms of q_{dB} and N:

$$\text{coherent integration: } Q_{dB} = q_{dB} + 10 \log N \text{ dB}, \tag{12.19h}$$

$$\text{coherent-on-receive: } Q_{dB} = q_{dB}(1.17 - 0.17N^{-0.282}) + 8 \log N \text{ dB}, \tag{12.19j}$$

$$\text{non-coherent: } Q_{dB} = q_{dB}(1.342 - 0.342N^{-0.282}) + 6 \log N \text{ dB}. \tag{12.19k}$$

When N is high Eq. (12.19k) is asymptotic to

$$Q_{dB} = 1.342q_{dB} + 6 \log N \text{ dB}. \tag{12.19l}$$

Figure 12.30 depicts these equations.

12.6.4 Swerling Case 2 targets

Because each Case 2 echo is completely decorrelated from its neighbours, coherent integration confers no advantage over non-coherent. Eq. (12.17c) for P_D (Section 12.5.6) contains a term N for the number of pulses integrated, coherently or otherwise. Figure 12.31 relates multipulse SNR, Q, with N for representative P_{FA} and P_D combinations. When $N = 1$, Q approximates the Case 1 value. Increasing N rapidly improves Q from this, the rate of improvement gradually falling until, when N is high, it approximates non-coherent integration of a Case 3 target, with slopes asymptotic to 4 dB/decade.

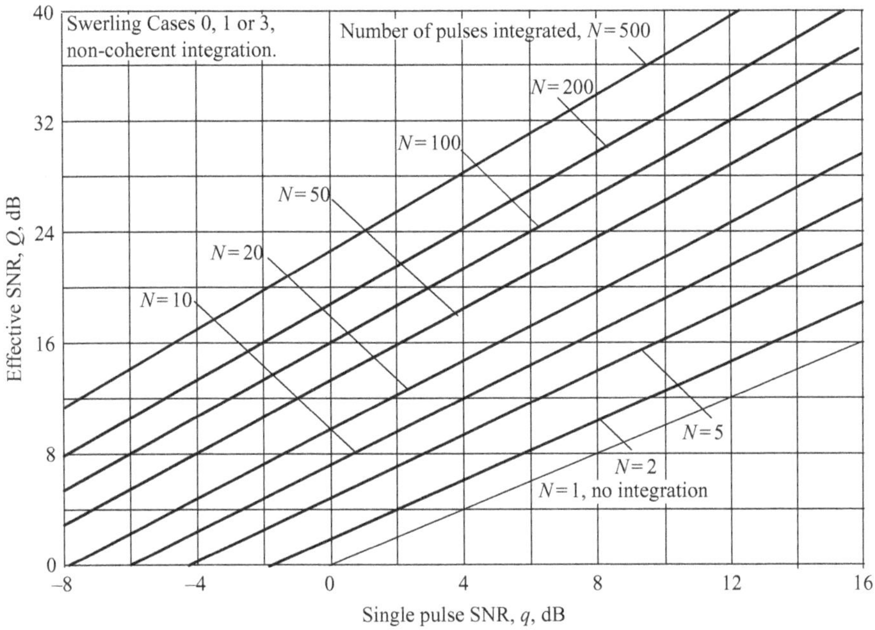

Figure 12.30 *Improvement of effective SNR by integration of N sweeps. The curves are linear but not quite parallel. For Case 2 see Figure 12.31*

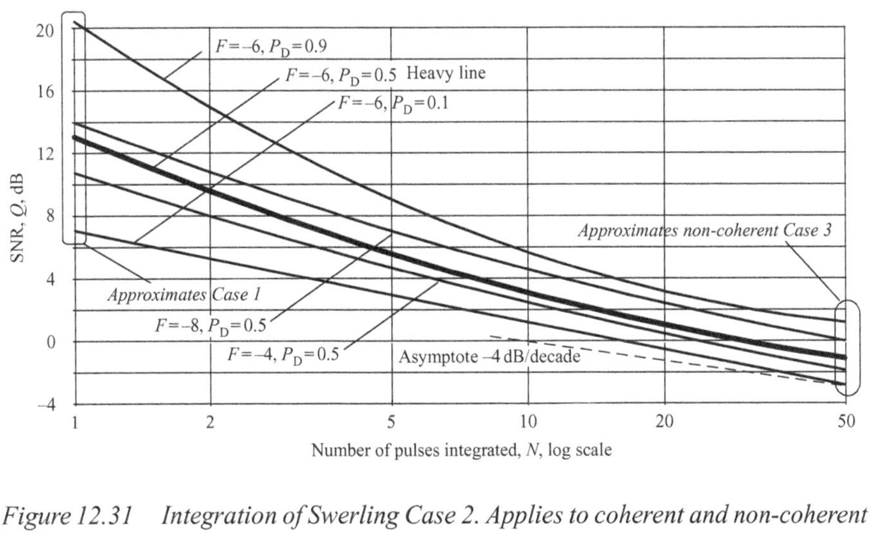

Figure 12.31 *Integration of Swerling Case 2. Applies to coherent and non-coherent systems*

12.6.5 M out of N integrators

Suppose there are N pulses per scan, M of which cross the primary detection threshold. For example, $N = 8$ and $M = 6$. If the single pulse probability of false alarm, P_{FA1}, is set low, say to 0.0574, $F = -1.24$, threshold $k = 2.400$ from Eq. (12.6) or Figure 12.4(*b*), the chance of M (here 6) consecutive false-alarm blips in the scan is $P_{FA1}^M = 0.0574^6 = 35.7 \times 10^{-9}$. There are $N!/[M!(N-M)!] = 28$ possible selections (the combination ${}^N C_M$), so the scan probability of false alarm, P_{FAs}, is

$$P_{FAs} = \frac{P_{FA1}^M N!}{[M!(N-M)!]}. \tag{12.20}$$

Here $P_{FAs} = 10^{-6}$. To get $P_{FA1} = 10^{-6}$, the threshold would have to be significantly higher at $k = 5.257$, so there is an integration gain, here $20\log(5.257/2.400) = 6.81$ dB. Digital counting techniques perform M out of N integration. As noted in Chapter 11, Section 11.9, 2 out of 2 integration is commonly employed to suppress running rabbits interference from distant or co-located radars.

12.6.6 Performance margin

Track-formers require some minimum P_D such as 0.5 on which reliably to form a target track. Using a spreadsheet of the form to be described in Chapter 14, alternative performance checks are possible.

- Calculate that actual P_D exceeds requirement at one or more spot ranges.
- Graph P_D to a base of range, checking that actual P_D exceeds requirement at all ranges of interest.
- Graph signal margin in hand, dB, relative to that necessary to achieve required P_D, to a base of range. Detection is assured when the margin is positive. This presentation is usually the most useful, for it directly indicates the system robustness against parameter changes such as increase of service loss.

12.6.7 Cursive displays

The long-persistence phosphor coating of the cursive CRT of the older radars has the useful property that spot brilliance is approximately proportional to the number of beam electrons striking the spot in a time frame of the order of a second, thereby integrating the individual returns within the sweep (and scan if there is insignificant movement in the scan period). For efficient integration, some feeble noise speckling is necessary, achieved by judicious setting of the gain control (in effect to the maximum false alarm rate tolerable to the observer). Brilliance has to be set rather low to bring echo paints within the rather small (~10 dB) dynamic range of the phosphor, necessitating a viewing hood to exclude daylight. Numerical integration gain is then approximately proportional to $\sqrt{N}$, that is 1.5 dB per doubling of number of pulses integrated (5 dB/decade) as indicated in Figures 12.28 and 12.29. The integration effect is partly psychological; alert and experienced operators doing slightly better, others somewhat worse.

$$\text{Cursive display plus operator integration gain} \sim 5 \log N \text{ dB}. \tag{12.21a}$$

Subtraction from Eq. (12.19b) gives the integration loss.

$$\text{Cursive display plus operator integration loss} \sim 5 \log N \text{ dB}. \tag{12.21b}$$

12.6.8 Analog integration

Modern radars use digital integration techniques, but radars may be encountered which use alternative forms of analogue integration to the long persistence tube, more recent responses receiving greater weighting than older responses. Summation methods include a resistor–capacitor or other form of low-pass filter device, a resistor–inductor–capacitor bandpass filter or a recirculating delay-line integrator. In general, the weight given to the current 1st pulse of voltage V_1 is $V_1 \times 1$, while the previous 2nd pulse is attenuated by weighting $V_2 \exp(-\gamma)$. Older pulse 3 is attenuated by $\exp(-2\gamma)$, weighted $V_3 \exp(-2\gamma)$, and so on.

$$\text{The weight of the } n\text{th pulse} = V_{\text{n}} \exp(1 - n)\gamma. \tag{12.21c}$$

For N pulses in the packet, the output voltage is

$$V = \Sigma_N = V_N \exp(1 - N)\gamma. \tag{12.21d}$$

If the pulse repetition interval (pri) is T s, RC low-pass filter timeconstant is RC s, or tube of persistence RC to $t/T = 1/\varepsilon$ (35 per cent brilliance, $\varepsilon = 2.7183\ldots$),

$$\gamma = \frac{T}{RC}. \tag{12.21e}$$

In an RLC resonant circuit form of bandpass filter,

$$\gamma = \frac{\pi\sqrt{L/C}}{R}. \tag{12.21f}$$

In a recirculating delay line,

$$\gamma = \text{loop gain}. \tag{12.21g}$$

12.6.9 Mitigation of losses in small scanners and wide bandwidth

For constant elevation beamwidth, as scanner gain is reduced, azimuth beamwidth rises, more clutter is illuminated and single-pulse SNR, q, falls. However, with more pulses in the packet N rises and integration partly restores sensitivity; wholly for fully coherent systems. Further, more clutter spikes arise within the detection cell and sea, land and ice clutters become more normal, Weibull weighting factor (W, Section 12.4.3) falling toward 0 dB.

Pulselength reduction on short range scales necessitates wider bandwidth and more thermal noise which tends to reduce q. If prf is raised, N rises and the performance loss is again mitigated.

12.6.10 Logarithmic receiver loss

A logarithmic receiver has output proportional to the logarithm of the input envelope, giving particularly high dynamic range, see Chapter 3, Section 3.5.7. For a single pulse, necessary SNR is as for a linear receiver. According to Skolnik [6, Section 9.6], multiple pulse SNR is slightly poorer, by $\sim$0.5 dB when $N = 10$ or $\sim$1.0 dB when $N = 100$, shown in Figure 12.29.

$$\text{Logarithmic loss} \sim 0.5 \log N \text{ dB.} \tag{12.22}$$

12.6.11 Detection at short range with ringing

Strength of feeder ringing clutter was established in Chapter 11, Section 11.8. On every pulse clutter occurs as one or more solid rings on the display at sub-kilometre ranges. This form of clutter does not fluctuate, so little or nothing is gained by integration of multiple pulses. For effective echo detection at a ring range, single pulse echo to ring ratio should well exceed 10 dB. Unfortunately, gain reduction to suppress the rings reduces sensitivity at other ranges also and does not improve short-range signal to ringing clutter ratio.

12.7 Setting the threshold

12.7.1 Interchangeability of receiver gain and threshold voltage

For convenience, we have regarded the receiver gain as fixed, and have seen that P_{FA} depends on the ratio k of threshold to noise plus clutter voltage at the amplifier output. Instead of adjusting threshold voltage, one could just as well change gain. P_D and P_{FA} can be controlled by adjustment of either gain or threshold voltage. In the following it is assumed that gain is varied because there is a control labelled gain, but as a matter of detail design this control might instead raise threshold voltage, or do a mixture of both – it is of no concern to the operator or to analysis of the system.

12.7.2 Inbuilt swept gain

High gain is required at long range to detect feeble echoes, but at short range the inverse fourth power range term in the radar range equation makes all navigationally significant targets relatively strong. To prevent receiver overload and to minimise the need for the operator to reset gain as target range changes, all radar receivers include an inbuilt swept gain or sensitivity–time control (STC) function. This increases gain from an initial low value during the first 100 μs or so after the transmitter fires, so reducing the necessary dynamic range. Figure 12.32 is a typical swept gain curve. For example, at range A, gain is -6 dB on maximum, halving noise voltage, doubling threshold/noise voltage ratio k and improving P_{FA} from say 10^{-6} to better than 10^{-12}. Noise is therefore always insignificant at short range.

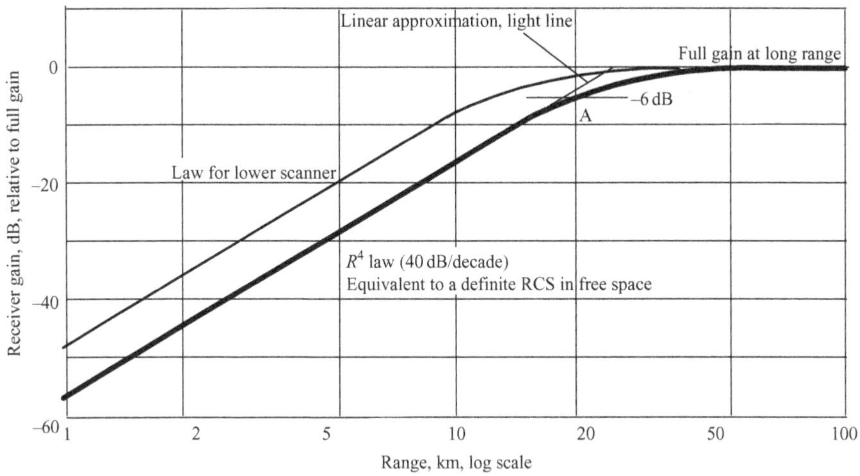

Figure 12.32 *Typical swept-gain law. Fan-beam scanner. Receiver gain is reduced (or threshold voltage is raised) at short range. The break point is set on installation to suit scanner height. Individual makes of radar vary*

STC tailors gain with range to match one of

- inverse fourth power law, to give nominally constant signal;
- inverse third power law, nominally constant sea clutter;
- inverse second power law, nominally constant precipitation clutter;
- some intermediate law or combination of laws.

Particularly when the scanner is mounted high, the optimum swept gain characteristic depends on the scanner elevation beamwidth, angle of depression and shape of polar diagram, differing between conventional fan-beam and inverse cosec^2 types. The break point of the swept-gain law should therefore be set during installation to suit scanner model and height. Radar manufacturers' swept gain curves doubtless differ in detail and are determined empirically from confidential trials results, seldom published. It may be sufficient to take the swept gain law as gain $\propto R^4$ below some critical range, as in the figure. Here, in free space signal strength becomes independent of range and the short-range detection threshold is equivalent to a definite RCS value, which we call the receiver swept gain threshold, expressed in dBm^2. The actual law might have a more rounded transition than shown. The swept gain sub-system partly utilises the TR cell attenuation reduction during recovery from the ionised state at instant of transmission.

12.7.3 Adaptive threshold

We have seen that one secret of successful detection of weak targets is to set the threshold as low as possible, to accept the maximum tolerable number of false alarms. Modern radar data processing systems measure the clutter environment at each part of the sea area within range and adapt the threshold voltage or receiver gain to hold P_{FA} constant at its optimum value. The facility is punningly called CFAR (constant false alarm rate) – it enables the radar to 'see far'. Bandwidth changes in response to range scale switching are accommodated by adjustment of

- threshold voltage;
- IF gain;
- detection cell count for Target Seen declaration.

12.7.4 Operator's gain control

The internal machine settings aim, under average conditions, to optimise operation in the view of the designer, who of course cannot predict all the vagaries of clutter in service. So a gain control is still provided, allowing the operator the following choices.

1. Set gain excessively high. Clutter and noise may exceed the receiver dynamic range, reducing SNR at the detector and spoiling P_D whether determined manually or by machine.
2. Set gain high. Gives high P_{FA}, shown as frequent noise speckles on unprocessed 'raw radar' display, but displays almost all weak echoes. Each speckle is a candidate target so the operator must first closely study the screen, mentally filter the brighter speckles, memorise their position, then compare with subsequent scans to pick out any persistent brightups at (or near, if the targets move) the original positions. The long persistence of a cursive display greatly assists. Even so, this tiring process demands undivided attention. Few operators can maintain concentration for more than a quarter-hour or so. Summarising, high gain maximises P_D of weak targets at expense of increased mental workload. On synthetic displays, processing may be overloaded.
3. Set gain moderately high. Of course, it is simpler to leave ARPA or ATA to get on with this boring tracking job on track-formed synthetic displays. The machine will not tire. But the high noise counts may occasionally meet the target track-formation criteria and generate and display phantom tracks. Beside confusing the operator, noise tracks clog up the track-former. If they reach its full capacity (often 24 tracks within maximum instrumented range, not the smaller sea area perhaps currently displayed), genuine targets will be dumped and not displayed, in effect risking arbitrary and dangerous reduction of P_D.
4. Set gain low. This certainly cleans up the display but risks missing weak targets. This setting is less tiring to observe so the operator is less likely to blunder. The conscientious operator resists the temptation to set low gain but restore the apparent quality by setting high screen Brilliance. Although giving a beautifully clear and clean display, low gain prevents detection of small targets. With advent of adaptive thresholds, the gain reduction available to the operator has been restricted, reducing the collision danger through lazy gain settings. Black coffee is definitely preferable to the fool's paradise of a low-gain/high-brilliance display combination!

12.8 Radar diversity

12.8.1 Principles

Chapter 6, Section 6.4.4, touched on the possibility of mitigating point target multipath null depths to a pair of scanners by attention to heights and operating frequencies. Chapter 7, Section 7.10, showed the resultant RCS of a pair of reflectors to be critically dependent on angle of look and on radar frequency. Figure 12.33(a) depicts one quadrant of the polar diagram of a rudimentary two-point-element target similar to Section 7.10.2, Figure 7.15(f). Initially it might be viewed on bearing $\theta 1$, where there is a null in the pattern. Shifting radar bearing a few degrees to $\theta 2$ yields a peak. Similarly, shifting frequency a few per cent from $f1$ to $f2$ without shift of bearing also moves RCS between a peak and a null. Clearly, viewing at diverse angles or by diverse frequencies partly fills nulls and reduces the probability that a target composed of several reflecting elements will continue to exhibit an extreme of RCS. Also the probability that two uncorrelated observations both yield an extreme value is less than for either radar alone. The probability distribution of RCS becomes more concentrated around the mean, moving part-way to Case 0, making detection possible at lower SNR.

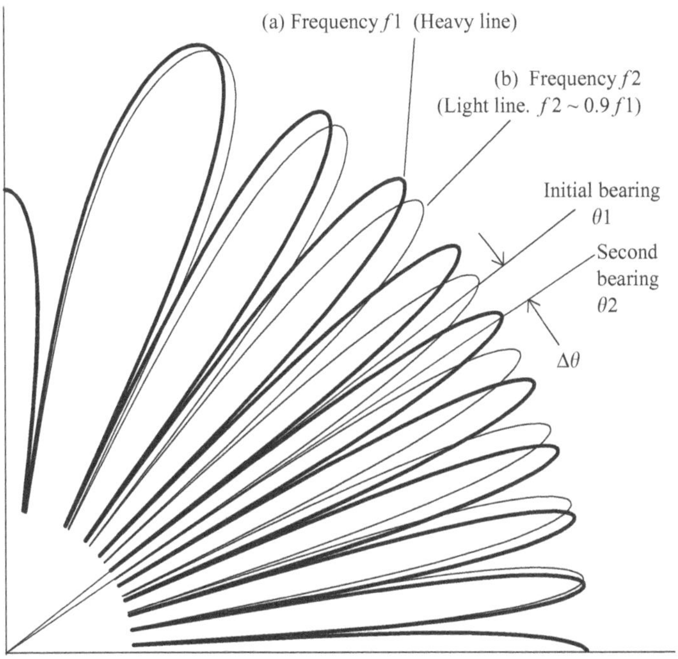

Figure 12.33 Bearing and frequency diversity. Part of target polar diagram at differing frequencies. Two equal reflecting elements; based on Figure 7.15(f)

Diversity also decorrelates precipitation clutter and partly decorrelates sea clutter, because the samples differ. Therefore improved detection in noise or clutter is obtained if a diversity technique combines the outputs of two (or more) radars, operated at differing frequencies (within a common band, or utilising two bands), and/or spaced apart laterally or in height. System reliability also improves; failure of either radar causes a moderate loss of performance (brownout or graceful degradation) rather than total blackout. We shall first indicate the separation criteria for dual diversity conditions to apply, then look at detection within a diversity system. The principles can be extended to multiple diversity, although we confine ourselves to twin systems. Switching the plane of polarisation may also give some diversity improvement, depending on target reflecting characteristics. Simultaneous use of frequency, space and polarisation diversity may give better results than the single-parameter diversity discussed below – the more the merrier!

12.8.2 Criterion for polar diagram decorrelation

For the target pattern map to be decorrelated, the separation, L, of the reflecting elements has to be less than $\frac{1}{2}c/\Delta f$, where c is the velocity of light, so

$$\Delta f \geqslant \frac{1}{2}\frac{c}{L} \tag{12.23}$$

The frequency separation, Δf, has to be sufficient for the smallest physical target size of interest. For example, to decorrelate echoes from a small craft, effective width between furthest reflection centres 1.0 m, requires $\Delta f > 150$ MHz, corresponding to 5 per cent frequency shift at 3 GHz but only 1.6 per cent at 9 GHz. A 3 GHz set could be twinned with a 9 GHz set, but single-band frequency diversity exchanges twin-scanner cost and alignment problems for simpler questions of differential squint and system bandwidth and is usually cheaper.

12.8.3 Criterion for precipitation clutter decorrelation

Twin identical radars with adjacent scanners receive identical target echoes, almost identical clutter returns and have similar but decorrelated receiver noise powers. Each radar's display would appear almost identical and little would be gained. Synchronising the transmitters (sometimes easier said than done) with a common scanner and receiver doubles signal but not noise, giving 3 dB more SNR, because transmitter power in effect has been doubled. But it is smarter to shift the frequency of the second radar away from the first by Δf to decorrelate its clutter returns. For effective decorrelation Δf must exceed the receiver bandwidth, B:

$$\Delta f > B. \tag{12.24a}$$

Some VTS sets operated on short range scales may use bandwidth as high as 20 MHz necessitating 20 MHz diversity. In practice, to allow for tolerances in the centre frequency of each set, higher separation might be chosen, perhaps 30–50 MHz. Any differential squint (Chapter 2, Section 2.2.7) can be corrected within the data processor. Even so, such diversity can readily be accommodated within either of the marine bands and within the design bandwidth of many scanners. Pairing 3 and 9 GHz radars always meets the criteria of Eqs (12.23) and (12.24), but of course demands twin scanners. Just as the reflection glints between differing parts of the target with time, it may also scintillate in position as frequency changes, reducing the benefit of diversity.

When co-sited radars observe a target such as a ship end-on which occupies an extended range bracket, d, their frequency difference must decorrelate the target ends by at least $\pi/2$ rad.

$$\Delta f > \frac{1}{2}\frac{c}{d}. \tag{12.24b}$$

For example, if $d = 10\,\text{m}$, $\Delta f > 15\,\text{MHz}$.

Separating the frequencies of a co-sited radar pair decorrelates sea clutter faster than echo. Signal to clutter ratio improves up to a maximum variously claimed to reach 6–12 dB at an optimum frequency diversity of the order of 100 MHz (9 GHz band). Often twin transceivers feed a common scanner, pulses being interleaved. Transmitter duplication is expensive, and some VTS system suppliers prefer a single transceiver, recouping performance with a bigger scanner. Most high-grade VTS schemes use one or other of these techniques, sometimes both.

12.8.4 Space diversity

Twin radars of the same frequency observing a target from different bearings receive decorrelated echoes when the angle subtended between the radars becomes comparable with the lobe-to-null angle of the target pattern map at the maximum range of interest. Chapter 7, Section 7.10.3, Eq. (7.26), gave the average lobe–lobe width, θ_{av}:

$$\theta_{\text{av}} = 45\frac{\lambda}{L}^{\circ}.$$

Lobe-null width $= \frac{1}{2}\theta_{\text{av}}$, so for diversity

$$\Delta\theta \geqslant 22.5\frac{\lambda}{L}^{\circ}. \tag{12.25}$$

For a pair of 3 cm radars and target width $L = 1\,\text{m}$, $\Delta\theta = 0.72^{\circ}$. To retain diversity to say 4 km range requires a baseline of 50 m. This exceeds the shipboard space generally available, particularly as the radars would need to be mounted athwartships to give good diversity in the important ahead direction. So space diversity is mainly confined to VTS systems, especially those including out-stations. Synthetic aperture radar, discussed in Chapter 16, extends the diversity principle by taking successive measurements along a baseline to simulate a single scanner whose aperture equals the baseline.

VTS occasionally uses height diversity. Usually the main high-gain scanner is mounted aloft on a mast, with an auxiliary scanner at its foot. The main scanner gives good long-range performance where the auxiliary may be horizon-limited or have inadequate signal to precipitation clutter ratio. The low scanner gives superior signal

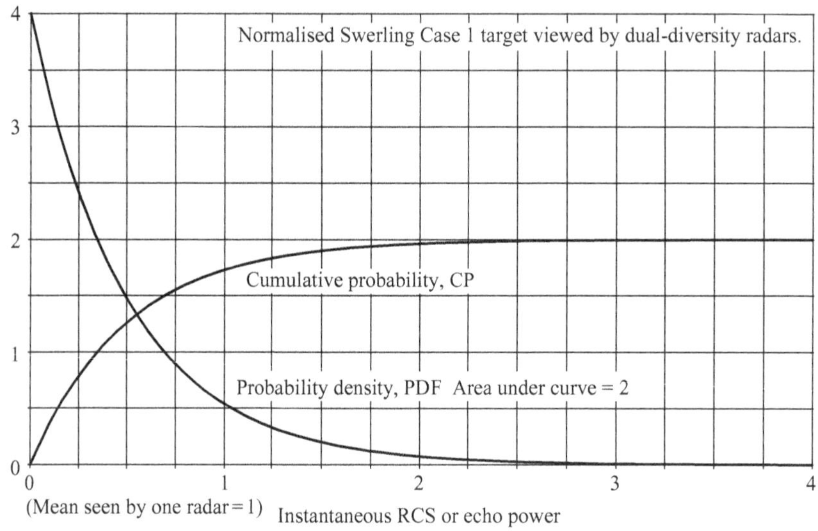

Figure 12.34 Probability distribution, Case 3b

to sea clutter ratio at short range. Intermediately, a diversity improvement is obtained. In the example, scanners at 10 and 60 m give diversity to 4 km.

12.8.5 Swerling Case 3b; Case 1 target observed by dual-diversity

When diversity criteria are satisfied, clutter spikes at the first radar are decorrelated from those at the second, and it is less likely both radars will simultaneously illuminate a TPM null. Extremes are partially smoothed out, changing distribution of Case 1 returns part way towards Case 0. The fluctuation of a system of twin radars plus Case 1 target with combined output is called Swerling Case 3b. The probability distribution function and cumulative probability are:

$$\text{PDF:} \quad P(\sigma) = \frac{4}{\bar{\sigma}^2} \exp\left(-2\frac{\sigma}{\bar{\sigma}}\right) \qquad [\sigma \geqslant 0]. \tag{12.26a}$$

$$\text{CP} = \frac{4}{\bar{\sigma}^2} \int_0^k \exp\left(-2\frac{\sigma}{\bar{\sigma}}\right) \, d\sigma. \tag{12.26b}$$

Figure 12.34 depicts these equations. CP is $1/\sigma$ that of Case 3a (Eq. (12.18b)), and is asymptotic to 2.0 rather than 1.0 because there are two radars.

12.8.6 Receiver combinations

The following analysis shows that alternative combinations of the radar outputs give differing performance. We express SNR, q, numerically. Subscripts (1) and (t) denote single-radar conditions and the overall performance of the twin system, respectively, so that the following apply.

Required system probability of detection $= P_{\mathrm{Dt}}$.

Required system probability of false alarm $= P_{\mathrm{FAt}}$, with exponent F_{t}.

1. Single channel alone requires (from Eq. (12.16d))

$$q_1 = \frac{F_1}{\log P_{\mathrm{D1}}} - 1 \text{ numerically.} \tag{12.27a}$$

2. In zero clutter, twin transmitters feeding a common scanner and single receiver double system transmitter power. Receiver noise is unaltered, so when clutter is much less than noise there is 3 dB reduction in necessary SNR, deriving solely from the extra power, with no diversity improvement as such. In principle a common frequency would give equal performance:

$$q_{\mathrm{t}} = \frac{q_1}{2} \text{ numerically.} \tag{12.27b}$$

3. In heavy clutter, twin transmitters as above collect doubled clutter, so when clutter is much larger than noise, using a common frequency there is no improvement:

$$q_{\mathrm{t}} = q_1. \tag{12.27c}$$

With frequency diversity the clutter returns are decorrelated and a Swerling Case 1 target behaves as Case 3b, giving an improvement when in the usual case where $P_{\mathrm{D}} > 0.3$ is required.

4. Diversity with OR output logic: twin transmitters and receivers with their outputs connected in OR configuration give a system false alarm when there is a false alarm in either or both channels:

$$P_{\mathrm{FA1}} = \tfrac{1}{2} P_{\mathrm{FAt}}, \quad \text{so } F_1 = F_{\mathrm{t}} - \log 2. \tag{12.27d}$$

In general, the probability of a target not being detected is $1 - P_{\mathrm{D}}$. Probability of both channels missing a target detection is the square of one alone doing so,

$$1 - P_{\mathrm{Dt}} = (1 - P_{\mathrm{D1}})^2.$$

Therefore

$$P_{\mathrm{D1}} = 1 - \sqrt{1 - P_{\mathrm{Dt}}}. \tag{12.27e}$$

Substituting in Eqs (12.27a), (12.27d) and (12.27e)

$$q_1 = \left\{ \frac{F_{\mathrm{t}} - \log 2}{\log[1 - \sqrt{1 - P_{\mathrm{Dt}}}]} \right\} - 1, \tag{12.27f}$$

from which

$$P_{\mathrm{Dt}} = 1 - \{1 - 10^{(F_{\mathrm{t}} - \log 2)/(q_1 + 1)}\}^2. \tag{12.27g}$$

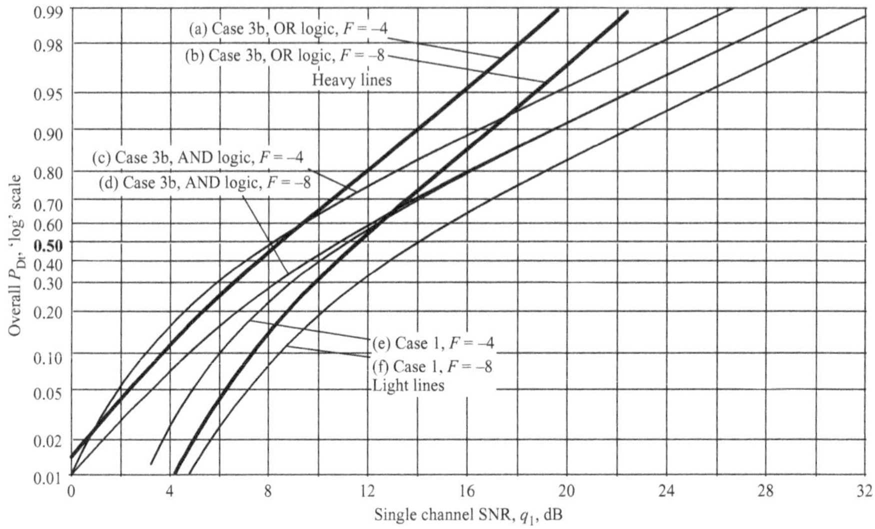

Figure 12.35 Swerling Case 3b for dual-diversity operation compared with Case 1

Figures 12.35(a) and (b) compare Eq. (12.27c) for Case 3b with two of the Case 1 curves of Figure 12.22 curves (e) and (f). Diversity improvement increases with P_D, exceeds 3 dB when $P_{Dt} > 0.64$ and is about 5 dB when $P_{Dt} = 0.9$.

5. Diversity with AND logic output: here there is a system false alarm only when simultaneous false alarms occur in both channels:

$$P_{FA1} = \sqrt{P_{FAt}}.$$

Therefore

$$F_1 = \tfrac{1}{2}F_t. \tag{12.27h}$$

If, for example, system false alarm probability is to be 10^{-6} or 10^{-8}, channel false alarm probability is set quite low at 10^{-3} or 10^{-4}, with low detection threshold and hence low q_1. The probability of the pair detecting a target is the square of one alone doing so, therefore

$$P_{D1} = \sqrt{P_{Dt}}. \tag{12.27i}$$

Therefore

$$q_1 = \frac{F_t}{(2 \log P_{Dt})} - 1 \tag{12.27j}$$

and

$$P_{Dt} = 10^{[F_t/(2q_1)]}. \tag{12.27k}$$

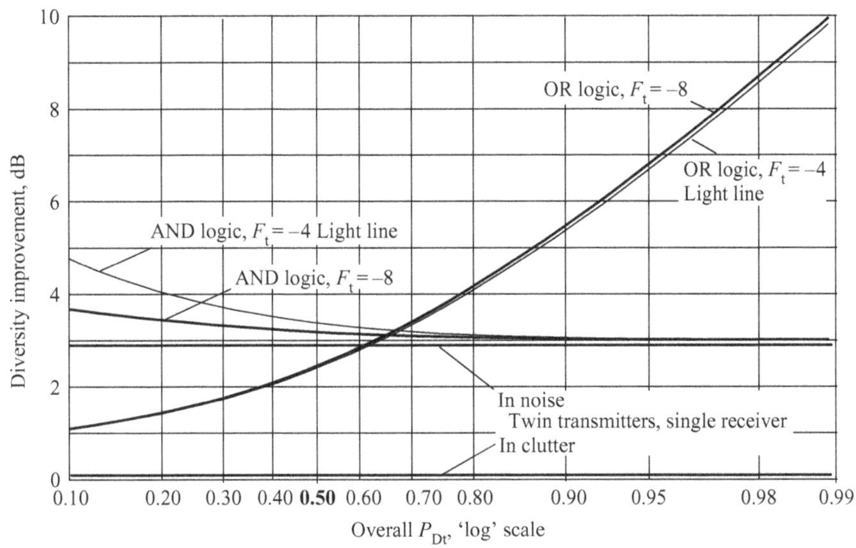

Figure 12.36 Diversity improvement. Detection improvement relative to single channel. Swerling Case 1 target. Confirms the superiority of OR logic when high P_D is required

Figures 12.35(c) and (d) illustrate the AND logic configuration. Although superior to a single radar alone, it is inferior to OR logic when P_{Dt} exceeds 0.6.

12.8.7 Combination performance

Figure 12.36 plots diversity gain to a base of system P_{Dt} for the above combinations, for the same two values of system P_{FA} exponent F_t, confirming the superiority of OR logic. Diversity is most beneficial when high P_D is required.

12.8.8 Practical problems

All radars contain sources of error which degrade the reported positions of targets. Each radar of a diversity pair will have its individual position error, so if both outputs are displayed on a PPI, in general two paints will be shown. Is there one target, seen (at slightly differing apparent positions) by both radars, or is the first radar viewing one target, perhaps a tug, while the second radar has a nearby second target, maybe a towed barge, the tug being in a TPM null? The problem of error in plot association is discussed in Chapter 13.

Combination of the outputs on a single display introduces circuit design problems, reduces operational flexibility (it would be difficult to optimise one radar for short range with the other set to a long range scale), and might compromise system integrity should one radar fail. Currently the pair of radars carried by ships operate almost independently, foregoing the potential benefits of diversity. Certain VTS stations do operate twin radars in diversity mode, usually both at 9 GHz, sharing a single scanner by interleaving transmission pulses.

12.9 Detection of active targets

12.9.1 RTEs and superhet racons

The discussion of active devices in Chapter 8 did not examine probability of detection. Unsaturated radar target enhancers act as passive point targets whose fluctuation characteristics are governed by the antenna radiation patterns, which are usually smooth enough for the RTE to become a Swerling Case 0 non-fluctuating target, even when not ground-fast. The RTE receiver noise component of output adds a generally insignificant component to system noise.

Receivers within saturated RTEs can be regarded as detecting all interrogations; $P_{D1} = 1$, subscript 1 denoting interrogate leg.

Racon receivers contribute noise but pick up no clutter, and see radar transmitters as non-fluctuating Swerling Case 0 sources. Because quite low sensitivity suffices to detect the one-way interrogation, receiver noise of superhet racons is small, their SNR usually being very high even when the interrogation is only marginally above receiver threshold, causing a rather sharp increase in P_{D1} from 0 to almost 1. Because of the numerous interrogating radars within range, pulse to pulse integration is not normally possible, or indeed necessary. These racons have Ricean probability distribution as for non-coherent radar receivers, Section 12.3. Unless the manufacturer's data sheet indicates otherwise, one may assume the racon receiver threshold is set for no more than one false interrogation per 20 s, with $P_{FA} \sim 10^{-8}$ (Eq. (12.1), $F = -8$, bandwidth ~5 MHz). The low false alarm rate provides a margin against deterioration in service, degradation at temperature extremes, and minimises both interference generation and battery drain. If a sidelobe suppression system adapts the threshold sensitivity according to interrogation strength or pulselength, P_{FA} is sometimes yet lower.

On the response path, the radar receiver picks up the usual noise and clutter. Active devices are point targets of rather smooth target pattern map and, when not combined with a platform of significant RCS, form Case 0 targets similar to passive point targets. Single-pulse SNR, q, is linked to P_D by Eqs (12.8) and (12.11a) and depicted by Figures 12.12–12.14. The benefits of integration remain available on the response leg. Observed at the radar display, overall P_D is the product of the interrogation and response leg probabilities of detection, P_{D1} and P_{D2}, respectively (unless the response transmitter becomes overloaded by an excessive number of interrogations), each having to exceed the required overall value. When the racon is muted, $P_{D1} = 0$.

$$P_D = P_{D1} \times P_{D2}. \tag{12.28}$$

The extended nature of racon response paints assists visual perception on the PPI at lower P_D than applicable to point targets. Indeed, the response tail may be displayed even when receiver swept gain has suppressed the initial portion; reliance should not however be placed on this effect, particularly as it causes range error. Considering say a 10 μs response as comprising 10 consecutive events when receiver bandwidth is say 1 MHz, the effective number of events integrated in the scan is multiplied by 10, improving SNR by at least 6 dB (non-coherent receiver, Figure 12.29) or 10 dB (coherent radar, Eq. (12.19b)). With this proviso, racons form Swerling Case 0 targets.

Offset frequency racons are received by the radar through a special receiver channel. The channel-splitting arrangement is likely to worsen the noise figure by a few decibels, but little or no clutter will be received if frequency offset is sufficient, enabling the racon channel threshold to be reduced. As the scanner may be operated away from its design frequency, its effective gain may be somewhat reduced and its sidelobes raised; appropriate values should be entered in the radar range equation.

Saturated RTE responses perform as Case 0 targets.

12.9.2 Racons, etc., with crystal-video receivers

A crystal-video receiver with square law demodulator can be used when it is unnecessary to gather interrogation frequency data, e.g. in swept frequency racons; Figure 12.37 shows the essentials. Radar proximity detector devices and the enabling receivers of more advanced RTEs have similar sensitivity to racon receivers and may also contain square-law crystal-video systems.

A 'crystal diode' demodulator when passing current generates several types of noise.

- Ordinary thermal noise in the spreading resistance.
- Shot noise in the p–n junction ('barrier noise'), caused by random electron emission, frequency distribution as for thermal noise.
- Thermal noise (noise figure ~2 dB) in the first stage of the following video amplifier.
- Flicker noise at low frequencies (below ~100 kHz). Noise density ∝ $1/f$.
- However, antenna and feeder losses and associated noise are negligible.

Crystal-video receivers have notoriously poor noise factors, caused by shot noise coupled with poor rectification efficiency. Luckily, the longest interrogation pulse of 1 MHz can be handled efficiently by a video amplifier whose bandpass filter's lower

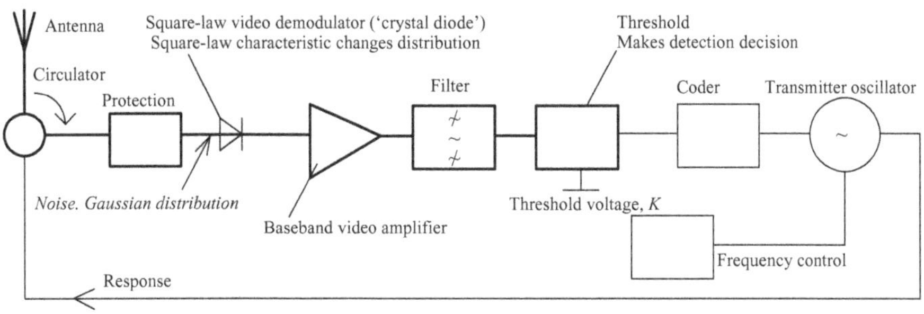

Figure 12.37 Crystal-video receiver. Shown as used in swept-frequency racon. Based on Figure 8.7, emphasising receiver components to detector

cut-off frequency is 500 kHz, keeping out flicker noise. The minimum pulselength of say 0.2 μs needing full sensitivity determines the upper cut-off frequency (base band operation, $B_n = 0.5/\tau$, see Chapter 3, Section 3.5.7), so upper cut-off frequency is ~2.5 MHz and noise bandwidth is ~2 MHz. Sufficient receiver performance is just achievable by direct microwave demodulation without resort to low noise microwave pre-amplifiers. (Crystal-video receivers are capable of −72 dBW sensitivity at 9 GHz for $F = -8$ and 3 dB pulselength 0.2 μs). Higher bandwidth may be preferred for proximity detectors, reducing sensitivity by a few decibels.

Swept frequency racons directly detect the demodulated microwave signal without an IF system. The original Gaussian distribution is retained, except that the detector has essentially a square law characteristic (voltage out ∝ power in), so Section 12.2 applies but with halved dB values. The small noise contribution from the succeeding video amplifier remains Gaussian but is insignificant.

Probability of detection of an interrogation, P_{D1}, is calculated on a single-pulse Swerling Case 0 basis as follows. As with superheterodyne racons, rated sensitivity, say −70 dBW, is usually taken at a fairly high P_{D1}, say 0.9.

For a linear system, from Eq. (12.4)

$$P_{D1} = \frac{1}{\sqrt{2\pi}} \exp\left[-\frac{1}{2}\left(K - \sqrt{2q}\right)^2\right].$$

The square law detector has the effect of squaring q, giving

$$P_{D1} = \frac{1}{\sqrt{2\pi}} \exp\left[-\frac{1}{2}\left(K - q\sqrt{2}\right)^2\right]. \tag{12.29a}$$

The racon receiver threshold, K, is found by putting $q = 0$, so $P_{D1} = P_{FA}$.

$$K = \sqrt{-4.605F - 1.8378}. \tag{12.29b}$$

Figure 12.8 applies with q_{dB} values halved (because square-law) and with rated sensitivity at say $P_{D1} = 0.9$, $F = -8$, as shown in Figure 12.38(b) for a rated sensitivity −70 dBW. Reducing interrogation strength by a mere 2 dB reduces interrogation P_{D1} from 0.9 to 0.1.

Alternatively, we may calculate the SNR at threshold sensitivity from Eq. (12.10a) (Case 0), Eq. (12.16d) (Case 1) or Eq. (12.18f) (Case 3a), add this value to the interrogation power to give the effective SNR and then calculate interrogation single-pulse P_D in the usual way, as for a radar receiver. For threshold $P_D = 0.9$ and $F = -8$, threshold SNRs are 14.21 dB (Case 0), 22.40 dB (Case 1) and 18.47 dB (Case 3a). These values are halved for a square law detector.

12.10 Practicalities

The accuracy of probability of detection calculations is detailed in Chapter 13. Meanwhile we note some practical factors which may modify the statistically determined probability. The list is doubtless incomplete and helps confirm the aphorism heading the chapter.

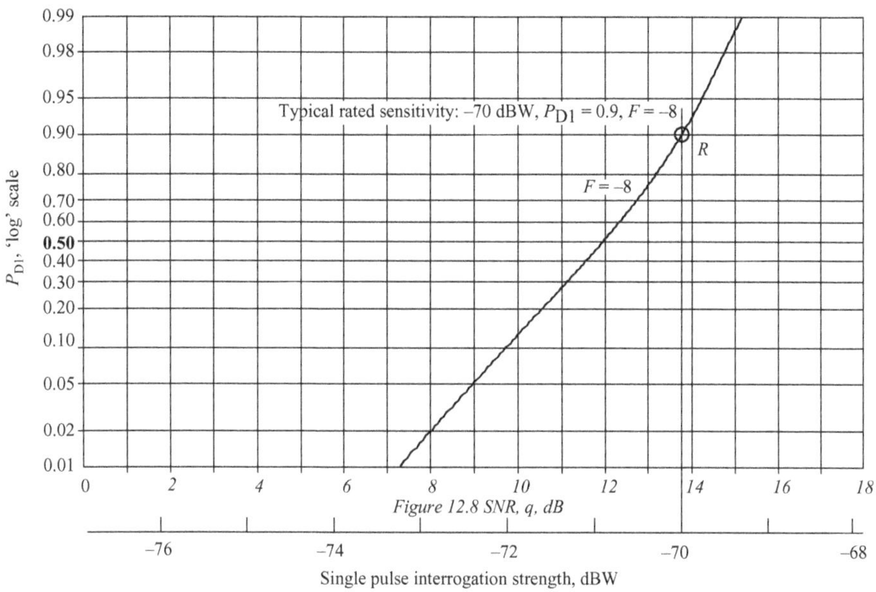

Figure 12.38 Probability of detection, crystal-video receiver. Based on Figure 12.8, assuming noise limited. Point R corresponds to R in Figure 12.8

12.10.1 Sidelobes and axial ghost echoes

Sidelobes of short-range targets can be a nuisance or navigationally dangerous even when displayed fairly infrequently, with quite low P_D. Returns received through sidelobes retain their usual fluctuation characteristics and the radar's pulse to pulse and scan to scan integration systems remains active. The radar threshold and false alarm exponent, F, of course remain as set for main-beam noise and clutter. Note the following.

1. *Passive targets*. The scanner gain term applies to transmit and receive legs. If, for example, the scanner main-beam gain is 30 dB and the sidelobes are 27 dB down, effective discrimination against sidelobes is $2 \times 27 = 54$ dB. Whether unwanted sidelobe echoes will be displayed is determined by recalculating the radar range equation for echo strength after resetting scanner gain, G, to $30 - 27 = 3$ dB and sidelobe P_D to, say, 0.1.
2. *Racons*. Because of the coding, a racon sidelobe response is unlikely to be mistaken for another racon, but its great length may mask adjacent echoes. Many modern racons include sidelobe suppression circuits which inhibit most or all responses to interrogations which the device calculates to be sidelobes. Not all racons have sidelobe suppression, and an unfortunate mix of interrogation frequencies and prfs may occasionally defeat those that do. Scanner gain occurs only once in the radar range equations for both the interrogation and response legs and

of course both legs must exceed their respective thresholds for sidelobe responses to be displayed. The example's discrimination would be 27 dB, so racon sidelobes are often relatively troublesome. Principal and secondary scanner sidelobes may display over a broad azimuth arc, in extreme cases approaching 360°.

On rare occasions racons are provided with directional antennas, usually approximating the simple ($\sin x/x$) radiation pattern of uniform aperture illumination. Here, radar main-beam may trigger responses through the racon sidelobes, extending operation beyond the intended sector. Calculation is merely a matter of replacing racon antenna gain by gain at its sidelobe, usually 13.3 dB down on main beam gain, similar to Chapter 2, Section 2.8.1, Figure 2.29.

3. *Radar target enhancers*. Interrogations via scanner sidelobes are unlikely to be sufficiently powerful to drive the RTE into saturation. Unsaturated, sidelobe performance is that of a passive target of the same RCS, sidelobe response being 54 dB down in the example. If saturation does occur, the device should be treated similarly to a racon, 27 dB down in the example.
4. *Ghost axial echoes* (Chapter 11, Section 11.8.3) can be calculated by an appropriate restatement of transmitter power in the range equation.

12.10.2 Roll and pitch

Figure 2.29 shows how roll in the radar-target plane as the scanner axis tilts alternately up and down reduces echo strength. When this reduction is included within the radar range equation, P_D fluctuates downwards from the even keel mean as the radar platform rolls (or pitches as appropriate) either way. Figure 12.39 restates Figure 2.17 in P_D terms for Swerling Case 1 targets having good even-keel P_Ds of 0.95 and 0.80. Figures 12.39(a) and (d) show that slight roll/pitch produces a modest near-sinusoidal echo strength fluctuation, the effect on overall detectability being slight, targets remaining readily detectable throughout the roll cycle. However, if roll or pitch peak amplitude reaches half of scanner elevation beamwidth, say from 10° to 12.5° for a typical marine scanner, the echo drops enough at roll/pitch peaks to spoil P_D. Figures 12.39(b) and (e), respectively, show that even-keel P_D of 0.95 falls to 0.81 and an initial P_D of 0.80 drops to less than 0.5 (the minimum for reliable detection) for 24 per cent of the roll cycle. When roll peak amplitude worsens further, to 0.75 beamwidths, the initial $0.95 P_D$ falls to 0.22, curve (c), and is less than 0.5 for 30 per cent of the cycle. Initial $0.8 P_D$ falls below 0.008, curve (f), and is less than 0.5 for 58 per cent of the time. Typical packets of echoes are shown as circles on curve (f). Scans near roll peaks receive very weak echoes and scan P_D collapses. If it becomes impossible to receive consistently strong pairs of scans, overall P_D will suffer particularly severely when scan to scan correlation is engaged. Swerling Case 0 or 3a targets have their fluctuation changed towards Case 1, because the echoes within individual scan packets retain their original correlation but the scans become decorrelated.

If roll time synchronises to a multiple of scan time (e.g. 24 rpm, scan time 2.5 s, roll period 10 s), and the scanner bears on the target at unfavourable instants X on the

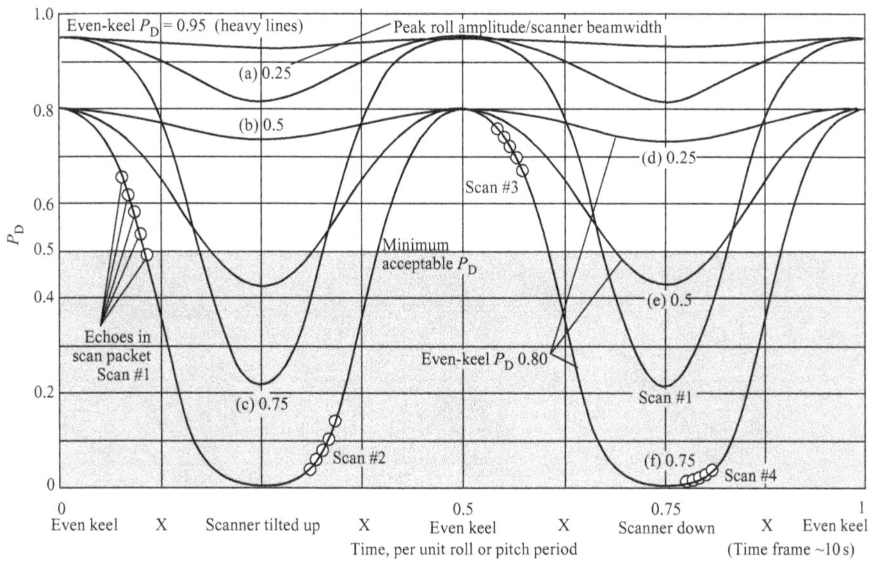

Figure 12.39 *Effect of roll or pitch on P_D. Moderate roll can severely reduce P_D at adverse points of the roll cycle. Swerling Case 1 target, uniform scanner aperture illumination. Echoes are correlated pulse to pulse but decorrelated scan to scan, biasing fluctuation characteristic of Swerling Case 0 or 3a targets towards Case 1*

figure, detection may remain weak for a minute or more. Often ascribed by operators to target fade, in reality the problem lies partly at the radar platform.

12.10.3 Wave screening

When a wave screens, masks or blocks a low target, the echo drops almost to zero, diffraction being very feeble. Long-term P_D is the product of (a) P_D during visible phases as determined by the methods set out in this chapter, and (b) the proportion of time that the target is visible. However high visible-phase P_D may be, if the target is screened for 10 per cent of the time, long-term P_D cannot exceed 0.9. Figures 12.40 and 12.41 restate Figures 12.22 and 12.23 for Swerling Case 1 and 3a echoes, respectively, having $P_{FA} = 10^{-6}$. Screening typically requires an additional few decibels SNR to assure 0.5 resultant P_D, although small targets may remained screened for so long that P_D remains below 0.5, however powerful the radar. Transition between screened and non-screened phases is not abrupt and screening may be incomplete, so P_D may oscillate roughly as Figure 12.39(e). Rhythmic screening may bias the fluctuation towards Swerling Case 1, as for roll in the previous section. If per-unit time screened is S

$$\text{screened } P_D = \text{unscreened } P_D \times (1 - S). \qquad (12.30a)$$

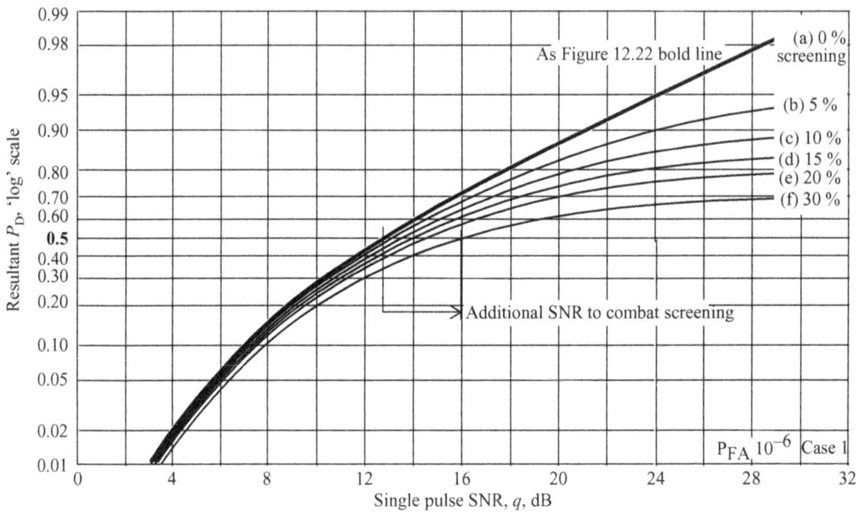

Figure 12.40 Reduction of effective P_D by screening, Swerling Case 1 target

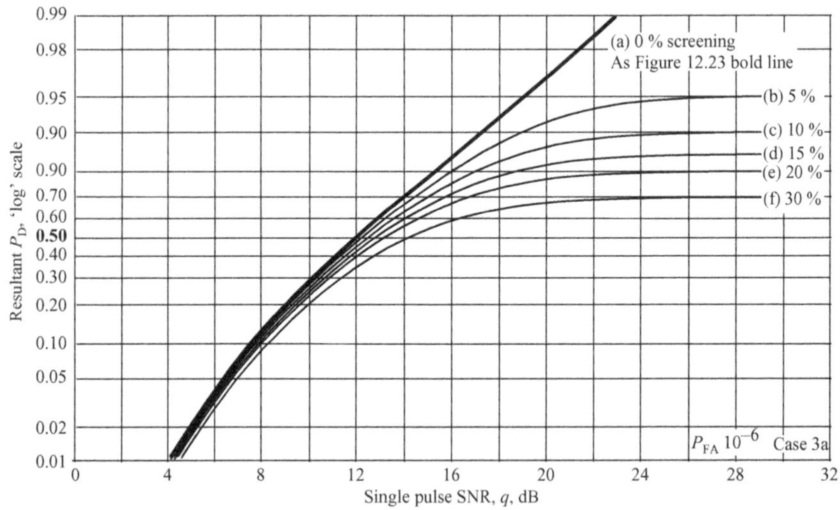

Figure 12.41 Reduction of effective P_D by screening, Swerling Case 3a target

With active devices, the paths of both legs are screened simultaneously or not at all, and the equation becomes

$$P_D = P_{D1} \times P_{D2} \times (1 - S). \tag{12.30b}$$

It is difficult to quantify S, which depends inversely on scanner and target effective heights (H and h or nj, respectively), significant wave height (h_s) and perhaps such secondary factors as sea/swell and obliquity of wave fronts to radar/target axis. The author tentatively suggests the following might be better than nothing:

$$S = \frac{h_s}{H} + \frac{h_s}{h \text{ or } nj} \text{ per unit.} \tag{12.30c}$$

12.10.4 Actual target fluctuation

The Swerling Cases assume echoes fluctuate with certain defined probability distributions, to which real targets may not conform; see for example, Chapter 10, Section 10.11, Figure 10.11, where the echo of a vessel on a constant heading remains consistently below the mean for many consecutive scans. It is probably best to calculate this condition by use of an artificially low RCS, say −5 or −10 dB on mean, retaining the Swerling case appropriate to the target class.

Fairly small ships may lie intermediate between Cases 1 and 3a, behaving almost as 3a to the 3 GHz band, migrating towards Case 1 at 9 GHz where the dimensions embrace more wavelengths. On the other hand, small Case 1 targets may change towards Case 3a when scanner rotation rate is slow, as in some VTS systems. Mounting a large passive or active reflector on a small ship might also bias it from Case 1 towards Case 3a, although the author does not know of any experimental backing for this hypothesis.

12.10.5 Losses

The text has noted several sources of performance loss within the radar. Individually usually rather small, in total these losses can significantly reduce the signal radiated towards the target, that received as echo or clutter, and the processing efficiency. Some of the losses degrade echoes but not noise or clutter. When calculating performance, rather than encumbering equations with numerous loss terms, it is convenient to group the transmission losses as a single term, L_t, reception losses as L_r and system processing losses as a third term L_p. Some of the hardware losses, for example, in the feeder, recur in L_t and L_r.

The values of many of the losses are not known with much certainty. Aggregate system loss, $L_t + L_r + L_p$, is around 13 dB and can rise to as high as 23 dB, so is by no means negligible. It is convenient to discuss losses as sources of error in detectability calculations in the next chapter.

12.10.6 Anomalous performance with small targets

The reason why ships carrying a pair of similar radars occasionally detect small yachts in clutter better at 3 than 9 GHz, seemingly contrary to the range equation, is under international investigation. Spreadsheet computation indicates that a small reflector producing the echo may be detected on the outer flank of the first 3 GHz multipath lobe, 9 GHz detection not occurring until the second lobe at shorter range.

12.11 Summary

Efficient target detection in clutter and noise is central to radar operation. Several strategies are used to maximise probability of detection (P_D). The following general summary omits a lot of secondary ifs and buts from the detailed treatments within the previous sections.

12.11.1 Targets

1. At given range, mean echo strength depends on RCS and environmental effects (atmospheric refraction and ducting at long range, multipath interference, atmospheric attenuation).
2. RCS of ships is difficult to measure and aspect-dependent. It conventionally lies within ±5 dB of numerical tonnage, gt, but is less for small, streamlined craft. When large ships and coastlines overflow scanner beamwidth or pulselength, effective RCS increases with range.
3. Range-dependent multipath interference alternately raises and reduces echo strength of point targets as indirect ray phasing changes and is most severe in calm water.
4. Extended targets show little multipath. Effective height is usually $\frac{1}{2}$ to $\frac{2}{3}$ tip height, affecting critical range where echo strength law changes from R^{-4} to R^{-8}.
5. Pulse–pulse or scan–scan echo fluctuation as the radar and/or target move in the seaway is approximated by a few Swerling case mathematical models. Large targets usually approximate Case 1, small ships Case 3 and fixed reflectors Case 0 (non-fluctuating).
6. Case 1 fluctuates slowly enough for a whole scan's echoes to be unusually small, needing exceptionally high mean SNR to assure high P_D. Case 3 is better, Case 0 best. Case differences become small when $P_D \sim 0.3$.

12.11.2 Noise

1. Receiver filters minimise noise, reducing effective bandwidth as far as possible without spoiling ranging accuracy. But big detection cells for narrow bandwidth collect more weather clutter; optimum bandwidth depends on clutter to noise ratio.
2. Noise is improved by minimising scanner and feeder losses and giving the receiver input stage a low-noise amplifier. Later stages, followed by less amplification, contribute insignificant noise. (Mismatched long feeders also introduce noise and ringing clutter; integration does not improve the latter.)
3. TR receiver protection cells recovering from the transmission are noisy for a few microseconds, but do not usually swamp strong short-range echoes. But insidious noise from faulty cells persists to long range, marring detection. Check performance frequently using built-in test equipment.

4. More transmitter power or scanner gain increase echo strength and signal to noise (but not clutter) ratio.
5. Noise limits long-range detection of small targets, but is soon swamped by moderate or strong clutter.

12.11.3 Precipitation

1. Precipitation along the radar-target path attenuates echoes, degrading SNR in a mildly temperature-dependent manner. To be conservative, assume storms extend through the whole path.
2. Precipitation clutter persists to long range.
3. Hydrometeor returns at the target enter the detection cells to compete with the echo, rising 5 dB for 2 : 1 equivalent rain rate increase.
4. Minimise precipitation clutter by decreasing scanner elevation and azimuth beamwidths, necessitating greater physical size. Roll and pitch limit shipborne elevation beamwidth to 20° minimum.
5. Precipitation clutter is proportional to (frequency)^4, favouring 3 GHz and limiting J (Ku) band to short range work.
6. Precipitation clutter distribution approximates Gaussian (normal), although scuds of rain may exhibit some organised structure.
7. Circular polarisation reduces rain (not snow) clutter more than echoes. The scanner must be switchable to linear polarisation in dry weather. Mechanical complexity and cost preclude shipborne use.
8. Precipitation clutter depends mildly on precipitation type and can be calculated with reasonable accuracy *if* the precipitation rate is known.

12.11.4 Sea-waves

1. Sea state is difficult to estimate and is imprecisely related to wave heights, which are variously defined.
2. Wave clutter rises about 5 dB per sea state number up to SS5, but is dependent on wind direction and whether the sea is fully developed or is a swell.
3. Minimise sea clutter by decreasing scanner azimuth beamwidth, needing wide physical aperture.
4. Sea clutter for given beamwidth depends on (frequency)^2, but for given aperture the advantage to 3 GHz is slighter.
5. Rough sea clutter becomes less Gaussian and perhaps better represented by Weibull distribution with varying shape factor; experimental data is sparse. Individual big waves may be resolved and displayed as 'targets'. Probability distribution may also vary between oceanic systems and the confused seas within a coastal bight.
6. Sea clutter rises when the radar looks down from a steep angle. It is often the limiting factor in detection of small distant targets. The sea clutter horizon is less than target horizon and sea clutter is a short to medium range problem.

12.11.5 Detection strategy

1. The usual magnetron radar is non-coherent. Some VTS radars are coherent or coherent-on-receive, preserving much of the phase information, improving integration efficiency.
2. Receiver dynamic range must be enough to prevent clipping of strong echo spikes which spoil signal to noise and clutter ratio. Logarithmic amplifier stages, good automatic gain control and clutter mapping can help. The operator optimises performance by the gain control.
3. Potential echoes are sorted from noise and clutter by an initial threshold, influenced by the gain setting.
4. The area within display scale range contains numerous detection cells, sized to match scanner azimuth beamwidth and chosen pulselength.
5. Digitised returns of a scan packet (two scans for scan–scan correlation) are allocated to cells according to scanner azimuth bearing and elapsed time from pulse transmission.
6. Each above-threshold return adds to the cell count. Counts exceeding M out of N possibles are declared as targets.
7. The thresholds adaptively provide the maximum number of false alarms tolerable by the processor or the operator.
8. Echoes correlate in specific cells: noise and clutter spikes have random distribution. Integration, especially of many returns, improves P_D.

12.11.6 Display accuracy

1. Steady plots, almost certainly representing echoes, demand high P_D and low P_{FA}. Weak plots jitter in strength and position, tiring the observer, who is unsure whether they are false alarms.
2. Jittery plots form very jittery tracks, giving imprecise future positions and closest points of approach (CPA). Details in Chapter 13.

12.11.7 System integration – diversity

- Using two (or more) scanner heights, geographic locations, frequencies or planes of polarisation can improve P_D – echoes correlate better than noise or clutter. Effective diversity demands sufficient spacings (in the widest sense). Correctly interconnected, performance exceeds that of a single double-power radar, but cost is high. Swerling Case 1 improves to Case 3, particularly beneficial at high overall P_D.

12.12 References

1 ROHAN, P.: ‘Surveillance radar performance prediction’ (Peter Perigrinus for The IEE, 1983)

2 RICE, S. O.: ‘Mathematical analysis of random noise’, *Bell Systems Technical Journal*, 1944, **23**, pp. 282–332

3 LEVANON, N.: 'Radar principles' (Wiley, New York, 1988)
4 SWERLING, P.: 'Probability of detection for fluctuating targets', *IRE Transactions*, 1960, **IT-6**, pp. 269–308 (Covered in the early pages of Skolnik, M. I.: 'Introduction to radar systems' (McGraw-Hill, 1985))
5 MEIKLE, H.: 'Modern radar systems' (Artech House, London, 2001), Section 12.25
6 SKOLNIK, M. I.: 'Introduction to radar systems' (McGraw-Hill, New York, 1985)

Chapter 13

Accuracy of position and track

'O let us never, never doubt what nobody is sure about.'

Hilaire Belloc, More Books for Worse Children, The Miracle

13.1 Introduction

13.1.1 The need to consider accuracy

Catastrophic marine accidents are rare. Of those collisions and strandings which do occur, many start as quite routine navigational situations; just another night, just another watch, just another target. Something then goes awry – perhaps the radar fails to display an islet or a target's calculated closest point of approach (CPA) is erroneous – jeopardising the crew, ship, cargo, passengers, other shipping or the environment, with losses running to a hundred million pounds. All significant targets have to be displayed to scale on the screen, often nowadays with chart features superimposed. We need to look in detail at the validity or accuracy of the plotted positions, for this accuracy, or lack of it, should always inform the navigator's judgement of the situation and the prudent action to take.

So far, we have concentrated on the process of detecting targets. Detection, although necessary, is not sufficient – the navigator would give little thanks to be told that there was another target somewhere or other within busy Tokyo Bay (Chapter 5, Figure 5.14). We have looked at how plots of target current positions are laid down on the display, showing where targets and coastal features are relative to own ship at the present time. This basic information is a useful start and must contain little error, but the navigator needs more: are targets moving or manoeuvring? What have they been doing? Are they hazardously close? Will they approach to close quarters? Where? When? Most of these questions involve movement, which is the rate of change of position with time, necessitating data collection during two scans at the very least. Movement is represented as a track, which is the target velocity vector either relative to own ship (relative motion display) or to the ground or the slowly moving water mass (true motion).

Target position and track have to be established with particularly high accuracy when extrapolating vectors into the future to find predicted CPA. This chapter therefore combines questions of accuracy, tracking and prediction. Tracking and prediction tasks are now automated, using digital processes based on present and past position data held within the frame stores. Modern radars often include as standard a plotting aid with the ability to track and predict several targets. Capabilities may be enhanced using either an automatic tracking aid (ATA, mandated for smaller cargo ships by IMO and supporting at least ten tracks) or the more comprehensive automatic radar plotting aid (ARPA, an IMO requirement for the bigger ships, 24 tracks minimum, often 50 or more being provided).

Accuracy is lack of error. We examine the several general kinds of error in Section 13.2. Good probability of detection is needed to get accurate plot positions, so we go on to analyse the accuracy of P_D calculations. Tracks are formed by joining successive plot positions, so plot inaccuracies cause track errors. Finally, we look at how extrapolation to obtain closest point of approach and similar predictions magnifies errors within the tracks. But first we look at the form of the display.

13.1.2 Display of target information

Chapter 3, Sections 3.8–3.10, discussed how targets can be displayed on an analog cursive long-persistence screen as raw radar plots showing current reported position, trails roughly indicating previous course and speed but outside the operator's control; or, after digital scan conversion, on a raster display as synthetic plots with trails available as needed.

Using a plotting aid adds capability. Additional processed target information can be displayed at will:

- track vectors showing courses made good to date;
- predicted tracks to some future instant;
- predicted closest point of approach to own ship, and time to CPA (TCPA);
- warning that a target has entered a guard zone surrounding own ship.

In the early days, the operator laboriously drew track lines on the convex face of the cursive tube for a few targets of interest, scan by scan, targets being identified by wax crayon colour up to a practical maximum of half a dozen. Advent of reflection plotters with concave semi-reflecting surface removed parallax error caused by the glass of the tube, which has to be thick to withstand the internal vacuum. But manoeuvre own ship – start again! Nowadays, tracking aids write tracks directly and automatically on the raster screen without parallax. Once the operator has assigned identity tags, they remain associated with their tracks even after switching to a short range scale throws them off the screen edge, and they do not smear as own ship manoeuvres.

Like any measuring instrument, radar is inherently subject to error. Cursive raw radar plots often jitter in position from scan to scan and occasionally fade in brilliance, irritating the operator trying to draw a reflection plotter track, but graphically indicating the errors inherent in the positioning process. The crisp and steady point of light denoting a synthetic target plot on a modern raster screen insidiously hides errors and uncertainties from view, but they are there just the same. It is easy to forget that all any plot really states is that noise, clutter or some reflecting part of an object, which might be a valid target, was fairly near the indicated position within the last few seconds.

Similarly, a track means that there is probably a target (but perhaps an unlucky set of successive noise or clutter spikes) whose historical progress has been approximately as shown, accuracy being degraded if, for instance, reflections happen to glint from one reflecting facet to another, or own ship is rolling heavily. Predicted track, CPA and TCPA, being forward extrapolations from historical data, magnify errors the further ahead they look, and of course are vitiated if the target consciously manoeuvres or cross-currents and wind change its leeway.

Explicitly or implicitly, the operator (or system designer) needs to know the following.

- Probability of detection – answered within Chapter 12 by reference to SNR and dependent on the accuracy with which the radar, target and environmental parameters are known.
- Target discrimination – minimum spacing before two targets merge as one.
- Range and bearing accuracy of the displayed plot.
- Speed, angle and position accuracy of the displayed track.
- Track, CPA and TCPA accuracy of the extrapolated track.

Most of these parameters depend on actual achieved SNR, so we shall try to evaluate the likely accuracy of SNR calculations before attempting answers to the operator's questions. But first we need to clarify the general concepts of error and accuracy.

13.1.3 Sources of error

Errors may be caused by combinations of factors within the radar, the target or the environment. Some error components indirectly depend on the operator's choice of control settings, for example, which affect receiver bandwidth and hence signal to noise ratio. Speaking broadly, errors may be systematic or random, both kinds being handled by rational processes discussed later. Not so blunders, which are operator's irrational mistakes, best detected by running a rough check, of the sort given in previous chapters, or by plotting a run of results to see whether any stand out suspect against the crowd. Never be fooled into thinking that the result of a complex calculation must be right because the computer gives a result to many significant figures – GIGO – garbage in, garbage out! Imperfect software can blunder also, as when some input parameter throws an algorithm outside its operational range.

Errors within calculations arise from the following.

- Uncertain mean echo and clutter strengths.
- Real-life radars falling short of theory, despite designers' best efforts.
- Simplifications and approximations within our analysis of the detection process.
- The operator having optimised the controls (range scale, gain, swept gain, etc.) for another target.

- Premature rounding of the values of individual terms in equations. It is best to retain one more significant figure than the term's accuracy warrants, dropping one significant figure from the result.

13.2 Forms of error

13.2.1 Absolute and relative error

We shall be using the words error and accuracy in their scientific senses whose meanings and properties must be understood before getting down to specifically radar problems. According to the dictionary,[1] error is 'the amount by which an observed or approximate numerical result differs from the true or exact one', accuracy being 'the degree of refinement in measurement or specification, as given by the extent of conformity with a standard or true value'. Accuracy is the complement of error.

Tolerance is the permissible error in an engineering component or a system parameter. For example, a scanner gain might be specified as 30 ± 1 dB, gain tolerance here being ±1 dB. The design aims for 30 dB, but acceptable limitations of slot-cutting accuracy may cause 1 dB performance error either way, the scanner designer having specified appropriate dimensional limits to which the machine shop must conform. The scanner would more likely nowadays be specified worst-case; 'Gain not less than 29 dB', ensuring the same minimum system performance as before. Although leaving the user unsure of average performance, it obviates the silly possibility of the supplier's test department rejecting scanners in which a fortunate combination of slot errors yields better performance than expected – it has happened!

A digital quantity – varying in steps – may be measured exactly, with complete accuracy and zero error. But to make such a measurement of an analog quantity – varying by infinitesimal amounts – is an ideal which cannot be attained. Such measurements are always in error by an amount which may or may not be significant for the purpose in hand. We can be sure we have twenty plums exactly, but not 1 kg exactly of plum jam, for all scales have some instrument error. Subsequent digitising does not remove the errors inherent in originally analog quantities, but may prevent accrual of further error in later processing. Subsequently putting the jam into fifty 20 g jars does not improve accuracy, for the jars have a manufacturing tolerance, but can prevent further error when serving one jar per diner.

Absolute error, δx, is the amount by which the measured value of a quantity, x, differs from the true value, X, all expressed in the same units:

$$\delta x = x - X. \qquad (13.1a)$$

Instrument reading minus absolute error gives the true value:

$$x - \delta x = X. \qquad (13.1b)$$

[1] Shorter Oxford English Dictionary definitions.

Eq. (13.1a) follows standard engineering practice and conforms with the dictionary, but alternatives exist: $\delta x = X - x$ (of opposite sign), or $\delta x = |x - X|$ (the modulus, always positive). If polarity is uncertain we can write absolute error as $\pm\delta x$.

Relative error, a pure number, is expressed as a fraction of the whole, and may be quoted as, say, 0.05 per unit.

$$\text{Relative error} = \frac{\delta x}{X} \text{ p.u.} \tag{13.1c}$$

Often only the measured value x is available, the true value being unknown. Relative error is then loosely referenced to x. The inaccuracy is small when $\delta x \ll X$.

$$\text{Relative error} \sim \frac{\delta x}{x} \text{ p.u.} \tag{13.1d}$$

$$\text{True value is:} \quad X = \frac{x}{1 - \text{relative error}}. \tag{13.1e}$$

Relative error may also be expressed as, say, 50 parts in 1000 or as 5 per cent. Relative error gives an immediate indication of the precision of the measurement. If X is power-related, decibels may be employed, here relative error is 0.13 dB. For example, SNR could be so expressed, but it is reckoned inelegant to express quantities such as range error in this way.

13.2.2 Systematic error

Systematic errors are those inherent in the construction and calibration of the measuring instrument, or its method of use, which apply to the whole set of measurements. For example, range measurements of a large target tanker may include systematic error if the time delay in a long feeder linking the scanner has not been allowed for, or the scanner position not been properly referenced to the ship's consistent common reference point (usually the conning position), or if the echo derives from accommodation-block scatterers aft – the bow might then be 250 m closer than the radar indicates. This absolute error of -0.25 km might be trivial at say 25 km range, where the relative error is -1 per cent. At 2.5 km range the absolute error remains -0.25 km, but the relative error rises to -10 per cent, which is navigationally significant. Constant systematic error is present on every scan of a set, but may change if circumstances alter; say the tanker turns to present its beam. If the feeder correction is known to have been omitted at time of installation, the range scale zero setting component of systematic error can be allowed for or calibrated out, in this case by subtracting (feeder length)/(relative velocity of propagation) from all measured range values, or by comparing radar range of a conspicuous object with range found independently, say from visual cross-bearings laid down on an accurate chart.

In this particular example, the raw display is an extended paint roughly representing the plan area of the scattering elements, giving some indication of target size and aspect. Glint shows as fluctuation of the area and is fairly obvious to the operator, who has to judge the target centre or bow when plotting on a reflection plotter. On the other hand, the synthetic plots usual on tracking aids show what the processor thinks is the centre of the target. Any glint is concealed by this centroiding and size is not usually indicated, perhaps lulling the operator into a false sense of the display accuracy.

There are several kinds of systematic error. The failure to measure range to the bow is a systematic error of method – there was insufficient radar sensitivity to detect the weak bow echo or the radar performed an unwanted centroiding function. The feeder introduced a systematic instrument error. Systematic operator error would result from habitual placement of the variable range marker on the furthest edge of the trace; mere sloppiness of placement, sometimes before and sometimes beyond being a random error. (Placing it on the wrong target is a blunder.)

These error components remain constant in absolute terms to form a systematic zero error, z m (meaning error in placement of the origin, rather than no error!). Such errors are summed arithmetically, taking due account of sign. Other systematic components of error might cause a constant relative error or systematic scaling error, s; for example if the clock driving the variable range marker ran slow, echoes would count too few clock pulses and appear at say 99 per cent of true range, a relative error of +1 per cent. Such error components would have to be calculated as the appropriate number of metres at the current range before addition to the zero error to give total systematic error.

If true range is R, measured range is r and systematic zero and scaling errors are z and s respectively, so $R = sr + z$, velocity or rate of change of true range is

$$\frac{\mathrm{d}R}{\mathrm{d}t} = s\frac{\mathrm{d}r}{\mathrm{d}t} + \frac{\mathrm{d}z}{\mathrm{d}t}\ \mathrm{m/s}. \tag{13.2a}$$

As z is independent of time, the last term is zero and rate of change of measured range gives a relative velocity independent of any zero error. If all measurements are made with the same instrument, say when determining relative movement from successive position measurements, systematic zero errors cancel out. There remains a scaling error, s, calculated velocity error being 1 per cent in the example. Mixing the readings from a pair of radars having differing scaling and zero errors would sharply increase velocity error.

When a measurement or calculation involves only products or quotients of the factors, which have per-unit [or percentage] *small* systematic scaling errors $s1$, $s2$, $s3, \ldots$, respectively, the maximum systematic error in the result approximates their sum:

$$\delta x \sim (s1 + s2 + s3 + \cdots)\ \text{per unit}. \tag{13.2b}$$

This rule would apply, for example, to the form of the radar range equation given in Chapter 4, Section 4.3.2:

$$S_{\mathrm{e(FS12)}} = PG^2\sigma\lambda^2(4\pi)^{-3}R^{-4}[L_\mathrm{t}L_\mathrm{r}]^{-1}\ \mathrm{W}. \tag{4.6a}$$

The dependence of $S_{\mathrm{e(FS12)}}$ on any particular term is found by partial differentiation with respect to that term. When the term is raised to power n in Eq. (4.6a) and ∂ represents the partial differential (without change of the other variables),

$$\partial(S_{\mathrm{e(FS12)}}) = n\partial(\text{term}). \tag{13.2c}$$

For example, the echo strength dependence on range, where $n = -4$, is -4 per cent increase in echo strength per 1 per cent increase in range.

The principle is applied to error components within the terms of the range equation as follows. If for example, per cent systematic scaling errors in the range equation components are: $P = -1$, $G = 2$, $\sigma = 3$, $\lambda = -4$, $R = 5$ and each L term $= 6$; total error $= -1 + 2 \times 2 + 3 - 2 \times 4 - 4 \times 5 - 2 \times 6 = -34$ per cent, much of which is contributed by the high exponent of the R^{-4} term.

If, however, the terms are thought to be subject to maximum systematic scaling errors of ± 1, ± 2, ± 3, ± 4 and ± 5 per cent, respectively, the total error always lies between $\pm(1+2\times 2+3+2\times 4+4\times 5+2\times 6) = \pm 48$ per cent. In simple scenarios having only two or three terms, we could argue that there is a significant chance of all terms taking their maximum negative or maximum positive values together and calculate as above, but where the terms are numerous, as in the range equation, the likelihood of all simultaneously taking the same limit values falls, and it is usual to assume a Gaussian statistical distribution along the lines of random errors in the next section if the upper and lower boundaries of each term's tolerances are known. Say it became critically necessary for safety reasons to ensure that error never exceeded some sum value, one would follow the summation of Eq. (13.2a). When examining the impact of errors within a formal safety assessment, one should always follow the methodology laid down for the assessment, or state the methodology adopted.

Repetition of a measurement within a system having only systematic errors does not refine the result. The errors apply equally to each measurement set, so the answers will be identical, all containing the same error.

13.2.3 Random error

Random errors arise from chance events such as clutter external to the radar, or random internal factors such as noise or granularity imparted to the data when digitised. Random errors vary in magnitude and polarity from instant to instant, some being less than the true value, others more. When the measurement is repeated, random errors tend to cancel, and the mean of n measurements approximates the true value.

$$X \sim \frac{1}{n}\sum_{0}^{n} x. \tag{13.3}$$

We saw this mechanism in action in Chapter 12, Section 12.6, where integration of multiple observations reduced the effects of noise and clutter, improving SNR.

When the various error components within one measurement are independent of one another, i.e. are decorrelated, resultant random error is best treated statistically, for it is unlikely that all its components will simultaneously have an extreme value. If the components are random (even if not individually having Gaussian distribution), it follows from the central limit theorem that the overall distribution is likely to approximate Gaussian (Chapter 11, Section 11.3.8; probability density function Figures 11.1(a) and 11.2). If the components are truly Gaussian, they have an infinitesimal chance of being infinity but distribution curves are often truncated in practice. This usually has little significance except when considering false alarm rates.

Professor Griffiths[2] comments that precision can be improved by reducing random errors by integrating multiple observations, but cannot reduce systematic errors (i.e. improve accuracy). For example, when measuring the length of a piece of string with a ruler, you can average many measurements to get a more precise answer, but if the ruler is wrongly made (a systematic error) the result will still be inaccurate.

The following assumes Gaussian distribution. The standard deviation, σ, is a measure of the amount of scatter and is the square root of the sum of the squares of the differences a, b, c, etc. from the mean:

$$\sigma = \sqrt{a^2 + b^2 + c^2 + \cdots}. \tag{13.4}$$

The squared function causes the largest of a set of errors to dominate. For example, if $a = 1$, $b = 2$ and $c = 3$, $\sigma = \sqrt{1 + 4 + 9} \sim 3.741$. Ignoring the smallest term would give $\sigma \sim 3.606$, only 3.6 per cent different. Ignoring all but the largest term gives $\sigma = 3.0$, 20 per cent different.

$$\text{Probable error} = 1 - \text{erf}\left(\frac{\sigma}{\sqrt{2}}\right). \tag{13.5a}$$

For the error function, erf (x), see Chapter 11, Section 11.3.8, Eq. (11.6). Probability and probable error are graphed in Figure 13.1. Figure 13.2 redraws probable error to a logarithmic scale to reveal behaviour at high σ values and Figure 13.3 shows probability logarithmically. Note that:

- 50 per cent of readings are less than, and 50 per cent exceed the median, which is 0.675σ,
- 31.6 per cent (approximately $\frac{1}{3}$) exceed 1 standard deviation, 1σ,
- 4.67 per cent exceed 2σ,
- 0.286 per cent exceed 3σ,
- 69.7×10^{-6} exceed 4σ.

The asymptote line of Figure 13.1 shows that when probability is low,

$$\text{Probability} \sim 0.82 \times \text{Error}. \tag{13.5b}$$

The probability of the error magnitude lying between x and $x + \delta x$ is

$$P(x) = \frac{1}{2\pi\sigma} \exp\left[\frac{-x^2}{2\sigma^2}\right]. \tag{13.6}$$

13.2.4 Latency

To make and report or display any measurement takes time. Considering the returns within a single scan, the sweeps in the packet occupy around 10 ms, and a similar time might be needed for analysis. The target is not then again scanned for ~2 s, so if two-scan correlation is employed, the result is not available until more than 2 s from

[2] H. Griffiths, personal communication, 2003.

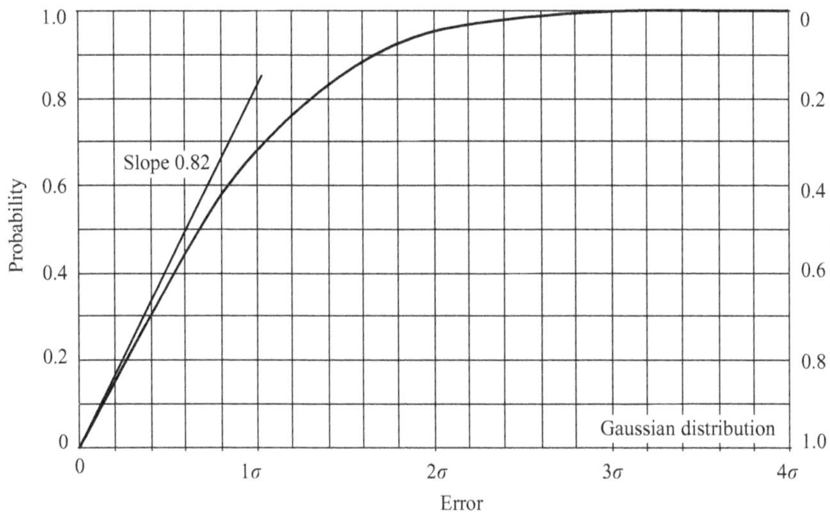

Figure 13.1 Probability and probable error. Probable error becomes small beyond the 2σ level

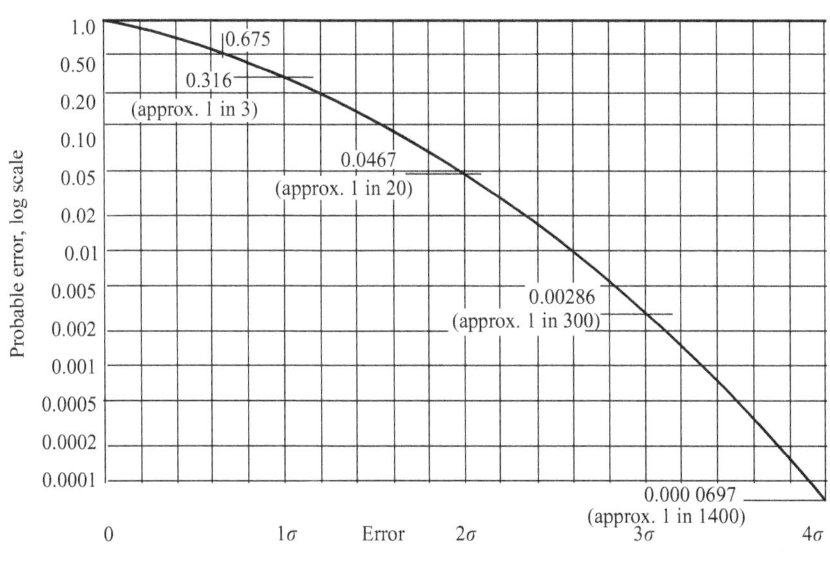

Figure 13.2 Probable error. Figure 13.1 redrawn to logarithmic scale

start of the measurement; longer if several scans are correlated. During this latency time the target may have moved several tens of metres, putting the displayed position in error by an amount proportional to relative velocity. In principle, latency errors can be calibrated out if velocities remain constant. Data should therefore be time tagged,

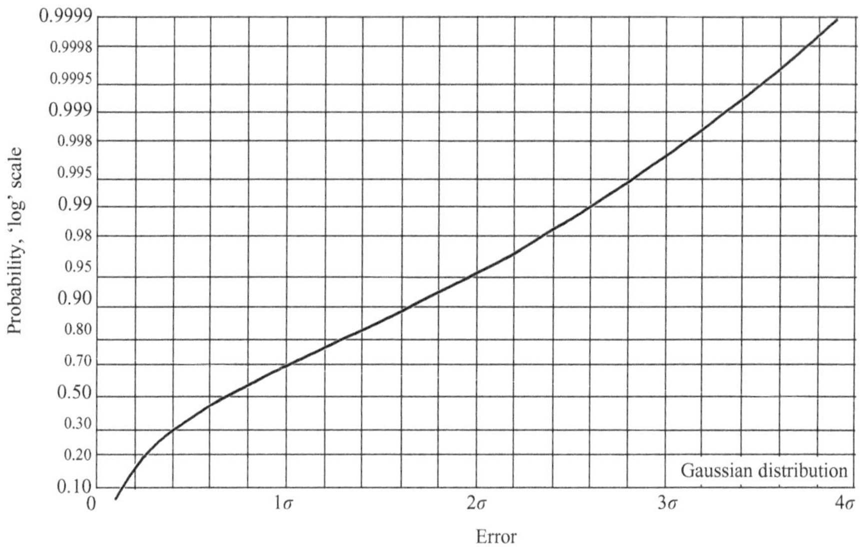

Figure 13.3 Probability

particularly when latency is unlikely to remain constant, for example when several out-stations feed a central VTS station over narrow-band or time-shared lines.

13.2.5 Quasi-random error

Some error components are not strictly random. The finite risetime of the receiver filter output causes a strong echo to be detected marginally faster (with less latency, giving shorter apparent range) than a weak echo; in Chapter 12, Section 12.3.1, Figure 12.10(*e*), compare instants of detection of strong echoes E8 and E9 with the preceding weaker ones. For given echo strength, the delay is least when receiver bandwidth is high, e.g., on short range scales. Latency from this cause is unlikely to exceed $0.1/B_{\mathrm{MHz}}$ μs ($15/B_{\mathrm{MHz}}$ m), where B_{MHz} is bandwidth in megahertz, equating to ~15 m maximum for 1 MHz bandwidth. Likewise, when SNR is good, plots may tend to be rotated anticlockwise by up to half a beamwidth towards the leading edge of the scanner beam.

13.3 Errors in terms within radar performance calculations

13.3.1 Introduction

This section summarises some sources of error within the calculations developed in earlier chapters.

Because losses are generally imperfectly known, they introduce error to calculation of SNR, which in turn affects accuracy of calculated probability of detection.

Atmospheric attenuation, discussed in Chapter 5, Section 5.9, is entered as a specific loss term L_A in the radar range equation. Many of the effects summarised below depend on wavelength, varying slightly within a single frequency band. Except when highest precision is required, it is usually acceptable to insert mid-band frequency or wavelength if actual values are unknown. It is never permissible to assume performance within one band describes performance in another; always re-calculate.

Most or all of the following loss components are systematic. They may improve as the radar warms up in the first few minutes of operation – or deteriorate if it overheats – and may depend somewhat on range scale.

Values ascribed to some losses may be arbitrary in absence of firm data. Relevant assumptions should always accompany SNR calculations.

13.3.2 Transmitter hardware losses

This group, L_t, includes all losses which reduce the radiated power below its nominal value.

1. *Short-pulse loss of transmitter power.* Modulator imperfections may reduce peak power by a couple of decibels on short pulselength.
2. *Duplexer.* Relative to the magnetron power output as reference, loss in the duplexer reduces power at the transceiver unit output by a few tenths of a decibel.
3. *Feeder resistive loss.* This depends on the length and type of feeder if there is one, see Chapter 2, Section 2.6.2, Table 2.2.
4. *Feeder mismatch loss.* This depends on the load mismatch (scanner, and rotating joint if a separate component), see Chapter 2, Section 2.6.2, Eq. (2.5b) and Figure 2.16. This loss remains if a transceiver is connected direct to a mismatched scanner with no intervening feeder and is 0.5 dB when VSWR $= 2$. Feeders are an unseen item, out in the weather and vulnerable to mechanical damage leading to additional mismatch and water ingress or condensation, raising loss.
5. *The scanner.* This contains several loss components, generally included within suppliers' quoted overall gain figures (Chapter 2, Section 2.7.16), so rarely needs further consideration. When VTS and range surveillance system parts are individually sourced from specialist suppliers, one must properly account for losses in rotating joints, etc. Scanner losses total about 2.5 dB excluding beamshape loss.

13.3.3 Service loss

Achieved in-service performance of new equipment depends on the amount of tender loving care bestowed by the installation team. Among other things, swept gain law and scanner tilt need matching to scanner height, range zero has to allow for feeder transit time and some data extraction settings depend on scanner azimuth aperture. Scanner bearing has to be collimated to the ship centreline. VTS and range surveillance sets have to be surveyed-in relative to chart datum, and it may be necessary to correct for latency in data transmission to the central processing system.

It is fair to assume that properly installed and maintained modern equipment in good working order meets the minimum values of transmitter power, noise figure, etc., promised in the manufacturer's data sheet with allowance for any installation-specific factors such as feeder loss. New equipment may exceed the minimum by a couple of dB, but this cannot be relied on. Any margin in hand gives a cushion against some of the uncertainties always surrounding environmental effects and target RCS. Provided there are no untruths, the supplier may legitimately highlight strengths without dwelling on weaknesses, so data sheets should be read forensically without jumping to unwarranted conclusions. For example, radar transmitter power may be stated at the magnetron flange rather than at the feeder input. If so, one should enquire the intermediate losses in the duplexer, or at least make an informed guess based on Chapter 2, Section 2.3.2.

It is prudent to assume system performance gradually deteriorates to an extent dependent on the roughness of operating service, quality of manufacture, use made of built-in test equipment and the servicing policy. For example, magnetrons slowly lose power and moisture may leak into feeders. Radomes and scanner windows are finished with self-cleaning slippery surfaces to help shed dirt, water and ice. Minor surface grime has little effect but thick buildup of ice or dirt, especially soot, may introduce attenuation and mismatch losses. Many ordinary paints are lossy and have rather high dielectric constant so should never be applied to radomes and windows, but sometimes are.

Digital technology makes modern radars far less prone than their analog ancestors to in-service deterioration as components drift, and offer few preset controls as hostage to those dubiously qualified servicing technicians who come aboard in foreign ports. Nevertheless, it is prudent to include a service loss term in the radar range equation to allow for minor shortfalls in performance parameters. The receiver service loss (Chapter 3, Section 3.2.2), say 1 dB, comprises scanner deterioration through dirt, etc.; feeder deterioration; loss of noise performance as the TR cell and mixer crystals age, and minor local oscillator tuning errors.

In absence of specific data, transceivers sheltered from the weather and not subject to temperature extremes may suffer 1 dB transmitter service loss. Where the transceiver is located at the mast-head, it is prudent to allow a couple of decibels additional service loss each way at temperature extremes. Service loss assumptions should always be stated with calculations.

13.3.4 Receiver hardware losses

The scanner losses recur on the receive leg, as does the feeder resistive loss. Mismatch loss is normally that of the transmit leg. Polarisation loss occurs if the target effective RCS is reduced by the chosen polarisation, for example, corner reflectors with circular. The duplexer and protection circuit (Chapter 3, Section 3.2.3) introduce loss, most conveniently expressed as raised noise temperatures, or incorporated within the system noise factor or noise figure.

The effective noise bandwidth depends primarily on the receiver bandwidth, which we have taken into account, but also to a small degree on the (usually unpublished) shape of the response curve. The effect is small, best handled within the service loss.

Published parameters naturally assume the radar is set to full sensitivity and the possibility of the operator desensitising the receiver using the differentiator control or by turning down control settings should be considered.

Receiver noise figure or factor is quoted in data sheets, either overall or for the first stage or LNA, in which case system noise figure is likely to be a decibel or so poorer. Noise figure deteriorates with age, especially if there are gas TR cells. Full use should be made of any performance check facilities; it is worthwhile to confirm which parts of the radar these embrace.

13.3.5 System processing losses

The following losses total about 8 dB in non-coherent modern marine radars.

- *Beamshape loss.* The assumption, made for ease of calculation, that the scanner beamshape is rectangular causes a beamshape loss (Chapter 3, Section 3.3.1) of about 1.6 dB (two-way) to non-coherent systems, about 2 dB to coherent systems and maybe as much as 4 dB when the platform yaws, rolls or pitches through a beamwidth.
- *Scanning loss.* The angular movement of the scanner between transmission of an interrogation and receipt of the echo was discussed in Chapter 2, Section 2.7.15 and is usually negligible.
- *Filter weighting loss.* For ease of calculation, we assume the receiver filter has rectangular frequency response, and the filter to be matched to the transmitted spectrum, together causing 1–3 dB loss, see Chapter 3, Section 3.5.2.
- *Quantising loss.* In general, the analogue voltage when digitised leaves a remainder which is ignored, causing a random error of 0.0834 × least significant bit power (−10.79 dB). Meikle [1, Section 10.4.1.1] quotes quantising losses for differing number of significant bits representing the noise as 0.35 dB (1 bit), 0.09 dB (2), 0.04 dB (3), 0.03 dB (4) and 0.01 dB (6 bits).
- *Quantising error.* This error arises when an analogue signal is digitised, because 1 bit may not quite equate to 2 : 1 voltage ratio. Where x is the least significant error expressed in error standard deviations, Meikle [1, equation 13.21] gives:

$$\text{Quantising error} = \sqrt{\frac{1+x^2}{12}}. \tag{13.7}$$

- *Straddling loss* of say 0.5 dB arises from the echo on average not sitting squarely within a single detection cell, see Chapter 3, Section 3.6.3.
- *Overflow* of large targets beyond a single detection cell in azimuth and/or range reduces cell RCS, see Chapter 10, Sections 10.7.1 and 10.7.2, respectively.
- *Integration loss* arises when the returns are imperfectly integrated, Chapter 12, Section 12.6.3. This loss is dependent on the radar integration scheme, the number of pulses and the fluctuation characteristics of the target, and includes:
- *Operator loss* if a cursive display is employed as the integrator (L_{op}, Chapter 3, Section 3.10.3, Eq. (3.5)).

13.3.6 Point target responses

The fairly uniform azimuth polar diagrams of most point reflectors can be severely degraded by interference effects with the host structure skin echo, Chapter 7, Section 7.10. Mean RCS is often only mildly frequency-dependent, but all active devices and a few passive reflectors such as the resonant patch type may have sharp band-edge cutoff. Here special attention should be paid to systems including radars working at or beyond the edges of the recognised marine bands. Echo strength depends critically on multipath interference (see Chapter 5, especially Section 5.3, Figure 5.4), which in turn depends on height and atmospheric refraction, which itself changes with weather or time of day.

As shown in Chapters 7, 8 and 12, RCS of point reflectors tends to be somewhat less uniform than that of active reflectors, but the fluctuation of the reflector alone probably remains approximately Swerling Case 0, the combination with the structure skin echo approximating Cases 3a or even 1. RCS of most passive point targets depends on the fourth power of frequency. The radar bands are about 2 per cent wide, so calculations assuming centre-band are likely to be in error by about ± 4 per cent (0.18 dB) at band edges.

As noted in Chapter 8, Section 8.4.3, poor interrogation strength at frequency agile racons causes response frequency jitter. Frequency error δf depends on interrogation-leg SNR, q, per Eq. (8.3b);

$$\delta f = \frac{1}{\tau\sqrt{2q}} \text{ Hz rms.}$$

If, say, radar receiver bandwidth is matched at $B = 1/\tau$ and $\delta f = B/4$ (giving $\sim$1 dB loss), $q = 8$ numerically, or 9 dB.

The distinctive responses of racons and SARTs are visually decoupled from any echoes of the host structure and are therefore non-fluctuating, Swerling Case 0. Although an RTE on its own may have an excellent radiation pattern and Case 0 response, in practice it usually operates with a host vessel having significant skin echo. As shown in Chapter 8, Section 8.15, the composite pattern is not uniform and may have Swerling Case 3a or Case 1 fluctuation characteristic, raising the necessary SNR for high P_D.

Saturation range of RTEs depends partly on the RTE itself, but partly on the radar and environmental parameters, so if these are changed system performance should always be re-calculated, rather than merely inserting the unsaturated RCS in the radar range equation. Heavy traffic may overload active devices, restricting the number of responses received by an interrogator and precluding achievement of high P_D.

Although antenna gains, receiver sensitivity and transmitter power of active devices may vary across the band, individual frequency dependencies are rarely quoted in data sheets, to reduce test cost and because there may well be device-to-device variations. Assuming the data sheet gives minimum values throughout the band, performance may turn out a couple of decibels better at a spot frequency. Performance may drop abruptly at band edges and it is never permissible to assume active devices work at all outside their declared bands.

The circuits of many active devices are poorly shielded from ambient temperature, perhaps raised by sunlight, and have to be designed to operate between say +55°C and −20°C for tropical and temperate locations, and lower for high latitudes or Eurasian or North American continental winters, where −20°C is not thought cold. Failing specific information, it is prudent to assume a couple of dB shortfall of receiver sensitivity and response power at temperature extremes.

13.3.7 Extended target RCS

Ships' RCS are rarely certainly known and are likely to vary significantly with aspect, radar band, deck cargo and other factors. Values may sometimes be deduced from observed detection range, otherwise RCS will have to be inferred from empirical data such as those in Chapter 10, Section 10.4. RCS usually rises somewhat with frequency. Depending on aspect, effective RCS may fall when the target dimensions overflow the detection cell at short range. Coast echo strengths are even less certain, and may vary with wet or dry weather or seasonally with vegetation growth.

13.3.8 Scanner rotation

The small rotation angle traversed between sweeps, $\delta\theta$, depends on prf and rotation rate, introducing an angular error which Meikle [1, p. 395] quotes as:

$$\text{error} \sim \frac{\delta\theta}{\sqrt{12}} = 0.289\delta\theta \text{ rad rms.} \tag{13.8}$$

Typically, for 1000 pps and 2.5 s scan time, error = 0.042° rms and is negligible compared with the beamwidth.

Centroiding is a computational method of reducing error of a number of azimuth measurements of varying amplitude, giving more weight to readings taken when SNR is high and error low. Range may be similarly centroided.

Centroid target bearing (or range)
= sum of the individual azimuth (or range) measurements
× their signal strength products divided by the sum of the
signal strengths. (13.9)

13.3.9 Environmental conditions

Atmospheric refraction plays little part at short range but may dominate long range detection, see Chapter 5, Sections 5.2 and 5.3. Not only do refraction index, n, and its height variation depend on weather and time of day, causing considerable variation of refraction factor, k, but ducting may be sufficiently severe to change maximum detectable range by a factor of two or more. These very important effects are generally not predictable, easily measured or readily inferred from meteorological instrument readings. It is best to calculate performance for a range of refraction values, including low values of k, which, although perhaps infrequent, may be allied to bad weather in which radar performance is vital.

Hydrometeors introduce attenuation and noise into the radar–target path. Attenuation depends on the path length and is negligible at short range or when the precipitation is localised. Performance calculations should therefore state the path length over which precipitation is assumed present, whole path being the worst case.

Precipitation also introduces clutter, which may be severe, see Chapter 11, Sections 11.4 and 11.5. Although it is rarely feasible to assess distant precipitation rates, at least the operator can adjust the radar to display and assess the severity of precipitation clutter, and in daylight it is often possible to see distant squalls. Clutter is chiefly significant when it surrounds targets. Clutter elsewhere may be significant if it causes the data extraction system or the operator to adopt suppression tactics which reduce in-clear target detectability, for example provoking short pulse operation which improves signal to clutter ratio for any targets lying within a squall at the expense of the signal to thermal noise ratio elsewhere. It may be necessary to recalculate for each available pulselength to determine optimum performance. Attenuation and clutter rise non-linearly with precipitation rate, and also depend on precipitation type and on atmospheric temperature, neither of which may accurately be known.

Fog can cause moderate attenuation but negligible clutter. It is not normally possible to detect fog banks by marine or VTS radar, making it difficult for the observer to forearm against performance loss.

The sea surface roughness at the grazing point affects strength of the forward-reflected indirect ray, and hence the resultant ray. This introduces moderate uncertainty of echo strength for extended targets, and very considerable uncertainty for point targets, active or passive. Surface roughness at the target also retro-reflects as sea clutter (Chapter 11, Sections 11.6 and 11.7) and is often the limiting factor in target detection. As with precipitation, clutter elsewhere may bias the operator into choice of short pulse, spoiling detection of any targets surrounded by more benign local clutter.

Alternative definitions of wave height are enumerated in Chapter 5, Section 5.7.4. Care is necessary to specify which is in use, a problem avoided when using sea state number. Sea roughness is somewhat difficult to assess, and may vary within the display area, particularly when a harbour VTS looks out toward the open sea. The effect of a given wave height also increases from a swell to a fully developed sea, driven by a local wind; whether the wind is rising, whether the radar is looking up-, down- or cross-wind and by plane of polarisation. The distribution of the clutter amplitude also becomes wider in heavy seas, see Chapter 11, Section 11.7.4.

Tide may vary effective scanner and target heights, depending whether the radar and its target are afloat or ground-fast. It should be remembered that the indirect ray grazing point may either be the water surface or terrain of some kind, possibly tidal mud-flats. Point target multipath null ranges can vary significantly with tide and it may be worth calculating for several tide states, or at least for low and high water ordinary spring tides. Coastal echoes can also vary considerably if, say, a beach covers at high tide.

13.4 Accuracy of calculations leading to SNR or P_D

13.4.1 Approximations within calculations

Previous chapters quantified environmental factors such as precipitation attenuation using algorithms, whose form may or may not be closely linked to the underlying physical process. Usually the algorithm matches some experimenter's observed results. The experimental conditions may not be quite identical to the current scenario, introducing error.

Published equations do not always lend themselves to calculation. Sometimes we have offered approximations, accompanied by indications of their error. For example, in Chapter 5, Section 5.8.4, the reflection coefficient of surface roughness, ρ_0, as given by the experimenters Brown and Miller, is represented by the algorithm of Eq. (5.41c). Its inconvenient Bessel function Eq. (5.41d) can be approximated by Eq. (5.41e), the error within the approximation being graphed in Figure 5.21.

The considerable uncertainty typical of radar data makes precise calculation impossible so it is misleading to express results to many significant figures. On the other hand, premature rounding of the individual terms in an equation introduces unnecessary error in the final result. It is best to insert each term to one more decimal place than its precision warrants, then round off the final result to one place less than the least certain important term. For example, we might judge that our antenna beamwidth of nominal 1.0° actually lies somewhere between 0.95° and 1.05° (0.01658–0.01833 rad). When calculating formulae, we would insert 0.0175 rad (1.0°) but round off the result to two significant figures. Computer spreadsheet methods of calculation facilitate analysis of the likely accuracy of results by repeat of calculations at upper and lower limit conditions. Rounding is deferred until the end of the calculation chain. The standard statistical methods for uncorrelated variables can also be used.

Beware the seductive influence of the computer printout containing umpteen significant figures. If the input is rough, the output cannot be better. Most of the digits will be dross. Often it is enough to round off results to the nearest decibel (26 per cent) and even that may overstate the real precision. This cavalier attitude may initially shock readers used to the exactitude of accountancy. Be assured that engineering budgets are never drafted on the same principle (well, hardly ever).

Throughout, we have stressed that system performance is hedged about with uncertainty, and we have just mentioned enough error sources for the pessimist to assert that all results of performance calculations must be meaningless. This is not in fact the case; most of the error sources are of moderate size and are uncorrelated, so the standard deviation of the resultant error is much less than the sum of the moduli of the individual component errors. Furthermore, depending on the task in hand, many of the error sources are inoperative or more or less cancel out, as described in the following task scenarios.

13.4.2 Radar comparisons

It is often necessary to answer practical questions such as: In a defined environmental scenario, is 9 GHz Radar A, with certain datasheet parameters, better than 3 GHz Radar B, having somewhat different parameters?

Here it is fair to take each radar's transmitter power, scanner gain, etc., at face value, or at least apply the same service loss to each. An arbitrary representative target type can be used, such as a small uniformly extended target of exactly 20 dB m^2 RCS (perhaps dropping 5 dB at 3 GHz per Chapter 10, Section 10.4.6, Eq. 10.5) and exactly 10 m effective height. Scanner height can be set at some exact value, such as the nominal height above sea level of the host ship's radar platform. Atmospheric refraction can be set in turn to several nominal values, say $k = 0.80$, 1.333 and 2.0. Precipitation rate can be set to a representative value, say heavy stratigraphic rain of 16 mm/h whole path at 20°C, so defining nominal atmospheric attenuation and precipitation clutter. Similarly, sea conditions would be set at a representative nominal value, so defining indirect ray reflection, multipath coefficient and sea clutter. It is immaterial that these nominal values may never simultaneously occur in practice. Small errors within the algorithms representing attenuation, clutter, etc., also almost cancel, leaving only negligible residual errors in the comparison.

Using a spreadsheet, to be described in the next chapter, we could plot SNR or even P_D to a base of range. The intercept on the individual radar's minimum detectable signal (or minimum acceptable P_D) then indicates maximum usable range. Both radars should be calculated with optimised control settings, e.g., pulse length. For example, Radar A might show 21.3 km and radar B 23.0 km, indicating the latter is more sensitive under the chosen conditions, giving 1.7 km more range. Put into actual service and measured with precipitation and sea states judged similar to the above paper figures, the above ranges are unlikely to be exactly observed, but it is very likely that Radar B would continue to exhibit about 1.5–2 km advantage. Similarly, if the scanner heights were actually 12 m rather than 10 m, the advantage would not change much.

To confirm Radar B's superiority, the calculations should preferably be repeated for a family of clutter scenarios and target sizes.

13.4.3 Mounting heights

Low scanner height minimises sea clutter, but great height gives longer horizon range and probably longer detection range except in heavy seas. Shipboard, height is likely to be constrained by the need for all-round visibility, but sometimes some limited choice is available. VTS and range surveillance systems are often bought against contractual requirements to detect or track specified small targets throughout a specified sea area in specified clutter, payment being subject to on-site acceptance trials. This can present bidders with severe problems. Should one go for a relatively high number of out-stations, mounted low, each having rather poor horizon range, or go for fewer stations with higher masts and narrower-beam scanners to bring the clutter back down? Out-stations incur high first and maintenance costs – access roads, security fences, buildings, power supplies, data links all have to be considered and land may be difficult to acquire. On the other hand, large scanners with stiff tall masts having safe maintenance access aloft are also expensive and may raise aesthetic objections. We shall return to this question in Chapter 15.

13.5 Plot and track accuracy

13.5.1 Instrument errors

Range is determined by the radar as the time delay between transmission and reception, and then measured by the operator using the display's electronic range marker or rings. Both these in essence rely on a quartz crystal oscillator and digital counter arrangement. The oscillator is inherently accurate and stable to better than 1 part in 10^6, so instrument range errors are small, systematic and largely self-cancelling when determining velocity or target-to-target separation.

Bearing on raster displays depends on digital retrieval of the initial R, θ information from the scanner, tagged with scanner azimuth, usually 1024–4096 positions per scan and measured against an electronic bearing marker, digitally generated and driven from the gyro or compass. Again, errors are small and mainly systematic. When displayed raw, the paint angular width approximates the target ship's projected width plus the scanner beamwidth. When the raw target is displayed:

$$\text{echo width} = \text{ship width} + \text{scanner subtended beamwidth}. \tag{13.10a}$$

In length, to minimise the risk of parts of the target lying closer than indicated, and to eliminate echo stretching when a target ship presents an open hold which acts as an echo box, sometimes only the leading edge of the echo pulse is used, discarding the pulselength-dependent body of the pulse. Ship length then becomes immaterial. Here:

$$\text{echo length} = \text{pulselength} \times c. \tag{13.10b}$$

Track formation is basically the generation of a straight line representing target velocity through a succession of plots laid down at successive instants, by use of computational algorithms. It is convenient to represent each target position as a point, drawing the best straight line among the points by regression (defined by Clapham [2] as a statistical procedure to determine the relationship between a dependent variable and one or more explanatory variables). Tracking aids tend therefore to show plots as synthetic points representing either the leading edge as just remarked, or the echo 'centre of gravity', rather than painting the whole raw echo area. In the latter case particularly, the resulting crisp display masks glint and fails to warn that some part of the target may lie closer to own ship.

13.5.2 Ship motions

Beside progressing smoothly along the course made good, ships also pitch, roll, yaw, surge, heave and sway to degrees dependent on sea state. The first three are oscillatory angular motions about the transverse, longitudinal and vertical axes respectively, the remainder being oscillatory bodily displacements about those axes. Ships may also carry trim or list/heel, semi-permanent components of pitch or roll, respectively. These eight displacements in effect move the ship's scanner about, introducing error when projected from the scanner to waterline level, the positioning plane of primary interest to the navigator.

Pitch, roll, surge and sway displace the scanner from its normal position, which is vertically above the nominal position relative to the average course made good. Pitch and surge cause the scanner to oscillate about the ship's mean forward velocity, while roll and sway introduce an oscillatory transverse component. Relative to own ship's waterline position, or the bridge, all targets appear to take on complementary oscillations. Motion periods ($\sim$10 s) are more or less constant for a given ship, but as they are decorrelated from scan rate, target positions are subjected to near-random errors on successive scans.

Severe roll or pitch swings targets lying near to the roll or pitch plane through the scanner elevation polar diagram, the effect becoming severe when peak roll exceeds half the scanner elevation beamwidth (IMO require sufficient beamwidth to cater for $\pm 10°$ roll). Chapter 12, Section 12.10.2, discusses resulting P_D fluctuation. The relationship connecting roll amplitude to sea state for a given ship is rarely published and varies with relative bearing of wave fronts, ship speed, condition of loading (in particular as it affects metacentric height), wave period and perhaps other factors. Insertion of inappropriate values is likely to introduce error into calculation of P_D in heavy seas. Roll or pitch can also introduce cross-polarisation, as discussed in Chapter 8, Sections 8.12 and 8.13, rarely with significant effect on echo strength.

Yaw, unless detected and allowed for, causes targets to oscillate in apparent bearing, introducing bearing error. The echo may spread among several detection cells, reducing hits per scan and spoiling P_D. Yaw may also blur actual changes of target bearing, making target manoeuvres difficult to spot.

Heave varies effective scanner height, which is usually only significant when it oscillates point target null ranges. Of course, targets afloat may also heave, modulating target height. Target roll and pitch introduce second-order and usually negligible height reductions.

13.5.3 Scan plane tilt errors

For analysis of tilt error, we assume a flat Earth (curvature merely introduces secondary errors at extreme range) and disregard all other instrument errors. The geometry is shown in Figure 13.4. With the radar platform on an even keel, the scanner axis is vertical with the scan plane horizontal. Surface targets at true azimuth Θ rad from tilt axis (for example the fore and aft line if the platform rolls) are displayed at azimuth θ, here coinciding with Θ. A target at range R has true polar coordinates R, Θ, which may be resolved into Cartesian coordinates $R \cos \Theta$ and $R \sin \Theta$ along and normal to the tilt axis, respectively. When the scanner plane is tilted at angle ν to the horizontal, the along-axis intercept on the scanner plane is unchanged, but the

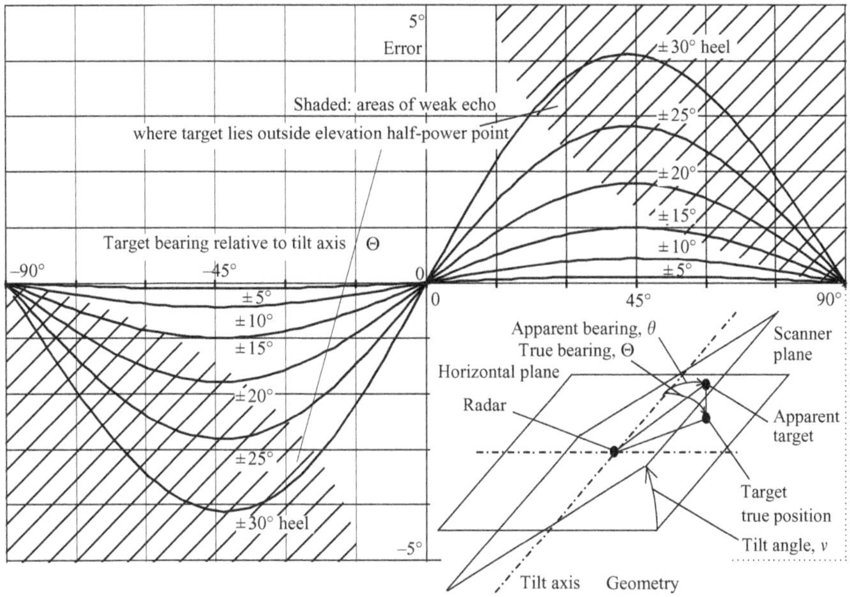

Figure 13.4 Scan plane tilt error. Severe pitch or roll introduces angular positioning error which depends on target bearing

normal intercept increases to $R \operatorname{cosec} \nu \sin \Theta$. Apparent azimuth bearing, θ, always lies further from the tilt plane and is given by

$$\tan \theta = \operatorname{cosec} \nu \tan \Theta. \tag{13.11a}$$

Figure 13.4 also plots the angular error, $\theta - \Theta$. Small when $\nu < 5°$, error rises in a square-law fashion with inclination, exceeding 4° when the platform rolls or pitches severely to 30° peak value and the target bears ±45 or ±135° to the tilt axis, for example, when the target is broad on the bow or stern quarters. In general, platforms roll or pitch asynchronously with scan rate, so quartering-target apparent bearings are subject to a quasi-random error which may be significant, sometimes exceeding the scanner beamwidth. In the hatched areas of the figure, the scanner boresight also inclines sufficiently to take the target outside the half-power elevation beamwidth, weakening the echo by more than 6 dB.

Figure 13.5 plots the bearing error fluctuation for a quartering target, seen from a sinusoidally rolling platform. The error is unidirectional, giving an angular mean bias proportional to the square of the peak tilt. In practice, mean errors may be reduced by the echo weakness or complete loss at tilt extremes. The target fluctuation characteristic, if not already Swerling Case 1, also migrates towards Case 1, because whole scan packets become correlated, see Chapter 12, Section 12.10.2 and Figure 12.39. The low SNR may reduce positioning accuracy, as shown in Section 13.5.4.

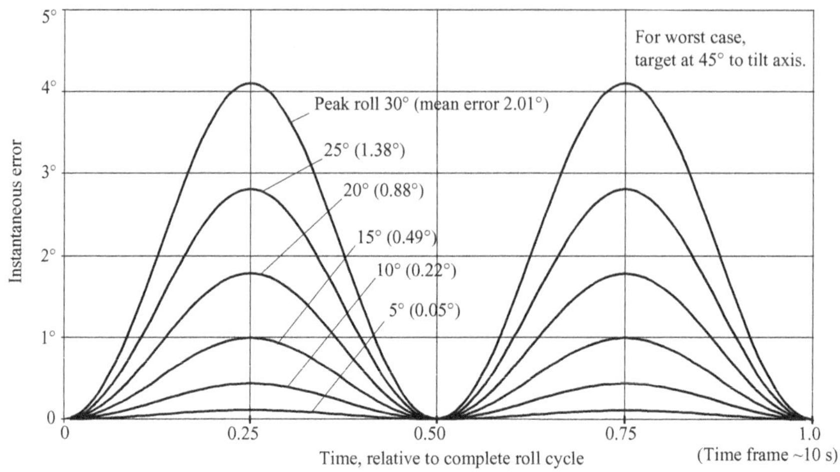

Figure 13.5 Bearing error fluctuation through platform roll cycle. Both systematic mean error and quasi-random errors are introduced by rolling or pitching

When the target is normal to the roll axis ($\Theta = 90°$), the range measurement is subject to $H \tan \nu$ range error. At other bearings,

$$\text{range error} = H \sin\Theta \tan \nu \text{ m.} \tag{13.11b}$$

For example, if $H = 35$ m, $\Theta = 90°$ and $\nu = \pm 15°$, range error $= \pm 9.4$ m.

This error is equally disposed about zero with no position bias. When roll is extreme, the target will again fall off the nose of the scanner elevation beam, reducing strength of some of the echoes. A saving grace is that high scanners tend to be associated with large ships having less roll and pitch. Expressed as linear displacement from true position, angular error is more important than range error except at sub-kilometre ranges. In principle, tilt-related errors may subsequently be corrected by algebraic processing, tilt being sensed by an orthogonal pair of clinometers.

A floating electronic bearing line is often provided to enable the operator to measure range and bearing between a pair of targets. Here systematic errors cancel but random errors accrue independently for each target and the readout error will approximate $\sqrt{2}$ the random error component of a single target relative to own ship, whose position within the system is of course accurately known.

13.5.4 Effects of SNR and bandwidth on plot accuracy

The following error expressions are approximations taking no account of detection cell size, straddling between adjacent range bins or some other secondary factors. Just as we tend to make more accurate length measurements by ruler when the light is good, the accuracy to which radar determines the position of a point target depends in part on numerical SNR, q. It is a fact of statistics that any measurement made with basic resolution x has error δx where:

$$\delta x \sim \frac{x}{\sqrt{2q}} \text{ m rms.} \tag{13.12a}$$

For range cell size R, receiver bandwidth B and velocity of propagation c, and if q is not very low, range error δR is

$$\delta R \sim \frac{c}{2B\sqrt{2q}} \text{ m rms.} \tag{13.12b}$$

If say $B = 1$ MHz and $q = 5$ (7 dB), $\delta R = 47$ m. Increasing q by a factor of 20, to 100 (20 dB), or increasing bandwidth to 4.5 MHz reduces δR to 11 m. Except at the shortest ranges, 47 m position uncertainty might seem trivial. But radar is not employed solely to give nowcasts; plotting aids predict future target positions, based on historical course and speed. Speed is computed as rate of change of apparent position from scan to scan. If, say, range is measured with errors of +47 m one scan and −47 m the next scan 2 s later, radial speed component will have 47 m/s = 169 km/h error. Even with good SNR of 20 dB ($q = 100$), and inspection time of five scans, error may be 16 km/h, introducing 8 km error in a position prediction made for 30 min hence.

Random bearing error $\delta\theta$ depends on the basic angular resolution $\Delta\theta$, which approximates scanner azimuth beamwidth, θ. Doubling aperture at a given operating frequency halves $\delta\theta$ and tripling wavelength by shifting from the 9 to the 3 GHz band, while retaining the same aperture length, triples $\delta\theta$.

$$\delta\theta \sim \frac{\theta}{\sqrt{2q}}. \tag{13.13}$$

13.5.5 Plotting aid prediction accuracy

Yaw, roll, etc., cause minor changes in aspect so echoes are received from successively different reflecting elements within a large target. This glint adds further error. As usual, range and bearing errors diminish as (a) more measurements are taken (b) through a longer inspection time. Unfortunately, the longer the inspection time, the less able is the predictor to cope with sharply manoeuvring targets and the longer it takes to realise that the target is manoeuvring. Accuracy of tracking aids such as ARPA or ATA (Figure 13.6) therefore demands higher SNR than necessary merely for reasonable P_D of echoes on the display screen. It is the need to take many measurements over a substantial inspection time which can make new target tracks so irritatingly slow to form. Manual plotting is no better; to allow reasonable screen displacement between one plot and the next, about 3 min should elapse, and a third confirmatory plot should be taken after a further 3 min. Allowing for appraisal, manual establishment of a reliable track line takes some 7 min, during which range may have closed by up to 10 km.

Figure 13.7 shows part of a display screen. Part (a) is a historic target plot at time $T0$, t s ago. The figure shows four alternative plot positions, a0, b0, c0, d0, each spaced around an error circle of radius e, 1σ from true position $p0$, where σ is the standard deviation. Because the error is random, the displayed plot could lie

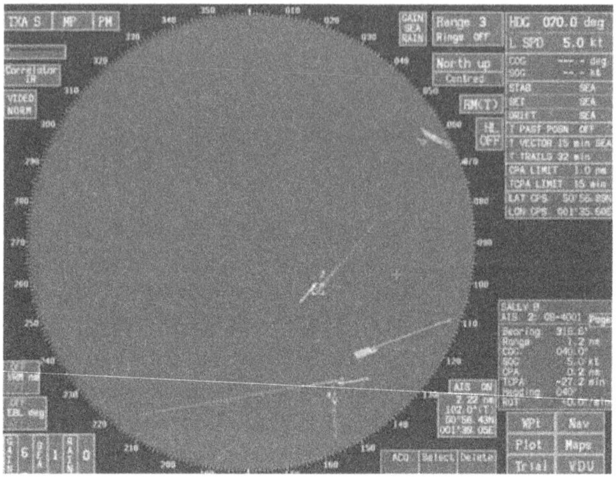

Figure 13.6 *Shipborne display with predicted vectors. Vector lengths ahead of the targets are proportional to the trail lengths behind them. Clutter speckles barely discernible. Data on the target marked 2 is presented on the alpha-numeric field to right of the display. Original in colour. Reproduced by permission of Kelvin Hughes Ltd, Ilford UK*

elsewhere, inside or outside the 1σ circle. Part (b) shows the current set of possible plots a1, ..., at time $T1$, circling true position $p1$ at radius e. Line $p0$, $p1$ represents the true target movement in time t, and is the target's velocity vector.

As noted earlier, the display must be able to show forward predictions of target movement, called vectors. Modern radars therefore can display predicted tracks for operator-selected forward times and deliver information of closest point of approach etc., as shown in Figure 13.6.

To predict where the target is likely to be mt seconds into the future at time $T2$, all that an operator or a machine computation can do is to extend velocity plot p0, p1 by drawing a straight line through the two observed positions, extended for a distance proportional to m, so that length a1, a2 $= m$(a0, a1). If there were no error in either displayed observation, the line would lie through p0 and p1 and extrapolate to p2, the true future position of the target. If however the observed positions happened to be a0 and a1, the predicted position would be at correct range but erroneous bearing, point a2. A similar result arises if the points were c0, c1, c2. If the points were b0, b1, b2 or d0, d1, d2 bearing would be correct with erroneous predicted range.

Figure 13.7(d) shows a set of possible predicted track vectors a1, a2, ..., d1, d2, obtained by extrapolating the existing vectors a0, a1, ..., d0, d1 in bearing and speed. The predicted vectors lie on a circle radius $\sqrt{2}me$, so it is fairer to draw a smaller circle of radius me to represent the 1σ locus, defining points a′2, ..., d′2. In the figure, $m = 2$, but in service a short set of observations, perhaps spanning 30 s, are often used as the basis for quite long-term predictions, up to perhaps half an hour, with m as high as 60. If the 1σ error in such a prediction is to be 1 km (about the roughest prediction ever navigationally useful), the 1σ error e of the two plots has to be as low as $1000/60\sqrt{2} = 12$ m, necessitating good SNR, wide bandwidth, wide scanner aperture and little rolling or pitching, especially if using a small scanner or the 3 GHz band where angular resolution is necessarily less.

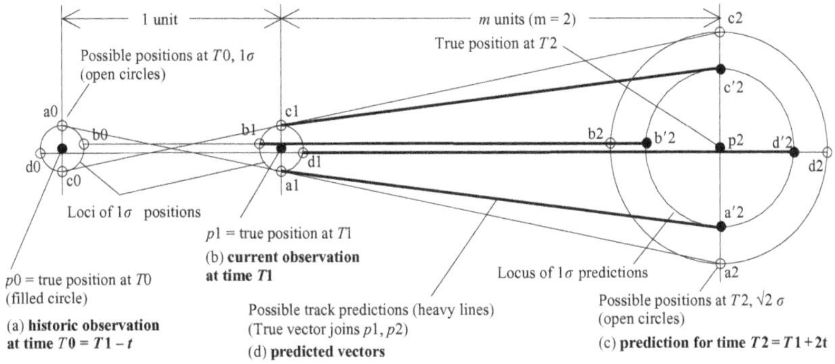

Figure 13.7 Prediction with random error. Target moving from left to right. Circular random-error loci

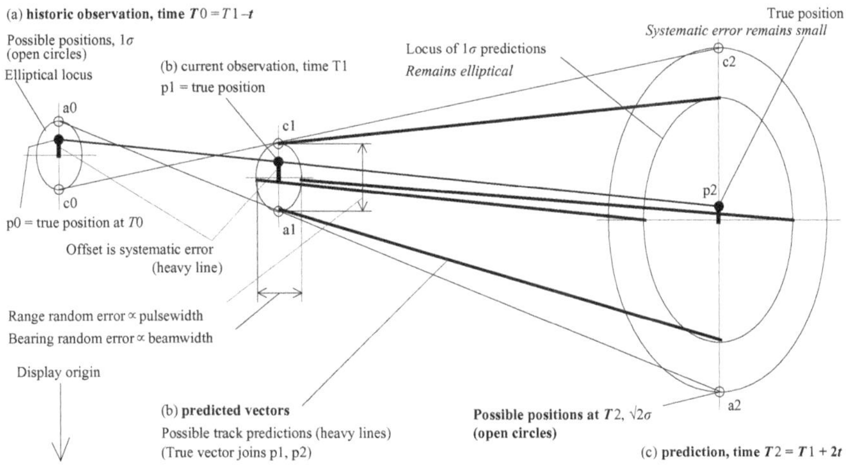

Figure 13.8 Prediction with systematic and random error. Systematic error component is not multiplied when predicting far ahead. Elliptical random-error loci

For simplicity, the above examples assume equal along- and cross-track errors, giving circular loci. Beside SNR, the 1σ plot error locus depends in range on pulse-length, and in bearing on scanner azimuth beamwidth × range, so in general each plot 1σ error locus is an ellipse with axes aligned on target bearing. Ellipticity varies with range, Figure 13.7 representing the special case of zero ellipticity. Systematic error (e.g. due to rolling) may bias the centres of all the ellipses from their true positions, as shown in Figure 13.8, which retains the previous notation. The locus of 1σ error in the prediction is also an ellipse. The prediction extrapolation does not magnify the systematic error component.

Of course, the radar makes not merely two but a dozen scans within a 30 s observation set, in effect integrating more echoes and considerably refining the prediction. Figure 13.9(*a*) shows a set of *n* (here 10) successive plots, each subject to equal

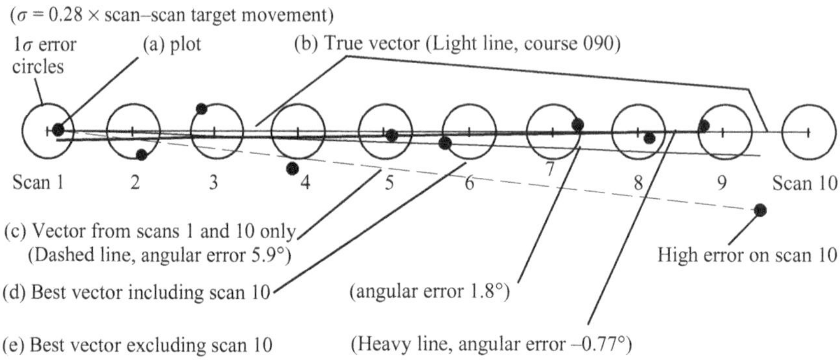

Figure 13.9 *Track error, several scans. Track error is reduced when multiple scans are available, and further reduced if high-error plots (scan 10) are ignored*

along-track and cross-track rms error. About $\frac{1}{3}$ of the plots lie outside the 1σ circles, as expected from Eq. (13.6). The plot angular errors are uniformly random distributed through the whole 360°. Plot 10 has 3σ error, to be expected once per 300 plots. The true target track is line (b), course 090°. Using only plots 1 and 10 as in Figures 13.7 and 13.8, the track vector is line (c), with error 5.9°. Taking the best straight lines among all 10 plots (d) considerably reduces the track error to 1.8°, a reduction factor of approximately $\sqrt{n}$ as might be expected from considerations of post-detection integration. Plot 10 (3σ) looks dubious to the eye. Discarding this plot gives track (e), with angular error again reduced (and changed in sign) to $-0.77°$. Presented with a set of plots of this sort, it is straightforward to arrange the track-forming algorithms to discard any plots differing 'significantly' from the mean track, reducing the speed and bearing errors in the formed track. This tends to bias the algorithm in favour of straight tracks, which is reasonable, for ships alter course relatively seldom.

Because noise plots are unlikely to persist at the same location for many scans together, or to progress in position in a rational manner, quite high false alarm rate can be tolerated, the trackformer rarely finding a set of noise plots which satisfies its algorithms to produce a false track. In effect, track formation reduces system bandwidth, evidenced by the longer time to form a track than a plot, so improves system signal to noise ratio. IMO requires ARPA to form tracks with 50 per cent blip/scan ratio. The author is always surprised how well modern ARPAs form and maintain tracks of weak and fading targets in unpromising clutter. They seem definitely superior to human performance, despite the brain's excellent pattern recognition skills. Of course these marginal tracks do not remain very steady and it would be imprudent to extrapolate far ahead from them.

13.5.6 Manoeuvres

The assumption of straight tracks should not be taken too far. Figure 13.10 shows how plot 10 might have been a valid, low-error plot from a sharp target manoeuvre,

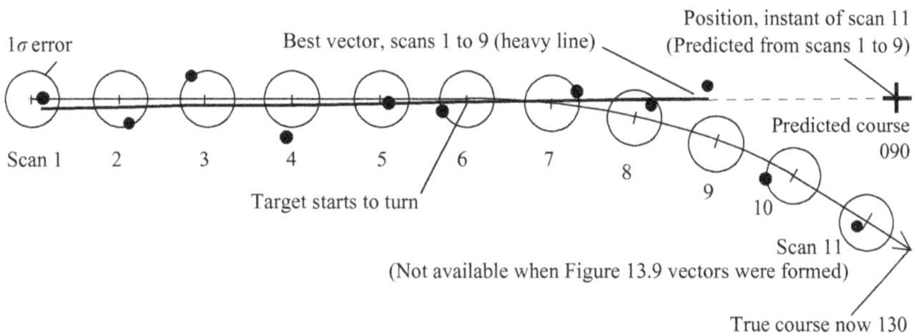

Figure 13.10 *Perhaps scan 10 was right after all! High-error plots must only be discarded with discretion*

plot 9 having had fairly large error. Early indication of manoeuvre is navigationally important, for the OOW needs as much time as possible to assimilate the changing situation, decide how to respond and turn or stop the ship if necessary – turning circles and stopping distances of VLCCs may exceed 8 km. Track-former algorithms must take account of the maximum manoeuvre capabilities of own ship and the nimblest targets likely to be encountered, including small high-powered HSC, setting a limit to the error reduction available from clever manipulation or smoothing of the data. As so often, it is possible to be too clever by half. An algorithm suiting a large vessel having low power to weight ratio such as a VLCC might be too sluggish for installation on a manoeuvrable tug. So although plotting aids and the more comprehensive track-formers used in groundfast VTS systems can reduce predicted vector errors, prudent navigators should take accuracy of predictions with a pinch of salt, especially when making long extrapolations, and continue to monitor any weak target whose predicted CPA is less than liberal. Sam Goldwyn spoke aright when he said that prediction is difficult, especially of the future!

It is navigationally important to spot quickly that a target has altered course or speed. Under Rule 8 of the Collision Regulations, alterations should be positive and large enough to be readily observed by another vessel, visually or by radar. Clearly, alterations of a few degrees or a few per cent speed reduction will take some time to show unmistakeably on another ship's radar, whether plotting manually or automatically, even disregarding vessel inertia. This is one reason why it is considered seamanlike only to make bold alterations, exceeding say 30° turn or 50 per cent speed reduction. Even so, ample time should be allowed for other ships to recognise that the manoeuvre has been made.

Echoes may sometimes go undetected, either because of the statistical fluctuation of strength, particularly with the usual Swerling Case 1 model, or because of wave screening if either own scanner or the target lies low in the water and the sea is rough. If a target which is being tracked fails to show, several scans are allowed to elapse before the target is declared lost. Meanwhile, the track is said to coast, continuing at the target's last-predicted course and speed. Naturally, this form of dead reckoning cannot warn of any manoeuvre started by the target until a few scans after echo

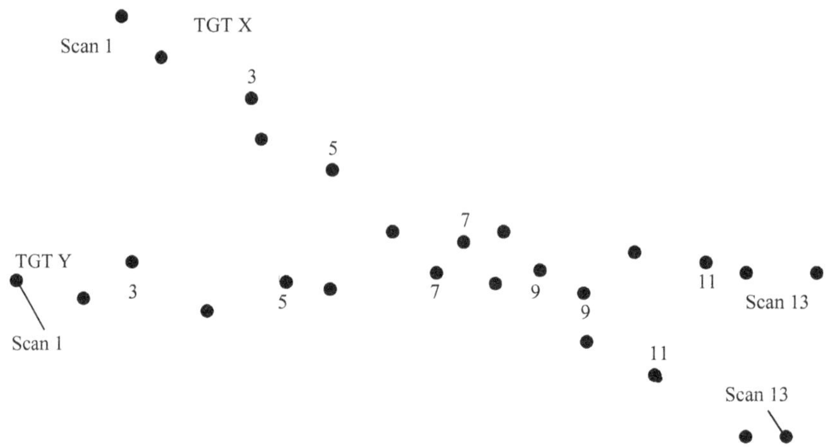

Figure 13.11 Plots on a VTS screen. From 13 successive scans containing random error

strength has recovered sufficiently for active tracking to resume. For example, if in Figure 13.10, the echo had faded out for scans 7–9, it would have been uncertain whether scan 10 represented the target after a manoeuvre, a strong clutter spike or a new target. Scan 11 would tend to indicate that either the original target had turned or a new target had been acquired with course 130 approximately. Not until further echoes had accrued on course 130 and none on 090 could either an observer or the machine gain confidence that the original target had turned 40° to starboard.

13.5.7 *Identity swap*

Figure 13.11 shows a series of plots on a VTS display, alternate scan numbers being indicated. Each plot is subject to random error, generally as Figure 13.10. SNR is assumed sufficiently high to exclude false noise plots. Unfortunately, the radar was not forming tracks of the targets. It seems clear that a target on course 130, which the operator has tagged TGT X, came to close quarters with a second, TGT Y, on 090 at about the time of plot 7. Two targets went in, two came out, but what really happened? Which is which? Did they both hold their courses or did each alter, to port and starboard respectively?

As track-forming algorithms are biassed towards presumption of non-manoeuvring target, the formed tracks would probably assume each target held its course, crossing as shown in Figure 13.12(a). (For clarity, the figure differentiates between the two ships' plots although of course neither the operator nor the track-former has the luxury of knowing which target is which.) Target X crosses Y's bow at time 8. This result is quite consistent with the data.

Figure 13.12(b) shows the alternative scenario in which each target starts to turn away from the other at time 6, giving substantially larger CPA. After the manoeuvres, X is on Y's former course and Y on X's. If the VTS operator or the algorithm does not

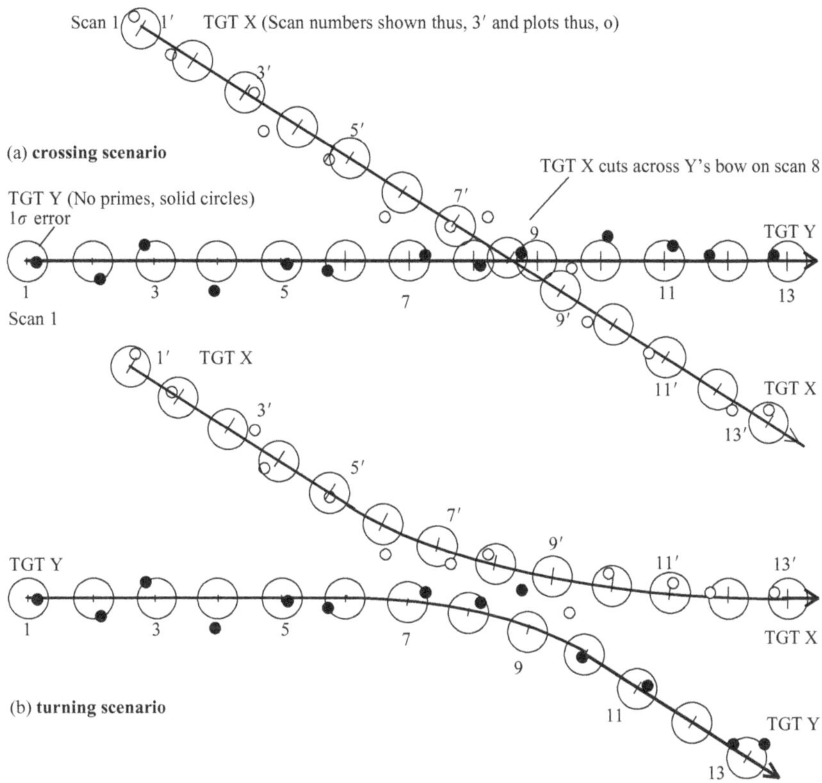

Figure 13.12 *Identity swap. It is impossible to tell from the data of Figure 13.11 which of the alternative interpretations A or B is correct*

spot this and assumes each held its course, the identification tags interchange, swap or seduce. This occurrence is clearly dangerous, for (a) the operator will not realise that each will likely soon make a further manoeuvre to get back onto its intended course and (b) any navigational advices issued by the VTS will be founded on a misapprehension and be erroneous. The figure uses perhaps a somewhat extreme scenario, but swap is a serious practical problem.

Swap is probable when targets converge within a few σ of one another, and is more likely when the radar random error is high – when SNR is low, pulses long and scanner beamwidth high. Swap is one of the main justifications for expensive large scanners in VTS service, married to short pulses. As usual, systematic errors tend to cancel and do not greatly affect seduction. However, targets changing aspect, as in Figure 13.12(*b*), may well start to echo from a different reflecting centre, causing a step change in the apparent track, making seduction more likely.

Identity swap can of course also occur on ships' tracking aids but tends to be of less consequence, own ship's primary task being to avoid collision with any target, be it X or Y.

Targets sometimes legitimately come close to one another, so are required by IMO to display as two entities down to very short spacing for navigational reasons. Cases in point are tugs or pilot boats alongside ships from which they may detach at any time. On the other hand, it is undesirable for a single target to register as two, which might happen with imperfect combination.

A pair of identical co-located targets gives a single video pulse, height 1.22× single-target echo, rounded by the restricted receiver bandwidth. As the targets start to separate in range, the video pulse develops a pair of rabbit's ears, height 1 unit, with a deepening saddle between them, finally splitting into two independent pulses each of unit height. The saddle height is 0.5 unit when the separation is twice the half-power pulselength, assuming the latter is matched to receiver bandwidth. For example for bandwidth 1 MHz, the half-power pulselength is $c/(2 \times 10^6) = 150$ m and the saddle height is 0.5 when the targets are 300 m apart. Range resolution will be of this order. Resolution of close-spaced targets therefore demands short pulse transmission with the attendant wide receiver bandwidth and possible squint difficulties of wide-aperture slotted scanners.

13.6 Combining data from multiple sensors

13.6.1 Shipborne radars

Primarily for reliability reasons IMO requires the bigger merchant ships to have at least two radars. The first is mandated to be 9 GHz, the second preferably being 3 GHz. The samples of clutter, especially sea clutter, and perhaps glint taken by such a diverse pair of radars are decorrelated, see Chapter 12, Section 12.8. In a perfect world a modest but definite improvement in detectability of small targets could be obtained by plot combination or plot association, combining their data in the signal processor at the average position of each plot pair, perhaps weighted towards the radar with the better current SNR or lower likely 1σ error. Combination would reduce systematic and random position errors, as well as improving the clutter. Perhaps of more practical importance, fades are partially decorrelated, reducing the probability of total target loss.

Digital scan conversion enables plots from two or more radars having different frequencies and unsynchronised prfs to be displayed on a single screen, the data source being identified if required by colour. More sophisticated processing enables the data streams to be electronically merged or combined before display. Because each radar plot contains position error, the combiner should include an algorithm requiring a target of one radar to be considered to relate to a target of the other when at a somewhat different reported position, only a single plot being displayed. But then it would be almost impossible to prevent the algorithm from merging a true pair of adjacent targets. Track-forming might reveal whether there really are two targets, but is only possible when there is sufficient relative velocity and does not help when the traffic is stationary, perhaps awaiting the tide at a harbour entrance.

In practice, plot combination is rarely attempted, partly to preserve the autonomy of each radar, facilitating operation on differing range scales and orientations (ship's head up, North up, etc.) for navigational convenience; and preventing a fault within either set from disabling the other, for example, by injecting excessive noise into the combiner. It has been found preferable to keep things simple, sacrificing the potential position error and clutter improvements, and displaying each radar's plots on its own separate screen, leaving it to the operator's judgement whether plots on the two screens represent the same target. The situation is analogous to the former problem of visual navigation in the Red Sea, whose East and West coast navigation marks were referenced to differing chart datum points several kilometres adrift. All went well as long as position fixes were confined to solely East or solely West coast marks with no attempt at combination!

Association of radar plots with the output plots of a different sensor system raises similar problems. Ship–shore (VTS) automatic identification systems (AIS) are being introduced and ship–ship AIS will follow. Targets (in radar parlance) determine their position, course and speed by analysis of radio signals received from GNSS, a constellation of navigational satellites such as dGPS. On demand or at preset intervals they transmit VHF messages to the VTS (or to other ships). The messages contain identity code, position, course, speed and status data. Position accuracy is generally an order of magnitude better than radar, with typical 1σ error 10 m. Positions are referenced to the datum of an international co-ordinate system such as WGS84. Velocity vectors are determined as rate of change of position relative to this ground-fast datum, whereas radars at sea may reference velocities to the local water mass whose currents affect all nearby targets equally. Although it is useful to assign AIS-derived identities to radar plots on the integrated navigation display, care must be taken that co-incident or nearly co-incident AIS and radar plots in fact do both refer to the same ship. It might be argued that the problem is eased by the good systematic accuracy usual on AIS. However, questions arise as to the part of the target delivering the position (for radar, some part of the plating, for GPS, the antenna), relative latencies in the reported positions and system position datum.

Some old sensors such as medium frequency radio beacon receivers were acknowledged to contain error comparable with that of the radar but were not displayed on a common screen. Navigators learned not to become too concerned about minor discrepancy and few problems arose. Now, electronic chart information (ECDIS) such as coastlines is routinely superimposed on many radar displays. Except when high magnification is engaged (when registration of ship's and chart datum may introduce insidious error), the very smudginess of the radar echo from many coastal features reminds the operator to guard against superimposition error, which may partly arise from tidal changes to the radar shoreline as well as from the usual radar error sources. The human mind seems to be more alert to imprecision in data presented in such analog format to the often spurious accuracy of a synthetic display. We all glance at the hands of a watch and register 'about ten past nine' whereas a digital display invites 'nine eleven thirteen' even when we know both instruments may be a minute slow or fast.

The following sections discuss currently used combination arrangements, primarily within VTS.

13.6.2 Coastal surveillance and VTS – simple system

The sea areas of some VTS systems are sufficiently compact and unobstructed for a single radar station to suffice. The VTS radar is then in essence similar to a conventional merchant ship installation, perhaps with narrower pulselength, say 50 ns, and a wider scanner to improve resolution and minimise random error (number of detection cells being suitably increased), so reducing risk of target swap and maximising detection of small craft in clutter. High resolution also enables the actual hull outline of the bigger targets to be displayed at short range, enabling the operator to see aspect, for example whether the ship has actually started to swing off the berth, without reliance on radio reports from the master, line handlers, etc., who sometimes have their private agendas. Transceivers are often duplicated for reliability, usually with cold standby, service being interrupted for a couple of minutes should the working unit fail. Hot standby eliminates this occasional hiatus but adds cost. Operation is usually in the 9 GHz band to maximise resolution, although occasionally 3 GHz is preferred for its low sea clutter where arrivals are the main interest. J (Ku) band has been used to maximise resolution, particularly in Japan, despite its poor rain performance.

Scanner height has to balance the requirement for low sea clutter (scanner low) against sufficient horizon without suffering screening by harbour installations. VTS masts rarely have to support more than a tonne or so of scanner, but should be stiff enough not to deflect more than about $\frac{1}{8}$ the scanner elevation beamwidth to minimise gain loss. The mast should not twist about the vertical axis by more than $\frac{1}{4}$ azimuth beamwidth otherwise echoes will be reallocated through a cluster of bearing cells. These rms values apply to the worst significant wind scenario and can drive up cost. Being a dynamic effect, moderate twist does not introduce significant mean bearing error. Luckily, the scanners are ground-fast, reducing latency problems somewhat – only the target, not the VTS system, can move in the time elapsing between a pair of scanners bearing on it.

Bigger systems have to cover a wider sea area than can be handled by a single radar head. Multiple-sensor systems with partially overlapping coverage must cater for target movement from sensor to sensor. With a few heads, perhaps at Control and a couple of outstations, it is simplest to operate each radar autonomously, with its own display screen. This may suffice when traffic is sparse and shipping activity is concentrated on a few more or less independent focal points, such as discrete quays, each in effect its own mini-port. The VTS operator has to move from screen to screen as a ship proceeds through the area, re-assigning its identity tag as the target transfers from one screen to the next. Each head may display plots on its own screen, plots from neighbouring heads being indicated in a different colour. However, this method needs a lot of data, requiring wide band links from the outstations. Fuller integration is achievable in several schemes of ascending complexity, as follows.

13.6.3 Autonomous radar heads with track-formers

Each remote station has its own plot extractor and track-former, the link to control reporting formed tracks. Each report in effect states a track exists on course A, speed B, passing through a point whose latitude and longitude are C, D, at time E. Data bandwidth is quite low and a conventional dedicated telephone channel may suffice, especially if permanent echoes have been suppressed. The out-station plot extractors take their radar's sweep by sweep data, set thresholds to the desired low P_{FA} and form tracks. Although there may be a general statement of prevailing clutter amplitude (derived from the detection threshold giving the current constant false alarm rate), clutter is discarded and cannot be displayed at the centre, rather decoupling the operator from actuality and making it difficult to optimise the radar controls via the reverse link.

Handover of tracks from station to station can be difficult. The central display of the whole system area ought to present seamless target progress without track discontinuity or loss of target identity. The central track combiner has to decide whether, say, the incipient track it is receiving on perhaps weak plots at the edge of outstation #2 coverage area represents an established track from its neighbour station #1, which is seeing SNR fall as the target recedes, the track accuracy deteriorating in course, speed and position. The problem is somewhat eased by provision of narrow beamwidth scanners. Outstation #1 views from a quite different aspect and glint may introduce a substantial position offset; radar #1 tracking the stern, say, and #2 the bow. Has #2 found a new target or do its plots refer to the established track, with different systematic and random error?

13.6.4 Central track-former or plot extractor

Here the out-station radars have local plot extractors but the trackformer is central. Plots are more numerous than tracks and should include some residual clutter if full system sensitivity is to be achieved, needing more data bandwidth than before. A semblance of clutter can be displayed, assisting the operator to assimilate the sea scenario. The track-former has to be large enough to handle the incoming data streams and is often duplicated to assure system reliability.

If wideband ($\sim$10 MHz) data links are available, each radar outstation can report the whole video bitstream to the centre, the latter becoming responsible for plot extraction and all subsequent activities. This solution probably gives maximum sensitivity, accuracy and flexibility, because no data is prematurely discarded, but tends to be expensive.

13.7 References

1 MEIKLE, H.: 'Modern radar systems' (Artech House, London, 2001)

2 CLAPHAM, C.: 'The concise Oxford dictionary of mathematics' (Oxford University Press, Oxford, 1996, 2nd edn.)

Chapter 14
Spreadsheet calculations

'Preserve me from unreasonable and immoderate sleep.'
Dr Samuel Johnson, Prayers and Meditations

14.1 Introduction

There are several well-known aids to calculation of surveillance radar performance. Blake's Worksheet [1] uses a pre-computer iterative pencil and paper approach. The CARPET [2] program runs on a PC. Both are slanted towards military and aviation applications and are not particularly straightforward to apply to civil marine problems.

To assist readers to compute the performance of their own systems, this chapter outlines a new family of spreadsheets for point passive, point active and extended passive targets, respectively. The spreadsheets themselves may be downloaded from the IEE Website (www.iee.org). Examples of the use of spreadsheets and their charts are included in Chapter 15. In this explanatory chapter, we assume basic familiarity with Microsoft Works® spreadsheet operation as well as the earlier chapters of this book. No programming skill is required; the user merely inserts the relevant system parameters and follows simple drills set out below. Results are presented numerically and can be charted (graphed) to several alternative bases.

The spreadsheets have uniform format. The first page (p1) contains cells in which the user enters details of the current task, date and so forth, followed by all necessary transceiver, scanner, environmental and target parameters, and minimum and maximum ranges of interest. Cells in this page display computed performance parameters of most frequent interest, including horizon range and probability of detection at two spot user-chosen ranges. Often p1 will provide all the information required, and is configured to fit a page of A4 paper. Grouping all the system parameters, together with calculated performance, on a single sheet avoids those irritating subsequent uncertainties about the exact parameters to which particular results relate. 'What ifs' are easily explored by changing parameters – to try the effect of weather or change of target size, as examples. Internal range calculations are in km, but the user may choose to operate in nautical miles (nmi). Other parameters use the same units and symbols as the rest of the book, with scanner and target height in metres. The spreadsheets primarily use decibel format, but numerical equivalents are given where important.

To the right of p1 are pre-loaded lookup tables for internal use, although scanner table S2 contains cells for entry of the azimuth polar diagram. With this exception, all table values are locked, visible but not adjustable. Also to the right of p1 is a user panel which facilitates charting (graphing) results to bases independent of range, such as scanner or target height. All detailed calculations follow, or are simply adapted from, the methods developed in the main text, using equations from Chapters 2 to 12, all calculated intermediate results, or the formulae used, being available for inspection if required.

The majority of calculations are directly or indirectly range-dependent, and are grouped within a matrix forming the lower pages of the spreadsheet. The matrix tabulates around 100 labelled rows of intermediate and final performance parameters such as clutter severities and SNR. The matrix columns are arranged in increasing range increments based on the minimum and maximum ranges of interest inserted in p1. The sheet first tentatively adopts these values unless it is necessary to adopt a wider range bracket, for example to provide sufficiently long range to obtain the diffraction region results essential for calculation of transition region multipath factor. The spreadsheet then computes a set of grazing angles roughly consistent with the tentative range set, from which final grazing point and target range sets are computed. Ranges increase in an approximately logarithmic manner, extending beyond the horizon where appropriate.

Charts can be displayed (and hard copies printed) of numerous parameters such as echo strength, noise, clutter strength, clutter cell size, all to a range base expressed as km, nmi, per cent of horizon, etc., with linear or logarithmic scaling, with minima and maxima chosen anywhere within the spreadsheet range bracket. Certain other chart bases are available, such as grazing angle. When a logarithmic range base is chosen, chart increments are approximately uniform.

Sections 14.2–14.8 detail the Point Passive Target spreadsheet, SS1, with operating instructions. Section 14.9 describes differences of the Extended Target (used for ships and coastlines) spreadsheet SS2 and its charts, and finally Section 14.10 gives the differences in the Active Point Target spreadsheet SS3 used for racons, RTEs, ramarks and radar warning receivers. Chapter 15 uses the spreadsheets to examine performance of some typical radar/target/environment/operator systems. This Chapter describes Version 1 of the spreadsheets.

14.2 Passive point targets: page 1

14.2.1 General arrangement

Figure 14.1 depicts page 1 as loaded for a 9 GHz radar. In general cells are arranged in pairs, cells in columns A, C, E, G and I labelling cells to their right within columns B, D, F, H and J, respectively. Pairs used for data entry by the user have ordinary roman type, page 1 pairs displaying internally computed calculations being identified by

	A	B	C	D	E	F	G	H	I	J
1	Point passive target spreadsheet				Spreadsheet ref	SS1 v1				
2	Date		User		Task	Chapter 15, Sec 15.2.8 Acceptance Tests.				
3	Transceiver		Scanner and feeder		Range bracket		Environment		Results	
4	Type		Type		Scaling, km/nmi	km	Type			
5	TYPICAL MARINE RADAR		12FT APERTURE		Max reqd *R*	13.5			*Scaling*	*km*
6	Frequency, MHz	9410	Az beamwidth°	0.5	*Achieved*	*14.475*	Refraction, *k*	3	Test R1	10
7	*Wavelength, m*	*0.0318*	El beamwidth°	20	Min reqd *R*	6.5	*Eff Earth rad E, km*	*19113.00*	*PD at R1*	*0.9890*
8	Tx P, kW	20	Gain, dBi	33.9	*Achieved*	*6.393*	Precip type: Stratiform =	3	Test R2	12
9	*Tx P,dBW*	*43.01*	*Efficiency, p.u.*	*0.98*	Target type		Precip rate, mm/hr	8	*PD at R2*	*0.1373*
10	Tx loss Lt, dB	0	Loss, dB	1	TRIHEDRAL REFLECTOR		*Equiv rain for atten*	*8.000*	*Min R for B15 PD*	*8.17*
11	Rx loss Lr, dB	0	Height H, *m*	28	Height h, m	3.5	*RCS, dBm²/m³*	*–56.34*	*Max R*	*11.13*
12	Service loss Ls, dB	2	El patt, 1 = sin, 2 = invc	1			*Loss lp, dB/km*	*0.117*	*R bracket*	*2.96*
13	Rx NF N, dB	3	Depression°	0	RCS, dB m²	20	Extent, p u	0.2	*Fill, p u*	*0.62*
14	RSG thld, dBm²	–10	Rotation, rpm	22	*RCS, m²*	*100*	Air temp, *C*	20	*Horizon R*	*44.28*
15	Reqd PD	0.6	CP improvement, dB				RH, %	90	*1st peak R*	*10.91*
16	*With screening*	*0.600*	Sidelobe below D8, dB	32			*Clr air loss, dB/km*	*0.012*	*1st null R*	*<F8*
17	Reqd Pfa expt, *F*	–6	Tolerable SL PD	0.1			Wave hgt, hs, m	0.125	*2nd peak R*	*<F8*
18	Proc loss, Lp dB	8	Feeder ohmic loss, dB	3.08	Swerling Case 0,	0	*Sea state*	*3*	*2nd null R*	*<F8*
19	Integ, n, c, cr, *p* = non	n	VSWR	1.5	Operator		Wind fctr C2, dB	0		
20	Scan/scan corrln y/n	y	*Refl coeff*	*0.200*	Pol, h, v, c	h	Screening, p.u.	0	*Max sidelobe R*	*<F8*
21			*Mismatch loss, dB*	*0.53*	Gain control, dB	0	Surface dielec, eta	81	*Ref echo, 1 km,*	*–81.33*
22			*EIRP, dBW*	*72.30*	Sw Gain control,	0	conductivity, *S*	0.24	*Sea clutter hori*	*32.72*
23			*System NF, dB*	*7.23*	SP = s, LP = 1	s	Tilt/pol loss 1-way dB	0		
24					Mode, Table S1	1				
25	TABLE S1, Mode	Max R	*Max R, km*	Plslgth, SP	LP, *µs*	Rx bw, MHz	PRF, pps	*Active mode*	*Selected, µs*	*Pulses inte*
26	*1*	2.78	2.78	0.05	0.1	25	5000	*1*	*0.05*	*37.88*
27	*2*	5.56	5.56	0.25	0.5	5	5000	*2*	*0.25*	*37.88*
28	*3*	11.11	11.11	0.5	1	3	2000	*3*	*0.5*	*15.15*
29	*4*	22.22	22.22	1	1	2	2000	*4*	*1*	*15.15*
30	*5*	88.9	88.9	1	1	2	1000	*5*	*1*	*7.58*
31	*6*							*6*	*0*	*0.00*

Figure 14.1 *Passive point target spreadsheet page 1. Cells A1 : J31. Contains cells in which system parameters are entered. Results panel displays often-required performance parameters. Maximum range, cell J11, is in this case 11.13 km. Shown set for a 100 m² passive reflector in rain, viewed by a hypothetical 9 GHz radar*

italics. Unused cells are stippled. Cells may contain a maximum of 256 characters. In general, user-entered values are displayed as entered, and must not contain commas, spaces or symbols, which would cause the spreadsheet to regard them as text. Letter entries are case-sensitive; use lower case as instructed below. Most computed results are shown rounded to two decimal places, but full values are retained internally, without introduction of rounding errors. In general, empty cells default to zero; where this precludes rational computation, internal stratagems usually avoid show of disconcerting error (ERR) messages. For convenience, we call the person using the spreadsheet the user, the person operating the real-life radar represented by the spreadsheet remaining the operator, as in earlier chapters. Unfortunately, around 75 parameters affect radar performance and all must be entered in page 1. This high number reflects the complexities of target detection and has been minimised as far as consistent with accuracy. The spreadsheet in effect calculates an equation having ~75 variables at 150 range increments. The task would be impossibly long by hand.

Equation references from previous chapters are included to aid understanding and to provide an 'audit trail' of the assumptions underlying the spreadsheet results. Font constraints have forced some symbol changes. A caution: computers work on the GIGO principle – garbage in, garbage out. Take care when entering data, 20 000 in the transmitter power kW cell gives *much* more performance than the (correct) 20! Page 1 is divided into several panels, as follows.

14.2.2 Title panel

Cells B1 : G1 contain the spreadsheet title. The date of compilation may be entered in cell B2 and is not automatically updated when the spreadsheet is opened. User may enter a name in cell B4, and the task in cells F2 : EX2, all using free text. If an entry overflows its cell it is not fully displayed, but can of course be read in full by clicking onto the cell.

14.2.3 Transceiver panel

Enter transceiver type if wished in cells B4 and A5 : B5, free text. Enter operating frequency, f MHz, in B6. Any radar frequency may be entered but those outside the main bands of 2900–3100 and 9300–9500 MHz may give less accurate results, because some environmental data, e.g. for sea clutter, then has to be extrapolated or approximated. The equivalent free-space wavelength, λ metres, per Eq. (2.3) shows in cell B7. Like all results cells, the formula can be read, in spreadsheet notation, by clicking onto the cell, here the formula is simple : = 299.7/B6.

Enter transmitter peak power, P kW, in cell B8. The equivalent dBW shows in cell B9. Enter transmitter loss, L_t dB, in cell B10 and receiver loss, L_r dB, in cell B11, e.g., respectively, if B8 represents power at the magnetron flange rather than the cabinet output port, or first-stage rather than overall receiver noise figure is used. If there is no loss, enter 0.

In cell B12 enter the two-way transceiver service loss, L_s dB. The spreadsheet divides this loss equally between transmit and receive legs.

Enter receiver noise figure, N dB, in cell B13. Receiver swept gain (RSG) threshold represents the smallest short-range target displayable in absence of clutter and characterises the receiver swept gain on the assumption of an R^{-4} law, see Chapter 12, Section 12.7.2. Enter RSG, dB m^2, in cell B14.

In cell B15 enter the required overall probability of detection (P_D, range 0 to 1, after any pulse–pulse or scan–scan integration or wave screening). A valid detection is declared when computed P_D exceeds B15 value. Cell B16 shows the (higher) required pre-screening P_D assuming target screening occurs per cell H20. (B15 $=$ 0.5 is about the lowest value for practical operation.) If B16 $>$ 1, the system cannot deliver the desired P_D with the entered screening value and shows screened. In cell B17 enter the required overall probability of false alarm (P_{FA}) exponent, F, typically −6. Include the minus sign.

Enter processing loss, applicable to targets but not clutter returns, L_P dB, in cell B18 (typically 8 dB). In cell B19 enter the radar processing type in lower-case characters; n for the usual non-coherent system, c for coherent, cr for coherent-on-receive or p for integration on the phosphor of a cursive PPI display tube. The spreadsheet defaults to n. In cell B20 enter y if the radar uses scan–scan correlation, otherwise n, the default setting.

14.2.4 Scanner and feeder panel, and Table S2

Enter scanner type details in cells D4 and C5 : D5 if desired, free text. Enter azimuth and elevation one-way half-power beamwidths (θ, ϕ degrees) in cells D6 : D7 and main-beam on-axis one-way nett gain, G dBi, in cell D8. Cell D9 indicates per-unit efficiency, expressed as the ratio between D8 and the rule of thumb gain $25\,000/\theta\phi$ after conversion to dB values (based on Eq. (2.7c)). The main use of D9 is to show *CHK GAIN* if nominal efficiency >1, warning that the entered gain and beamwidths may be mutually inconsistent; check also if efficiency shows a very low value, say less than about 0.5. D9 is not used within the spreadsheet. Enter the scanner loss, typically 1 dB, in cell D10. This is solely used in calculation of system noise figure and is not subtracted from D8 in radar range equation calculations.

Make any necessary tidal correction, then enter cell D11 with scanner height, H m above water level. The minimum value for proper spreadsheet operation with practical system parameters is 0.25 m, so if a lower value is entered, intermediate cell AJ29 adopts 0.25 m as the working value. Enter cell D12 according to the scanner elevation radiation pattern: 1 for the usual $(\sin x)/x$ uniform illumination, 2 for inverse cosecant squared, 3 for other illuminations. In the last case, go to Table S2 (cells M22 : BT28) and enter the appropriate one-way gain loss (dB) relative to beam axis in P27 : BT27 for each of the angles off axis (degrees, positive downwards, maximum 30) of P23 : BT23. The working relative gain (dB) appropriate to cell D12 shows in P28 : BT28. The lowest angle, cell P23, is set to the extreme value of $-89°$ to prevent overflow problems.

There is no specific upper limit to H and h, but the geometrical approximations of Chapter 5 may introduce significant error if heights of the order of a kilometre or more are entered.

Enter the boresight depression (degrees, positive downwards) in cell D13. Effect of platform roll or pitch can be explored by successive entry of differing angles in this cell. Enter rotation rate (rpm) in cell D14 and the circular polarisation improvement factor, dB, in cell D15 if applicable. Otherwise leave blank or enter 0. (Choice of polarisation is contained within the operator panel, Section 14.2.7.)

Enter the principal azimuth sidelobe level in cell D16, as (positive) dB below main beam gain one-way of cell D8, typically 27 dB. In cell D17 enter the tolerable probability of detection (0 to 1, say 0.1) of the resulting sidelobe display of echoes at false azimuths. This will generally be much lower than the minimum required echo P_D, cell B15.

Enter the one-way feeder ohmic loss, dB, in cell D18 (Chapter 2, Section 2.6.1, Table 2.2), and VSWR (1 upwards, typically 2, assumed the same whether viewed from the transceiver or scanner) in cell D19. Cell D20 shows reflection coefficient, ρ per Eq. (2.5a), and cell D21 the mismatch loss, dB one-way, per Eq. (2.5b).

Cell D22 shows the EIRP, dBW, calculated by subtraction of transmitter loss (cell B10), half service loss ($\frac{1}{2}$ cell B12) and feeder losses (cells D18, D21) from the sum of transmitter power (cell B9) and scanner gain (cell D8). Cell D23 computes system noise figure, N_S as the sum of the receiver numerical noise factor, feeder ohmic loss and scanner loss, converted to dB form.

14.2.5 Range bracket panel

The spreadsheet computes performance at 150 spot ranges. The lowest and highest internal range values approximate where possible the user's desired lowest and highest ranges, except where lower minimum and/or higher maximum values have to be adopted for internal reasons. The range bracket should be chosen to match the ranges of practical interest to minimise rounding errors and to deliver the smoothest available charts.

In cell F4, scaling, enter nmi if results are desired in nautical miles, otherwise enter km, the default setting. Cell AA29 carries the appropriate scaling constant to convert ranges to km for internal use and restore results for display. Enter maximum and minimum desired spreadsheet ranges (km or nmi as chosen, *not* metres) in cells F5, F7; cells AG29 and AE29 provide the km equivalent.

The spreadsheet must operate either wholly within the interference region, at relatively short ranges; or must span transition and diffraction region boundaries R_A and R_B to compute m_t as a necessary preliminary to finding the effective multipath factor, M. The user cannot operate wholly in the diffraction region, where ducting tends to introduce unquantifiable error. Using approximation Eq. (6.2b), cell AM29 decides whether F5 is less than the likely R_A, and if so enters 1. Then cell BE29 approximates the km equivalent of desired maximum range unless the latter falls in the attenuation region, overcome by setting to 1.5 times horizon range (which will almost always exceed the range of the R_B criterion, initially $m_d = -20$ dB; this is necessary because R_B is yet unknown). BE29 may exceed maximum detectable range.

The initial tentative minimum range, cell BM29, has to be:

(a) $<16Hh/(R\lambda)$,
(b) $<0.9 \times$ the initial maximum range, to give a sensible range bracket, and

(c) high enough to give low enough sea clutter grazing angle, $\beta(\beta < 0.18\,\text{rad})$ for the algorithm of Eq. (11.11) to apply, so $R > H/180$ from the geometry of Figure 5.12(a).

These are computed in cell BM29, repeated in cell E35. Condition (c) precludes achievement of zero range, but entering 0 in cell F7 yields short enough range for practical purposes, usually a few tens of metres.

Cell BK29 computes the logarithmic range increment to give a geometric progression from tentative minimum range E35 to tentative maximum range AM29, reached in cell EX35. The progression multiplies the previous cell value by 10^{BK29}, where $\text{BK29} = (\log(\text{BE29}/\text{BM29}))/(\text{BI29} - 1)$; cell BI29 is set at the number of active columns, 150.

Cells F8 and F6, respectively, show the achieved minimum and maximum ranges from the first and last actual range cells in the computation matrix (cells E38 and EX38 as described in Section 14.3.2, converted from km). Results are computed at 150 points within these boundaries and charts may be drawn anywhere within them.

14.2.6 Target panel

Target type may be entered in cells F9 and E10 : F10, free text. After making any tidal correction, enter point target height above waterline, *h* metres, in cell F11. Cell AK29 sets the minimum working value of 0.25 m. Enter target average radar cross section (RCS, dB m^2) for the conditions in question (frequency, aspect, polarisation, etc.) in cell F13; cell F14 shows the RCS in m^2. Enter the Swerling Case in cell F18 as 0, 1 or 3 (although treated as Case 3a, do not enter 3a). Default is Case 1, which gives poorest performance. Cases 2, 3b and 4 are not available.

14.2.7 Operator panel and Table S1

The polarisation field is located within the operator panel since circularly polarised scanners have an operator's switch to select a linear polarisation when there is no rain. Enter cell F20 as h (horizontal, default setting), v (vertical), or c (circular). If the target is polarisation-sensitive, enter a polarisation (or tilt) two-way loss, dB, in cell H23 of the environment panel.

Cells F21, F22 replicate the operator's gain and swept gain controls, expressed as dB below full radar sensitivity; enter positive values or 0 for full sensitivity. If the radar has a short/long pulse (SP/LP) switch, enter s (default) or l in cell F23.

Mode cell F24 and Table S1 (cells A25 : I31) replicate the operator's Range switch. Most radars change pulselength (τ μs), receiver bandwidth (*B* MHz) and pulse repetition frequency (prf, pulses per second) in several steps as the Range Scale control is advanced. The conditions pertaining to the shortest ranges we call mode 1. There is provision for a maximum of six modes. In Table S1 cell B26 enter the maximum range (units of cell F4) to which the shortest-range parameters apply. For example, if the parameters change at 1.5, 3, 6 and 12 nmi and the maximum instrumented range is 48 nmi (88.9 km), Table S1 appears as in Figure 14.1. Cells C26 : C31 give km equivalents. If the scale changes are at certain nmi, and the spreadsheet is operated in km, the appropriate km values must be entered in cells B26 : B31. Follow with the appropriate pulselengths, bandwidth and prf in D26 : G26. Repeat for the longer modes in the lower rows of the table. If the radar has, say, only five modes as in Figure 14.1, cells B31 : G31 may be left blank. Table S1 maximum range is not interlocked with maximum range set in cell F5 and some spreadsheet results beyond maximum instrumented range (MIR) may show ERR (error), others showing >MIR.

Enter the operator-selected mode in cell F24. In the example, if F24 = 1, the spreadsheet would have assumed mode 1 parameters when computing for ranges <1.5 nmi, mode 2 parameters for ranges between 1.5 and 3 nmi and modes 3 upwards for longer ranges; in other words, it is assumed that the operator sets range scale to the shortest which will display the target without offcentring. If the operator is using the 6 nmi scale, enter mode 3. The parameters of this mode will then apply at all ranges up to 6 nmi, with automatic changeover to mode 4 parameters when target range lies between 6 and 12 nmi and mode 5 parameters beyond. Charts of, say, SNR to a base of range may have steps at the mode boundary ranges as, say, receiver bandwidth changes; examples are included in Chapter 15. If F24 = 3 and no higher modes have been entered, mode 3 parameters are applied at all ranges. The mode facility allows short range performance to be computed with short pulse, etc., without wrongly allowing the user to assume conditions not available to the operator, such as short pulses and high prf at long range. If F24 is set to, say, 3, then medium-range parameters will remain in use at the shorter ranges, mimicking operator retention of a long range scale when viewing a close-in target.

Enter the pulselength used on each range scale in column D26 : D31; if the radar has SP/LP facility (cell F23), enter the long pulse settings in column E26 : E31. Enter appropriate receiver bandwidths in cell F26 : F31 and pulse repetition frequencies in column G26 : G31; it is assumed that bandwidth and prf are not changed by the SP/LP switch; if the radar changes these parameters, enter the new parameters manually. Cells H26 : H31 display the active mode. Cells I26 : I31 show the selected pulselength and J26 : J31 show the number of pulses integrated for the scan rate, azimuth beamwidth, prf and is doubled for correlation over two scans if B20 = y.

14.2.8 Environment panel

Enter environment type in cells H4 and G5 : H5, free text, say ‘typical winter North Atlantic’. Enter atmospheric refraction coefficient, k (>0) in cell H6. The ‘standard’ value is 1.333 but lower values occur in bad weather, higher in good (Chapter 5, Sections 5.2 and 5.3). The effective Earth radius, E km, shows in cell H7 per Eq. (5.1). Cell BC29 uses the effective values of H, h and E (cells AI29, AK29, H7, respectively) to compute horizon range, km, from Eq. (5.23a), independently of the main matrix. It is shown in the chosen units within cell J14.

In cell H8 enter the precipitation or fog type: Stratiform rain = 1, Orographic rain = 2, Thunder rain = 3, Ice crystals = 4, Wet snow = 5, Dry snow = 6, Advection fog = 7, Radiation fog = 8. Cell G9 shows Fog visibility, km, if appropriate, otherwise Precipitation rate, mm/h. As prompted, enter in cell H9 either fog optical visibility, V km, or precipitation rate, r mm of rain equivalent per hour. Cell H10 shows the equivalent stratiform rain rate for attenuation calculations, per Eq. (5.46d) if either of the fogs was selected, otherwise the precipitation rate set in cell H9.

Cell H11 shows the appropriate precipitation clutter RCS per m^2 suiting the precipitation type in cell H8, obtained from Table S4, (cells BV22 : CD28 reproducing Table 11.1 and Eq. (11.7)). In the table, an artificial 0.000001 is added to the H10 rain rate to avoid $-\infty$ problems when H10 $= 0$. Fogs cause no clutter.

Cell H12 shows the precipitation loss, l_p dB/km one-way per Eq. (5.43), appropriate to the weather and frequency; if cell B6 $<$ 5000 MHz, 3 GHz values are used; between 5000 and 12 400 MHz, 9 GHz values and above that J band values; intermediate frequencies such as 6000 MHz may therefore suffer some error. Precipitation and fog do not always extend over the whole radar – target path. Enter the extent (from 0 to 1 p.u.) of the path subject to the precipitation or fog in cell H13, guided by Eq. (5.44). Cell H13 defaults to 1 (whole path) even when H13 $= 0$. The spreadsheet assumes the target always to lie within the clutter.

Enter air temperature (degree Celsius) and relative humidity (RH, from 0 to 100 per cent) in cells H14, H15, respectively; Cell H16 shows the clear air loss, L_c dB/km one-way per Eq. (5.49b).

Enter significant wave height, h_s m, in cell H17. Equivalent sea state number appears as an integer 0–5 in cell H18, obtained by lookup of cells CG24 : CG29 within Table S5 (cells CF22 : CM29), which is based on the modified values within Table 5.3. This is admittedly not entirely satisfactory, with steps from one sea state to the next (it would also be possible to connect sea state and wave height by Eqs (5.38) of Chapter 5, Section 5.7.5), but is the best available in absence of a reliable function smoothly linking h_s to sea clutter. Enter in cell H19 the sea clutter wind factor (C2 $= -2.5$ to $+2.5$ dB for looking downwind to upwind, respectively, and several dB negative for a swell rather than a fully developed sea). The spreadsheet extracts working values for factors A, B and C of Eq. (11.11) from Table S5 cells CM23 : CM25, adding factor C2 to factor C1 (from Table 11.3 according to the scanner polarisation, cell F20). Weibull exponent, c, is computed from sea state in cell CM26 per Eq. (11.19) and Weibull weighting factor, W dB Eq. (12.13b), in cell CM27. Table S5 recognises only the 3 and 9 GHz bands, with changeover at 5000 MHz, and does not interpolate for other frequencies per Eq. (11.11b).

Enter, in cell H20, the average time proportion (from 0 to 1) that most of the target is screened from the scanner by sea waves, default 0. For the water or land surface at the grazing point, enter dielectric constant ε in cell H21 and conductivity, S, in cell H22, guided by Table 5.4 (for sea water $\varepsilon = 81$ and $S \sim 4$).

In cell H23 enter, as a positive dB value, the loss of effective RCS or one-way echo strength caused by target tilt or polarisation.

14.2.9 Results and user panels

The results panel (km/nmi units of cell F4, repeated in cell J5) shows: probabilities of detection at the spreadsheet range next below two user-selected ranges, the minimum and maximum ranges giving the required P_D, the range bracket and fill through which that P_D is achieved, horizon range, ranges of the first two (longest range) multipath peaks and nulls, maximum sidelobe range and the sea clutter horizon range for $\beta = 0$. Cell J21 gives the reference free-space echo strength (figure of merit, F_{12}) at 1 km for a 0 dB m^2 RCS target (always km; not scaled according to cell F4; reduced to suit the Gain setting, cell F20; but not swept gain, cell F21), calculated using Eq. (4.12), including physical losses such as L_t and feeder losses affecting echoes and clutter returns, but excluding processing loss L_p. Multipath and atmospheric loss terms are excluded. Horizon and J21 are always calculated, even if the horizon and 1 km are outside the spreadsheet range or maximum instantaneous range (MIR). Horizon uses free-space range equation Eq. (5.22b) (including k).

Most of the result cells J7 : J21 are derived from the main calculation matrix, see Sections 14.6.7 and 14.7. Logical IF functions show >F6 or <F8 if ranges in question lie outside the spreadsheet range, and >MIR if beyond the maximum instrumented range, the maximum range in cells B26 : B31. Cell J13, per-unit fill, calculates, unless insufficient data is available, the proportion of the range interval between minimum and maximum detection ranges at which detection occurs. Cell J20 gives the maximum range to which sidelobes are detected. Cells AU29, AW29, AY29, BA29 and BC29 perform preliminary range calculations.

The User Panel, cells K3 : EX21 to the right of p1, facilitates charting results to bases independent of range. For example, to chart how performance varies with scanner height, H, proceed as follows. First enter in cell I4 the name of the independent variable, here H. Enter the lowest value of interest, say 1 (m) in cell J4. In the cell in which the independent variable is usually entered, here D12, enter = J4, so the spreadsheet adopts the value entered in J4. Cell J5 reminds of the scaling km/nmi and cells J7 : J21 compute results of the dependent variables as usual. Use the Copy and Paste Special, Insert Value procedure to place J4 : J21 values into cells K4 : K21. Reset J4 to a new value of the independent variable, say 2 (m). Again Copy and Paste Special, Insert Value the new contents of J4 : J21 into the next column, L4 : L21. Repeat as often as desired. A chart may then be constructed to a base of the chosen independent variable, here H row K4 onwards as abscissa, with up to six quantities chosen from rows K11 : K21 as ordinates. Pasted quantities in column K onwards are simply numbers and *do not* respond to subsequent page 1 parameter changes. Chapter 15, Section 15.2.3, Figures 15.17, 15.21 and 15.22 below is an example.

It is prudent to double-click function key f9 after entry of new values to ensure the spreadsheet recalculates properly. Readers of Chapter 13 will need no reminding that display of a result to numerous significant figures does *not* infer great accuracy. Most of the trailing digits have no practical significance, but are retained to show trends.

14.3 Geometry panel

14.3.1 Layout

The body of the spreadsheet calculates each parameter leading to detection as a row, with each of a set of ranges as a column. Column A names the parameter, B the main text equation(s) on which it is based, C its symbol and unit. Column D sometimes contains miscellaneous items. Columns E to EX are devoted to a set of 150 range values, rising approximately logarithmically from km equivalents of cell F8 value to cell F6 value. It is not normally necessary for users to inspect the Geometry panel, except to scrutinise trends when an unexpected result is delivered. Values are not rounded, so trends are apparent even when a cell scarcely differs from its neighbour.

The spreadsheet refers to a number of internal parameters which are not range-dependent and are named and calculated in Table S3 cells M29 : BU30, or the other tables. The column E cell of a row is occasionally calculated differently from the remainder and is then identified in the spreadsheet by **bold type**. In the following, # represents the letter(s) of the column under discussion. Factors of 1000 and $\pi/180$ frequently appear, reflecting the mixed m/km and radians/degrees units in page 1.

Row 32 is not used.

14.3.2 Establishment of α and R series

To establish a rising, nominally logarithmic, series of range values (numerically in geometric progression), the working set of columns are first numbered, cells E34 : EX34 being labelled serially from 1 to 150. The next row contains the tentative geometric range series, described in Section 14.2.5. This indirect approach avoids the difficult direct calculation of grazing angle from the system geometry.

The grazing angle, α rad, for range E35 is calculated in cell D36 per Eq. (5.14d), which is accurate only at short range. The remaining cells are calculated similarly: #36 = **D36** × **BC29** × **E35** × (**BC29** − **E35**)/(1/#35 − 1/BC29). For convenience the terms in bold are pre-calculated in cell D35. This empirical formula yields a set of falling α; incremented per cell BK29, going through zero at the horizon, with shallower slope at longer range, from a maximum consistent with the adopted minimum range to a value algebraically suiting the adopted maximum range. Beyond the horizon, row 36 range increments are exactly logarithmic so ranges rise in geometric progression; meaningless negative α values are displayed but not used.

Row 37 calculates grazing point range, $D1$ km, from Eq. (5.12a); again meaningless post-horizon values are not used. Row 38 calculates the final set of target ranges, R km, from Eqs (5.10) and (5.12b) out to the horizon, and in geometric progression beyond. These ranges may differ slightly from row 35, but their progression remains substantially geometric throughout. They are accurate for the α values of row 36 and form the main charting abscissa. Cell E38 and EX38 values are scaled for entry as minimum and maximum achieved ranges in cells F8 and F6, respectively. The following two rows compute log R and R/R_{HOR}. Row 52 gives $\log(R/R_{\mathrm{HOR}})$, range in nautical miles is in row 73 and log R_{nmi} is in row 82. They are available as bases for charts having range-related abscissa.

14.3.3 Scanner and target heights

Rows 41 and 42 compute effective scanner and target heights H', h', respectively, from Eq. (5.9a), falling from near the cell D11 and F11 actual values to zero at the horizon. Meaningless negative post-horizon values are not used. Row 43 computes the indirect/direct ray path length difference, Δ m, from Eq. (5.16a). Again this gives meaningless and unused post-horizon results (here positive).

14.3.4 Angles and effective scanner gain

Rows 44–48 compute the geometrical angles summarised in Table 5.1, using the equations there listed. The grazing angle at the target foot, β rad, row 48, is made to shows zero instead of negative values when the foot is below its horizon, so giving zero sea clutter.

The direct and indirect rays to the target, and the sea-clutter ray to the target foot, are in general below the scanner beam axis (unless the axis is depressed) so the effective scanner gain is reduced below cell D8 value by an extent dependent on: (a) the elevation beamwidth, cell D7; (b) scanner depression angle, δ degrees D13; (c) angles κ and χ (Figure 5.8(*b*)); (d) the elevation pattern, cell D12. The gain reductions, $\Gamma_{\eta-\delta}, -\Gamma_{\kappa-\delta}, \Gamma_{\chi-\delta}$ dB, for the range in question are computed in rows 49–51, respectively, by reference to Table S2.

14.4 Environmental effects

14.4.1 Diffraction region

Row 53 computes $-\Phi$ (Eq. 5.16b) (to get a set of rising values to suit HLOOKUP syntax), row 54 repeats target range, row 38, for internal HLOOKUP purposes and 55 contains the negative of diffraction region multipath factor m_{d}, computed by Eq. (6.7b), using values for L, U, Z, z, $f(Z)$ and $f(z)$ per Eqs (6.8)–(6.11) in Table S3 cells N30 : Y30. Row 56 gives a monotonically rising set of values from the inverted bell-shaped row 55, returning the artificial value −999 when the slope of row 55 is falling. Cell D55 gives an artificial value which assists calculation of cell E56. Cell D56 gives the minimum of $-m_{\mathrm{d}}$, used in cell AC30 to bias the working diffraction multipath threshold down from −20 dB when working with low H and h.

14.4.2 Interference region multipath

Row 57 repeats column number for Lookup purposes. Divergence, d, is computed in row 58 using Eq. (5.22); d is not used beyond the horizon and an IF function sets to zero to prevent disconcerting show of ERR (error).

Rows 59–65 compute the parameters C to K used in sea forward reflection calculations, Eqs (5.40), and ρ_0 itself is computed in row 66 from Eq. (5.40f), followed by the associated phase angle, ψ, row 67, from Eq. (5.40g).

In row 68 comes the total interference region indirect ray phase shift ϕ rad (Eq. (5.16c)), given as $-\phi = -(\Phi + \psi)$ to get a rising series of values.

Row 69 computes roughness term Y used when finding coefficient of surface roughness, ρ_{s}, per Eq. (5.41a), followed by ρ_{s} itself in row 70, using Eqs (5.41c) and (5.41e). Differential gain loss, g_{dif}, numerical between direct and indirect rays $\Gamma_{\kappa-\delta}, -\Gamma_{\eta-\delta}$, rows 50 and 49 is computed in row 71 and under most circumstances is close to 1.00. Forward reflection coefficient, ρ numerical, is then computed in row 72 using Eq. (5.39). Row 73 gives range in nmi; it is inserted here for convenience of HLOOKUP functions in Table S2. Row 74 gives Φ for charting use. Row 76 shows the multipath region for the range in question. These preliminaries clear the way to computation in row 77 of interference region multipath factor, m_p dB, using Eq (6.4). For row 75 see Section 14.5.1.

14.4.3 Transition region multipath

The transition multipath factor, m_t, is found by the curve-fitting method of Chapter 6, Section 6.6.2. The lower boundary of the transition region, R'_A km, is set by cell AE30, the criterion being the path length phase shift in cell AA30, set to $\Phi = \pi/2$ rad. The upper boundary, R''_B km, is set by cell AI30, the criterion being $m_d = -20$ dB, cell AC30 unless a lower value (−cell D56) is needed for low H and h. Cells AG30, AK30 looks up the column numbers for R'_A, R''_B. Ranges R''_A, R'_B (cells AM30, AO30) are found by stepping forward 1 column from R'_A, R''_B and using Lookup functions against the column number, row 57. Multipath values at these ranges are obtained by Lookup and shown in AP30 : AW30, with slopes S_A and S_B in cells BD30 : BG30. For convenience, various functions of the general form $R'_B - R'_A$ (Section 6.6.2, factors c, d and e) are computed in cells AX30 : BC30 to facilitate computation of factors s, r, q, p in cells BH30 : BO30 per Eqs (6.22), (6.18), (6.21) and (6.20), respectively. Row 78 uses these factors to compute m_t per Eq. (6.15). If the whole spreadsheet lies in the interference region, some of the Table S3 cells show ERR. This is not significant.

14.4.4 Overall multipath factor

Row 80 chooses, by IF function, the overall multipath factor, M, from the interference (row 77), transition (row 78) and diffraction (row 79, = −row 55) multipath factors according to row 76 value. If the whole spreadsheet lies in the interference region, transition and diffraction region multipath factors show ERR and row 80 equals row 77. In this spreadsheet, M applies both to interrogate and return legs.

14.4.5 Atmospheric loss

The range bracket through which precipitation occurs is the row 38 range, multiplied by H13 unless per-unit precipitation extent is set <1 in cell H13. Total precipitation loss component is this extent times the loss per km l_p, cell H12. To this is added range times clear air loss per km, cell H16, to give total atmospheric loss, L_A dB/km one-way, row 81.

14.5 Signals at the radar receiver, single pulse

14.5.1 Effective mode

Strengths of noise, etc., at the receiver depend on bandwidth, etc., and in turn on mode, which is range-dependent. The active mode is computed in row 75 by nested IF functions relating the maximum mode ranges in Table S1 column C25 : C31 to row 38 range. If row 38 range exceeds maximum instrumented range, mode 1 is assumed to prevent appearance of unwanted ERR in the later spreadsheet columns.

14.5.2 Noise and swept gain floor

Row 83 computes system noise power, dBW, from Eq. (3.2b), using the system noise factor of cell D23, active bandwidth selected from Table S1 and operating temperature per cell H14 (assuming the receiver, scanner and feeder are all at ambient air temperature).

The spreadsheet assumes the swept gain law is R^{-4}, so that in free space the threshold without noise is equivalent to the definite RCS set in cell B14, modified by the swept gain control, cell F22, see Chapter 12, Section 12.7.2. The residual swept gain (T_{sg}, dBW) threshold, row 84, equals the reference echo J21 + B14 + F22 – 40 log R (km). Adding row 83 gives row 85, the Noise + RSG floor, converted to dBW.

14.5.3 Precipitation clutter

The spreadsheet assumes the target always lies within the selected precipitation even if precipitation extent, cell H13, is set less than 1 (including 0). Row 86 computes the mean precipitation clutter RCS dB m^2 in the detection cell, per Eq. (11.9) from the unit-volume clutter (σ_p, cell H11 using Eq. (11.7a) and Table 11.1, valid for all frequencies) and the cell volume, which depends on the scanner beamwidths (cells D6, D7), pulselength (Table S1) and range, row 38, per Eq. (11.8).

The precipitation clutter mean return, C_{p0} dBW, is computed in row 87 by Eq. (11.10a) assuming one-way multipath = 1.5 dB as in Chapter 11 by doubling the effective cell volume in row 86, for simplicity using the reference echo, cell J21. If the scanner is circularly polarised (cell F20 = c), the return is depressed by the CP improvement entered in cell D15. Two-way atmospheric loss, 2 × #81, is subtracted.

14.5.4 Sea clutter

Row 88 computes the mean sea clutter RCS in the detection cell, σ_{s0} dB m^2 for the 3 or 9 GHz bands, from the unit-area clutter (Table S5 using Eq. (11.11a) and Table 11.3, based on the sea-state number computed in cell H18 from the h_s in cell H17) and the cell area (Eq. 11.13), which depends on the scanner azimuth beamwidth (cell D6), pulselength (Table S1) and range, row 38. If frequency $<$5000 MHz, Table 11.2.3 GHz data are used, corrected for frequency by Eq. (11.11b); higher frequencies are based on the 9.4 GHz data, facilitating use at C and J bands. Factor $C1$, dependent on polarisation, is set automatically per Table 11.3, but factor $C2$, dependent on wind direction, must be set by the user in cell H19. Factors A, B, C (including H19 value), and Weibull shape and weighting factors, c and W, are computed in Table S5 cells CM23 : CM27, respectively, using Eqs (11.11), (11.19) and (12.13b).

Row 89 computes the sea clutter RCS within the detection cell area per Eq. (11.15). Row 90 computes the sea clutter mean return, C_S dBW, per Eq. (11.16b) modified by the two-way sea clutter ray axis loss, $2\Gamma_{\chi-\delta}$, row 51 and atmospheric loss, row 81.

14.5.5 *Total noise and clutter*

Row 91 converts the residual swept gain threshold, precipitation and sea clutter components to power, sums them and reconverts to dBW. Short-range ringing clutter (Chapter 11, Section 11.8) is excluded. Rows 92–94 are not used in this spreadsheet.

14.5.6 *Echo*

Row 95 calculates the free-space echo by adding target RCS (dB m^2, as entered in cell F13, less tilt/polarisation loss, cell H23) to the reference echo (cell J21, dB m^2, target at 1 km) and subtracting 40 log R km, generally per Eq. (4.8) and assuming the target lies on the nose of the scanner beam.

Row 96 computes the target actual echo by subtracting the two-way beam axis loss (2× row 50) two-way atmospheric loss (2× row 81) and adding two-way multipath (2 × row 80). Row 97 gives the single-pulse SNR, q dB, by subtracting clutters and noise from the echo, also subtracting processing loss per cell B18; row 98 gives q numerical equivalent. These q values are approximate; they include all processing and integration losses relative to theoretical integration gain, which strictly should not be applied until row 102.

14.6 Main beam detection, multiple pulses

14.6.1 *Pulses integrated*

Row 100 performs a HLOOKUP to obtain the number of pulses integrated per scan from the active mode row 75 and Table S1 column J.

14.6.2 *Integration gain*

Row 101 computes integration gain, dB, per Eq. (12.19c) for a non-coherent system. Row 102 selects this, or the integration gain relative to other values entered in cell B19 (reminder displayed in cell D102) per Eqs (12.19b), (12.19d) or (12.19f). This operation is split into two rows to avoid memory overload. Default is n, non-coherent, as this is commonest and gives poor performance. Addition to row 97 gives SNR with integration, Q, given in rows 103 (dB) and 104 (numerical).

14.6.3 *Swerling Case 0*

For this echo fluctuation, row 105 calculates factor $D = \log\,[P_D/(1 - P_D)]$ per the modified Levanon approximation of Eq. (12.11a) with modification constant $C = \frac{1}{3}$.

Cell D106 computes the Q (dB) needed for the required screened P_D of cell B16, using the false alarm exponent F entered in cell B17 and the D value computed in cell W29 from cell B16. The unmodified Levanon approximation is used, Eq. (12.10a), enabling results to be obtained at all P_D values. Cells E106 : EX106 then compute achieved probability of detection for the prevailing SNR per Eq. (12.10c), allowing for screening, cell H20, Eq. (12.28a). IF functions block attempts at calculation of row 105 extreme values.

Row 107 calculates the equivalent single pulse SNR, q dB, which would achieve required cell B15 P_D, by subtraction of the integration gain, row 102, from D106.

14.6.4 *Swerling Case 1*

Cell D108 computes Q for the required screened P_D, using Eq. (12.16b). Row 108 computes achieved screened P_D for the Q of row 104, using Eq. (12.16b). Row 109 calculates the equivalent required single pulse SNR, q dB, by subtraction of the integration gain, row 102 (Eq. 1219g).

14.6.5 *Swerling Case 3a*

Row 110 computes achieved P_D per Eq. (12.18e), row 111 giving the required equivalent q dB.

14.6.6 *Chosen case performance*

The required case is entered in cell F18, defaulting to Case 1 in this spreadsheet, with a reminder in cell D112. Using IF functions, this selects the appropriate screened P_D in row 112 from rows 106, 108 or 110. Row 113 shows its factor D, to facilitate charting to 'log' P_D base. The range is limited to $-6 < D < 6$ to prevent extreme-value problems when $P_D \sim 0$ or 1. Row 114 indicates the signal in hand (dB) above the value giving the required unscreened P_D, cell B15, as a guide to robustness of detection with possible system performance variations and accuracy of tracks formed on the target (Chapter 13); #114 = #97 − #107, 109 or 111. If selected P_D equals or exceeds the value entered in cell B15, detection is declared by 1 in row 115. Row 116 carries event ranges at which detectability changes, showing detection-start ranges as negative and detection-cease ranges as positive values. Row 117 sets row 116 positive values to very high number 9999 and changes sign of row 117 negative values, necessitating presetting cell D115 = 0. These are used in the Results panel, Section 14.6.8.

14.6.7 *Event labels*

Row 118 generates data labels in the column in which important events occur, for optional attachment to chart series where the software will support this. The labels are: RA, RB, MAX SL, HOR, MIR; signifying R_A, R_B (lower and upper bounds of multipath transition region), maximum sidelobe detection range, horizon range and maximum instrumented range, respectively. The decision tree IF functions use cells AS29, AE30, AI30, J20 and BG29.

14.6.8 Results panel

Most of the following result quantities use IF functions to show <F8 or >F6 if outside the spreadsheet range limits, or >MIR if beyond maximum instrumented range. Results are delivered in cell F4 (=J5) units, km or nmi.

For the test ranges entered in cells J6 and J8, cells J7 and J9 look up achieved P_D from row 112. The ranges may lie anywhere within the spreadsheet range bracket, F8 to F6; the P_Ds shown (0 to 1, rounded to four decimal places) relate to the row 38 range next below the km version of the chosen value.

Cell J10 shows the minimum range delivering the required screened P_D per cell B16 as the minimum within row 117. Cell J11 shows the maximum range giving the required P_D, as the maximum number in row 116. Cell J12 shows the total number of km or nmi bracketed between J10 and J11 where the echo was detected, by addition of the row 116 entries but showing 0 if the sum is negative. For example, if required P_D is equalled or exceeded between 1 and 2 km, has a null between 2 and 4 km, with a second peak between 4 and 8 km, J12 shows $2 - 1 + 8 - 4 = 5.00$. Cell J13 shows this range bracket as per-unit fill of the range bracket F8 to J11. Less than 1 in J13 warns there are multipath nulls having less than specified minimum P_D at intermediate ranges, the lower the J13 value, the more extensive the loss of detection. J13 shows N/A if J12 indicated >F6 or <F8.

As previously mentioned, cell J14 computes horizon range independently of the matrix, from cell BC29, this range may not coincide with a range entry in rows 38 (km) or 56 (nmi).

Cells J15 : J18 show ranges of the first two peaks and nulls, by looking up via cells AT29 : BA29 the ranges (row 73 expressed in km) at which $-\phi$ row 68, is -2π, -3π, -4π or -5π, respectively, per Eqs (5.24) and (5.25), assuming curved Earth.

14.7 Sidelobes

Rows 120–128 replicate some of rows 103–116 for the principal sidelobe as follows, with scanner one-way gain reduced per cell D16. Row 120 is 2 × D16 dB below row 103, row 121 being the numerical equivalent. Rows 122 and 123 replicate rows 105 and 106 for Case 0. Row 124 replicates row 108 for Case 1, likewise row 125 replicates row 110 for Case 3a. Rows 126–128 replicate rows 112, 115 and 116. Row 126 is based on the maximum tolerable probability of detection of the principal sidelobe, cell D18. The user is seldom likely to require to find q for a given sidelobe P_D or to find minimum sidelobe range, so the other main-beam detectability rows have not been replicated. Results panel cell J20 shows the maximum range, km or nmi, at which sidelobe P_D exceeds the value entered in cell D18, following the method of cell J11.

14.8 Graphs

14.8.1 Chart construction

Graph sheets (called charts by Microsoft) may contain up to six curves (called series), having a common abscissa and one or two ordinate scales. Up to eight charts are supported. Charts may show, say, exactly where a curve cuts some value, perhaps where P_D cuts the 0.5 probability gridline intermediate between two spreadsheet range increments, enabling closer estimation of range for $P_D = 0.5$ than available from the results panel. However, this additional 'accuracy' is probably illusory, since the whole spreadsheet is subject to numerous small errors as warned in Chapter 13. Series values change in response to subsequent changes of page 1 parameters, unless a Copy and Paste Special, Insert Value procedure has been used. Chapter 15 contains numerous examples of 'raw' Charts, and similar Charts form the basis of some figures in previous chapters: for example, Figures 5.4, 5.9, 5.10 and 6.3. These inserted a chart as an Object into Microsoft Draw, then added annotations, altered line thickness, etc., to give figures of 'report' quality.

Often the user will wish to plot to a base of range, expressed linearly in km using row 38 or nmi using row 73. When a logarithmic scaling is required, the user may command the spreadsheet charting function to use logarithmic scaling of row 38 or 73 quantities, the scale then always extending from the power of 10 next below the lowest selected range cell to the power of 10 next above the highest. For example, it may be desired to plot between 5 and 50 km, but the chart will extend between 1 and 100 km, cramping the range bracket of interest. Choosing as base row 39 or 82 gives log R in km or nmi, respectively, improving flexibility. In the example, row 39 values extend between 0.7 and 1.7 log R units without cramping, but the reader has to convert values to km, etc., manually. An alternative is to plot to a base of per unit radar horizon range, say from 0.2 to 1.5 p.u., equivalent to 6–45 km if R_{hor} is 30 km; here row 40 gives linear scaling and row 52 logarithmic.

Charts may also be plotted to a base of grazing angle, α row 36 if all events occur within horizon range, where α falls to zero (post-horizon, row 36 negative quantities have no physical meaning).

The relationship between R and α depends on f, H, h, and k, cells B6, D11, F11 and H6, respectively. If these remain unaltered, any of the quantities represented by rows 41 onwards can be used as chart ordinates. For example, it would be possible to plot effective scanner and target heights, H and h, rows 41 and 42, to a left-hand ordinate of metres, with system noise, precipitation clutter, sea clutter and target echo, rows 83, 87, 90 and 96, respectively, to a right-hand ordinate scaled in dBW, all to a common abscissa such as range or α.

To construct a chart having, say, six series (plots) of P_D to a base of R, for dependent variables which do not affect the α/R relationship, say two values of precipitation rate, r, each for 3 values of wave height h_s, one sets the first pair of dependent r, h_s, values in cells H9, H17, respectively. Using the Copy and Paste Special, Insert Value facility, transfer the results row(s) of interest, here P_D row 112, to a spare row, say 130, at the foot of the spreadsheet. Reset the dependent variables and again Copy and Paste Special, Insert Value, this time to row 131. Continue for all six dependent variables and construct a chart with range (row 38 for linear scaling) as abscissa, with 'series' rows 130, 131, ... The ordinate in this case would be scaled from 0 to 1. Examples are included in Chapter 15, Section 15.1.1, Figures 15.7 and 15.8.

To plot a family of values of one or more of f, H, h and k, which affect the α/R relationship, the procedure is extended as the following example. Suppose curves of P_D are to be plotted for six scanner heights of $H = 1$, 2, 4, 8, 16 and 32 m at 9400 MHz, $h = 4$ m, $k = 0.8$, and the range bracket of interest is 1–50 km. Enter all the other system parameters, with F11 at 4, F4 at km and H6 at 0.8. Next, enter F7 and F5 as 1 and 50 (km), respectively, and enter H at its minimum value, 1 (m), in cell D11. Check that achieved minimum and maximum ranges, cells F8 and F6, 1.002 and 52.789 km, respectively, cover the whole bracket of interest, if necessary adjusting F5 or F7 for full coverage.

Using the Copy and Paste Special, Insert Value facility, transfer the range row of interest, for example, row 38, as well as the results row(s) of interest, here row 112, to unused rows 130 and 131 at the foot of the spreadsheet. Repeat the whole procedure for the next H value. Accumulate a family of up to six row pairs of range/P_D results and use the HLOOKUP function to transfer the P_Ds to a common range base. All six curves linking P_D with range may now be charted in the usual way, taking the common range base as abscissa. HLOOKUP may introduce minor discontinuities in the curves.

Charts can be annotated or embellished for reports by importing the chart as an object in the Microsoft Draw facility. To assist readers to familiarise themselves with the spreadsheet, two X–Y scatter charts are preloaded, as follows.

14.8.2 Chart 1, detectability

Preloaded Chart 1 plots signal and clutter strengths and P_D to a base of range. Users can readily change series colours, line patterns and markers to suit preferences.

Series 1 (black line) is row 85, noise + RSG floor; series 2 (red) is row 87, precipitation return; series 3 (turquoise) is sea clutter return, row 90; series 4 (blue) is target echo strength, row 96; series 5 (bright green) is P_D, row 112, and series 6 (pink) is signal in hand, row 114. The range base is row 38, km, linear scale.

The left ordinate, used for all except series 5, is dB, scaled linearly. The series 5 right ordinate is P_D, also linear. The user can format the minimum, maximum and interval of the three axes to suit the task in hand.

14.8.3 Chart 2, geometry

The six series of this chart are coloured as before and, respectively plot: grazing angle, α; grazing point range (right-hand axis), D1 km; effective scanner and target heights, H' and h' metres, indirect ray path-length phase shift, Φ rad (row 74), and multipath factor, M. The base is log R, km (respectively, rows 36, 37, 41, 42, 74, 80 versus 39).

14.9 Extended passive targets

Multipath of extended targets was discussed in Chapter 9, and target RCS in Chapter 10. Spreadsheet SS2 for extended passive targets such as vessels and coastlines is similar to that described above for passive point targets, with the following significant differences.

14.9.1 Spreadsheet page 1

Figure 14.2 shows page 1. Layout closely follows the point target version described in detail above. The following entries are made in the Target panel.

1. Enter target type in cell F9, free text.
2. Enter in cell F10 target kind: coast for a coastline target of great radial extent, ship for a target of finite radial extent (default setting).
3. Enter tip height, j metres, in cell F11. In cell F12, enter height factor, n, typically 0.5 (see Chapter 9, Section 9.3). Cell AC29 constrains the working value between 0.01 and 1. Cell AK29 constrains the value of nj to >0.25 m. Horizon range (cell J14) is computed to target tip. The target height used to compute target range, R, in row 38 is AC29 × AK29.
4. Cell E13 shows RCS, dB m^2/m^2 if coast was entered in cell F10; enter RCS, dB m^2 per square metre of coastline in cell F13, typically -14. If F10 was entered other than coast, cell E13 shows Total RCS, dB m^2. Enter the target total RCS, dB m^2, in cell F13. Cell F14 shows the RCS m^2/m^2 (i.e. the numerical reflectivity) or total RCS m^2, respectively.
5. If F10 = coast, cell E15 shows N/A and cell F15 need not be entered. Otherwise, E15 shows Extent, radial, m; enter the radial extent of the target echoing surfaces, (the transverse width seen by the radar, metres). In cell F16 enter the axial extent of the echoing surfaces (this applies to both coast and ship). Width of a ship head-on to the radar would be entered in F15 and length in F16, interchanged for a beam-on ship.
6. Enter the target Swerling Case in cell F17 as for a point target (0, 1 or 3; most extended targets are Case 1, the default setting).

In the Results panel, instead of multipath peak and null ranges, which do not apply to extended targets, cell J15 shows azimuth overspill range for ship targets or N/A for coast, obtained from the geometry via cell BQ29, the overspill range, km. Cell J16 shows rough-sea critical range, R_c, from Eq. (9.14b). Cells I17 : J19 are not used.

14.9.2 Remainder of spreadsheet

Row 42 shows target effective height, h', falling below nj as range increases. Row 74 computes r'/h' (Eq. (9.11d)) preliminary to computation of m_p in row 77 using Eq. (9.12a). Rows 78–80 compute m_t, m_d, and M as before. Meanwhile row 82 computes the sea-state dependent version of the critical range, R'_c km, by Eq. (9.14c). This is not used in the spreadsheet, but is available for plotting.

	A	B	C	D	E	F	G	H	I	J
1										
2	Extended passive target spreadsheet				Spreadsheet Ref	SS2 v1				
3	Date		User		Task	Chapter 15, Sec 15.3.1.				
4	Transceiver		Scanner and feeder		Range bracket		Environment		Results	
5	Type		Type		Scaling, km/nmi	nmi	Type			
6	YACHT RADAR				Max reqd *R*	25			*Scaling*	*nmi*
7	Frequency, MHz	9450	Az beamwidth°	4	*Achieved*	*25.000*	Refraction, *k*	2	Test R1	4
8	*Wavelength, m*	*0.0317*	El beamwidth°	20	Min reqd *R*	0.5	*Eff Earth rad E, km*	*12742.00*	*PD at R1*	*0.6029*
9	Tx P, kW	1	Gain, dB	24.5	*Achieved*	*0.367*	Precip type: Stratiform =	7	Test R2	5
10	*Tx P,dB W*	*30.00*	*Efficiency, p.u.*	*0.90*	Target type		Fog vis, km	0.05	*PD at R2*	*0.0304*
11	Tx loss Lt, dB	1	Loss, dB	1	Kind: ship, coast	coast	*Equiv rain for atten*	*2.266*	*Min R for B15 PD*	*<F8*
12	Rx loss Lr, dB	0	Height H, m	4	Tip height j, m	15	*RCS, dBm²/m³*	*−9999.00*	*Max R*	*4.0847*
13	Service loss Ls, dB	2	El patt, 1 = sin, 2 = invc	1	Height factor, *n*	0.66	*Loss lp, dB/km*	*0.024*	*R bracket*	*3.72*
14	Rx NF N, dB	3	Depression°	0	RCS, dBm²/m2	−15	Extent, p u	1	*Fill, p u*	*1.00*
15	RSG thld, dB m²	−10	Rotation, rpm	30	*RCS, num*	*0.06*	Air temp, *C*	4	*Horizon R*	*16.01*
16	Reqd PD	0.6	CP improvement, dB		N/A	5	RH, %	100	*N/A*	*0.04*
17	*With screening*	*0.600*	Sidelobe below D8, *d*	27	Extent, axial, m	1	*Clr air loss, dB/km*	*0.009*	*R crit*	*18.31*
18	Reqd Pfa expt, *F*	−6	Tolerable SL PD	0.1			Wave hgt, hs, m	0.025		
19	Proc loss, Lp dB	8	Feeder ohmic loss, *d*	0	Swerling Case 0,1,3	0	*Sea state*	*2*		
20	Integ, n, c, cr, *p* = non	n	VSWR	2	Operator		Wind fctr C2, dB	−2.5		
21	Scan/scan corrln y/n	y	*Refl coeff*	*0.333*	Pol, h, v, c	h	Screening, p.u.	0	*Max sidelobe R*	*<F8*
22			*Mismatch loss, dB*	*1.53*	Gain control, dB	0	Surface dielec, eta	81	*Ref echo, 1 km, FS*	*−109.02*
23			*EIRP, dBW*	*51.97*	Sw Gain control,	0	conductivity, *S*	0.24	*Sea clutter horizon*	*5.45*
24			*System NF, dB*	*6.29*	SP = s, LP = 1	s	Tilt/pol loss 1-way dB	0		
25					Mode, Table S1	1				
26	TABLE S1, Mode	Max R	*Max R, km*	Plslgth, SP	LP, *µs*	Rx bw, MHz	PRF, pps	*Active mod*	*Selected, µs*	*Pulses inte*
27	*1*	3	*5.555555556*	0.05		25	5000	*1*	*0.05*	*222.22*
28	*2*	6	*11.11111111*	0.25		5	5000	*2*	*0.25*	*222.22*
29	*3*	30	*55.55555556*	0.5		3	2000	*3*	*0.5*	*88.89*
30	*4*		*0*					*4*	*0*	*0.00*
31	*5*		*0*					*5*	*0*	*0.00*
	6		*0*					*6*	*0*	*0.00*

Figure 14.2 Extended target spreadsheet SS2 page 1. Cells A3 : J31. Layout is similar to the point target spreadsheet, Figure 14.1. Set for detection of a cliff by a 9 GHz band yacht radar in fog

Rows 92 and 93 compute effective target RCS width, height and depth components (m^2, allowing for effective pulselength) and row 94 gives the total RCS in dB m^2, using the geometry of Eqs (10.14) and (10.15). For ship targets, rows 93 and 94 are the proportion of the F13/F14 RCS within the radar detection cell radial and axial dimensions, assuming the total target RCS is uniformly distributed within the rectangle F15 × F16 m^2; Lookup of effective pulselength from Table S1 and IF functions give full target RCS when its dimensions are small or range is long. The coast case assumes the radial detection cell dimension is filled at all ranges.

Target free space and actual echoes, probabilities of detection, etc., are then computed as for point targets. Tables S1–S5 are as before, with a few additions; working n is added in cell AK29 as already noted. The User panel remains available. Row 118 includes RC, critical range.

Graphs may be constructed as Section 14.8. Chart 1 uses identical series. Chart 2 drops H' (row 41) in favour of effective RCS, dB m^2, row 94, which is usually more informative. Φ is in row 99.

14.10 Active point targets

14.10.1 Target types

Multiplicity of active target types, and the sometimes differing multipath factors, M_1, and M_2, between the interrogation and response legs make this spreadsheet SS3 considerably more complicated than the others. To minimise its size, no direct provision has been made for entry of passive targets, use spreadsheet GDC14S1 described in Section 14.2. For comparative purposes, a passive point target may be simulated by ascribing very high saturated output power to a RTE of the desired unsaturated RCS. Figure 14.3 shows page 1 of the spreadsheet, which caters for the following devices, entered by number in cell F9.

1. Search and rescue transponder (SART).
2. In-band racon with square-law demodulator (crystal-video receiver, e.g. swept-frequency racon).
3. Offset-frequency racon, square-law demodulator.
4. In-band racon, linear-law demodulator (the usual modern type).
5. Offset-frequency racon, linear-law demodulator.
6. Ramark.
7. Radar Target Enhancer (RTE) without economy circuit.
8. RTE with economy circuit having square-law demodulator.
9. Radar warning receiver with square-law demodulator.
10. Radar warning receiver with linear-law demodulator.

Enter the interrogation fluctuation Case as 0, 1 or 3. The default is Case 0. Cell L28 shows 1 if the demodulator is square law and 0 for linear, the default setting. Cell BS30

	A	B	C	D	E	F	G	H	I	J	K	L
1	Active point target spreadsheet				Spreadsheet ref	SS3 v1						
2	Date		User		Task			Target type	Offset frequency racon			
3	Transceiver		Scanner and feeder		Range bracket		Environment		Results		User panel	
4	Type		Type		Scaling ,km/nmi	nmi	Type					
5					Max R reqd	50			*Scaling*	nmi		
6	Frequency, MHz	3050	Az beamwidth°	3	*Achieved*	*50.000*	Refraction, *k*	1.3333	Test R1	13		
7	*Wavelength, m*	*0.0983*	El beamwidth°	20	Min R reqd	1	*Eff Earth rad E, km*	*8494.45*	*PD at R1*	*0.9957*		
8	Tx P, kW	30	Gain, dB	26	*Achieved*	*1.002*	Precip type: Stratiform =	5	Test R2	16		
9	*Tx P,dBW*	*44.77*	*Efficiency, p.u.*	*0.96*	Target type: 1 =	5	Precip rate, mm/hr	10	*PD at R2*	*0.0857*		
10	Tx loss Lt, dB	0	Loss, dB	1	Ae gain dB	5	*Equiv rain for atten*	*10.000*	*Min R for B15 PD*	*1.03*		
11	Rx loss Lr, dB	0	Height H, m	30	Rx ae hgt h, m	30	*RCS, dBm²/m²*	*–68.82*	*Max R*	*14.37*		
12	Service loss Ls, dB	2	El patt,1 = sin,2 = invco	1	Tx ae ofst, m	0	*Loss lp, dB/km*	*0.002*	*R bracket*	*10.92*		
13	Rx NF N, dB	2	Depression°	0	N/A	–70	Extent, p u	1	*Fill, p u*	*0.82*		
14	RSG thld, dBm²	–10	Rotation, rpm	22	N/A		Air temp, *C*	–2	*Horizon R*	*24.38*		
15	Reqd PD	0.8	CP improvement, dB		*N/A*	*N/A*	RH, %	100	*1st peak R*	*11.67*		
16	*With screening*	*0.800*	Sidelobe below D9, *d*	27	Resp P, dBW	0	*Clr air loss, dB/km*	*0.016*	*1st null R*	*7.84*		
17	Reqd Pfa expt, *F*	–6	Tolerable SL PD	0.1	*Watts*	*1.00*	Wave hgt, hs, m	0.300	*2nd peak R*	*5.84*		
18	Proc loss, Lp dB	8	Feeder ohmic loss, *d*	1	Sw Case 0,1,3	0	*Sea state*	*4*	*2nd null R*	*4.48*		
19	Integration: n, c, cr, p	n	VSWR	2	Operator		Wind fctr C2, dB	2	*N/A*	*N/A*		
20	Scan/scan corlm y/n	y	*Refl coeff*	*0.333*	Pol, h, v, c	h	Screening, p.u.	0	*Max sidelobe R*	*<F8*		
21	Racon ch oset, MHz	–50	*Mismatch loss, dB*	*1.53*	Gain control, dB	0	Surface dielec, eta	81	*Ref echo, 1 km, F*	*–83.43*		
22	*Resp wavelength*	*0.0999*	*EIRP, dBW*	*67.24*	Sw Gain control, *d*	0	conductivity, *S*	0.24	*Sea clutter horizo*	*12.19*		
23	NF, dB	9	*System NF, dB*	*10.20*	SP = s, LP = 1	s	Tilt/pol loss 1-way dB	0	*Interrog thld R*	*14.01*		
24	Rx Bw MHz	8			Mode, Table S1	3					Target sec'y features	
25	TABLE S1, Mode	Max R	*Max R, km*	Plslgth, SP	LP, *µs*	Rx bw, MHz	PRF, pps	*Active mod*	*Selected, µs*	*Pulses inte*	Sw Case 0,	0
26	*1*	3	*5.555555556*	0.05	0.05	25	5000	*3*	*0.05*	*227.27*	N/A	–8
27	*2*	6	*11.11111111*	0.25	0.25	5	5000	*3*	*0.25*	*227.27*	N/A	0.9
28	*3*	12	*22.22222222*	0.5	0.5	3	2000	*3*	*0.5*	*90.91*	Demod law,	*0*
29	*4*	96	*177.7777778*	1	1	1.5	750	*4*	*1*	*34.09*	N/A	0.2
30	*5*		*0*					*5*	*0*	*0.00*	N/A	40
31	*6*		*0*					*6*	*0*	*0.00*	El bw° 1way	25

Figure 14.3 Active point target spreadsheet SS3 page 1. Cells A3 : J31. Layout is similar to the point target version, Figure 14.1, with additional cells. Shown for an experimental offset frequency racon with linear demodulator giving good detection range while working with a ship's 3 GHz radar in a snowstorm

shows 1 for a SART, racon or ramark (which have fixed response power); cell AC29 shows 1 for devices including a demodulator, and cell BQ30 shows 1 for devices giving a response, that is all except the warning receivers. Cell AO29 calculated $D = \log(P_D/(1 - P_D))$ for a racon interrogation (for the usual threshold $P_D = 0.9$, $D = 0.9542$) and cell BU30 gives the threshold SNR as described in Chapter 12, Section 12.9.2.

14.10.2 Radar auxiliary racon channel

The radar needs a special auxiliary receiver channel if it is to receive offset frequency racons. When cell BO29 = 1 the following apply.

1. Cell A21 shows racon channel offset. Enter in cell B21 the racon response frequency offset, MHz above the radar transmitter frequency of cell B6. Approximations within the spreadsheet introduce unacceptable errors if the interrogation and response frequencies differ by of the order of 1 GHz or more; for example, the spreadsheet is unsuited to cross-band racon and similar transponder devices interrogating in say C band (5 GHz) and responding at say 9 GHz. Cell B22 shows the response wavelength, m.
2. Cell A23 shows Noise factor, dB. In cell B23 enter the noise factor of the beacon channel, which is often worse than the main echo channel. System noise factor, cell D23, changes to suit.
3. Cell A24 shows Rx Bw MHz. Enter in cell B24 the bandwidth of the auxiliary channel.

Other entries in cell F9 cause cells A21 : A24 to show N/A; cells B21 : B24 may then be occupied or left empty. It is assumed the residual swept gain threshold, gain and swept gain controls (cells B14, F21, F22, respectively) operate as usual.

14.10.3 Device antenna

Some active devices, particularly RTEs, have two antennas. Enter the gain (assumed the same for both), dB, in cell F10. Using Table 8.1, enter any slant, polarisation or tilt loss, dB positive number one-way, in cell H23 of the environment panel; all radar range equation calculations use the effective gain, F10 − H23. Any feeder losses are assumed to be included within the antenna gain term. In cell F11 enter the height of the receive antenna, metres after making any tidal correction, and in cell F12 enter the height offset of the response antenna above the receive antenna, metres. If there is only one antenna, as for most racons, enter 0. Enter any tilt or polarisation loss in cell H23, positive dB.

14.10.4 Device characteristics

If cell AC29 = 1, cell E13 shows Threshold sensitivity, dBW. In cell F13, enter the threshold signal strength, S_{thsL} dBW. This is the datasheet interrogation strength at the device receiver which causes $0.9P_D$ for a single 1 μs interrogation pulse. If cell AC29 = 1, cell K29 shows Video TC, μs. In cell L29 enter the receiver timeconstant, τ (μs) the interrogation pulselength which causes 3 dB reduction of sensitivity; cell K30 shows dB/decade rolloff. In cell L30 enter the nominal rolloff of the receiver, N, to shorter pulses, usually 20 or 40 dB/decade. Enter 'datasheet' interrogation P_{FA} exponent F in cell L26, typically −8, and threshold P_D in cell L27, typically 0.9. Enter the one-way antenna elevation bandwidth, degrees, in cell L31.

If cell BQ30 = 1, cell E16 shows Tx dBW. Enter the rated response power, dBW, in cell F16. For RTEs use the saturated power. Cell F17 gives the response power in watts. Enter the Swerling Case in cell F18 as 1, 2 or 3. The same Case is assumed to apply to interrogation and response; many active devices are Case 0, the default setting.

If applicable, cell AO29 shows the value of $D = \log(P_D/(1 - P_D))$ at datasheet sensitivity. Cells W29, Y29 show the values of factors s and t used in calculation of Case 0 SNR, Eq. (12.11b) and cell BU29 computes the SNR, q dB, for the F, P_D and demodulator law set in cells L26, L27 and L28, respectively, using IF functions to double the slope if square law demodulation; Eqs (12.11b), (12.16d) and (12.18f), respectively, are used according to chosen case.

The Operator, Environmental effects and Results panels are generally as spreadsheet GDC14S1. The Geometry and Environmental Effects panels of the matrix are as for point passive targets and apply in particular to the interrogation leg. They are followed by a new panel.

14.10.5 Device interrogation panel

This new panel (A82:E293) calculates the detectability of interrogations reaching racons, SARTs, RTEs with auxiliary detection facilities and radar warning receivers, then calculates the response power where applicable. Results of this panel are not used for ramarks, which are not interrogated. The ordinary point target multipath factor is used.

Row 74 states effective antenna gain allowing for tilt and polarisation loss. Row 83 calculates the interrogation strength under free space conditions, $S_{tgt(FS1)}$, from the radar EIRP (cell D22) and Eq. (4.11a) less multipath and atmospheric attenuation terms. Row 84 includes these terms to give the actual interrogation, S_{tgt}, dBW, at the device receiver input. Row 85 calculates the receiver floor for the radar pulselength (obtained using Eq. (8.2c) from mode row 75 and Table S1) and the device long-pulse threshold sensitivity (cell F13). Row 86 calculates the effective SNR dB as (row 84 − row 85 + BU29 [SNR for the datasheet reference point]), the SNR then being doubled if the demodulator is square-law (cell L28). The next row gives the numerical equivalent. Noise modulated on the interrogation by feeder ohmic loss is unlikely materially to affect detection of interrogations and is ignored.

Row 88 gives the non-screened D value for Case 0. The following row gives the equivalent P_D including a screening term (cell H20). Rows 90 and 91 give the respective screened P_Ds for Case 1 per Eq. (12.16b) and Case 3a per Eq. (12.18e). Row 92 gives the P_D of the chosen interrogation Case, default being Case 0.

To assist consideration of the number of radars the device will support, row 93 shows, except for ramarks, the approximate average number of valid interrogations per second accepted by the device, being P_D × scanner azimuth beamwidth/360 × active prf per Table S1 and row 75, assuming the scanner beamshape to be rectangular.

This completes the computations for radar warning devices. Their probability of detection is given by row 92. They will probably give adequate warning at quite low P_D.

14.10.6 Device response panel

Offsetting the response in frequency or antenna height (cells B21 and F12, respectively) changes the indirect ray phase, affecting the response leg multipath factor. Cells N30 : BO30 of Table S3 contain the interrogation-leg transition and diffraction region intermediate parameters, as for the point passive target. Cells N31 : BO31 repeat these for the response leg. Row 94 recalculates path length phase shift $-\Phi$ by addition to the interrogation leg value (row 53) of rates of change of phase with: (a) frequency × frequency offset, per Eq. (5.16f) and (b) target height × height offset, per Eq. (5.16e). Row 96 then recalculates m_{p2} for the response leg, retaining the row 67, 71 and 72 interrogation leg values for ψ, g_{dif} and ρ, because they will not have changed significantly for the relatively small offsets in question. Row 96 calculates m_t similarly to row 77; row 97 restates row 38 for Lookup purposes, row 98 calculates m_{t2} as for row 78, row 99 calculates m_{d2} as for row 79 and row 100 chooses the appropriate row to adopt as response leg multipath factor M_2. Three interrogation-leg region boundaries are retained, row 76.

If applicable, enter the RTE amplifier gain, dB, in cell F14. Cell F15 then gives the unsaturated RCS in dB m^2, using Eq. (8.18a). Row 101 sums row 84 and the gain, to give the RTE prospective response power (results are only used if the device is an RTE); if greater than the saturated power, cell F16, row 102 gives 1, flagging that the RTE is saturated. IF functions also give 1 for racons, SARTs and ramarks. Row 103 gives the actual response power; as row 101 if row 102 = 0, otherwise as cell F16, the device rated power. Row 104 adds the device antenna gain to give the response EIRP component applicable to the scanner polarisation, dBW. Row 105 gives the saturated and unsaturated change ranges, used to give maximum unsaturated range, if applicable, in results cell J19.

14.10.7 Remaining matrix panels

The next part of the matrix (to row 129) is generally as for a point target, with row numbers increased by 24. Differences are as follows.

Row 107, system noise factor, uses the offset channel bandwidth (cell B24) if an offset response, (cell BO29 = 1).

If the response is offset in frequency (cell BO29 = 1), rows 111 (Precipitation clutter return) and 114 (Sea clutter return) are labelled N/A and row 115 is labelled Noise + RSG (receiver swept gain), dBW. For in-band devices, row 115 remains labelled Noise, RSG + clutters and sums the noise and clutter powers in the usual way. Rows 116–118 are not used.

Row 119 computes the free space response at the radar receiver input using the one-way range equation Eq. (4.11) and the response wavelength per cell B22, less atmospheric attenuation and multipath terms. Row 120 introduces these terms, one-way, to give the actual response, dBW (using interrogation path $\Gamma_{\kappa-\delta}$ and atmospheric attenuation L_A values, rows 50 and 81, respectively, as the changes with response offsets are small). Cell D121 contains the standard SART return sweep rate, $Z = 27.04$ MHz/μs, and row 121 computes the single pulse SNR, q dB, with numerical equivalent in the next row. If the target is a SART, Eq. (8.8) is used to give the sweep filter loss term L_s using Z. The forward sweep has higher loss and is ignored.

Rows 123–128 are as GDC14S1 equivalent rows 99 to 104. Rows 130, 133 and 136 give the Case 0, 1 and 3a response P_Ds and new rows 131, 134 and 137 give the overall P_Ds, which are response P_D, multiplied, if the device includes a threshold (cell BS30 = 1), by the interrogation P_Ds of rows 89–91, respectively, and otherwise by the screening, (1 − cell H20). Rows 132, 135 and 138 then give the required single-pulse SNR, q dB. Rows 139 (defaulting to Case 0) gives the chosen case response P_D. Rows 140–145 give overall (interrogate and response) results for the chosen case, as rows 112–117 of the passive point target spreadsheet. The event labels of row 146 are as that spreadsheet.

Calculation of sidelobe detection again requires separate consideration of the interrogate (rows 147–154) and response (rows 155–169) paths.

14.10.8 Results panel

Horizon, and first and second peak and null ranges (cells J14 : J18) are calculated for the interrogation leg. Event ranges for response legs of offset devices will differ somewhat. If the device is an RTE, cell I19 shows Max sat R and J19 gives the maximum range (cell F4 units) at which the RTE transmitter saturates, from row 105. Other devices show N/A in both cells.

If appropriate, cell I23 shows Interrogation threshold range; cell J23 then shows the maximum range at which interrogation P_D exceeds the required overall screened P_D of cell B16, taken as the maximum in row 94, rescaled if necessary. The J23 range gives a rough indication of whether the interrogate path limits overall performance; if the response leg is more sensitive, overall maximum range (cell J11) will approximate J23. If J23 ≫ J11, the response leg is weak.

14.10.9 Charts

Chart 1, Detectability, Interrogation at device, dBW, row 84: Precipitation and Sea clutter returns, rows 111 and 114; Target response, row 120: overall P_D (right-hand scale) and Response signal in hand, rows 140 and 142; row 140 uses the right-hand ordinate. Base is log Range, nmi units, row 82. Rows 111 and 114 only apply if the response is not offset.

Chart 2, Geometry, has the same quantities as for the point passive target, Section 14.8.3, Φ (Row 106) and M are for the interrogation leg. Range base is as Chart 1. Chart colours are as Section 14.8.2.

14.11 References

1 BLAKE, L. V.: 'A guide to basic pulse-radar maximum range calculations, Parts 1 and 2.' Figure 8-3 Pulse-radar range calculation worksheet. NRL (USA) Reports 5868 and 7010, 1969
2 CARPET 2.0.: 'Computer aided radar performance evaluation tool', TNO Physics and Electronics, The Netherlands.
Website: TNO.Observation.Systems@fel.tno.nl

Chapter 15

Worked examples

'... intended to give artistic verisimilitude to an otherwise bald and unconvincing narrative.'

W. S. Gilbert, The Mikado

This chapter presents hypothetical case studies of radar system detectability, using the spreadsheets of Chapter 14 to illustrate some of the factors meriting consideration when designing, specifying or analysing the detection aspects of typical systems. Likely importance of the major factors in play are indicated, but, rather than relying on the chapter's results as a rule of thumb, readers should take the plunge and run the spreadsheets for their specific systems. For brevity, this chapter excludes some of the re-iterations and alternatives necessary to real-life situations. None of the equipment represents any particular make or model.

Ranges are usually stated to two places of decimals. Readers of Chapter 13 will appreciate that the calculations contain so many uncertainties and approximations that range accuracy is really much poorer, with perhaps a kilometre or more error. However, if one of a pair of closely related calculations gives a range which is say 1.00 km more than the second, the true range difference will probably lie within say ± 20 per cent of 1 km.

15.1 Deep-sea ship viewing ships

15.1.1 Nine gigahertz band, small craft target

The radar is a typical shipboard 9 GHz installation, the target being a small craft having RCS 15 dB m^2 (31.62 m^2), tip height 3 m, height factor $n = 0.5$, 10 per cent screened by waves and Swerling Case 3a fluctuation. We are primarily interested in the detectability at 3.5 and 5 nmi range, which for many ships are the safe minimum and optimum first detection ranges. We operate extended target spreadsheet SS2, choosing to work in nmi. The system parameters are shown in Page 1 of the spreadsheet,

	A	B	C	D	E	F	G	H	I	J
1	Extended passive target spreadsheet				Spreadsheet Ref	SS2 v1				
2	Date		User		Task	Chapter 15, Sec 15.1.1.				
3	Transceiver		Scanner and feeder		Range bracket		Environment		Results	
4	Type		Type		Scaling, km/nmi	nmi				
5					Max reqd R	10			*Scaling*	*nmi*
6	Frequency, MHz	9410	Az beamwidth°	0.9	*Achieved*	*25.531*	Refraction, k	1.333	Test R1	3.5
7	*Wavelength, m*	*0.0318*	El beamwidth°	20	Min reqd R	0.1	*Eff Earth rad E, km*	*8492.54*	*PD at R1*	*0.7412*
8	Tx P, kW	20	Gain, dB	31	*Achieved*	*0.100*	Precip type: Stratiform =	1	Test R2	5
9	*Tx P, dBW*	*43.01*	*Efficiency, p.u.*	*0.91*	Target type		Precip rate, mm/h	4	*PD at R2*	*0.0205*
10	Tx loss Lt, dB	0	Loss, dB	1	Kind: ship, coast	ship	*Equiv rain for atten*	*4.000*	*Min R for B15 PD*	*<F8*
11	Rx loss Lr, dB	0	Height H, m	35	Tip height j, m	3	*RCS, dB m²/m³*	*–62.94*	*Max R*	*3.87*
12	Service loss Ls, dB	2	El patt, 1 = sin, 2 = invco	1	Height factor, n	0.5	*Loss lp, dB/km*	*0.049*	*R bracket*	*3.77*
13	Rx NF N, dB	3	Depression°	0	Tot RCS, dB m²	15	Extent, p.u.	1	*Fill, p.u.*	*1.00*
14	RSG thld, dBm²	–10	Rotation, rpm	22	*RCS, num*	*31.62*	Air temp, C	20	*Horizon R*	*17.02*
15	Reqd PD	0.5	CP improvement, dB		Extent, radial, m	2	RH, %	70	*Az ospill R*	*0.07*
16	*With screening*	*0.5*	Sidelobe below D8, d	29	Extent, axial, m	10	*Clr air loss, dB/km*	*0.011*	*R crit*	*7.12*
17	Reqd Pfa expt, F	–6	Tolerable SL PD	0.1			Wave hgt, hs, m	0.025		
18	Proc loss, Lp dB	8	Feeder ohmic loss, d	2	Swerling Case 0,	3	*Sea state*	*2*		
19	Integ, n, c, cr, p = non-	n	VSWR	1.5	Operator		Wind fctr C2, dB	0		
20	Scan/scan corrln y/n	y	*Refl coeff*	*0.200*	Pol, h, v, c	h	Screening S, p.u.	0	*Max sidelobe R*	*<F8*
21			*Mismatch loss, dB*	*0.53*	Gain control, dB	0	Surface dielec, eta	81	*Ref echo, 1 km, FS*	*–84.97*
22			*EIRP, dBW*	*70.48*	Sw Gain control, d	0	conductivity, S	0.24	*Sea clutter horizon*	*13.17*
23			*System NF, dB*	*6.85*	SP = s, LP = 1	s	Tilt/pol loss 1-way dB	1		
24					Mode, Table S1	1				
25	TABLE S1, Mode	Max R	*Max R, km*	Plslgth, SP	LP, µs	Rx bw, MHz	PRF, pps	*Active mod*	*Selected, µs*	*Pulses inte*
26	*1*	1.5	*2.777777778*	0.05	0.1	25	5000	*1*	*0.05*	*68.18*
27	*2*	3	*5.555555556*	0.25	0.5	5	5000	*2*	*0.25*	*68.18*
28	*3*	6	*11.11111111*	0.5	1	3	2000	*3*	*0.5*	*27.27*
29	*4*	12	*22.22222222*	1	1	1.5	2000	*4*	*1*	*27.27*
30	*5*	48	*88.88888889*	1	1	1.5	1000	*5*	*1*	*13.64*
31	*6*		*0*					*6*	*0*	*0.00*

Figure 15.1 Spreadsheet SS2 page 1 set for deep-sea ship with small target vessel in clutter, 9 GHz band. Extended target spreadsheet. Cells A3:J31 shown

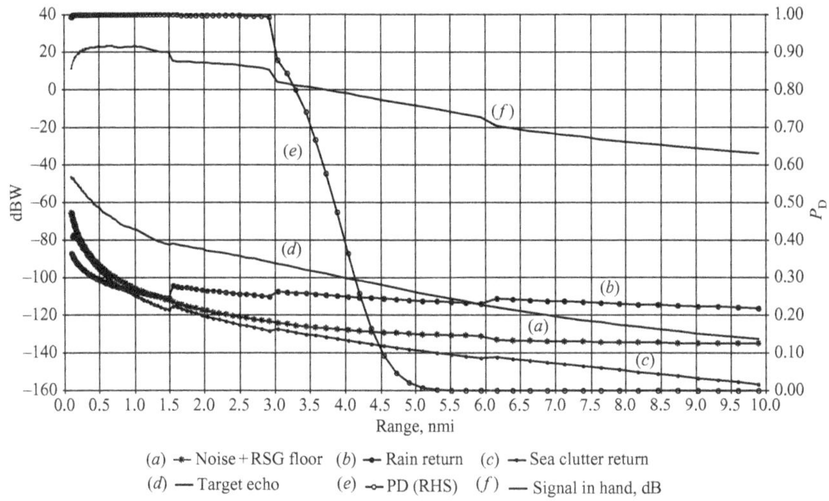

Figure 15.2 Echo and clutter returns. Conditions of Figure 15.1, moderate clutter ($r = 4$ mm/h, SS2). Craft detectable to 3.87 nmi for $P_D = 0.5$, limited by rain clutter, which well exceeds sea clutter or noise at that range. No wave screening

Figure 15.1, set for modest clutter comprising stratiform rain 4 mm/h whole path, sea state 2. Initially we assume the radar is set to a short range scale, entering cell F24 as mode 1 and using its short-pulse transmission. Maximum range (cell J9) is 3.70 nmi. The working range bracket has to extend to 25.531 nmi to embrace R_B, enabling calculation of transition region multipath beyond $R_A = 5.72$ nmi.

Figure 15.2 shows Chart 1 of the spreadsheet. The target echo well exceeds the noise and clutter components at short range, giving P_D of 0.9, limited by the wave screening. At longer range, the echo falls more rapidly than the rain clutter return and P_D starts to fall, reaching the minimum acceptable value of 0.5 (cell B15) at 3.87 nmi, where the signal in hand curve crosses 0 dB. Setting maximum range to say 5 nmi, keeping the whole spreadsheet in the interference region would give closer spacing, with cell J9 range closer to 3.87 nmi. At short range noise plus the receiver swept gain are the dominant competitors to the echo, but at longer range the rain clutter predominates. The steps at 1.5, 3 and 6 nmi arise as the radar receiver bandwidth changes when the operator changes range scale. At ranges near 3.87 nmi, the sea clutter return is fully 20 dB below the rain clutter, suggesting that maximum detectable range would hold up well against a rougher sea state.

Figure 15.3 shows conditions at sea state 4. Maximum detectable range falls slightly, to the scaling step at 3.0 nmi, but P_D becomes poor at short range. Cell J13 shows that the fill has fallen from 1 to 0.71 p.u., meaning that the target remains detectable at only 71 per cent of ranges between 0.1 and 3 nmi. This would make it difficult for ARPA to maintain track.

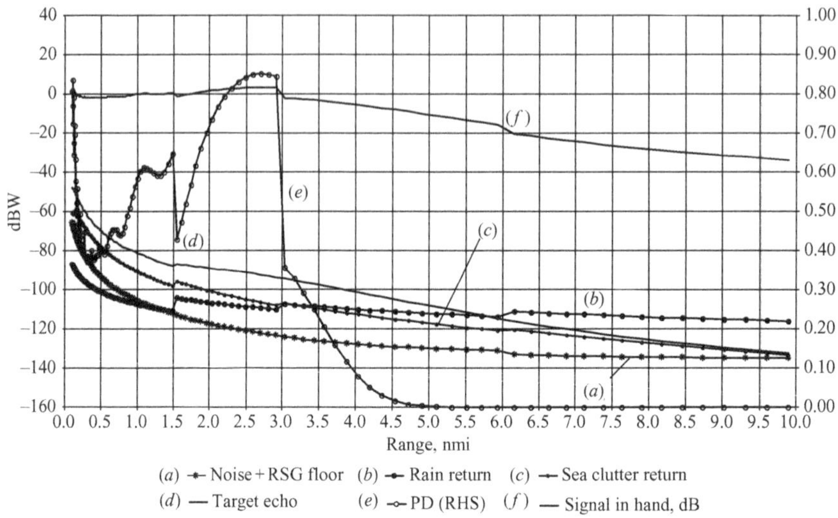

Figure 15.3 *Rough sea. Sea state 4, giving increased clutter, otherwise as Figure 15.2 (r = 4 mm/h). Although maximum detectable range is not much reduced (to 3.0 nmi), detectability becomes poor at short range, confirming that sea clutter is chiefly a short-range problem*

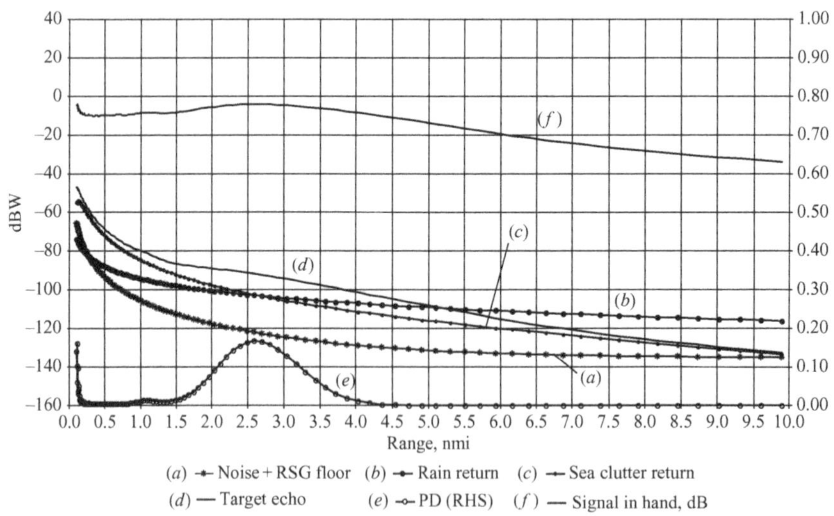

Figure 15.4 *Radar controls adverse. Long pulse setting, with 12 nmi range scale retained at all target ranges, otherwise as Figure 15.3. Target undetectable at any range*

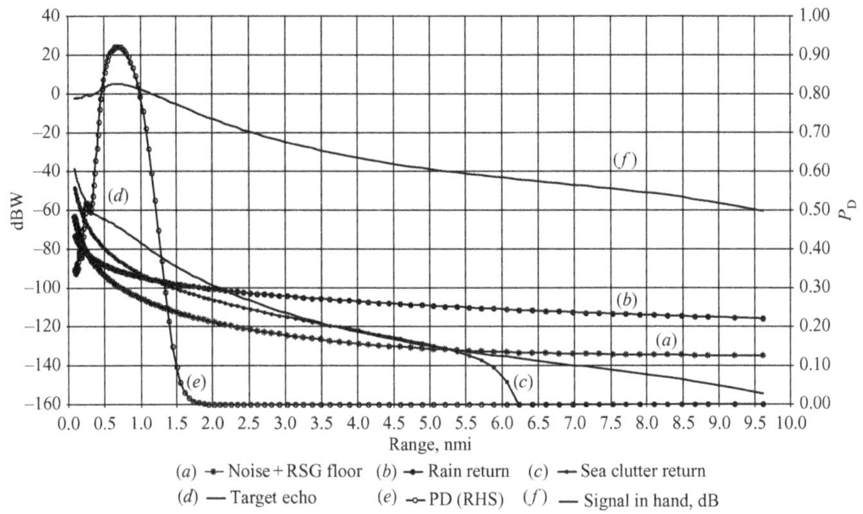

Figure 15.5 *Low scanner. Height reduced from 35 to 8 m, otherwise as Figure 15.4. Reduction of sea clutter return permits detection between 0.3 and 1.2 nmi*

Figure 15.4 shows how badly detectability may suffer when the radar controls are not optimised for the target in question. Here (using Mode 5) the 12 nmi range scale is retained at all target ranges – perhaps the operator is concerned about the movement of other targets at several miles range – and the long pulse setting is engaged, perhaps to maximise detectability of a distant target in noise, beyond the rainy area. The longer pulse covers a bigger clutter footprint so nowhere does the SNR reach that necessary for P_D of 0.5 and the target is never detected. The highest P_D is 0.17 at 2.6 nmi range. It would be quite unrealistic to expect the most vigilant operator or the best data extraction system to spot that the target existed.

Figure 15.5 shows the beneficial effect of a low scanner under these conditions. Reducing H from 35 to 8 m renders the target detectable between 0.20 and 1.24 nmi, although target horizon range falls from 17.02 to 10.15 nmi. The sea clutter horizon falls from 13.17 to 6.29 nmi and short-range sea clutter falls by some 10 dB.

Figures 15.6 and 15.7 show the effect of precipitation clutter on detectability for high and low scanners respectively. For the high scanner and $P_D = 0.5$, maximum detectable range in the clear is 7.2 nmi. Range falls as rain rate increases. With excessive rain of 40 mm/h, P_D is 0.27 at 1.5 nmi when the radar uses the 3 nmi range scale, but rises to 0.88 when switching to the 1.5 nmi scale; because the pulselength is reduced, the corresponding detection cell area reduction collecting 3 dB less clutter. Using a low scanner halves low-clutter detection ranges, but has little effect in high clutter where ranges are so low that the horizon change is almost immaterial. Figure 15.8 depicts part of the spreadsheet matrix for the different values of r. The curves of Figure 15.7 represent rows E130:EX130 through to E135:EX135.

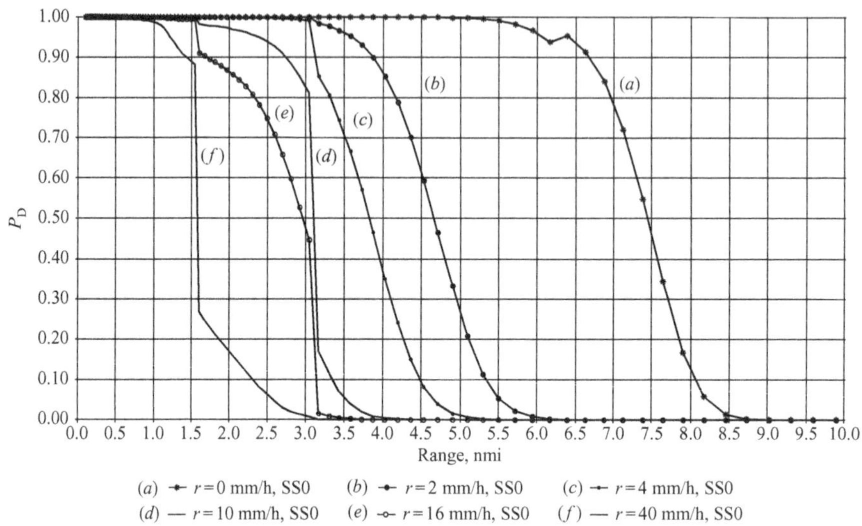

Figure 15.6 *Rain clutter. System of Figure 15.1, sea state 0, SP, Mode 1, scanner height 35 m. Rain for full pathlength. Detection range 7.45 nmi in clear, reduced to 1.5 nmi in 40 mm/h excessive rain*

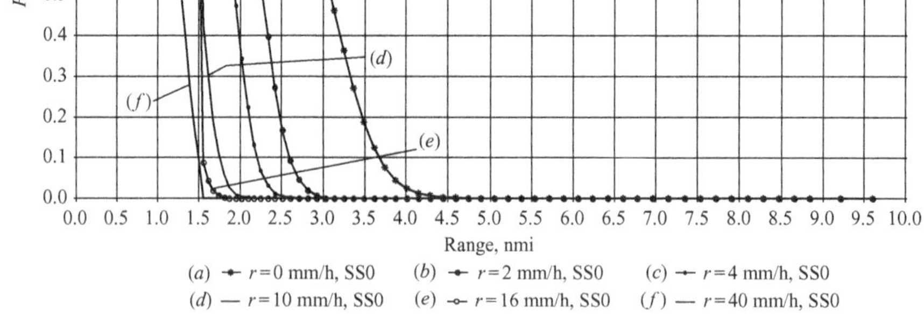

Figure 15.7 *Rain clutter, low scanner. Scanner height reduced to 8 m, otherwise as Figure 15.6. Maximum detectable ranges fall, particularly in the lighter rain rates, because the closer horizon reduces target RCS*

	A	B	C	D	E	F	G
129	Various rain clutters, SS0, H=35 m, SP, Mode 1, no screening Ch3						
130	Row 112		r=	0	0.99	0.99	1.00
131			r=	2	0.99	0.99	1.00
132			r=	4	0.994064614	0.9940859	0.996875135
133			r=	10	0.993624511	0.9936212	0.996611305
134			r=	16	0.992987581	0.9929459	0.99622547
135			r=	40	0.988468104	0.9881021	0.993403515

Etc.

Figure 15.8 Spreadsheet detail for Figure 15.7, using Copy, Paste Special, Values Only facility

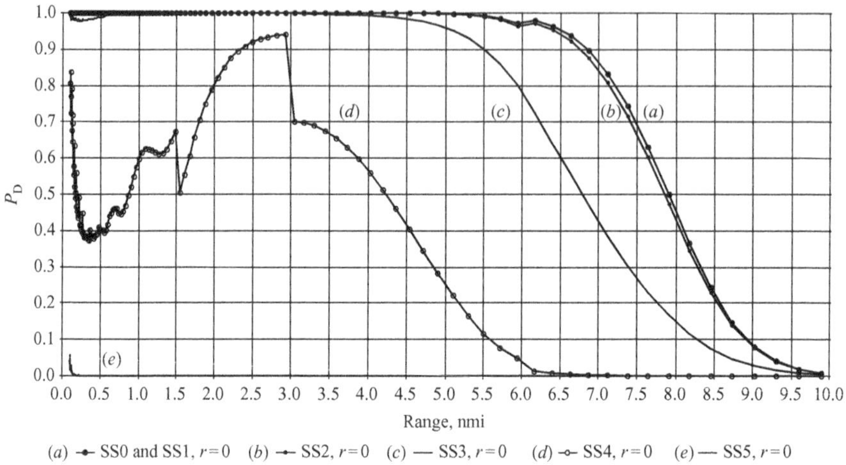

Figure 15.9 *Sea clutter. Scanner height 35 m, no precipitation, otherwise as Figure 15.1. No screening*

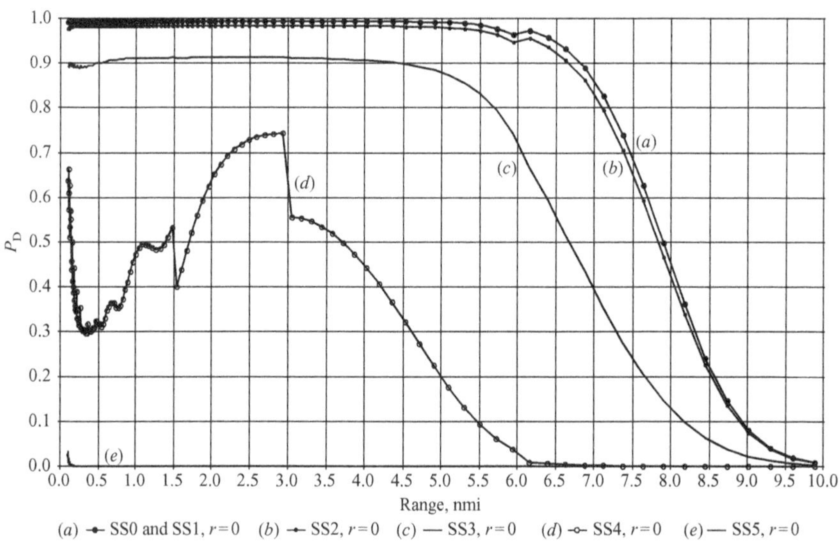

Figure 15.10 *Wave screening. As Figure 15.9 but wave screening as Chapter 12, Section 12.10.3, Eq. (12.29c) reduces maximum available P_D in high sea states*

Figures 15.9–15.11 show detectability in sea clutter but no precipitation. At low height, detectability falls in light clutter because target horizon falls. This effect is overcome in heavy clutter by the shallower angle at which the surface is illuminated, returning less clutter and improving P_D.

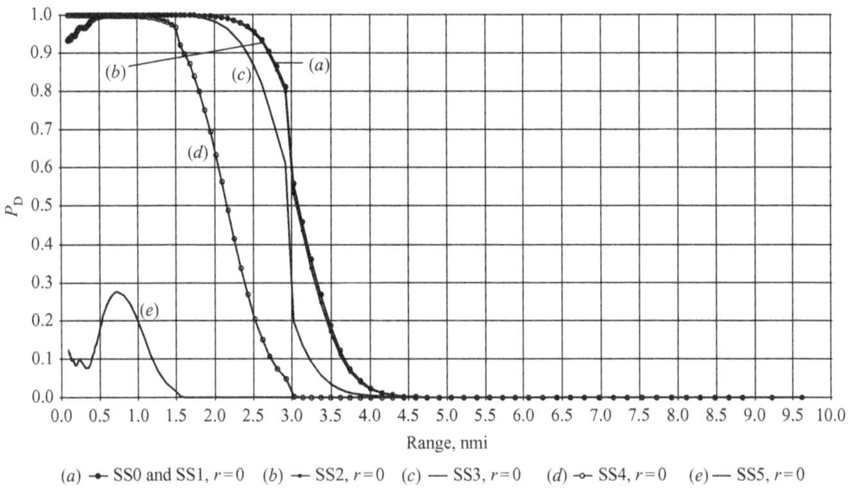

Figure 15.11 *Sea clutter, low scanner. Scanner height 8 m, otherwise as Figure 15.9. No screening*

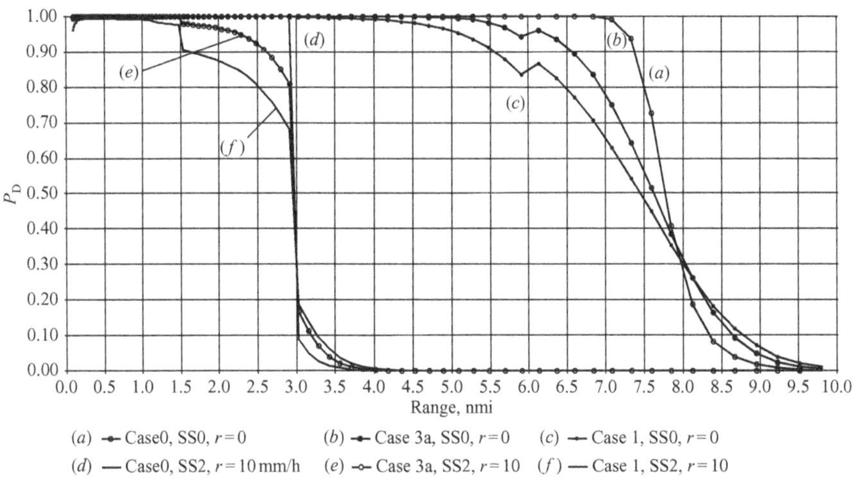

Figure 15.12 *Swerling case change. Conditions of Figure 15.1. Longer-range curves assume no clutter; lower-range curves assume rain 10 mm/h whole path, sea state 2*

So far, we have used Swerling Case 3a target fluctuation, as we were modelling a small craft. Figure 15.12 shows the effect of fluctuation case on detectability for clear and moderate clutter conditions. As expected from Chapter 12, case has most effect when high P_D is required and can change detection range by ~30 per cent at $P_D = 0.9$. The most common Case 1 gives the poorest performance.

15.1.2 Three gigahertz band, small craft target

Figure 15.13 shows spreadsheet SS2 Page 1 reconfigured for a typical deep-sea ship's 3 GHz radar and scanner. Transmitter power and receiver noise figure are assumed somewhat better but scanner gain is worse than the 9 GHz case. Scanner height remains 35 m but feeder loss has been reduced to a typical 3 GHz value. The target RCS has been reduced by 2 dB to 13 dB m^2, a reduction with frequency typical of many real-life targets. The environment remains as before. Figures 15.14–15.18 depict performance. Precipitation clutter's dependence on the fourth power of frequency makes it much less significant. Sea clutter falls because it is approximately proportional to f^2; however, both clutters are partially restored by the wider scanner azimuth beamwidth, which is raised by the longer wavelength from 0.9° to 2.0°.

15.2 VTS installation

15.2.1 Scenario

A Harbour Authority needs a VTS radar of moderate performance. The area is tropical semi-desert, with little rain but high humidity and frequent heavy swell. They prefer the 9 GHz band for its better angular resolution than 3 GHz, without the high cost of J band. Their two main detection requirements are (a) small tugs of RCS 15 dB m^2 to 10 km range in swell of SS3, with a rather nominal 1 mm/h whole-path rain and (b) deep-sea ships approaching the Port at 10 km in occasional thundery squalls of 16 mm/h heavy rain and fully developed SS4 incoming sea; each at $P_D > 0.6$ for good trackforming performance. The Port is occasionally visited by 50 000 gt cruise ships, but the 150 000 tonners currently coming into service are too big to enter. The scanner will have to be tower-mounted, for the land lies low. The minimum tower height to clear local buildings is 8 m. The Port hopes a relatively inexpensive standard deep-sea ship's radar might suffice. The tug target is examined first, assuming these squat vessels have tip height $j = 5$ m with height factor $n = 0.5$ and Swerling Case 1 fluctuation characteristics. In the first instance no allowance is made for possible screening by waves and it is assumed the transceiver is mounted at the scanner, obviating feeder loss. The chosen scaling is kilometres.

15.2.2 P_D variation with range; effect of scanner height

Figure 15.19 shows the settings of extended target spreadsheet SS2 Page 1. Wind factor C2 (cell H19) is set to −3 dB to suit swell rather than sea. The Results panel shows the tug is detectable only to 5.60 km at the $P_D = 0.6$ level, confirmed by Figure 15.20, which indicates that precipitation clutter dominates at long range where the echo falls away towards $-\infty$ at the horizon (20.87 km) and the scaling step at 3 nmi causes a jump in P_D from 0.24 to 0.74. Detectability falls slightly at short range because of

	A	B	C	D	E	F	G	H	I	J
1	Extended passive target spreadsheet				Spreadsheet Ref	SS2 v1				
2	Date		User		Task	Chapter 15, Sec 15.1.2.				
3	Transceiver		Scanner and feeder		Range bracket		Environment		Results	
4	Type		Type		Scaling, km/nmi	nmi				
5					Max reqd R	10			*Scaling*	*nmi*
6	Frequency, MHz	3050	Az beamwidth°	2	*Achieved*	*25.531*	Refraction, k	1.333	Test R1	3.5
7	*Wavelength, m*	*0.0983*	El beamwidth°	20	Min reqd R	0.1	*Eff Earth rad E, km*	*8492.54*	*PD at R1*	*0.6377*
8	Tx P, kW	25	Gain, dB	27.5	*Achieved*	*0.100*	Precip type: Stratiform =	1	Test R2	5
9	*Tx P,dBW*	*43.98*	*Efficiency, p.u.*	*0.90*	Target type		Precip rate, mm/h	4	*PD at R2*	*0.0097*
10	Tx loss Lt, dB	0	Loss, dB	1	Kind: ship, coast	ship	*Equiv rain for atten*	*4.000*	*Min R for B15 PD*	*<F8*
11	Rx loss Lr, dB	0	Height H, m	35	Tip height j, m	3	*RCS, $dB/m^2/m^3$*	*–82.51*	*Max R*	*3.72*
12	Service loss Ls, dB	2	El patt, 1 = sin, 2 = invc	1	Height factor, n	0.5	*Loss lp, dB/km*	*0.001*	*R bracket*	*3.62*
13	Rx NF N, dB	2	Depression°	0	Tot RCS, dB m^2	13	Extent, p.u.	1	*Fill, p.u.*	*1.00*
14	RSG thld, dBm^2	–10	Rotation, rpm	20	*RCS, num*	*19.95*	Air temp, C	20	*Horizon R*	*17.02*
15	Reqd PD	0.5	CP improvement, dB		Extent, radial, m	2	RH, %	70	*Az ospill R*	*0.03*
16	*With screening*	*0.500*	Sidelobe below D8, d	29	Extent, axial, m	10	*Clr air loss, dB/km*	*0.007*	*R crit*	*9.31*
17	Reqd Pfa expt, F	–6	Tolerable SL PD	0.1			Wave hgt, hs, m	0.600		
18	Proc loss, Lp dB	8	Feeder ohmic loss, d	0	Swerling Case 0, 1	3	*Sea state*	*5*		
19	Integ, n, c, cr, p = non	n	VSWR	1.5	Operator		Wind fctr C2, dB	0		
20	Scan/scan corrln y/n	y	*Refl coeff*	*0.200*	Pol, h, v, c	h	Screening S, p.u.	0	*Max sidelobe R*	*<F8*
21			*Mismatch loss, dB*	*0.53*	Gain control, dB	0	Surface dielec, eta	81	*Ref echo, 1 km, FS*	*–77.21*
22			*EIRP, dBW*	*69.95*	Sw Gain control,	0	conductivity, S	0.24	*Sea clutter horizon*	*13.17*
23			*System NF, dB*	*5.85*	SP = s, LP = 1	s	Tilt/pol loss 1-way dB	1		
24					Mode, Table S1	1				
25	TABLE S1, Mode	Max R	*Max R, km*	Plslgth, SP	LP, μs	Rx bw, MHz	PRF, pps	*Active mod*	*Selected, μs*	*Pulses inte*
26	*1*	1.5	*2.777777778*	0.05	0.1	25	5000	*1*	*0.05*	*166.67*
27	*2*	3	*5.555555556*	0.25	0.5	5	5000	*2*	*0.25*	*166.67*
28	*3*	6	*11.11111111*	0.5	1	3	2000	*3*	*0.5*	*66.67*
29	*4*	12	*22.22222222*	1	1	1.5	2000	*4*	*1*	*66.67*
30	*5*	48	*88.88888889*	1	1	1.5	1000	*5*	*1*	*33.33*
31	*6*		*0*					*6*	*0*	*0.00*

Figure 15.13 Spreadsheet SS2 for 3 GHz corresponding to Figure 15.1. The same target, RCS assumed reduced by 2 dB by the longer wavelength. Cells A3:J31

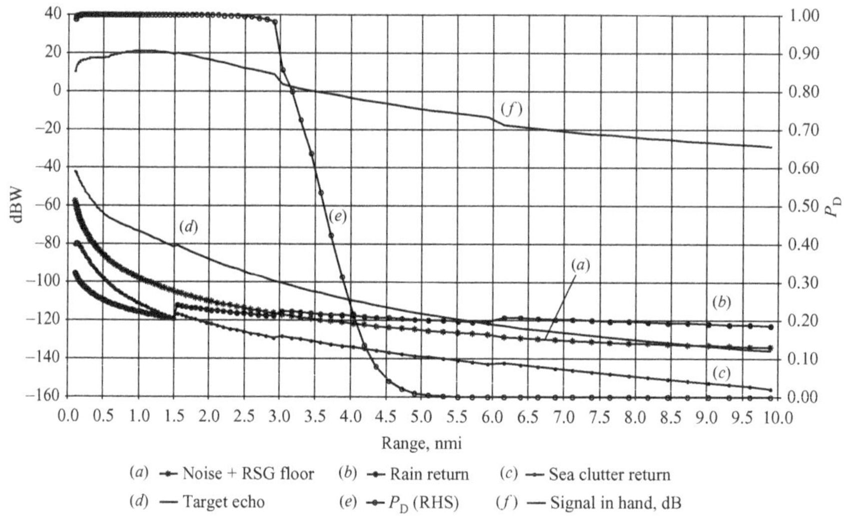

Figure 15.14 *Echo and clutter returns. System of Figure 15.13. Compared with Figure 15.3, clutter returns are reduced by the longer wavelength, offset by the wider scanner azimuth beamwidth and a small drop in target RCS. The effects combine to reduce detection range by 4 per cent, from 3.87 to 3.72 nmi*

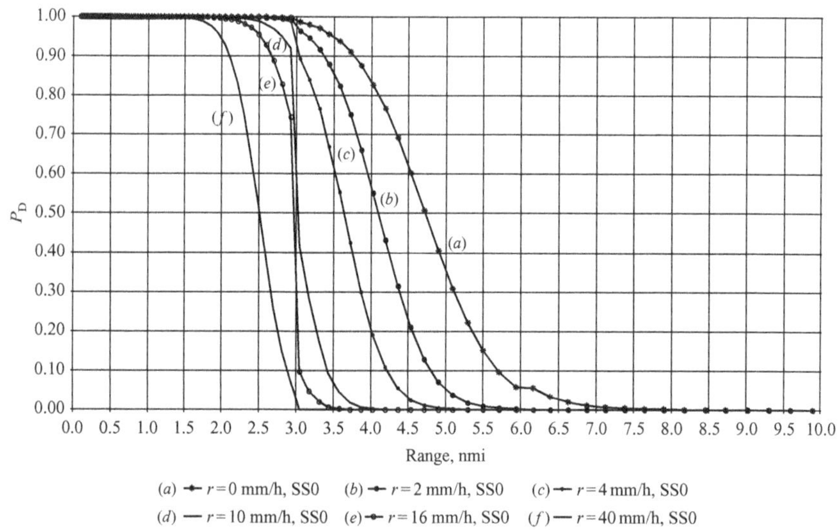

Figure 15.15 *Rain clutter, 3 GHz band. System of Figure 15.13, sea state 0, SP, Mode 1, scanner height 35 m. Rain occupies full pathlength. Compared to the 9 GHz band system of Figure 15.6, maximum detection range is shorter in the clear but significantly better in heavy precipitation*

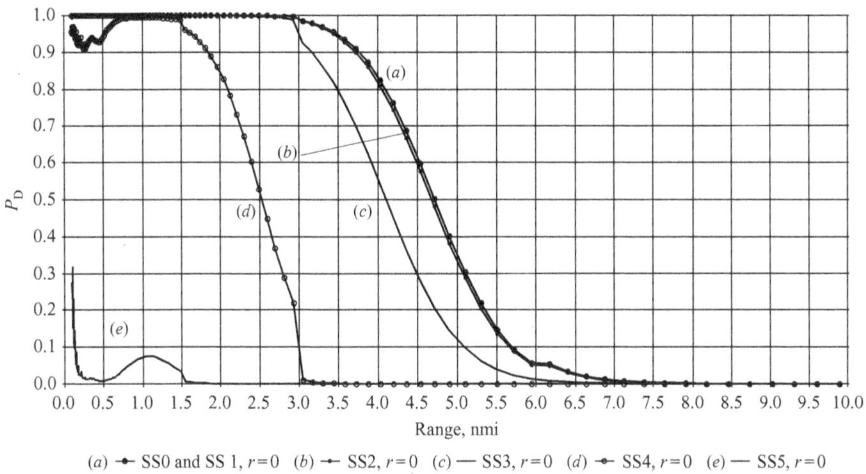

Figure 15.16 Sea clutter, 3 GHz band. Performance is somewhat inferior to the similar 9 GHz system of Figure 15.9. The craft is undetectable in SS5

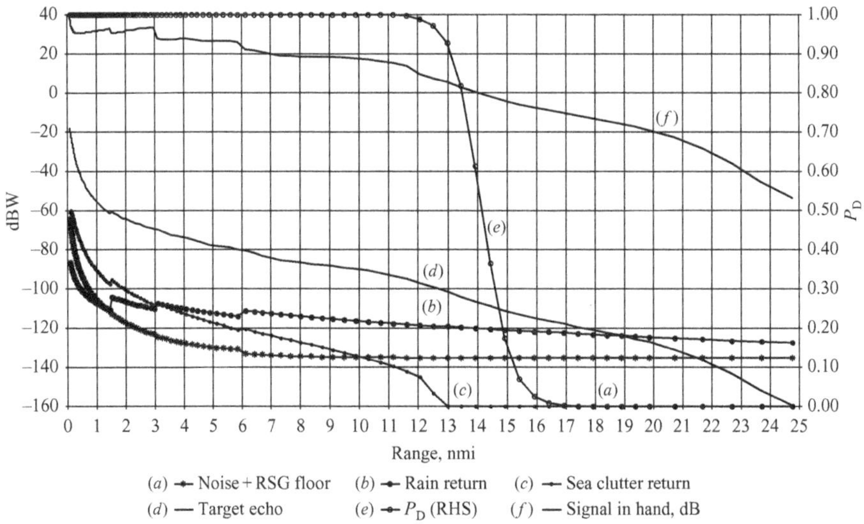

Figure 15.17 Ship target, 9 GHz band. Moderate clutter, r = 4 mm/h, SS4. Compare with Figure 15.3. Maximum detectable range 13.9 nmi. Sea clutter horizon 13.2 nmi, target horizon range 21.8 nmi

sea clutter. Sea clutter falls toward its horizon ($\beta = 0$), which is at 11.66 km. Would a taller mast to increase horizon range give better detection range? Quick successive insertions of increasing H values in cell D12 show maximum detection range does in fact increase. For more detailed analysis we proceed as follows.

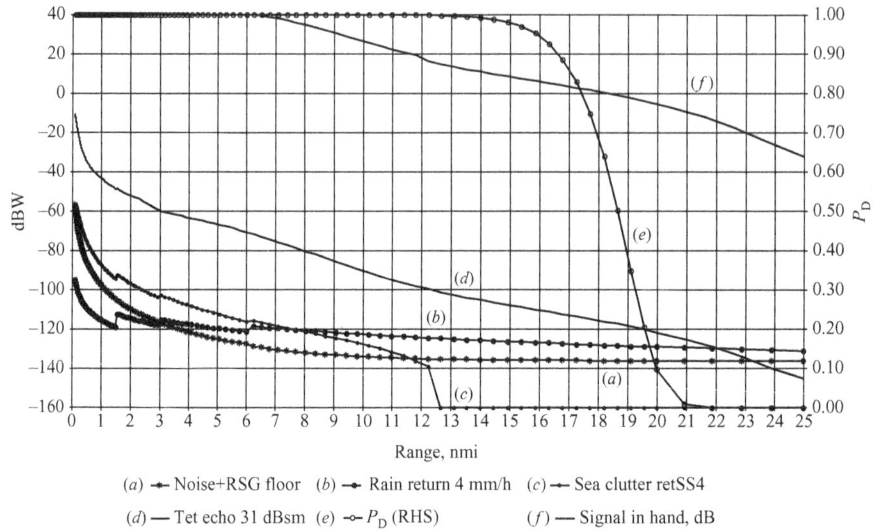

Figure 15.18 Ship target, 3 GHz band. Compare with the small target of Figure 15.14. Clutter conditions as Figure 15.17, but maximum detectable range increased to 18.6 nmi

Successive *H* values are entered in cell D12, giving a set of maximum and horizon range values in cells J11 and J14, respectively, which are copied by the Copy, Paste Special function to rows 4, 11 and 14, respectively, Figure 15.21. Figure 15.22 is constructed from these values; it would of course be easy to plot additional curves such as P_D at 10 km. The detection range curve is useful in cost/benefit analysis. As mast cost rises very sharply with height, it is important not to over-specify. In practice, the Harbour ought to run the spreadsheet a number of times to explore fully the interaction between height, target and environmental factors, but on the basis of Figure 15.22, $H = 28$ m is the minimum giving the required 10 km range.

15.2.3 Scanner aperture

Would it be preferable to use a better radar to enable a lower mast? The Harbour now explores the performance improvement available from a wider scanner, azimuth beamwidth reduced from 0.9° to 0.5° and gain up from 31 to 33.5 dB. The new scanner promises 9.72 km range at 24 m height and the Harbour decides to settle for this, the lower mast offsetting the additional mast stiffness and scanner costs, with the additional benefit of better angular resolution. Compromises of this nature are commonplace, much of the art and skill of all system engineering being to find the best cost/benefit ratio for the task in hand. The systems engineer needs to retain some nominal performance in reserve against the inevitable uncertainties of life, but Management needs to detect how far the engineer is listing to starboard with the weight of expensive dBs ‘in the hip pocket’.

	A	B	C	D	E	F	G	H	I	J
1	Extended passive target spreadsheet				Spreadsheet Ref	SS2 v1				
2	Date		User		Task	Chapter 15, Sec 15.2.1.				
3	Transceiver		Scanner and feeder		Range bracket		Environment		Results	
4	Type		Type		Scaling, km/nmi	km				
5	STANDARD MARINE RADAR		9FT APERTURE		Max reqd R	60			*Scaling*	*km*
6	Frequency, MHz	9410	Az beamwidth°	0.9	*Achieved*	*60.000*	Refraction, k	1.333	Test R1	5
7	*Wavelength, m*	*0.0318*	El beamwidth°	20	Min reqd R	0.5	*Eff Earth rad E, km*	*8492.54*	*PD at R1*	*0.8667*
8	Tx P, kW	20	Gain, dB	31	*Achieved*	*0.405*	Precip type: Stratiform =	3	Test R2	10
9	*Tx P,dBW*	*43.01*	*Efficiency, p.u.*	*0.91*	Target type	TUG	Precip rate, mm/h	2	*PD at R2*	*0.0000*
10	Tx loss Lt, dB	0	Loss, dB	1	Kind: ship, coast	ship	*Equiv rain for atten*	*2.000*	*Min R for B15 PD*	*<F8*
11	Rx loss Lr, dB	0	Height *H*, m	8	Tip height j, m	5	*RCS, dB m²/m³*	*–64.59*	*Max R*	*5.60*
12	Service loss Ls, dB	2	El patt, 1 = sin,2 = invco	1	Height factor, n	0.5	*Loss lp, dB/km*	*0.021*	*R bracket*	*5.19*
13	Rx NF N, dB	3	Depression°	0	Tot RCS, dB m²	17	Extent, p.u.	1	*Fill, p.u.*	*1.00*
14	RSG thld, dBm²	–10	Rotation, rpm	22	*RCS, num*	*50.12*	Air temp, °C	30	*Horizon R*	*20.87*
15	Reqd PD	0.6	CP improvement, dB		Extent, radial, m	5	RH, %	90	*Az ospill R*	*0.32*
16	*With screening*	*0.600*	Sidelobe below D8, d	29	Extent, axial, m	15	*Clr air loss, dB/km*	*0.016*	*R crit*	*5.35*
17	Reqd Pfa expt, F	–6	Tolerable SL PD	0.1			Wave hgt, hs, m	0.300		
18	Proc loss, Lp dB	8	Feeder ohmic loss, d	0	Swerling Case 0, 1, 3	1	*Sea state*	*4*		
19	Integ, n, c, cr, p = non	n	VSWR	1.5	Operator		Wind fctr C2, dB	–3		
20	Scan/scan corrln y/n	y	*Refl coeff*	*0.200*	Pol, h, v, c	h	Screening S, p.u.	0	*Max sidelobe R*	*<F8*
21			*Mismatch loss, dB*	*0.53*	Gain control, dB	0	Surface dielec, eta	81	*Ref echo, 1 km, FS*	*–80.97*
22			*EIRP, dBW*	*72.48*	Sw Gain control, d	0	conductivity, S	0.24	*Sea clutter horizon*	*11.66*
23			*System NF, dB*	*6.29*	SP = s, LP = 1	s	Tilt/pol loss 1-way dB	0		
24					Mode, Table S1	1				
25	TABLE S1, Mode	Max R	*Max R, km*	Plslgth, SP	LP, µs	Rx bw, MHz	PRF, pps	*Active mod*	*Selected, µs*	*Pulses inte*
26	*1*	2.78	*2.78*	0.05	0.1	25	5000	*1*	*0.05*	*68.18*
27	*2*	5.56	*5.56*	0.25	0.5	5	5000	*2*	*0.25*	*68.18*
28	*3*	11.11	*11.11*	0.5	1	3	2000	*3*	*0.5*	*27.27*
29	*4*	22.22	*22.22*	1	1	1.5	2000	*4*	*1*	*27.27*
30	*5*	88.9	*88.9*	1	1	1.5	1000	*5*	*1*	*13.64*
31	*6*		*0*					*6*	*0*	*0.00*

Figure 15.19 Spreadsheet SS2, tug on VTS radar. Cells A3:J31

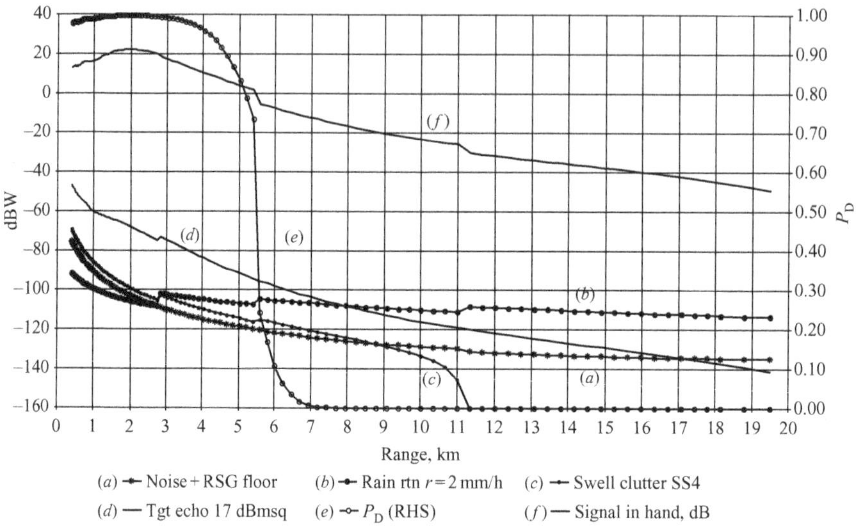

Figure 15.20 *Tug detectability, conditions of Figure 15.19, $H = 8$ m. Steps at 1.5, 3 and 6 nmi caused by Mode 1 operation*

15.2.4 Feeder

So far we have assumed an up-mast transceiver (Feeder ohmic loss, cell D18 = 0 dB). For easy access, service engineers prefer all the electronics to be at ground level, despite the performance loss in the lengthy feeder to the high scanner. (Shipboard radars often also have long feeders for similar reasons and the following analysis can readily be applied to the example of Section 15.1.) Entering loss 0.11 dB/m × 24 m = 2.64 dB in cell D18 indicates that detection range is unimpaired. This is because performance is clutter-limited rather than noise-limited and the feeder attenuates clutter as much as echo, leaving the signal to clutter ratio unaltered. With no clutter, noise-limited detection range is reduced from 17.35 to 15.07 km by the feeder. This price may be thought worth paying for the improvement in serviceability.

The Port settles for this configuration. Figure 15.23 shows the tug's detectability.

15.2.5 Atmospheric refraction

So far, we have used a standard atmosphere. How much will variation in refraction affect detection range? It is straightforward to repeat the Copy, Paste Special, procedure in the User Panel to give a curve relation detection range with k-factor. It is slightly less straightforward to plot several P_D curves to a Range abscissa, because change of k varies the spreadsheet range base. The procedure is to duplicate the Range and P_D rows of the spreadsheet for the required k, as rows 130, 131 of Figure 15.24. For each k, the range and P_D are Copy, Paste Special Values Only into rows 133–134, 135–136, etc. The P_Ds are then brought to row 133 range base in rows 145–149 using

	I	J	K	L	M	N	O	P	
3	Results		User panel						
4	Scr height H	40	8	10	12	14	16	18	Etc.
5	*Scaling*	*km*	km	km	km	km	km	km	
6	Test R1	5	5	5	5	5	5	5	
7	*PD at R1*	*0.9361*	0.8666845	0.9381502	0.9717483	0.9807306	0.9849058	0.9867374	
8	Test R2	10	10	10	10	10	10	10	
9	*PD at R2*	*0.8121*	7.923E–06	3.853E–05	0.0002026	0.0027808	0.0135166	0.0505374	
10	*Min R for B15 PD*	*<F8*	<F8	<F8	<F8	<F8	<F8	<F8	
11	*Max R*	*11.37*	5.5968298	5.6959622	6.1939646	6.7107149	7.2514004	7.8199745	
12	*R bracket*	*10.89*	5.1914751	5.2780121	5.7664025	6.2755815	6.8101537	7.3736928	
13	*Fill, p.u.*	*1.00*	1	1	1	1	1	1	
14	*Horizon R*	*35.28*	20.872287	22.248184	23.49209	24.635979	25.700686	26.70068	
15	*Az ospill R*	*0.32*	0.3182837	0.3182837	0.3182837	0.3182837	0.3182837	0.3182837	
16	*R crit*	*19.81*	5.3501676	6.5246141	7.6185764	8.8474578	9.8859031	10.65057	

Figure 15.21 *Construction of Figure 15.22, using Copy, Paste Special technique in User Panel*

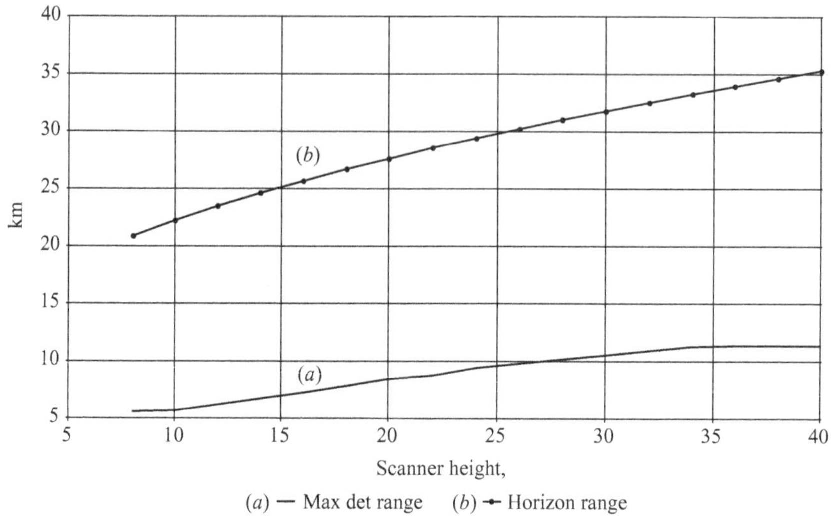

Figure 15.22 *Scanner height variation. System of Figure 15.19. No feeder loss. Shows how horizon and detection ranges increase with scanner height*

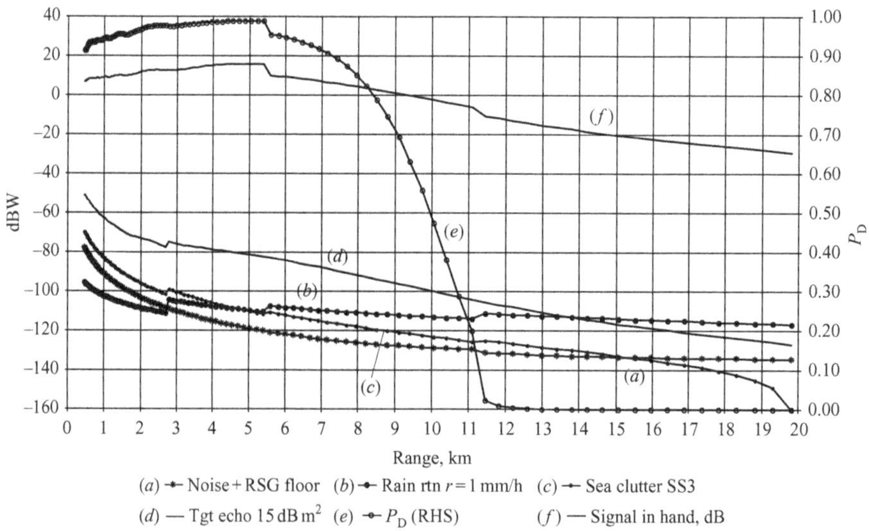

Figure 15.23 *Detectability, chosen system. 24 m mast with transceiver at foot, 33.5 dB scanner, 0.5° azimuth beamwidth. Tug target. Compare with Figure 15.20. Sea clutter dominant below 4.5 km, rain clutter beyond. Detection range 10.0 km*

the HLOOKUP facility. Finally a set of curves is constructed using rows 134, 144, 145, ..., 149, to a common row 133 range abscissa, giving Figure 15.25. This figure shows that, for $P_D = 0.6$, the tug detection range falls from 18.1 km when $k = 10$ to 13.9 km when $k = 0.6$. The percentage variation would be less in clutter where ranges are lower, giving the atmosphere less opportunity to curve the rays.

15.2.6 Coaster

We re-enter the extended target spreadsheet (Chapter 14, Figure 14.2) for the worst-case scenario of a small (1600 gt) coaster whose RCS is assumed from rule of thumb Eq. (10.5b) to be $10 \log 1600 = 32.04\,\text{dB}\,\text{m}^2$, tip height $j = 10$ m, height factor $n = 0.66$, length 60 m and beam 15 m. Screening is set to 0.1. Weather is assumed bad: atmospheric refraction factor very low ($k = 0.8$), thunder rain rate $r = 16$ mm/h half path, SS5 with wind blowing towards the radar to give wind factor C2 $= +2.5$ dB. We assume feeder loss as before. The Results panel shows that maximum range is 13.61 km under these difficult conditions. In fair weather (SS2, $r = 0$, $k = 2$), the coaster is detected to 33.10 km.

Modifying the radar to retain short pulselength to long range reduces the clutter entering the detection cell (although many more cells are required) but only gives an additional 8 per cent range in severe clutter, so the Port decides not to take this non-standard option.

15.2.7 Sidelobes

Occasionally a 50 000 gt cruise ship visits the Port. Obviously RCS is going to be fully adequate for detection in all circumstances. But will sidelobes be troublesome? We raise the spreadsheet target RCS to the average likely from Eq. (10.5b): $10 \log 50\,000 = 47\,\text{dB}\,\text{m}^2$, $j = 35$ m, $n = 0.66$, length 300 m, beam 40 m, retaining fair weather conditions. The standard scanner has gain to principal sidelobes of -29 dB on main beam. We set permissible sidelobe $P_D = 0.1$. The Results panel shows sidelobe P_D exceeds 0.1 at $R = 27.7$ km. This is undesirable, since it will be important for the VTS station to check that no small craft stray into the ship's path. A superior scanner with principal sidelobes -33 dB on main beam almost eliminates display of sidelobes (the improvement works on both interrogate and response legs). Figure 15.26 shows main beam and sidelobe P_Ds with each scanner, and how operation on mode 1 reduces the sidelobe problem at the shorter ranges. Logarithmic range scaling reveals the short-range performance. Here the ambit is log R from -0.3 to $+1.9$, corresponding to 0.5 to ~80 km. This presentation is sometimes useful when needing to show specific range brackets without cramping short ranges.

15.2.8 Purchase specification

The Port should now mull over scanner specifications and decide what expense is justified. A thorough knowledge of the importance to operations of occasional sidelobe interference, the Port budget and likely traffic developments in the years after

	A	B	C	D	E	F	G	H	
129	Variation of k factor.								
130	Range, km, row 38		k=	10	0.46	0.47	0.49	0.51	
131	Chosen case PD, row 112		k=	10	0.99	0.99	0.99	0.99	
132									
133	Range, km, row 38		k=	0.6	0.458062916	0.4734277	0.489317657	0.5057515	Etc.
134	Chosen case PD, row 112		k=	0.6	0.988748775	0.98918	0.989870085	0.9898779	
135	Range, km, row 38		k=	0.9	0.46	0.47	0.49	0.50	
136	Chosen case PD, row 112		k=	0.9	0.99	0.99	0.99	0.99	
137	Range, km, row 38		k=	1.3333	0.457138413	0.472353	0.488080191	0.5043377	
138	Chosen case PD, row 112		k=	1.3333	0.988699807	0.9891078	0.989837199	0.9899056	
139	Range, km, row 38		k=	2	0.45688675	0.4720474	0.487715909	0.5039097	
140	Chosen case PD, row 112		k=	2	0.988687446	0.9890873	0.98982581	0.9899145	
141	Range, km, row 38		k=	5	0.45658506	0.4727922	0.489577387	0.5069615	
142	Chosen case PD, row 112		k=	5	0.988673227	0.9891247	0.989870244	0.9900908	
143	Range, km, row 38		k=	10	0.456484565	0.4737565	0.491683393	0.5102904	
144	Chosen case PD, row 112		k=	10	0.988668641	0.9891767	0.989905332	0.9900707	
145	PD, row 133 range base		k=	0.9	0.988718439	0.9891366	0.989851622	0.9898939	
146	PD, row 133 range base		k=	1.3333	0.988699807	0.9891078	0.989837199	0.9899056	
147	PD, row 133 range base		k=	2	0.988687446	0.9890873	0.98982581	0.9899145	
148	PD, row 133 range base		k=	5	0.988673227	0.9891247	0.989124652	0.9898702	
149	PD, row 133 range base		k=	10	0.988668641	0.9886686	0.989176657	0.9899053	

Figure 15.24 Construction of Figure 15.25. Details in text

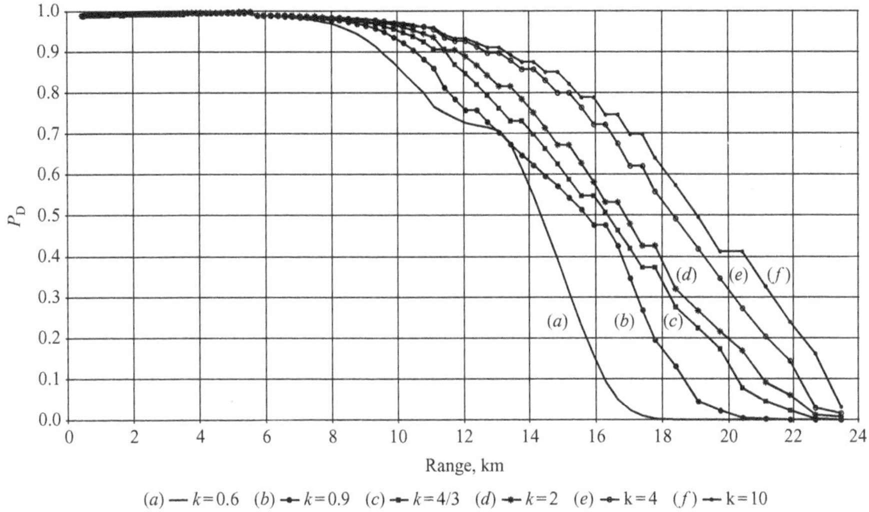

Figure 15.25 Atmospheric refraction k factor. Tug, no clutter. This unseen factor considerably affects detection range of distant targets

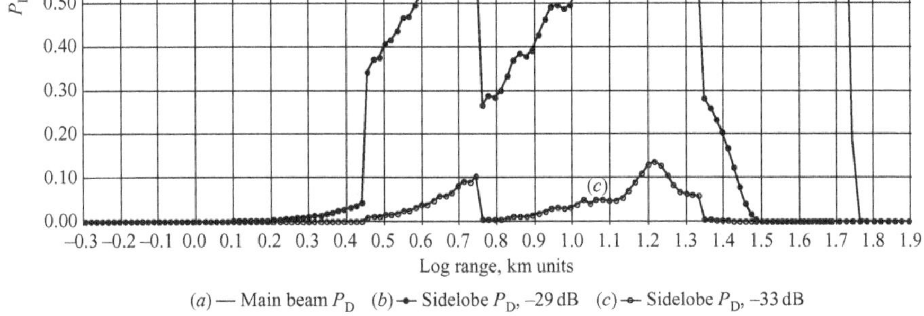

Figure 15.26 Liner. Main beam and sidelobe P_Ds. RCS 47 dB m^2, clear weather. Reduction of scanner sidelobes by 4 dB sharply reduces display of sidelobes. Abscissa is 0.5–80 km, scaled logarithmically

system installation are all necessary before reaching the final decision. We shall not whisper that the answer might even be to forego VTS and buy a new tug!

It is decided that $H = 22$ m with transceiver at the mast foot connected by feeder to the bigger scanner having superior sidelobes probably gives the best cost/benefit ratio. Decisions of this sort are often difficult and in practice many more scenarios would be, or ought to be, explored before making the decision. Time spent in reconnaissance is rarely wasted and a couple of days could profitably be employed by injecting a wide variety of scenarios into the spreadsheet to give confidence in the robustness of predictions, printouts preserving an audit trail of the decision-making process, demonstrating to all that the choice was thoroughly researched and that there is a reasonable margin of performance in hand.

The Port eventually overcomes the Finance Director's ritual plea that funding is especially difficult this fiscal year and issues an international call for tenders for the VTS radar system. Formerly this might have been a fairly brief statement of the Operational Requirement – what targets must be seen at what range in what weather – together of course with delivery time, installation requirements, payment terms, etc. This approach minimised the Port's bidding cost and relied in part on the reputation of the chosen supplier to provide an operationally satisfactory system within Sale of Goods legislation. Suppliers tended to be left some freedom to choose, say, whether to go for a low mast with big scanner, or higher mast with smaller scanner, and carried the risk of mis-guessing target RCS, prevailing atmospheric refraction coefficient and so forth.

Nowadays it is usual to prepare a fully detailed technical specification, which bidders are to meet in all respects, perhaps with a clause permitting alternatives which can be demonstrated to provide equivalent or better performance. Indeed, many State-controlled organisations are required by law to procure from the bottom-dollar supplier offering full compliance with the published specification. It is a moot point whether this approach is really an improvement. It may eliminate graft but certainly increases buyer's and bidders' costs, which latter must eventually be reflected in prices if bidders are to remain in business. The tendency is to specify such factory testable parameters as transmitter power and scanner gain. Purchasers may not fully realise that this transfers much of the onus of system design onto them. Suppliers should be especially wary of customers who try to have their cake and eat it, specifying equipment parameters as well as operational requirements, not invariably consistently.

The specification should reflect a formally agreed internal Operational Requirement rather than a vague wish list. Grief may result if the specification is not tightly drafted. Writing and interpretation of specifications has become a lawyers' art form and the supplier may claim carte blanche over any undefined detail. The author remembers quite legitimately offering a coherent-on-receive system where the purchaser of a Missile Firing Range surveillance radar wrongly thought it had specified the more expensive fully coherent alternative, and a Government customer being prepared to accept completely untested spares, having omitted the relevant clause from the very detailed contract! Acceptance tests requirements must also be carefully spelt out.

Ports purchase VTS systems at such rare intervals that in-house expertise cannot be accumulated. They are well advised to engage a specialist consultancy to draft the specification and perform technical bid analysis. But the wise Port Engineer will want to understand the basis of the analyses. Likewise bidders, especially if foreign, generally employ a specialist commercial Agent to guide them through the thickets of local procurement requirements, often dismissed as 'boiler plate' but poised to drop on the unwary in tonnage quantity from great height.

15.2.9 Site acceptance tests

The Port and its Consultant formalise the Purchase Specification, call tenders, analyse their technical content, weed out the no-hopers, make quite sure that one particularly low bidder has forgotten no vital aspect, finesse the requirement to match tendered prices to budget, get best and final offers from the leading three candidates and place an order on the chosen supplier, perhaps the one offering the easiest credit terms or the most attractive long-term training and maintenance package, rather than the very best technical solution.

The supplier finally sets the system to work. Before handover, the Port naturally requires a demonstration of compliance with the agreed specification, based on detection range of a test target. Rather than the vagaries of a ship's RCS, it has been agreed to use a calibrated trihedral corner reflector, measuring echo margin in hand by insertion of calibrated attenuation into the receiver intermediate frequency amplifier until P_D is observed to be 0.6 at range 10 km, on a day with little or no rain and moderate sea clutter; measured margin to be transformed to standard clutter conditions using point target spreadsheet SS1. Note that this design of acceptance test puts the risk of tug RCS turning out lower than thought with the Port. With the tug specified as test target, this risk would lie with the supplier, which might impact on price.

A 20 dB m^2 reflector was chosen to raise RCS well above that of the rigid inflatable boat carrying it, and to give a moderate margin over threshold if the radar is operating to specification. As atmospheric refractive index, k, is so difficult to determine, target height, h, was chosen as 3.5 m to sit the first multipath peak near 10 km to minimise the effect of refraction changes on echo strength – the aim being to test equipment, not the environment! The spreadsheet is similar to that of Chapter 14, Section 14.2.1, Figure 14.1. The upper set of curves of Figure 15.27 graph signal margin for a just-compliant system to a base of range for several k values. Margin, 25.4 dB, varies only ± 0.2 dB for variation of $k =$ between 1.0 and 4.0.

After an irritatingly calm spell, eventually a day of thunder squalls gives an opportunity to test the system under some clutter at least. The reflector and mast are rigged on a rigid inflatable boat (RIB), which tracks rain showers outwards from 7 km under radiotelephone direction from the radar, aligning the reflector toward the radar at all times. The receiver attenuator is adjusted to maintain $P_D = 0.6$. The solid points inserted in Figure 15.27 show dB in hand, read off the attenuator, versus range. Observed rain rate at the target is 8 mm/h average, the rain patch occupying 0.2 of the radar to target path, with an active sea, SS2. Afterwards, the boat crew agree that the

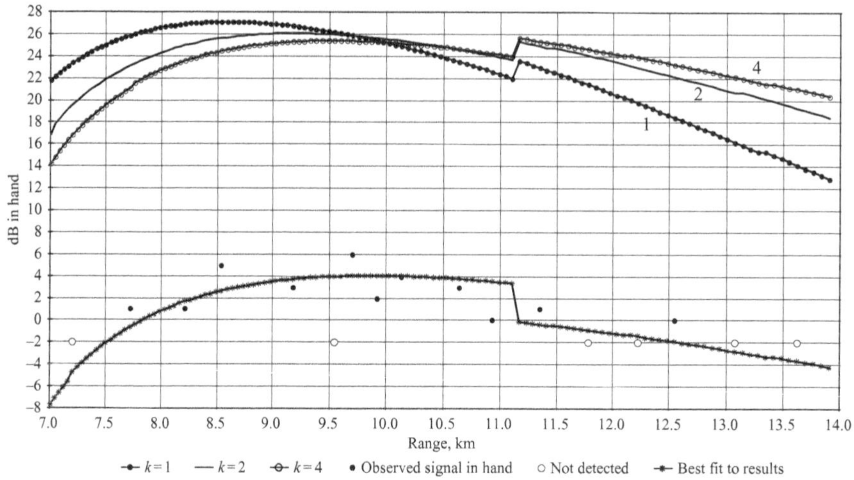

Figure 15.27 Acceptance test results. Point target near first multipath peak. RCS 20 dB m², Upper curves: variation of dB in hand with k factor is minimal near 10 km range with h = 3.5 m. SS2, r = 0. Lower curves: test results (dots) and reconciliation

low 9.5 km result is probably from temporary reflector misalignment. Discarding this result, at first glance performance looks about 18 dB below specification. However, after inserting the higher rain rates and sea states in the spreadsheet, it is found possible to fit a curve to the results, with $k = 3$ and 1.5 dB margin in hand on specification. (The sense of the range-change step at 11.1 km = 6 nmi arises because long-range performance is now limited by clutter rather than noise.) Performance being satisfactory in all other respects, the Port accepts the system.

15.3 Small craft radar

15.3.1 Detection of cliffs

Small craft in particular need to know likely detection range of coastal features for passage planning. At what range will a line of 15 m high cliffs be detected if dense advection fog occurs, optical visibility 0.05 km (Chapter 5, Section 5.9.4, Eq. (5.46d)), swell of SS2, no precipitation, air temperature 4°C, relative humidity (RH) 100 per cent? Cliff reflectivity is assumed $-15\,\mathrm{dB\,m^2/m^2}$ (Chapter 10, Section 10.12.2, Eq. (10.18)). Fluctuation is assumed Swerling Case 1 because the vessel will roll. The radar is a typical small-craft installation having the parameters shown in Chapter 14, Section 14.9.1, Figure 14.2. The results panel shows maximum range in clear conditions (visibility 10 km, RH 50 per cent) of 4.08 nmi, with no detectable change in dense fog despite the additional atmospheric attenuation of

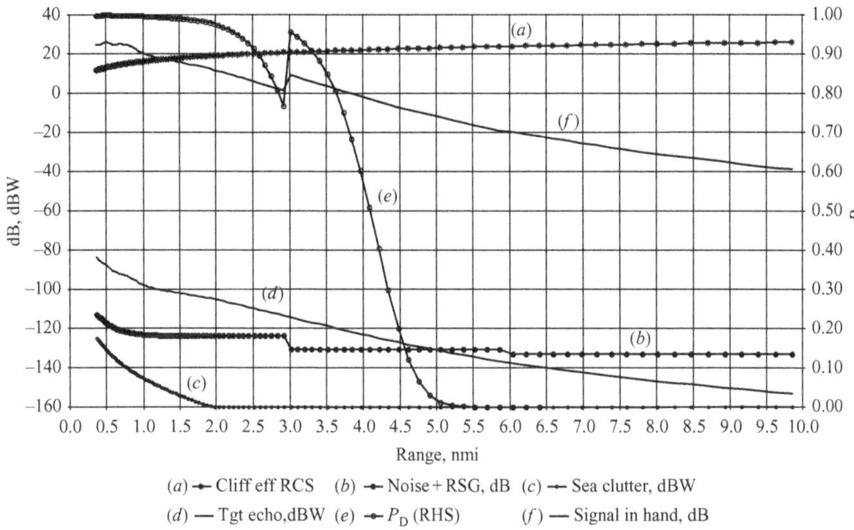

Figure 15.28 Cliff in fog. Cliff height 15 m. Clear weather barely affects results

0.024 dB/km. Observed fog ranges often decrease, probably because of reduction of atmospheric refraction factor and prevalence of ducting which the spreadsheet cannot account for.

The test range results in the Results panel indicate that fog P_D drops sharply at the detectability limit; P_D 0.6029 at 4.0 nmi and 0.0304 at 5.0 nmi, because the cliff has been modelled as a uniform structure. In practice, the cliff would be irregular and RCS would not vary so smoothly with range changes, so P_D would fall less steeply.

Figure 15.28, for performance in fog, shows the following.

- Cliff effective RCS initially rises with range (at 3 dB/octave) because cliff illuminated width is proportional to range.
- Detection range is noise-limited. (Operation on a long range scale reduces receiver bandwidth insufficiently to improve range significantly.)
- Sea clutter at SS2 is always insignificant, the scanner being low.

15.3.2 Cliff height

How does detection range vary with cliff height, *j*? Variation of this parameter affects the spreadsheet range base. We elect to Copy, Paste Special, Values Only the Results panel into the User panel, a similar process to Section 15.2.2, Figure 15.21, for a succession of cliff heights rising in nominally geometric progression. The electronic engineering E12 series (for resistor values, etc.) is useful: 1.0, 1.2, 1.5, 1.8, 2.2, 2.7, 3.3, 3.9, 4.7, 5.6, 6.8, 8.2, [10, 12, ...]. We insert Tip height, *j* m, in cell J4 and enter = J4 in cell F13. We enter RCS, dB m^2/m^2 in cell J17 and enter = J17 in cell F13. The exercise could be repeated for clear or other environmental conditions.

Figure 15.29 shows results. A 10 m cliff is detectable at 3.12 nmi, rising to 11.3 nmi for a 100 m cliff, these values varying between $\frac{1}{4}$ and $\frac{1}{3}$ of horizon range. Of course, actual detection ranges depend on real rather than assumed RCS per unit area of the cliff and might differ considerably from the figure.

15.3.3 Encounter with a coaster

The approaching rhythmic thud of a ship's engine in fog is one of the yachtsman's worst nightmares. Where is she? Am I in her path? Will she see me? Where can I go? Would I see her if I had bought a radar?

'She' may be a 1600 gt coaster, bows on. Chapter 10, Section 10.4.9, Eq. (10.6) gives minimum likely RCS as 21.72 dB m^2. Many such vessels are low-built to traverse bridges on the Rhine or other river systems, overall height about 6 m and often notoriously lightly manned for their punishing schedules, Figure 10.4. The ship will have Swerling Case 1 fluctuation. We suppose the yacht has ventured into open waters, with SS4. Entry of the appropriate parameters, including wave screening, in Page 1 of the extended target spreadsheet gives the yacht radar first detection range of 3.20 nmi (5.92 km), allowing the skipper a quarter of an hour to start engine, track the target and form a course out of danger.

If the ship has a low-performance 10 kW, 9 GHz radar with a 6 ft (1.83 m aperture) scanner, and the yacht has the target characteristics shown on Figure 15.1, the ship may not see the yacht beyond 1.72 nmi, leaving little time for it to assess the situation and manoeuvre. Although the ship has the more powerful radar, mounted higher, the yacht has by far the bigger target. The yacht radar here well justifies its cost by nearly doubling the available escape time from a potentially lethal situation.

15.4 Active targets

15.4.1 Detecting a buoy racon

Many important navigational buoys carry racons or sometimes radar target enhancers. Others generally sport a reflector of nominally 10 dB m^2 RCS, these devices usually being 2–5 m above sea level, let us say a high focal plane buoy, $h = 4$ m, Figure 7.9. We will see how well the ship's radar of Section 15.1 picks up all these devices, using the active point target spreadsheet SS3. We assume stratiform rain of 4 mm/h whole path, SS3 and atmospheric refraction factor $\frac{4}{3}$ and that the ship uses a long range scale (Mode 5). Figure 15.30 plots the probability of detection of a typical racon of antenna gain 5 dB, height 4 m, long-pulse threshold sensitivity −70 dBW, rolling off at −40 dB/decade below the −3 dB pulselength of 0.2 μs, linear detector.

Interrogate leg detection range is 13.5 nmi. The response leg is less sensitive in the prevailing precipitation clutter, with $P_{\mathrm{D}} = 0.5$ at 10.61 nmi, which defines the overall performance. Without any clutter, response leg detection range would be 14.23 nmi and overall detectable range would be set by the interrogation leg at 13.5 nmi.

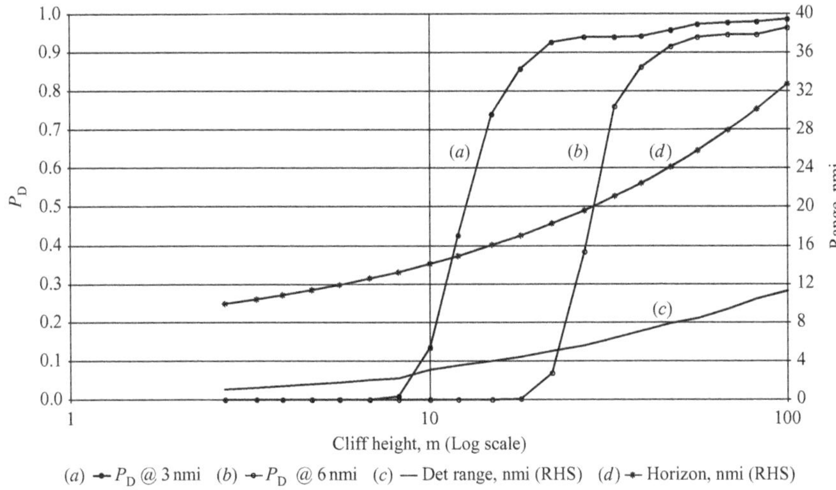

Figure 15.29 *Cliff height. Shows variation of P_D at spot ranges of 3 and 6 nmi, maximum detectable range ($P_D = 0.6$) and horizon range of the cliff top. Abscissa 1 to 100 m, logarithmically scaled*

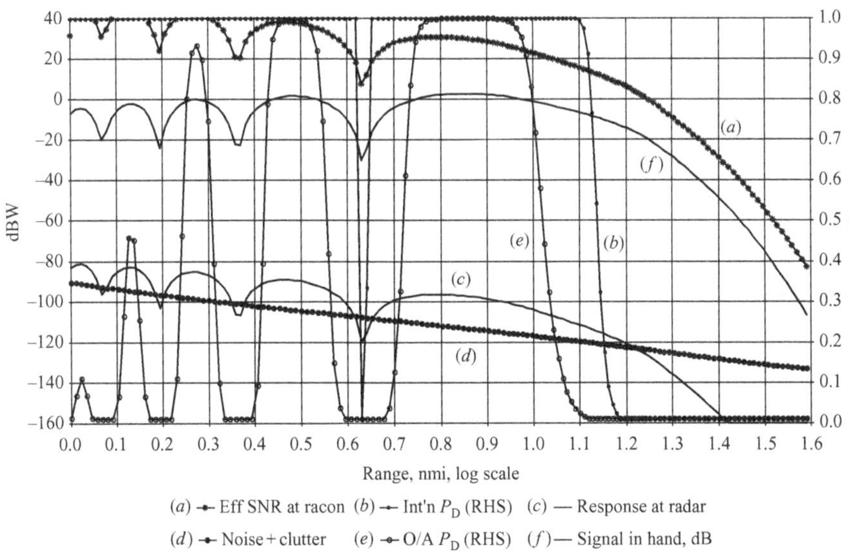

Figure 15.30 *Ship detecting racon. Range scale 1–40 nmi, logarithmic. The overall P_D is set by the weak response channel. Maximum range 10.61 nmi but multipath nulls at shorter ranges and no detection below 1.79 nmi due to the inverse square response power law*

	A	B	C	D	E	F	G	H	I	J	K	L
1	Active point target spreadsheet				Spreadsheet ref	SS3 v1						
2	Date		User		Task	Chapter 15, Sec 15.4.2		Target type	RTE			
3	Transceiver		Scanner and feeder		Range bracket		Environment		Results		User panel	
4	Type		Type		Scaling ,km/nmi	nmi	Type					
5	SHIP'S RADAR				Max R reqd	40			*Scaling*	nmi		
6	Frequency, MHz	9410	Az beamwidth°	0.9	*Achieved*	*40.000*	Refraction, *k*	1.3333	Test R1	3.5		
7	*Wavelength, m*	*0.0318*	El beamwidth°	20	Min R reqd	1	*Eff Earth rad E, km*	*8494.45*	*PD at R1*	*0.0195*		
8	Tx P, kW	20	Gain, dB	31	*Achieved*	*1.001*	Precip type: Stratiform = 1	1	Test R2	5		
9	*Tx P,dBW*	*43.01*	*Efficiency, p.u.*	*0.91*	Target type: 1 =	7	Precip rate, mm/h	4	*PD at R2*	*0.3092*		
10	Tx loss Lt, dB	0	Loss, dB	1	Ae gain dB	5	*Equiv rain for atten*	*4.000*	*Min R for B15 PD*	*1.61*		
11	Rx loss Lr, dB	0	Height H, m	35	Rx ae hgt h, m	4	*RCS, dB m²/m³*	*−62.94*	*Max R*	*9.03*		
12	Service loss Ls, dB	2	El patt, 1 = sin,2 = invco	1	Tx ae ofst, m	−0.5	*Loss lp, dB/km*	*0.049*	*R bracket*	*5.04*		
13	Rx NF N, dB	3	Depression°	0	N/A	−70	Extent, p.u.	1	*Fill, p u*	*0.63*		
14	RSG thld, dBm²	−10	Rotation, rpm	22	RTE amp gain, dB	55.93	Air temp, C	20	*Horizon R*	*17.62*		
15	Reqd PD	0.5	CP improvement, dB		*Unsat RCS dB m²*	*25.00*	RH, %	70	*1st peak R*	*>F6*		
16	*With screening*	*0.500*	Sidelobe below D8, d	29	Sat P, dBW	0	*Clr air loss, dB/km*	*0.027*	*1st null R*	*4.26*		
17	Reqd Pfa expt, F	−6	Tolerable SL PD	0.01	*Watts*	*1.00*	Wave hgt, hs, m	0.125	*2nd peak R*	*2.99*		
18	Proc loss, Lp dB	8	Feeder ohmic loss, d	2	Sw Case 0,1,3	0	*Sea state*	*3*	*2nd null R*	*2.28*		
19	Integration: n, c, cr, p	n	VSWR	1.5	Operator		Wind fctr C2, dB	0	*Max sat R*	*8.20*		
20	Scan/scan corlrn y/n	y	*Refl coeff*	*0.200*	Pol, h, v, c	h	Screening, p.u.	0	*Max sidelobe R*	*<F8*		
21	N/A	0	*Mismatch loss, dB*	*0.53*	Gain control, dB	0	Surface dielec, eta	81	*Ref echo, 1 km, FS*	*−84.97*		
22	*Resp wavelength*	*0.0318*	*EIRP, dBW*	*70.48*	Sw Gain control, d	0	conductivity, S	0.24	*Sea clutter horizon*	*13.17*		
23	N/A	9	*System NF, dB*	*6.85*	SP = s, LP = 1	s	Tilt/pol loss 1-way dB	0	*N/A*	*N/A*		
24	N/A	8			Mode, Table S1	5						
25	TABLE S1, Mode	Max R	*Max R, km*	Plslgth, SP	LP, μs	Rx bw, MHz	PRF, pps	*Active mod*	*Selected, μs*	*Pulses inte*	Target sec'y features	
26	*1*	1.5	*2.777777778*	0.05	0.1	25	5000	*5*	*0.05*	*68.18*	N/A	0
27	*2*	3	*5.555555556*	0.25	0.5	5	5000	*5*	*0.25*	*68.18*	N/A	−8
28	*3*	6	*11.11111111*	0.5	1	3	2000	*5*	*0.5*	*27.27*	N/A	0.9
29	*4*	12	*22.22222222*	1	1	1.5	2000	*5*	*1*	*27.27*	N/A	*0*
30	*5*	48	*88.88888889*	1	1	1.5	1000	*5*	*1*	*13.64*	N/A	0.2
31	*6*		*0*					*6*	*0*	*0.00*	N/A	40
											El bw° 1way	25

Figure 15.31 Spreadsheet SS3 for ship observing RTE

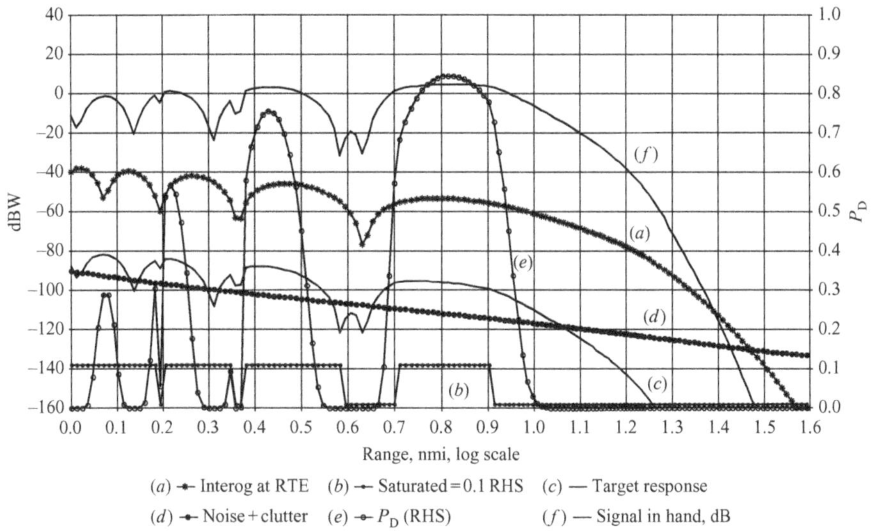

Figure 15.32 Ship observing RTE. Maximum detectable range 9.03 nmi, but poor detection at many shorter ranges (minimum detectable range 1.61 nmi, fill 0.63 p.u.) due to multipath interference, coupled with RTE saturation, as with the racon of the same response power

Had the radar been the small craft type of Section 15.3, but retaining $H = 35$ m, the feeble interrogation would restrict detection range to 8.82 nmi, showing the desirability of retaining good racon receiver sensitivity.

15.4.2 Detecting a radar target enhancer

The RTE is assumed to have its upper antenna at 4 m with the other antenna 0.5 m below, unsaturated RCS 25 dB m^2 (316 m^2, quite high), and saturated response power 0 dBW, as for the racon. Figures 15.31 and 15.32 show the spreadsheet Page 1 and the performance, which is generally similar to the racon because both have the same saturated response power. The nicks in the target response curve are formed by the differing multipath null ranges of the interrogation and response legs because of the differing antenna heights.

Chapter 16
Future possibilities

C. J. Baker

'There's is no use trying,' she said: 'one can't believe impossible things.' 'I dare say you haven't had much practice,' said the Queen. . . . 'Why, sometimes I've believed as many as six impossible things before breakfast.'

Lewis Carroll, Through the Looking-Glass

16.1 Introduction

This chapter reviews a number of potential future developments that may lead to changes in civil marine radar system design. As ever, changes are likely to arise from alterations to the conditions that currently prevail. These may be factors internal to the marine radar industry such as technology improvements, or they may be external factors such as regulatory and legislative alterations. It is always impossible to predict the future with a high degree of confidence, but a look back at recent history does tell us that one thing is certain, change will occur. Indeed it is a trend in the World today for change to occur increasingly rapidly. At first sight civil marine radar has not altered greatly since its inception. Closer inspection reveals that (a) regulatory legislation has imposed certain performance demands while somewhat inhibiting variation in design. Additionally (b) there has always been strong competition in the market place which has resulted in much emphasis on price as a discriminator between systems. Consequently, new technology developments that do not reduce system costs have not always been adopted unless prescribed by legislative changes. As a result, the design of marine radar systems has not hugely altered from the original adaptations of military technology developed during the Second World War. These are well described in the early chapters of *The Use of Radar at Sea*, first published 1952, while Chapter 2 of this book describes current practice. It might be argued, therefore, that the time is long overdue for marine radar to be subject to quite radical change. However, before speculating whether and how civil marine and VTS radar may develop, we need first to examine the developing requirements that would drive such changes. Some may be evolutionary and continue trends already in place. Others, if they materialise, could be called revolutionary and may call forth major changes in system design and implementation.

In the remainder of this chapter we examine the likely drivers for change and the possible technology responses and developments that either are being researched or are likely to occur. This allows us to examine options able to satisfy emerging future requirements for the performance and specification of civil marine radar systems.

16.2 The drivers for change

The drivers likely to lead to changes and developments in future systems fall into five principal categories: customer requirements, regulatory change, cost-effectiveness, environment and technology. We now consider them in more detail to determine their potential.

16.2.1 Customer requirements

It might be argued that the strongest requirement expressed by customers is price reduction. The market place today is truly global and highly price sensitive. It is likely that price will continue to be a key purchasing criterion. This may restrict introduction of new technologies to those areas that reduce costs, possibly at the expense of capability improvements. An exception is cost increases resulting from legislative change, usually accepted by the industry provided a level competitive playing field is maintained.

One area where there has been a consistent demand for enhanced capability is improved detection of small targets in sea clutter, particularly in high sea-states and with minimal false alarms. For example, improved detection of yachts may help avert potential collision disasters, a very real hazard. Clearly detection of small targets is desirable, instilling in the mariner a greater sense of confidence that his equipment is providing an accurate picture of the surrounding environment, especially in the foulest weather. Alternatively small craft could carry reflectors or some other device that would show on the operator's display. Indeed the performance of such reflectors is being prescribed by the International Marine Organisation (IMO), Chapter 7, Section 7.6.1, which may encourage development of novel passive and active reflectors.

It is a feature of the conventional magnetron transmitter that it generates very short pulses with particularly fast rise and fall times, helping to minimise the clutter return by permitting small range cells. Coupled with intelligent clutter filtering, resulting performance of current marine radar is extremely good. Whilst there is still room for improvement, the point of diminishing returns is, perhaps, being approached. Other relatively recent improvements, at least part customer driven, include colour displays, integration with GPS and much more fully integrated bridge systems. Overall, though, consumer pressure for performance and capability improvements has not been as strong a driver as it tends to be in other technology markets, a feature likely to continue. Indeed performance advantage is probably not the strongest differential in the market place but it certainly plays a significant role. This may change. Indeed there is some evidence of customers desiring a degree of imaging capability to more accurately map out coastlines, other fixed features and perhaps even to obtain imagery of other vessels. It should also be remembered that there are numerous classes of ship, faced with differing navigation problems. These range from container ships that have difficulty in seeing nearby traffic due to cargo obscuration to fishing vessels wishing to detect flocks of gulls indicating shoals of fish, and from coastal to open sea operation.

It is also important to look outside the current application of radar for purely marine navigation. For example, the adoption of low-cost vehicle collision management and obstacle avoidance radar systems for tasks such as harbour navigation or as part of an intruder alerting system may generate a consumer demand for new products, albeit of a different type to traditional navigation radar systems. Indeed specialist Doppler radar is already sometimes used as a docking aid.

16.2.2 Regulatory change

Alteration to the legal requirements for marine radar systems forms a second and perhaps more powerful driver likely to have a significant impact on future marine radar. Current systems have to be compliant with IMO regulations and have emissions controlled within limits set by the International Telecommunications Union (ITU-R). One example of such a regulatory description is expressed by the IEC/British Standards Institution (BSI) document 'Maritime navigation and radio-communication equipment and systems-radar' (BS EN 60936-1:2000; fully compliant with IMO regulations, see Chapter 1, Section 1.5). The various IMO regulations specifying equipment for detection, surveillance and navigation are a fairly prescriptive mixture of technical and capability specifications, perhaps necessary and desirable in the early days of the adoption of radar when specifications were evolved on a pragmatic and safety conscious basis. However, this prescription has removed a number of design freedoms from the radar engineer, in part explaining why the marine radars of today have largely similar designs to those of yesteryear, albeit incorporating more modern technology in a number of instances. They do, however, offer a vastly improved performance. To be fair, the resulting standardisation makes it much easier for officers transferring from ship to ship.

The IMO radar regulations have not been fundamentally altered for many years. Indeed they refer to items such as the persistence of targets painted on to the operator's screen. Whilst this harks back to cursive cathode ray tube (CRT) displays it is a useful feature and still required today. It is quite likely, though, that the regulations will change both in the near- and the long-term future. The upsurge in the use of radio telecommunications (both mobile and point to point) has placed increasing pressure on the way in which the electromagnetic spectrum is used. Indeed bands currently allocated to both civil and military radar services may have quite severe restrictions placed on the way their edges roll off. Current frequency allocations (e.g. the whole marine 3 GHz band) could be reduced even more severely or in the extreme be completely reallocated to telecommunications services. There has already been a tightening of frequency allocations and there are likely to be further re-definitions, perhaps in two or more steps with the first being a further tightening of the roll-off of the transmitter spectral output, to reduce the possibility of interference with adjacent allocations to other users. The second may well be a very severe re-definition of spectral roll-off requiring much sharper spectral emission cut-off to enable more efficient use of the spectrum. The first case will probably require adaptation of the magnetron, a notoriously dirty instrument in spectral terms as it inherently transmits significant amounts of power beyond the 3 dB roll-off points of its nominal operating bandwidth. Magnetron output filters are the subject of research and development. It is quite likely that better filters will quickly find their way into operational equipment if the regulations are changed. The more severe re-definition of spectral roll-off may well be beyond the limits of filter designs, necessitating a move to a different form of transmitter technology altogether. If this happens, it will lead to very significant changes to current equipment either through the development of a 'coherent magnetron' or the universal adoption of more conventional coherent transmitter technology, such as travelling wave tubes (TWT) or solid-state amplifiers.

Currently, the regulations prescribe the spectral roll-off in relative terms compared to the peak power in the spectrum, permitting very high absolute radiated powers (high EIRP). Thus it is quite likely that the problem of mutual interference will be little affected unless the regulatory definitions are altered.

Were a radar band to be completely re-allocated to another service, marine radar systems would be required to operate in a different band. For example, there seems to be less pressure on the 9 GHz band than the 3 GHz. If 3 GHz were lost, there would be implications for sensitivity as rain clutter will be more of a problem at the higher frequency. It may be possible to use much more speculative techniques. These might include modes of operation that are compatible with telecommunication applications, permitting co-existence in the same region of spectrum thus preserving the 3 GHz band.

Alternatively, and more speculatively, passive systems exploiting emissions of opportunity might provide a surveillance and navigation capability that avoids the need for the radar system to transmit at all. One such example is the US Lockheed-Martin Silent Sentry system developed for airspace surveillance. It seems doubtful whether useful emissions of opportunity exist in the high seas, so this technology is not a good candidate for general marine navigation applications. Nevertheless in coastal areas one may imagine a 'radar lighthouse' as an all weather aid to local navigation exploiting bistatic target reflections.

Currently, the greatest pressure is being exerted is on the 3 GHz band, which lies close to current mobile telephony equipment bands. In addition, there is intense speculation that future adoption of wireless local area network (WLAN) technology would bring equal pressure to bear at C band (4–8 GHz), pre-empting its marine radar use as an alternative to 3 GHz. Frequency re-allocations of frequencies are agreed at triennial World Administrative Radio Conferences (WARC). This generally means that it takes 6 years (sometimes longer) for a recommendation to be implemented and rolled out, because so much work has to be undertaken to determine the feasibility and impact of any proposed changes. However, it must remain a very real possibility that in the relatively near future there will be alterations to the bands allocated to radar. A seemingly small change to the regulations could lead to a very big impact. For example, it could force all future radars to switch from magnetron transmitters to an alternative. The exact changes to the regulations and their resulting impact is as yet unclear, however, it is an aspect that the marine radar industry must be alert to in its strategic planning.

16.2.3 Cost effectiveness

The third driver for change is a need to reduce system and manufacturing costs. This is also a customer driven requirement but is something that all manufacturers strive for to remain competitive. Costs will also be a function of the highly regulated operating environment. Users are required by international maritime law to have a compliant radar system and thus there is a tendency to reduce their own costs by purchasing the least expensive compliant system available. At the very least this tends to place a premium on costs with systems costing significantly above the market norm being non-competitive. In the future it is likely that this situation will remain largely unchanged and cost will remain an important factor in customer choice. This tends to focus the ingenuity of the radar designer into areas other than achievement of improved technical performance via the adoption of new technology, innovative system designs and smart production techniques. Cost of conventional marine radar has been vigorously attacked by the industry for many years, particularly fuelled by the introduction of very low cost systems by the Japanese radar industry (particularly in the leisure part of the market). Cost minimisation has been aided by technology developments such as the advent of cheap analogue to digital converters, digital signal processors and liquid crystal displays (LCDs), bought about by developments in the mass PC market. In addition manufacturing costs have been reduced by outsourcing sub-system production to countries having much lower labour costs, the manufacturer concentrating on system integration, testing, maintenance and sales. This is likely to be a continuing trend with the industry perpetually in search of ways of minimising costs.

16.2.4 Environment

Environmental factors are also likely to continue to influence developments. Marine radar has always had to operate in the harshest of environments and this is unlikely to change. However, coastal waters are becoming an increasingly busy and important area of operation, so the radar has to cope not only with sea clutter but also the more extreme dynamic ranges associated with land clutter. The clutter environment could become even worse in some areas with the growing number of offshore and coastal wind farms. These consist of large (~50 m) towers supporting rotating ~40 m turbine blades placed so high efficiently to couple wind energy. There will be significant radar reflection from both the tower and the blades. The blade echoes, in particular, will be spread over a range of Doppler frequencies. This would also reduce sensitivity of magnetron based systems. The impact of wind farms on current systems is not clear and remains to be fully analysed but should be confined to quite localised areas.

16.2.5 Technology

Technology change is the last of our categories that may influence future system design. Two distinct types of change may be envisaged; adoption of new technologies and the adaptation of existing technologies. New technology could improve performance and provides manufacturers with a means of differentiating their products in a very competitive market place. Examples are clutter rejection algorithms able to detect smaller targets and provide more robust performance in the highest sea-states, and emerging technology such as micro engineered machined (MEM) circuit components that lead to significant reductions in cost. Complete MEMs radar systems have already been fabricated. Alternatively new technology may help create new products and demand. The example of adopting car collision management radar for harbour navigation was mentioned earlier.

There are certainly a number of technology developments that may impact civil marine radar. The marine radar industry is too small itself to generate much new technology but will look to harness developments emerging from other areas. Fast-paced technology development in the mobile communications arena such as high power, high efficiency amplifiers and low cost electronically scanned antennas may have radar applications. LCDs are fast becoming the norm and new display technologies such as plasma screens are entering the market. Second, existing technologies developed for advanced military radar may find their way into civil systems should the appropriate conditions come about. Just a few examples include sophisticated electronic (rather than mechanical) scanning, adaptive beam forming for clutter mitigation and improvements to digital signal processing.

As stated earlier, the only certainty about the future is that things will change. Thus, while we have introduced a number of anticipated drivers for change likely to influence marine radar developments, it is only possible to speculate on the resulting changes that may occur. Whether or not they will come about is dependent on many interrelated diverse and unpredictable factors.

The next section considers some of the developments taking place in other radar application areas, particularly military radar, where systems tend to be the most technologically advanced, helping us to hypothesis the relationships between drivers for change and possible technical developments. The technology types considered are divided into four categories: hardware, processing, systems and infrastructure.

16.3 Hardware developments

16.3.1 Transmitters

The magnetron transmitter has been the subject of much improvement since the adoption of post Second World War units in the early days of marine radar. It is now a very compact, low cost and relatively reliable device. Nevertheless its output tends to spread over quite a wide range of spectral frequencies and is not completely stable. As noted in the previous section, this is becoming increasingly incompatible with the more efficient use of the spectrum, for which the global explosion in wireless telecommunications makes a compelling commercial case. Two main approaches have been and are being considered to reduce spurious emissions from magnetrons.

The first is to filter the output. A number of approaches are under development based upon narrow band filtering or the use of diplexers. Co-axial magnetrons may also be an alternative as they offer a cleaner starting spectrum than conventional magnetrons. A co-axial magnetron typically consists of a conventional magnetron with a co-axial output cavity added. Energy is fed to this external cavity via slots cut into every second (co-phased) inner cavity of the conventional magnetron to improve the output stability.

The second approach is to feed a small part of the magnetron output back into the cavity in an attempt to 'injection lock' it to a chosen (crystal controlled) frequency. Although some successful research has been published this form of 'coherent' magnetron has failed to find its way to manufacture. This is most probably due to a combination of (a) not all the problems having been solved and until recently (b) insufficient regulatory pressure. However, if such a device were available it could potentially provide a source of low cost, very high power transmit energy for a wide variety of other applications as well as marine navigation. These might include electronically scanned weather radar or military systems where stealth technology at the target requires higher sensitivities than can otherwise be generated at affordable cost.

For many applications other than marine radar the need to detect movement with a high degree of sensitivity has resulted in the development of coherent transmitters such as the Travelling Wave Tube (TWT). The TWT was developed in the 1960s and enabled wideband, moderate power pulses to be transmitted with a high degree of precision. These coherent radar systems (see Chapter 2, Section 2.2) retain knowledge of phase as well as amplitude and the rate of change of phase provides a direct measure of target velocity. The TWT is a linear beam vacuum amplifier that converts the kinetic energy of an electron beam into microwave energy. As shown in Figure 16.1, it consists of four elements, an electron gun, a helical coil, a collector and an electromagnetic cavity. A microwave signal is input at one end of the helix. Its effective speed is reduced by the extra propagation length represented by the helix to

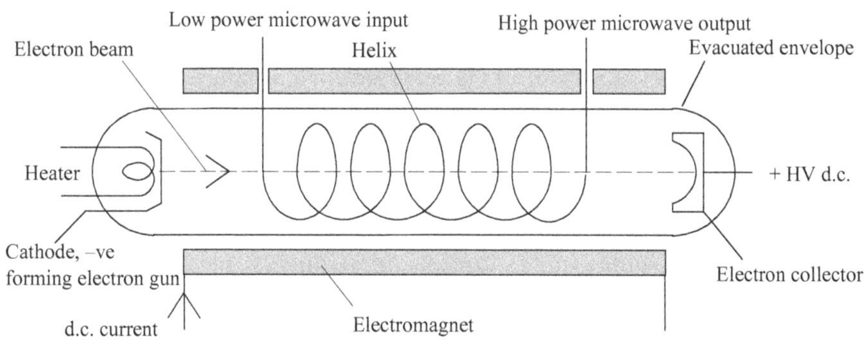

Figure 16.1 Travelling wave tube

a little below that of the electrons emitted by the gun. The fields induced encourage bunching of electrons and the strong fields they generate transfer energy to the more slowly moving microwave signal, which is amplified. There are quite a number of variations on this theme such as coupled cavities able to handle higher powers. TWTs are quite sophisticated, complex devices produced in relatively small numbers and consequently are very costly. This, coupled with the fact that targets can be quite adequately detected with the magnetron transmitter has precluded adoption by the marine radar industry. It is difficult to see this situation changing in the future.

However, relatively low power solid state transmitters may offer a viable alternative to the magnetron. These are, spectrally, inherently much cleaner, have reasonably high efficiencies and if mass produced for the telecommunications industry are likely to be affordable. Indeed if the legislation governing emissions in the radar band is significantly tightened then it is possible only a solid state transmitter solution will be compliant. This would effectively enforce adoption and ultimately they could supplant the magnetron. Their relatively low peak powers mean that a number of devices would probably have to be used. These could be distributed along the length of the antenna itself or combined before feeding the antenna. Current technology is based on gallium arsenide (GaAs) substrates; in the future gallium nitride (GaN) and silicon carbide (SiC) offer greater power and efficiency combinations.

16.3.2 Scanners

The simplest scanner design exploiting low power solid state transmitters is the linear array. Here radiating elements may be thought of as taking the place of slots in a conventional slotted array antenna (Chapter 2, Sections 2.7 and 2.8). This has the advantage of forming the antenna from the transmitter elements directly, obviating feeder loss. However, it will mean that the antenna will need careful protection from foul weather; also this scheme does have the disadvantage of requiring extra servicing and maintenance at the scanner which could incur additional costs. An alternative might be to combine transmitters at deck level and feed the power to the antenna using a conventional feeder.

If the phase to each element forming a linear array can be individually controlled then the antenna can, in principle, be made to scan electronically. This would have the advantage of removing the need for the turning gear necessary for mechanical scanning. It also facilitates adaptation of the beam pattern to help reject clutter, stabilise the beam and enable non-linear scanning (for more effective target detection, e.g. in the forward direction). Quite a number of problems remain to be overcome before electronically scanned arrays are likely to be adopted for marine radar. Typically linear electronically scanned arrays have a maximum scan angle of approximately $\pm 60°$ from broadside (perpendicular to the linear dimension of the array). This limitation is due to a combination of beam spreading and higher sidelobes. Beam spreading results from the projection of the linear antenna onto the scanned angle resulting in a foreshortening effect which reduces the equivalent aperture and hence increases the beamwidth by an inverse function of the cosine of the scan angle. Thus, either multiple antenna faces or a circular array would be necessary.

A circular array would occupy a volume equivalent to that swept by a similar sized mechanically scanned antenna. Clearly, this has quite severe consequences for system and platform integration as it necessitates a radome not unlike that carried by an AWACS aircraft. For leisure craft where the antenna sizes are not so large this may prove less of a problem. Indeed current mechanically scanned antennas are often housed in a closed circular radome of similar size. The other and perhaps much more significant factor likely to hold back the development of electronically scanned array radar for marine applications is cost. Typically, integrated monolithic microwave integrated circuits (MMICs) hosting both RF transmit and receive (T/R) functions cost in excess of £700 per module. Assuming an antenna length of 1 m, nearly seventy modules would be required at half-wavelength spacing (to avoid grating lobes), resulting in a cost of £50 000 before taking into account the remaining sub-systems comprising the radar. A great deal of on-going research is attempting to find ways to dramatically reduce costs. It is possible that technology developed for the telecommunications market, where devices are produced in much greater quantities, giving economies of scale, may help to bring down the unit cost of T/R modules. However, the different operating frequencies and technical requirements of radar will still result in a cost premium for devices for application in marine navigation.

In addition, phase shifters are required for electronic scanning. These have two main impacts. Firstly, they are lossy (a few decibels) and this must be taken into account in any system design. Secondly they represent an additional cost. Currently phase shifters typically comprise diode switched line lengths or are ferrite based. There is emerging research on MEM phase shifters that may offer a lower cost solution. However, experimental devices are quite lossy and are not able to achieve very high switching speeds. In the short term, at least, it unlikely that electronically scanned array technology will be adopted for civil marine radar unless the cost benefit equation changes dramatically. If it is ever to do so then it will certainly have to compete with mechanically scanned antennas on a cost basis. Perhaps one possible exception is the circular antenna that uses simple beam switching rather than relying on phase control. This does not have the adaptability of a phased array but does represent a complete 'no moving parts' radar. This may be more feasible for leisure craft where smaller sized antennas are often enclosed in a protective radome anyway. Even so, performance and cost will have to equate or better that of the magnetron based systems.

16.3.3 Digitisation

Another key technology is analog to digital converters (ADCs). These have been available for many years and their performance in terms of digitisation speed, linearity and dynamic range has been improving year on year and is set to improve further still. Their wide use in the PC market has brought cost down and each year the same money buys a higher specification device. Digitisation speeds vary from a few tens of kilohertz to several gigahertz, cost tending to be a direct function of the product of digitisation speed and dynamic range. Low-cost modest performance devices have been finding their way into marine radar systems for some time, making subsequent data processing much more flexible. The trend for higher performance at lower price makes it likely that, in the near term, they will replace part of the IF circuitry, offsetting their cost. RF digitisation up to about 2 GHz is available now and in the longer term fully digital radar may become a reality. Whether this becomes affordable or truly has enough advantage is much less clear.

Fortunately, the advances being made in analog to digital (A/D) technology are largely matched by those taking place in commercial-off-the-shelf (COTS) digital processors and digital memory, again fuelled by the PC market. High-speed ADCs coupled with high capacity digital COTS processors make a tremendously powerful platform upon which radar data can be processed and targets detected. Indeed today's PCs have processing speeds equalling the marine radar 3 GHz band. This will facilitate the use of more sophisticated signal processing algorithms such as fully adaptive thresholding and even inverse synthetic aperture radar (ISAR) imaging.

It is now possible to see how a number of technologies could come together to form the next generation of marine radar systems and that this will only happen if provoked by legislation, cost savings or some other external factor. For example, moving to a distributed solid state transmitter will require new waveforms and additional data processing. This, in turn, implies a need for high performance ADCs and high-capacity processors. Generally, technologies exist and will be subjected to even more development in the future. A combination of these could form the core for next generation marine radar surveillance and navigation systems.

16.4 Processing enhancements

16.4.1 Moving target indication

In all radar systems the technology alone provides little useful capability. It must be married to suitable processing algorithms (simple examples are detection and tracking) to enable the system to deliver the required performance. As discussed earlier, replacing the magnetron with a distributed coherent solid state power source would support a much wider series of waveforms. In particular moving target indication (MTI) techniques could provide more sensitivity in high clutter.

Coherent solid state transmitters enable the phase and amplitude of a backscattered signal to be measured with reference to a fixed point in time. The Doppler frequency of moving targets is a function of the rate of change of phase of the received signal, in turn directly related to the radial velocity component. Fourier processing creates a series of frequency filters in which moving targets may be detected provided they can be distinguished from a dynamic clutter background. This may be done using a sequence of pulses over the dwell of the scanned beam and a series of frequency filters for each range bin to distinguish movers from clutter before painting onto the operator's screen much as echoes are today.

16.4.2 Long pulses

Different waveform designs will be necessary to ensure equivalent performance to today's magnetron-based systems. Both to be feasible and affordable, available peak power from a solid state transmitter will likely be considerably less than that of the magnetron. Thus, the first waveform design change will be an enforced one to increase the pulse length or even move to continuous wave (CW) operation just to maintain range sensitivity. Avoidance of range ambiguities places an upper limit on the pulse duration. For example, if the maximum unambiguous detection range is set at 89 km (the standard 48 nmi range scale) then the maximum pulse repetition frequency will be 1.68 kHz (i.e. an inter pulse period of 595 μs). If the minimum range is set to be 24 nmi then the receiver will have to be switched on after 297 μs. In effect this implies a realistic maximum pulse length of about 270 μs, because of finite rise and fall times. Any longer would result in the radar transmitting whilst receiving, reducing system dynamic range and limiting overall sensitivity.

Assuming a pulse duration of 270 μs, the power generated by the solid state transmitter can be compared with that of the magnetron. A magnetron will typically have a peak power in the region of 10 kW and pulse duration of, say, 0.1 μs with pulse repetition frequency (prf) 1 kHz, providing an average power of 1 W. To give the same mean power on target and hence system sensitivity, for the same scanner gain the solid state version must also produce 1 W average (higher power systems perhaps average 10 W). Affordable solid state transmitters are likely to have a peak power in the region of 0.1 W. With pulse length 270 μs and also 1 kHz prf, this gives an average power per transmitter of 0.025 W. Therefore approximately 40 transmitter elements will provide an average power of 1 W and give equal range sensitivity under unambiguous operation.

To avoid grating lobes, element spacing has to be less than half a wavelength; for a 1 m antenna this implies 66 transmitters (assuming the transit elements are distributed on the antenna). Thus alternatively we could have (a) 66 transmitters with a peak power of only 0.06 W, or (b) five rows of 66 transmitters each with an output power of 0.12 W would provide 10 W average as well as some elevation gain. Clearly the intimate relationship amongst cost, sensitivity requirements and design parameters has to be carefully examined to ensure an optimum configuration. For example, in a practical system it is desirable to minimise the overall pulse length to reduce sea clutter. In addition, the more efficient adoption of coherent processing techniques should lead to improvements in overall sensitivity enabling lower power amplifiers to be used.

16.4.3 Pulse compression

The short pulse duration and fast rise and fall times characteristic of the output waveform of a magnetron have the valuable effect of helping to reduce the quantity of sea clutter from which targets have to be detected as well as providing accurate ranging. The much longer pulses necessary for the solid state solution provide no such help. Indeed they are so long that they provide little in the way of target location in radial range. A 0.1 μs pulse gives 15 m range resolution, whereas 270 μs provides 40.5 km! This makes the solid state solution initially seem quite unattractive (if not impossible). To avoid such effects, some form of pulse compression will have to be employed. Here the waveform is coded such that matched filtering on reception maximises signal to noise and produces an equivalent pulse length to the modulation rate of the code used. An example is simple frequency modulation of the carrier across a predefined frequency bracket. The bracket swept is inversely proportional to range resolution. A modulation of 10 MHz is equivalent to a pulse length of 0.1 μs, restoring the resolution. In this way compressed pulse lengths with properties similar to those of the magnetron can be generated from much longer real pulses. In general, these techniques are very well known and have been in existence for many years. Digital processors allow a wide variety of codes, from which those with the most useful properties can be selected. However, ensuring adequate sensitivity and sufficient suppression of clutter present new challenges for the marine radar designer. In particular, sea clutter may prove more difficult to suppress. In high sea states, clutter may interact with the transmitted pulse to alter the form of return such that the pulse compression features high range sidelobes, which in part may be balanced by the superior MTI performance offered by coherent processing.

16.4.4 Continuous wave transmission

Another alternative waveform consistent with the use of a solid state transmitter is CW operation. Potentially this would have the advantage of only requiring, say, a single 1 W transmitter to provide equivalent average power to our example magnetron-based system (although several lower power transmitters may be more cost effective). A disadvantage is that the radar, in effect, will be 'deafening' itself by transmitting and receiving at the same time. This raises the noise floor against which targets have to compete, reducing system sensitivity and dynamic range. Pure CW operation will also be completely range ambiguous, providing no target ranging information. However, CW radar systems have found useful application and there are methods for helping to overcome the problems of simultaneous transmission and reception. A simple approach uses separate transmit and receive antennas, perhaps augmented by radar absorbing baffles to prevent the transmitted signal from leaking directly into the receiver. In addition, any close-by objects could reflect powerfully into the receiver, having the same effect as transmitter leakage, demanding careful siting and perhaps further radar absorbing baffles. Indeed, if two antennas are necessary, cost may preclude take up of CW operation.

As a pure CW tone provides no range information (or time stamp) and is completely range ambiguous it is, in raw form, of little use to the mariner. To get target range, the waveform must be coded in some way. Techniques too numerous to describe here have been established and successfully applied. Perhaps the simplest of these and illustrative of the principles is the frequency modulated CW waveform (FMCW). The CW carrier is modulated in frequency across a predetermined bracket over a chosen period of time as for frequency modulated pulse compression. The resulting waveform therefore has a value of frequency that can be associated with a particular transmission instant, creating the necessary time stamp against which targets may be ranged. On reception the signal is de-modulated, leading to the establishment of a series of range intervals (or cells) that can be correctly associated with target position. The resolution is determined by the total ambit of the frequency sweep (e.g. a

50 MHz sweep gives 3 m range resolution). The repetition rate of the frequency sweep has equivalence to the prf of a pulsed system and determines the Doppler ambiguity properties. Frequency modulation is very often the choice of the CW radar designer as it is cheap and easy to generate using Gunn diode oscillators. Great care has to be taken to ensure linearity of the frequency modulation; any deviations result in cross products which themselves generate ambiguities and false targets. The performance of CW waveforms for marine applications are, at best, unproven and it may well be that again sea clutter is a significant problem. However, CW radar systems designed for land applications have been operated over the sea and found to work successfully, although their performance is unquantified. Other waveform codes, such as Barker, Costas and Phase codes, have a number of differing properties. They are alternatives to FMCW, having differing ambiguity characteristics, but are compatible with digital techniques although generally more complex to implement.

Some existing CW radar systems have been developed with marine navigation as one of their applications. Two notable examples are PILOT developed originally by Philips Research (now Thales) and the CRM-100 produced by PIT.[1]

16.4.5 Target profiling

A further advantage of waveform coding (whatever code scheme is employed) is that very high range resolutions can be achieved. Current technology will support total bandwidths of over 2 GHz, either instantaneously or in a series of steps on a per-pulse basis. For example ten pulses each with 100 MHz bandwidth could be reassembled to provide a total bandwidth of 1 GHz, equivalent to a range resolution of 15 cm. A resolution of the order of 1 m is quite routine and helps to improve the detection of small targets by spatially filtering-out clutter. However, at these high resolutions sea clutter deviates significantly from that of well-behaved Gaussian noise. In particular, there is a much higher probability of high clutter returns. These 'sea spikes' (Chapter 5, Section 5.7.1; Chapter 11, Section 11.7.2) tend to persist for several milliseconds. They have a very target-like appearance, thus increasing false alarms, necessitating sophisticated adaptive signal detection processing. Transmission of wide bandwidths is counter to current ITU regulations that seek to improve efficient spectrum use.

The radar may observe a head-on ship target of projected length 30 m as being made up of 30 values of target reflectivity, one per 1 m range resolution cell. This is known as a range-profile, whose form provides information about the ship type and leads to the possibility of classification as well as detection. Whilst the utility of such information is unproven, this does illustrate the sort of additional capability that new radar designs might offer. For example, when coupled with aspect (derived from briefly tracking the ship's heading) a fairly accurate measure of true length can be made, instantly giving a degree of basic classification. A measured length of 100 m is unlikely to be a yacht. The range profile also provides more detailed information. It must be remembered, however, that the range profile is a measure

[1] Przenysłowy Instytut Telekomunikacji (Telecommunications Research Institute), Warsaw, Poland. Quiet Naval Radar CRM-100.

of the backscatter of electromagnetic radiation from each separate range cell. Thus, simplistically, an area containing significant superstructure is likely (on average) to be more reflective than flat decking. Although the range profile can be quite variable this general behaviour can be used as an aid to further classification. For example, a ferry has little but superstructure and will tend to generate a range profile of quite high magnitude in most range cells. An oil tanker tends to have its main superstructure elements predominantly at one end and has a range profile with low values in most range cells except those coinciding with the accommodation block. This difference between the two profiles helps to classify them. However, there are a great many vessels with a huge variety of designs and the route to routine classification is by no means straightforward. Although automatic classification techniques have and are being developed, they tend to be specific to military interests. Even these are reputed not to provide consistently reliable and robust classification. Nevertheless, there is a growing body of research on this topic that could be exploited for the civil marine community.

Range profiles may be considered as one-dimensional images, the resolution increment in the cross range dimension being determined by the scanner beamwidth and therefore can be considered to be infinitely large when illuminating ship targets. Inverse Synthetic Aperture Radar (ISAR) is a technique enabling two-dimensional high resolution, the additional detail generated providing a much more informative representation of the target. ISAR uses the motion of the target ship to map out or synthesise an aperture over a period of time (e.g. using 32–64 pulses and a prf of 1–2 kHz). The synthetic aperture results from the change in viewing angle between the illuminating radar and the target. The synthesised aperture is much larger than the real aperture, giving high resolution. Ideally, the aperture is formed in a direction that results in a resolution in a plane orthogonal to that in which the range resolution acts, to provide a standard two-dimensional image. For example, a ship viewed side on and rolling with an angular range of $\pm 0.5°$ will produce a synthetic aperture in a plane that resolves the ship in height. Essentially the higher parts of the ship move at faster velocities than lower parts and this property is used in the aperture synthesis to provide the height resolution. Figure 16.2 shows a typical ISAR image. It is clearly ‘ship-like’ but is actually a measure of electromagnetic backscatter and should not be

Figure 16.2 ISAR image and photographic image of a ship. Courtesy of Thales Ltd

confused with the optical image. Similarly, yaw would generate resolution in length. Note, for this geometry a pitching motion has no component towards the illuminating radar and thus no useful aperture can be formed in this plane. In addition any linear motion of the target is likely to de-focus the image; this must be corrected if the image is to be further exploited for target classification. Lastly, self-motion of the ship hosting the radar will also cause a de-focusing effect that requires corrective processing.

Re-focusing can be carried out in two stages. First, bulk linear motion effects are removed by calculation of average phase compensation. The second stage is to refocus using a trial and error method that attempts to 'guess' the correct imaging parameters. If the first guess makes things worse then the next guess should change the parameters in the opposite sense. Iteration leads to a best-focused image (which can still exhibit a degree of defocus). In general, real ship motion will enable imagery to be generated in an arbitrary plane, as it will exhibit all types of motion. In fact there are likely to be changes in the way the ship moves resulting in variable image projection planes and even varying resolution. In effect, the generated image will have an unknown plane of projection and unknown scaling. Further interrogative processing is required to determine these parameters before it can be used as a basis for classifying the target; the processing methods are still the subject of research. However, many experimental observations exist and high quality imagery can be generated, usually by dwelling on the target ship until a 'good' image is formed. This normally takes a few seconds, sometimes longer. The procedure would therefore be to detect a target in the normal surveillance mode and then change to a 'staring' mode with the antenna trained on the target ship. It would be a matter of judgement as to whether or not this is a 'good' use of radar resource.

The classification process, whether range profiles or ISAR images are used, can be manual, automatic or a combination of both. Manual classification is accomplished by selecting the best match after comparing a range profile of the target with a library set of known ship range profiles. Numerous schemes for automatic classification have appeared in the research literature, many mimicking the manual approach. More advanced methods attempt to learn to recognise radar signatures based upon their experience. That is to say, the library set is built up from measurements of known ships and is improved all the time as and when new data can be added. In this way the radar system should become better and better at recognising ships. Techniques such as neural networks and genetic algorithms underpin this approach. However, no existing methods yield truly robust and reliable results. One problem is that a complete library generation means measuring the World's shipping fleet, an unrealistically large task. Also, even a small refit or change in deck cargo distribution could have a large effect on the radar signature. Finally, the time-variant and sea state dependent multipath component of illumination will also add variability to the observed radar signature (i.e. the range profile or the ISAR image) which will reduce the performance of the classification scheme.

Motion of the ship carrying the radar system can also be used to attempt to improve resolution of a scene to be surveyed (a coastline for example). To generate 1 m resolution imagery, an angular change of the order of one degree is required. Linear

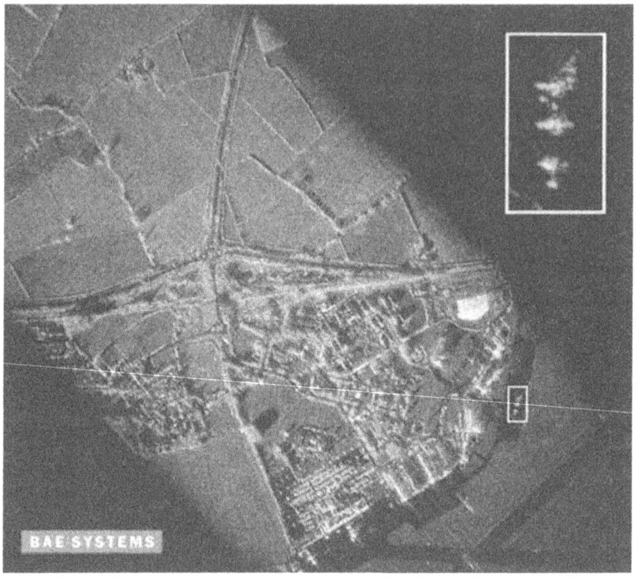

Figure 16.3 *SAR image of Edinburgh Airport. The inset shows a magnified area of the runway containing aircraft. Courtesy of BAE Systems*

forward motion of the ship could be used. The ship would have to travel approximately 200 m at a range of 10 km from the zone to be imaged in order to subtend the necessary one degree angle and thus must stare for many seconds. Synthesising an aperture in this way is known as SAR. Here the aperture is synthesised from motion of the platform containing the radar system and the target should be stationary. Figure 16.3 shows an example of a high resolution SAR image in which quite finely detailed features may be observed. In ISAR, as seen earlier, the aperture is synthesised via motion of the target whilst the radar is, hopefully, stationary. In SAR, as in ISAR, any motion of the radar antenna from an ideal case requires 'motion compensation' and focusing using techniques similar in principle to those which correct unwanted motions in ISAR image formation.

It is unclear whether or not imagery of ships, the coast and immediate surrounding land to 1 m resolution has any commercial purpose, except possibly for coastguard operations such as drug interdiction. However, it does illustrate the sort of additional capability offered if coherent transmitters are employed in future radar system designs. In short, a number of new design freedoms are presented to the designer of future marine radar.

16.4.6 Monopulse

Another well-established radar technique that could be adopted is monopulse. Monopulse can improve the angular location accuracy from an antenna by approximately a factor of ten relative to the conventional use of beamwidth alone.

A distributed planar antenna is well suited to this task. To form a monopulse beam two overlapped beam components are required on azimuth (for example the antenna could be split into two halves). The echo from each is received in two separate channels that can be compared in either amplitude or phase. For example, if amplitudes are compared and the left beam signal is found to be stronger than the right, the target must be offset to the left of boresight. The magnitude of the difference is related to the angular offset, so the true target bearing can be evaluated more accurately. If desired, the antenna could be trained to track a target, effectively staring at it, as for ISAR. More generally, monopulse may be used to track targets more accurately, tracks being updated on each scan whilst in the surveillance mode (track-while-scan). A track processor will also initiate new tracks and delete old ones as they are observed. This mature technique has found application in a wide variety of roles, particularly in air traffic control and military radar systems.

Unsurprisingly, additional processing tasks such as those described above will require additional processing capacity. Fortunately, as pointed out earlier, this is quite consistent with trends in processor technology and features selected from those outlined above will be chosen via a complex balance of cost, complexity and need. Perhaps more significantly, is implied a growing role for software supporting the radar and its operating modes, representing a particularly fast growing cost component. Increasing concern across the entire radar industry is resulting in some interesting developments. For example, Gedae is a commercially available radar software development tool, already developed and now being taken up by the military radar industry. It is possible to speculate that, at least within a single company, a strategy will emerge of developing non-hardware specific operating systems upon which applications can be designed and built. This clearly has parallels with the approaches adopted within the PC and mobile phone industries. Whether or not this is a worthwhile investment for a single marine radar manufacturer is unclear. However, as digitisation increasingly dominates the design of the radar systems, it will become increasingly important to tackle software development in a cost optimal fashion. Equally, increasing digitisation should provide an environment in which software upgrades and improved detection algorithms can be relatively easily incorporated.

16.5 Integrated systems

A very clear, relatively recent, trend is a move towards fully integrated bridge systems (IBS) (see Chapter 2, Section 2.1.3, Figure 2.4). These incorporate navigation sensors, steering, engine room monitoring and control functions into a single unit, with the advantages of providing a more comprehensive solution to navigation and surveillance and allowing the industry to take a more overarching (and commercially beneficial) system-integration role rather than being a component or sub-systems supplier. It does of course mean that the radar system will tend to become a component of the larger integrated system, although still needing to operate in a stand-alone mode that maintains compliance with the IMO regulations. Nevertheless, IBS does open up potential capability improvements to the ship's ability to sense its environment.

For example, combined inputs from the radar, GPS, AIS, sonar, infrared camera, electronic maps (ECDIS) and gyro systems could be 'fused' into a single display. Digital processing helps to facilitate this data fusion and should lead to a more reliable picture of activity that can be tailored to the mariner's needs. Having done this, it may be possible to relax the requirements on individual sensors, provided that the overall capability for navigation remains consistent with that demanded by IMO regulations. The use of data fusion for marine navigation is very much in its infancy with only radar data and chart information currently being placed onto the same display. For example, the 3 and 9 GHz band radar outputs are not combined, although doing so could improve both sensitivity and reliability of detection.

16.6 Infrastructure and implementation

Earlier in this chapter alterations to the design of future marine radar systems were postulated in a fairly cavalier manner. However, in practice, the radar cannot be considered in isolation as it must interact with a number of external equipments, including racons, search and rescue transponders (SARTs) and possibly other forms of transponder or beacon. Transponder-based systems react to illumination from a radar by transmitting either their own signal or re-transmitting a facsimile of the illuminating signal back to the radar. Currently, the frequency, power and waveform type of marine radar systems triggers them; that is, they are designed to react to illumination provided by magnetron based systems and do not necessarily react to other forms of electromagnetic radiation. In this way they avoid creating an unnecessary electromagnetic hazard that could interfere with other RF systems. The much longer pulse lengths and associated lower peak powers that would characterise a solid state alternative radar will not necessarily be able to trigger the transponders, particularly at longer ranges where the power densities received will be that much lower. The point at which this happens will be determined by the exact design of the solid state variant. Simplistically, from the one-way inverse square law, a reduction in peak power by the factor of 10 000 (from 10 kW to 1 W) used in the earlier example would reduce the maximum 'trigger range' by a factor of approximately 100.

Transponder and beacon outputs are also designed to have characteristic waveforms mandated to suit magnetron radars. A new generation solid state radar must receive these waveforms and detect them as beacon sources without corrupting surveillance data. Recognising a signal as a racon or SART ought to be quite straightforward, although representing another task for the processor to absorb. Some care will be needed to avoid inadvertent reception by ensuring that the range and antenna sidelobes are sufficiently well controlled, otherwise the radar will be 'jammed' by the transponder. This aspect of performance will have to be thoroughly investigated, with demonstrations that any new design can be integrated into existing infrastructure. This need for compatibility could adversely affect the costs of either the radar directly or through the financing of necessary modifications to transponder units. The area is critically important as it represents a combination of both safety and potentially very significant extra expense. However, marine radar systems must be compliant with the marine infrastructure to form a complete operating system and this aspect must form part of any overall re-design. An alternative approach under discussion for the 3 GHz band is to drop the necessity for operation with transponders altogether and in the longer term marine radar system designs might not need remain compatible with racons; SARTs are confined to 9 GHz.

Training, reliability and maintainability should be fully taken into account. Training and re-training of radar operators to keep pace with technology advances is a costly and time-consuming business. Any new radar design should aim to minimise the necessity for re-training. Indeed, in principle, the flexibility of digital processing should make operation easier; this should be part of the design goal. At the very least, the new radar design should be able to mimic the old quite closely to ease the transition. However, if new and more complicated modes such as ISAR imaging and range-profiling are incorporated then the radar operator will be (a) faced with new tasks and (b) have additional mode selection decisions to make. A high degree of re-skilling might then be required to ensure that the correct decision is taken at the right time for the prevailing conditions.

Although magnetrons have improved in reliability markedly over the years, solid state technology is, in general, even more reliable. Failure of a single power module within a distributed solid state transmitter would not lead to complete radar failure but instead to a reduction in performance, particularly detection range. The system could be programmed to detect module failure and not only report it but also adjust itself to minimise the impact on resulting performance, eg through control of scanner side-lobes. More reliable systems naturally lead to cost savings in maintenance, improving customer cost effectiveness of any new designs. As indicated above, the increased use of digital processing facilitates a high degree of self-testing and even limited self-correction. There are, however, areas where reliability and maintenance may not change greatly. Perhaps the most obvious is in the antenna turning gear which is already a highly developed component of current marine radar systems. Should electronic scanning ever become affordable, the need for any mechanical steering would be obviated, leading to further improvements in overall reliability. At the moment this is some way off.

16.7 Other uses of radar for commercial and leisure shipping

Earlier in this chapter, a number of developments in other branches of the radar industry were highlighted that may lead to marine applications. In particular, in recent years, there has been tremendous progress in the development of very short range, very low cost automotive radar. The current trend is towards using groups of sensors (TV, acoustic, radar, etc.) to provide a local 'picture' of the immediate environment around the vehicle, enabling it to set up an 'exclusion zone'. If the sensor system detects anything entering the zone, evasive action may be taken. Eventually it is hoped that this can be done automatically, bringing about significant improvements in car safety, although the human will retain overall responsibility for the safe passage of the vehicle. Such a system could well have application in the marine environment, particularly as an aid during difficult harbour manoeuvres in poor visibility. The technology for the front-end radar component costs only a few tens of pounds and is based on the FMCW principle. If ships move to a central processing unit then the sensors could use spare capacity rather than requiring their own dedicated processors – an extension of the integrated bridge system concept. More speculatively, there may be a trade-off between localised and centralised data processing that offers more optimum data fusion performance. The same technology could also be used to sense intruders. Again an 'electronic fence' is created around the ship and any movement detected crossing the fence triggers an alarm. This is not dissimilar to the passive infrared (PIR) detector found in most homes. However, the radar allows a tailored definition of the electronic fence and reacts to movement rather than changes to ambient thermal radiation.

Piracy is still a very real danger in certain regions around the globe. The craft used are often small, capable of relatively high speeds and quite highly manoeuvrable. The coherent radar technology resulting from the use of solid-state transmitters should enable such craft to be detected in velocity and angle by the primary surveillance radar, enabling them to be tracked. Sophisticated signal interpretation could be developed to infer hostile patterns of activity and provide sufficient threat warning. An alternative would be to use a dedicated system with a medium range capability. This could also double up as an intruder alerter by operating it on a short range-mode. Overall, the technology exists to make a wide range of very versatile radar sensing systems. Whether or not they find application will depend on the requirements of the market place.

For many years, ground radar systems have been used to measure properties of the Earth's atmosphere to enable more accurate predictions of weather phenomena. Most of the instruments are large, operate in C-band (approximately 5 GHz), have very high power and require extremely low antenna sidelobe levels. Whilst their performance exceeds that available within the confines of a ship, the value of being able to sense the local weather conditions makes weather radar a potentially very useful tool. It would enable storms to be circumnavigated and would improve course setting and safety, minimising heavy weather cargo damage. Relatively small weather radar systems do exist; indeed, they are carried by most large aircraft. Their migration to the marine environment, enabling at least a crude estimate of weather a few tens of kilometres distant, should be possible but will depend on cost-effectiveness.

Radar shares many common properties and technologies with RF communication systems, leading, potentially, to systems able to combine both applications. For example, a communication mode built into a radar system would provide high bandwidth for a data link. Ship to shore links might be capable of supporting TV, video, etc. In the same way, high data rate ship to ship communications could also be supported. The precise drivers for such a capability are not clear but certainly this type of operation could be supported by today's technology. It is conceivable that the boundary between communication and radar systems could become increasingly blurred.

There are other techniques that could, potentially, have a place in maritime applications. Passive coherent location (PCL) 'borrows' transmissions of opportunity and thus only consists of a receiver, which looks for a direct transmission of opportunity (such as a radio broadcast), which it correlates with indirect scattered radiation from the area of interest. Thus it is in effect a bistatic radar, the transmitter and receiver being at differing sites. Target range and bearing information are deduced from knowledge of site positions. Well away from the coast, the only transmissions of opportunity likely to be present are those of communications or navigation (GPS) orbiting satellites. Target detection range would be primarily determined by the power density of the radiation emitted by the satellite. Detailed calculations are required to determine what this detection range may be to evaluate the utility of the surveillance and navigation performance. If feasible, this would certainly offer a low cost solution, as no transmitter is required. Indeed it might be possible to operate with a magnetron-based system out to sea and just use PCL for short-range coastal operations where the wide magnetron spectrum becomes troublesome to telecommunications services. Again, detailed studies are required to determine whether or not this is remotely possible.

16.8 In conclusion

Clearly, there are a number of compelling reasons likely to lead to changes in marine radar at some point in the future. The strongest of these seems to be regulatory. Equally, there is a great deal of existing and emerging technology that can support a wide range of new design options. In this chapter we have postulated a number of possible developments and system designs. It is impossible to predict the future with any certainty and thus the developments described may or may not find application. Much will depend on how the World changes. The only sure prediction is that the status quo will not remain. Regulatory change will initially spark active evaluation of solid state radar designs as an alternative to new filter designs for magnetrons. Either or both could find their way into future systems. Certainly existing technology could be used to migrate to quite differing designs to those used currently and generally these design changes will offer a range of new operating modes. It should be remembered that current marine radar systems provide extremely good performance that alternatives will struggle to match. Technology developments in military radar, computing and communications may well provide further advances that can migrate into marine radar systems as the cost-effectiveness equation becomes appropriately balanced.

Conventional marine radar has long given good service and continues to do so. This book has tried to set out its strengths and limitations when detecting targets. In this chapter we have outlined some of the numerous technical possibilities for improvement to meet changing circumstances and to provide new facilities. Because shipping is international and operates in a competitive environment, the regulators have to balance the undoubted need for standardisation and maintenance of safety with the freedom to explore these new opportunities. Possibly the first take up of some of the possibilities described may be VTS and coastguard users, who are more lightly regulated and have rather different operational requirements from those of merchant shipping.

Although the marine industry does not have the research and development facilities and budgets available to the military – IMO has no technical development remit – the regulatory agencies of leading maritime States have demonstrated their interest in fostering and part-funding navigational innovations showing promise of becoming worthwhile contributors to safe navigation. A good example is the recent international programme of studies, calculations, laboratory testing, sea trialling and debate leading to IMO's adoption of a better standard for radar reflectors.

Previous major radar advances, such as ARPA, racons and RTEs, owe a good deal to leading shipping companies, Lighthouse Authorities and Harbour Boards. Over the years they have willingly made trials facilities available, encouraging and complementing the efforts of the manufacturing firms. Unfortunately, the users – Officers of the Watch and VTS operators – have not always played as full a part in developments as might be wished. Beside being away at sea or busy in the ports, the technical options have not always been presented to them in intelligible terms. The authors hope this book may go some way towards redressing this problem and that radar will continue to evolve as an important aid to the free and safe passage of marine traffic.

Appendix A1

Glossary

This glossary gives the meanings usually borne in the text of the main technical terms, especially the less self-evident. Shades of meaning might differ in other radar contexts, for example, air traffic control. Source references are [S] = New Shorter Oxford English Dictionary; [C] = Clapham C, Oxford Concise Dictionary of Mathematics. Glossary cross-entries are *italicised*.

Active device (active reflector) Device such as a *racon* connected to a source of power and able to respond at greater power than that within the *interrogation*. (*Secondary radar* reflector, for example, a radar target enhancer, where an onboard power supply augment response power.)

Algorithm A precisely described routine procedure that can be applied and systematically followed through to a conclusion. [C].

Amplifier Device increasing output power of input signal using power from an external source. Linear amplifiers do not distort the waveform. See *gain*.

Analog(ue) Designating, pertaining to, or operating with signals or information represented by a continuously variable quantity such as spatial position, voltage, etc. [S]. See *digital*.

Anaprop Anomalous propagation caused by atmospheric refraction effects.

Antenna Converts between current in a conductor and electromagnetic radiation in space. Usually a *reciprocal device*. See *scanner.*

Aperture width, area The effective electrical width, area of an *antenna* or other *radiator*.

Attenuation (1) Scattering or absorption of electromagnetic energy as heat, usually expressed in *dB*. (2) Proportion of output energy entering a device or network such as a *filter*.

Band(width) *Spectrum* width in the frequency domain occupied by a signal or handled by a component, for example, a *filter*.

Baseband Signal spectrum after stripping off *RF* or *IF* carriers. Means video frequency, applicable to the bright-up of a *cursive PPI.*

Beam(width) (The angular width between half-power points of) the bundle of rays forming the main beam or *lobe* of a directional *antenna* or *scanner*.

Boresight Antenna electrical axis.

Capacitance Ability to store electrical charge; the ratio between the change in the electrostatic charge in a body to the corresponding change in its potential [S]. At a.c., current leads voltage by $\pi/2$ rad.

Capillary wave Small sea-wave impressed by breeze, the basic *scatterer* governing sea clutter and reflection at the *grazing point*. See *gravity wave*, *sea*.

Carrier An *RF* bearer of information carried as a *modulation.*

Clutter (map) Unwanted *signals*, *echoes* or *noise* from sea-waves or precipitation which tend to degrade reception of wanted echoes, whether or not displayed. Clutter maps store locations of significant clutter to facilitate adaptive processing.

Coherent (oscillator, COHO) Receiver system preserving the phase information component of the echo (in such receivers: Local oscillator whose frequency is locked to a transmitter carrier sub-harmonic).

Correlation Between two random variables, … is a measure of the extent to which a change in one tends to correspond to a change in the other … Independent random variables have zero correlation … [C].

Crystal-video receiver Receiver used in some *secondary radars* with direct *RF demodulation* rather than the usual superheterodyne of primary radars.

Cursive *Raw radar PPI* display built up in real time from radial *sweeps*, using long persistence phosphor.

Datasheet value The value of a parameter quoted in the supplier's published datasheet.

Decade In ratio 10 to 1. See *octave*.

Decibel (dB) [dBW, dB m^2, dBi] A logarithmic expression of power-related ratios. [dBW are decibels relative to 1.0 W; dB m^2 relative to a reference *RCS* of 1.0 m^2, dBi denotes *antenna gain* relative to an *isotropic radiator*.]

Declare Finally decide that candidate signals associated with a location constitute a *target*, to be displayed.

Demodulate Extract *baseband* signals from the *RF* (*crystal-video receivers*) or *IF* carrier, using a rectification process. A first step towards *detection*.

Detect The whole process of determination that a signal contains a valid *target echo*. Many authors use detect in the narrower sense of *demodulate*.

Detection cell One of a large number of digitised locations covering the sea area under surveillance within which *detection* decisions are made. Sometimes called a range bin.

Dielectric (constant) A vacuum or insulator storing electric charge after a voltage is applied. Dielectric constant or permittivity is the charge stored relative to a vacuum having $\varepsilon = 1.0$.

Differentiation (1) The mathematical process of determination of the rate of change of a quantity to find the gradient or slope of its curve. (2) Extraction of the changes within a solid block of signal, for example, to aid *detection* of weak *echoes* within an area of strong clutter by use of a fast time constant circuit.

Diffraction (region) The bending of *rays* by a nearby obstruction. (The region illuminated by the bending near and beyond the obstruction.)

Digital Computation and signal processing using data in the form of discrete on-off bits rather than *analog* elements.

Diode Device passing current only when the anode is positive to the cathode, used to rectify or switch a.c. signals.

Distribution In statistics. The distribution of a random variable is concerned with the way in which the probability of it taking a certain value ... varies ... The distribution ... of a continuously random variable is given by its probability density function. [C].

Divergence Beam spreading when rays are reflected from the spherical face of the curved Earth.

Diversity Combination of two or more radars having differing attributes (height, location, frequency, polarisation, etc.) to improve *detectability* through the decorrelation between their data outputs.

Domain Consideration of events, for example, a pulse train, from the standpoint of spectrum (frequency domain) or waveform (time domain). Domains are interchangeable using mathematical transforms.

Duct(ing) A horizontal atmospheric layer having temperature and moisture gradient boundary changes sufficient to form a waveguide which can trap signals between the duct boundaries, helping or hindering point to point transmissions. (Ducting, also called trapping, is the resulting *anaprop*.)

Dynamic range Amplitude bracket which a device, for example, an *amplifier*, will faithfully handle without output distortion.

Echo The *signal* from a passive *target.* Secondary radars reply with responses rather than echoes.

Effective Earth radius Fictitious radius straightening *rays* which have been curved by atmospheric *refraction*, simplifying geometrical calculation.

Envelope detection *Detection* of a signal at the stage when it is mixed with a carrier (usually at *IF*).

Environment The sea surface, atmosphere, fog, ice and precipitation, which modify *target detectability* and are absent from *free space*.

Equivalent isotropic radiated power (EIRP) Product of transmitter power and *antenna* or *scanner gain*.

Extended target One too big to be treated as a geometrical *point target*, for example, a ship or coastline.

Far field The Fraunhofer region, used for ordinary radar target detection, within which an antenna or scanner beam is fully formed. See *field*, *near field*.

Feeder Transmission line. The coaxial cable or *waveguide* connecting *scanner* with transmit/receive unit. May significantly degrade system performance.

Field The a.c. electric or magnetic potential or *flux* existing within a vacuum or material.

Filter Frequency-dependent device, for example, used to control a *receiver bandwidth*. May have bandpass, bandstop, low-pass, high-pass or other defined characteristics.

Flat-Earth approximation Assumption that the Earth's surface is flat and that *rays* travel in straight lines, not curved by atmospheric *refraction*, simplifying geometrical analysis.

Flux (density) *RF* power flow radiated from an *antenna*, etc. (Flux density is flux per unit cross-sectional area.)

Free space Hypothetical *propagation* condition unbounded by the Earth's surface, without other *environmental* constraints.

Frequency The number of complete cycles per second of a uniformly repetitive signal, expressed in Hz. Multipliers are kHz, MHz, GHz.

Gain (1) Of an *amplifier*, etc: ratio of output to input power. (2) Of an *antenna*: ratio of output flux density on *boresight* to flux density from an *isotropic radiator* fed by the same source.

Geometrical optics Analysis tool assuming rays behave as ballistic particles obeying Newtonian laws of motion.

Gravity wave Conventional sea wave formed by wind. See *capillary wave*.

Grazing point Point on Earth's surface at which oblique indirect *ray* from *scanner* specularly reflects to reach *point target*, causing *multipath interference* with direct ray

Harmonic A *frequency* which is an exact multiple (twice for second harmonic) of another.

Impedance The opposition to the passage of an alternating current provided by the vector sum of the series *resistance*, *capacitance* and *inductance*. Vector quantity, expressed in ohms.

Inductance Ability to store magnetic energy. At a.c., current lags voltage by $\pi/2$ rad.

Integration (1) The mathematical process of determination of the area under a curve; the inverse of *differentiation*. (2) Summation or averaging of a packet of signals, for example to aid *detection* of correlated weak echoes within uncorrelated clutter.

Interference (region) Unwanted interaction of one signal with another, especially when both are at the same *frequency*. Region at relatively short ranges subject to *multipath interference* between direct and indirect rays, the latter arriving via the *grazing point*).

Intermediate frequency (IF) In superheterodyne receivers: *frequency* between *RF* and *baseband*, typically 50 MHz, at which received signals are chiefly amplified and filtered.

Interrogation A primary radar's transmitter pulse received at a *secondary radar*.

Isotropic Of a property, for example, *antenna gain*, not varying with direction. An isotropic antenna radiates through the whole 4π steradians (sr) and has unity *gain*, 0 dBi.

Latency Elapsed time from collection of data to its availability as usable information.

Leg The radar to target or target to radar path.

Lobe A solid angle within which a directional antenna radiates strongly, either directly or as a result of constructive (co-phased) *multipath interference* with indirect rays forward reflected from the *grazing point*. The main beam is surrounded by unwanted *sidelobes*. Regions of strong echo from *point targets* are also called lobes. Between the lobes are *nulls*.

Local oscillator (LO) Low power *RF* oscillator whose *frequency* beats (is *modulated*) with an incoming signal at transmitter frequency to give the *IF* difference frequency in ordinary superheterodyne receivers.

Loss (Usually undesired) reduction of signal strength during passage through a medium or a component, the energy often being dissipated as heat. Ohmic loss is caused by *resistivity* of the conductor. See *mismatch*.

Macro (micro) structure/geometry Structural features much larger (smaller) than a wavelength, or the resulting geometrical analysis. For example, *gravity* and *capillary waves*.

Magnetron High power *RF oscillator* which generates radar transmitter power when *pulsed* by the *modulator*.

Match See *mismatch*.

Maximum instrumented range The maximum range at which the radar system is designed to display, constrained by *pulse repetition frequency*, etc., rather than sensitivity or target *RCS*.

Minimum detectable signal (MDS) The minimum echo power which is *detectable* in the prevailing *clutter* and *noise*.

(Mis)match Match: the efficiency with which a device accepts power presented to it. Mismatch is quantified as voltage standing wave ratio (VSWR) or *reflection coefficient* in its sense 1. Mismatch is lack of match, some energy being rejected.

Mixer Device (often based on *diodes*) converting *RF* to *IF* by injection of the *local oscillation*.

Modulator/modulation Device/process impressing one signal on another, for example, *pulses* on an *RF* carrier.

Multipath interference Interference between direct and indirect scanner/target *rays*, the latter via the *grazing point*.

Near field The Fresnel region, within which an antenna beam is not fully formed and the *radiation pattern* is range-dependent. See *far field*.

Noise (factor/figure) [margin] Clutter-like random disturbances within the receiver, akin to hiss in *radio* receivers. Noise factor is the ratio by which actual noise power exceeds the theoretical minimum; noise figure is this ratio expressed in dB. Margin is the excess of *signal to noise ratio* over that required for target detection.

Non-coherent Receiver system, usual to marine radars, which discards the phase information component of the echo. See *coherent*.

Null A point of minimum or zero signal caused by destructive interference between two competing near-equal signal components of opposing phase. See *lobe*.

Object A *target*, element of a target or item of *clutter*.

Octave Ratio of 2 to 1. See *decade*.

Oscillator Device generating an a.c. signal, continuously (e.g. *LO*) or when pulsed by a *modulator* (e.g. *magnetron*).

Packet The group of transmissions, or their echoes, illuminating a small target during a single *scan*.

Phase The displacement, expressed in radians, of one cyclically repetitive quantity, for example, current, relative to another, for example, voltage, of the same or an exact multiple *frequency*.

Phase front The plane, normal to *boresight*, formed by the *wavelets* comprising a propagating *RF beam*.

Plan position indicator (PPI) Viewing screen or 'scope' depicting *plots*, *tracks* and ancillary data mapwise, laid down as a *raster* or *cursive*ly.

Platform The ship, tower, buoy, etc., carrying a primary or secondary radar or a reflector.

Plot (association) A *declared detection* painted on the *PPI*. (Association is machine examination of successive plots to determine whether they form a *track* of a single target in motion relative to the radar *platform*.)

Point target Having sufficiently small dimensions to be regarded as a geometrical point when *ray tracing*.

Polar diagram Depicts *radiation pattern* of an antenna or variation of target *RCS* with angle (azimuth or elevation) on polar graph paper. See *target pattern map*.

Polarisation Plane of the electrical *field* component of a *ray* from or to an *antenna*, viewed in the direction of energy flow (IEEE convention). May be linear (horizontal, vertical or slant) or circular (left- or right-hand).

Port An input or output terminal of a device.

Post-detection integration *Integration* (sense 2) at *baseband* after the *demodulator* (the usual marine radar arrangement).

Precipitation clutter Unwanted echoes from hydrometeors such as raindrops, snowflakes, etc.

Probability of detection (false alarm) Statistical probability that a genuine *target* will be *detected* or *declared* under stated circumstances, for example single pulse. (That a non-existent *target* will be wrongly *detected* or *declared*.)

Propagate Transmit *RF* energy a long distance through space or a *dielectric*, direction of travel being mutually perpendicular to the *phase fronts*.

Pulse repetition frequency (interval, length) The prf is rate of pulse transmission, pulses per second (pps). The pri is the time interval between pulse starts. Pulselength is the pulse duration, expressed in microseconds or metres range equivalent.

Pulse (train) A short episode of energy interspersed with silence (repeated indefinitely).

Racon Radar beacon. *Secondary radar* marking an aid to navigation (A to N) station with distinctive response on the *PPI*.

Radar cross section (RCS) The ability of an *object* to retro-reflect. Expressed in m^2 or dB m^2, relative to the reflection of a sphere of silhouette area 1 m^2, whose RCS is 0 dB m^2. The usual monostatic RCS is in the direction of the illuminator, bistatic is at any other specified solid angle.

Radar range equation Describes the range or signal strength performance of a radar, target and environment system.

Radiator (radiation pattern) Element of an *antenna*, for example, a slot in a slotted array *scanner* (radiation pattern is its *polar diagram* or part thereof, drawn in Cartesian coordinates).

Radio (frequency, RF) (1) The transmission and reception of sub-infrared electromagnetic waves through space for wireless communication and radar purposes. (2) Electromagnetic waves of length exceeding ~1 m in contradistinction to the centimetric waves used for marine radar, or to optics. (3) RF designates signals or components working at or near radar transmitter frequency, rather than *baseband* or *IF*.

Ramark Continuously transmitting *secondary* radar beacon giving bearing but not range.

Raster *PPI* display laid down zigzag fashion as in TV receivers. See *cursive*.

Raw (radar) Radar received signals in relatively unprocessed form, usually analog, especially when displayed.

Ray (tracing) The path taken by a component of radiated energy of infinitesimal width, perpendicular to the *wavefront*. Ray tracing describes performance in terms of the interplay of rays using the principles of *geometrical optics*.

Reciprocal device Device having similar characteristics when its input and output *ports* are interchanged, for example, a *resistor* or an *antenna* whose *radiation pattern* and *gain* are the same whether transmitting or receiving.

Rectify Convert an a.c. signal to unidirectional d.c., usually using a *diode*.

Reflection coefficient (1) Proportion of incident power rejected by a *mismatch*. (2) Proportion of incident voltage reflected by the sea surface, etc.

Reflector (scanner) Scatterer, usually metal, from which *rays* bounce in a defined manner. (A *scanner* using a reflector to define *polar diagram* and *gain*.)

Refractive index, gradient Refraction is bending of *rays* due to variation of *dielectric constant* (refractive index) of the medium. Refractive gradient is the rate of change of index with height.

Resistance (resistor) Ability to impede energy flow by conversion to heat, voltage drop per unit current being expressed in ohms (Ω). At a.c., current is in phase with voltage (device providing resistance).

Saturation (range) Driven beyond the *dynamic range*, so small input changes no longer provoke correct output power changes. (Range at which saturation occurs.)

Scan(ner) Cause an area to be systematically and repeatedly traversed by a *beam* (using a narrow-beam continually rotating *antenna* illuminating the sea surface.)

Scatterer Anything which re-radiates incident *RF* energy in any direction.

Sea (1) The body of salt (or, in this book fresh) water floating the targets, including lakes and rivers. (2) The system of sea-waves in dynamic equilibrium with the currently prevailing wind.

Sea clutter Unwanted returns from sea waves.

Search and rescue transponder (SART) *Secondary radar* akin to a *racon*, carried for distress purposes.

Secondary radar A cooperative radar with no display which only responds when interrogated; a *racon* or *SART*. (Navigation radars are primary radars.)

Seduction Capture of one *target's* series of *plots* or its *track*, by a close-by second target.

Service loss Loss accounting from the likely shortfall in radar performance from the *datasheet* as-new state when running in ordinary service.

Sideband *Spectrum* component in *frequency domain* lying to one side of the *carrier*.

Sidelobe Unwanted off-axis *lobe* of antenna *radiation pattern*; echo received through such a lobe.

Signal Electrical event carrying information on a target.

Signal to noise ratio (SNR) (1) Ratio between signal and noise powers in the system bandwidth. Determines detectability. (2) Signal to (noise plus clutter) power ratio.

Single pulse P_D *Probability of detection* of a single echo pulse, without benefit of *integration* with its fellows within the *packet*.

Skin (echo) The surface of a *platform* carrying a *secondary radar*; (the platform's unaided echo as a passive *reflector*).

Spectrum The distribution with *frequency* of the power within a *signal*, often shown as a graph of power per unit bandwidth to a base of frequency.

Squint The angular displacement between *antenna* electrical (*boresight*) and mechanical beam axes. May vary with frequency, differential squint being the change of angle for a given frequency change.

Standard atmosphere Having defined *refraction* typical of the air mass over the continental United States.

Stealthed In this book, a vessel whose shape or construction wittingly or otherwise minimises *RCS*.

Sweep The sequence of events comprising a single pulse transmission, its reflection by an *object*, reception, processing and (for cursive systems) display. Compare with *scan*.

Swept frequency Form of *racon* whose response transmitter *frequency* repetitively sweeps the frequency *band*, whether *interrogated* or not.

Swerling Case One of several mathematical models describing fluctuation of target *RCS* or *echo*.

Target Object of current interest to the radar operator.

Target pattern map Map-like representation of variation of an *object's RCS* through a large solid angle. See *polar diagram*.

Threshold(ing) Subject a waveform to a d.c. threshold voltage, using, for example, a *diode*, events exceeding threshold being candidate *detections*.

Tonnage Ship's capacity per a mercantile formula only loosely related to displacement or maximum cargo weight. Expressed in gt (gross tons).

TR cell *Waveguide* device which short-circuits to protect receiver when the *transmitter* fires and is a component of the *duplexer*.

Track (1) The course made good by a *target*. (2) Its representation on the PPI, formed from successive *plots*; alternatively vector.

Trackformer Digital subsystem outputting a vector statement of target course and speed from successive *plots*.

Trail Display of previous target *plots* at successively lower brilliance similar to the wake to indicate approximate relative velocity.

Transformer An *impedance*-changing passive component. A step-down transformer input port has high voltage, low current; the output port having low voltage, high current.

Transition region Region between *interference* and *diffraction regions* where both effects are in play.

Tune Align in frequency a *signal* and a *filter* or another signal.

Unambiguous Having only one *sweep* in play at any instant, so there is unlikely to be ambiguity whether an echo is from the current sweep at short range, or from a previous sweep at longer range. Antonym of *ambiguous*.

Vessel traffic service (VTS) Monitors and advises shipping in port or coastal waters, from data gathered by sensors including shore-based radar.

Video See *baseband*.

Wavefront Imaginary surface, perpendicular to direction of *propagation*, within which all *rays* in *beam* have same phase.

Waveguide Metal pipe, usually rectangular in section, within which *RF* travels with low loss by a process of total internal reflection.

Wavelet Conceptual spherical *wavefront* of radiation from an element of a larger *radiator*.

Appendix A2
Statistics details

This appendix expands on material in Chapters 11 and 12.

A2.1 Log-normal distribution

Further to Chapter 11, Section 11.7.3, probability is

$$p(x) = \frac{1}{x\sigma\sqrt{2\pi}} \exp\left[-\frac{1}{2}\left(\frac{\ln x - x_{\mathrm{m}}}{\sigma}\right)^2\right]$$

where

σ = standard deviation

x_{m} = median x (half events are < median and half > median)

$$\text{mean/median} = \exp\left(\frac{\sigma^2}{2}\right).$$

Using logarithmic abscissa

$$\sigma_{\mathrm{dB}}^2 = \frac{20}{[\ln 10\ \ (\text{mean : median ratio, dB})]} = 0.686\ (\text{mean : median ratio, dB})$$

The mean, σ, and median, σ_{m}, are related to the standard deviation, σ_0, by

$$10 \log \sigma = 10 \log \sigma_{\mathrm{m}} + 0.115\sigma_0.$$

A2.2 Rayleigh distribution

Further to Chapter 12, Section 12.3.3, the mean [1] $= \sqrt{(\pi/2)}\sigma = 1.2533\sigma$ and lies at 54.4 per cent of the cumulative distribution; median $= \sqrt{(2\ln 2)}\sigma = 1.1774\sigma$. The mode is the most frequently occurring observation and equals σ.

Shearman[1] points out that if we put $p(v)$ in terms of the instantaneous power, $w = v^2$, then $p(w)\,\mathrm{d}w$ must equal $p(v)\,\mathrm{d}v$, whence

$$p(w) = p(v)\frac{\mathrm{d}w}{\mathrm{d}v} = \frac{p(v)}{2v}.$$

Hence,

$$p\{w\} = \frac{1}{w_0}\exp\left[-\frac{w}{w_0}\right]$$

where w_0 is the mean value of the power, w, which in the notional 1 Ω is the same as the variance of v, which is σ^2. The above equation is the exponential probability density distribution of w. Hence, this is sometimes called the Rayleigh-power probability density function (PDF) [2]. However, the Rayleigh PDF is not the same as the exponential PDF, which is:

$$p(v) \propto \exp\left(-\frac{|v|}{\sigma}\right).$$

Rayleigh distribution is the probability of each of the two parts of a bivariate Gaussian distribution and is a chi-squared distribution with two degrees of freedom. The standard deviation is equal to the mean. Rayleigh distribution is also a special case of Weibull distribution, Chapter 11, Section 11.7.4, Eq. (11.17c), with shape parameter $c = 1$.

A2.3 Ricean distribution

Further to Chapter 12, Section 12.3.4, the Ricean probability distribution is:

$$p(r) = \frac{r}{\sigma^2}\exp\left[-\frac{r^2 + s^2}{2\sigma^2}\right] I_0\left(\frac{rs}{\sigma^2}\right)$$

where $I_0(z)$ is the Bessel function of zero order and imaginary argument, given by

$$I_0(z) = \frac{1}{2\pi}\int_0^{2\pi} \exp[z\cos\theta]\,\mathrm{d}\theta.$$

The following notes, provided by Shearman, explain the reason why this probability applies to non-coherent radar receivers carrying noisy signals.

A2.3.1 Noise

When data is amplitude modulated on to a carrier to form an echo pulse, the effect of superimposed noise is more complicated than in baseband systems because the signal

[1] E. D. R. SHEARMAN, personal communication, January 2003.

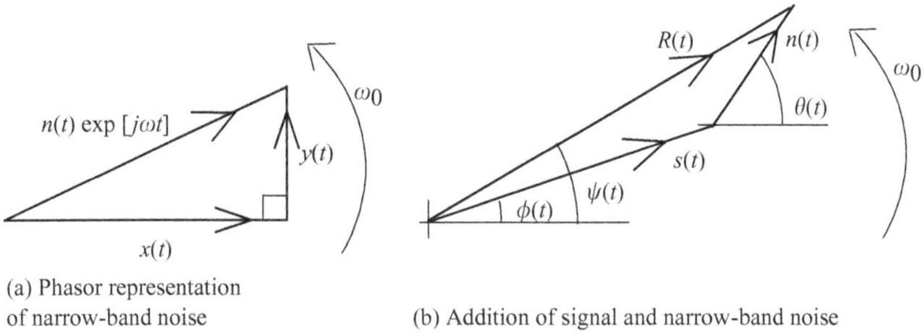

Figure A2.1 Noise and signal vectors

and noise are both phasors. In marine radar receivers the bandpass filter has small fractional bandwidth and the noise voltage can be represented by

$$n(t)\cos(\omega_0 t + \theta(t)) = x(t)\cos\omega_0 t - y(t)\sin\omega_0 t$$

where

$$x(t) = n(t)\cos\theta(t) \qquad y(t) = n(t)\sin\theta(t).$$

Functions $x(t)$, $n(t)$, $\theta(t)$ and $y(t)$ are time-varying quantities fluctuating with a timescale of the order of the inverse of the filter bandwidth, while ω_0 is the angular IF centre frequency. The phasors are shown in stationary form in Figure A2.1(a).

It can be shown that for narrow band noise $x(t)$ and $y(t)$ are independent and that both have Gaussian PDFs with zero mean and the same rms value, say σ.

Thus,

$$p(x) = \frac{1}{\sigma\sqrt{2\pi}}\exp\left[-\frac{x^2}{(2\sigma)^2}\right]$$

and

$$p(y) = \frac{1}{\sigma\sqrt{2\pi}}\exp\left[-\frac{y^2}{(2\sigma)^2}\right].$$

The mean noise power, N, is given by (bold type denotes mean values)

$$N = \frac{\mathbf{n^2(t)}}{2} = \frac{\mathbf{x^2(t)}}{2} + \frac{\mathbf{y^2(t)}}{2}.$$

Since the rms values of $x(t)$ and $y(t)$ are the same and both are equal to σ,

$$N = \mathbf{x^2(t)} = \mathbf{y^2(t)} = \sigma^2.$$

Substituting,

$$p(x) = \frac{1}{\sqrt{2\pi N}}\exp\left[-\frac{x^2}{2N}\right]$$

and

$$p(y) = \frac{1}{\sqrt{2\pi N}} \exp\left[-\frac{y^2}{2N}\right].$$

The PDF of $\theta(t)$ is uniform over 2π (i.e. may equally be of any phase) and the envelope, $n(t)$, has Rayleigh distribution:

$$p(\theta) = \frac{1}{2\pi} \qquad 0 < \theta < 2\pi$$

and

$$p(n) = \left(\frac{n}{N}\right) \exp\left[\frac{1}{2}\frac{r^2}{N}\right].$$

Having obtained representations of the noise, we next consider modulated signals with superimposed noise.

A2.3.2 Signal

The signal can be represented by

$$V_c s(t) \exp[j\phi(t)]$$

where $s(t)$ and $\phi(t)$ represent the time-varying amplitude and angle modulation, respectively. Both $s(t)$ and $\phi(t)$, like the corresponding noise quantities, fluctuate with a time scale which is the inverse of the order of the IF bandwidth. The input to the demodulator is thus the phasor resultant, $R(t)\cos(\omega_0 t + \psi(t))$, of the two quantities, one varying in a random fashion in both phase and amplitude and the other with meaningful amplitude modulation.

The phasor sum is shown in Figure A2.1(b), and in algebraic form can be written as

$$R(t)\exp[j\phi(t)] = V_c s(t)\exp[j(t)] + n\phi(t)\exp[j\phi(t)]$$

$$= V_c s(t)\exp[j\phi(t)]\mathbf{x}\left[\left\{1 + \frac{\boldsymbol{n(t)}}{\boldsymbol{V_c\, s(t)}}\right\}\mathbf{exp}[\,\boldsymbol{j\theta(t) - \phi(t)}]\right].$$

In this form the resultant voltage appears as the product of two factors: (a) the uncorrupted signal; (b) the term shown bold, which weights the amplitude and rotates the phase, both depending on SNR. If the term $s(t)V_c/n(t)$ is large (high SNR), the bold term $\rightarrow 1$ and noise hardly affects the signal; if the term is small the resultant PDF approximates that of noise alone. We are mainly interested in the intermediate case where both signal and noise are comparable, with SNR of the order of 5.

We start with an unmodulated carrier plus noise, which was analysed by Rice [3]. We take $V_c s(t)$ as a constant amplitude sinusoid, $A\cos\omega t$, to which are added the in-phase and quadrature noise components $x(t)\cos\omega_0 t$ and $y(t)\sin\omega_0 t$. Thus,

$$R(t)\exp[j\psi(t)] = (A + x(t)) + jy(t)$$

$$R^2 = (A + x)^2 + y^2 = (x')^2 + y^2, \text{ say}$$

and

$$\psi = \arctan\left[\frac{y}{A+x}\right] = \arctan\left(\frac{y}{x'}\right).$$

Thus, given the above Gaussian PDFs, $p(x)$ and $p(y)$, for $x(t)$ and $y(t)$, the joint probability density $q(R,\psi)$ may be found. From this, the probability density $q(R)$ is found by integrating over all ψ:

$$q(R) = \left(\frac{R}{N}\right)\exp\left[-\frac{R^2+A^2}{2N}\right] I_0\left(\frac{RA}{N}\right).$$

Normalised,

$$q(R) = RI_0(RA)\exp[-\tfrac{1}{2}(R^2+A^2)].$$

This is Eq. (12.7), whose PDF is plotted in Figure 12.11 curves (a), (c) and (e). I_0 is the modified Bessel function of the first kind and zero order. The parameter $A^2/2N = (A/\sqrt{2})^2/N$ is the ratio of carrier to noise power, P_c/N. As expected, it degenerates to Rayleigh distribution for $A \to 0$, shown by Figure 12.11(a). For A large, the PDF becomes a Gaussian distribution centred about the carrier amplitude, converging to this condition for increasing carrier/noise ratio.

A2.4 Solution of Eq. (12.8)

Further to Section 12.3.4, the single pulse P_D for Ricean distribution is

$$P_D = \int_k^\infty \frac{R(1+1/(8RA))}{\sqrt{2\pi RA}}\exp(RA)\exp\left[-\frac{1}{2}(R^2+A^2)\right]\mathrm{d}R. \qquad (12.8)$$

When $RA \gg 1$, $A \gg |R-A|$ and terms in A^3 and beyond are negligible; P_D is approximated by the first few terms of an infinite series which re-introduces the error function erf(Z) (encountered when we looked at Gaussian noise in Chapter 11, Section 11.3.8 and defined in Eq. (11.6)):

$$P_D \sim \frac{1}{2}(1-\mathrm{erf})\left[\frac{k-A}{\sqrt{2}}\right] + \frac{1}{2A\sqrt{2\pi}}\exp\left[-\frac{1}{2}(k-A)^2\right]\cdots.$$

When $A \gg 1$, this equation is approximated by the following. The terms in bold type are part of an infinite series giving a second-order correction:

$$P_D \sim \frac{1}{2}\left[1-\mathrm{erf}\,\frac{k-A}{\sqrt{2}}\right] + \left[\mathbf{1} - \frac{\boldsymbol{k-A}}{\mathbf{4}\boldsymbol{A}} + \frac{\mathbf{(1+(\boldsymbol{k-A})^2)}}{\mathbf{8}\boldsymbol{A}^2} - \cdots\right]$$
$$\times\left[\frac{1}{2\sqrt{2\pi}}A\right]\exp\left(-\frac{1}{2}(k-A)^2\right).$$

Substituting $A = \sqrt{2q}$ from Eq. (12.3), where q is the single-pulse signal power to mean noise power ratio introduced in Section 12.2.2, and tidying up:

$$P_\mathrm{D} = \frac{1}{2}\left[1 - \mathrm{erf}\left(\frac{k}{\sqrt{2}} - \sqrt{q}\right)\right] + \left[\frac{11}{8} - \frac{k}{2\sqrt{2q}} + \frac{1+k^2}{16q} \times \frac{1}{4\sqrt{\pi q}}\right]$$
$$\times \exp\left[k\sqrt{2q} - q - \frac{k^2}{2}\right].$$

This expression is passably convenient (the error function term demands a lookup table) when finding P_D from known values of k (substituted for F per Eq. (12.6c)), but is grim when q has to be found from a given P_D.

A2.5 Weibull distribution

The median, σ_m, is when $\mathrm{CP} = \frac{1}{2}$: $\sigma_\mathrm{m} = (\sigma \ln 2)^{1/2c}$. The instantaneous RCS is related to the cumulative probability and shape parameter by

$$\sigma = \sigma_\mathrm{m}(\ln 2)^{-1/2c}\left\{\ln\left[\frac{1}{1-\mathrm{CP}}\right]\right\}^{1/2c}$$

A set of events, such as clutter power within a radar receiver, can be tested for Weibull probability by plotting on graph paper scaled proportional to $\log\{\ln[1/(1 - P_\mathrm{D})]\}$, to a base of RCS in dB. A straight line indicates Weibull distribution. The slope of the line is $1/2c$.

A2.6 References

1 MEIKLE, H.: ‘Modern radar systems’ (Artech House, London, 2001), Section 2.7

2 SKOLNIK, M. I. (Ed.): ‘Introduction to radar systems’ (McGraw-Hill, New York, 1985), p. 23

3 RICE, S. O.: *Bell System Technical Journal*, 1948, **27**, pp. 109–57

Index